copyright © 1980 Marie Tharp

Introductory
OCEANOGRAPHY

Introductory
OCEANOGRAPHY
Tenth Edition

Harold V. Thurman
Emeritus Professor Mt. San Antonio College

Alan P. Trujillo
Associate Professor Palomar College

PEARSON

Prentice Hall

Upper Saddle River, New Jersey 07458

Library of Congress Cataloging-in-Publication Data

Thurman, Harold V.
   Introductory oceanography.—10th ed. / Harold V. Thurman, Alan P. Trujillo.
      p. cm.
   Includes bibliographical references and index.
   ISBN 0-13-143888-3
    1. Oceanography. I. Trujillo, Alan P. II. Title.

GC16.T45 2004
551.46—dc22

2003063235

> *To my wife,*
> *Sandra Ryan Trujillo,*
> *with all my love*

Executive Editor: *Patrick Lynch*
Editor in Chief, Science: *John Challice*
Associate Editor: *Melanie Cutler*
Executive Managing Editor: *Kathleen Schiaparelli*
Assistant Managing Editor: *Beth Sweeten*
Production Editor: *Kim Dellas*
Vice President ESM Production and Manufacturing: *David W. Riccardi*
Manufacturing Manager: *Trudy Pisciotti*
Manufacturing Buyer: *Alan Fischer*
Creative Director: *Carole Anson*
Art Director: *John Christiana*
Director of Creative Services: *Paul Belfanti*
Managing Editor, AV Management & Production: *Patty Burns*
AV Art Editor: *Jessica Einsig*
Assistant Managing Editor, Science Media: *Nicole Bush*
Assistant Managing Editor, Science Supplements: *Becca Richter*
Senior Marketing Manager: *Christine Henry*
Editorial Assistant: *Sean Hale*
Interior Design: *Jonathan Boylan*
Cover Design: *Jonathan Boylan*
Cover Photos: Satellite view of the edge of Bahama Bank courtesy of NASA's Earth Observatory. Sand Pattern Image of Bahama Bank acquired by the Landsat 7 ETM+ satellite. Processed by Institute for Marine Remote Sensing, University of South Florida, courtesy of *Serge Andrefouet/Frank Muller-Karger.*
  Starfish photo courtesy of *Jack and Sue Drafahl*
  Illustration of chemistry in ocean courtesy of *Quade Paul*
  Big Sur Coastline photo courtesy of *David Muench; Corbis Bettmann*
Photo Research: *Yvonne Gerin*
Manager of Formatting: *Jim Sullivan*
Assistant Manager of Formatting: *Allyson Graesser*
Electronic Production Specialist: *Clara Bartunek*
Electronic Page Makeup: *Clara Bartunek, Beth Gschwind & Karen Stephens*

© 2004, 2002, 2001, 1999, 1996 Pearson Education Inc.
Pearson Prentice Hall
Pearson Education, Inc.
Upper Saddle River, New Jersey 07458

Printed in the United States of America

10 9 8 7 6 5 4

ISBN 0-13-143888-3

Pearson Education Ltd., *London*
Pearson Education Australia Pty., Limited, *Sydney*
Pearson Education Singapore, *Pte. Ltd.*
Pearson Education North Asia Ltd., *Hong Kong*
Pearson Education Canada, *Ltd., Toronto*
Pearson Educación de Mexico, *S.A. de C.V.*
Pearson Education—Japan, *Tokyo*
Pearson Education Malaysia, *Pte. Ltd.*

Brief Contents

Introduction   1

1   Planet Ocean: A Historical Perspective   8

2   Origins: Beginnings of the Universe, Earth, and Life   40

3   Plate Tectonics and the Ocean Floor   64

4   Marine Provinces   110

5   Marine Sediments   134

6   Water and Seawater   163

7   Air–Sea Interaction   193

8   Ocean Circulation   229

9   Waves and Water Dynamics   264

10   Tides   294

11   The Coast: Beaches and Shoreline Processes   319

12   Coastal Waters and Marginal Seas   248

13   The Marine Habitat   371

14   Biological Productivity and Energy Transfer   394

15   Animals of the Pelagic Environment   425

16   Animals of the Benthic Environment   457

17   Marine Resources   491

18   Marine Environmental Concerns   518

Afterword   544

Appendix I   546

Appendix II   550

Appendix III   552

Appendix IV   555

Appendix V   559

Appendix VI   568

Glossary   565

Credits and Acknowledgments   592

Index   597

Contents

About the Authors xiii

Preface   xiv

  To the Student   xiv
  To the Instructor   xiv
  What's New in this Edition?   xiv
  The New Instructional Package   xv
  Acknowledgements   xv

Introduction   1

  What Is Oceanography?   2
  Earth's Oceans   4
  Rational Use of Technology?   5
  The Nature of Scientific Inquiry   5
    Observations   6
    Hypothesis   6
    Testing   6
    Theory   6
    Theories and the Truth   6

1   Planet Ocean: A Historical Perspective   8

Key Questions   9
Geography of the Oceans   9
  The Four Principal Oceans, Plus One   10
  The Seven Seas?   11
  Comparing the Oceans to the Continents   11
Early Exploration of the Oceans   12
  Pacific Navigators   12
Box 1–1   How Do Sailors Know where They Are at Sea?:
From Stick Charts to Satellites   14
  Advancements by Mediterranean Cultures: The
    Phoenicians, Greeks, and Romans   13
  The Middle Ages and the Ming Dynasty   16
  European Exploration during the Renaissance   18
The Beginnings of Ocean Science   20
  European Contributions   20
Box 1–2   Charles Darwin and the Voyage of
HMS Beagle   23
  American Contributions   25
Twentieth-Century Oceanography and Beyond   27
  Voyage of the Meteor   28
  World War II and the Expansion of Oceanography   28
  Oceanography Gets Institutionalized   28
  International Cooperation: Research in the World Ocean

Satellite Oceanography   30
A Human Presence in the Ocean   31
  Submersibles: A History   31
Box 1–3   Diving Into the Marine Environment   32
  ROVs and AUVs   33
  Living under the Sea   34
Chapter in Review   38
Key Terms   39
Questions and Exercises   39

2   Origins: Beginnings of the Universe, Earth, and Life   40

Key Questions   41
Origin of the Universe   41
  Light-Year: An Astronomical Distance   42
  Moving Galaxies and the Redshift   42
  The Big Bang   43
Origin of the Solar System and Earth   43
  The Nebular Hypothesis   43
  Protoearth   44
  Density Stratification   46
Origin of the Atmosphere and the Oceans   46
  Origin of the Atmosphere   46
  Origin of the Oceans   47
  Oceans on Other Worlds?   48
  The Development of Ocean Salinity   49
Cycling and Mass Balance   49
  Source of Ocean Water   49
  Sources of Salts in the Ocean   50
  Element Mass Balances   50
Origin of Life   51
  A Working Definition of Life   51
  The Importance of Oxygen to Life   51
Box 2–1   Life on Mars?   52
  The First Organic Substances   53
  The First Organisms   54
  The First Autotrophs   54
  Multicellular Life and Symbiosis   57
  Evolution and Natural Selection   57
  Changes to Earth's Environment   57
Radiometric Dating and the Geologic Time Scale   58
Box 2–2   "Deep" Time   60

Chapter in Review   62
Key Terms   62
Questions and Exercises   62

3 Plate Tectonics and the Ocean Floor 64

Key Questions 65
Evidence for Continental Drift 66
  Fit of the Continents 66
  Matching Sequences of Rocks and Mountain Chains 66
  Glacial Ages and Other Climate Evidence 67
  Distribution of Organisms 68
  Objections to the Continental Drift Model 69
Evidence for Plate Tectonics 70
  Earth's Magnetic Field and Paleomagnetism 70
Box 3–1   Do Sea Turtles (and Other Animals) Use Earth's Magnetic Field for Navigation?   72
  Sea Floor Spreading and Features of the Ocean Basins 74
  Other Evidence from the Ocean Basins 75
  Worldwide Earthquakes 77
  The Acceptance of a Theory 79
Earth Structure 79
  Chemical Composition versus Physical Properties 79
  Near the Surface 81
  Isostatic Adjustment 83
Plate Boundaries 84
  Divergent Boundaries 84
  Convergent Boundaries 89
  Transform Boundaries 84
Testing the Model: Some Applications of Plate Tectonics 94
  Mantle Plumes and Hotspots 94
  Seamounts and Tablemounts 96
Box 3–2   Ophiolites: A Gift from the Sea Floor to the Bronze Age   98
  Coral Reef Development 100
  Detecting Plate Motion with Satellites 102
  The Past: Paleoceanography 102
  The Future: Some Bold Predictions 106
Plate Tectonics … To Be Continued 107
Chapter in Review 108
Key Terms 108
Questions and Exercises 109

4 Marine Provinces 110

Key Questions 111
Bathymetry 111
  Bathymetric Techniques 111
Box 4–1   Sea Floor Mapping from Space   114
The Hypsographic Curve 116
Provinces of the Ocean Floor 117
  Features of Continental Margins 117
  Features of the Deep-Ocean Basin 121

Box 4–2   A Grand "Break": Evidence for Turbidity Currents   122
  Features of the Midocean Ridge 125
Chapter in Review 132
Key Terms 133
Questions and Exercises 133

5 Marine Sediments 134

Key Questions 135
Lithogenous Sediment 135
Box 5–1   Collecting the Historical Record of the Deep-Ocean Floor   136
  Origin 136
  Composition 136
  Sediment Texture 136
  Distribution 140
Biogenous Sediment 144
  Origin 144
  Composition 145
  Distribution 145
Box 5–2   Diatoms: The Most Important Things You Have (Probably) Never Heard Of   147
Hydrogenous Sediment 153
  Origin 153
  Composition and Distribution 153
Cosmogenous Sediment 155
  Origin, Composition, and Distribution 155
Box 5–3   When the Dinosaurs Died: The Cretaceous–Tertiary (K–T) Event   156
Mixtures 157
Distribution of Neritic and Pelagic Deposits: A Summary 158
  Neritic Deposits 158
  Pelagic Deposits 158
Chapter in Review 161
Key Terms 162
Questions and Exercises 162

6 Water and Seawater 163

Key Questions 164
Atomic Structure 164
The Water Molecule 164
  Geometry 165
  Polarity 165
  Interconnections of Molecules 165
  Water: The Universal Solvent 166
Water's Thermal Properties 167
  Heat, Temperature, and Changes of State 167
  Water's Freezing and Boiling Points 167

Water's Heat Capacity   168
Water's Latent Heats   169
Global Thermostatic Effects   170
Water Density   170
Seawater   174
Salinity   174
Salinity Variations   174
Box 6–1   How to Avoid Goiters   176
Determining Salinity   176
Dissolved Components Added and Removed from Seawater   177
Residence Time   179
Ocean Salinity Change through Time   179
Dissolved Gases in Seawater   180
Time to Grab a Soda   180
Gas Exchange between the Atmosphere and Ocean   180
Acidity and Alkalinity of Seawater   181
The pH Scale   182
The Carbonate Buffering System   182
Processes Affecting Seawater Salinity   183
Processes That Decrease Seawater Salinity   183
Processes That Increase Seawater Salinity   183
The Hydrologic Cycle   184
Surface and Depth Salinity Variation   185
Surface Salinity Variation   185
Depth Salinity Variation   185
Seawater Density   187
Pycnocline and Thermocline   188
Comparing Pure Water and Seawater   190
Chapter in Review   190
Key Terms   191
Questions and Exercises   191

7   Air–Sea Interaction   193

Key Questions   194
Uneven Solar Heating on Earth   194
Distribution of Solar Energy   194
Earth's Seasons   195
Oceanic Heat Flow   197
The Atmosphere: Physical Properties   197
An Example: A Nonspinning Earth   199
The Coriolis Effect   200
Example 1: Perspectives and Frames of Reference on a Merry-Go-Round   200
Example 2: A Tale of Two Missiles   201
Changes in the Coriolis Effect with Latitude   202
Atmospheric Circulation Cells on a Spinning Earth   203
Circulation Cells   203
Pressure   203
Wind Belts   203
Boundaries   203

Circulation Cells: Idealized or Real?   205
The Oceans, Weather, and Climate   205
Winds   206
Storms   207
Tropical Cyclones (Hurricanes)   209
Box 7–1   The Storm of the Century: Galveston, Texas (1900)   213
The Ocean's Climate Patterns   214
Sea Ice   215
Sea Ice Formation   215
Sea Ice in the Arctic Ocean   216
Sea Ice near Antarctica   217
Icebergs   217
The RMS *Titanic* Disaster   217
Shelf Ice   219
The Atmosphere's Greenhouse Effect   219
Which Gases Contribute to the Greenhouse Effect?   221
What Changes Will Occur as a Result of Increased Global Warming?   222
The IPCC and the Kyoto Protocol   223
The Ocean's Role in Reducing the Greenhouse Effect   224
What Should We Do About the Increasing Greenhouse Gases?   224
Box 7–2   The ATOC Experiment: SOFAR So Good?   225
Chapter in Review   226
Key Terms   227
Questions and Exercises   228

8   Ocean Circulation   229

Key Questions   230
Measuring Ocean Currents   230
Box 8–1   Running Shoes as Drift Meters: Just Do It   231
Surface Currents   233
Equatorial Currents, Boundary Currents, and Gyres   233
Ekman Spiral and Ekman Transport   235
Geostrophic Currents   237
Western Intensification   237
Equatorial Countercurrents   239
Ocean Currents and Climate   239
Upwelling and Downwelling   240
Diverging Surface Water   241
Converging Surface Water   241
Coastal Upwelling and Downwelling   241
Other Upwelling   242
Surface Currents of the Oceans   242
Antarctic Circulation   242
Atlantic Ocean Circulation   244
Pacific Ocean Circulation   247
Box 8–2   El Niño and the Incredible Shrinking Marine Iguanas of the Galápagos Islands   252
Indian Ocean Circulation   256

Deep Currents   257
   Origin of Thermohaline Circulation   257
   Sources of Deep Water   258
   Worldwide Deep-Water Circulation   259
Chapter in Review   261
Key Terms   262
Questions and Exercises   263

9   Waves and Water Dynamics   264

Key Questions   265
What Causes Waves?   265
How Waves Move   267
Wave Characteristics   268
   Circular Orbital Motion   268
   Deep-Water Waves   269
   Shallow-Water Waves   271
   Transitional Waves   271
Wind-Generated Waves   272
   "Sea"   272
   Swell   276
   Rogue Waves   278
Box 9–1   Yachting in Monster Seas: The Sydney to
Hobart and Fastnet Racing Disasters   279
   Surf   281
   Wave Refraction   283
   Wave Diffraction   284
   Wave Reflection   284
Tsunami   285
   Coastal Effects   286
   Historic Tsunami   287
Box 9–2   The Biggest Wave in Recorded History: Lituya
Bay, Alaska (1958)   288
   Tsunami Warning System   290
Chapter in Review   292
Key Terms   292
Questions and Exercises   293

10   Tides 294

Key Questions   295
Generating Tides   295
   Tide-Generating Forces   295
Equilibrium Theory of Tides   300
   Tidal Bulges: The Moon's Effect   300
   Tidal Bulges: The Sun's Effect   301
   Earth's Rotation   302
   The Monthly Tidal Cycle   302
   Declination of the Moon and Sun   305
   Effects of Elliptical Orbits   306
   Prediction of Equilibrium Tides   306

Dynamic Theory of Tides   308
   Amphidromic Points and Cotidal Lines   308
   Effect of the Continents   309
   Other Considerations   309
Tidal Patterns   309
Tidal Phenomena   311
Box 10–1   Tidal Bores: Boring Waves These Are Not!
312
   Tides in Lakes   313
   Tides in Narrow Basins Connected to the Ocean   313
   An Example of Tidal Extremes: The Bay of Fundy   313
Box 10–2   Grunions: Doing What Comes Naturally on
the Beach   314
   Coastal Tidal Currents   316
Chapter in Review   317
Key Terms   318
Questions and Exercises   318

11   The Coast: Beaches and Shoreline Processes   319

Key Questions   320
The Coastal Region   320
   Beach Terminology   320
   Beach Composition and Slope   321
   Movement of Sand on the Beach   321
Box 11–1   Warning: Rip Currents . . . Do You Know
What to Do?   323
Erosional- and Depositional-Type Shores   325
   Features of Erosional-Type Shores   325
   Features of Depositional-Type Shores   327
Classification of Coasts   333
Emerging and Submerging Shorelines   335
   Tectonic and Isostatic Movements of Earth's Crust   335
   Eustatic Changes in Sea Level   336
   Sea Level and the Greenhouse Effect   336
Characteristics of U.S. Coasts   337
   The Atlantic Coast   338
   The Gulf Coast   338
   The Pacific Coast   338
Hard Stabilization   340
   Groins and Groin Fields   340
   Jetties   341
   Breakwaters   341
   Seawalls   343
   Alternatives to Hard Stabilization   343
Box 11–2   The Move of the Century: Relocating the
Cape Hatteras Lighthouse   345
Chapter in Review   346
Key Terms   347
Questions and Exercises   347

12 Coastal Waters and Marginal Seas   348

Key Questions   349
Coastal Waters   349
  Salinity   349
  Temperature   349
  Coastal Geostrophic Currents   350
Estuaries   351
  Origin of Estuaries   351
  Water Mixing in Estuaries   352
  Estuaries and Human Activities   354
Coastal Wetlands   355
  Serious Loss of Valuable Wetlands   357
Lagoons   357
  Laguna Madre   357
Marginal Seas   358
  Marginal Seas of the Atlantic Ocean   358
  Marginal Seas of the Pacific Ocean   358
Box 12–1   When a Sea Was Dry: Clues from the
  Mediterranean   359
  Marginal Seas of the Indian Ocean   366
Chapter in Review   369
Key Terms   370
Questions and Exercises   370

13 The Marine Habitat   371

Key Questions   372
Classification of Living Things   372
Classification of Marine Organisms   374
  Plankton (Floaters)   374
  Nekton (Swimmers)   375
  Benthos (Bottom Dwellers)   376
Distribution of Life in the Oceans   378
  Why Are There So Few Marine Species?   378
Adaptations of Organisms to the Marine Environment   379
  Need for Physical Support   379
  Water's Viscosity   380
  Temperature   381
  Salinity   382
  Dissolved Gases   385
  Water's High Transparency   385
  Pressure   387
Divisions of the Marine Environment   388
  Pelagic (Open Sea) Environment   388
Box 13–1   A False Bottom: The Deep Scattering Layer
  (DSL)   389
  Benthic (Sea Bottom) Environment   391
Chapter in Review   392
Key Terms   393
Questions and Exercises   393

14 Biological Productivity and Energy Transfer   394

Key Questions   395
Primary Productivity   395
  Photosynthetic Productivity   395
  Measurement of Primary Productivity   396
  Availability of Nutrients   396
  Availability of Solar Radiation   397
  Margins of the Oceans   398
  Light Transmission in Ocean Water   398
Photosynthetic Marine Organisms   402
  Seed-Bearing Plants (Anthophyta)   402
  Macroscopic (Large) Algae   403
  Microscopic (Small) Algae   404
Regional Productivity   405
Box 14–1   Red Tides: Was Alfred Hitchcock's The Birds
  Based on Fact?   406
Box 14–2   Pfiesteria: A Morphing Peril to Fish and
  Humans   407
  Productivity in Polar Oceans   410
  Productivity in Tropical Oceans   411
  Productivity in Temperate Oceans   412
Energy Flow   413
  Energy Flow in Marine Ecosystems   413
  Symbiosis   414
Biogeochemical Cycling   415
  Carbon, Nitrogen, and Phosphorous Cycles   416
  The Silicon Cycle   418
Feeding Relationships   418
  Trophic Levels   419
  Transfer Efficiency   419
  Food Chains, Food Webs, and the Biomass Pyramid   420
  Microbes in the Marine Environment   421
Chapter in Review   423
Key Terms   423
Questions and Exercises   424

15 Animals of the Pelagic Environment   425

Key Questions   426
Staying Above the Ocean Floor   426
  Gas Containers   426
  Floating Heterotrophs (Zooplankton)   427
  Swimming Organisms (Nekton)   432
Box 15–1   Some Myths (and Facts) About Sharks   437
Adaptations for Seeking Prey   438
  Lungers versus Cruisers   438
  Speed and Body Size   438
  Cold-Blooded versus Warm-Blooded   438
  Circulatory System Modifications   439

Adaptations to Avoid Being Prey   439
  Schooling   439
Marine Mammals   440
  Order Carnivora   441
  Order Sirenia   442
  Order Cetacea   442
  An Example of Migration: Gray Whales   451
Box 15–2   Killer Whales: A Reputation Deserved?   453
Chapter in Review   454
Key Terms   455
Questions and Exercises   455

16 Animals of the Benthic Environment   457

Key Questions   458
Rocky Shores   458
  Spray (Supralittoral) Zone   459
  High Tide Zone   462
  Middle Tide Zone   462
  Low Tide Zone   464
Sediment-Covered Shores   465
  The Sediment   465
  Intertidal Zonation   466
  Life in the Sediment   466
  Sandy Beaches   466
  Mud Flats   469
Shallow Offshore Ocean Floor   469
  Rocky Bottoms (Sublittoral)   470
  Coral Reefs   473
Box 16–1   How White I Am: Coral Bleaching and Other Diseases   475
The Deep-Ocean Floor   479
  The Physical Environment   479
  Food Sources and Species Diversity   480
  Deep-Sea Hydrothermal Vent Biocommunities   480
Box 16–2   How Long Would Your Remains Remain on the Sea Floor?   481
  Low-Temperature Seep Biocommunities   485
  The Deep Biosphere   488
Chapter in Review   488
Key Terms   489
Questions and Exercises   490

17 Marine Resources   491

Key Questions   492
Laws and Regulations   492
  Mare liberum and the Territorial Sea   492
  Law of the Sea   492
Ecosystems and Fisheries   493
  Fish Recruitment and Survival   494

  Primary Productivity Effect on Fisheries   494
  Upwelling and Fisheries   494
  World Fishery   494
  Incidental Catch   496
  Fisheries Management   497
Box 17–1   A Case Study in Fisheries Mismanagement: The Peruvian Anchoveta Fishery   498
  Seafood Choices   499
Mariculture   500
  Fish   500
  Crustaceans   502
  Bivalves   502
  Algae   502
Energy Resources   503
  Power from Offshore Winds   503
  Power from Currents   503
  Power from Waves   503
  Power from Tides   505
  Power from Thermal Energy   507
Geologic Resources   509
  Petroleum   509
  Gas Hydrates   510
  Sand and Gravel   510
  Phosphorite (Phosphate Minerals)   511
  Metal Sulfides   511
  Manganese Nodules and Crusts   511
Chemical Resources   513
  Fresh Water   513
  Evaporative Salts   514
  Drugs from the Sea   515
Chapter in Review   516
Key Terms   516
Questions and Exercises   517

18 Marine Environmental Concerns   518

Key Questions   519
What Is Pollution?   519
  Predicting the Effects of Pollution on Marine Organisms   520
Main Types of Marine Pollution   520
  Petroleum   520
  Sewage Sludge   525
Box 18–1   The *Exxon Valdez* Oil Spill: Not the Worst Spill Ever   526
  Radioactive Waste   528
  DDT and PCBs   530
  Mercury and Minamata Disease   531
  Non-Point-Source Pollution and Trash   533
Box 18–2   From A to Z in Plastics: The Miracle Substance?   534

Waste Heat   535
Other Concerns   535
   Non-Native Species   535
   Eutrophication   535
   Ozone Depletion and Phytoplankton Production   537
   Coastal Population Increase and Habitat Destruction   538
Laws Governing Marine Waters   539
   Marine Pollution Control in the United States   539
   International Efforts   539
   Current Laws on Ocean Dumping   540
Chapter in Review   541
Key Terms   542
Questions and Exercises   542

Afterword   544

   Marine Sanctuaries and Marine Reserves   544
   What Can I Do?   544
   Box Aft-1   Ten Simple Things You Can Do to Prevent
   Marine Pollution   546

Appendix I   546

Appendix II   550

Appendix III   552

Appendix IV   555

Appendix V   559

Appendix VI   568

Glossary   565

Credits and Acknowledgements   592

Index   597

About the Authors

Harold V. Thurman

Hal Thurman retired in May 1994, after 24 years of teaching in the Earth Sciences Department of Mt. San Antonio College in Walnut, California. Interest in geology led to a bachelor's degree from Oklahoma A & M University, followed by seven years working as a petroleum geologist, mainly in the Gulf of Mexico, where his interest in the oceans developed. He earned a master's degree from California State University at Los Angeles and then joined the Earth Sciences faculty at Mt. San Antonio College. Other books that Hal has coauthored include *Essentials of Oceanography* (with Alan Trujillo) and a marine biology textbook. He has also written articles on the Pacific, Atlantic, Indian, and Arctic Oceans for the 1994 edition of *World Book Encyclopedia* and served as a consultant on the National Geographic publication, *Realms of the Sea*. He still enjoys going to sea on vacations with his wife Iantha.

Alan P. Trujillo

Al Trujillo teaches at Palomar Community College in San Marcos, California, where he is co-Director of the Oceanography Program and Chair of the Earth Sciences Department. He received his bachelor's degree in geology from the University of California at Davis and his master's degree in geology from Northern Arizona University, afterward working for several years in industry as a developmental geologist, hydrogeologist, and computer specialist. Al began teaching in the Earth Sciences Department at Palomar in 1990 and in 1997 was awarded Palomar's Distinguished Faculty Award for Excellence in Teaching. He has coauthored *Essentials of Oceanography* with Hal Thurman and is a contributing author for other Earth science textbooks, including *Earth*, 7th edition and *Earth Science*, 10th edition. In addition to writing and teaching, Al works as a naturalist and lecturer aboard natural history expedition vessels in Alaska and the Sea of Cortez/Baja California. His research interests include beach processes, sea cliff erosion, and computer applications in oceanography. Al and his wife, Sandy, have two children, Karl and Eva.

Preface

To the Student

Welcome! You're about to embark on a journey that is far from ordinary. Over the course of this term, you will discover the central role the oceans play in the vast global system of which you are a part.

The book's content was carefully developed to provide a foundation in science by examining the vast body of oceanic knowledge. This knowledge includes information from a variety of scientific disciplines—geology, chemistry, physics, and biology—as they relate to the oceans. However, no formal background in any of these disciplines is required to master the subject matter contained within this book. Our desire is to have you take away from your oceanography course much more than just a collection of facts. Instead, we want you to develop a fundamental understanding of *how the oceans work.*

Taken as a whole, the components of the ocean—its sea floor, chemical constituents, physical components, and life forms—comprise one of Earth's largest interacting, interrelated, and interdependent systems. Because humans are beginning to impact Earth systems, it is important to understand not only how the oceans operate, but also how the oceans interact with Earth's other systems (such as its atmosphere, biosphere, and hydrosphere) as part of a larger picture. Thus, this book uses a systems approach to highlight the interdisciplinary relationship among oceanographic phenomena and how those phenomena affect other Earth systems.

To that end—and to help you make the most of your study time—we focused the presentation in this book by organizing the material around three essential components:

1. **Concepts:** General ideas derived or inferred from specific instances or occurrences (for instance, the concept of density can be used to explain why the oceans are layered)
2. **Processes:** Actions or occurrences that bring about a result (for instance, the process of waves breaking at an angle to the shore results in the movement of sediment along the shoreline).
3. **Principles:** Rules or laws concerning the functioning of natural phenomena or mechanical processes (for instance, the principle of sea floor spreading suggests that the geographic positions of the continents have changed through time).

Interwoven within these concepts, processes, and principles are hundreds of photographs, illustrations, real-world examples, and applications that make the material relevant and accessible (and maybe sometimes even *entertaining*) by bringing the science to life.

Ultimately, it is our hope that by understanding how the oceans work, you will develop a new awareness and appreciation of all aspects of the marine environment and its role in Earth systems. So enjoy and immerse yourself! You're in for an exciting ride.

Alan Trujillo
Harold Thurman

To the Instructor

The tenth edition of *Introductory Oceanography* is designed to accompany an introductory college-level course in general or physical oceanography taught to students with no formal background in mathematics or science. Like previous editions, the goal of this edition of the textbook is to present clearly the relationships of scientific principles to ocean phenomena in an engaging and meaningful way. This edition, like its predecessors, emphasizes fundamental marine processes and the basic issues of interactions among humans, the ocean, and the atmosphere.

The text has benefited from the addition of a new coauthor, Alan Trujillo. As a result, the writing style has been improved, several sections have been reorganized, and material has been updated and refined throughout the text. You'll also notice many new features, which are detailed in the "What's New in this Edition?" section below. This edition has also benefited from extensive review by knowledgeable reviewers.

The 18-chapter format of this textbook is designed for coverage of the material in a 15- or 16-week semester. For courses taught on a 10-week quarter system, the instructor may need to select those chapters that cover the topic and concepts of primary relevance to their course. The text also contains enough depth of material to be used for a year-long course of study.

Chapters in the text have been designed to be self-contained and thus can be covered in any order. Following the introductory chapters on the history of oceanography (Chapter 1) and the origin of Earth, the atmosphere, the oceans, and life itself (Chapter 2), the four major academic disciplines of oceanography are represented in the following chapters:

- Geological oceanography: Chapters 3, 4, 5, and parts of Chapters 11, 12, and 17
- Chemical oceanography: Chapters 6 and parts of Chapters 14, 17, and 18
- Physical oceanography: Chapters 7–10 and parts of Chapters 11, 12, and 17
- Biological oceanography: Chapters 13–16 and parts of Chapters 17 and 18

We believe that oceanography is at its best when it links together several scientific disciplines and shows how they are interrelated in the oceans. Therefore, this interdisciplinary approach is a key element of every chapter. In addition, the final two chapters on marine resources and marine environmental concerns illustrate human interaction with the oceans.

What's New in This Edition?

Changes in this edition are designed to increase the readability, relevance, and appeal of this book. The major changes include the following:

- Reorganization, revision, and/or additions in nearly all chapters; the two most prominent improvements are
  - Combining the chapters "The Physical Properties of Water and Seawater" and "The Chemistry of Seawater" from the previous edition into one chapter, "Water and Seawater" (Chapter 6)
  - Splitting the chapter "Exploitation and Pollution of Marine Resources" from the previous edition and adding additional information to create two new chapters, "Marine Resources" (Chapter 17) and "Marine Environmental Concerns" (Chapter 18)
- Addition of a new Introduction, which includes a description of the scientific method
- Addition of Key Questions at the beginning of each chapter that are linked to new highlighted Concept Statements within the text of the chapter
- Explanation of word etymons (*etumon* = the true sense of a word) as new terms are introduced in an effort to demystify scientific terms by showing what the terms actually mean
- Addition of "Students Sometimes Ask . . ." questions within each chapter that contains actual student questions, along with the authors' answers
- Inclusion of 31 new feature boxes that focus on some of the most recent discoveries in oceanography
- Feature boxes within each chapter, which are organized around three new themes:
  - Research methods in oceanography
  - People and the ocean environment
  - Historical features
- Addition or modification of over 35 tables, which organize and summarize important data
- Inclusion of over 100 new photos and illustrations
- Modification or redrawing of over 150 existing figures to add clarity and improve the illustration package
- A revised "Chapter in Review" summary feature at the end of each chapter
- Addition of a new appendix, "Careers in Oceanography"
- Continued effort to refine the style and clarity of the writing

Additionally, this edition continues to offer some of the previous edition's most popular features, including the following:

- Extensive rigor and depth of material, particularly in the coverage of tides and biological processes
- Use of the international metric system (Système International or SI units) with comparable English system units in parenthesis
- Notation of key terms with bold print, which are defined when they are introduced and are included in the glossary
- The end-of-chapter questions and exercises

The New Instructional Package

For the Student

- The New York Times *Themes of the Times–Oceanography* is a unique newspaper-format supplement featuring recent articles about oceanography culled from the pages of the *New York Times*. This supplement, available for wrapping with the text at no charge, encourages you to make connections between what you're learning in the classroom and recent news events as reported in the media.

For the Instructor

- A **Transparency Set** of over 150 acetates of key illustrations from the text, selected by the authors and enlarged for excellent classroom visibility. (*Note*: All figures are available on the Instructor's Resource CD-ROM.)
- The **Instructor's Resource CD-ROM**, which includes all illustrations, tables, and photographs (for which we could acquire permission) from the text in high-resolution, 16-bit JPEG files. The JPEG files are organized by chapter and can be imported easily into lecture presentation software (such as Microsoft PowerPoint). In addition, the IRCD includes animations of key oceanographic processes for use in your lecture. These Flash animations are preloaded into PowerPoint files to make it easier to project them. Also included on the IRCD are the Instructor's Manual and Test Item File, in Microsoft Word format.
- Three types of **PowerPoint files**, which are included on the IRCD: (1) A file for each chapter of all figures, tables, and photographs, in sequence, preloaded into slides; (2) All of the animations, preloaded into slides; and, (3) Customizable lecture presentations, containing text outlines and figures, created by Darlene Richardson, of the Indiana University of Pennsylvania.
- A new **Instructor's Manual**, authored by Darlene Richardson of Indiana University of Pennsylvania, which includes a new Test Item File authored by Catherine A. Teare-Ketter of University of Georgia

Acknowledgments

The authors are indebted to many individuals for their helpful comments and suggestions during the revision of this book. Al Trujillo is particularly indebted to his colleagues at Palomar Community College, Patty Deen and Lisa DuBois, for their keen interest in the project and for allowing him to use some of their creative ideas in the book. They are simply the finest colleagues imaginable.

Many individuals at Scripps Institution of Oceanography have been particularly helpful, especially the staff at *Explorations* magazine. Thanks also go to the many people around the world who helped locate images or willingly donated photographs for use in the text. Without the kind help of these people, the job of tracking down experts and finding images would have been a much larger task.

Many people were instrumental in helping the text evolve from its manuscript stage. In particular, Patrick Lynch, Geoscience Editor at Prentice Hall, expertly guided the project and offered many insightful suggestions to improve the text. Kim Dellas, Production Editor, deserves special recognition for her hard work in turning the manuscript into the book you see today.

Al Trujillo would also like to thank his students, whose questions provided the material for the "Students Sometimes Ask ..." questions and whose continued input has proved invaluable for improving the text. Since scientists (and good teachers) are always experimenting, thanks also for allowing yourselves to be a captive audience with which to conduct my experiments.

Al Trujillo also thanks his patient and understanding family for putting up with his absence during the long hours of preparing this book. Lastly, appreciation is extended to the chocolate manufacturers Hershey, See's, and Ghiradelli for providing inspiration. A heartfelt thanks to all of you!

Many other individuals (including several anonymous reviewers) have provided valuable technical reviews for this and previous works. The following reviewers are gratefully acknowledged:

Franz E. Anderson	*University of New Hampshire*
Tsing Bardin	*San Francisco City College*
John Barlow	*Marine Maritime Academy*
Harold Cones	*Christopher Newport University*
Clay Harris	*Middle Tennessee State University*
Joseph Holliday	*El Camino Community College*
Mary Anne Holmes	*University of Nebraska, Lincoln*
Ron Johnson	*Old Dominion University*
Bjorn Kjerfve	*University of South Carolina*
Charles E. Knowles	*North Carolina State University*
Michael Lane	*Kapiolani Community College*
Stephen A. Macko	*University of Virginia, Charlottesville*
James M. McWhorter	*Miami-Dade Community College*
Darrell A. Milburn	*Tulane University*
John Mylroie	*Mississippi State University*
Bruce Railsback	*University of Georgia*
Arthur Snoke	*Virginia Polytechnic Institute*
Richard W. Spinrad	*George Mason University*
Piyush Srivastav	*University of Nebraska, Lincoln*
Chellie Teal	*Wichita State University*
Katryn Weise	*San Francisco City College*
John H. Wormuth	*Texas A&M University*

Although this book has benefited from careful review by many individuals, the accuracy of the information rests with the authors. If you find errors or have comments about the text, please contact us at

Al Trujillo
Department of Earth Sciences
Palomar College
1140 W. Mission Rd.
San Marcos, CA 92069
(760) 744-1150 ext. 2734
atrujillo@palomar.edu
Web: http://daphne.palomar.edu/atrujillo

Hal Thurman
17580 SE 88th Covington Circle
The Villages, FL 32162
(352) 751-0822
ilhvt@thevillages.net

Alan Trujillo
Harold Thurman

Introduction

"Oceanography is not so much a science as a collection of scientists who find common cause in trying to understand the complex nature of the ocean. In the vast salty seas that encompass the earth, there is plenty of room for persons trained in physics, chemistry, biology, and engineering to practice their specialties. Thus, an oceanographer is any scientifically trained person who spends much of his [or her] career on ocean problems."

—Willard Bascom, oceanographer and explorer (1980)

Welcome to a book about the oceans. As you read this book, we hope that it elicits a sense of wonder and a spirit of curiosity about our watery planet. The oceans represent many different things to different people. To some, it is a wilderness of beauty and tranquility, a refuge from hectic civilized lives. Others see it as a vast recreational area that inspires either rest or physical challenge. To others, it is a mysterious place that is full of unknown wonders. And to others, it is a place of employment unmatched by any on land. To be sure, its splendor has inspired artists, writers, and poets for centuries (Figure I-1). Whatever your view, we hope that understanding the way the oceans work will increase your appreciation of the marine environment. Above all, take time to admire the oceans.

Introductory Oceanography was written to help students develop an *awareness about the marine environment*—that is, develop an appreciation for the oceans by learning about oceanic processes (how the oceans behave) and their interrelationships (how physical entities are related to one another in the oceans). In this tenth edition, our goal is the same: to give the reader the scientific background to understand the basic principles underlying oceanic phenomena. In this way, one can then make informed decisions about the oceans in the years to come. We hope that some of you will be inspired so much by the oceans that you will continue to study them formally or informally in the future (for those who may be considering a life-long career in oceanography, see Appendix V, "Careers in Oceanography," for some tips and practical advice).

What Is Oceanography?

Oceanography (*ocean* = the marine environment, *graphy* = the name of a descriptive science) is literally the description of the marine environment. Unfortunately, this definition does not fully portray the extent of what oceanography encompasses: oceanography is much more than just *describing* marine phenomena. Oceanography could be more accurately called the scientific study of all aspects of the marine environment. Hence, the field of study called oceanography could (and maybe *should*) be called oceanology (*ocean* = the marine environment, *ology* = the study of) However, the science of studying

Figure I–1 The ocean environment at Jalama Beach, California.

the oceans has traditionally been called oceanography. It is also called *marine science* and includes the study of the water of the ocean, the life within it, and the (not so) solid earth beneath it.

Since prehistoric time, people have used the oceans as a means of transportation and as a source of food. However, the importance of ocean processes has been studied technically only since the 1930s. The impetus for this study began with the search for petroleum, continued with the emphasis on ocean warfare during World War II, and more recently has been expressed in the concern for the well-being of the ocean environment. Historically, those who make their living fishing in the ocean go where the physical processes of the oceans offer good fishing. But how marine life interrelates with ocean geology, chemistry, and physics to create good fishing grounds has been more or less a mystery until only recently when scientists in these disciplines began to investigate the oceans with new technology.

Oceanography is typically divided into different academic disciplines (or subfields) of study. The four main disciplines of oceanography that are covered in this book are as follows:

- *Geological oceanography,* which is the study of the structure of the sea floor and how the sea floor has

changed through time; the creation of sea floor features; and the history of sediments deposited on it.
- *Chemical oceanography,* which is the study of the chemical composition and properties of seawater; how to extract certain chemicals from seawater; and the effects of pollutants.
- *Physical oceanography,* which is the study of waves, tides, and currents; the ocean-atmosphere relationship that influences weather and climate; and the transmission of light and sound in the oceans.
- *Biological oceanography,* which is the study of the various oceanic life forms and their relationships to one another; adaptations to the marine environment; and developing ecologically sound methods of harvesting seafood.

Other disciplines include ocean engineering, marine archaeology, and marine policy. Since the study of oceanography often examines in detail all the different disciplines of oceanography, it is frequently described as being an *interdisciplinary* science, or one covering all the disciplines of science as they apply to the oceans (Figure I-2). The content of this book includes the broad range of interdisciplinary science topics that comprises the field of oceanography. In essence, this is a book about *all* aspects of the oceans.

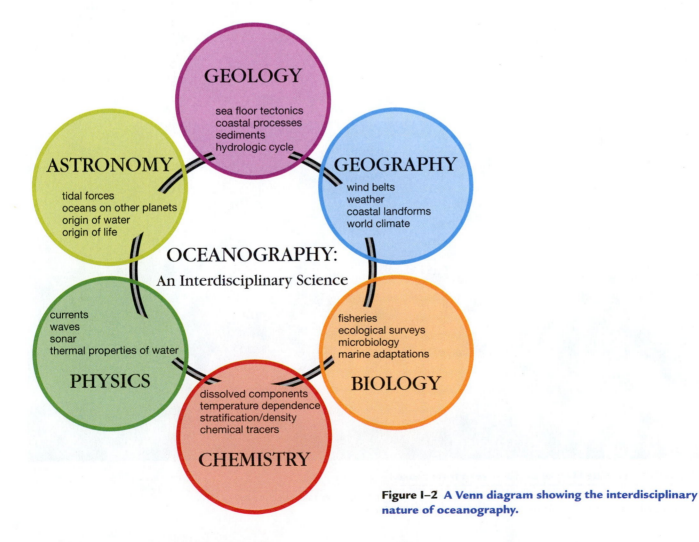

Figure I–2 **A Venn diagram showing the interdisciplinary nature of oceanography.**

Earth's Oceans

The oceans are the largest and most prominent feature on Earth. In fact, they are the single most defining feature of our planet. As viewed from space, our planet is a beautiful blue, white, and brown globe (Figure I-3). It is our oceans of liquid water that sets us apart in the solar system. No other planet has an ocean (however, a recent discovery of fluid-filled cracks on some of Jupiter's moons—most notably Europa—has lead to speculation that there may be an ocean of liquid water beneath the ice!). The fact that our planet has so much water, *and in the liquid form*, is unique in the solar system.

The oceans determine where our continents end, and thus have shaped political boundaries and human history many times. The oceans conceal many features; in fact, the majority of Earth's geographic features are on the ocean floor and 80% of Earth's volcanic activity takes place underwater. Remarkably, there was once more known about the surface of the moon than about the floor of the oceans! Fortunately, over the past several decades, our knowledge of both has increased dramatically.

The oceans influence weather all over the globe, even in continental areas far from any ocean. The oceans are also the lungs of the planet, taking carbon dioxide gas (CO_2) out of the atmosphere and replacing it with oxygen gas (O_2). Some scientists have estimated that the oceans supply as much as 70% of the oxygen humans breathe.

The oceans are in large part responsible for the development of life on Earth, providing a stable environment in which life could evolve over millions of years. Today, the oceans contain the greatest number of living things on the planet, from microscopic bacteria and algae to the largest life form alive today (the blue whale). Interestingly, water is the major component of nearly every life form on Earth, and our own body fluid chemistry is remarkably similar to the chemistry of seawater.

Figure I–3 The Blue Marble: Earth as seen from space.

This photo of Earth, showing its interrelated atmosphere, oceans, and land, was taken by the *Apollo 17* crew on humankind's last trip to the Moon in December 1972.

The oceans hold many secrets waiting to be discovered, and new discoveries about the oceans are made nearly every day. The oceans are a source of food, minerals, and energy that remain largely untapped. Over half of the world population lives in coastal areas near the oceans, taking advantage of the mild climate, an inexpensive form of transportation, and vast recreational opportunities. And unfortunately, the oceans are also the dumping ground for many of society's wastes.

Rational Use of Technology?

Many stresses have been put on the oceans by an ever-increasing human population. For instance, population studies reveal that over 50% of world population—some 3.2 billion people—live along the coastline and over 80% of all Americans live within an hour's drive from an ocean or the Great Lakes. In the future, these figures are expected to increase (in the United States, eight of the 10 largest cities are in coastal environments and 3600 people move to the coast every day). This migration to the coasts will further mar the delicate natural balance that exists in the coastal ocean (Figure I-4). Specifically, the migration is resulting in more harbor and channel dredging, industrial waste and sewage disposal at sea, chemical spills, cooling of power plants with seawater, and the destruction of wetlands that are vital to the cleansing of runoff waters and to the maintenance of coastal fisheries.

Although it may seem as if humans have severely and irreversibly damaged the oceans, our impact has been felt mostly in coastal areas. The world's oceans are a vast resource that has not yet been lethally damaged. Humans have been able to inflict only minor damage here and there along the margins of the oceans. However, as our technology makes us more powerful, the threat of irreversible harm becomes greater. For instance, in the open ocean (those areas far from shore), deep-ocean mining and nuclear waste disposal have been proposed. How do we as a society deal with these increased demands on the marine environment? How do we regulate the ocean's use?

If used wisely, our technology can actually reduce the threat of irreversible harm. Which path will we take? We all need to evaluate carefully our own actions and the effects those actions have on the environment. In addition, we need to make conscientious decisions about those we elect to public office. Some of you may even have direct responsibility for initiating legislation that affects our environment. It is our hope that you, as a student of the marine environment, will gain enough knowledge while studying oceanography to help your community (and perhaps even your nation) make rational use of technology in the oceans.

The Nature of Scientific Inquiry

In modern society, scientific studies are increasingly being used to substantiate the need for action. However, there is often little understanding of how science operates. For instance, how certain are we about a particular scientific theory? How are facts different from theories?

The overall goal of science is to discover the underlying patterns in the natural world and then to use this knowledge to make predictions about what should or should not be expected to happen given a certain set of circumstances. Scientists develop explanations about the causes and effects of various natural phenomena (such as why Earth has seasons or what the structure of matter is).

Figure I–4 Heavily developed coastline of Miami Beach, Florida.

This work is based on an assumption that all natural phenomena are controlled by understandable physical processes and the same physical processes operating today have been operating throughout time. Consequently, science has demonstrated remarkable power in allowing scientists to describe the natural world accurately, to identify the underlying causes of natural phenomena, and to better predict future events that rely on natural processes.

Science supports the explanation of the natural world that best explains all available observations. Scientific inquiry is formalized into what is called the **scientific method**, which is used to formulate scientific theories (Figure I-5).

Observations

The scientific method begins with *observations*, which are occurrences we can measure with our senses. They are things we can manipulate, see, feel, hear, touch, or smell, often by experimenting with them directly or by using sophisticated tools (such as a microscope or telescope) to sense them. If an observation is repeatedly confirmed—that is, made so many times that it is assumed to be completely valid—then it can be called a *scientific fact*.

Hypothesis

As observations are being made, the human mind attempts to sort out the observations in a way that reveals some underlying order or pattern in the objects or phenomena being observed. This sorting process—which involves a great deal of trial and error—seems to be driven by a fundamental human urge to make sense of our world. This is how **hypotheses** (*hypo* = under, *thesis* = an arranging) are made. A hypothesis is sometimes labeled as an informed or educated guess, but it is more than that. A hypothesis is a tentative, testable statement about the general nature of the phenomena observed. In other words, a hypothesis is an initial idea of how or why things happen in nature.

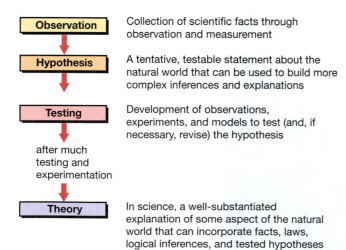

Observation	Collection of scientific facts through observation and measurement
Hypothesis	A tentative, testable statement about the natural world that can be used to build more complex inferences and explanations
Testing	Development of observations, experiments, and models to test (and, if necessary, revise) the hypothesis
after much testing and experimentation	
Theory	In science, a well-substantiated explanation of some aspect of the natural world that can incorporate facts, laws, logical inferences, and tested hypotheses

Figure I–5 The scientific method.

Suppose we wanted to understand why whales *breach* (that is, why whales sometimes leap entirely out of water). After scientists observe breaching many times, they can organize their observations into a hypothesis. For instance, one hypothesis is that a breaching whale is trying to dislodge parasites from its body. Scientists often have multiple working hypotheses (for example, whales may use breaching to communicate with other whales). If a hypothesis cannot be tested, it is not scientifically useful no matter how interesting it might seem.

Testing

Hypotheses are used to predict certain occurrences that lead to further research and the refinement of those hypotheses. For instance, the hypothesis that a breaching whale is trying to dislodge its parasites suggests that breaching whales have more parasites than whales that don't breach. Analyzing the number of parasites on breaching versus nonbreaching whales would either support that hypothesis or cause it to be recycled and modified.

In science, the validity of any explanation is determined by its coherence with observations in the natural world and its ability to predict further observations. Only after much testing and experimentation—usually done by many experimenters using a wide variety of repeatable tests—does a hypothesis gain validity where it can be advanced to the next step.

Theory

If a hypothesis has been strengthened by additional observations and if it is successful in predicting additional phenomena, then it can be advanced to what is called a **theory** (*theoria* = a looking at). A theory is a well-substantiated explanation of some aspect of the natural world that can incorporate facts, laws (descriptive generalizations about the behavior of an aspect of the natural world), logical inferences, and tested hypotheses. A theory is not a guess or a hunch. Rather, it is an understanding that develops from extensive observation, experimentation, and creative reflection.

In science, theories are only formalized after many years of testing and verifying predictions. Thus, scientific theories are those that have been rigorously scrutinized to the point where most scientists agree that they are the best explanation of certain observable facts. Examples of prominent, well-accepted theories that are held with a very high degree of confidence include geology's theory of plate tectonics (which will be covered in Chapter 3) and biology's theory of evolution.

Theories and the Truth

We've seen how the scientific method is used to develop theories, but does science ever arrive at the undisputed "truth"? Science never reaches an absolute truth because we can never be certain that we have all the observations, especially considering that new technology will be available in the future to examine phenomena in different ways. New observations are always possible, so the nature of scientific truth is subject to change. Therefore, it is more accurate to say that science arrives at that which is *probably* true, based on the available observations.

It is not a downfall of science that scientific ideas are modified as more observations are collected. In fact, the opposite is true. Science is a process that depends on re-examining ideas as new observations are made. Thus, science progresses when new observations yield new hypotheses and modification of theories. As a result, science is littered with hypotheses that have been abandoned in favor of later explanations that fit new observations. One of the best known is the idea that Earth was at the center of the universe, a proposal that was supported by the apparent daily motion of the Sun, Moon, and stars around Earth.

The statements of science should never be accepted as the "final truth." Instead, over time, they generally form a sequence of increasingly more accurate statements. Theories are the end-points in science and do not turn into facts through the accumulation of evidence. Nevertheless, the data can become so convincing that the accuracy of a theory is no longer questioned. For instance, the *heliocentric* (*helios* = sun, *centric* = center) theory of our solar system states that Earth revolves around the Sun rather than vise versa. Such concepts are supported by such abundant observational and experimental evidence that they are no longer questioned in science.

Is there really such a formal method to science as the scientific method suggests? Actually, the work of scientists is much less formal and is not always done in a clearly logical and systematic manner. Like a detective analyzing a crime scene, a scientist uses ingenuity and serendipity, visualizes models, and sometimes follows hunches in order to unravel the mysteries of nature.

"Our environment, our health, our economic prospects, our national defense, the foods we eat, and the air that we breathe— even our genetic future—will depend upon how wisely we apply the technologies that become available. And to do this we need a population of scientists, but also of citizens, of workers, of administrators, of policy makers ... who can grasp the scientific way of thinking."

—Leon M. Lederman, Nobel Laureate and President of the American Association for the Advancement of Science (2000)

CHAPTER 1
Planet Ocean: A Historical Perspective

John Harrison's chronometer H1. In the late 1700s, the development of accurate timepieces for use at sea was a major advancement in new technology that allowed more precise determination of position at sea and opened further exploration of the world ocean. The first of Harrison's chronometers was driven by a helical balance spring that remained horizontal and independent of ship motion.

Key Questions

- What are the four principal oceans on Earth?
- Where is the deepest part of the oceans, and have humans ever visited there?
- How was early exploration of the oceans achieved?
- What was accomplished during the Age of Discovery?
- When were the first voyages for scientific knowledge?
- What tools allow researchers to study the oceans?

The answers to these questions (and much more) can be found in the highlighted concept statements within this chapter.

"Strange and beautiful things were brought to us from time to time which seemed to give us a glimpse of the edge of some unfamiliar world."

—C. Wyville Thomson, *The* Challenger *Expedition* (1876)

In spite of the ocean's huge extent over the surface area of Earth, it has not prevented humans from exploring its furthest reaches. Since early times, humans have developed technology that has allowed civilizations to travel across large stretches of open ocean. Today, we can cross an ocean in a day by plane. Even so, much of the deep ocean remains out of reach and woefully unexplored. In fact, the surface of the Moon has been more accurately mapped than most parts of the sea floor. Yet satellites at great distances above Earth are being used to gain knowledge about our watery home.

It seems perplexing that our planet is called "Earth" when 70.8% of its surface is covered by oceans. Many early human cultures that lived near the Mediterranean (*medi* = middle, *terra* = land) Sea envisioned the world as composed of large landmasses surrounded by marginal bodies of water (Figure 1–1). How surprised they must have been when they ventured into the larger oceans of the world. Our planet is misnamed "Earth" because we live on the land portion of the planet. If we were marine animals, our planet would probably be called "Ocean," "Water," "Hydro," "Aqua," or even "Oceanus," to indicate the prominence of Earth's oceans.

Geography of the Oceans

A world map (Figure 1–2) shows how extensive our oceans really are. Notice that *the oceans dominate the surface area of the globe*. If you have ever traveled by boat across an ocean (or flown across one in an airplane), you soon realize that the oceans are enormous. Notice, also, that *the oceans are interconnected* and form a single continuous body of seawater, which is why the oceans are commonly referred to as a "world ocean" (singular, not plural). For instance, a vessel at sea can travel from one ocean to another, whereas it is impossible to travel on land from one continent to most others without crossing an ocean. In addition, the oceans contain 97.2% of all the water on or near Earth's surface, so *the volume of the oceans is immense*.

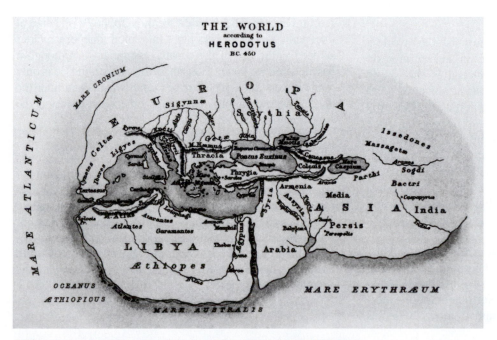

Figure 1–1 An early map of the world by Herodotus (450 B.C.).

The world according to the Greek Herodotus in 450 B.C., showing the prominence of the Mediterranean Sea surrounded by the continents of Europe, Libya (Africa), and Asia. Seas are labeled "mare" and are shown as a band of water encircling the land.

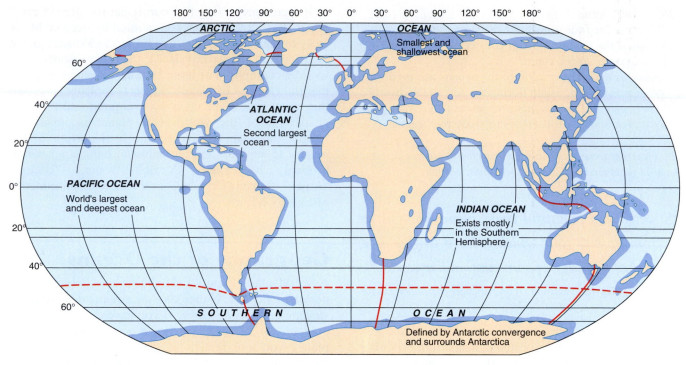

Figure 1–2 **Earth's oceans.**

Map showing the four principal oceans, plus the Southern or Antarctic Ocean. Dark blue shading represents shallow areas.

The Four Principal Oceans, Plus One

Our world ocean can be divided into four principal oceans (plus an additional ocean), based on the shape of the ocean basins and the positions of the continents (Figure 1–2).

Pacific Ocean The **Pacific Ocean** is the world's largest, covering over half of the ocean surface area on Earth (Figure 1–3b). The Pacific Ocean is the single largest geographic feature on the planet, spanning over one-third of Earth's entire surface. The Pacific Ocean is so large that *all* of the continents could fit into the space occupied by it—with room left over! Although the Pacific Ocean is also the deepest ocean in the world (Figure 1–3c), it contains many small tropical islands. It was named in 1520 by explorer Ferdinand Magellan's party, in honor of the fine weather they encountered while crossing into the Pacific (*paci* = peace) Ocean.

Atlantic Ocean The **Atlantic Ocean** is about half the size of the Pacific Ocean and is not quite as deep (Figure 1–3). It separates the Old World (Europe, Asia, and Africa) from the New World (North and South America). The Atlantic Ocean was named after the Atlas Mountains in northwest Africa.

Indian Ocean The **Indian Ocean** is slightly smaller than the Atlantic Ocean and has about the same average depth (Figure 1–3). It resides mostly in the Southern Hemisphere (south of the Equator, or below 0 degrees latitude in Figure 1–2). The Indian Ocean was named for its proximity to the subcontinent of India.

Arctic Ocean The **Arctic Ocean** is about 7% the size of the Pacific Ocean and is only a little more than one-quarter as deep as the rest of the oceans (Figure 1–3). Although it has a permanent layer of sea ice at the surface, the ice is only a few meters thick. The Arctic Ocean was named after the northern constellation Ursa Major, otherwise known as the Big Dipper, or the Bear (*arktos* = bear).

Southern Ocean or Antarctic Ocean Oceanographers recognize an additional ocean near the continent of Antarctica in the Southern Hemisphere (Figure 1–2). Defined by the meeting of currents near Antarctica called the Antarctic Convergence, the **Southern Ocean** or **Antarctic Ocean** is really the portions of the Pacific, Atlantic, and Indian Oceans south of about 50 degrees south latitude. This ocean was named for its location in the Southern Hemisphere.

The four principal oceans are the Pacific, Atlantic, Indian, and Arctic Oceans. An additional ocean, the Southern or Antarctic Ocean, is also recognized.

The Seven Seas?

"Sailing the seven seas" is a familiar phrase in literature and songs, but are there really seven seas? If there are four principal oceans (plus one), what is the difference between a sea and an ocean? In common use, the terms "sea" and "ocean" are often used interchangeably. For instance, a *sea* star lives in the *ocean*; the *ocean* is full of *sea*water; *sea* ice forms in the *ocean*; and one might stroll the *sea*shore while living in *ocean*-front property. Technically, however, seas are defined as follows:

- Smaller and shallower than an ocean (this is why the Arctic Ocean might be more appropriately considered a sea)
- Composed of salt water (many "seas," such as the Caspian Sea in Asia, are actually large fresh water lakes)
- Somewhat enclosed by land (but some seas, such as the Sargasso Sea in the Atlantic Ocean, are defined by strong ocean currents rather than by land)

Although the definition of a sea leaves some room for interpretation, it does allow us to answer the question about the seven seas. If we count the oceans as seas and split the Pacific Ocean and the Atlantic Ocean arbitrarily at the Equator, then we have seven seas: the North Pacific, the South Pacific, the North Atlantic, the South Atlantic, the Indian, the Arctic, and the Southern or Antarctic.

Comparing the Oceans to the Continents

Figure 1–3d shows that the average depth of the world's oceans is 3729 meters[1] (12,234 feet). This means that there must be some extremely deep areas in the ocean to offset the shallow areas close to shore. Figure 1–3d also shows that

---

[1]Throughout this book, metric measurements are used (and the corresponding English measurements are in parentheses). See Appendix I, "Metric and English Units Compared," for conversion factors between the two systems of units.

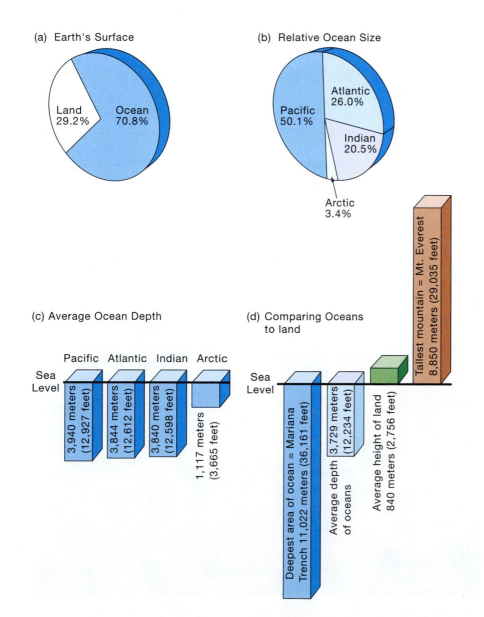

(a) Earth's Surface

Land 29.2%
Ocean 70.8%

(b) Relative Ocean Size

Pacific 50.1%
Atlantic 26.0%
Indian 20.5%
Arctic 3.4%

(c) Average Ocean Depth

Sea Level

Pacific — 3,940 meters (12,927 feet)
Atlantic — 3,844 meters (12,612 feet)
Indian — 3,840 meters (12,598 feet)
Arctic — 1,117 meters (3,665 feet)

(d) Comparing Oceans to land

Sea Level

Deepest area of ocean = Mariana Trench 11,022 meters (36,161 feet)
Average depth of oceans 3,729 meters (12,234 feet)
Average height of land 840 meters (2,756 feet)
Tallest mountain = Mt. Everest 8,850 meters (29,035 feet)

Figure 1–3 Ocean size and depth.

(a) Relative proportions of land and ocean on Earth's surface. **(b)** Relative size of the four principal oceans. **(c)** Average ocean depth. **(d)** Comparing average and maximum depth of the oceans to average and maximum height of land.

the deepest depth in the oceans (the Challenger Deep region of the Mariana Trench, which is near Guam) is a staggering 11,022 meters (36,161 feet) below sea level.

Humans have indeed visited the deepest part of the oceans—where there is crushing high pressure, complete darkness, and near-freezing water temperatures—over 40 years ago! In January 1960, U.S. Navy Lt. Don Walsh and explorer Jacques Piccard descended to the bottom of the Challenger Deep region of the Mariana Trench in the *Trieste*, a deep diving bathyscaphe (*bathos* = depth, *scaphe* = a small ship) (Figure 1–4). At 9906 meters (32,500 feet), the men heard a loud cracking sound that shook the cabin. They were unable to see that a 7.6-centimeter (3-inch) Plexiglas viewing port had cracked (miraculously, it held for the rest of the dive). More than five hours after leaving the surface, they reached the bottom at 10,912 meters (35,800 feet)—a record depth of human descent that has not been broken since. They did see some life forms that are adapted to life in the deep: a small flatfish, a shrimp, and some jellyfish.

By comparison, Figure 1–3d shows that the average height of the continents is only 840 meters (2756 feet). The average height of the land is not very far above sea level. The highest mountain in the world (the mountain with the greatest height above sea level) is Mount Everest in the Himalaya Mountains of Asia at 8850 meters (29,035 feet). Even so, it is a full 2172 meters (7126 feet) shorter than the Mariana Trench is deep. The mountain with the *greatest total height* from base to top is Mauna Kea on the island of Hawaii in the United States. It measures 4206 meters (13,800 feet) above sea level and 5426 meters (17,800 feet) from sea level down to its base, for a total height of 9632 meters (31,601 feet). The total height of Mauna Kea is 782 meters (2566 feet) higher than Mount Everest, but it is still 1390 meters (4560 feet) shorter than the Mariana Trench is deep. No mountain on Earth is taller than the Mariana Trench is deep.

> The deepest part of the ocean is the Mariana Trench in the Pacific Ocean. It is 11,022 meters (36,161 feet) deep and was visited by humans in 1960 in a specially designed deep diving bathysphere.

Early Exploration of the Oceans

Humankind probably first viewed the oceans as a source of food. Later, vessels were built to move upon the ocean's surface and transport ocean-going people to new fishing grounds. The oceans also provided an inexpensive and efficient way to move large and heavy objects, facilitating trade and interaction between cultures.

Pacific Navigators

It is not known who first developed navigation, but it may have been ancestors of Pacific Islanders. The peopling of the Pacific Islands (Oceania) is somewhat perplexing because there is no evidence that people actually evolved on these islands. Their presence required travel over hundreds or even thousands of miles of open ocean from the continents, probably in small vessels of that time (double canoes, outrigger canoes, or balsa rafts). The islands in the Pacific Ocean are

Figure 1–4 The U.S. Navy's bathyscaphe *Trieste*.

The *Trieste* suspended on a crane before its record-setting deep dive in 1960. The 1.8-meter (6-foot) diameter diving chamber (round ball below the float) accommodated two people and had steel walls 7.6 centimeters (3 inches) thick.

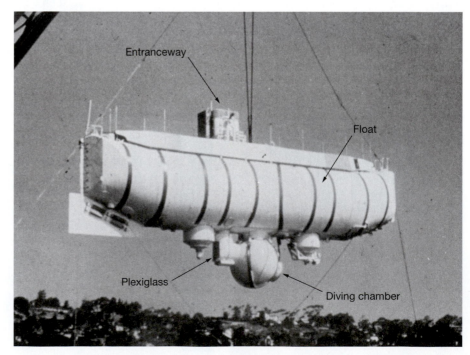

Entranceway

Float

Plexiglass

Diving chamber

widely scattered, so it is likely that only a fortunate few of the voyagers made landfall and that many others perished during the voyage. Figure 1–5 shows the three major island regions in the Pacific Ocean: Micronesia (*micro* = small, *nesia* = islands), Melanesia (*melan* = black, *nesia* = islands), and Polynesia (*poly* = many, *nesia* = islands), which covers the largest area.

No written records of Pacific human history exist before the arrival of Europeans in the 16th century. Nevertheless, the movement of Asian peoples into Micronesia and Melanesia is easy to imagine because distances between islands are relatively short. In Polynesia, however, large distances separate island groups, which must have presented great challenges to ocean voyagers (Box 1–1). Easter Island, for example, at the southeastern corner of the triangular-shaped Polynesian islands region, is over 1600 kilometers (1000 miles) from Pitcairn Island, the next nearest island. Clearly, a voyage to the Hawaiian Islands must have been one of the most difficult because they are 3000 kilometers (1860 miles) from the nearest inhabited islands, the Marquesas Islands (Figure 1–5).

Archeological evidence suggests that humans from New Guinea may have occupied New Ireland as early as 4000 or 5000 B.C. However, there is little evidence of human travel farther into the Pacific Ocean before 1100 B.C. By then, pottery makers called the *"Lapita people"* had traveled on to Fiji, Tonga, and Samoa. From there, Polynesians sailed on to the Marquesas (30 B.C.). The Marquesas appear to have been the starting point for voyages to Easter Island (400 A.D.), the Hawaiian Islands (500 A.D.), and New Zealand (900 A.D.).

Archeological evidence from the eastern Pacific confirms human occupation of the Marquesas Islands by 300 A.D. Although the Maori of New Zealand, the Hawaiians, and the Easter Islanders of the 16th century were Polynesian, there is no clear evidence to explain how these peoples arrived at these destinations.

Thor Heyerdahl, an adventurous biologist/anthropologist, proposed that voyagers from South America may have reached islands of the South Pacific before the coming of the Polynesians. In 1947, he sailed the ***Kon Tiki***—a balsa raft designed like those that were used by South American navigators at the time of European discovery—from South America to the Tuamotu Islands, a journey of over 11,300 kilometers (7000 miles) (Figure 1–5). Although the voyage of the *Kon Tiki* demonstrated that early South Americans could have traveled to Polynesia just as easily as early Asian cultures, no anthropological evidence exists of such a migration. Further, comparative DNA studies show a strong genetic relationship between the peoples of Easter Island and Polynesia, but none between these groups and natives in coastal North or South America.

Advancements by Mediterranean Cultures: The Phoenicians, Greeks, and Romans

The first humans from the Western Hemisphere known to have developed the art of navigation were the *Phoenicians*, who lived at the eastern end of the Mediterranean Sea, in the present-day area of Egypt, Syria, Lebanon, and Israel. As early as 2000 B.C., they explored the Mediterranean Sea, the Red Sea, and the Indian Ocean. The first recorded circumnavigation of Africa, in 590 B.C., was made by the Phoenicians, who had also sailed as far north as the British Isles.

Herodotus mapped what the ancient *Greeks* knew of the world in 450 B.C. (see Figure 1–1). His map shows the Mediterranean Sea surrounded by three continents that, in turn, are surrounded by a continuous ocean.

The Greek astronomer-geographer **Pytheas** sailed northward to Iceland in 325 B.C. using a simple yet elegant method for determining latitude (one's position north or south) in the Northern Hemisphere. His method involved measuring the angle between an observer's line of sight to the North Star and line of sight to the northern horizon (see Appendix III, "Latitude and Longitude on Earth"). Despite Pytheas's method for determining latitude, it was still impossible to accurately determine longitude (one's position east or west).

One of the key repositories of scientific knowledge at the time was the **Library of Alexandria** in Alexandria, Egypt, which was founded in the 3rd century B.C. by

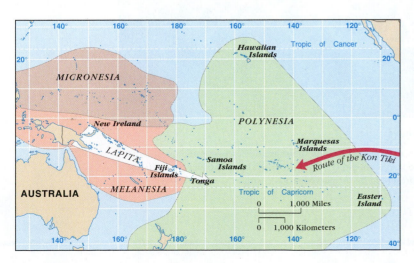

Figure 1–5 The peopling of the Pacific islands.

The major island groups of the Pacific Ocean are Micronesia (*brown shading*), Melanesia (*red shading*), and Polynesia (*green shading*). The "Lapita people" present in New Ireland 5000–4000 B.C. can be traced to Fiji, Tonga, and Samoa by 1100 B.C. The route of Thor Heyerdahl's balsa raft *Kon Tiki* is also shown.

How do you know where you are in the ocean without roads, signposts, or any land in sight? How do you determine the distance to a destination? How do you find your way back to a good fishing spot or where you have discovered sunken treasure? Sailors have relied on a variety of navigation tools to help answer questions such as these by locating where they are at sea.

Some of the first navigators were the Polynesians. Remarkably, the Polynesians were able to successfully navigate to small, distant islands located at great distances across the Pacific Ocean. These early navigators must have been very aware of the marine environment and been able to read subtle differences in the ocean and sky. The tools they used to help them navigate between islands included the Sun and Moon, the nighttime stars, the behavior of marine organisms, various ocean properties, and an ingenious device called a *stick chart* (Figure 1A). These stick charts accurately depicted the consistent pattern of ocean waves. By orienting their vessels relative to this regular ocean wave direction, sailors could successfully navigate at sea. The bent wave directions let them know when they were getting close to an island—even one that was located beyond the horizon.

The importance of knowing where you are at sea is illustrated by a tragic incident in 1707, when a British battle fleet was over 160 kilometers (100 miles) off course and ran aground in the Isles of Scilly near England, with the loss of four ships and nearly 2000 men. **Latitude** (location north or south) was relatively easy to determine at sea by measuring the position of the Sun and stars using a device called a *sextant* (*sextant* = sixth, in reference to the instrument's arc, which is one-sixth of a circle) (Figure 1B). The accident occurred because the ship's crew had no way of keeping track of their **longitude** (location east or west; see Appendix III, "Latitude and Longitude on Earth"). To determine longitude, which is a function of time, it was necessary to know the time difference between a reference meridian and when the Sun was directly overhead of a ship at sea (noon local time). The pendulum-driven

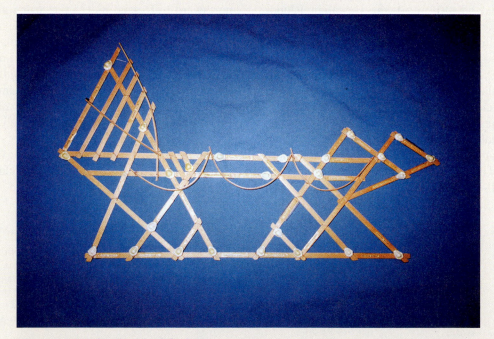

Figure 1A Navigational stick chart.

This bamboo stick chart of Micronesia's Marshall Islands shows islands (represented by shells at the junctions of the sticks), regular ocean wave direction (represented by the straight strips), and waves that bend around islands (represented by the curved strips). Similar stick charts were used by early Polynesian navigators.

Figure 1B Using a sextant.

This hand-held sextant is similar to the ones used by early navigators to determine latitude.

clocks in use in the early 1700s, however, would not work for long on a rocking ship at sea. In 1714, the British Parliament offered a £20,000 prize (about $20 million today) for developing a device that would work well enough at sea to determine longitude within half a degree or 30 nautical miles (34.5 statute miles) after a voyage to the West Indies.

A cabinetmaker in Lincolnshire, England, named **John Harrison**, began working on such a timepiece in 1728, which was dubbed the *chronometer* (*chrono* = time, *meter* = measure). Harrison's first chronometer, H1 (see chapter-opening figure), was successfully tested in 1736, but he received only £500 of the prize because the device was deemed too complex, costly,

and fragile. Eventually, his more compact fourth version, H4—which resembles an oversized pocket watch—was tested during a trans-Atlantic voyage in 1761. Upon reaching Jamaica, it was so accurate that it had lost only *five seconds* of time, a longitude error of only 0.02 degree or 1.2 nautical miles (1.4 statute miles)! Although Harrison's chronometer greatly exceeded the requirements of the government, the committee in charge of the prize withheld payment, mostly because the astronomers on the board wanted the solution to come from measurement of the stars. Because the committee refused to award him the prize without further proof, a second sea trial was conducted in 1764, which confirmed his success. Harrison was reluctantly granted £10,000. Only when King George III intervened in 1773 did Harrison finally receive the remaining prize money and recognition for his life work—at an age of 80 years old.

Today, navigating at sea relies on the **Global Positioning System (GPS)**, which was initiated in the 1970s by the United States Department of Defense when they began launching a $13 billion system of satellites (Figure 1C). Initially designed for military purposes but now available for a variety of civilian uses, GPS relies on 24 satellites that send continuous radio signals to the surface. Position is determined by very accurate measurement of the time of travel of radio signals from the satellites to receivers on board a ship (or on land). Thus, a vessel can determine its exact latitude and longitude to within a few meters—a small fraction of the length of most ships. Navigators from days gone by would be amazed at how quickly and accurately a vessel's location can be determined, but they might say that it has taken all the adventure out of navigating at sea.

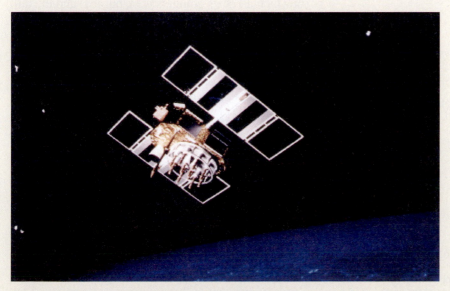

Figure 1C Global Positioning System (GPS) satellite.

Currently, 24 of these satellites orbit about 20,000 kilometers (12,400 miles) above Earth, each taking 12 hours to go around Earth once. The satellite measures 5.2 meters (17 feet) across with its solar panels unfurled.

Alexander the Great. It housed an impressive collection of written knowledge that attracted scientists, poets, philosophers, artists, and writers, who studied and researched there. The Library of Alexandria soon became the intellectual capital of the world, featuring history's greatest accumulation of ancient writings.

Like many scholars of this time who considered Earth to be a sphere, the Greek **Eratosthenes** (276–192 B.C.), the second librarian at the Library of Alexandria, cleverly used geometric relationships to determine Earth's circumference (Figure 1–6). He had heard that at noon on the longest day of the year the Sun shone directly onto the waters of a deep vertical well in Syene (Aswan), which was 800 kilometers (500 miles) to the south. At the same time in Alexandria, he noticed that a vertical pole cast a slight shadow. He accurately measured the angle of the shadow, which was 7.2 degrees, or about one-fiftieth of a circle. Using algebra, he determined Earth to be 40,000 kilometers (24,855 miles) in circumference, which compares well with the true value of 40,032 kilometers (24,875 miles) known today.[2]

By 250 B.C., the Greeks had reached Great Britain and Ireland. A few centuries later, the *Romans* had conquered the indigenous peoples of the British Isles and built extensive settlements and fortifications there.

The Roman **Strabo** (63 B.C.–24 A.D.), by observing volcanic activity in what is now Italy, concluded that the land periodically sank and rose, causing the sea to invade and recede from the continents. He also recognized the role of streams in eroding the continents and depositing sediment in the seas.

An Egyptian-Greek named **Claudius Ptolemy** (c. 165 B.C.–c. 127 B.C.) produced a map of the world in about 150 A.D. that represented the extent of Roman knowledge at that time (Figure 1–7). The map included the continents of Europe, Asia, and Africa, as did earlier Greek maps, but it also included vertical lines of longitude and horizontal lines of latitude, which had been developed by Alexandrian scholars. Moreover, Ptolemy showed the known seas to be surrounded by land, much of which was as yet unknown and proved to be a great enticement to explorers.

Ptolemy also introduced an (erroneous) update to Eratosthenes's surprisingly accurate estimate of Earth's circumference. Unfortunately, Ptolemy wrongly depended on flawed calculations and an overestimation of the size of Asia, so he determined Earth's circumference to be 29,000 kilometers (18,000 miles), which is about 28% too small. Remarkably, nearly 1500 years later, Ptolemy's error caused explorer Christopher Columbus to believe he had encountered parts of Asia rather than a new world.

> In spite of large ocean distances and primitive sailing vessels, early exploration of the oceans was achieved by Pacific navigators and the Phoenicians, Greeks, and Romans.

The Middle Ages and the Ming Dynasty

After the destruction of the Library of Alexandria in 415 A.D. (in which all of its contents were burned) and the fall

---

[2]There is continuing debate about the accuracy of the distance between Alexandria and Syene known by Eratosthenes. Even assuming a large error in this value (and other errors about the exact positions of the two cities, which tended to correct each other), his calculations were not off by more than a few percent.

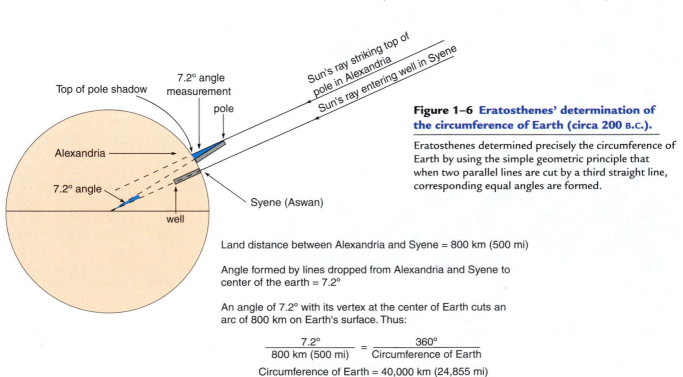

Figure 1–6 Eratosthenes' determination of the circumference of Earth (circa 200 B.C.).

Eratosthenes determined precisely the circumference of Earth by using the simple geometric principle that when two parallel lines are cut by a third straight line, corresponding equal angles are formed.

Land distance between Alexandria and Syene = 800 km (500 mi)

Angle formed by lines dropped from Alexandria and Syene to center of the earth = 7.2°

An angle of 7.2° with its vertex at the center of Earth cuts an arc of 800 km on Earth's surface. Thus:

$$\frac{7.2°}{800 \text{ km (500 mi)}} = \frac{360°}{\text{Circumference of Earth}}$$

Circumference of Earth = 40,000 km (24,855 mi)
Value known today = 40,032 km (24,875 mi)

Figure 1–7 Ptolemy's map of the known world (circa 150 A.D.).

Ptolemy's map of the world shows a reasonable representation of the Mediterranean region, Europe including the British Isles, northern Africa, and western and central Asia. The lack of accurate longitudinal measurements gives the map a stretched look in the east-west direction.

of the Roman Empire in 476 A.D., the achievements of the Phoenicians, Greeks, and Romans were mostly lost. Some of the knowledge, however, was retained by the *Arabs*, who controlled northern Africa and Spain. The Arabs used this knowledge to become the dominant navigators in the Mediterranean Sea area and to trade extensively with East Africa, India, and southeast Asia. The Arabs were able to trade across the Indian Ocean because they had learned how to take advantage of the seasonal patterns of monsoon winds. During the summer, when monsoon winds blow from the southwest, ships laden with goods would leave the Arabian ports and sail eastward across the Indian Ocean. During the winter, when the trade winds blow from the northeast, ships would return west.[3]

Meanwhile, in the rest of southern and eastern Europe, Christianity was on the rise. Scientific inquiry counter to religious teachings was actively suppressed and the knowledge gained by previous civilizations was either lost or ignored. As a result, the Western concept of world geography degenerated considerably during these so-called Dark Ages. **Cosmas**, a 6th-century navigator, drew a map of Earth as a flat rectangle 10,000 by 20,000 kilometers (6200 by 12,400 miles) (Figure 1–8). Another notion envisioned the world as a flat disk with Jerusalem at the center.

The Vikings During the Middle Ages in northern Europe, the *Vikings* of Scandinavia, who had excellent ships and good navigation skills, actively explored the

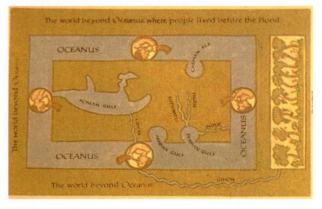

Figure 1–8 Cosmas's map of the known world (6th century A.D.).

Cosmas's map of the world depicts a rather square landmass surrounded by a single ocean (*Oceanus*). The land beyond Oceanus is labeled as "The world beyond Oceanus where people lived before the Flood."

Atlantic Ocean (Figure 1–9). Late in the ninth century, aided by a period of worldwide climatic warming, the Vikings colonized Iceland. In about 981, **Erik "the Red" Thorvaldson** sailed westward from Iceland and discovered Greenland. He may also have traveled further westward to Baffin Island. He returned to Iceland and led the first wave of Viking colonists to Greenland in 985. **Bjarni Herjolfsson** sailed from Iceland to join the colonists, but he sailed too far southwest and is thought to be the first Viking to have seen what is now called Newfoundland. Bjarni did not land but instead returned to the new colony at Greenland. **Leif Eriksson**, son of Erik the Red, became intrigued by Bjarni's stories about the new land Bjarni had seen. In 995, Lief bought Bjarni's ship and set out from Greenland

---

[3]More details about Indian Ocean monsoons can be found in Chapter 8, "Ocean Circulation."

Figure 1–9 Viking colonies in the North Atlantic.

Map showing the routes and dates of Viking explorations and the locations of the colonies that were established in Iceland, Greenland, and parts of North America.

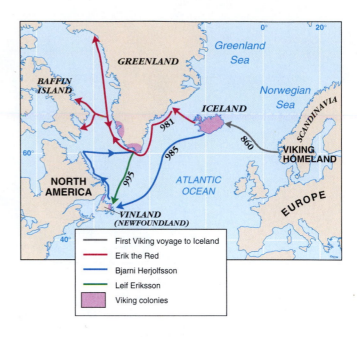

for the land that Bjarni had seen to the southwest. Leif spent the winter in that portion of North America and named the land *Vinland* (now Newfoundland, Canada) after the grapes that were found there. Climatic cooling and inappropriate farming practices for the region caused these Viking colonies in Greenland and Vinland to struggle and die out by about 1450.

Exploration by the Chinese   Near the end of Viking exploration in the North Atlantic, the Chinese *Ming Dynasty* sent large convoys of ships out on what were apparently diplomatic missions. From 1405 to 1433, seven voyages were made, involving a total of 317 ships and 37,000 men. These ships were larger and more technologically advanced than anything in Europe: They had as many as five masts, hulls divided into separate watertight compartments, and were equipped with magnetic compasses and navigational charts. The highly successful Ming voyages reached as far as Africa.

European Exploration during the Renaissance

The *Renaissance* (from the 14th through the 17th centuries) was a time of great enlightenment in Europe. In 1410, Ptolemy's map of the world was republished in Europe, complete with the erroneously small circumference of Earth. Other records of Greek and Roman accomplishments probably also became accessible to the Christian Europeans about this time because Christians gained control of Spain from the Arabs, and in doing so gained access to the knowledge stored in Spanish libraries. These sources, viewed anew, provided strong impetus to seek out new lands and push the boundaries of the known world.

The Age of Discovery   The 30-year period from 1492 to 1522 is known as the *Age of Discovery*. During this time, Europeans explored the continents of North and South America and the globe was circumnavigated for the first time. As a result, Europeans learned the true extent of the world's oceans and that human populations existed elsewhere on newly "discovered" continents and islands with cultures vastly different from those familiar to European voyagers.

Why was there such an increase in ocean exploration during the Age of Discovery? One reason was that Sultan Mohammed II had captured Constantinople (the capital of eastern Christendom) in 1453, which isolated Mediterranean port cities from the riches of the India, Asia, and the East Indies (modern-day Indonesia). The Western world had to search for new Eastern trade routes by sea.

The Portuguese, under the leadership of **Prince Henry the Navigator** (1392–1460), led a renewed effort to explore outside Europe. The prince established a marine institution at Sagres to improve Portuguese sailing skills. The treacherous journey around the tip of Africa was a great obstacle to an alternative trade route. Cape Agulhas (at the southern tip of Africa) was first rounded by **Bartholomeu Diaz** in 1486. He was followed in 1498 by **Vasco da Gama**, who continued around the tip of Africa to India, thus establishing a new Eastern trade route to Asia.

Meanwhile, the Italian navigator and explorer **Christopher Columbus** began a series of famous voyages, which were financed by Spanish monarchs. After years of difficulties in initiating his first voyage, he finally set sail with 88 men and three ships (the *Niña*, the *Pinta*, and the *Santa María*) on August 3, 1492. He sailed west from Spain in 1492, hoping to find a new route to the East Indies (today the country of Indonesia) across the Atlantic Ocean (Figure 1–10). During the morning of October 12, 1492, the first land was sighted, which is generally believed to have been Watling Island in the Bahama Islands southeast of Florida. Earth's circumference, however, had been substantially underestimated, leading Columbus to believe he had arrived in the East Indies somewhere near India. Consequently, he called the inhabitants "Indians," and the area is known today as the West Indies.

Upon his return to Spain and the announcement of his discovery, additional voyages were planned. During the next 10 years, Columbus made three more trips across the Atlantic Ocean, exploring various islands in and around the Caribbean and the coast of Central and South America. Although today he is considered a master mariner, he died in neglect in 1506, still convinced that he had explored islands near India.

Even though Christopher Columbus is widely credited with discovering North America, he never actually set foot on the continent. The lands he did "discover" were already populated with many natives, and the Vikings predated his voyage to North America by about 500

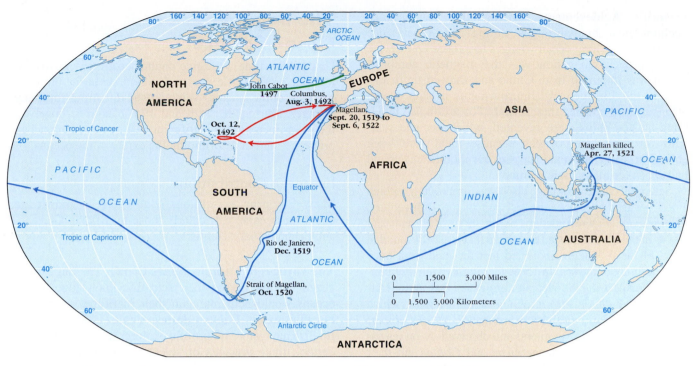

Figure 1–10 Voyages of European explorers during the Age of Discovery (1492–1522).

Christopher Columbus's voyage in 1492 (*red*) opened the New World to Europe. John Cabot's exploration took him from England to the New World in 1497 (*green*). Ferdinand Magellan's party circumnavigated the globe from 1519 to 1522 (*blue*).

years. Still, his journeys inspired other navigators to explore the "New World." For example, in 1497, only five years after Columbus's first voyage, Englishman **John Cabot** landed somewhere on the northeast coast of North America (Figure 1–10). Later, Europeans first saw the Pacific Ocean in 1513, when **Vasco Núñez de Balboa** attempted a land crossing of the Isthmus of Panama and sighted a large ocean to the west from atop a mountain.

The culmination of the Age of Discovery was a remarkable circumnavigation of the globe initiated by **Ferdinand Magellan** (Figure 1–10). Magellan left Spain in September 1519 with five ships and 280 sailors. He crossed the Atlantic Ocean, sailed down the eastern coast of South America, and traveled through a passage to the Pacific Ocean at 52 degrees south latitude, now named the Strait of Magellan in his honor. After landing in the Philippines on March 15, 1521, Magellan was killed in a fight with the inhabitants of these islands. **Juan Sebastian del Caño** completed the circumnavigation by taking the last of the ships, the *Victoria*, across the Indian Ocean, around Africa, and back to Spain in 1522. After three years, just one ship and 18 men completed the voyage.

> During the Age of Discovery (1492–1522) Europeans explored the continents of North and South America and the globe was circumnavigated for the first time.

The English Defeat the Spanish Armada   With the considerable knowledge gained during the Age of Discovery, the Spanish and Portuguese ruled the seas. During this time, the Spanish initiated many voyages to take gold and other treasures from Aztec and Inca cultures in Mexico and South America. The English and Dutch, meanwhile, used smaller, more maneuverable ships to rob the bulky Spanish galleons of their treasures. For example, Englishman **Sir Francis Drake**, who was knighted by Queen Elizabeth I but viewed as a notorious pirate by Spain and Portugal, circumnavigated the world from 1577 to 1580, looting settlements and treasure-laden Spanish and Portuguese galleons as he went. Looting such as this resulted in many confrontations at sea.

In 1588, Spain had had enough of English raids on Spanish ships and attempted to punish England for openly supporting such piracy. The Spanish Armada, consisting of over 100 fighting and support vessels, set sail to invade England. As the fleet lay anchored the night before the invasion in Dunkirk Harbor, France, the English sent frigates they had set on fire into the formation. Most of the remaining Spanish ships, fleeing into the treacherous waters of the North Sea and Atlantic, were wrecked by storms along the Scottish and Irish coasts. With the defeat of the Spanish Armada, the English had control of the seas and thus became the dominant world power—a status they retained until early in the 20th century.

Scientific Achievements during the Renaissance
Technological breakthroughs during the Renaissance helped to make long ocean voyages less risky. One device that improved navigation was the magnetic compass, which was imported from China and showed up in Europe in the 13th century. Improvements in ship design also surfaced during this period. For example, larger three-masted ships were invented, which became standard for long sea voyages because they were large enough to carry adequate supplies for months at sea and had space for many crew members.

Scientific understanding also was improving, which aided voyages at sea. In the last part of the 15th century, *Leonardo da Vinci* recorded his observations of currents and waves in the Mediterranean and noted that seashells found in rocks in the mountains indicated that the land must once have been under the sea. In 1569, *Gerhardus Mercator* constructed maps using the Mercator projection that showed latitude and longitude lines crossing at right angles, which proved particularly useful for navigation at sea. *Robert Boyle* published his experiments and observations on seawater chemistry in 1674. *Copernicus* and *Galileo* did their work on the motions of planetary bodies and the structure of the solar system. *Isaac Newton* and *Leibnitz* invented calculus, which allowed Newton to quantify the fundamental laws of physics, thereby laying the foundation in 1687 for the development of modern tidal theory.

The Beginnings of Ocean Science

Up until this time, voyaging on the oceans was mostly for trade or profit. Little was understood about the physical behavior of the oceans, which proved disastrous for many voyages. During the 18th and 19th centuries, however, there was increased scientific interest in measuring ocean properties and in determining the extent of the oceans.

European Contributions

Many European expeditions contributed to knowledge about the oceans, particularly those by British, German, and Scandinavian scientists.

The Voyages of Captain James Cook   The English realized that increasing their scientific knowledge of the oceans would help maintain their maritime superiority. For this reason, Captain **James Cook** (1728–1779), an English navigator and prolific explorer (Figure 1–11), undertook three voyages of scientific discovery with the ships *Endeavour*, *Resolution*, and *Adventure* between 1768 and 1779. He searched for the continent Terra Australis ("Southern Land," or Antarctica) and concluded that it lay beneath or beyond the extensive ice fields of the southern oceans, if it existed at all. Cook also mapped

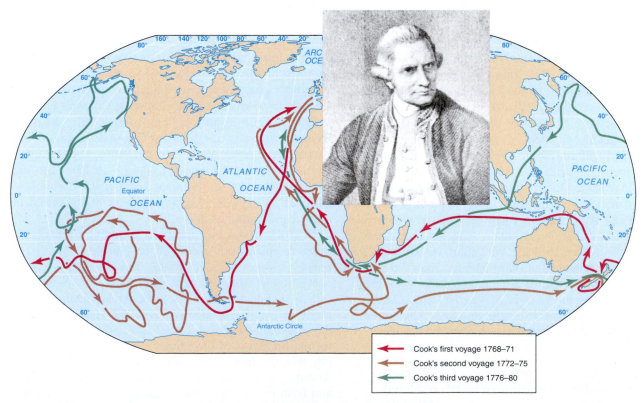

←	Cook's first voyage 1768–71
←	Cook's second voyage 1772–75
←	Cook's third voyage 1776–80

Figure 1–11 Captain James Cook (1728–1779) and his voyages of exploration.
Routes taken by Captain James Cook (*inset*) on his three scientific voyages. Cook was killed in 1779 in Hawaii during his third voyage.

many previously unknown islands, including the South Georgia, South Sandwich, and the Hawaiian Islands. During his last voyage, Cook searched for the fabled "northwest passage" from the Pacific Ocean to the Atlantic Ocean and stopped in Hawaii, where he was killed in a skirmish with native Hawaiians.

Cook's expeditions added greatly to the scientific knowledge of the oceans. He determined the outline of the Pacific Ocean and was the first person known to cross the Antarctic Circle in his search for Antarctica. Cook initiated systematic sampling of subsurface water temperatures, measuring winds and currents, taking **soundings** (which are depth measurements that, at the time, were taken by lowering a long rope with a weight on the end to the sea floor), and collecting data on coral reefs. Cook also discovered that a shipboard diet containing the German staple sauerkraut prevented his crew from contracting scurvy, a disease that incapacitated sailors. Scurvy is caused by a vitamin C deficiency, and the cabbage used to make sauerkraut contains large quantities of vitamin C. In addition, by proving the value of John Harrison's newly invented timepiece as a means of determining longitude (see Box 1–1), Cook made possible the first accurate maps of Earth's surface, some of which are still in use today.

The Rosses, Edward Forbes, and Life in the Deep Sea

One of the first major controversies in marine science developed around the work of **Sir John Ross** (1777–1856), his nephew **Sir James Clark Ross** (1800–1862), and **Edward Forbes** (1815–1854)—all British naturalists and scientists. Sir John Ross explored Baffin Bay in Canada in 1817 and 1818, where he made depth measurements. He also collected samples of bottom-dwelling organisms and sediments with a device of his own design called a *deep-sea clamm*, which resembled a pair of metal jaws that retrieved "bites" of the bottom mud. He obtained worms in mud samples from a depth of 1.8 kilometers (1.1 miles) and even retrieved a starfish from this depth, which had become entangled in the line that was used to lower the clamm to the bottom.

Sir James Clark Ross extended the soundings to greater depths on voyages to the Antarctic during 1839–1843 (Figure 1–12). Despite using a 7-kilometer (4.3-mile) sounding line, it was insufficient at times to reach the bottom of the ocean. The animals recovered from the cold waters of the Antarctic were the same species his uncle had recovered from the Arctic, and they were found to be very sensitive to temperature increase. Sir James concluded that the waters of the deep ocean must have a uniformly low temperature and that life exists at all levels in the ocean, including the deep-ocean floor.

During this same time, Edward Forbes was conducting oceanographic shipboard experiments to study the vertical distribution of life in the ocean. He observed that plant life (marine algae) is limited to the sunlit surface waters. Forbes also found that the concentration of animals was greater near the surface and decreased with increasing depth until only a small trace of life—if any—remained in the deepest waters.

Forbes's followers may have misinterpreted his findings because they concluded that no life whatsoever existed in the deep ocean. They reasoned that life would be impossible in the deep ocean because the pressure is so high and there is no light and oxygen. Their conclusions, moreover, contradicted the findings of the Rosses, who had discovered many types of life in the deep ocean. Today, it is well established that life exists at all levels in the ocean, and those organisms that are found in the deep ocean are well adapted to its dark, cold, high pressure environment.

The HMS *Challenger* Expedition: Birth of Oceanography

Oceanography as a scientific discipline is widely regarded to have begun in 1872 with the **HMS *Challenger* Expedition**. The Ross-Forbes controversy over the distribution of life in the ocean helped stimulate public and political support in Britain for an expedition with the express purpose of studying the ocean for scientific purposes.

Figure 1–12 The Ross Expedition in Antarctica (circa 1840).

Based on marine life brought up by dredges in Baffin Bay, Canada (by Sir John Ross in 1817–1818) and off Antarctica (by Sir James Clark Ross in 1839–1843), Sir James concluded that life existed at all levels in the ocean, including the deep-ocean floor.

In 1871, the Royal Society recommended that the British government provide funds for an expedition to help answer many unresolved questions about the oceans, such as, Did life forms live in the deep ocean? If so, what were the physical and chemical conditions there? What was the nature of sea floor deposits? The British government agreed to sponsor an expedition to explore the world's deep oceans and conduct scientific investigations. In 1872, a reserve warship was refitted to support scientific studies and renamed the HMS *Challenger* (Figure 1–13, *inset*). It contained a staff of six scientists under the direction of **C. Wyville Thompson** (1830–1882).

From time to time during the course of the voyage, the ship would stop to measure the water depth using a sounding line, the bottom temperature with newly developed thermometers that could withstand the high pressure at depth, and atmospheric and meteorologic conditions. In addition, a sample of the bottom water was collected and the bottom sediment was dredged. Other measurements included trawling the bottom for life using a net, collecting organisms at the surface, determining temperature at various depths, gathering samples of seawater from certain depths, and recording surface and deep-water currents.

Challenger returned in May 1876 after nearly three and a half years of circumnavigating the world (Figure 1–13).

During the 127,500-kilometer (79,200-mile) voyage, the scientists performed 492 deep-sea soundings, dredged the bottom 133 times, trawled open water 151 times, took 263 water temperature readings, and collected water samples from as deep as 1830 meters (6000 feet). As with most oceanographic expeditions, the real work of analyzing the data was just beginning. In fact, it took nearly 20 years to compile the expedition results into 50 volumes.

Two of the most remarkable contributions of the voyage were the discovery and classification of 4717 new species of marine life and the measurement of a then-record water depth of 8183 meters (26,847 feet) in the Mariana Trench. In 1884, chemist **William Dittmar** analyzed 77 of the seawater samples collected by the expedition. These were the first detailed analyses of ocean water and contributed greatly to understanding ocean dissolved components and their constancy over great distances.[4]

Victor Hensen Coins the Term *Plankton*   In the late 19th century, there was great international concern about fluctuations in the abundance of commercial fish. Although scientific committees were formed to investigate the problem,

---

[4]More information about seawater salinity and its constancy of composition is contained in Chapter 6, "Water and Seawater."

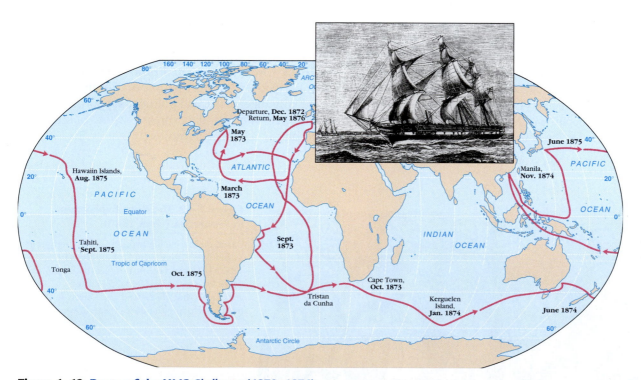

Figure 1–13 Route of the HMS *Challenger* (1872–1876).

Map showing the route traveled by HMS *Challenger* (*inset*) during its voyage of December 1872 to May 1876, which was the first large-scale voyage with the express purpose of studying the ocean for scientific purposes.

BOX 1–2   Research Methods in Oceanography

CHARLES DARWIN AND THE VOYAGE OF HMS *BEAGLE*

In the early 19th century, the English naturalist **Charles Darwin** (Figure 1D, *right inset*), whose interest was investigating the whole of nature, led him to propose the *theory of evolution* by natural selection. More than any other, Darwin has shaped scientific understanding of the underlying processes operating in nature. Most of the observations upon which he based his theory were made aboard the vessel HMS *Beagle* during its famous expedition from 1831 to 1836 that circumnavigated the globe (Figure 1D).

The *Beagle* sailed from Devonport, England, on December 27, 1831, under the command of *Captain Robert Fitzroy*. The major objective of the voyage was to complete a survey of the coast of Patagonia (Argentina) and Tierra del Fuego and to make chronometric measurements. The voyage al-

lowed Darwin—who was unpaid and often seasick during the voyage—to study the plants and animals throughout the world, particularly the 14 species of finches in the Galápagos Islands. These finches differ greatly in the configuration of their beaks (Figure 1D, *left inset*), which are suited to their diverse feeding habitats. After his return to England, he noted the adaptations of finches and other organisms living in different environments and concluded that all organisms change slowly over time as a product of their environment.

Darwin also realized that birds and mammals must have evolved from reptiles and that the similar skeletal framework of the human, the bat, the horse, the giraffe, the elephant, the porpoise, and other vertebrates required that they be grouped together.

Darwin suggested that the superficial differences between populations were the result of adaptation to different environments and modes of existence.

In 1858, Darwin and *Alfred Russel Wallace* independently and simultaneously published summaries of their ideas about natural selection. A year later Darwin set forth the structure of his theory and the evidence to support it in *The Origin of Species*, which dealt not so much with the origin of life but with the evolution of living things into their many forms. Darwin's ideas were highly controversial at the time because they stood in stark conflict with what most people believed. Today, most of Darwin's ideas have been so thoroughly embraced by scientists and so widely accepted by the general public that they are now the underpinnings of the modern study of biology.

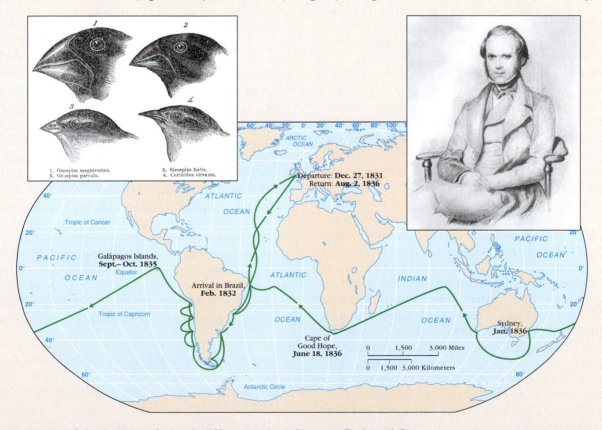

Figure 1D **Route of the HMS *Beagle*, beak differences in Galápagos finches (*left inset*), and naturalist Charles Darwin (*right inset*).**

German **Victor Hensen** (1835–1924) solved the problem alone by pioneering an ecological approach, studying the ocean habitat and distribution of the fishes' food supply instead of the fish themselves. His work was the first in what would become the field of *marine ecology.*

In 1887, Hensen coined the term **plankton** (*planktos* = wandering) for floating microscopic marine organisms and developed equipment and quantitative methods for their study. Plankton are important because practically all animal life in the sea is dependent on them, either directly or indirectly.[5] At that time, most biologists expected to find more plankton in the tropics than in colder waters. Hensen's "Plankton Expedition" aboard the *National* in 1889 showed that the reverse was true. Later work showed that the warm surface waters of the tropics become depleted of nutrients whereas cold, high-latitude surface waters maintained higher nutrient levels. It is well established today that nutrients determine plankton abundance, which, in turn, affects fish populations.

The Voyage of the *Fram* and Polar Oceanography
Norwegian explorer **Fridtjof Nansen** (1861–1930) was interested in a voyage to determine the existence of land in the unexplored Arctic. A previous expedition that piqued Nansen's curiosity was the ill-fated voyage of American *George Washington DeLong* and his crew on the *Jeanette* in 1879. DeLong had attempted to sail through the Bering Strait between North America (Alaska) and Asia (Russia) to Wrangell Island. From there, he planned to trek overland to the North Pole. At the time, it was believed that Wrangell Island might be the tip of a peninsula extending southward from an as-yet-undiscovered Arctic continent that lay beneath Arctic Ocean ice.

The *Jeanette* became stuck in the polar ice pack on September 6, 1879, and drifted north of Wrangell Island. This proved that the island was not a peninsula of an Arctic continent, because the ship could not have drifted past it in the ice. After two years of drifting, the *Jeanette* was crushed while still embedded within the ice off the New Siberian Islands. Unfortunately, DeLong and many of his crew perished in the Lena Delta region of eastern Siberia.

Five years later, a number of items from the *Jeanette* were found frozen in the pack ice off the southwestern coast of Greenland, over 4800 kilometers (3000 miles) from where she sank. Based on this unusual evidence, Nansen reasoned that the articles must have drifted with Arctic ice from the wreck of the *Jeanette* to Greenland by following a path that would have taken them near the North Pole. If Arctic ice could be verified to drift that far, it would help prove that an Arctic continent did not exist. Arctic ice could also provide a means of transporting an expedition to the North Pole, which had never been visited before. Nansen envisioned that a ship, once locked in the ice, might be able to follow a path similar to the transported articles, carrying the vessel near the North Pole.

Nansen eventually raised enough money to build the *Fram*, a 39-meter- (128-foot)-long wooden ship (Figure 1–14). The *Fram* was designed with extra reinforcing in its hull so that the expanding ice would not crush it but rather would force it up to the surface, free of the grip of the growing ice. Provisions for 13 men for five years were stored on the small ship, and the crew set sail from Oslo, Norway, on June 24, 1893. On September 21, after an arduous voyage along the northern coast of Siberia, the ship became entrapped in ice at 78.5 degrees north latitude, 1100 kilometers (683 miles) from the North Pole. The crew had begun the long and lonely endeavor of drifting slowly with the pack ice as it was moved by winds and currents (Figure 1–14, *map*). During their time locked in the Arctic ice, the explorers were able to collect depth soundings, seawater samples, and ice measurements. On November 15, 1895, the *Fram* reached its northernmost point, only 394 kilometers (244 miles) from the pole, but the explorers were unable to get to the North Pole.

On August 13, 1896, after being locked in ice for almost three years, the *Fram* broke free and floated again in the open sea. She had drifted 1658 kilometers (1028 miles) during an amazingly successful journey that proved no continent existed under the Arctic Ocean. It also showed that the polar ice was not of glacial origin but formed by freezing seawater. During the voyage, the crew had found that the depth of the Arctic Ocean exceeded 3000 meters (9840 feet). They also discovered a body of relatively warm water with temperatures as high as 1.5°C (35°F) between the depths of 150 and 900 meters (500 and 3000 feet), which Nansen correctly described as being a mass of Atlantic Ocean water that had sunk below the less saline Arctic Ocean water.

Nansen's accomplishments include laying the groundwork for future Arctic exploration and developing a sampling apparatus called the *Nansen bottle* for collecting water samples at depth to determine water temperature and salinity. Further, his observations of the direction of ice drift relative to the wind direction helped **V. Walfrid Ekman** (1874–1954), a Swedish physicist, develop the mathematical explanation of this phenomenon, known today as the Ekman spiral.[6] Nansen (who was awarded the Nobel Peace Prize in 1922), Ekman, and other Scandinavian scientists led the way in developing the discipline of physical oceanography, which is the study of physical phenomena in the ocean, including waves, tides, and currents.

Some of the first voyages for scientific knowledge were the voyages of Captain James Cook (1768–1779), the HMS *Challenger* Expedition (1872–1876), and the voyage of the *Fram* (1893–1896).

[5]More information about plankton and food webs can be found in Chapter 14, "Biological Productivity and Energy Transfer."

[6]Further discussion of the Ekman spiral is in Chapter 8, "Ocean Circulation."

Figure 1–14 **The voyage of the *Fram* (1893–1896).**

Map showing the route traveled by the *Fram* (*inset*) after becoming locked in ice near the New Siberian Islands, September 21, 1893. On August 13, 1896, the little wooden ship was safely released from the ice off Spitsbergen after drifting 1658 kilometers (1028 miles), proving no continent existed under the Arctic Ocean.

American Contributions

During the time that Europeans were exploring and studying the oceans, Americans contributed to knowledge about the oceans, too.

Benjamin Franklin and the Gulf Stream **Benjamin Franklin** is well known as an inventor, a statesman, and one of the founding fathers of our country. He even held the position of deputy postmaster general of the American colonies from 1753 to 1774. Remarkably, he also became known as one of the first physical oceanographers because he contributed greatly to the understanding of the Gulf Stream, a North Atlantic Ocean surface current. Why would a postmaster general be interested in an ocean current?

Franklin became interested in North Atlantic Ocean circulation patterns because he needed to explain why mail ships coming from Europe to New England took about two weeks less time when they took a longer, more southerly route than when they took a more direct, northerly route. In about 1769 or 1770, Franklin mentioned this dilemma to his cousin, a Nantucket whaling captain named **Timothy Folger**. Folger told Franklin that a strong current with which the mail ships were unfamiliar was impeding their journey because it flowed against them. The captains of the whaling ships were familiar with the current because they often hunted whales along its boundaries. The whalers often met the mail ships within the current and told their crews they would make swifter progress if they avoided the current. The British captains of the mail ships, however, would not accept advice from simple American fishers, so they continued to make slow progress within the current. If the winds were light, their ships were actually carried *backward*!

Folger sketched the current on a map for Franklin, including directions for avoiding it by taking a more southerly route when sailing from Europe to North America. Franklin then asked other ship captains for information concerning the movement of surface waters in the North Atlantic Ocean. Franklin inferred that there was a significant current moving northward along the eastern coast of the United States, which then headed east across the North Atlantic. He concluded that

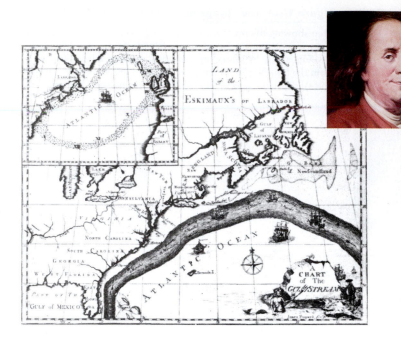

Figure 1–15 Benjamin Franklin's chart of the Gulf Stream (1777).

In 1777, Benjamin Franklin (*inset*) published this chart of the Gulf Stream, which originated from a drawing made by his cousin, Timothy Folger, a whaling captain. The strong Gulf Stream current helped explain why British mail ships took two weeks longer sailing eastward to the U.S. colonies than in the opposite direction.

this current was responsible for aiding the progress of ships traveling through the North Atlantic to Europe and slowing ships traveling in the reverse direction. He subsequently published a map of the current in 1777 based on these observations (Figure 1–15) and distributed it to the captains of the mail ships (who initially ignored it). This strong current is named the *Gulf Stream* because it carries warm water from the Gulf of Mexico and because it is narrow and well defined—similar to a stream, but in the ocean.

Matthew Fontaine Maury: The "Father of Oceanography" Matthew Fontaine Maury (1806–1873) (Figure 1–16), a career officer in the U.S. Navy, was placed in charge of the Depot of Naval Charts and Instruments after being crippled in a stagecoach accident early in his career. Maury turned his mundane job of cataloging ship records into an incredible source of knowledge that contributed to safe navigation and the international standardization of methods for marine data collection and reporting. He studied and compiled a huge and neglected treasure trove of information from ship's logs, which included regular readings of temperature and wind directions. Maury organized the first International Meteorological Conference held in Brussels in 1853 to establish uniform methods for collection of these variables at sea. In 1855, Maury published a summarized version of his compiled data and other ocean knowledge in the first textbook on oceanography, *The Physical Geography of the Sea*. Because of his enormous contribution to safe seafaring, Maury is often referred to as the "father of oceanography."

Alexander Agassiz: Advancements in Ocean Sampling Biological oceanography was strongly supported in America by the activities of **Alexander Agassiz** (Figure 1–17). The son of the great Swiss scientist Louis Agassiz, he immigrated to the United States in 1846 with his parents. He graduated from Harvard University in 1855, became superintendent of a large copper mining operation in 1866, and was a multimillionaire by age 40. Agassiz was a friend of C. Wyville Thompson, of the British *Challenger* Expedition, and his successor *Sir John Murray*,

Figure 1–16 U.S. Navy Lieutenant Matthew F. Maury (1806–1873).

Photograph of an engraving by Lemuel S. Punderson after a daguerreotype, autographed and inscribed by Lt. Maury himself, who is considered the "father of oceanography" for his contributions to safe seafaring, including his text *The Physical Geography of the Sea*.

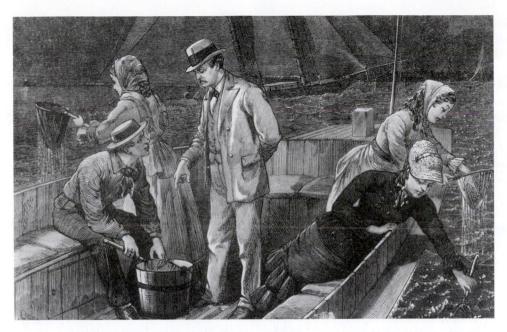

Figure 1–17 *Alexander Aggasiz (1835–1910) and associates collecting specimens.*

Alexander Aggasiz (*standing*), the son of famous Swiss naturalist Louis Aggasiz, contributed substantial money, effort, and ideas for creative sampling devices in the development of U.S. ocean study.

who helped prepare the reports from the voyage of the *Challenger*. In fact, Agassiz helped process the organisms collected during the *Challenger* Expedition.

Agassiz contributed substantial money and effort to the development of U.S. ocean study and was strongly opinionated—but not always right. For example, he disagreed with Darwin's hypothesis of coral reef development[7] and with the idea that there was an extensive midwater plankton community, both of which became well accepted. These controversies, however, did not stop Agassiz from financing scientific voyages and supporting the Museum of Comparative Zoology at Harvard, where valuable analysis of marine organisms has been conducted ever since.

Agassiz is widely acknowledged as the driving force that brought oceanography recognition as a science. He initiated the system that first brought oceanographic research strong financial and institutional support. He was also noted for his ability to organize efficient and effective research voyages, many of which occurred in the Caribbean Sea, the Gulf of Mexico, and off the Atlantic Coast of Florida.

In addition, Agassiz's training as a mining engineer led him to develop ingenious oceanographic sampling devices—the prototypes for many devices in use on research vessels today—that improved the quantitative value of the biological samples recovered. One such device, a dredge used to collect organisms living on the bottom, was designed to work equally well no matter which way it landed on the ocean floor. It worked so well, in fact, that one haul from

---

[7]Darwin's hypothesis of the stages of progression of coral reef development is further explained in Chapter 3, "Plate Tectonics and the Ocean Floor."

3219 meters (10,560 feet) brought up more specimens of deep-sea fishes than the *Challenger* had collected in its entire three-and-a-half-year expedition!

Twentieth Century Oceanography and Beyond

During the 20th century, vast technological improvements built on the successes of earlier scientific expeditions. In addition, there was continued interest in cataloging and displaying collections of marine organisms. In 1903, for example, **Prince Albert I** of Monaco established the Musée Océanographique (Figure 1–18),

Figure 1–18 *Musée Océanographique, Monaco.*

The Musée Océanographique, an oceanographic museum and aquarium, was established in 1903 by Prince Albert I of Monaco to house the extensive collections of marine organisms collected during his research ships between 1886 and 1922.

which is an oceanographic museum and aquarium that houses extensive samples of marine life collected by his research ships from 1886 to 1922.

Ironically, both World War I and World War II were of great benefit to oceanography. In particular, the development of German U-boat submarines in World War I led to the creation of the **echo sounder**, which uses sound at specific frequencies to probe ocean depths. After the war, this technology (dubbed **sonar**, which stands for *sound navigation and ranging*) was adapted to make depth surveys of the ocean floor, allowing the ability to "see" into the ocean using sound. With sonar, researchers were able to conduct depth surveys much more quickly and cheaply than they had previously by using individual soundings (depth measurements), which were collected with a long weighted line.

Voyage of the *Meteor*

In 1925, the German ship *Meteor* set the 20th century standard for detailed multidisciplinary studies of the ocean. For 25 months, scientists studied the topography, currents, and chemistry of the South Atlantic, setting up a series of 310 sampling stations (Figure 1–19). The ship also carried two echo sounders, which were used to map the depth and shape of the ocean floor. This detailed mapping led to the discovery of a large mountain range running through the middle of the South Atlantic Ocean, which was observed to influence deep-water currents. As a result of this expedition, its chief scientist, **George Wüst**, introduced the four-layer structure of vertical ocean layering that is still in use today.[8]

World War II and the Expansion of Oceanography

During World War II, anti-submarine warfare inspired many improvements in sonar technology, which led to increased knowledge of sound transmission in the ocean. In addition, the U.S. military performed and supported many studies on physical properties of the ocean such as waves, currents, and sea floor topography. This developing body of oceanographic knowledge greatly aided the military during battles at sea and during amphibious warfare.

After World War II, there was an impressive expansion in oceanography, which continues today. What spurred this expansion was the realization of the importance of marine problems to society as a whole, and so governments made moneys available for research that made it feasible to study the ocean on a scale and to a degree of complexity never before possible. Particularly, the U.S. government recognized the strategic advantage of ocean science and saw the need to maintain its lead in oceanographic research. As a result, the Sea Grant program was established to fund oceanographic research at a greater number of universities. Later, funding

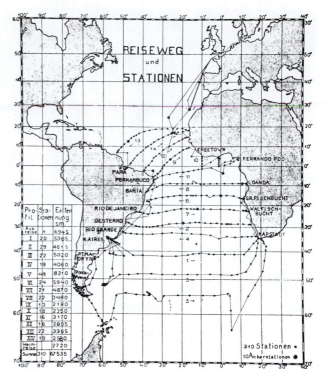

Figure 1–19 Voyage of the *Meteor* (1925–1927).
Route and transect stations of the German vessel *Meteor* during its 1925–1927 research cruise that crisscrossed the South Atlantic Ocean. *Reiseweg* means "sea voyage route" and *stationen* means "stations."

for U.S. oceanographic research shifted from the Office of Naval Research to the National Science Foundation, which supports much oceanographic research today.

Oceanography Gets Institutionalized

Figure 1–20 shows some of the most prominent oceanographic research institutions in the United States. The first oceanographic institution in the U.S., founded in 1912 in La Jolla, California, is now called the **Scripps Institution of Oceanography** and is part of the University of California at San Diego. In the 1930s, **Woods Hole Oceanographic Institution** in Woods Hole, Massachusetts, grew out of the summer seaside laboratory that Ellen Swallow Richards had established in the late 19th century to teach science to women; it is now affiliated with both Harvard University and the Massachusetts Institute of Technology. The University of Miami's **Rosenstiel School of Marine and Atmospheric Sciences** in Miami, Florida, was founded in the 1930s. The **Lamont-Doherty Earth Observatory**, associated with Columbia University, was founded at Torrey Cliffs, Palisades, New York, in 1949 as the Lamont Geological Observatory. Today, numerous schools both in the United States and abroad conduct oceanographic research and offer undergraduate and graduate degrees in oceanography or marine science.

---

[8]For more information about this and other details on ocean currents, see Chapter 8, "Ocean Circulation."

(a)

(b)

(c)

Figure 1–20 Leading U.S. oceanographic institutions.
(a) Scripps Institution of Oceanography, La Jolla, California.
(b) Woods Hole Oceanographic Institution, Woods Hole, Massachusetts. **(c)** Rosenstiel School of Marine and Atmospheric Sciences, Miami, Florida.

International Cooperation: Research in the World Ocean

Recognizing the interconnection of all the world's oceans, scientists began to ask questions that could only be answered by studying the oceans on a global scale. As a result, many oceanographic research projects became global in scope. This also necessitated larger and more capable oceanographic research vessels to carry out extended research in even the most remote areas of the world ocean.

The first large-scale, cooperative research effort between academic institutions and both U.S. and international governments was the *Deep Sea Drilling Project* (*DSDP*), which was initiated in 1968 and later was supervised by *Texas A & M University*. The initial goal of DSDP was to prove or disprove the newly emerging theory of plate tectonics.[9] To accomplish this, a specially outfitted ship drilled and retrieved cores from the sea floor. Additional details of the history and development of DSDP is contained in Box 5–1 in Chapter 5, "Marine Sediments."

At about this same time, the first computers became available. As compared to computers today, they were expensive, fragile, bulky, and limited in scope. Their usefulness was proved, however, by aiding scientists with complicated data analysis, and, as a result, they began to be used in oceanographic research and on oceanographic research vessels.

There are many other major national and international programs that have contributed (or are contributing) to ocean science. Among the most notable are the following (including their acronyms and dates of operation):

- International Geophysical Year (IGY, 1957–1958)
- International Decade of Ocean Exploration (IDOE, 1970s)
- International Indian Ocean Expedition (IIOE, 1959–1965)
- Geochemical Ocean Sections (GEOSECS, 1972–1978)
- Joint Global Ocean Flux Study (JGOFS, 1987–present)
- Ridge Inter-Disciplinary Global Experiments (RIDGE/RIDGE 2000, 1989–present)

STUDENTS SOMETIMES ASK...
I've heard of the U.S. Geological Survey and the National Biological Service. Is there a U.S. Oceanographical Survey that serves as the branch of the U.S. government dedicated to researching oceanographic phenomena?

Not quite. The branch of the U.S. government that oversees oceanographic research is the Commerce Department's National Oceanic and Atmospheric Administration (NOAA, pronounced "*noah*"). Scientists at NOAA work to ensure wise use of ocean resources through the National Ocean Service, the National Oceanographic Data Center, the National Marine Fisheries Service, and the National Sea Grant Office branches. Other U.S. government agencies that work with oceanographic data include the U.S. Naval Oceanographic Office, the Office of Naval Research, the U.S. Coast Guard, and the U.S. Geological Survey (coastal processes and marine geology).

[9]For more details on the development of the theory of plate tectonics, see Chapter 3, "Plate Tectonics and the Ocean Floor."

- World Ocean Circulation Experiment (WOCE, 1990–1998)
- Climate Variability and Predictability (CLIVAR, 1999–present)
- Census of Marine Life (CoML, 1999–present)

Satellite Oceanography

Surprisingly, one of the best ways to see the ocean is to observe it from space. Instruments aboard satellites can measure various ocean properties such as temperature, ice cover, roughness of the surface (waves), and water color (which indicates plankton abundance) on a global scale every few days. Sensitive instruments can also accurately measure the topography of the sea surface, which allows scientists to indirectly map the shape of the sea floor below (see Box 4–1 in Chapter 4, "Marine Provinces").

The first dedicated oceanographic satellite was **Seasat A** (Figure 1–21), which was launched in 1978. Unfortunately, it shorted out after only three months. During its time of operation, however, it collected more data about the oceans than was collected in the previous 200 years of ship-based ocean science, clearly demonstrating the benefit of satellite oceanography. Recent U.S. satellite missions are shown in Table 1–1.

> Development of sonar for military use allowed exploration of the sea floor. Today, international cooperation and ocean-observing satellites facilitate research in the world ocean.

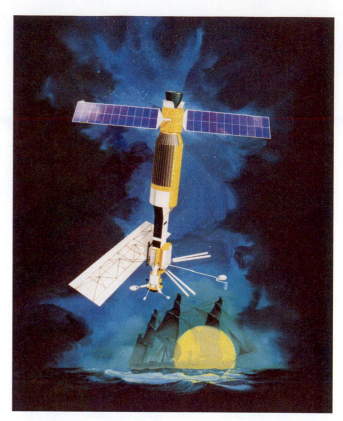

Figure 1–21 *Seasat A.*

Launched in 1978, *Seasat A* ushered in satellite oceanography. Although it shorted out only three months later, it collected a wealth of data and proved the value of satellites in oceanography. The ghostly image below *Seasat A* is the HMS *Challenger*.

TABLE 1–1 **Recent oceanographic satellite missions sponsored by the United States.**

Satellite	Dates	Mission objective
Seasat A	1978 (3 months)	Measure ocean surface temperature
Nimbus-7/Coastal Zone Color Scanner (CZCS)	1978–1986	Measure ocean color
Geosat	1985–1989	Measure sea level heights
TOPEX/Poseidon (joint mission with France)	1992–	Measure sea level heights
NASA Scatterometer (NSCAT)	1996–1997	Measure atmospheric winds and ocean waves
SeaStar/SeaWiFS	1997–	Measure ocean color
QuikSCAT/SeaWinds	1999–	Measure atmospheric winds and ocean waves
Terra (flagship of NASA's Earth Observing System)	1999–	Monitor Earth's climate, including measuring ocean, temperature productivity, and ice cover
Jason-1 (joint mission with France)	2001–	Replace TOPEX/Poseidon; measure sea level heights
GRACE (joint mission with Germany)	2002–	Study ocean currents; improve sea level height data; measure climate change
Aqua (joint mission with Japan and Brazil)	2002–	Measure water cycle to quantify climate change
ICESat	2003–	Measure ice sheet retreat/growth in response to climate change

A Human Presence in the Ocean

Despite our current ability to conduct extensive ship-based ocean research and observe the ocean by satellite, there is still a need to explore the deep ocean, an area known as "inner space." However, the inaccessibility of the deep ocean and its harsh conditions have limited human exploration (Box 1–3). One way to explore the deep ocean is in specially designed vessels called **submersibles** that can transport surface conditions to the intense pressure at depth.

Submersibles: A History

History credits Alexander the Great with the first descent in a sealed waterproof container, which reportedly took place in 332 B.C. Unfortunately, there is no record of what Alexander's submersible looked like. Much later, the *submarine* was developed, which can submerge, propel itself underwater, and surface under its own power. Because submarines are difficult to detect, they can be used to sink enemy ships. The earliest report of a submarine used in warfare is from the early 16th century, when Greenlanders used sealskins to waterproof a three-person, oar-powered submarine, which was used to drill holes in the sides of Norwegian ships.

In 1934, reaching even deeper depths for scientific exploration was the goal for naturalist **William Beebe** and his engineer-associate **Otis Barton**. They used a submersible called a **bathysphere** (*bathos* = depth, *sphere* = ball) to observe marine life in the clear waters off Bermuda. The bathysphere (Figure 1–22)—a heavy steel ball with small windows—was suspended from a ship and lowered to a then-record depth of 923 meters (3028 feet), allowing the first descriptions of the deep. Prior to this historic dive, the farthest down a living human had descended was 160 meters (525 feet)!

In 1964, the research submersible *Alvin* (Figure 1–23) from Woods Hole Oceanographic Institution began to explore the deep ocean. At 7.6 meters (25 feet) long, *Alvin* can carry a crew of one pilot and two scientists to a depth of 4000 meters (13,120 feet) and maneuver independently along the sea floor. Since it was commissioned, *Alvin* has made numerous dives to allow oceanographers to explore the deep ocean floor and retrieve samples. Under the direction of oceanographer **Robert Ballard**, some notable accomplishments of *Alvin* include discovering unique life forms along hot water springs at the mid-ocean ridge in 1977 and locating the sunken wreck of the RMS *Titanic* in 1985.

Other notable deep-diving submersibles include *Sea Cliff II*, *Shinkai 6500*, and *Trieste* (which carried two passengers to the deepest depth in the ocean; see Figure 1–4) (Figure 1–24). Currently, the deepest-diving manned submersible is *Shinkai 6500*, a Japanese research vessel

Figure 1–22 William Beebe (1877–1962) and his bathysphere.

In 1934, William Beebe (exiting the bathysphere) and Otis Barton (standing at left) descended to a record depth of 923 meters (3028 feet) in this steel bathysphere, which weighed 2268 kilograms (5000 pounds). To combat the high pressure at depth, the bathysphere had walls that were 0.5 meter (1.5 feet) thick and small windows made of fused quartz.

BOX 1–3 People and Ocean Environment
DIVING INTO THE MARINE ENVIRONMENT

Throughout history, humans have submerged themselves in the marine environment to observe it directly for scientific exploration, profit, or adventure (Figure 1E). As early as 4500 B.C., brave and skillful divers reached depths of 30 meters (98 feet) on one breath of air to retrieve red coral and mother-of-pearl shells. Later, diving bells (bell-shaped structures full of trapped air) were lowered into the sea to provide passengers or underwater divers with an air supply. In 360 B.C., Aristotle, in his *Problematum*, recorded the use by Greek sponge divers of kettles full of air lowered into the sea. However, technology to move around freely while breathing underwater was not developed until 1943, when *Jacques-Yves Cousteau* and *Émile Gagnan* invented the fully automatic, compressed-air *Aqualung*. The equipment was later dubbed "**scuba**," an acronym for *s*elf-*c*ontained *u*nderwater *b*reathing *a*pparatus, and is used by millions of recreational divers today. By using scuba, divers can experience the ocean first hand, leading to a fuller appreciation of the wonder and beauty of the marine environment.

Those who venture underwater must contend with many obstacles inherent in ocean diving, such as low temperatures, darkness, and the effects of greatly increased pressure. To combat low temperatures, specially designed clothing is worn. Waterproof, high-intensity diving lights are used to combat darkness. To combat the deleterious effects of pressure, depth and duration of dives must be limited. As a result, most scuba divers rarely venture below a depth of 30 meters (98 feet)—where the pressure is three times that at the surface—and they stay there less than 30 minutes.

It is relatively dangerous for humans to enter the marine environment because our bodies are adapted to living in the relatively low pressure of the atmosphere. In water, pressure increases rapidly with depth, to which anyone who has been to the bottom of the deep end of a swimming pool can attest. The increased pressure at depth in the ocean can cause problems for divers. For instance, higher pressure causes more nitrogen to be dissolved in a diver's body and may cause a disorienting condition known as *nitrogen narcosis*, or "rapture of the deep." Further, if a diver surfaces too rapidly, expanding gases within the body can catastrophically rupture cell membranes (a condition called *barotrauma*).

In addition, when divers return to the surface, they may experience *decompression illness* (the "bends"). The "bends" affects divers that ascend to the lower pressure at the surface too rapidly, causing nitrogen bubbles to form in the bloodstream and other tissues (analogous to the bubbles that form in a carbonated beverage when the container is opened). Various symptoms can result, from nosebleed and joint pain (which causes divers to stoop over, hence the term the "bends") to permanent neurological injury and even fatal paralysis. To avoid it, divers must ascend slowly, allowing time for excess dissolved nitrogen to be eliminated from the blood via the lungs.

Despite these risks, divers venture to greater and greater depths in the ocean. In 1962, Hannes Keller and Peter Small made an open-ocean dive from a diving bell to a then record-breaking depth of 304 meters (1000 feet). Although they used a special gas mixture, Small died once they returned to the surface. Presently, the record ocean dive is 534 meters (1752 feet), but researchers who study the physiology of deep divers have simulated a dive to 701 meters (2300 feet) in a pressure chamber using a special mix of oxygen, hydrogen, and helium gases. Researchers believe that humans will eventually be able to stay underwater for extended periods of time at depths below 600 meters (1970 feet).

Figure 1E **Oceanographer and explorer Willard Bascom.**

that can dive to 6500 meters (21,320 feet). Only the unmanned robotic vessel *Kaiko* can reach the deepest trenches. Table 1–2 lists some important achievements in the history of human descent into the ocean.

ROVs and AUVs

Some of the negative factors associated with manned submersibles are the risk to human life, the high cost of the systems required to accommodate humans, and on deep dives, the relatively short time that can be spent making observations. Of 12 hours of launch time for a typical submersible, for example, only four hours can be spent on the bottom before the crew must return to the surface. Consequently, unmanned **remotely operated vehicles (ROVs)** armed with cameras and a variety of measurement and manipulative devices have become popular for research and industry applications.

There are many advantages to ROVs. For example, they can stay on the bottom as long as the ship to which they are tethered can stay on location, so they can explore vast areas of the sea floor with only infrequent trips to the surface. They can make computer-assisted maps based on sonar data they gather, record what they see with video and still cameras, and collect specimens with remotely operated arms. They also operate at a much lower cost than manned submersibles. They are so useful that research teams often combine a manned submersible with an ROV deployed by the submersible. In 1985–1986, for instance, a team consisting of the ROV *Argo-Jason* (Figure 1–25a) and the submersible *Alvin* provided the first images of the RMS *Titanic* (Figure 1–26), which sank in 1912 after colliding with a huge iceberg off Newfoundland.

Autonomous underwater vehicles (AUVs) are also being developed for underwater surveys. These vehicles are not tethered to a ship at the surface, so they have increased mobility and can carry out specific data gathering missions of long duration and in the deepest parts of the ocean without human intervention or the risk to human life. In the past, AUVs have not received the wide usage that ROVs have because of technical limitations and problems with reliability. Recently, however, many of these problems have been resolved and AUVs are beginning to make their mark on marine science. For example, AUVs have been successfully used in a variety of applications, such as making detailed sea floor maps, studying algal blooms, tracking fish stocks, and exploring marine life under Antarctic sea ice. Also, there is growing interest in using AUVs by the military (for hunting mines and for surveying the sea floor in coastal waters before amphibious assaults) and the oil industry (for inspecting deep-water oilfields).

Figure 1–25b shows *ABE* (*Autonomous Benthic Explorer*), an AUV developed at Woods Hole Oceanographic Institution. Hovering independently just 40 meters (131 feet) above the sea floor, *ABE* recently surveyed the sea floor near the Galápagos Islands and discovered an important new hot spring that would have been much too time-consuming for manned craft and much too deep for tethered craft to locate.

Figure 1–25c shows a *Slocum glider*, an AUV developed at Rutgers University and named after *Joshua Slocum*, who sailed alone around the world in a small vessel from 1895 to 1898. The Slocum glider resembles a torpedo with wings but operates on much the same principle as an air glider. Lacking bulky motors, the glider's wings allow it to cover long distances while gradually sinking through the water column, collecting data as it descends. On reaching the bottom, it uses a pump to change its buoyancy and rise to the surface, where it downloads data and receives new instructions by satellite for its next glide downward. Thus, a glider can saw-tooth its way back and forth across an area of interest for days or even weeks at a time.

Still, there is a need for direct human observation of the deep ocean floor that robotic craft cannot duplicate. There is no substitute for having an adaptable, resourceful, inventive human for certain tasks on the deep ocean floor. In addition, manned submersibles keep alive the chance for human travel to the great ocean depths in the spirit of exploration.

Figure 1–23 The deep-diving submersible *Alvin*.

Since it was commissioned in 1964, the research submersible *Alvin* from Woods Hole Oceanographic Institution has safely carried thousands of scientists to the sea floor and back.

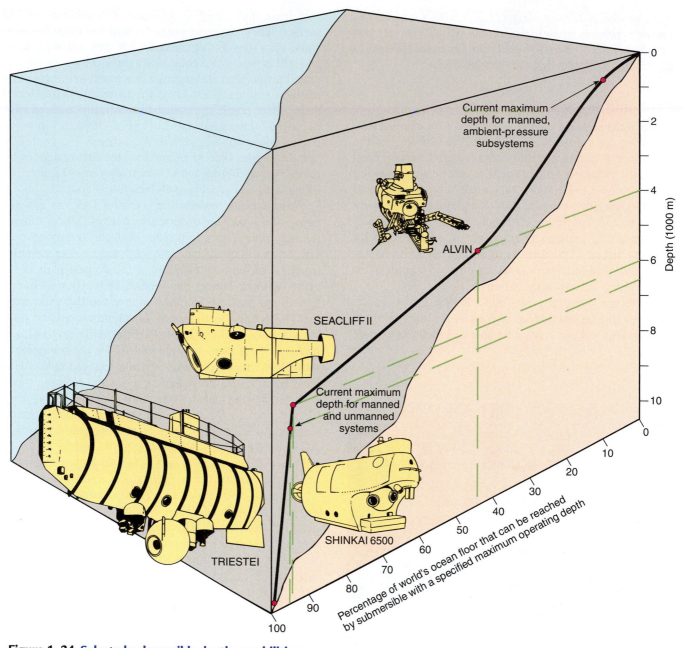

Figure 1–24 Selected submersible depth capabilities.
Graph showing the percentage of ocean floor reachable by four manned submersibles. *Alvin* has a maximum operating depth of 4000 meters (13,120 feet) and can explore about 44% of the ocean floor. *Sea Cliff II* has an operations limit of 6000 meters (19,680 feet) and can explore about 93% of the ocean floor. *Shinkai 6500*, the world's deepest-diving manned research submarine, has a maximum operational depth of 6500 meters (20,865 feet) and is capable of exploring 97% of the ocean floor. The 1960 dive of *Trieste* to a depth of 10,912 meters (35,800 feet) demonstrated its capability to explore any part of the deep ocean, but it is no longer in operation.

Living under the Sea

The shallow sea floor of the continental shelf, which adjoins the continents, covers about 26 million square kilometers (10 million square miles), an area almost as large as Africa. Because of its shallow depth and abundant food, minerals, and other resources, the continental shelf has often been considered an ideal place to establish permanent underwater living habitats.

Many attempts have been made to advance humankind's settlement of the continental shelf. One of the first trials in underwater living took place in 1962, when two "aquanauts" lived at a depth of 26 meters (85 feet) for one week in Jacques-Yves Cousteau's Continental Shelf Station (*Conshelf*) offshore Marseilles, France in the Mediterranean Sea. Cousteau's Conshelf project culminated with five aquanauts staying a month at 11 meters (36 feet) below the surface in the Red Sea.

TABLE 1–2 **Some important achievements in the history of human descent into the ocean.**

360 B.C.	In his *Problematum*, Aristotle records the use of air trapped in kettles lowered into the sea by Greek divers.
330 B.C.	Alexander the Great descends in a glass, berrylike bell during the siege of Tyre.
1620	Dutch inventor Cornelius van Drebbel tests the first submarine, descending to a depth of 5 meters (16 feet) in the Thames River with King James I of England on board.
1690	Sir Edmond Halley descends into the Thames River in a lead-weighted diving bell with air replenished from barrels lowered into the water.
1715	Englishman John Lethbridge tests a leather-covered barrel of air with viewing ports and waterproof armholes to a depth of 18 meters (59 feet).
1776	David Bushnell's *Turtle* attempts the first military submarine attack on the British ship HMS *Eagle*. The attempt failed, but the British moved their fleet to less accessible waters.
1800	Robert Fulton builds the submarine *Nautilus*, which was powered by a hand-driven screw propeller.
1837	Augustus Sieve invents a prototype of the modern helmeted diving suit.
1913	The German company Neufeldt and Kuhnke manufactures an armored diving suit with articulated arms and legs.
1934	William Beebe and Otis Barton descend in a bathysphere (see Figure 1–22) to a depth of 923 meters (3028 feet) off Bermuda and describe never-before-seen marine life.
1943	Jacques-Yves Cousteau and Émile Gagnan invent the fully automatic, compressed air aqualung (self-contained underwater breathing apparatus, or SCUBA).
1952	Hugh Bradner develops a prototype of the neoprene wet suit, which becomes universally adopted by divers because it enables them to go deeper and stay underwater longer.
1960	Jacques Piccard and Donald Walsh descend into the Challenger Deep of the Mariana Trench in the bathyscaphe *Trieste* to a depth of 10,912 meters (35,800 feet), a record of human descent that still stands today.
1962	Hannes Keller and Peter Small make an open-sea dive from a diving bell to a depth of 304 meters (1000 feet) using a special gas mixture. Small dies during decompression.
1964	Research submersible *Alvin* (see Figure 1–23), which can carry a pilot and two scientists to a depth of 4000 meters (13,120 feet), begins missions to explore the deep ocean.
1966	Submersibles *Alvin*, *Aluminaut*, and *Cubmarine*, with the aid of an unmanned remote-controlled vehicle, *CURV*, locate and recover an atomic bomb from water over 800 meters (2624 feet) deep off Palomares, Spain.
1969	Pressurized diving suit, *Jim*, is built to operate at depths below 304 meters (1000 feet).
1975	U.S. Navy divers use a pressurized, tethered sphere as transport to and from an open water dive at 350 meters (1148 feet).
1976	Jacques Mayol free dives to 100 meters (328 feet) while holding his breath for 3 minutes, 40 seconds.
1977	Submersible *Alvin* discovers hydrothermal vents in the Galápagos Rift.
1979	Sylvia Earle makes a dive in *Jim* to 381 meters (1250 feet) off Hawaii, the first dive made in the suit without a tether to the surface.
1981	Diving scientists at Duke University simulate a dive to 686 meters (2250 feet) in a pressure chamber.
1995	*Keiko*, a small unmanned Japanese robotic vessel, sets a new depth record of descent at 10,978 meters (36,017 feet) in the Challenger Deep of the Mariana Trench.
1996	Francisco Ferreras free-dives to 130 meters (428 feet) while holding his breath for 2 minutes, 11 seconds.
1996	Graham S. Hawkes successfully tests his single-passenger mini-sub, *Deep Flight I*, which has a depth capability of 1000 meters (3280 feet).
2002	Submersible *Alvin* completes its 3835th dive, having successfully transported 11,300 people to the sea floor and back.

(a)

(b)

(c)

Figure 1–25 ROVs and AUVs.

(a) *Jason* is a remotely operated vehicle (ROV) used in conjunction with the manned submersible *Alvin* that can be operated on a tether or from a surface ship. (b) *ABE* (Autonomous Benthic Explorer) is an untethered autonomous underwater vehicle (AUV) that can remain beneath the ocean surface for months, allowing it to collect data continuously. (c) Launching a Slocum glider, which is an AUV that resembles a torpedo with wings. Lacking bulky motors, it glides through the water column collecting data and has the ability to resurface and glide downward multiple times.

Figure 1–26 RMS *Titanic* discovered in 1985 by an ROV.

RMS *Titanic* (circa 1912), which sank 600 kilometers (373 miles) south of Newfoundland after it collided with an iceberg at night. It was discovered at the bottom of the North Atlantic during sea trials of the deep-sea investigative instrument *Argo* in 1985 (*inset*).

In 1964, the **Sealab** project was initiated by the U.S. Navy to study the feasibility of living underwater for long periods. A U.S. Navy team of four aquanauts remained 11 days at a depth of 59 meters (193 feet) off Bermuda in Sealab I. In 1965, Sealab II (Figure 1–27a) provided a habitat for three 10-person teams to stay 15 days each at a depth of 62 meters (205 feet) off La Jolla, California. This project generated much public interest because it included *M. Scott Carpenter*, the second U.S. astronaut to orbit Earth, and a trained Navy dolphin named Tuffy who, by responding to signals, could swim errands and reach lost divers.

In all these habitats, the aquanauts made daily dives without incident because the habitats were at the same pressure as the surrounding water. How could these aquanauts stay at these high pressures for so long? Experiments using high-pressure chambers in the late 1950s revealed that certain mixtures of gases increased the length of time divers could stay at depth. It was estimated that divers could stay at high pressure almost indefinitely once they became saturated with these mixtures. In the Sealab project, for instance, the deleterious effects of nitrogen were avoided by replacing it with helium gas within the habitat. However, the helium-rich atmosphere distorted the aquanauts' voices, making them high-pitched and cartoon-like. This provided some interesting moments, such as when President Lyndon Johnson called to congratulate the aquanauts in Sealab II.

Unfortunately, the Sealab project was hampered by unrealistic goals, poor administration, and questionable safety practices. After many lengthy delays, Sealab III

(a)

(b)

Figure 1–27 Underwater habitats Sealab II and Aquarius.

(a) Experimental underwater living habitat Sealab II, which has since been retired.
(b) Aquarius, one of only three underwater habitats in use today.

came to a halt in 1969 when aquanaut *Berry Cannon* died while establishing the habitat at a depth of 186 meters (610 feet) off San Clemente Island, California.

Today, only three small underwater habitats remain functional, all near Key Largo, Florida. They are the Jules' Undersea Lodge, which at a depth of 9 meters (30 feet) is the world's only underwater hotel; the MarineLab Undersea Laboratory, a research and education lab; and Aquarius, a research facility operated by NOAA and the National Undersea Research Program at the University of North Carolina at Wilmington. Aquarius (Figure 1–27b) accommodates a crew of six at a depth of 19 meters (63 feet) on 10-day missions to study coral reef communities.

> Submersibles transport researchers to areas of the ocean too deep to be reached by divers, while ROVs and AUVs are robotic craft used to sample the oceans.

Yes, the technology exists, but it won't be readily available for some time. The concept of breathing an inert liquid filled with a large amount of dissolved oxygen gas is called "liquid breathing" or "liquid ventilation." The technique uses a chemical called perfluorocarbon, which is a colorless, odorless, nontoxic fluid that readily absorbs oxygen and was initially used to cool electronic equipment. The fluid is infused with oxygen and breathed into the lungs, where it exchanges oxygen directly into the lung tissue. The first experiments were conducted successfully in 1966 to save ill animals. In 1989, it was used on a prematurely born infant (the first human) in an unsuccessful effort to save her life. Since then, it has been used as an emergency medical technique with good results to help children and adults who are in danger of respiratory failure. Although still in its experimental stage, the use of liquid breathing may enable people to dive to record depths in the future. This technology was featured in the 1989 movie *The Abyss*.

Chapter in Review

- *Water covers 70.8% of Earth's surface.* The world ocean is a *single interconnected body of water*, which is large in size and volume. It can be divided into *four principal oceans* (the Pacific, Atlantic, Indian, and Arctic Oceans) plus an additional ocean (the Southern or Antarctic Ocean). Even though there is a technical distinction between a sea and an ocean, the two terms are used interchangeably. In comparing the oceans to the continents, it is apparent that *the average land surface does not rise very far above sea level* and that *there is not a mountain on Earth that is as tall as the ocean is deep*.

- In the Pacific, *people who populated the Pacific Islands may have been the first great navigators*. The *Phoenicians* were the first civilization in the Mediterranean region to undertake journeys as far as the Indian Ocean for trading purposes. The *Greeks* and *Romans* who succeeded them also were successful traders, explorers, and conquerors of regions around the Mediterranean and extending to the British Isles. Their scientific accomplishments included establishment of the *Library at Alexandria, latitude determination,* and *calculation of Earth's circumference*.

- After the fall of the Roman Empire, *the scientific knowledge of the Greeks and Romans was lost* to Europe until the 15th century. While the rest of Europe languished with respect to ocean exploration during the Middle Ages, the *Vikings* settled Iceland and Greenland and established a colony in Newfoundland. The Chinese Ming Dynasty sent diplomatic ships as far as Africa in the early 15th century. When Europe emerged from the Dark Ages, *Spain and Portugal led the way*, intent on finding new trading routes to India and China.

- The *Age of Discovery* during the Renaissance in Europe renewed the Western world's interest in exploring the unknown. It began with the voyage of *Christopher Columbus* in 1492 and ended in 1522 with the first circumnavigation of Earth by a voyage initiated by *Ferdinand Magellan*. The *defeat of the Spanish Armada* by the English in 1588 changed the balance of power on the seas. Scientific achievements during the Renaissance improved the ability to undertake long ocean voyages.

- The *beginnings of ocean science* are marked by the *voyages of exploration of Captain James Cook* (1768–1779); the *HMS* Challenger *Expedition* (1872–1876), which was the first major voyage to collect oceanographic information; and the voyage of the *Fram* (1893–1896), which initiated the study of polar oceanography. European scientists who made important contributions to knowledge of marine life included *Charles Darwin, Sir John Ross, Sir James Ross, Edward Forbes,* and *Victor Hensen*.

- Americans who contributed to knowledge about the oceans during the 18th and 19th centuries included *Benjamin Franklin*, who mapped the Gulf Stream; *Matthew Fontaine Maury*, who developed international standards for reporting oceanographic data and wrote the first oceanography textbook; and *Alexander Agassiz*, who made major advancements in ocean sampling.

- *Twentieth century oceanography is characterized by the use of new technologies*, many of which were developed in World Wars I and II. *Institutions dedicated to oceanography* also were established in this century, leading to *international cooperation* between research institutions today. *Ocean-observing satellites* have greatly improved our ability to study large-scale phenomena of the ocean and to map ocean features.

- Specially designed *submersibles* can transport researchers to areas of the ocean too deep to be reached by divers. Less expensive, safer, *unmanned robotic remote vessels* have made it possible for oceanographers to remotely observe the deep ocean. *Underwater habitats* have allowed researchers to live under the sea for extended periods.

Key Terms

Key people

Agassiz, Alexander (p. 26)
Ballard, Robert (p. 30)
Barton, Otis (p. 30)
Beebe, William (p. 30)
Cabot, John (p. 19)
Columbus, Christopher (p. 18)
Cook, James (p. 20)
Cosmas (p. 17)
da Gama, Vasco (p. 18)
Darwin, Charles (p. 23)
de Balboa, Vasco Núñez (p. 19)
del Cão, Juan Sebastian (p. 19)
Diaz, Bartholomeu (p. 18)
Dittmar, William (p. 22)
Drake, Sir Francis (p. 19)
Ekman, V. Walfrid (p. 24)
Eratosthenes (p. 16)
Eriksson, Leif (p. 17)
Folger, Timothy (p. 25)
Forbes, Edward (p. 21)

Franklin, Benjamin (p. 25)
Harrison, John (p. 15)
Hensen, Victor (p. 22)
Herjolfsson, Bjarni (p. 17)
Herodotus (p. 13)
Heyerdahl, Thor (p. 13)
Magellan, Ferdinand (p. 19)
Maury, Matthew Fontaine (p. 26)
Nansen, Fridtjof (p. 24)
Prince Albert I (p. 27)
Prince Henry the Navigator (p. 18)
Ptolemy, Claudius (p. 16)
Pytheas (p. 13)
Ross, Sir James Clark (p. 21)
Ross, Sir John (p. 21)
Strabo (p. 16)
Thompson, C. Wyville (p. 22)
Thorvaldson, Erik "the Red" (p. 17)
Wüst, George (p. 28)

Key places and things

Alvin (p. 30)
Antarctic Ocean (p. 10)
Arctic Ocean (p. 10)
Atlantic Ocean (p. 10)
Autonomous underwater vehicle (AUV) (p. 33)
Bathysphere (p. 30)
Echo sounder (p. 28)
Fram (p. 24)
Global Positioning System (GPS) (p. 15)
HMS *Challenger* Expedition (p. 21)
Indian Ocean (p. 10)
Kon Tiki (p. 13)
Lamont-Doherty Earth Observatory (p. 28)
Latitude (p. 14)
Library of Alexandria (p. 13)
Longitude (p. 14)

Pacific Ocean (p. 10)
Plankton (p. 24)
Remotely operated vehicle (ROV) (p. 33)
Rosenstiel School of Marine and Atmospheric Sciences (p. 28)
Scripps Institution of Oceanography (p. 28)
Scuba (p. 32)
Sealab (p. 37)
Seasat A (p. 30)
Sonar (p. 28)
Sounding (p. 21)
Southern Ocean (p. 10)
Submersible (p. 30)
Woods Hole Oceanographic Institution (p. 28)

Questions and Exercises

1. How did the view of the ocean by early Mediterranean cultures influence the naming of planet Earth?

2. What is the difference between an ocean and a sea? Which ones are the seven seas?

3. Describe the development of navigation techniques that have enabled sailors to navigate in the open ocean far from land.

4. Construct a time line that includes the major events of human history that have resulted in a greater understanding of our planet in general and the oceans in particular.

5. Using a diagram, illustrate the method used by Eratosthenes to calculate the circumference of Earth.

6. While the Arabs dominated the Mediterranean region during the Middle Ages, what were the most significant ocean-related events taking place in northern Europe?

7. Describe the important events in oceanography that occurred during the Age of Discovery in Europe.

8. List some of the major achievements of Captain James Cook.

9. List some major achievements of the voyage of HMS *Challenger*.

10. Describe the voyage of the *Fram* and how it helped prove that there was no continent beneath the Arctic ice pack.

11. Why did Benjamin Franklin want to know about the surface current pattern in the North Atlantic Ocean?

12. What was Matthew Fontaine Maury's major contribution to an increased knowledge of the oceans?

13. What important oceanographic inventions and data came out of World Wars I and II?

14. List the features of the oceans that could be studied remotely by use of satellite-borne sensors.

15. Discuss what problems the human body can experience as a result of diving underwater.

CHAPTER 2

Origins: Beginnings of the Universe, Earth, and Life

Dense cluster of galaxies. The small circular swirls in this image represent distant galaxies, each of which is composed of large numbers of individual stars. The deepest view of the universe so far, astronomers use such images to study the history and development of the universe. To collect this image, the Advanced Camera for Surveys aboard NASA's Hubble Space Telescope peered for more than 13 hours through the center of one of the most massive galaxy clusters known, Abell 1689, which acted as a gravitational lens.

Key Questions

- How did the universe originate?
- What sequence of events led to the origin of Earth?
- Where did Earth's oceans come from?
- Did life on Earth begin in the oceans?
- How old is Earth?

The answers to these questions (and much more) can be found in the highlighted concept statements within this chapter.

"Beginnings; are apt to be shadowy."

—Rachel Carson, *The Sea Around Us* (1956)

Why does Earth have such an abundance of water—especially in the liquid form—when it is so scarce on most other bodies in the solar system? Where did Earth's water come from? How has the composition of our planet changed over time? What distribution of energy and mass were needed to create the right conditions for life to originate? Is an ocean essential to the origin of life? Unfortunately, some of these questions don't have definitive answers. In this chapter, we explore what is known about the origin of our universe, Earth, and life.

Origin of the Universe

Humans have always been fascinated by the stars, and some who took their stargazing seriously gradually figured out that not all the points of light visible in the night sky were the same. In fact, eight of the closest ones are not even stars; they are *planets* (*planetes* = wanderer), which were so named because they appeared to wander relative to the fixed background of distant stars. Earth and its eight neighboring planets orbit around an average sized star, the Sun, and comprise our **solar system**.

Our solar system lies about two-thirds of the way from the center of a huge spiral of stars called the **Milky Way galaxy** (*galaxias* = milky) (Figure 2–1a). The Milky Way galaxy contains more than 100 billion stars and gets its name because it is a broad band of faint light, giving it a milky appearance in the night sky. In fact, all the stars that can be seen at night with the naked eye belong to the Milky Way galaxy. Still, *Proxima Centauri*, the star nearest to the Sun, is more than 4×10^{13} (40 trillion) kilometers [2.5×10^{13} (25 trillion) miles] away.[1] Figure 2–1b is an image of galaxy NGC 4414, a distant spiral **galaxy** that resembles our own, taken by NASA's Hubble Space Telescope.

The Milky Way is only one of numerous galaxies in the much larger **universe** (*universum* = whole), and it is smaller than most others are. On a clear night, if you look very carefully from a vantage point in the Northern Hemisphere, you might see a hazy patch of light within the constellation Andromeda. That patch is our nearest spiral neighbor, the Andromeda galaxy,[2] which

---

[1]See Appendix I, "Metric and English Units Compared," for a discussion of scientific notation.

[2]The distant Andromeda galaxy is named after the constellation through which it can be seen from Earth.

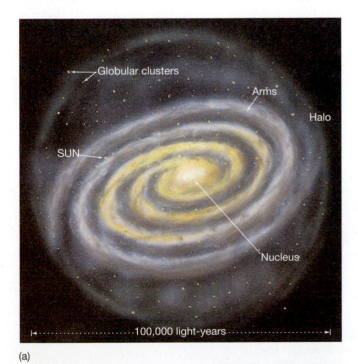

(a)

(b)

Figure 2–1 The Milky Way and NGC 4414 galaxies.

(a) Artist's conception of the Milky Way galaxy, which is about 100,000 light-years across and has a central nucleus with spiral arms. Our solar system is located about two-thirds the distance from the center of the Milky Way galaxy.
(b) NASA Hubble Space Telescope image of galaxy NGC 4414, which is a spiral galaxy similar to the Milky Way.

is about 3×10^{19} kilometers (1.9×10^{19} miles) away. Powerful new telescopes have enabled astronomers to observe even more distant galaxies (see chapter-opening photo). Within 1×10^{22} kilometers (0.6×10^{22} miles) of our galaxy, in fact, there are at least 100 million others, and astronomers estimate that there are some 100 billion galaxies in the universe, each harboring an enormous number of stars.

Light-Year: An Astronomical Distance

To deal with the large distances involved in astronomy, astronomers have developed a handy unit for measuring distance: the **light-year**, which is equal to the distance light travels in one year. Measuring distance with a unit of time might seem illogical because most moving objects can move at a variety of speeds. Light, however, always travels at exactly the same speed of 299,792 kilometers (186,282 miles) per second, so each light-year is a constant distance. Astronomers have found that using light-years makes the expression of large (astronomical) distances much easier. For example, since light travels almost 10 trillion kilometers (6.2 trillion miles) in one year, Proxima Centauri is about 4 light-years from the Sun. For the rest of this chapter, light-years will be used for astronomical distances.

The Milky Way galaxy is 100,000 light-years in diameter (Figure 2–1a). Stars are much closer together near the center of the galaxy and much farther apart toward the edges. The central *nucleus* (*nucleos* = a little nut) of the galaxy is only about 10,000 to 15,000 light-years across and, within the Milky Way, there are approximately 100 billion stars. From statistical considerations, astronomers estimate that tens of millions of these stars have families of planets, and perhaps millions of these planets could be inhabited by intelligent creatures.

Moving Galaxies and the Redshift

By observing changes in the light radiating from these distant galaxies, astronomers have determined that nearly all of them are moving away from us. It has also been determined that the most distant galaxies are traveling more rapidly than the closest ones. In fact, speeds of more than 250,000 kilometers (155,000 miles) per second—which is 80% of the speed of light—have been deduced for these galaxies.

How can the speed of a distant galaxy be determined? A galaxy's speed can be calculated by measuring its shift in the pattern of light emission toward the red end of the spectrum—called a *redshift*. This effect is similar to how the pitch of an emergency vehicle siren changes as it approaches and passes you. Moreover, the greater the observed redshift, the faster the galaxy is moving, and as a result, the more distant it is from Earth (Figure 2–2).

It is important to remember that as the motion of celestial objects is observed from Earth, Earth is not standing still in space. Earth orbits the Sun at about 30 kilometers (18.6 miles) per second, and the solar system spins around at 220 kilometers (137 miles) per second as the Milky Way rotates in space. The Milky Way is also moving out from the center of the universe at high speed. We don't feel any of this motion, however, because everything that we perceive is moving at the same speed we are.

To illustrate this concept, imagine traveling in a car on a straight stretch of freeway. If another car is traveling nearby at the same speed as your car, you cannot perceive any change in distance between the vehicles. If you ignore the scenery passing by, you would be unable to determine that you are moving by observing the other car. However, if you see a car overtake your vehicle, you know it is traveling faster than you are. If the ride is perfectly smooth, you can't determine whether you are going

Constellation in which galaxy may be seen	Velocity km/s (mi/s)	Distance from sun (light-years)	Shift of absorption lines		
			Violet		**Red**
			(short wavelength)	(long wavelength)	
Virgo	1,200 (745)	43,000,000	shift		
Corona Borealis	21,500 (13,351)	728,000,000	shift		
Hydra	61,000 (37,881)	1,960,000,000	shift		

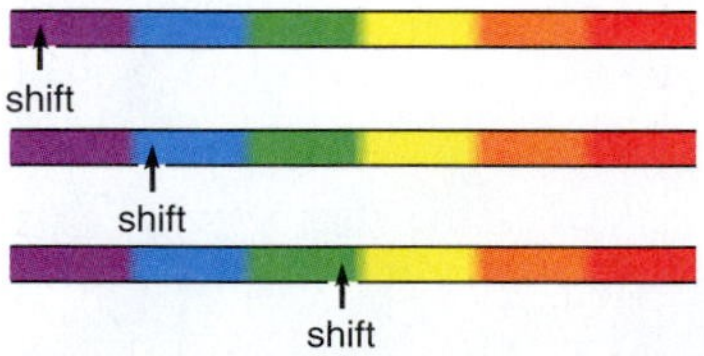

Figure 2–2 Motion of selected galaxies and the redshift.

Three galaxies moving away from the Sun show characteristic relationships between velocity, distance from the Sun, and the apparent shift of absorption lines toward the red end of the light spectrum (redshift) for a wavelength of violet light (*arrows*). For the closest galaxy (*top*), the redshift is minimal and the spectrum still represents violet light. However, in the more distant galaxies (*center* and *bottom*), the redshift increases to the blue and green portions of the visible spectrum—indicating that these galaxies are traveling at higher velocities.

100 kilometers (62 miles) per hour and being passed by a car going 140 kilometers (87 miles) per hour, or whether you are standing still and being passed by a car going 40 kilometers (25 miles) per hour. Note that *the change in the relative positions* of the two vehicles is the same in both instances.

The Big Bang

From our point of view on Earth, it appears that most of the galaxies in the universe are moving away from us. But if we remember that we are moving too, a more reasonable explanation is that we are riding along within an expanding universe. Astronomers have concluded that all galaxies are moving away from a center and from one another as if they were fragments from some ancient explosion. Unlike a bomb explosion, however, where pieces lose speed due to friction and gravity as they move away from the center, the pieces of the universe are actually *gaining* speed as they move away from the center. That's why the most distant galaxies are moving the fastest.

When did this explosion occur? If all of these galaxies are moving away from a central point, it might be reasoned that they all originated from a single large mass. If so, this origination time can be calculated from their speeds how long it has taken them to reach their present positions. In this way, astronomers estimate the age of the universe to be about 13.7 billion years old.

The idea of an exploding universe is called the **big bang theory**, which suggests that all matter originated as a dense, hot, supermassive ball with extremely high temperature and pressure conditions that underwent a cataclysmic explosion. According to the theory, elementary particles formed within the first one-billionth of a second; after one-hundredth of a second, neutrons, protons, and electrons formed. In about 25 minutes, the temperature dropped to 1 billion degrees, allowing neutrons and protons to combine to form atomic nuclei. At about 3000 degrees, nuclei attracted electrons to form the first *atoms* (*a* = not, *tomos* = cut). All matter began as atoms of hydrogen and helium, the two lightest elements.

After the blinding flash of the big bang, the universe plunged into a darkness called the Dark Ages, when there were no stars, no galaxies, and no light. After about 200 million years of darkness, stars began to form from the clouds of helium and hydrogen that moved outward from the central explosion. When these clouds became large enough, they started to contract and increase in temperature. As temperatures within the clouds became high enough, a process known as a **fusion** (*fusus* = to melt) **reaction** was initiated. A fusion reaction occurs when temperatures reach tens of millions of degrees and hydrogen atoms are converted to helium atoms, releasing large amounts of energy in the form of light.

At some point in its history, every star undergoes another contraction, and the helium is burned, producing carbon. With successive contractions, heavier elements such as oxygen, silicon, and iron are produced. When a star develops an iron core and contracts, it explodes, producing a *supernova* (*super* = great, *novus* = new). The material from this explosion contains many of the newly produced heavier elements. When this material is ejected into space, it forms interstellar clouds of gas and space dust that recombine to form new stars.

> The big bang, which occurred about 13.7 billion years ago, created the raw materials in the universe to produce galaxies, including the Milky Way galaxy, of which our solar system is a part.

? STUDENTS SOMETIMES ASK ...

I have a hard time buying into the idea of the universe starting as a "big bang." Did it really happen?

You're not the first to have this doubt. In fact, the name *big bang* was originally coined by cosmologist Fred Hoyle as a sarcastic comment on the believability of the theory. The big bang theory proposes that our universe began as a violent explosion, from which the universe continues to expand, evolve, and cool. Through decades of experimentation and observation, scientists have gathered substantial evidence that supports this theory. Despite this fact, the big bang theory, like all other scientific theories, can never be proved. It is always possible that a future observation will disprove a previously accepted theory. Nevertheless, the big bang has replaced all alternative theories and remains the only widely accepted scientific model for the origin of the universe.

Origin of the Solar System and Earth

Earth is the third of nine planets in our solar system that revolve around the Sun (Figure 2–3). Evidence suggests that the Sun and the rest of the solar system formed about 5 billion years ago from a huge cloud of gas and space dust called a **nebula** (*nebula* = a cloud). Astronomers base this hypothesis on the orderly nature of our solar system and the consistent age of meteorites (pieces of the early solar system). Using sophisticated telescopes, astronomers have also been able to observe distant nebula in various stages of formation (Figure 2–4).

The Nebular Hypothesis

According to the **nebular hypothesis** (Figure 2–5), all bodies in the solar system formed from an enormous cloud composed mostly of hydrogen and helium, with only a small percentage of heavier elements. As this huge accumulation of gas and dust revolved around its center,

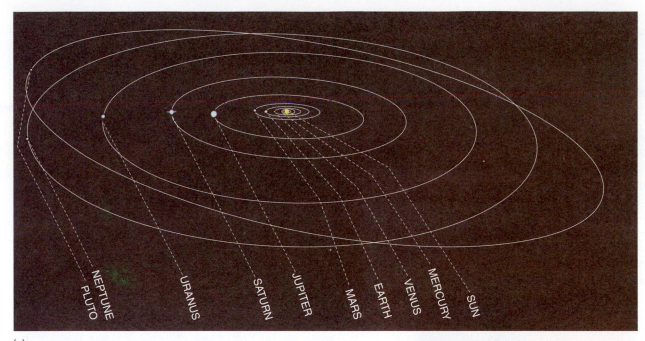

(a)

Figure 2–3 The solar system.

(a) Orbits of the planets of the solar system, drawn to scale. (b) Relative sizes of the Sun and the planets. Distance not to scale.

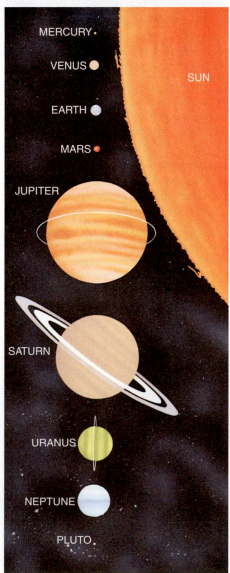

the Sun began to form as the force of gravity concentrated particles. In its early stages, the diameter of the Sun may have equaled or exceeded the diameter of our entire solar system today.

As the nebular matter that formed the Sun contracted, small amounts of it were left behind in eddies, similar to whirlpools in a stream. The material in these eddies was the beginning of the **protoplanets** (*proto* = original, *planetes* = wanderer) and their orbiting satellites, which later consolidated into the present planets and their moons.

Protoearth

Protoearth looked very different from Earth today. It was a huge mass, perhaps 1000 times greater in diameter than Earth today and 500 times more massive. There were neither oceans nor any life on the planet. In addition, the structure of the deep Protoearth is thought to have been *homogenous* (*homo* = alike, *genous* = producing), which means that it had a uniform composition throughout. The structure of Protoearth changed, however, when its heavier constituents migrated toward the center to form a heavy core.

Throughout this process, meteorites from space bombarded Protoearth. Late in the stage of planetary formation, a large body about the size of Mars struck Earth. Planetary scientists believe that the object's rocky outer layer was propelled into orbit around Earth as the Moon, while its metallic interior remained with Earth.

(b)

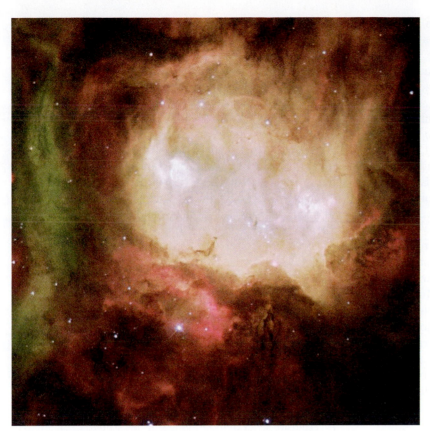

Figure 2–4 The Ghost Head Nebula.

NASA's Hubble Space Telescope image of the Ghost Head Nebula (NGC 2080), which is a site of active star formation.

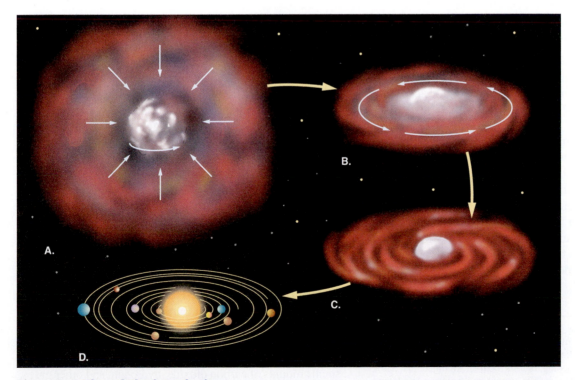

B.

A.

C.

D.

Figure 2–5 The nebular hypothesis.

(a) A huge cloud of dust and gases (a nebula) contracts. (b) Most of the material is gravitationally swept toward the center, producing the Sun, while the remainder flattens into a disk. (c) Small eddies are created by the circular motion. (d) In time, most of the remaining debris forms the planets and their moons.

do the same elements of debris collect together? the planets in our solar system are all different.

During this early formation of the protoplanets and their satellites, the Sun condensed into such a hot, concentrated mass that forces within its interior began releasing energy through atomic fusion. Recall that during a fusion reaction, hydrogen is converted to helium and large amounts of energy are released.

In addition to light energy, the Sun also emits ionized (electrically charged) particles that make up the solar wind. In the early stages of our solar system, the solar wind blew away the nebular gas that remained from the formation of the planets and their satellites. Eventually, these light gases were literally boiled away from the four inner planets as the planetary atmospheres were heated up by the Sun.

Meanwhile, the four protoplanets closest to the Sun (Mercury, Venus, Earth, and Mars) were heated so intensely by solar radiation that their initial atmospheres (mostly hydrogen and helium) boiled away. Much of this lost gas was captured by the larger planets beyond Mars: Jupiter, Saturn, Uranus, and Neptune. Additionally, the combination of ionized solar particles and internal warming of these protoplanets caused them to drastically shrink in size. As the protoplanets continued to contract, heat was produced deep within their cores from the spontaneous disintegration of atoms, called *radioactivity* (*radio* = radiation, *acti* = a ray).

> The nebular hypothesis suggests that Earth and all the bodies of the solar system were created by contraction of a cloud of gas and space dust.

Density Stratification

The release of internal heat became so intense that Earth's surface became molten. Once Earth became a ball of hot liquid rock, the elements were able to segregate according to their **densities**[3] in a process called **density stratification** (*strati* = a layer, *fication* = making). The highest-density materials (primarily iron and nickel) concentrated in the core, whereas lower and lower–density components (primarily rocky material) formed concentric spheres around the core. If you've ever noticed how oil-and-vinegar salad dressing settles out into a lower-density top layer (the oil) and a higher-density bottom layer (the vinegar), then you've seen how density stratification causes separate layers to form. Thus, Earth became a layered sphere based on density.

---

[3]Density is defined as mass per unit volume. An easy way to remember this is that density is a measure of *how heavy something is for its size*. For instance, an object that has a low density is light for its size (like a dry sponge, foam packing, or a surfboard). An object that has a high density is heavy for its size (like cement, most metals, or a large container full of water). Note that density has nothing to do with the *thickness* of an object: Some objects (like a stack of foam packing) can be thick but have low density.

The cutaway view of Earth's interior in Figure 2–6 shows that the high-density **core** can be divided into a solid *inner core* and a liquid *outer core*, both of which are composed of iron and nickel. Surrounding the core is the lower-density rocky **mantle** (a solid zone), and surrounding the mantle is the even lower-density **crust**. There are two kinds of crust—oceanic and continental. *Oceanic crust* underlies the ocean basins, has a higher density than continental crust, and is 4 to 10 kilometers (2.5 to 6.2 miles) thick. *Continental crust* underlies the continents, has a lower density than oceanic crust, and is 35 to 60 kilometers (22 to 37 miles) thick. The internal structure of Earth is discussed in more detail in Chapter 3, "Plate Tectonics and the Ocean Floor."

?
STUDENTS SOMETIMES ASK ...
How do we know about the internal structure of Earth?

You might suspect that the internal structure of Earth has been sampled directly. However, humans have never penetrated beneath the crust! The internal structure of Earth is determined by using indirect observations. Every time there is an earthquake, waves of energy (called *seismic waves*) penetrate Earth's interior. Seismic waves change their speed and are bent and reflected as they move through zones having different properties. An extensive series of monitoring stations around the world detects and records this energy. The data are analyzed and used to work out the structure of Earth's interior.

Origin of the Atmosphere and the Oceans

Earth's geologic history began about 4.6 billion years ago, when the planet had cooled sufficiently for the crust to become solid. At that time, only a small fraction of Earth's original hydrogen and helium atmosphere remained, and there was still a lot of volcanic activity. Intense meteorite bombardment of the surface also continued until about 3.9 billion years ago.

Origin of the Atmosphere

Where did the atmosphere come from? Most likely, an early atmosphere was expelled from inside Earth by a process called **outgassing**. During the period of density stratification, the lowest-density material contained within Earth was composed of various gases. These gases rose to the surface and were expelled to form Earth's early atmosphere. What was the composition of these gases? They are believed to have been similar to the gases emitted from volcanoes, geysers, and hot springs today; mostly water vapor (steam), with small amounts of carbon dioxide, hydrogen, and other gases. The composition of this early atmosphere was not, however, the same composition as today's atmosphere. There was probably little free oxygen and nitrogen, but large amounts

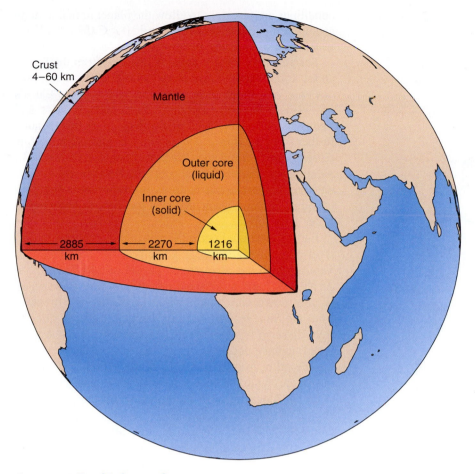

Crust
4–60 km

Mantle

Outer core
(liquid)

Inner core
(solid)

← 2885 →
km

← 2270 →
km

1216
km

Figure 2–6 Earth's internal structure.

A cutaway view of Earth's interior, showing the major subdivisions of Earth structure.
The inner core (highest density), outer core, and mantle are drawn to scale, but the
thickness of the crust (lowest density) is exaggerated about five times.

of carbon dioxide, water vapor, sulfur dioxide, and methane
were present. The composition of the atmosphere changed
over time because of the influence of life and possible
changes in the mixing of material in the mantle.

Origin of the Oceans

Where did the oceans come from? Their origin is directly
linked to the origin of the atmosphere. Figure 2–7 shows
that as Earth cooled, the water vapor released to the atmosphere during outgassing condensed and fell to Earth.
Evidence suggests that by at least 4 billion years ago,
most of the water vapor from outgassing had accumulated to form the first permanent oceans on Earth.

> Originally, Earth had no oceans. The oceans (and
> atmosphere) came from inside Earth as a result of
> outgassing and were present by at least 4 billion
> years ago.

STUDENTS SOMETIMES ASK ...

*You mentioned that the oceans came from inside Earth.
However, I've heard that the oceans came from
outer space as icy comets. Which one is true?*

Comets, being about half water, were once widely held to be
the source of Earth's oceans. During Earth's early development, space debris left over from the origin of the solar system
bombarded the young planet, and there could have been plenty of water supplied to Earth. However, spectral analyses of the
chemical composition of three comets—Halley, Hyakutake,
and Hale-Bopp—during near-Earth passes they made in 1986,
1996, and 1997, respectively, revealed a crucial chemical difference between the hydrogen in comet ice and that in Earth's
water. If comets supplied large quantities of water to Earth,
much of Earth's water would still exhibit the telltale type of hydrogen identified in comets. Instead, this type of hydrogen is
exceedingly rare in water on Earth. Assuming that the compositions of these three comets are representative of all comets, it
seems unlikely that comets supplied much water to Earth.

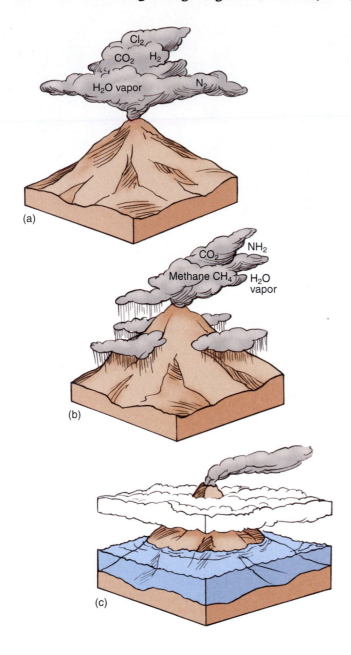

Figure 2–7   Formation of Earth's early atmosphere and oceans.

Early in Earth's history, widespread volcanic activity released large amounts of water vapor (H_2O *vapor*) and smaller quantities of various gases such as carbon dioxide (CO_2), chlorine (Cl_2), hydrogen (H_2), and nitrogen (N_2). This produced an atmosphere containing water vapor, carbon dioxide, methane (CH_4), and ammonia (NH_2). As Earth cooled, the water vapor **(a)** condensed into clouds and **(b)** fell to Earth's surface, where it accumulated to form the oceans **(c)**.

Oceans on Other Worlds?

Since our neighboring planets had a similar origin, it seems surprising that they do not have oceans while Earth does. What is unique about Earth that allowed it to develop an ocean? It turns out that a planet's average distance from the Sun and its rotational period are the critical factors in determining a planet's surface temperature. If both conditions are just right to allow the planet to maintain an average temperature between 0 and 100° C (32 to 212° F), water at its surface will be liquid.

For example, Earth's average distance from the Sun is about 150,000,000 kilometers (93,000,000 miles); this does not vary much throughout the year because Earth's orbital path is nearly circular.[4] Even though Mars and Venus also have nearly circular orbits, Mars orbits much further from the Sun and so is much colder; Venus orbits closer to the Sun and so is much warmer (see Figure 2–3). Although astronomers have confirmed the presence of ice on Mars and erosional features on the planet suggest it once had running water, Mars is too cold to have liquid water at its surface.

Earth's rotational period (one rotation every 24 hours) is also relatively rapid, especially when compared to Venus, which has a rotational period of 244 Earth-days. As a result, Earth's surface does not have time to warm up or cool down very much while facing toward or away from the Sun, respectively. In essence, Earth is just the right distance from the Sun (and rotates at just the right speed) to have liquid oceans.

One other important factor is the presence of an atmosphere, which acts like an insulating blanket around a planet, blocking both incoming solar energy and escaping re-radiated energy. Even with Earth's fortuitous distance from the Sun and rotational period, its average surface temperature would be a chilly −21°C (−5.8°F) if not for the atmosphere. The heat-trapping effectiveness of an atmosphere depends on the gases it contains and the wavelengths of the solar and re-radiated energy. This warming phenomenon is referred to as the *greenhouse effect*.[5] For example, greenhouse warming caused by the carbon dioxide-rich atmosphere on Venus further raises the temperature there, so any liquid water would have boiled away long ago. Greenhouse warming caused by Earth's atmosphere raises the average surface temperature to 14°C (57°F), ensuring that the oceans will neither freeze nor boil away.

Remarkably, some of the best prospects of finding liquid water within our solar system lie beneath the icy surfaces of some of Jupiter's moons. For instance, Europa is suspected to have an ocean of liquid water hidden under its outer covering of ice. Detailed images sent back to Earth from the *Galileo* spacecraft have revealed that Europa's icy surface is quite young and exhibits cracks apparently filled with dark fluid from below. This suggests that under its icy shell, Europa must have a warm, mobile interior—and perhaps an ocean. Because the presence of water in the liquid form is a necessity for life as we know

---

[4]For more detailed information about the effects of Earth's elliptical orbit, see Chapter 10, "Tides."

[5]For more about the atmosphere's greenhouse effect, see Chapter 7, "Air–Sea Interaction."

it, there has been much interest in sending an orbiter to Europa—and eventually a lander capable of launching a robotic submarine—to determine if it too may harbor life.

The Development of Ocean Salinity

The relentless rainfall that landed on Earth's rocky surface dissolved many elements and compounds and carried them into the newly forming oceans. Even though Earth's oceans have existed since early in the formation of the planet, its chemical composition must have changed. This is because the high carbon dioxide and sulfur dioxide content in the early atmosphere would have created a very acidic rain, capable of dissolving grater amounts of minerals in the crust than occurs today. In addition, volcanic gases such as chlorine became dissolved in the atmosphere. As rain fell and washed to the ocean, it carried some of these dissolved compounds, which accumulated in the newly forming oceans.[6] Eventually, a balance between inputs and outputs was reached, producing an ocean with a chemical composition similar to today's oceans. Further aspects of the oceans' salinity are explored in Chapter 6, "Water and Seawater."

? STUDENTS SOMETIMES ASK ...

*Are the oceans growing more or less salty
or has their salinity remained constant over time?*

We can answer this question by studying the proportion of water vapor to chloride ion, Cl^-, using ancient marine rocks. Chloride ion is important because it forms part of the most common salts in the ocean (e.g., sodium chloride, potassium chloride, and magnesium chloride). Also, chloride ion is produced by outgassing, like the water vapor that formed the oceans. Currently, there is no indication that the ratio of water vapor to chloride ion has fluctuated throughout geologic time, so it can be reasonably concluded that the oceans' salinity has been relatively constant through time.

Cycling and Mass Balance

If input from outer space is discounted, the elements that make up the atmosphere and the ocean must have come from within Earth's interior and been brought to the surface by volcanic activity. Can a whole ocean full of water come from volcanic eruptions? Is there enough chloride in volcanic emissions to account for all the chloride in the sea? These questions can be answered using simple mass balance calculations.

---

[6]Note that some of these dissolved components were removed or modified by chemical reactions between ocean water and rocks on the sea floor.

Source of Ocean Water

Studies have shown that the material brought to Earth's surface by volcanoes comes from the lower crust or the upper mantle. Let's start our examination of the source of ocean water by determining the mass of Earth's mantle.

Earth's mantle has a volume of 1.0×10^{27} cubic centimeters and an average density of 4.5 grams per cubic centimeter. The general equation is

$$\text{volume} \times \text{density} = \text{mass} \qquad (2-1)$$

To determine the mass of material in the mantle, we plug values into Equation (2–1):

$$(1.0 \times 10^{27} \text{ cm}^3) \times (4.5 \text{ g/cm}^3) = \\ 4.5 \times 10^{27} \text{ grams} \qquad (2-2)$$

The same method can be used to determine the mass of water in the present-day oceans.

If all of the ocean's water came from the mantle, how much mass has been lost from the mantle? To answer this, we need to compare the mass of the ocean to the mass of the mantle before water loss (which equals present-day mantle mass plus the mass of ocean water). We calculate:

$$\frac{1.4 \times 10^{24} \text{ g}}{(4500 \times 10^{24} \text{ g}) + (1.4 \times 10^{24} \text{ g})} = \\ 0.00031 \textit{ or } 0.031\% \qquad (2-3)$$

Therefore, the mantle would need to have lost only 0.031% of its mass as water to produce Earth's oceans.

The next question to ask is how much water (by weight percent) Earth's mantle could have contained. To determine this, we need to find some analogous material that is similar in composition to the original mantle material. Probably the best analogs are silicate-containing stony meteorites, which are fragments of the early solar system that have remained largely unchanged. The average water content of these meteorites is about 0.5% by weight, which is about 16 times more than the 0.031% that was calculated as being necessary to account for the present oceans.

Additionally, recent laboratory studies of certain minerals subjected to conditions that simulate those deep within the planet suggest that there was enough water released from minerals in the mantle as Earth cooled to supply about five times the amount of water present in Earth's oceans. Therefore, the mantle could very easily have served as the source for water in Earth's oceans if there was a sufficient rate of escape from the mantle to the surface.

Recall that there have been oceans on Earth for about 4 billion years. If the amount of water in volcanic emissions is considered and an average rate of discharge over the last 4 billion years is assumed, calculations show that volcanoes have produced enough water vapor to fill the oceans more than 100 times. Even if 99% of

EXAMPLE 2–1

Given that the volume of seawater is 1.4 × 10²⁴ grams and its average density is 1.0 grams per cubic centimeter, what is the ocean's mass?

We can plug values into Equation (2–1) to determine the answer:

$$(1.4 \times 10^{24} \text{ cm}^3) \times (1.0 \text{ g/cm}^3) = 1.4 \times 10^{24} \text{ grams}$$

The ocean's mass is 1.4 × 10²⁴ grams.

this water were recycled, there would still be enough to account for the present-day oceans. Clearly, there is strong evidence that suggests Earth's mantle is a likely source of ocean water.

Sources of Salts in the Ocean

To understand the sources and processes that control the amount of salt in the ocean, let's look more closely at the chemical and physical weathering processes that break down rock. *Chemical weathering* releases elements contained in rock by dissolving them. *Physical weathering* breaks down rocks by various natural processes that crack, split, smash, pulverize, and grind rocks into smaller pieces.

Water carries both dissolved materials and solid particles from source areas toward the oceans. Although most dissolved materials make it to the oceans, larger solid particles often do not. Moreover, water flowing from the mountains toward the oceans encounters increasingly more gentle slopes, thereby losing velocity along the way. Because the size of particles that water can carry depends on its velocity, the larger particles are deposited relatively close to their source and may not make it to the ocean, whereas the finest particles are carried all the way to the ocean where they finally settle out in deep, quiet water.

Volcanic gases emitted into the atmosphere may also end up in the ocean. All gases dissolve to some extent in water. Chlorine, sulfur gases, and carbon dioxide in the atmosphere all dissolve in rain and thereby enter the ocean.

Element Mass Balances

In the preceding section, it was implied that all the material that is now sediment, all the dissolved salt in the ocean, and all the gases in the atmosphere are thought to have come from *primary crystalline rocks* of the solid Earth. If this is true, then one should be able to add up the masses for each of the elements in sediments, the oceans, and the atmosphere to see whether the total balances with what could have been weathered from primary crystalline rocks (Figure 2–8). In this context, the primary rocks, sediments, oceans and atmosphere are referred to as *reservoirs*.

Estimates for the total element masses in each reservoir are obtained by multiplying the volume of the reservoir by the average element concentration in the reservoir (similar to the calculation in Example 2–1). Mass estimates are most accurate for those reservoirs for which the volumes are known and in which the element concentrations are not highly variable, such that the average value is representative. Furthermore, ocean and atmosphere element mass estimates are much more accurate than the estimates for primary crystalline rocks and sediments because rocks and sediments are composed of a variety of different materials, so their properties are more difficult to estimate. Another difficulty in making these estimates is separating recycled material from primary material.

Nonetheless, for the most common elements in crystalline rocks (Na, Ca, K, Si, Mg, and Fe), the balance sheets agree well, within the error of the estimates. For certain elements (Cl, S, C [as CO_2], and N), it appears that large quantities have come from somewhere other than primary crystalline rock. These elements and compounds are found in the atmosphere, ocean, and sedimentary rocks in far greater amounts than would have been made available by chemical weathering of primary crystalline rocks and are called **excess volatiles** (*volatilis* = flying) (Table 2–1). This imbalance suggests that volcanic activity rather than surface weathering is the source for these elements. It also implies that volcanic activity is the source for the water in the oceans and atmosphere.

Figure 2–8 Geochemical balances.

All the material eroded from primary crystalline rocks (*left*) is accounted for by corresponding material in the atmosphere, oceans, and sediments (*right*), creating a balance of geochemical components.

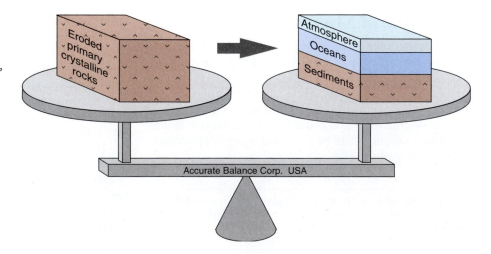

TABLE 2–1 **Excess volatiles (values are in grams $\times 10^{20}$).**

Compound or element:	H$_2$O	C as CO$_2$	Cl	N	S	H, B, Br, Ar, F, etc.
Found in:						
a. Atmosphere and ocean	14,600	1.5	276	39	13	1.7
b. Buried in sedimentary rocks	2,100	920.0	30	4	15	15.0
Total	16,700	921.5	306	43	28	16.7
Supplied by:						
a. Weathering of crystalline rocks	130	11.0	5	0.6	6	3.5
Difference of (found in) − **(supplied by), which are** **known as** *excess volatiles*	16,570	910.5	301	42.4	22	13.2

?
STUDENTS SOMETIMES ASK ...

Does Earth's continuing volcanic activity
(and outgassing) mean that the oceans
will gradually cover more and more of the surface?

Not necessarily, because the surface area of the oceans depends directly on the volume of the basin in which they form. During the initial solidification of Earth's crust, there may have been no distinct continents or ocean basins. Instead, both may have formed gradually as the oceans themselves formed. If the continents formed gradually, then the capacity of the oceans' basins to hold water may have gradually increased, too. It is likely the surface area of the oceans has been relatively constant for a long period of geologic time and the only major change in the ocean's character has been an increase in depth.

Origin of Life

How did life begin on Earth? One recent hypothesis is that the organic building blocks of life may have arrived embedded in meteors, comets, or cosmic dust. Alternatively, life may have originated around **hydrothermal** (*hydro* = water, *thermo* = heat) **vents**—hot springs— deep in the ocean. Yet another idea is that life originated in rock material deep below Earth's surface.

According to the fossil record on Earth, the earliest known life forms were primitive bacteria that lived about 3.5 billion years ago. This record indicates that the basic building blocks for the origin of life came from materials already present on Earth. The oceans were the most likely place for these materials to interact and produce life.

A Working Definition of Life

What is life? It might seem easy to differentiate those that are living from the non-living, but the unusual nature of some life forms makes defining life a challenging task. Also, both living and nonliving things are composed of the same basic building blocks: atoms, which move continuously in and out of living and nonliving systems. This free exchange of identical components between life and nonlife is one of the factors that complicates attempts at formally defining life.

A simple definition of life is that it consumes energy from its environment. Using this definition, a car engine could probably be classified as alive. An engine, however, cannot self-replicate or otherwise reproduce itself, which is another key component of life.

Several other qualities are crucial in defining life. Water probably needs to be a part of a living organism because living things need a solvent for biochemical reactions—though ammonia might also work. A living thing probably has to have some sort of a membrane to distinguish itself from its environment. In addition, most living things tend to respond to stimuli or adapt to their environment. Lastly, life as we know it is carbon-based, since carbon is so useful in making chemical compounds. Because NASA's definition of life must encompass the potential of extraterrestrial life (Box 2–1), NASA have been using a fairly simple working definition of life: "Life is a self-sustained chemical system capable of undergoing Darwinian evolution." However, even this definition is problematic in that it would likely require observation of several successive generations over a considerable length of time to verify evolution in a life form.

A good working definition of life, then, should incorporate most of these ideas: that living things can capture, store, and transmit energy; they are capable of reproduction; they can adapt to their environment; and they change through time.

The Importance of Oxygen to Life

Oxygen, which comprises almost 21% of our present atmosphere, is essential to human life for two reasons. First, our bodies need oxygen to "burn" (*oxidize*) food, releasing energy to our cells. Second, oxygen in the form of **ozone** (*ozon* = to smell[7]) protects Earth's surface from most of the Sun's harmful ultraviolet radiation (which is why there is such concern over the development of an ozone hole over Antarctica).

---

[7]Ozone gets its name because of its pungent, irritating odor.

BOX 2–1   Research Methods in Oceanography

LIFE ON MARS?

Some material that has fallen to Earth is thought to have originated as fragments blasted off the surface of Mars from collisions with meteoroids. One of these fragments, which was discovered in Antarctica in 1984 and named meteorite ALH 84001, caused a great scientific controversy in 1996 when researchers from NASA reported they had found what appeared to be fossilized microbes within the meteorite. During study of the meteorite, scientists found unusual structures resembling those that are formed by terrestrial bacteria (Figure 2A), although on a much smaller scale. If the structures could be confirmed as being produced by a life form, it would provide compelling evidence for life on Mars.

One group of skeptics immediately countered with the fact that the meteorite has been on Earth for thousands of years and had plenty of time to become contaminated by terrestrial life. Another group was skeptical as to the biologic origin of the evidence; they claimed it could have been created by inorganic processes. What tools can scientists use to tell the difference?

One of the signatures of terrestrial life is its preference for light *isotopes* (*iso* = same, *topos* = place) of elements. If an organism needs to use carbon, for example, it will preferentially use ^{12}C instead of ^{13}C; if it needs sulfur, it will use ^{32}S instead of ^{34}S, and so on. Thus, organic matter and other compounds generated by metabolic processes will be enriched in light isotopes relative to the surrounding environment. When an isotopic analysis of the meteorite was conducted, it did not show enrichment in light isotopes. But what if the life forms that evolved on Mars didn't preferentially use light isotopes?

The NASA scientists claimed unique chemical signatures found in the meteorite are indicative of production by a life form. Particularly, the presence of complex organic molecules called PAHs (polycyclic aromatic hydrocarbons) in the meteorite suggests the presence of a life form. PAHs, however, can result from nonbiologic processes and are even abundant in interstellar space. Other researchers analyzed the meteorite for the presence of amino acids, the building blocks of life, but could not find any. In addition, the extremely small size of the structures limits the possibility that they were produced by a life form. But could an entirely different type of life—one much smaller than those known on Earth that does not use amino acids—be present on Mars?

Although it appears unlikely that the unusual structures identified in Martian meteorite ALH 84001 were created by a life form, it highlights the question of whether scientists will be able to recognize life on other worlds. Mars seems to be a prime candidate for life because there is strong evidence that Mars once had liquid water at its surface, thereby fulfilling one of the most important conditions for life. Did life begin on Mars but die out when the water became trapped at the poles? Or, were other conditions never just right for life to form? Are Martian and Earth life forms both seeds from space—Earth's seed flourished, but Mars's seed died?

The existence of complex organic molecules across space, combined with the recent discovery of planets around other stars, makes it likely that the conditions conducive to life have developed in other solar systems as well. If so, perhaps there are many other worlds with life, some of them even evolving into intelligent civilizations

with which our civilization can communicate. To test this hypothesis, "listening" telescopes are being monitored, radio messages have been sent, and probes containing messages have been launched into space in hopes that another civilization on a different world might be looking and listening, too. One prominent effort in this regard is the Search for Extraterrestrial Intelligence in the Universe (SETI) Institute in Mountain View, California, which was established in 1984 to explore the origin, nature, and prevalence of life in the universe.

Although life has existed on Earth very nearly since its beginnings, the fossil record suggests it can survive here only temporarily. In the history of life on Earth, there have been many mass extinction events (one of the most famous is the Cretaceous–Tertiary event that wiped out the dinosaurs; see Box 5-3). Nonetheless, at least some life forms have managed to avoid demise and carry on, proving that life always seems to find a way to survive. The unanswered question is whether life could have originated and survived on a different world.

Figure 2A Close-up of Martian meteorite ALH 84001.

The tube-like structure in the middle of this photomicrograph is one of the controversial features of Martian meteorite ALH 84001. The structure is 0.5 micron long, where one micron equals one-millionth of a meter.

Fortunately, much of the energy represented by ultraviolet radiation is exhausted in the layer of the upper atmosphere called the *stratosphere* (*stratus* = to extend, *sphere* = a ball). Here, ozone (O_3) is created when ultraviolet light bombards oxygen molecules (O_2) and knocks some of them apart. Chemically, the reaction is:

$$O_2 + \text{ultraviolet light} \rightarrow O + O \qquad (2\text{--}4)$$

Each oxygen atom produced is very reactive and readily combines with another oxygen molecule to produce a molecule of ozone:

$$O_2 + O \rightarrow O_3 \qquad (2\text{--}5)$$

These two reactions occur so rapidly that oxygen molecules are essentially undetectable in the stratosphere.

In the process of forming ozone, oxygen molecules absorb ultraviolet radiation. Once ozone is created, it has the ability to block incoming ultraviolet radiation. Oxygen—either as O_2 or O_3—that is not bonded to other atoms is known as *free oxygen*.

Evidence suggests that Earth's early atmosphere (the product of outgassing) was different from Earth's initial hydrogen-helium atmosphere and different from the mostly nitrogen-oxygen atmosphere of today. The early atmosphere probably contained large percentages of water vapor and carbon dioxide and smaller percentages of hydrogen, methane, and ammonia, but very little free oxygen.

Why was there so little free oxygen in the early atmosphere? Oxygen may well have been outgassed, but oxygen and iron have a strong affinity for each other.[8] Iron occurs in two forms: ferrous iron (Fe^{2+}) and ferric iron (Fe^{3+}), with ferrous iron reacting readily with oxygen to produce ferric iron. Most of the iron in volcanic rocks at Earth's surface is ferrous, and there was probably sufficient iron in early volcanic rocks to chemically bind most of the oxygen that was outgassed. Thus, any oxygen released by volcanic activity was quickly used up in the conversion of ferrous iron to the ferric state, effectively removing it from the early atmosphere.

Without oxygen in Earth's early atmosphere, moreover, there would have been no ozone layer to block most of the Sun's harmful ultraviolet radiation. In fact, the lack of a protective ozone layer may have been a key component in influencing the development of life on Earth.

The First Organic Substances

The main elements in organic compounds and in living things are hydrogen, carbon, and, to a lesser extent, nitrogen. *Amino acids* and *nucleotides* are the two types of organic compounds that are the building blocks of living tissue on Earth. Only 20 different amino acids exist; there are only five nucleotides. From these, all of the more complex organic molecules are formed: *Proteins* form chains of amino acids in different combinations, and the nucleic acids *DNA* and *RNA* form chains of the five nucleotides.

One advantage of the absence of oxygen in the early atmosphere was the chemical stability of gases containing reduced carbon and nitrogen, such as methane and ammonia, as well as the more complex molecules that are produced from them. These gases formed by combination of the hydrogen, nitrogen, and carbon outgassed from the mantle.

In 1952, a 22-year old graduate student of chemist Harold Urey at the University of Chicago named **Stanley Miller** (Figure 2–9b) conducted a laboratory experiment that had profound implications about the development of life on Earth. In Miller's experiment, he exposed a mixture of carbon dioxide, methane, ammonia, hydrogen, and water (the components of the early atmosphere and ocean) to ultraviolet light (from the Sun) and an electrical spark (to imitate lightning) (Figure 2–9). After a few weeks, the clear water turned pink and then brown, indicating the formation of a large assortment of organic molecules including amino acids, which are the basic components of life. Perhaps, Miller suggested, this was how organic compounds were made on the ancient Earth before life existed.

Throughout the 1960s, scientists continued to work with equipment similar to that used by Miller and, using either electricity or ultraviolet radiation, succeeded in synthesizing all 20 types of amino acids and the five nucleotides. All of these compounds must have been present in the "primordial soup" of ocean water within which life arose.

Miller's now-famous laboratory experiment of a simulated primitive Earth in a bottle demonstrated that vast amounts of organic molecules could have been produced in Earth's early oceans. What is unclear, however, is precisely how this organic material developed into more complex molecular structures—such as proteins and DNA—that are intrinsic to life. For example, many of these small organic molecules occur as dissolved compounds in water and are separated from other molecules by layers of attached water molecules (called *hydration spheres*). In this setting, it would be incredibly difficult for small molecules to form complex chains. However, experiments have shown that evaporating or heating the primordial soup to remove the water—or allowing the organic compounds to settle out onto clay mineral surfaces—are ways in which these individual molecules could become linked into larger molecules.

Alternatively, some scientists have recently moved the primordial soup pot from the ocean surface to the deep sea floor, where there exist hot springs called hydrothermal vents. These hydrothermal vents spew murky, mineral-rich clouds of fluids into the ocean that could also have generated the right conditions to produce life's precursor molecules.

Whether created at the surface or in the deep ocean, the organic material must have become chemically

---

[8]As an example of the strong affinity of iron and oxygen, consider how common rust—a compound of iron and oxygen—is on Earth's surface.

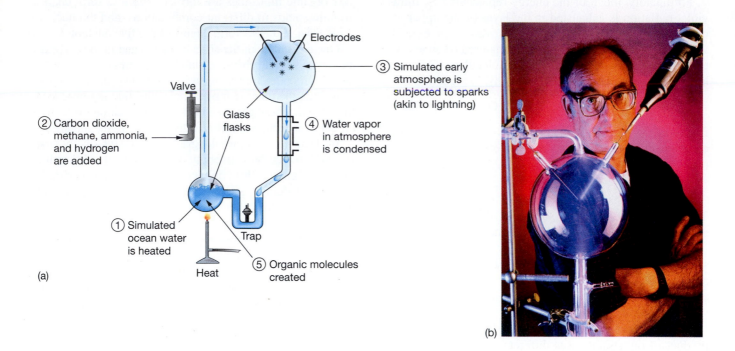

Figure 2–9 Creation of organic molecules.

(a) Laboratory apparatus used by Stanley Miller to simulate the conditions of the early atmosphere and the oceans. The experiment produced organic molecules and suggests that the basic components of life were created in the oceans. **(b)** Stanley Miller in 1999, with his famous apparatus in the foreground.

self-reproductive at some point. Further, it must have developed the ability to actively metabolize food and grow toward a characteristic size and shape dictated by internal molecular codes.

> Organic molecules were produced in a simulation of Earth's early atmosphere and ocean, suggesting that life most likely originated in the oceans.

The First Organisms

From a practical standpoint, there is an enormous increase in complexity from large organic molecules to a self-organized, replicating, living organism that is able to control and regulate its internal environment. One consideration is the amino acids that comprise proteins and the nucleotides that make up nucleic acids in organisms are linked in very specific orders. Under experimental conditions, chains of organic molecules can be produced, but they contain randomly ordered molecules.

Another consideration is that self-organization requires an organism to be physically separated from the external environment so that its internal chemistry can remain within certain limits, regardless of changes in the environment. The first living organisms were probably little more than simple membranes surrounding internal fluids that were very similar in composition to the pri-

mordial soup. The membrane had to allow needed molecules to enter the organism, waste products to leave the organism, and regulatory molecules (proteins) to be retained. The membrane also had to be easily mended after being split in two when the organism divided to form new individuals.

The very earliest forms of life were probably **heterotrophs** (*hetero* = different, *tropho* = nourishment). Heterotrophs require an external food supply, which was abundantly available in the form of nonliving organic matter in the ocean around them. *Fermenting* (*fermentum* = to boil) *bacteria* are modern-day examples of heterotrophic organisms. These bacteria obtain energy by breaking down complex organic molecules such as sugars and carbohydrates, which are obtained from their environment. Their waste products are simpler molecules such as ethyl alcohol and acetic acid. Humans utilize fermenting bacteria to create a variety of foods, including wine, beer, cheese, and vinegar.

The First Autotrophs

The **autotrophs** (*auto* = self, *tropho* = nourishment), which can manufacture their own food supply, evolved later and had a distinct advantage over heterotrophs. The first heterotrophs could only survive and increase their numbers if organic molecules were created rapidly enough by ultraviolet radiation or lightning strikes to keep pace with their expanding populations. They were also limited to living only within the nutrient-rich parts of the primordial soup that was the early oceans. Autotrophs, on the other hand, can prosper simply by making what they need wherever they can find raw materials in the presence of an energy source. Possible energy sources include chemical reactions that release energy, electricity (as was used in the Miller experimental apparatus), and sunlight.

The most abundant energy source available during the early stages of life was—and still is—sunlight. Although damaging short wavelength ultraviolet radiation provided the energy that synthesized the first organic molecules out of the primordial soup, organisms could not harness that energy without sustaining damage to their existing molecules. Since water naturally blocks ultraviolet radiation, the first organisms to use sunlight must have lived far enough below the water's surface to escape damage, yet close enough to the surface to take advantage of the non-damaging, longer wavelength visible light that penetrates to greater depths in the ocean.

The first autotrophs were probably similar to our present-day **anaerobic** (*an* = without, *aero* = air) **bacteria**, which live without atmospheric oxygen. They may have been able to derive energy from inorganic compounds at deep-water hydrothermal vents using a process called **chemosynthesis** (*chemo* = chemistry, *syn* =with, *thesis* = an arranging). In fact, the recent discovery of 3.2 billion year old microfossils of bacteria from deep-water marine rocks found in what is now

Australia supports the idea of a high-temperature origin of life on the deep ocean floor in the absence of light.

Photosynthesis and Respiration Eventually, more complex single-celled autotrophs evolved. They developed a green pigment called *chlorophyll* (*chloro* = green, *phyll* = leaf), which captures the Sun's energy through **photosynthesis** (*photo* = light, *syn* = with, *thesis* = an arranging). In photosynthesis (Figure 2–10, *top*), plants capture light energy and store it as sugars. A chemical reaction in which energy is captured or absorbed is said to be *endothermic* (*endo* = inside, *thermo* = heat). In **respiration** (*respir* = to breathe) (Figure 2–10, *middle*), the sugars are oxidized with oxygen so their stored energy can be used to carry on the life processes of the plant or the animal that eats the plant. A chemical reaction that releases energy is said to be *exothermic* (*exo* = outside, *thermo* = heat).

Not only are photosynthesis and respiration chemically opposite processes, but they are also complementary because the products of photosynthesis (sugars and oxygen) are used during respiration and the products of respiration (water and carbon dioxide) are used in photosynthesis (Figure 2–10, *bottom*). Thus, autotrophs (algae and plants) and heterotrophs (most bacteria and animals) have developed a mutual need for each other.

The first photosynthesizers were probably similar to modern sulfur bacteria that use hydrogen sulfide as a source of hydrogen and carbon dioxide as a carbon source. Both of these compounds were abundant in Earth's early atmosphere and in the primordial soup (early ocean). The waste product of these photosynthe-

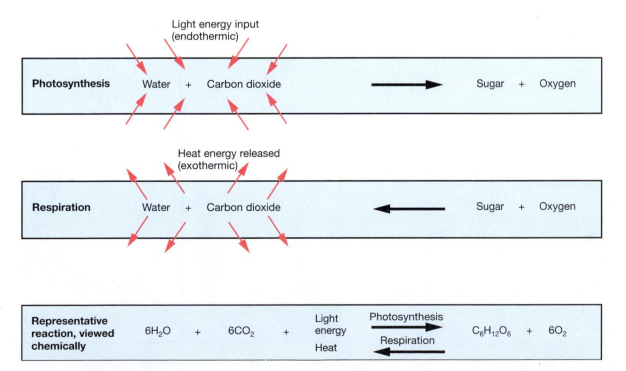

Figure 2–10 Photosynthesis (*top*), respiration (*middle*), and representative reactions viewed chemically (*bottom*).

sizing reactions is pure solid sulfur, which is excreted harmlessly by the organism. At some point, however, autotrophs evolved that could use water as a source of hydrogen. These organisms, ancestors of modern **cyanobacteria** (*kuanos* = dark blue), had a great advantage in that water was abundant everywhere. Just as the splitting of hydrogen sulfide produced sulfur as a waste product, the waste product of those reactions was free oxygen, which was toxic to many organisms.

What is the oldest evidence on Earth of photosynthesis? Rocks found in Greenland and dated at 3.8 billion years contain somewhat ambiguous chemical evidence that could have belonged to photosynthetic bacteria. Unfortunately, fossils from these rocks are not present, either because the rocks initially lacked fossils or because rocks this old have been so extensively altered. The oldest unequivocal evidence for primitive photosynthetic bacteria comes from fossilized remains of organisms recovered from rocks in northwestern Australia that formed on the sea floor 3.465 billion years ago (Figure 2–11). Some of these fossils resemble cyanobacteria, others resemble *coccoidal* (*coccus* = berry, *eidos* = like) *bacteria*, and some even look as if they were preserved in the act of dividing! Certainly, by 3.5 billion years ago, photosynthetic life was flourishing.

However, the oldest rocks containing iron oxide (rust)—an indicator of an oxygen-rich atmosphere—do not appear until about 2 billion years ago. Thus, photosynthetic organisms took at least 1.5 billion years to develop and begin producing abundant free oxygen in the atmosphere (Figure 2–12). At the same time, when a large amount of oxygen-rich (ferric) iron sank to the base of the mantle, it may have been heated by the core, risen as a plume to the ocean floor, and began releasing large amounts of oxygen about 2.5 billion years ago.

The Oxygen Crisis For anaerobic bacteria that had grown successfully in an oxygen-free world, all this oxygen was nothing short of a catastrophe! The increased atmospheric oxygen caused the ozone concentration in the upper atmosphere to build up, thereby shielding Earth's surface from ultraviolet radiation—and effectively eliminating anaerobic bacteria's food supply of organic molecules in the primordial soup. In addition, oxygen (particularly in the presence of light) is highly reactive with organic matter. When anaerobic bacteria are exposed to toxic oxygen and light, they are killed instantaneously. By 1.8 billion years ago, the atmosphere's oxygen content had increased to such a high level that it began causing the extinction of many anaerobic organisms. Nonetheless, descendants of such bacteria survive on Earth today in isolated microenvironments that are dark and free of oxygen, such as deep in soil or rocks, in garbage, and inside other organisms.

Cyanobacteria, however, evolved to exploit this new high oxygen environment. A metabolic pathway evolved that enabled these organisms to use oxygen to release energy from organic matter. They were the first organisms with the capacity for **aerobic** (*aero* = air) **respiration**; up until that time, respiration was strictly anaerobic. Because oxygen is so reactive with organic matter compared with the compounds metabolized in anaerobic respiration, it also yields much more energy. Compared with anaerobic respiration, aerobic respiration yields nearly *20 times* more energy—a fact that has led to its widespread use on Earth.

In early cyanobacteria, the same structures were used in photosynthesis and respiration, with photosynthesis occurring during the day and respiration occurring at night. Modern algae and plants can carry on both processes simultaneously because their cells have separate structures for the two processes: **mitochondria** (*mitos* = thread, *khondros* = grain) for respiration and **plastids** (*plastos* = molded) for photosynthesis. Mitochondria are also the site of respiration in animal cells, but those cells lack plastids and cannot synthesize organic matter. Although modern cells are much more complex than the cells of the cyanobacteria and their early contemporaries, the first critical steps in their design likely occurred at this time in Earth's history as a result of the oxygen crisis.

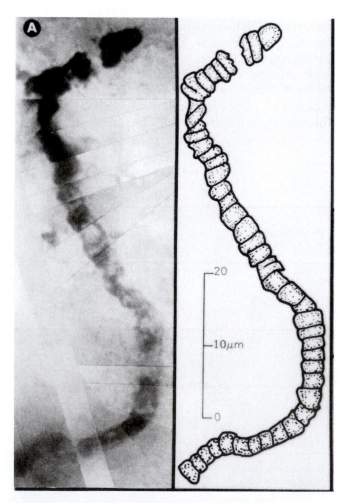

Figure 2–11 Earth's oldest fossilized life form.

Photomicrograph of fossilized bacteria (*left*) with interpretive drawing (*right*) from 3.465 billion year old rocks in northwestern Australia. Scale is in microns.

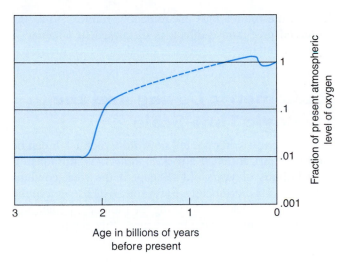

Figure 2-12 Oxygen concentration in Earth's atmosphere.

Concentration of atmospheric oxygen over the past 3 billion years based on data from geochemical and fossil evidence. The steep rise in oxygen about 2 billion years ago indicates the advent of photosynthesis. Dashed part of the curve indicates uncertainty.

Multicellular Life and Symbiosis

The earliest living organisms were **prokaryotic** (*pro* = before, *karuotos* = nuts) **cells**, which consisted of a single cell with no central nucleus and no internal membrane; as a result, the genetic material was loose within the cell. On the other hand, **eukaryotic** (*eu* = good, *karuotos* = nuts) **cells**, including modern plant and animal cells, contain membrane-bound nuclei, have complex membrane systems, and possess other intracellular bodies such as mitochondria and plastids. The central nucleus contains the genetic material of the cell, which is tightly coiled into structures called *chromosomes* (*khroma* = color, *soma* = a body). The oldest preserved eukaryotic cells are from single–celled organisms that lived about 1.4 to 1.6 billion years ago.

Remarkably, there seem to be no transitional forms between prokaryotic and eukaryotic cells. This fact and other chemical and genetic evidence have led evolutionary biologists to suggest that eukaryotic cells began as cooperative interactions between groups of bacteria. Mitochondria, for example, may have begun as predatory, oxygen-respiring bacteria that ate their hosts from the inside out. Eventually, these bacteria survived within their host without killing it and, at the same time, the host had the advantage of utilizing their metabolic byproducts as food. Similarly, plastids may have been photosynthesizing bacteria that were ingested by larger organisms and ended up surviving within the host and providing it with new organic matter as food. A relationship in which two or more organisms associate in a way that benefits at least one of them is called **symbiosis** (*sym* = together, *bios* = life). Symbiosis is such a successful adaptation that it is still employed today by many organisms.[9]

How did symbiosis begin? Prokaryotic cells undergo simple division in order to reproduce. Sometimes, the division is incomplete and two cells remain attached to each other.[10] Scientists think that "mistakes" of this sort led to prokaryotic and eukaryotic multicellular organisms. With these colonial arrangements, member cells could become specialized for specific tasks, such as locomotion, sensing, photosynthesis, respiration, or reproduction. Another possibility is that primitive cells took up other cells initially as food but, instead of digesting them, the food provided other benefits to the host, so the host and the "food" entered into a symbiotic relationship. In addition, the swapping of single or multiple genes between organisms—called *lateral gene transfer*—could have provided organisms with new specialized traits that enabled them to succeed in their environment.

Moreover, eukaryotic cell colonies proved to be such an advantageous lifestyle that by about 1 billion years ago, all sorts of new eukaryotic organisms emerged. By 700 million years ago, numerous types of complex, multicellular, soft-bodied animals existed, which are the predecessors of all modern life that exists today.

Evolution and Natural Selection

Every living organism that inhabits Earth today is the result of **evolution** by the process of **natural selection** that has been going on since these early times. Evolution is the theory that groups of organisms adapt and change with the passage of time. Descendants differ morphologically and physiologically from their ancestors. Certain advantageous traits are naturally selected and passed on from one generation to the next. Evolution is the process by which various **species** (*species* = a kind) have been able to inhabit increasingly numerous environments on Earth.[11]

As species adapted to Earth's various environments, they also modified the environments in which they lived. For example, when plants emerged from the oceans and inhabited the land, they changed it from a harsh and bleak landscape (much like the Moon's surface today) to one that was green and lush. The ocean changed, too, as vast quantities of dead organisms accumulated on the sea floor. Some of these accumulations have been turned into rock and uplifted onto continents, sometimes at high elevations. Because these rocks formed on the sea floor, there is much to be learned about the oceans of the past by studying them.

Changes to Earth's Environment

The development and successful evolution of photosynthetic organisms is greatly responsible for the world as

---

[9]Symbiosis will be discussed in more detail in Chapter 14, "Biological Productivity and Energy Transfer."

[10]Life forms in which a number of similar cells live attached together are called *colonies*.

[11]See Box 1–2 for further discussion of Darwin's theory of evolution.

we know it today (Figure 2–13). These organisms reduced the amount of carbon dioxide in the early atmosphere from relatively high levels to 0.037% today. At the same time, they increased the amount of free oxygen in the early atmosphere from very low levels to 21% today. The increasing level of oxygen, in turn, made it possible for most present-day animals to develop.

The remains of ancient plants and animals buried in oxygen-free environments have become the oil, natural gas, and coal deposits of today. These so-called *fossil fuels* provide over 90% of the energy currently used to power modern society. Humans depend not only on the food energy stored in today's plants but also on the energy stored in plants during the geologic past.

Because of increased burning of fossil fuels for home heating, industry, power generation, and transportation during the industrial age, the atmospheric concentration of carbon dioxide and other gases that help warm the atmosphere has increased, too. Many people are concerned that increased global warming will cause serious prob-

lems in the future. This phenomenon, referred to as the increased greenhouse effect, is discussed in Chapter 7, "Air–Sea Interaction."

Radiometric Dating and the Geologic Time Scale

How can Earth scientists tell how old a rock is? It can be a difficult task to tell if a rock is thousands, millions, or even billions of years old—except if the rock contains telltale fossils. Fortunately, Earth scientists can determine how old most rocks are by using the radioactive materials contained within rocks. In essence, this technique involves reading a rock's internal "rock clock."

Most rocks on Earth (as well as those from outer space) contain small amounts of radioactive materials such as uranium, thorium, and potassium. These radioactive materials spontaneously break apart or decay into atoms of other elements at predictable rates.

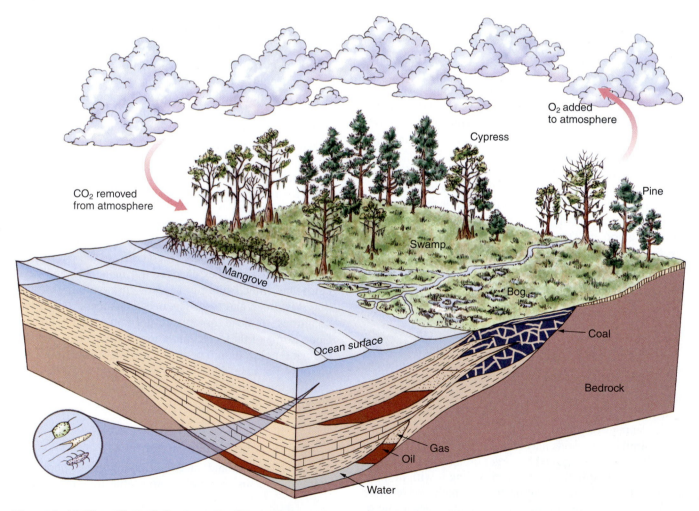

Figure 2–13 The effect of plants on Earth's environment.

As microscopic photosynthetic cells (*inset*) became established in the ocean, Earth's atmosphere was enriched in oxygen and depleted in carbon dioxide. As organisms died and accumulated on the ocean floor, some of their remains were converted to oil and gas. The same process occurred on land, sometimes producing coal.

Radioactive materials have a characteristic *half-life*, which is the time required for one-half of the atoms in a sample to decay to other atoms. The older the rock is, the more radioactive material will have been converted to decay product. Of course, this method works only if there is a *closed system*: that is, no decay product atoms are lost or gained, and the decay product atoms are clearly discernable from other atoms of the same element. Analytical instruments can accurately measure the amount of radioactive material and the amount of resulting decay product in rocks. By comparing these two quantities and knowing the rate of decay for a radioactive element, the age of the rock can thus be determined. Such dating is referred to as **radiometric** (*radio* = radioactivity, *metri* = measure) **age dating** and is an extremely powerful tool for determining the age of rocks.

For example, Figure 2–14 shows two locations where rocks contain small amounts of a particular type of radioactive uranium, which has a half-life of 713 million years and decays to a stable form of lead. Measurements at Location A indicate that half of the radioactive uranium has been converted to lead, its decay product. Thus, one half-life has elapsed and the rock is 713 million years old. At Location B, three-quarters of the original radioactive material has converted to its decay product, so the rock here must be older than that at Location A.

Two half-lives have elapsed and the rock is 1426 million (1.4 billion) years old. Using this method, hundreds of thousands of rock samples have been age dated from around the world.

The ages of rocks on Earth and the names of the geologic time periods are shown in the **geologic time scale** (Figure 2–15; see also Box 2–2). Initially, the divisions between geologic periods were based on major extinction episodes as recorded in the fossil record. As radiometric age dates became available, they were also included on the geologic time scale.

The time scale indicates Earth is 4.6 billion years old. The oldest known intact rocks on Earth, located in northwestern Canada, are 3.96 billion years old. In western Australia, however, geologists have recently discovered a 4.4 billion-year-old mineral fragment trapped within another rock. Starting with the origin of the first single-celled organisms about 3.5 billion years ago, the geologic time scale also shows important advances in the development of life forms on Earth.

> Earth scientists can accurately determine the age of most rocks by analyzing their radioactive components, some of which indicate that Earth is 4.6 billion years old.

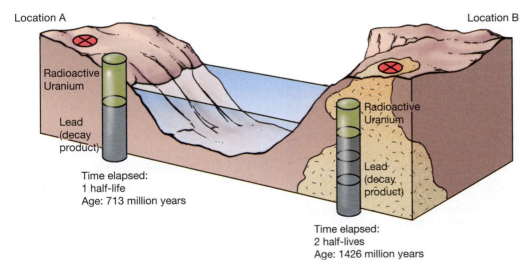

Figure 2–14 Radiometric age dating.

Locations A and B contain rocks with radioactive uranium that has a half-life of 713 million years and decays to lead. At Location A, one-half of the radioactive uranium has converted to lead, indicating that one half-life has elapsed so the rocks are 713 million years old. At Location B, three-quarters of the radioactive uranium has converted to lead, indicating that two half-lives have elapsed and that the rocks are 1426 million (1.4 billion) years old.

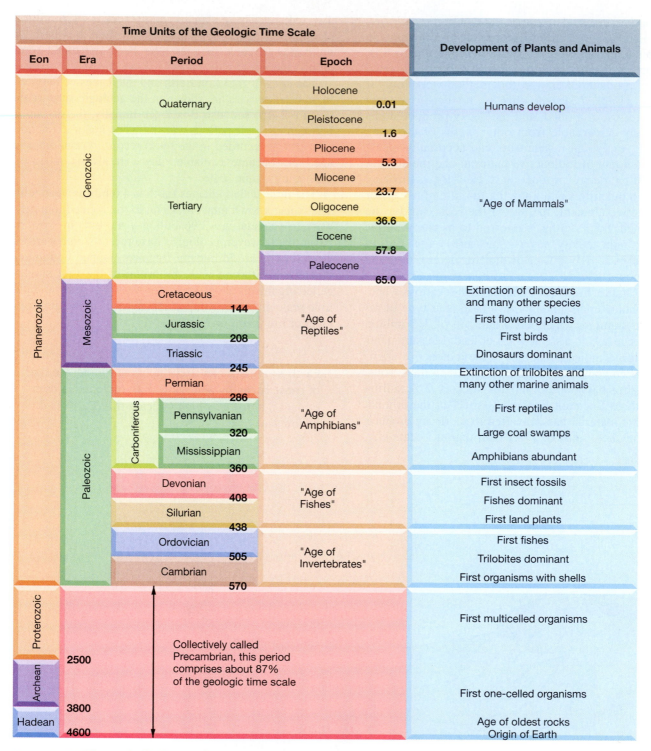

Figure 2–15 The geologic time scale.

Numbers on the time scale represent time in millions of years before the present; significant advances in the development of plants and animals on Earth are also shown.

BOX 2–2 Historical Feature

"DEEP" TIME

The time line of Earth history shown in Figure 2B indicates that Earth is 4.6 billion years old. It is difficult to comprehend how old Earth is, however, because 4.6 billion (that's 4600 million, or 4,600,000,000) is so enormous. To gain some idea of the immensity of a number in the billions, how high would a stack of one billion fresh one-dollar bills be (Table 2A)?

The problem with an example such as this is that the mind becomes immune to comprehending such large amounts of money—or any such large numbers. For instance, how long would it take to count to 4.6 billion if you counted one number every second for 24 hours a day, seven days a week, 365 days a year? The answer can be easily calculated, remembering that there are 60 seconds in a minute and 60 minutes in an hour. The rather surprising answer is given at the end of the Box if you need help.

TABLE 2A **Thickness of stacks of fresh one-dollar bills.**

Thickness of a stack of . . .	Would be . . .
100 fresh one-dollar bills	1 centimeter (0.4 inch) high
1000 fresh one-dollar bills	10 centimeters (4 inches) high
1,000,000 (one million) fresh one-dollar bills	100 meters (330 feet) high
1,000,000,000 (one billion) fresh one-dollar bills	100 kilometers (62 miles) high

Another way to try to visualize the enormity of geologic time is by representing its entirety with a roll of toilet paper.[12] If you use a standard 500-sheet roll and round off the age of Earth to 5 billion years, then each sheet on the roll represents 10 million years. As you unroll the paper, keep in mind that you are progressing through all of geologic time and that the last sheet on the roll (the last 10 million years) is about four times longer than the time humans have existed on Earth. In fact, all of recorded human history is represented by the last $^1/2000$th of a sheet, and a long human lifetime of 100 years is only $^1/100,000$th of a sheet! It is indeed humbling to realize how insignificant the length of time human existence on Earth has been.

Can one really imagine the space taken up by $1 billion, or the lifetimes it would take to count to 4.6 billion, or how many sheets of toilet paper have gone by in 4.6 billion years of Earth history? The same scale problem exists with visualizing millions or billions of years, often called "deep" time or geologic time. But keep trying, and maybe someday you will get a glimpse into the huge expanse of time represented by the geologic time scale. It will amaze you.

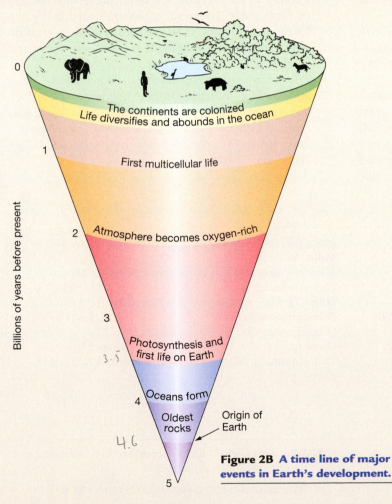

Billions of years before present

0 — The continents are colonized
Life diversifies and abounds in the ocean

1 — First multicellular life

2 — Atmosphere becomes oxygen-rich

3

3.5 — Photosynthesis and first life on Earth

Oceans form

4 — Oldest rocks — Origin of Earth

4.6

5

Figure 2B A time line of major events in Earth's development.

[12]Toilet paper has many advantages: It is something with which most people are familiar, it is inexpensive and readily available, it is long and linear (like geologic time), and it is even perforated into individual sheets.

Chapter in Review

- *Our galaxy, the Milky Way galaxy, is composed of numerous stars and is one of countless galaxies in the universe.* Galaxies are thought to be accumulations of debris from the *big bang*, the explosion that formed the universe about 13.7 billion years ago. Stars result from accumulation of gaseous masses that collapsed in on themselves, resulting in internal temperatures and pressures extreme enough to set off *fusion reactions*, burning hydrogen to produce helium. Stars give rise to heavier elements by successive expansions and contractions.

- *Our solar system, consisting of the Sun and nine planets, most likely formed from a huge cloud of gas and space dust called a nebula.* According to the *nebular hypothesis*, the nebular matter contracted to form the Sun, and the planets were formed from eddies of material that remained. The Sun, composed of hydrogen and helium, was massive enough and concentrated enough to emit large amounts of energy. The Sun also emitted ionized particles that swept away any nebular gas that remained from the formation of the planets and their satellites.

- *The Protoearth, more massive and larger than Earth today, was molten and homogenous.* The *initial atmosphere*, composed mostly of hydrogen and helium, was later driven off into space by intense solar radiation. The Protoearth began a period of rearrangement, forming a layered structure of core, mantle, and crust through *density stratification*. Also during this period, *outgassing* produced an early atmosphere rich in water vapor and carbon dioxide. Once Earth's surface cooled sufficiently, the water vapor condensed and accumu-

lated to give Earth its oceans. Rainfall on the surface dissolved compounds that, when carried to the ocean, made it salty.

- *Life is thought to have begun in the oceans.* Ultraviolet radiation from the Sun and hydrogen, carbon dioxide, methane, ammonia, and inorganic molecules from the oceans may have combined to produce carbon-containing molecules. Certain combinations of these molecules eventually produced *heterotrophic organisms* (which cannot make their own food) that were probably similar to present-day anaerobic bacteria. Eventually, *autotrophs evolved* that had the ability to make their own food through *chemosynthesis*. Later, some cells developed chlorophyll, which *made photosynthesis possible* and led to the development of plants.

- *Photosynthetic organisms altered the environment* by extracting carbon dioxide from the atmosphere and also by releasing free oxygen, which was lethal to many existing organisms. Eventually, *organisms evolved* that could use oxygen for respiration, thereby thriving in the oxygen-rich atmosphere that remains on Earth today. *Symbiotic relationships* led to the evolution of eukaryotic cells, and from these, a great diversity of plants and animals evolved into forms that inhabit Earth today.

- *Radiometric age dating is used to determine the age of most rocks.* Information from extinctions of organisms and from age dating rocks comprises the *geologic time scale*, which indicates that *Earth has experienced a long history of changes since its origin 4.6 billion years ago.*

Key Terms

Aerobic respiration (p. 56)
Anaerobic bacteria (p. 55)
Autotroph (p. 54)
Big Bang theory (p. 43)
Chemosynthesis (p. 55)
Core (p. 46)
Crust (p. 46)
Cyanobacteria (p. 56)
Density (p. 46)
Density stratification (p. 46)

Eukaryotic cell (p. 57)
Evolution (p. 57)
Excess volatile (p. 50)
Fusion reaction (p. 43)
Galaxy (p. 41)
Geologic time scale (p. 59)
Heterotroph (p. 54)
Hydrothermal vent (p. 51)
Light-year (p. 42)
Mantle (p. 46)

Milky Way galaxy (p. 41)
Miller, Stanley (p. 53)
Mitochondria (p. 56)
Natural selection (p. 57)
Nebula (p. 43)
Nebular hypothesis (p. 43)
Outgassing (p. 46)
Ozone (p. 51)
Photosynthesis (p. 55)
Plastid (p. 56)

Prokaryotic cell (p. 57)
Protoearth (p. 44)
Protoplanet (p. 44)
Radiometric age dating (p. 59)
Respiration (p. 55)
Solar system (p. 41)
Species (p. 57)
Symbiosis (p. 57)
Universe (p. 41)

Questions and Exercises

1. How do the observed motions of galaxies support the big bang theory?
2. Discuss the origin of the solar system using the nebular hypothesis.
3. How was the Protoearth different from today's Earth?
4. What is density stratification, and how did it change the Protoearth?

5. What is the origin of Earth's oceans and how is it related to the origin of Earth's atmosphere?
6. Have the oceans always been salty? Why or why not?
7. Describe some basic characteristics of living things and list the order in which they evolved.

8. How does the presence of oxygen (O_2) in our atmosphere help reduce the amount of ultraviolet radiation that reaches Earth's surface?

9. What was Stanley Miller's experiment, and what did it help demonstrate?

10. Discuss photosynthesis and respiration, and explain their relationship to the chemical processes of storing and releasing energy by organisms.

11. As plants evolved on Earth, great changes in Earth's environment were produced. Describe some of the major changes caused by plants.

12. Earth has had three atmospheres (initial, early, and present). Describe the composition and origin of each one.

13. What events must have occurred for life to evolve?

14. Construct a representation of the geologic time scale, using an appropriate quantity of any substance (other than dollar bills or toilet paper). Be sure to indicate some of the major changes that have occurred on Earth since its origin.

CHAPTER 3
Plate Tectonics and the Ocean Floor

Tall mountains created by tectonic uplift. Tall coastal mountains such as these in Glacier Bay National Park in southeast Alaska have been uplifted by plate tectonic processes, creating a large amount of relief. Some of the uplifted rocks here have come from distant areas and include parts of the sea floor.

Key Questions

■ What evidence did Alfred Wegener use to formulate his idea of continental drift?

■ How did the early idea of continental drift differ from the more modern version of plate tectonics?

■ What are the lines of evidence that support the theory of plate tectonics?

■ How do structure, composition, and physical properties vary within the deep Earth?

■ What types of features are found at the three main types of plate boundaries?

■ How do mantle plumes and hotspots fit into the plate tectonic model?

■ How have the features on Earth looked in the past and how will they look in the future?

The answers to these questions (and much more) can be found in the highlighted concept statements within this chapter.

"It is just as if we were to refit the torn pieces of a newspaper by matching their edges and then check whether the lines of print run smoothly across. If they do, there is nothing left but to conclude that the pieces were in fact joined in this way."

—Alfred Wegener, *The Origin of Continents and Oceans* (1929)

Several thousand earthquakes and dozens of volcanic eruptions occur on land and under the oceans each year, indicating that our planet is very dynamic. These events have occurred throughout history, constantly changing the surface of our planet, yet only a few decades ago most scientists believed the continents were stationary over geologic time. Since then, however, a bold new theory has been advanced that helps explain, for the first time, the following surface features and phenomena on Earth:

• The worldwide locations of volcanoes, faults, earthquakes, and mountain building

• Why mountains on Earth haven't been eroded away

• The origin of most landforms and ocean floor features

• How the continents and ocean floor formed, and why they are different

• The continuing development of Earth's surface

• The distribution of past and present life on Earth

This revolutionary new theory is called **plate tectonics** (*plate* = plates of the **lithosphere**; *tekton* = to build), or "the new global geology." According to the theory of plate tectonics, the outermost portion of Earth is composed of a patchwork of thin, rigid plates[1] that move hor-

izontally with respect to one another, like icebergs floating on water. The interaction of these plates as they move builds features of Earth's crust (such as volcanoes, mountain belts, and ocean basins). As a result, the continents are mobile and move about on Earth's surface, controlled by forces deep within Earth.

STUDENTS SOMETIMES ASK ...

How long has plate tectonics been operating on Earth? Will it ever stop?

It's difficult to say how long plate tectonics has been operating on Earth with much certainty because our planet has been so dynamic over the past several billion years. Evidence to support any conclusion about plate tectonics before 180 million years ago has been largely eliminated by the ongoing destruction of sea floor, which gets recycled on average about every 110 million years. However, very old marine rock sequences uplifted onto continents show certain characteristics of plate motion and suggest that plate tectonics has been operating for at least the last 3 billion years of Earth history.

Looking into the future, the forces that drive plates will likely decrease until plates no longer move. This is because plate tectonic processes are powered by heat released from within Earth (which is of a finite amount). However, the erosional work of water will continue to erode Earth's features. What a different world it will be then—an Earth with no earthquakes, no volcanoes, and no mountains. Flatness will prevail!

In this chapter, we examine the early ideas about the movement of plates, the evidence for those ideas, and how they led to the theory of plate tectonics. Then we look at Earth structure, plate boundaries, and some applications of plate tectonics, including what our planet may look like in the future.

Evidence for Continental Drift

Alfred Wegener (Figure 3–1), a German meteorologist and geophysicist, was the first to advance the idea of mobile continents in 1912. He envisioned that the continents were slowly drifting across the globe and called his idea **continental drift**. Let's examine the evidence that Wegener compiled that led him to formulate the idea of drifting continents.

Fit of the Continents

The idea that continents—particularly South America and Africa—fit together like pieces of a jigsaw puzzle originated with the development of reasonably accurate world maps.

---

[1]These thin, rigid plates are pieces of the lithosphere (*lithos* = rock, *sphere* = ball) that comprise Earth's outermost portion.

Figure 3–1 Alfred Wegener, circa 1912–1913.

As far back as 1620, Sir Francis Bacon wrote about how the continents appeared to fit together. However, little significance was given to this idea until 1912, when Wegener used the shapes of matching shorelines on different continents as a supporting piece of evidence for continental drift.

Wegener suggested that during the geologic past, the continents collided to form a large landmass, which he named **Pangaea** (*pan* = all, *gaea* = Earth) (Figure 3–2). Further, a huge ocean surrounded Pangaea, called **Panthalassa** (*pan* = all, *thalassa* = sea). Panthalassa, the ancient precursor of the Pacific Ocean, included several smaller seas, including the **Tethys** (*Tethys* = a Greek seagoddess) **Sea**. Wegener's evidence indicated that about 200 million years ago, Pangaea began to split apart, and the various continental masses started to drift toward their present geographic positions.

Wegener's attempt at matching shorelines revealed considerable areas of crustal overlap and large gaps. Some of the differences could be explained by material deposited by rivers or eroded from coastlines. What Wegener didn't know was the shallow parts of the ocean floor close to shore are closely related to the continents. In the early 1960s, Sir Edward Bullard and two associates used a computer to fit the continents together (Figure 3–3). Instead of using the shorelines of the continents as Wegener had done, Bullard achieved the best fit (i.e., with minimal overlaps or gaps) by using a depth of 2000 meters (6560 feet) below sea level. This depth corresponds to halfway between the shoreline and the deep ocean basins and represents the true edge of the continents. By using this depth, the overall fit of the continents was even better than expected.

Matching Sequences of Rocks and Mountain Chains

If the continents were once together as Wegener had hypothesized, then evidence should appear in rock sequences that were originally continuous but now separated by large distances. To test the idea of drifting continents, geologists began comparing the rocks along the edges of continents with rocks found in adjacent positions on matching continents. They wanted to see if the

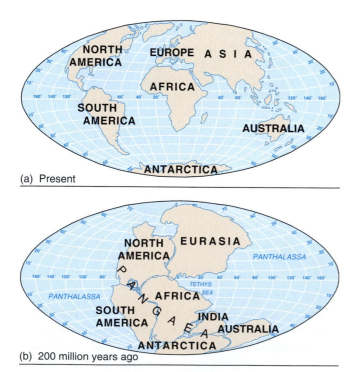

(a) Present

(b) 200 million years ago

Figure 3–2 Reconstruction of Pangaea.

The positions of the continents about 200 million years ago, showing the supercontinent of Pangaea and the single large ocean, Panthalassa.

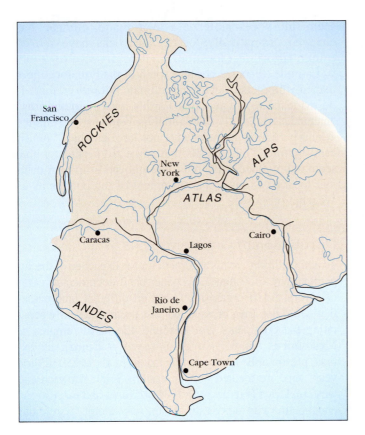

Figure 3–3 Computer fit of the continents.

In 1965, Sir Edward Bullard used a depth of 2000 meters (6560 feet) for a best-fit match of the continents, which shows few gaps and minimal overlap.

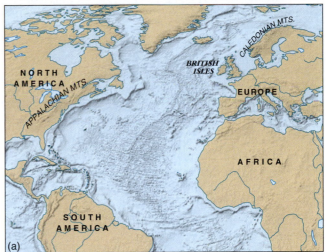

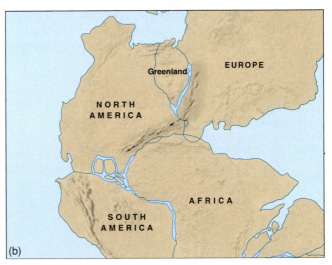

Figure 3–4 Matching mountain ranges across the North Atlantic Ocean.

(a) Present-day positions of continents and mountain ranges.
(b) Positions of the continents about 300 million years ago, showing how mountain ranges with similar age, type, and structure form one continuous belt.

rocks had similar types, ages, and structural styles (the type and degree of deformation). In some areas younger rocks had been deposited during the millions of years since the continents separated, covering the rocks that held the key to the past history of the continents. In other areas, the rocks had been eroded away. Nevertheless, in many other areas the key rocks were present.

These studies showed that many rock sequences from one continent matched up with identical rock sequences on a matching continent—although the two were separated by an ocean. In addition, mountain ranges that terminated abruptly at the edge of a continent continued on another continent across an ocean basin, with identical rock sequences, ages, and structural styles. Figure 3–4 shows, for example, how similar rocks from the Appalachian Mountains in North America match up with identical rocks from the British Isles and the Caledonian Mountains in Europe.

Wegener noted the similarities in rock sequences on both sides of the Atlantic and used the information as a supporting piece of evidence for continental drift. He suggested that mountains such as those seen on opposite sides of the Atlantic formed during the collision when Pangaea was formed. Later, when the continents split apart, once-continuous mountain ranges were separated. Confirmation of this idea exists in a similar match with mountains extending from South America through Antarctica and across Australia.

Glacial Ages and Other Climate Evidence

Wegener also noticed the occurrence of past glacial activity in areas now tropical and suggested that it, too,

provided supporting evidence for drifting continents. Currently, the only places in the world where thick continental *ice sheets* occur are in the polar regions of Greenland and Antarctica. However, evidence of ancient glaciation is found in the lower latitude regions of South America, Africa, India, and Australia.

These deposits, which have been dated at 300 million years old, indicate one of two possibilities: (1) There was a worldwide **ice age** and even tropical areas were covered by thick ice, or (2) some continents that are now in tropical areas were once located much closer to one of the poles. It is unlikely that the entire world was covered by ice 300 million years ago because coal deposits from the same geologic age now present in North America and Europe originated as vast semitropical swamps. Thus, a reasonable conclusion is that some of the continents must have been closer to the poles than they are today.

There is another type of glacial evidence that indicates certain continents have moved from more polar regions during the last 300 million years. When glaciers flow, they move and abrade the underlying rocks, leaving grooves that indicate the direction of flow. The blue arrows in Figure 3–5a show how the glaciers would have flowed away from the South Pole on Pangaea 300 million years ago. The direction of flow is consistent with the grooves found on many continents today (Figure 3–5b), providing additional evidence for drifting continents.

Many examples of plant and animal fossils indicate very different climates than today. Two such examples are fossil palm trees in Arctic Spitsbergen and coal deposits in Antarctica. Earth's past environments can be interpreted from these rocks because plants and animals need specific environmental conditions in which to live. Corals, for example, generally need seawater above 18 degrees Centigrade (°C) or 64 degrees Fahrenheit (°F) in order to survive. When fossil corals are found in areas that are cold today, two explanations

seem most plausible: (1) Worldwide climate has changed dramatically; or (2) the rocks have moved from their original location.

Latitude (distance north or south of the Equator), more than anything else, determines climate. Moreover, there is no evidence to suggest that Earth's axis of rotation has changed significantly throughout its history, so the climate at any particular latitude must not have changed significantly either. Thus, fossils that come from climates that seem out of place today must have moved from their original location through the movement of the continents as Wegener proposed.

Distribution of Organisms

To add credibility to his argument for the existence of the supercontinent of Pangaea, Wegener cited documented cases of several fossil organisms found on different landmasses that could not have crossed the vast oceans presently separating the continents. For example, the fossil remains of **Mesosaurus** (*meso* = middle, *saurus* = lizard), an extinct, presumably aquatic reptile that lived about 250 million years ago, are located only in eastern South America and western Africa (Figure 3–6). If *Mesosaurus* had been strong enough to swim across an ocean, why aren't its remains more widely distributed?

Figure 3–5 Ice age on Pangaea.

(a) Reconstruction of the supercontinent Pangaea, showing the area covered by glacial ice about 300 million years ago. Arrows indicate direction of ice flow.
(b) The positions of the continents today.

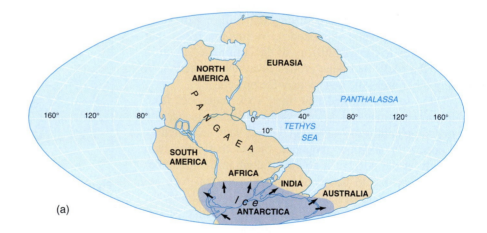

(a)

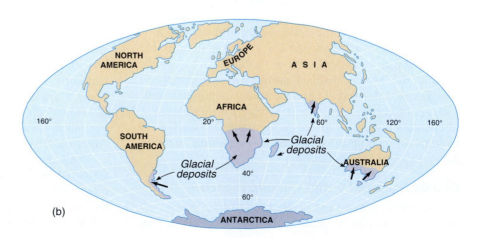

(b)

Wegener's idea of continental drift provided an elegant solution to this problem. He suggested that the continents were closer together in the geologic past, so *Mesosaurus* didn't have to be a good swimmer to leave remains on two different continents. Later, after *Mesosaurus* became extinct, the continents moved to their present-day positions, and a large ocean now separates the once-connected landmasses. Other examples of similar fossils on different continents include those of plants, which would have had a difficult time traversing a large ocean.

Before continental drift, several ideas were proposed to help explain the curious pattern of these fossils, such as the existence of island stepping stones or a land bridge. It was even suggested that at least one pair of land-dwelling *Mesosaurus* survived the arduous journey across several thousand kilometers of open ocean by rafting on floating logs. However, there is no evidence to support the idea of island stepping stones or a land bridge and the idea of *Mesosaurus* rafting across an ocean seems implausible.

Wegener also cited the distribution of present-day organisms as evidence to support the concept of drifting continents. For example, modern organisms with similar ancestries clearly had to evolve in isolation during the last few million years. Most obvious of these are the Australian marsupials (such as the kangaroos, koalas, and wombats), which have a distinct similarity to the marsupial opossums found in the Americas.

Objections to the Continental Drift Model

In 1915, Wegener published his ideas in *The Origins of Continents and Oceans*, but the book did not attract much attention until it was translated into English, French, Spanish, and Russian in 1924. From that point until his death in 1930,[2] his drift hypothesis received much hostile criticism—and sometimes open ridicule—from the scientific community because of the mechanism he proposed

for the movement of the continents. Wegener suggested the continents plowed through the ocean basins to reach their present day positions and that the leading edges of the continents deformed into mountain ridges because of the drag imposed by ocean rocks. Further, the driving mechanism he proposed was a combination of the gravitational attraction of Earth's equatorial bulge and tidal forces from the Sun and Moon.

Scientists rejected the idea as too fantastic and contrary to the laws of physics. Material strength calculations showed that ocean rock was too strong for continental rock to plow through it and analysis of gravitational and tidal forces indicated that they were too small to move the great continental landmasses. Even without an acceptable mechanism, many geologists who studied rocks in South America and Africa accepted continental drift. North American geologists—most of whom were unfamiliar with these Southern Hemisphere rock sequences—remained highly skeptical.

As compelling as Wegener's evidence may seem today, he was unable to convince the scientific community as a whole of the validity of his ideas. Although his hypothesis was correct in principle, it contained several incorrect details, such as the driving mechanism for continental motion and how continents move across ocean basins. In order for any scientific viewpoint to gain wide acceptance, it must explain all available observations and have supporting evidence from a wide variety of scientific fields. This supporting evidence would not come until more details of the nature of the ocean floor were revealed, which, along with new technology that enabled scientists to determine the original positions of rocks on Earth, provided additional observations in support of drifting continents.

---

[2]Wegener perished in 1930 during an expedition in Greenland while collecting data to help support his idea of continental drift.

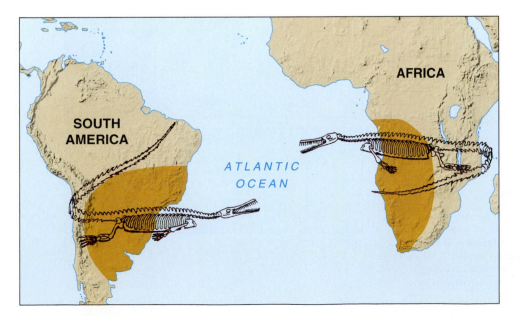

Figure 3–6 Fossils of *Mesosaurus.*

Mesosaurus fossils are found only in South America and Africa and appear to link these two continents.

Alfred Wegener used a variety of interdisciplinary information from land to support continental drift. However, he did not have a suitable mechanism or any information about the sea floor.

Evidence for Plate Tectonics

Very little new information about Wegener's continental drift hypothesis was introduced between the time of Wegener's death in 1930 and the early 1950s. However, bathymetric studies of the sea floor using sonar that were initiated during. World War II and continued after the war provided critical evidence in support of drifting continents. In addition, technology unavailable in Wegener's time enabled scientists to begin analyzing the way rocks retained the signature of Earth's **magnetic field**. These developments caused scientists to reexamine continental drift and advance it into the more encompassing theory of plate tectonics.

Earth's Magnetic Field and Paleomagnetism

Earth's magnetic field is shown in Figure 3–7. The invisible lines of magnetic force that originate within Earth and travel out into space resemble the magnetic field produced by a large bar magnet.[3] Similar to Earth's magnetic field, the ends of a bar magnet have opposite polarities (labeled either + and − or N for north and S for south)

that cause magnetic objects to align parallel to its magnetic field. In addition, notice in Figure 3–7b that Earth's geographic north pole (the rotational axis) and Earth's magnetic north pole (magnetic north) do not coincide.

STUDENTS SOMETIMES ASK...
What causes Earth's magnetic field?

Studies of Earth's magnetic field and research in the field of *magnetodynamics* suggest that convective movement of fluids in Earth's liquid iron-nickel outer core is the cause of Earth's magnetic field. The most widely accepted view is that the core behaves like a self-sustaining *dynamo*, which converts the energy in convective motion into magnetic energy. Interestingly, most other planets (and even some planet's moons) have magnetic fields. The Sun has a strong magnetic field, which reverses its orientation about every 22 years and is closely tied to the Sun's 11-year sunspot cycle.

---

[3]The properties of a magnetic field can be explored easily enough with a bar magnet and some iron particles. Place the iron particles on a table and place a bar magnet nearby. Depending on the strength of the magnet, you should get a pattern resembling Figure 3-7a.

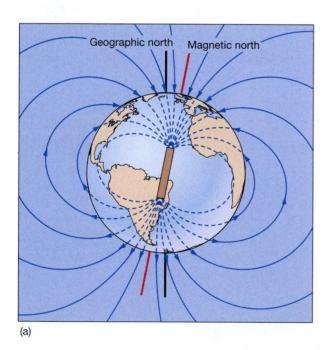

(a)

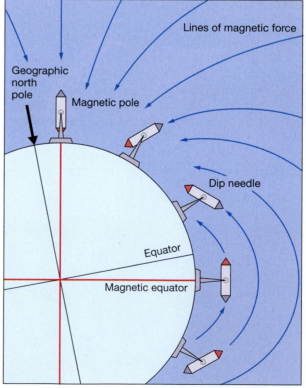

(b)

Figure 3–7 Earth's magnetic field.

(a) Earth's magnetic field generates invisible lines of magnetic force similar to a large bar magnet. Note that magnetic north and true north are not in the same exact location. **(b)** Earth's magnetic field causes a dip needle to align parallel to the lines of magnetic force and change orientation with increasing latitude. Consequently, the latitude can be determined based on the dip angle.

Rocks Affected by Earth's Magnetic Field Recall from Chapter 2 that *igneous* (*igne* = fire, *ous* = full of) *rocks* solidify from molten *magma* (*magma* = a mass) either underground or after volcanic eruptions at the surface, which produce *lava* (*lava* = to wash). Nearly all igneous rocks contain **magnetite**, a naturally magnetic iron mineral. Particles of magnetite in magma align themselves with Earth's magnetic field because magma and lava are fluid. Volcanic lavas such as basalt are high in magnetite and solidify from molten material at temperatures in excess of 1000°C (1832°F); however, the magnetic signatures are not set until the rock cools below 600°C (1112°F), the temperature called the **Curie point** after Pierre Curie, a French physical chemist known for his work, with his wife Marie, on radioactivity.

At the Curie point, magnetite particles in igneous rocks become fixed in the direction of Earth's magnetic field and record the angle of Earth's magnetic field at that place and time. In essence, grains of magnetite serve as tiny compass needles that record the strength and orientation of Earth's magnetic field. Unless the rock is again heated to the Currie point, which would allow magnetite grains to be mobile, these magnetite grains contain information about the magnetic field where the rock originated regardless of where the rock subsequently moves.

Magnetite is also deposited in sediments. As long as the sediment is surrounded by water, the magnetite particles can align themselves with Earth's magnetic field. As discussed in Chapter 2, sediment can be buried and solidified into *sedimentary* (*sedimentum* = settling) rock. As this occurs, the particles are no longer able to realign themselves if they are subsequently moved. Thus, magnetite grains in sedimentary rocks also contain information about the magnetic field where the rock originated. Although other rock types have been successfully used to reveal information about Earth's ancient magnetic field, the most reliable ones are igneous rocks that have high concentrations of magnetite such as basalt.

Paleomagnetism The study of Earth's ancient magnetic field is called **paleomagnetism** (*paleo* = ancient). The scientists who study paleomagnetism analyze magnetite particles in rocks to determine not only their north–south direction but also their angle relative to Earth's surface. The degree to which a magnetite particle points into Earth is called its **magnetic dip**, or **magnetic inclination**.

Magnetic dip is directly related to latitude. Figure 3–7b shows that a dip needle does not dip at all at Earth's magnetic equator. Instead, the needle lies horizontal to Earth's surface. At Earth's magnetic north pole, however, a dip needle points straight into the surface. A dip needle at Earth's south magnetic pole is also vertical to the surface, but it points out instead of in. Thus, magnetic dip increases with increasing latitude, from 0 degrees at the magnetic equator to 90 degrees at the magnetic poles. Because magnetic dip is retained in magnetically oriented rocks, measuring the dip angle reveals the latitude at which the rock initially formed. For instance, rocks found today near the Equator in southern India with a high magnetic dip angle suggest that they were not formed at that location but at a higher latitude. Done with care, paleomagnetism is an extremely powerful tool for interpreting where rocks first formed. Based on paleomagnetic studies, convincing arguments could finally be made that the continents had drifted relative to one another.

Apparent Polar Wandering When magnetic dip data for rocks on the continents were used to determine the apparent position of the magnetic north pole over time, it appeared that the magnetic pole was wandering. Figure 3–8a, for example, shows the **polar wandering curves** for North America and Eurasia. Both curves have a similar shape but, for all

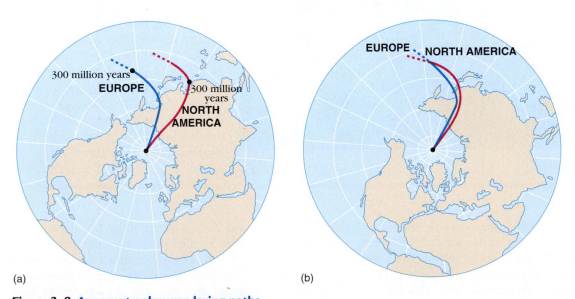

(a) (b)

Figure 3–8 Apparent polar wandering paths.

(a) Apparent polar wandering paths for North America and Eurasia resulted in a dilemma because they were not in alignment. (b) The positions of the polar wandering paths when the landmasses are assembled.

BOX 3–1 Research Methods in Oceanography

DO SEA TURTLES (AND OTHER ANIMALS) USE EARTH'S MAGNETIC FIELD FOR NAVIGATION?

Sea turtles travel great distances across the open ocean so they can lay their eggs on the island where they themselves were hatched. How do they know where the island is located and how do they navigate at sea during their long voyage? Studies have indicated that during their migration, green sea turtles (*Chelonia mydas*; Figure 3A) often travel in an essentially straight-line path to reach their destination. One hypothesis suggests that, like the Polynesian navigators, the sea turtles use wave direction to help them steer. However, sea turtles have been radio-tagged and tracked by satellites, which reveals that they continue along their straight-line path independent of wave direction.

Research in *magnetoreception*, the study of an animal's ability to sense magnetic fields, suggests that sea turtles use Earth's magnetic field for navigation. For instance, turtles can distinguish between different magnetic inclination angles, which in effect would allow them to sense latitude. Sea turtles can also distinguish magnetic field intensity, a rough indication of longitude. By sensing these two magnetic field properties, a sea turtle could determine its position at sea and relocate a tiny island thousands of kilometers away. Like any good navigator, sea turtles may also use other tools, such as olfactory (scent) clues, Sun angles, local landmarks, and oceanographic phenomena.

Other animals may also use magnetic properties to navigate. For example, some whales and dolphins may detect and follow the magnetic stripes on the sea floor during their movements, which may help to explain why whales sometimes beach themselves. In addition, certain bacteria use the magnetic mineral magnetite to align themselves parallel to Earth's magnetic field. Subsequently, magnetite has been found in many other organisms that have a "homing" ability, including tuna, salmon, honeybees, pigeons, turtles, and even humans. What has been unclear is how these animals detect—and potentially use—Earth's magnetic field. Recent findings by a research team studying rainbow trout (close relatives of salmon) have traced magnetically receptive fibers of nerves back to the brain, more closely linking a magnetic sense with an organism's sensory system.

Do humans have an innate ability to use Earth's magnetic field for navigation? Studies conducted on humans indicate that the majority of people can identify north after being blindfolded and disoriented. Interestingly, many people point *south* instead of north, but this direction is along the lines of magnetic force. Similarly, migratory animals that rely on magnetism for navigation will not be confused by a reversal in Earth's magnetic field and will still be able to get to where they need to go. The detection of a directional sense in animals seems likely to remain an intriguing and elusive mystery in animal behavior.

Figure 3A Green sea turtle.

rocks older than about 70 million years, the pole determined from North American rocks lies to the west of that determined from Eurasian rocks. There can be only one north and one south magnetic pole at any given time, however, and it is unlikely that their positions change with time. This discrepancy implies that magnetic poles remained stationary while North America and Eurasia moved relative to the pole and relative to each other. Figure 3–8b shows that when the continents are moved into the positions they occupied when they were part of Pangaea, the two wandering curves match up, providing strong evidence that the continents have moved throughout geologic time.

Magnetic Polarity Reversals Magnetic compasses on Earth today follow lines of magnetic force and point toward magnetic north. It turns out, however, that the **polarity** (the directional orientation of the magnetic field) has reversed itself periodically throughout geologic time. Thus, the north magnetic pole has become the south magnetic pole and vice versa. Figure 3–9 shows how rocks have recorded the switching of Earth's magnetic polarity through time.

The time during which a particular paleomagnetic condition existed ("normal" or "reversed") can be determined by radiometric dating. Over the last 76 million years, magnetic polarity has switched irregularly at the rate of about once or twice each million years. It takes a few thousand years for a change in polarity to occur and is identified in rock sequences by a gradual decrease in the intensity of the magnetic field of one polarity, followed by a gradual increase in the intensity of the magnetic field of opposite polarity. Earth's present magnetic field has been weakening during the past 150 years, which suggests Earth's current "normal" polarity might reverse itself within the next 2000 years.

Paleomagnetism and the Ocean Floor Paleomagnetism had certainly proved its usefulness on land, but, up until the mid-1950s, it had only been conducted on continental rocks. Would the ocean floor also show variations in magnetic polarity? To test this idea, the United States Coast and Geodetic Survey in conjunction with scientists from Scripps Institution of Oceanography undertook an extensive deep-water mapping program off Oregon and Washington in 1955. Using a sensitive instrument called a *magnetometer* (*magneto* = magnetism, *meter* = measure), which is towed behind a research ship, the scientists spent several weeks at sea moving back and forth in a regularly spaced pattern, measuring Earth's magnetic field and how it was affected by the magnetic properties of rocks on the ocean floor.

When the scientists analyzed their data, it revealed that the entire surveyed area had a pattern of north–south stripes in a surprisingly regular and alternating pattern of above-average and below-average magnetism. What was even more surprising was that the pattern appeared to be symmetrical with respect to a long mountain range that was fortuitously in the middle of their survey area.

Detailed paleomagnetism studies of this and other areas of the sea floor confirmed that a similar pattern of alternating stripes of above-average and below-average magnetism. These stripes are called **magnetic anomalies** (*a* = without, *nomo* = law; an anomaly is a departure from normal conditions). The ocean floor had embedded in it a regular pattern of alternating magnetic stripes unlike anywhere on land. Researchers had a difficult time explaining why the ocean floor had such a regular pattern of magnetic anomalies. Nor could they explain how the sequence on one side of the underwater mountain range matched the sequence on the opposite side—in essence, they were a mirror image of each other. To understand how this pattern could have formed, more information about ocean floor features and their origin was needed.

Figure 3–9 Paleomagnetism preserved in rocks.

The switching of Earth's magnetic polarity through time is preserved in rocks like these lava flows.

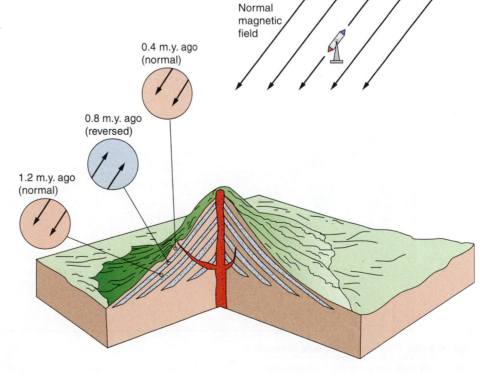

Sea Floor Spreading and Features of the Ocean Basins

Geologist *Harry Hess* (1906–1969), when he was a U.S. Navy captain in World War II, developed the habit of leaving his depth recorder on at all times while his ship was traveling at sea. After the war, compilation of these and many other depth records showed extensive mountain ridges near the centers of ocean basins and extremely deep, narrow trenches at the edges of ocean basins. In 1960, Hess published the idea of **sea floor spreading** with **convection** (*con* = with, *vect* = carried) **cells** as the driving mechanism (Figure 3–10). He suggested that new ocean crust was created at the ridges, split apart, moved away from the ridges, and later disappeared back into the deep Earth at trenches. Mindful of the resistance of North American scientists to the idea of continental drift, Hess referred to his own work as "geopoetry."

As it turns out, Hess's initial ideas about sea floor spreading have been confirmed. The **mid-ocean ridge** (Figure 3–10) is a continuous underwater mountain range that winds through every ocean basin in the world and resembles the seam on a baseball. It is entirely volcanic in origin, wraps one-and-a-half times around the globe, and rises over 2.5 kilometers (1.5 miles) above the surrounding deep ocean floor. It even rises above sea level in places such as Iceland. New ocean floor forms at the crest, or axis, of the mid-ocean ridge. By the process of sea-floor spreading, new ocean floor is split in two and carried away from the axis, replaced by the upwelling of volcanic material that fills the void with new strips of sea floor. Sea floor spreading occurs along the axis of the mid-ocean ridge, which is

referred to as a **spreading center**. One way to think of the mid-ocean ridge is as a zipper that is being pulled apart. Thus, Earth's zipper (the mid-ocean ridge) is becoming unzipped!

At the same time, ocean floor is being destroyed at **ocean trenches**. Trenches are the deepest parts of the ocean floor and resemble a narrow crease or trough (Figure 3–10). Some of the largest earthquakes in the world occur near these trenches, caused by a lithospheric plate that bends downward and slowly plunges back into Earth's interior. This process is called **subduction** (*sub* = under, *duct* = lead), and the sloping area from the trench along the downward plate is called a **subduction zone**.

In 1963, geologists **Fredrick Vine** and **Drummond Matthews** of Cambridge University combined the seemingly unrelated pattern of magnetic sea floor stripes with the process of sea floor spreading to explain the perplexing pattern of alternating and symmetric stripes on the sea floor (Figure 3–11). They proposed that the above-average magnetic stripes represented "normal" polarity sea floor rocks that enhanced Earth's current normal magnetic polarity and the below-average magnetic stripes represented the presence of "reverse" polarity sea floor rocks that subtracted from Earth's current polarity. The pattern could be created when newly formed rocks at the mid-ocean ridge are magnetized with whichever polarity exists on Earth during their formation. As those rocks are slowly moved away from the crest of the mid-ocean ridge, the periodic switches of Earth's magnetic polarity are recorded in subsequent rock. The result is an alternating pattern of magnetic polarity stripes that are symmetric with respect to the mid-ocean ridge.

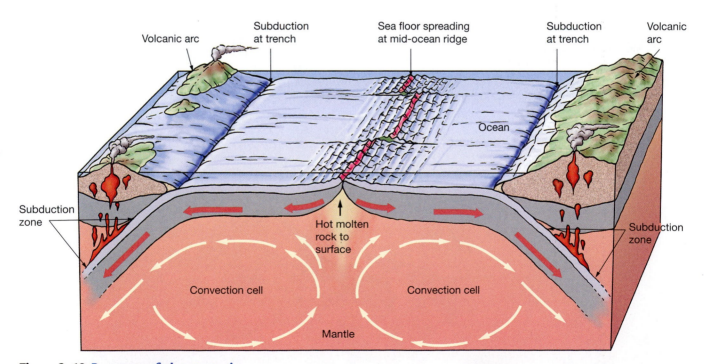

Figure 3–10 Processes of plate tectonics.

Hot molten rock comes to the surface at the mid-ocean ridge and moves outward by the process of sea floor spreading. Eventually, sea floor is destroyed at the trenches, where the process of subduction occurs. Convection of material in the mantle produces convection cells.

The pattern of alternating reversals of Earth's magnetic field as recorded in the sea floor was the most convincing piece of evidence set forth to support the concept of sea floor spreading—and, as a result, continental drift. However, the continents weren't plowing through the ocean basins as Wegener had envisioned. Instead, the ocean floor was a conveyer belt that was being continuously formed at the mid-ocean ridge and destroyed at the trenches, with the continents just passively riding along on the conveyer. By the late 1960s, most geologists had changed their stand on continental drift in light of this new evidence.

> The plate tectonic model states that new sea floor is created at the mid-ocean ridge where it moves outward by the process of sea floor spreading and is destroyed by subduction into ocean trenches.

Other Evidence from the Ocean Basins

Even though the tide of scientific opinion had indeed switched to favor a mobile Earth, additional evidence from the ocean floor would further support the ideas of continental drift and sea floor spreading.

Age of the Ocean Floor In the late 1960s, an ambitious deep-sea drilling program was initiated to test the existence of sea floor spreading. One of the program's primary missions was to drill into and collect ocean floor rocks for radiometric age dating. If sea floor spreading does indeed occur, then the youngest sea floor rocks would be atop the mid-ocean ridge and the ages of rocks would increase on either side of the ridge in a symmetric pattern.

The map in Figure 3–12, showing the age of the ocean floor beneath deep-sea deposits, is based on the pattern of magnetic stripes verified with thousands of radiometrically age dated samples. It shows the ocean floor is youngest along the mid-ocean ridge, where new ocean floor is created, and the age of rocks increases with increasing distance in either direction away from the axis of the ridge. The symmetric pattern of ocean floor ages confirms that the process of sea floor spreading must indeed be occurring.

The Atlantic Ocean has the simplest and most symmetric pattern of age distribution in Figure 3–12. The pattern results from the newly formed Mid-Atlantic Ridge that rifted Pangaea apart. The Pacific Ocean has the least symmetric pattern because many subduction zones sur-

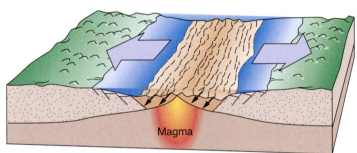

Figure 3–11 Magnetic evidence of sea floor spreading.

As new basalt is added to the ocean floor at mid-ocean ridges, it is magnetized according to Earth's existing magnetic field.

(a) Period of normal magnetism

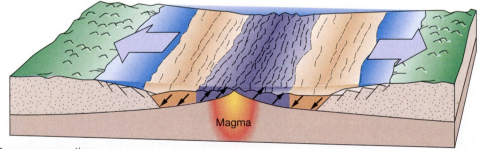

(b) Period of reverse magnetism

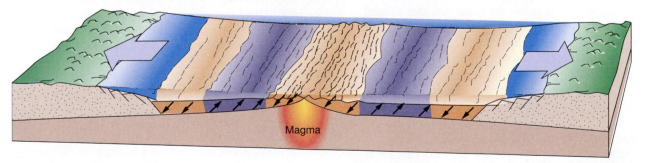

(c) Period of normal magnetism

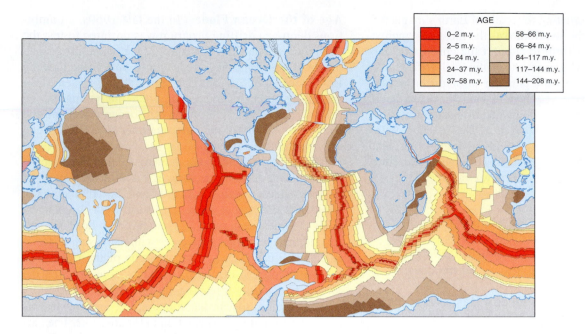

AGE

0–2 m.y.	58–66 m.y.
2–5 m.y.	66–84 m.y.
5–24 m.y.	84–117 m.y.
24–37 m.y.	117–144 m.y.
37–58 m.y.	144–208 m.y.

Figure 3–12   Age of the ocean crust beneath deep-sea deposits.

The youngest rocks (*bright red areas*) are found along the mid-ocean ridge. Farther away from the mid-ocean ridge, the rocks increase linearly in age in either direction. Age shown in millions of years before present.

round it. For example, ocean floor east of the East Pacific Rise that is older than 40 million years old has already been subducted. The ocean floor in the northwestern Pacific, about 180 million years old, has not yet been subducted. A portion of the East Pacific Rise has even disappeared under North America. The age bands in the Pacific Ocean are wider than those in the Atlantic and Indian Oceans, which suggests the rate of sea floor spreading is greatest in the Pacific Ocean.

Recall from Chapter 2 that the ocean is at least 4 billion years old. However, the oldest ocean floor is only 180 million years old (or 0.18 billion years old), and the majority of the ocean floor is not even half that old (see Figure 3–12). How could the ocean floor be so incredibly young while the oceans themselves are so phenomenally old? According to plate tectonic theory, new ocean floor is created at the mid-ocean ridge by sea floor spreading and moves off the ridge to eventually be subducted and remelted in the mantle. In this way, the ocean floor keeps regenerating itself. The floor beneath the oceans today is not the same one that existed beneath the oceans 4 billion years ago.

If the rocks that comprise the ocean floor are so young, why are continental rocks so old? Based on radiometric age dating, the oldest rocks on land are about 4 billion years old. Many other continental rocks approach this age, implying that the same processes that constantly renew the sea floor do not operate on land. Rather, evidence suggests that continental rocks do not get recycled by the process of sea floor spreading and thus remain at Earth's surface for long periods of time.

STUDENTS SOMETIMES ASK ...

How fast do plates move, and have they always moved at the same rate?

Currently, plates move an average of 2 to 12 centimeters (1 to 5 inches) per year, which is about as fast as a person's fingernails grow. A person's fingernail growth is dependent on many factors, including heredity, gender, diet, and amount of exercise, but averages about 8 centimeters (3 inches) per year. This may not sound very fast, but the plates have been moving for millions of years. Even an object moving slowly will eventually travel a great distance over a very long time. For instance, fingernails growing at a rate of 8 centimeters (3 inches) per year for 1 million years would be 80 kilometers (50 miles) long!

Evidence shows the plates were moving faster millions of years ago. Geologists can determine the rate of plate motion in the past by analyzing the width of new oceanic crust produced by sea floor spreading since fast spreading produces more sea floor rock. (Using this relationship and by looking at Figure 3–12, you should be able to determine whether the Pacific Ocean or the Atlantic Ocean has a faster spreading rate.) Recent studies using this same technique indicate that about 50 million years ago, India attained a speed of 19 centimeters (7.5 inches) per year. Other research indicates that about 530 million years ago, plate motions may have been as high as 30 centimeters (1 foot) per year! What caused these rapid bursts of plate motion? Geologists are not sure why plates moved more rapidly in the past, but increased heat release from Earth's interior is a likely mechanism.

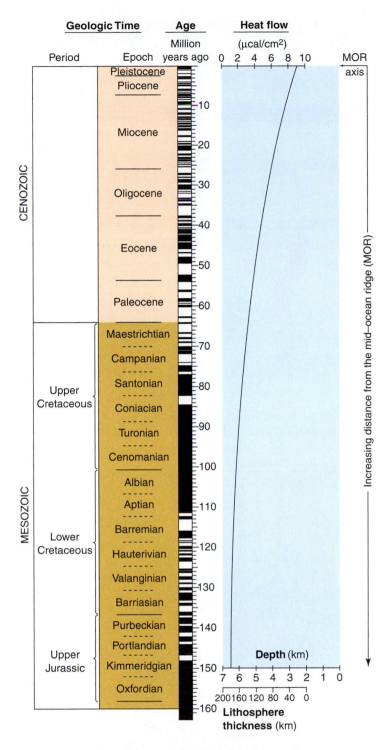

Geologic Time | **Age** | **Heat flow**

Figure 3–13 Relationships relative to the mid-ocean ridge.

From left, geologic time scale; magnetic polarity reversal scale (*black bands* = normal polarity; *white bands* = reversed polarity); geologic age in millions of years before present; and graph showing decreasing heat flow, increasing ocean depth, and increasing lithosphere thickness with increasing distance from the mid-ocean ridge.

as little as one-tenth the average. Increased heat flow at the mid-ocean ridge and decreased heat flow at subduction zones is what would be expected based on thin crust at the mid-ocean ridge and a double thickness of crust at the trenches (see Figure 3–10).

A summary of various physical relationships relative to the mid-ocean ridge is shown in Table 3–1. Moreover, many of these relationships exist because as sea floor spreading brings new crustal material to the surface along the mid-ocean ridge, it cools; as it cools, it contracts, and as it contracts, it subsides. Figure 3–13 is a graphical representation of many of these same relationships but also shows specific geologic ages and the pattern of magnetic polarity reversals through time.

Worldwide Earthquakes

Earthquakes are sudden releases of energy caused by fault movement or volcanic eruptions. The map in Figure 3–14a shows that most large earthquakes occur along ocean trenches, reflecting the energy released during subduction. Other earthquakes occur along the mid-ocean ridge, reflecting the energy released during sea floor spreading. Still others occur along major faults in the sea floor and on land, reflecting the energy released when moving plates contact other plates along their edges. The two maps in Figure 3–14 show that the distribution of worldwide earthquakes closely matches the locations of plate boundaries.

Heat Flow The heat from Earth's interior is released to the surface as **heat flow**. Current models indicate this heat moves to the surface with magma in convective motion. Most of the heat is carried to regions of the mid-ocean ridge spreading centers (see Figure 3–10). Cooler portions of the mantle descend in deep-sea trenches to complete each circular-moving convection cell.

Heat flow measurements show the amount of heat flowing to the surface along the mid-ocean ridge can be up to eight times greater than the average amount flowing to other parts of Earth's crust. Additionally, heat flow at deep-sea trenches, where ocean floor is subducted, can be

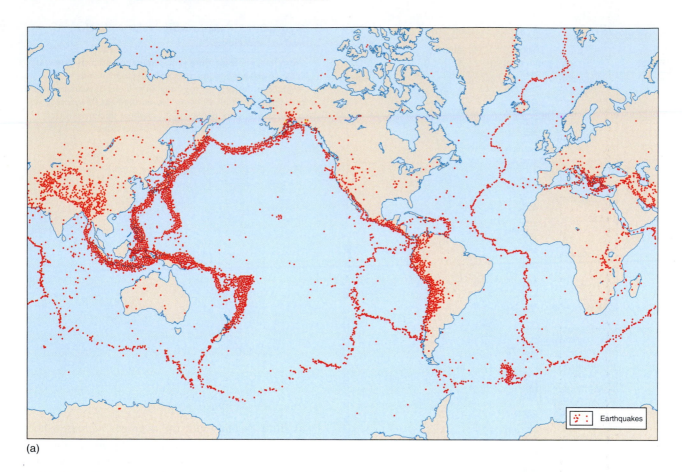

(a)

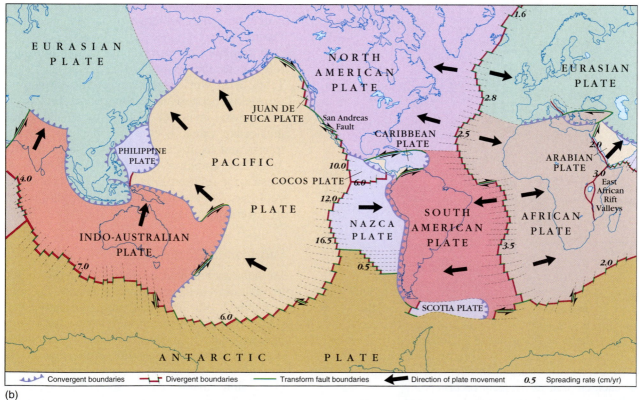

(b)

Figure 3–14 Earthquakes and lithospheric plates.

(a) Distribution of earthquakes with magnitudes equal to or greater than $M_w = 5.0$ for the period 1980–1990. **(b)** Plate boundaries define the major lithospheric plates (*shaded*), with arrows indicating the direction of motion and numbers representing the rate of motion in centimeters per year. Notice how closely the pattern of major earthquakes follows plate boundaries.

TABLE 3–1 **Relationships relative to increasing distance in either direction from the axis of a mid-ocean ridge.**

Relationship	Reason
1. Volcanic activity decreases	Magma chambers are concentrated only along the crest of the mid-ocean ridge
2. The age of ocean crust increases	New ocean floor is created at the mid-ocean ridge
3. The thickness of sea floor deposits increases	Older ocean floor has more time to accumulate a thicker deposit of debris due to settling
4. The thickness of the lithospheric plate increases	As plates move away from the spreading center, they are cooler and gain thickness as material beneath the lithosphere attaches to the bottom of the plate
5. Heat flow decreases	As the thickness of the lithospheric plate increases (see Relationship 3), less heat can escape because conduction heat transfer through the lithosphere is poor
6. Water depth increases	The mid-ocean ridge is a topographically high feature, and as plates move away from there, they move into deeper water because plates undergo thermal contraction and isostatic adjustment downward

Evidence for plate tectonics includes many types of information from land and the sea floor, including the symmetric pattern of magnetic stripes relative to the mid-ocean ridge.

The Acceptance of a Theory

The accumulation of these and many other lines of evidence in support of moving continents has convinced scientists of the validity of continental drift. Since the late 1960s, the concepts of continental drift and sea floor spreading have been united into a much more encompassing theory known as plate tectonics, which describes the movement of the outermost portion of Earth and the resulting creating of continental and sea floor features.

Although several mechanisms have been proposed for the force or forces responsible for driving this motion, none of them are able to explain all aspects of plate motion. Nevertheless, it is clear that the unequal distribution of heat within Earth is the underlying driving force for this movement. The various properties of Earth's internal layers can help account for the movement of plates.

Earth Structure

Earth's internal structure consists of a series of nested spheres (similar to the layers of an onion) that differ in density. Let's examine these layers and discover their importance to plate tectonic processes.

Chemical Composition versus Physical Properties

Earth is a layered sphere based on density, with the highest-density material found near the center of Earth and the lowest-density material located near the surface. The cross-sectional view of Earth in Figure 3-15 shows that Earth's inner structure can be subdivided according to its chemical composition (the chemical makeup of Earth materials) or its physical properties (how the rocks respond to increased temperature and pressure at depth).

Chemical Composition Based on chemical composition, Earth consists of three layers: the **core**, the **mantle**, and the **crust** (Figure 3–15). If Earth were an apple, then the crust would be its thin skin. It extends from the surface to an average depth of 30 kilometers (20 miles). The crust is composed of relatively low-density rock, consisting mostly of various *silicate minerals* (common rock-forming minerals with silicon and oxygen that form silicate tetrahedra, as discussed in Chapter 2). There are two types of crust, oceanic and continental, which will be discussed in the next section.

Immediately below the crust is the mantle. It occupies the largest volume of the three layers and extends to a depth of 2900 kilometers (1800 miles). The mantle is composed of relatively high-density iron and magnesium silicate rock.

Beneath the mantle is the core. It forms a large mass from a depth of 2900 kilometers (1800 miles) to the center of Earth at 6370 kilometers (3960 miles). The core is composed of even higher-density metal (mostly iron and nickel).

Physical Properties Based on physical properties, Earth is composed of five layers: the **inner core**, the **outer core**, the **mesosphere** (*mesos* = middle, *sphere* = ball), the **asthenosphere** (*asthenos* = weak, *sphere* = ball), and, as previously mentioned, the lithosphere (Figure 3–15). The lithosphere is Earth's cool, rigid, outermost layer. It extends from the surface to an average depth of about 100 kilometers (62 miles) and includes the crust plus the topmost portion of the mantle. The lithosphere is *brittle* (*brytten* = to shatter), meaning that it will fracture when force is applied to it. The plates involved in plate tectonic motion are the plates of the lithosphere.

Figure 3–15 Comparison of Earth's chemical composition and physical properties.

A cross-sectional view of Earth with Earth's layers classified by chemical composition shown along the left side of the diagram. For comparison, Earth's layers classified by physical properties are shown along the right side of the diagram.

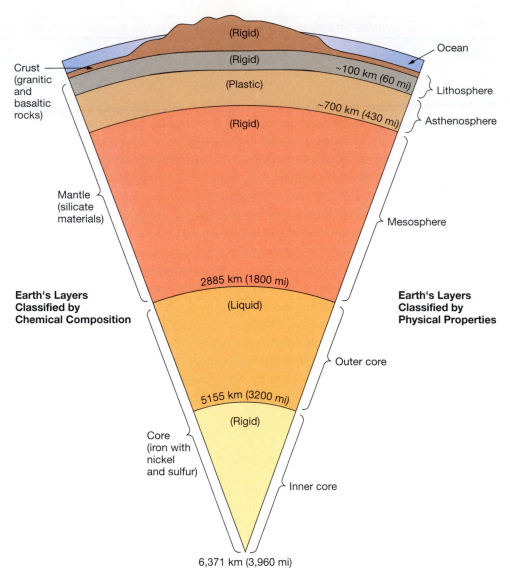

Figure 3–15 Comparison of Earth's chemical composition and physical properties.

A cross-sectional view of Earth with Earth's layers classified by chemical composition shown along the left side of the diagram. For comparison, Earth's layers classified by physical properties are shown along the right side of the diagram.

Beneath the lithosphere is the asthenosphere. The asthenosphere is *plastic* (*plasticus* = to mold), meaning it will flow when a gradual force is applied to it. It extends from about 100 kilometers (62 miles) to 700 kilometers (430 miles) below the surface, which is the base of the upper mantle. At these depths, it is hot enough to partially melt portions of most rocks.

Beneath the asthenosphere is the mesosphere. The mesosphere extends to a depth of 2900 kilometers (1800 miles), which corresponds to the middle and lower mantle. Although the asthenosphere deforms plastically, the meso-sphere is rigid due to increased pressure at these depths.

Beneath the mesosphere is the core. The core consists of the outer core, which is liquid and capable of flowing; and the inner core, which is rigid and does not flow. Again, the increased pressure at the center of Earth keeps the inner core from flowing.

Knowledge of Earth's Inner Structure How have scientists determined the inner structure of Earth? Remarkably, the deepest well in the world, which was drilled in the Kola Peninsula of Russia, reached a depth of 12,266 meters (40,478 feet, or nearly 8 miles) in 1992 but still never came close to penetrating the crust. Instead of directly sampling the deep Earth, Earth scientists have had to rely on *indirect* methods to understand what lies deep below the surface. For example, determining the gravitational attraction that Earth exerts on other bodies of the solar system gives scientists information about the density of deep Earth layers.

Another powerful indirect sampling method is used by *seismologists* (*seismo* = earthquake, *ologist* = one who studies), who analyze the pattern of seismic waves bouncing around within Earth (similar to listening through a doctor's stethoscope) to decipher the chemical composition and physical properties of Earth's internal layers. All earthquakes generate low-frequency *seismic waves*, which travel at different speeds dependant on the density of the materials they pass through. After an earthquake is generated, a host of waves travels outward from the earthquake's source at different speeds. Two of the most important of these waves for determining Earth structure are the P and S waves.

Motion toward and
away from wall

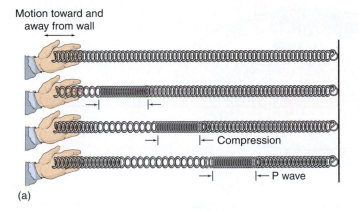

(a)

Motion up
and down

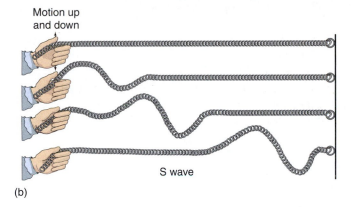

S wave

(b)

P wave

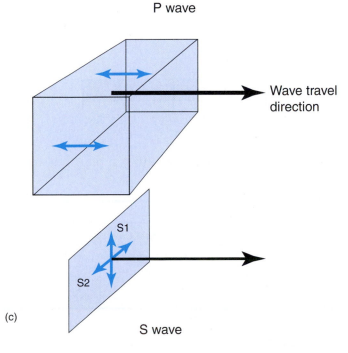

Wave travel
direction

S1

S2

(c)

S wave

Figure 3–16 P and S waves.

(a) P waves are created when a spring is compressed or stretched, causing back-and-forth compression and extension along the length of the spring. **(b)** S waves are created when a spring is shaken up and down (S1) [or back and forth (S2)], causing side-to-side shearing of the spring. **(c)** Taken together, P, S1, and S2 waves record motion of materials in three dimensions.

The *P* (primary) *wave* is a compressional wave that compresses and extends the material through which it travels. A spring attached to a wall, for example, exhibits this type of phenomenon when you pull it, which causes extension, or push it, which causes compression (Figure 3–16a). The *S* (secondary) *wave* is a shear wave that creates transverse motions like those created by moving the end of the spring up and down or back and forth (Figure 3–16b). In fact, there are two S waves, called S1 and S2, that move at right angles to each other. At seismic listening stations, speedy P waves arrive first, followed by more slowly moving S waves (which is how they received their names). Together, P and S waves create motion in three dimensions (Figure 3–16c).

Solid rocks transmit P and S waves, but liquids transmit only P waves because liquids cannot be sheared. The patterns of S wave arrivals at Earth's surface are a major line of evidence that helps Earth scientists determine that the outer core of Earth must be liquid (see Figure 3–15).

The wealth of seismic information has recently allowed seismologists to create three-dimensional maps of Earth that can be viewed as vertical slices (similar to CAT-scan images used in medical applications). This technique is called *seismic tomography* (*tomos* = section, *graphy* = the name of a descriptive science). Based on small variations in seismic shear wave velocities, the slices display temperature differences within Earth's mantle (Figure 3–17). These images show hot plumes of magma rising to the surface as well as the colder slabs of lithosphere descending during subduction, adding to the evidence in support of mantle convection.

Near the Surface

The top portion of Figure 3–18 shows an enlargement of Earth's layers closest to the surface.

Lithosphere The lithosphere is a relatively cool, rigid shell that includes all of the crust and the topmost part of the mantle. In essence, the topmost part of the mantle is attached to the crust and the two act as a single unit, approximately 100 kilometers (62 miles) thick. The expanded view in Figure 3–18 shows that the crust portion of the lithosphere is further subdivided into oceanic crust and continental crust, which are compared in Table 3–2.

Oceanic crust is composed of the igneous rock **basalt**, which is dark-colored and has a relatively high density of about 3.0 grams per cubic centimeter.[4] The average thickness of the oceanic crust is only about 8 kilometers (5 miles). Basalt originates as molten magma beneath Earth's crust. This magma comes to the surface mostly along the mid-ocean ridge, where it cools and hardens to form new oceanic crust.

---

[4]As a comparison, water has a density of 1.0 grams per cubic centimeter. Thus, basalt, which has a density of 3.0 grams per cubic centimeter, is three times the density of water.

Figure 3–17 Shear wave tomography reveals mantle structure.

(a) View of Earth's mantle at 1500 kilometers (932 miles) depth showing hot, low-velocity (*red*) and cold, high-velocity (*blue*) regions. **(b)** Views of two slices of the mantle beneath the eastern (*upper*) and western (*lower*) Pacific Ocean, with various features labeled.

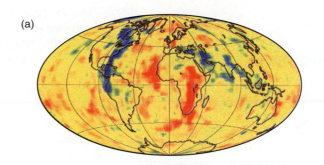

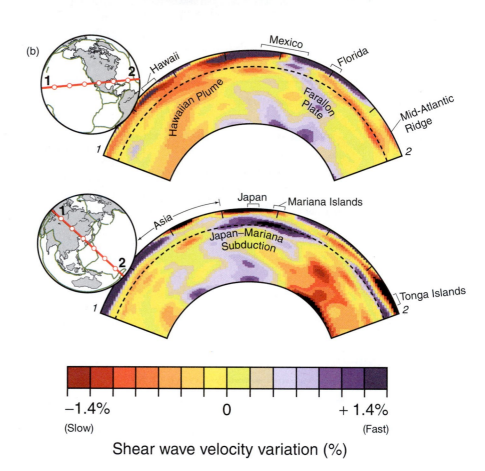

Shear wave velocity variation (%)

−1.4% (Slow) 0 + 1.4% (Fast)

TABLE 3–2 Comparing oceanic and continental crust.

	Oceanic Crust	Continental Crust
Main rock type	Basalt (dark-colored igneous rock)	Granite (light-colored igneous rock)
Density (grams per cubic centimeter)	3.0	2.7
Average thickness	8 kilometers (5 miles)	35 kilometers (22 miles)

Continental crust is composed mostly of a lower-density and lighter-colored igneous rock **granite**.[5] It has a density of about 2.7 grams per cubic centimeter. The average thickness of the continental crust is about 35 kilometers (22 miles) but may reach a maximum of 60 kilometers (37 miles) beneath the highest mountain ranges. Most granite originates beneath the surface as molten magma that cools and hardens within Earth's crust. No matter which type of crust is at the surface, it is all part of the lithosphere.

---

[5]At the surface, continental crust is often covered by a relatively thin layer of surface sediments. Below these, granite can be found.

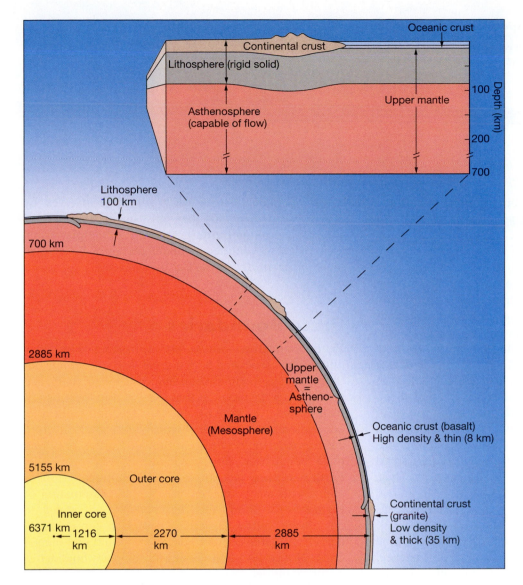

Figure 3–18 **Internal structure of Earth.**

Enlargement (*top*) shows that the rigid lithosphere includes the crust (either continental or oceanic) plus the topmost part of the mantle to about 100 kilometers (60 miles) depth. Beneath the lithosphere, the plastic asthenosphere extends to a depth of 700 kilometers (430 miles).

Asthenosphere The asthenosphere is a relatively hot, plastic region beneath the lithosphere. It extends from the base of the lithosphere to a depth of about 700 kilometers (430 miles) and is entirely contained within the upper mantle. The asthenosphere can deform without fracturing if a force is applied slowly. It has the ability to flow but has high **viscosity** (*viscos* = sticky). Viscosity is a measure of a substance's resistance to flow.[6] Studies indicate that the high-viscosity asthenosphere is flowing slowly through time, which has important implications in plate tectonic processes.

> Earth has differences in composition and physical properties that create layers such as the brittle lithosphere and the plastic asthenosphere, which is capable of flowing slowly over time.

Isostatic Adjustment

Isostatic (*iso* = equal, *stasis* = standing) **adjustment**— the vertical movement of crust—is the result of the buoyancy of Earth's lithosphere as it floats on the denser, plastic-like asthenosphere below. Isostatic adjustment of Earth's crust accounts for the difference in heights

---

[6]Substances that have high viscosity (a high resistance to flow) include toothpaste, honey, tar, and silly putty; a common substance that has low viscosity is water. A substance's viscosity often changes with temperature. For instance, as honey is heated, it flows more easily.

Figure 3–19 **A container ship experiences isostatic adjustment.**
A ship will ride higher in water when it is empty and will ride lower in water when it is loaded with cargo, illustrating the principle of isostatic adjustment.

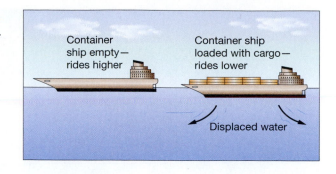

of the continental and oceanic crusts and contributes to the mechanism of global plate tectonics.

The container ship in Figure 3–19 provides an example of isostatic adjustment. An empty ship floats high in the water. Once the ship is loaded with cargo, though, the ship undergoes isostatic adjustment and floats lower in the water (but hopefully won't sink!). When the cargo is unloaded, the ship isostatically adjusts itself and floats higher again.

Similarly, both continental and oceanic crust float on the denser mantle beneath. Oceanic crust is denser than continental crust, however, so oceanic crust floats lower in the mantle because of isostatic adjustment. Oceanic crust is also thin, which creates low areas for the oceans to occupy. Areas where the continental crust is thickest (in large mountain ranges on the continents) float higher than continental crust of normal thickness, also because of isostatic adjustment. Thus, tall mountain ranges on Earth are composed of a great thickness of crustal material that in essence keeps them buoyed up.

Areas that are exposed to an increased or decreased load experience isostatic adjustment. For instance, during the most recent Ice Age (which occurred during the Pleistocene Epoch between 2 million and 10,000 years ago), massive ice sheets covered far northern regions such as Scandinavia and northern Canada. The additional weight of ice several kilometers thick caused these areas to isostatically adjust themselves lower in the mantle. Since the end of the Ice Age, the reduced load on these areas caused by the melting of ice caused these areas to rise and experience **isostatic rebound**, which continues today. The rate at which isostatic rebound occurs gives scientists important information about the properties of the upper mantle.

Isostatic adjustment also affects some areas of Earth that have higher elevations than would normally be expected, such as eastern Africa and the Colorado Plateau of the southwestern United States. In these locations, hot, low-density mantle material provides additional buoyancy that is usually associated with deep roots of continental crust and causes them to rise to anomalously high levels.

Further, isostatic adjustment provides additional evidence for the movement of Earth's tectonic plates. Because continents isostatically adjust themselves by moving *vertically*, then they must not be firmly fixed in one position on Earth. If this is true, the plates that contain these continents should certainly be able to move *horizontally* across Earth's surface.

Plate Boundaries

Plate boundaries—where plates interact with each other—are associated with a great deal of tectonic activity, such as mountain building, volcanic activity, and earthquakes. In fact, the first clues to the locations of plate boundaries were the dramatic tectonic events that occur there. For example, Figure 3–14 shows the close correspondence between worldwide earthquakes and plate boundaries. Further, Figure 3–14b shows that Earth's surface is composed of seven major plates along with many smaller ones. Close examination of Figure 3–14b shows that the boundaries of plates do not always follow coastlines and, as a consequence, nearly all plates contain both oceanic and continental crust. Notice also that about 90% of plate boundaries occur on the sea floor.

There are three types of plate boundaries, as shown in Figure 3–20. **Divergent** (*di* = apart, *vergere* = to incline) **boundaries** are found along oceanic ridges where new lithosphere is being added. **Convergent** (*con*= together, *vergere* = to incline) **boundaries** are found where plates are moving together and one plate subducts beneath the other. **Transform** (*trans* = across, *form* = shape) **boundaries** are found where lithospheric plates slowly grind past one another. Table 3–3 summarizes characteristics, tectonic processes, and examples of these plate boundaries.

Divergent Boundaries

Divergent plate boundaries occur where two plates move apart, such as along the crest of the mid-ocean ridge where sea floor spreading creates new oceanic lithosphere (Figure 3–21). A common feature along the mid-ocean ridge is a central downdropped linear **rift valley** (Figure 3–22). Pull-apart faults located along the central rift valley show that the plates are being *continuously*

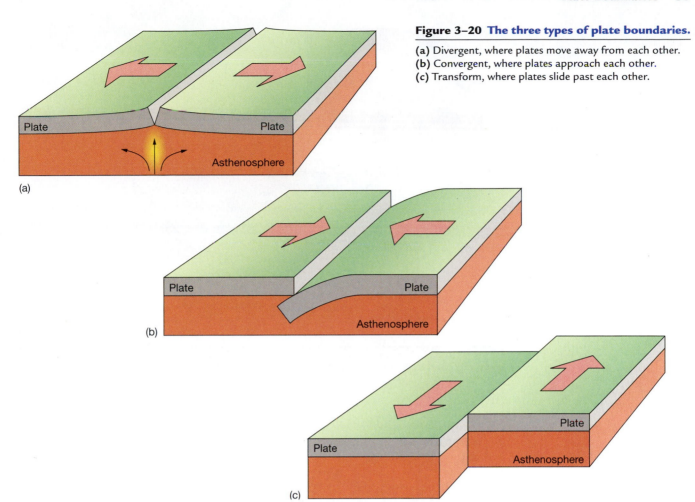

Figure 3–20 **The three types of plate boundaries.**
(a) Divergent, where plates move away from each other.
(b) Convergent, where plates approach each other.
(c) Transform, where plates slide past each other.

TABLE 3–3 **Characteristics, tectonic processes, and examples of plate boundaries.**

Plate boundary	Plate movement	Crust type(s)	Sea floor created or destroyed?	Tectonic process	Geographic examples
Divergent plate boundaries	Apart ← →	Ocean–ocean	New sea floor created	Sea floor spreading	Mid-Atlantic Ridge, East Pacific Rise
		Continent–continent	As continent splits apart, new sea floor created	Continental rifting	East Africa Rift Valleys, Red Sea, Gulf of California
Convergent plate boundaries	Together → ←	Ocean–continent	Old sea floor destroyed	Subduction	Andes Mountains, Cascade Mountains
		Ocean–ocean	Old sea floor destroyed	Subduction	Aleutian Islands, Mariana Islands
		Continent–continent	N/A	Collision	Himalaya Mountains, Alps
Transform plate boundaries	Past each other → ←	Oceanic	Sea floor neither created nor destroyed	Transform faulting	Mendocino fault, Eltanin fault (between mid-ocean ridges)
		Continental	Sea floor neither created nor destroyed	Transform faulting	San Andreas fault, Alpine fault (New Zealand)

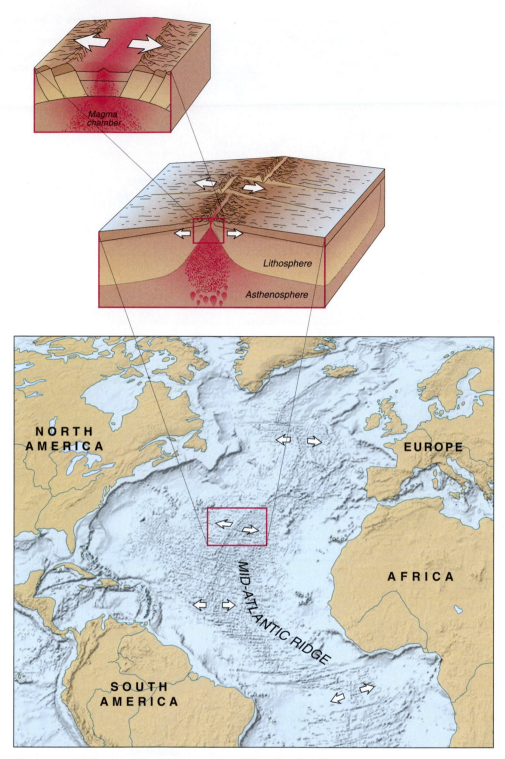

Figure 3–21 Divergent boundary at the Mid-Atlantic Ridge.

Most divergent plate boundaries occur along the crest of the mid-ocean ridge, where sea floor spreading creates new oceanic lithosphere.

pulled apart rather than being pushed apart by the upwelling of material beneath the mid-ocean ridge. Upwelling of magma beneath the mid-ocean ridge is simply filling in the void left by the separating plates of lithosphere. In the process, sea floor spreading produces

about 20 cubic kilometers (4.8 cubic miles) of new ocean crust worldwide each year.

Studies of the geometry of *magma chambers* along mid-ocean ridges indicate that magma rises to the surface in discrete blobs at intervals along the mid-ocean ridge

Figure 3–22 Rift valley on Iceland.

View along a rift valley looking south from Laki volcano in Iceland, which sits atop the Mid-Atlantic Ridge. Note bus to the left of the rift valley in the middle of the picture for scale.

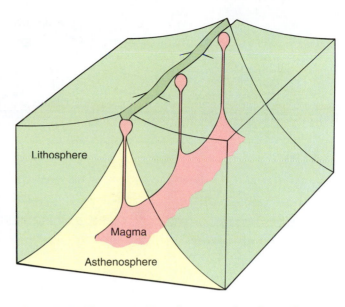

Figure 3–23 Segmentation of magma chambers along the mid-ocean ridge.

Beneath the mid-ocean ridge, heat-bearing asthenosphere (*yellow*) rises between separating lithospheric plates (*green*). A zone of partially molten asthenosphere (*orange*) supplies blobs of low-density magma as shallow magma chambers along the ridge axis.

axis (Figure 3–23). These shallow magma chambers supply molten material to the central rift valley as sea-floor spreading occurs.

Figure 3–24 shows how the development of a mid-ocean ridge creates an ocean basin. Initially, molten material rises to the surface, causing upwarping and thinning of the crust. Volcanic activity produces vast quantities of high-density basaltic rock. As the plates begin to move apart, a linear rift valley is formed and volcanism continues. Further **rifting** of the land and more spreading cause the area to drop below sea level. When this occurs, the rift valley eventually floods with seawater and a young linear sea is formed. After millions of years of sea floor spreading, a full-fledged ocean basin is created with a mid-ocean ridge in the middle.

Two different stages of ocean basin development are shown in the map of East Africa in Figure 3–25. First, the rift valleys are actively pulled apart and are at the rift valley stage of formation. Second, the Red Sea is at

the linear sea stage. It has rifted apart so far that the land has dropped below sea level. The Gulf of California in Mexico is another linear sea. The Gulf of California and the Red Sea are two of the youngest seas in the world, having been created only a few million years ago. If plate motions continue rifting the plates apart in these areas, they will eventually become large oceans.

The rate at which the sea floor spreads apart varies along the mid-ocean ridge and dramatically affects its appearance. Faster spreading, for instance, produces broader and less rugged segments of the global mid-ocean ridge system. This is because fast-spreading segments of the mid-ocean ridge produce vast amounts of rock, which move away from the spreading center at a rapid rate and consequently undergo less thermal contraction and subsidence than slower spreading segments do. In addition, central rift valleys on slow-spreading segments tend to be larger and better developed.

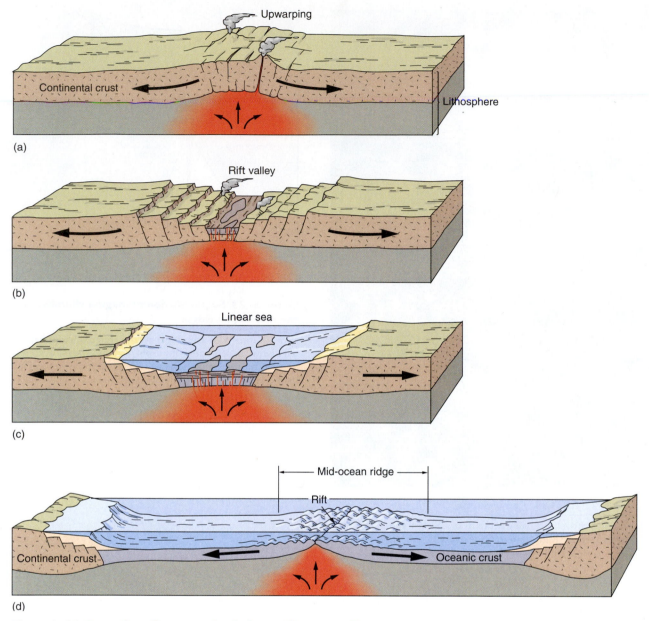

Figure 3–24 Formation of an ocean basin by sea floor spreading.

Sequence of events in the formation of an ocean basin. **(a)** A shallow heat source develops under a continent, causing initial upwarping and volcanic activity. **(b)** Movement apart creates a rift valley. **(c)** With increased spreading, a linear sea is formed. **(d)** After millions of years, a full-fledged ocean basin is created, separating continental pieces that were once connected.

The gently sloping and fast-spreading parts of the mid-ocean ridge are called **oceanic rises**. For example, the **East Pacific Rise** (Figure 3–26, *bottom*) between the Pacific and Nazca Plates is a broad, low, gentle swelling of the sea floor with a small central rift valley and has a spreading rate as high as 16.5 centimeters (6.5 inches) per year.[7] Conversely, steeper-sloping and slower-spreading areas of the mid-ocean ridge are called **oceanic ridges**. The **Mid-Atlantic Ridge** (Figure 3–26, *top*) between the South American and African Plates is an example of a tall, steep, rugged oceanic ridge with a prominent rift valley and has a spreading rate of only 2 to 3 centimeters (0.8 to 1.2 inches) per year.

The amount of energy released by earthquakes along the divergent plate boundaries is closely related to the spreading rate. The faster the sea floor spreads, the less energy released in each earthquake. Earthquake intensity

---

[7]The spreading rate is the total widening rate of an ocean basin resulting from motion of both plates away from a spreading center.

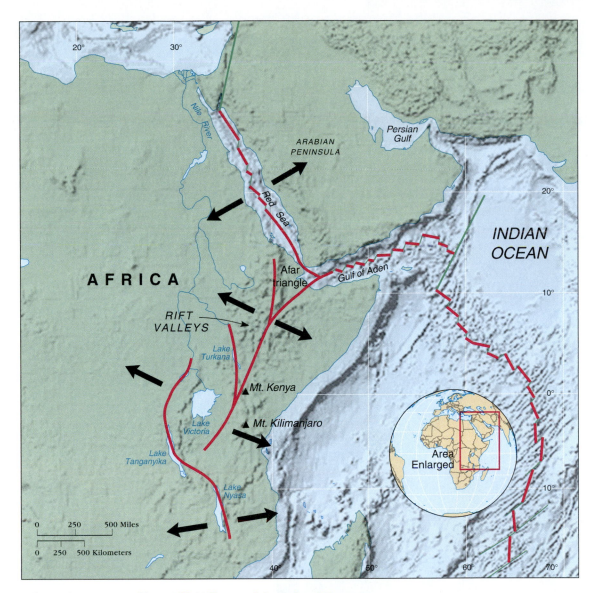

Figure 3–25 East African Rift Valleys and Associated Features

Parts of east Africa are splitting apart (*arrows*), creating a series of linear downdropped rift valleys (*red lines*) along with prominent volcanoes (*triangles*). Similarly, the Red Sea and Gulf of Aden have split apart so far they are now below sea level. The mid-ocean ridge in the Indian Ocean has experienced similar stages of development.

is usually measured on a scale called the **seismic moment magnitude**, which reflects the energy released to create very long-period seismic waves. Because it more adequately represents larger magnitude earthquakes, the moment magnitude scale is increasingly used instead of the well-known Richter scale and is represented by the symbol M_w. Earthquakes in the rift valley of the slow-spreading Mid-Atlantic Ridge reach a maximum magnitude of about $M_w = 6.0$, whereas those occurring along the axis of the fast-spreading East Pacific Rise seldom exceed $M_w = 4.5$.[8]

Convergent Boundaries

Convergent boundaries—where two plates move together and collide—result in the destruction of ocean crust as one plate plunges below the other and is remelted in the mantle. The physiographic ocean floor feature associated with a convergent plate boundary is a deep-ocean trench. Trenches are deep linear scars where subduction occurs. Melting in the subduction

---

[8]Note that each one-unit increase of earthquake magnitude represents an increase of energy release of about 30 times.

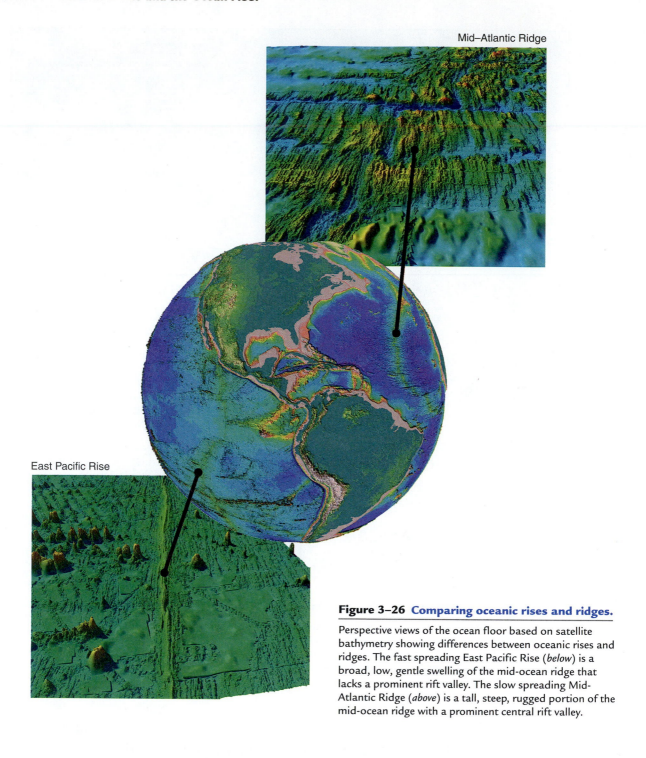

Mid–Atlantic Ridge

East Pacific Rise

Figure 3–26 **Comparing oceanic rises and ridges.**

Perspective views of the ocean floor based on satellite bathymetry showing differences between oceanic rises and ridges. The fast spreading East Pacific Rise (*below*) is a broad, low, gentle swelling of the mid-ocean ridge that lacks a prominent rift valley. The slow spreading Mid-Atlantic Ridge (*above*) is a tall, steep, rugged portion of the mid-ocean ridge with a prominent central rift valley.

zone causes an arc-shaped row of highly active and explosively erupting volcanoes that parallel the trench called a **volcanic arc**.

Figure 3–27 shows the three subtypes of convergent boundaries that result from interactions between two different types of crust (oceanic and continental).

Oceanic–Continental Convergence When an oceanic and a continental plate converge, the denser oceanic plate

is subducted (Figure 3–27a). The oceanic plate becomes heated as it is subducted into the asthenosphere and some of the basaltic rock is melted. This molten rock mixes with superheated gases (mostly water and other volatiles) and begins to rise to the surface through the overriding continental plate. The rising basalt-rich magma mixes with the granite of the continental crust, producing lava in volcanic eruptions at the surface that is intermediate in composition between basalt and granite. One type of vol-

Figure 3–27 Three sub-types of convergent plate boundaries.

(a) Oceanic–continental convergence.
(b) Oceanic–oceanic convergence. **(c)** Continental–continental convergence.

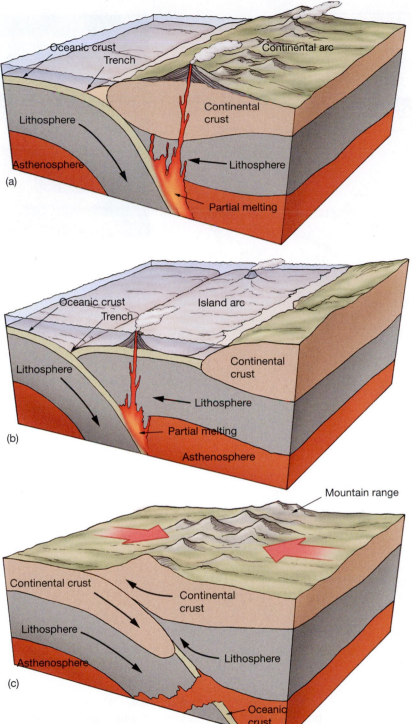

canic rock with this composition is called **andesite**, named after the Andes Mountains of South America because it is so common there. Andesitic volcanic eruptions are usually quite explosive and have historically been very destructive because andesite contains high silica content and is quite viscous. The result of this volcanic activity on the continent above the subduction zone produces a type of volcanic arc called a **continental arc**, which occurs along the edge of a continent. Continental arcs are created by andesitic volcanic eruptions and by the folding and uplifting associated with plate collision.

If the spreading center producing the subducting plate is far enough from the subduction zone, an oceanic trench becomes well developed along the margin of the continent. The Peru-Chile Trench is an example, and the Andes Mountains are the associated continental arc produced by melting of the subducting plate. If the spreading center producing the subducting plate is close to the subduction zone,

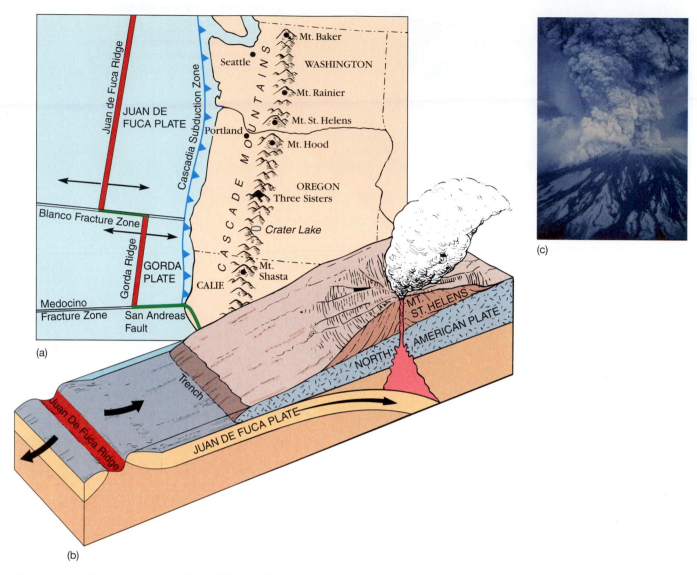

Figure 3–28 Convergent tectonic activity produces the Cascade Mountains.

(a) Tectonic features of the Cascade Mountain Range and vicinity. (b) The volcanoes of the Cascade Mountains are created by the subduction of the Juan de Fuca Plate beneath the North American Plate. (c) The eruption of Mount St. Helens in 1980.

however, the trench is not nearly as well developed. This is the case where the Juan de Fuca Plate subducts beneath the North American Plate off the coasts of Washington and Oregon to produce the Cascade Mountains continental arc (Figure 3–28). Here, the Juan de Fuca Ridge is so close to the North American Plate that the subducting lithosphere is less than 10 million years old and has not cooled enough to become very deep. In addition, the large amount of sediment carried to the ocean by the Columbia River has filled most of the trench with sediment. Many of the Cascade volcanoes of this continental arc have been active within the last 100 years. Most recently, Mount St. Helens erupted in May 1980, killing 62 people.

Oceanic–Oceanic Convergence When two oceanic plates converge, the denser oceanic plate is subducted (Figure 3–27b). Typically, the older oceanic plate is denser because it has had more time to cool and contract. This type of convergence produces the deepest trenches in the world, such as the Mariana Trench in the western Pacific Ocean. The subducting oceanic plate becomes heated and melts in the asthenosphere. Similar to oceanic–continental convergence, molten material mixes with superheated gases and rises to the surface to produce volcanoes. The molten material is mostly basaltic because there is no mixing with granitic rocks from the continents. Basalt contains less silica and is less viscous than andesite—thus the eruptions are not nearly as destructive. The result of this volcanic activity is a type of volcanic arc called an **island arc**, which usually is separated from the nearest continent by a marginal sea. Examples of island arc/trench systems are the West Indies' Leeward and Windward Islands/Puerto Rico Trench in the Caribbean Sea and the Aleutian Islands/Aleutian Trench in the North Pacific Ocean.

Continental–Continental Convergence When two continental plates converge, which one is subducted? You might expect the older of the two (which is probably the denser one) will be subducted. Continental lithosphere forms differently than oceanic lithosphere, however, and old continental lithosphere is no denser than young continental lithosphere. It turns out that neither subducts because both are too low in density to be pulled very far down into the mantle. Instead, a tall uplifted mountain range is created by the collision of the two

plates (Figure 3–27c). These mountains are composed of folded and deformed sedimentary rocks originally deposited on the sea floor that previously separated the two continental plates. The oceanic crust itself may subduct beneath such mountains. A prime example of continental–continental convergence is the collision of India with Asia (Figure 3–29). It began 45 million years ago and has created the Himalaya Mountains, presently the tallest mountains on Earth.

Earthquakes Associated with Convergent Boundaries Both spreading centers and trench systems are characterized by earthquakes, but in different ways. Spreading centers have shallow earthquakes, usually less than 10 kilometers (6 miles) deep. Earthquakes associated with trenches, on the other hand, vary from near the surface down to 670 kilometers (415 miles) deep, which are the deepest earthquakes in the world. These earthquakes are clustered in a band about 20 kilometers (12.5 miles)

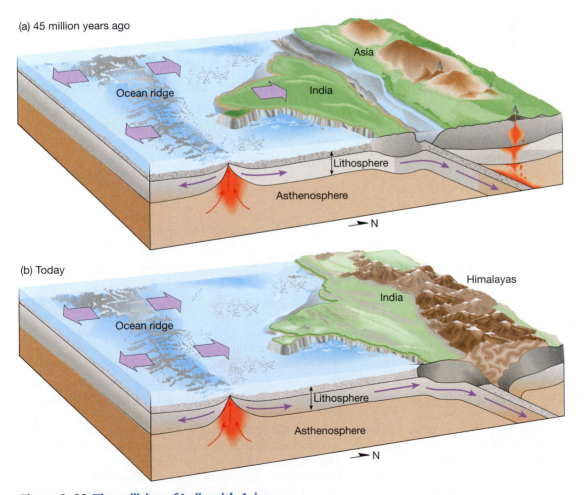

Figure 3–29 The collision of India with Asia.

(a) Sea floor spreading along the mid-ocean ridge south of India caused the collision of India with Asia, which began about 45 million years ago. **(b)** The collision closed the shallow sea between India and Asia, crumpled the two continents together, and is responsible for the continued uplift of the Himalaya Mountains.

thick of successively deeper earthquakes extending from the trench that dips at an angle of approximately 45 degrees, becoming steeper with depth. This band is called a *Wadati-Benioff seismic zone*, which represents a subducting plate in a convergent plate boundary.

Many factors combine to produce large earthquakes at convergent boundaries. The forces involved in convergent-plate boundary collisions are enormous. Huge lithospheric slabs of rock are relentlessly pushing against each other, and the subducting plate must actually bend as it dives below the surface. In addition, thick crust associated with convergent boundaries tends to store more energy than the thinner crust at divergent boundaries. Also, mineral structure changes occur at the higher pressures encountered deep below the surface, which are thought to produce changes in volume that lead to some of the most powerful earthquakes in the world. In fact, the largest earthquake ever recorded was the 1960 Chilean earthquake near the Peru-Chile Trench, which had a magnitude of $M_w = 9.5$!

Transform Boundaries

A global sea floor map (such as the one inside the front cover of this book) shows that the mid-ocean ridge is offset by many large features oriented perpendicular (at right angles) to the crest of the ridge. What causes these offsets? They occur because spreading at a mid-ocean ridge only occurs perpendicular to the axis of a ridge and all parts of a plate must move together. As a result, offsets are oriented perpendicular to the ridge and parallel to each other to accommodate spreading of a linear ridge system on a spherical Earth. In addition, the offsets allow different segments of the mid-ocean ridge to spread apart at different rates. These offsets—called **transform faults**—give the mid-ocean ridge a zigzag appearance. There are thousands of these transform faults, some large and some small, which dissect the global mid-ocean ridge.

There are two types of transform faults. The first and most common type occurs wholly on the ocean floor and is called an **oceanic transform fault**. The second type cuts across a continent and is called a **continental transform fault**. Regardless of type, though, transform faults *always* occur between two segments of a mid-ocean ridge, as shown in Figure 3–30.

The movement of one plate past another—a process called **transform faulting**—produces shallow but often strong earthquakes in the lithosphere. Magnitudes of $M_w = 7.0$ have been recorded along some oceanic transform faults. One of the best studied faults in the world is California's **San Andreas Fault**, a continental transform fault that runs from the Gulf of California past San Francisco and beyond into northern California. Because the San Andreas Fault cuts through continental crust, which is much thicker than oceanic crust, earthquakes are considerably larger than those produced by oceanic transform faults, sometimes up to $M_w = 8.5$.

Because California experiences large periodic earthquakes, many people are mistakenly concerned that it will "fall off into the ocean" during a large earthquake along the San Andreas Fault. These earthquakes occur as the Pacific Plate continues to move to the northwest past the North American Plate at a rate of about 5 centimeters (2 inches) a year. At this rate, Los Angeles (on the Pacific Plate) will be adjacent to San Francisco (on the North American Plate) in about 18.5 million years—a length of time for about 1 million generations of people to live their lives. Although California will never fall into the ocean, people living near this fault should be very aware they are likely to experience a large earthquake within their lifetime.

> The three main types of plate boundaries are divergent (plates moving apart such as at the mid-ocean ridge), convergent (plates moving together such as at an ocean trench), and transform (plates sliding past each other such as at a transform fault).

Testing the Model: Some Applications of Plate Tectonics

One of the strengths of plate tectonic theory is how it unifies so many seemingly separate events into a single consistent model. Let's look at a few examples that illustrate how plate tectonic processes can be used to explain the origin of features that, up until the acceptance of plate tectonics, were difficult to explain.

Mantle Plumes and Hotspots

Although the theory of plate tectonics helped explain the origin of many features near plate boundaries, it did not seem to explain the origin of *intraplate* (*intra* = within, *plate* = plate of the lithosphere) *features* that are far from any plate boundary. For instance, how can plate tectonics explain volcanic islands near the middle of a plate?

According to the plate tectonic model, volcanism in the middle of a plate is caused by the presence of mantle plumes that are most likely related to the positions of convection cells in the mantle. **Mantle plumes** (*pluma* = a soft feather) are columnar areas of hot molten rock that arise from deep within the mantle. The areas where mantle plumes come to the surface are called **hotspots** and are marked by an abundance of volcanic activity. For example, the continuing volcanism in Yellowstone National Park and Hawaii is caused by hotspots.

Note that hotspots are different from either a volcanic arc or a mid-ocean ridge, even though all three are marked by a high degree of volcanic activity. Rela-

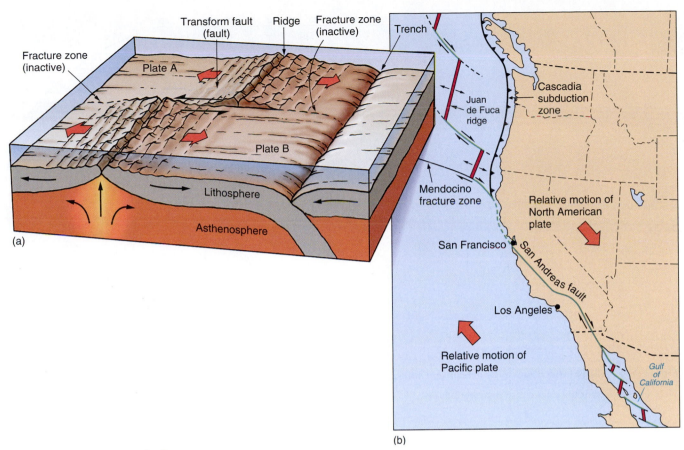

Figure 3–30 Transform faults.

The Juan de Fuca Ridge is offset by several oceanic transform faults. Also shown is the San Andreas Fault, a continental transform fault connecting the Juan de Fuca Ridge and the spreading center in the Gulf of California.

tive to basalts typical of hotspots, those produced at the mid-ocean ridge are chemically different, containing reduced concentrations of the elements potassium, rubidium, cesium, uranium, and thorium. Based on this difference, Earth scientists have suggested that mid-ocean ridge basalt is probably derived from the upper mantle, whereas hotspot basalt is likely from the lower mantle, perhaps as deep as the core-mantle boundary. If this is true, it implies that mantle plumes may have a very deep Earth source.

Worldwide, more than 100 hotspots have been active within the last 10 million years and several dozen remain active today. Figure 3–31 shows the global distribution of prominent hotspots. In general, hotspots do not coincide with plate boundaries. Notable exceptions are those that are near divergent boundaries where the lithosphere is thin, such as at the Galápagos Islands and Iceland. Not only does Iceland straddle the Mid-Atlantic Ridge (a divergent plate boundary), it also sits directly over a 150-kilometer (93-mile)-wide

mantle plume, which accounts for its remarkable amount of volcanic activity. In fact, the volume of volcanic rock generated at Iceland is so large that it has caused Iceland to be one of the few places along the global mid-ocean ridge that rises high above sea level.

Throughout the Pacific Plate, many island chains are oriented in a northwestward–southeastward direction. The most intensely studied of these is the **Hawaiian Islands–Emperor Seamount Chain** in the northern Pacific Ocean (Figure 3–32). What created this chain of over 100 intraplate volcanoes that stretch over 5800 kilometers (3000 miles)? Further, what caused the prominent bend in the overall direction that occurs in the middle of the chain?

To help answer these questions, look at the ages of the volcanoes in the chain. Every volcano in the chain has long since become extinct, except the volcano Kilauea on the island of Hawaii, which is the southeast-ernmost island of the chain. The age of volcanoes progressively increases northwestward from Hawaii (Figure 3–32). To the northwest, the volcanoes in-

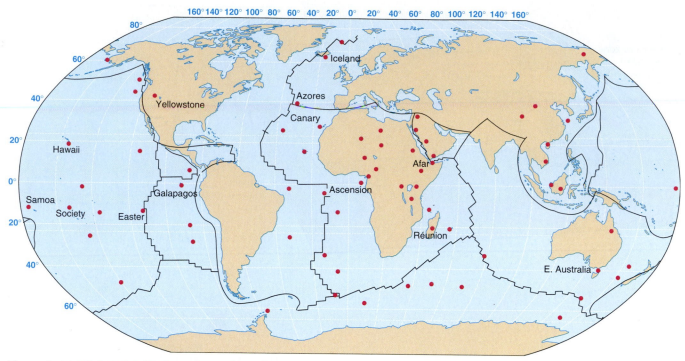

Figure 3–31 Global distribution of prominent hotspots.

Black lines represent the locations of plate boundaries.

crease in age past Suiko Seamount (65 million years old) to Detroit Seamount (81 million years old) near the Aleutian Trench.

These age relationships suggest that the Pacific Plate has moved steadily northwestward while the underlying mantle plume remained relatively stationary. The resulting Hawaiian hotspot created each of the volcanoes in the chain. As the plate moved, it carried the active volcano off the hotspot and a new volcano began forming, younger in age than the previous one. A chain of extinct volcanoes that is progressively older as one travels away from a hotspot is called a **nematath** (*nema* = thread, *tath* = dung or manure). Evidence suggests that about 40 million years ago, the Pacific Plate shifted from a northerly to a northwesterly direction with a rearrangement of plate boundaries along western North America. This change in plate motion accounts for the bend in Figure 3–32 about halfway through the chain, separating the Hawaiian Islands from the Emperor Seamounts. Alternatively, recent research suggests that the change in direction of the chain seen in Figure 3-32 could be related to movement of the Hawaiian hotspot itself.

If Hawaii is directly above the hotspot now, what will become of it in the future? It will be carried to the northwest off the hotspot, become inactive, and eventually be subducted into the Aleutian Trench like all the rest of the volcanoes in the chain to the north of it. In turn, other volcanoes will build up over the hotspot. In fact, a 3500-meter (11,500-foot) volcano already exists 32 kilometers (20 miles) southeast of Hawaii, named **Loihi**. Still 1 kilometer (0.6 mile) below sea level, Loihi is volcanically active and should reach the surface sometime between

30,000 and 100,000 years from now at its current rate of activity. As it builds above sea level, it will become the newest island in the long chain of volcanoes created by the Hawaiian hotspot.

> Mantle plumes create hotspots at the surface, which produce volcanic chains called nemataths that record the motion of plates.

Seamounts and Tablemounts

Many areas of the ocean floor (most notably on the Pacific Plate) contain tall volcanic peaks that are called **seamounts** if conical on top, and **tablemounts**—or **guyots**, after Princeton University's first geology professor Arnold Guyot[9]—if flat on top. Until the theory of plate tectonics, how seamounts and tablemounts formed was unclear. The theory explained why tablemounts were flat on top and why the tops of some tablemounts had shallow-water deposits despite being located in very deep water.

The origin of many seamounts and tablemounts is related to the volcanic activity occurring at hotspots; others are related to processes occurring at the mid-ocean ridge (Figure 3–33). Because of sea floor spreading, active volcanoes (seamounts) occur along the crest of the mid-ocean ridge. Some may be built up so high they rise above sea level and become islands, at which point wave erosion becomes

---

[9]Guyot is pronounced "GEE-oh" with a hard g as in "give."

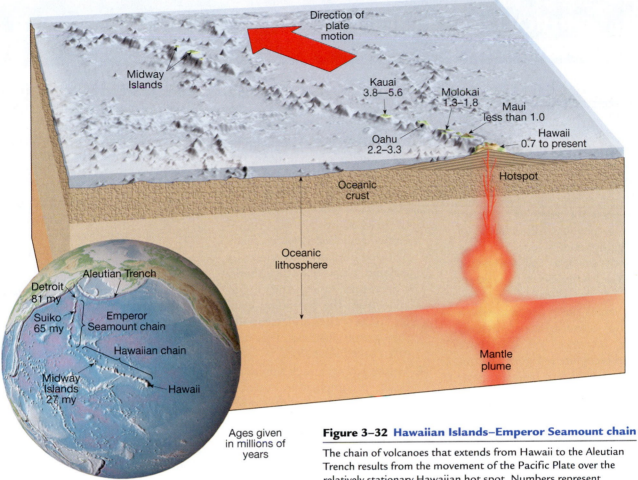

Figure 3–32 **Hawaiian Islands–Emperor Seamount chain**

The chain of volcanoes that extends from Hawaii to the Aleutian Trench results from the movement of the Pacific Plate over the relatively stationary Hawaiian hot spot. Numbers represent radiometric age dates in millions of years before present.

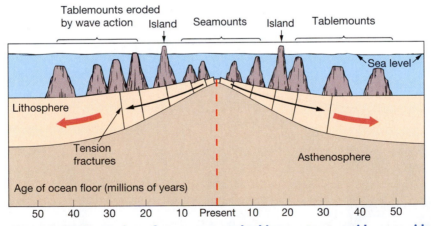

Figure 3–33 **Formation of seamounts and tablemounts at a mid-ocean ridge.**

Seamounts are formed along a mid-ocean ridge. If they are tall enough to reach the surface, their tops get eroded flat by wave activity and become tablemounts. Through sea floor spreading, seamounts and tablemounts are transported into deeper water, sometime carrying with them evidence of their tops reaching shallower water.

important. When sea floor spreading has moved the seamount off its source of magma (whether it is a mid-ocean ridge or a hot spot), the top of the seamount can be flattened by waves in just a few million years. This flattened seamount—now a tablemount—continues to be carried away from its source and, after millions of years, is submerged deeper into the ocean. Frequently, tops of tablemounts contain evidence of shallow-water conditions (such as ancient coral reef deposits) that were carried with it into deeper water.

BOX 3–2   Historical Feature

OPHIOLITES: A GIFT FROM THE SEA FLOOR TO THE BRONZE AGE

As a result of plate tectonic processes, most oceanic lithosphere that is created at spreading centers is ultimately subducted and returned to the underlying mantle. However, a very small amount—only about 0.001%—of oceanic lithosphere does not get subducted; instead, it is uplifted onto the edge of a continent in a process called *obduction* (*ob* = against, *duct* = lead).   Fragments of the sea floor that are hoisted onto land are composed of a unique sequence of rocks called *ophiolites* (*ophis* = snake, *lite* = stone), which have long, slender, curvy shapes. Ophiolite sequences on land have long been recognized (Figure 3B) but only recently have they been connected to their environment of formation on the sea floor.

Figure 3C (*part a, left*) shows an idealized ophiolite sequence, which contains the following rock types vertically from top to bottom:

1. Layered *marine sediments*, which get progressively older with depth.

2. *Pillow basalts*, which indicate the lava flows that produced them flowed out on the ocean floor and cooled quickly during contact with water.

3. *Vertical basalt dikes*, which form by being injected into cracks in oceanic crust as lithospheric plates pull apart and move away from the spreading center. Because they are essentially thin planar features, they are often called *sheet dikes*.

4. *Massive gabbro*, a rock similar to basalt in composition but of a coarser texture due to slower cooling. This gabbro may represent the cooling of magma near the roof of the magma chamber that was the source of the overlying basalts.

5. *Layered gabbro* that may have crystallized within the upper region of the magma chamber.

6. *Layered peridotite*, a mantle rock composed mostly of iron- and magnesium-rich minerals called pyroxene and olivine. The boundary between the peridotite and the overlying gabbro is thought to represent the contact between ocean crust and the underlying mantle.

The main features of an idealized ophiolite sequence can also be observed in cores drilled into the sea floor near mid-ocean ridges. Such is the case for cores recovered from the Deep Sea Drilling Project (DSDP) Drill Hole No. 504B, which is located on the southern flank of the Costa Rica Rift (Figure 3C). The hole was drilled to a depth of 1350 meters (4429 feet) beneath the ocean floor, making it the deepest hole ever drilled in ocean crust. Although it did not

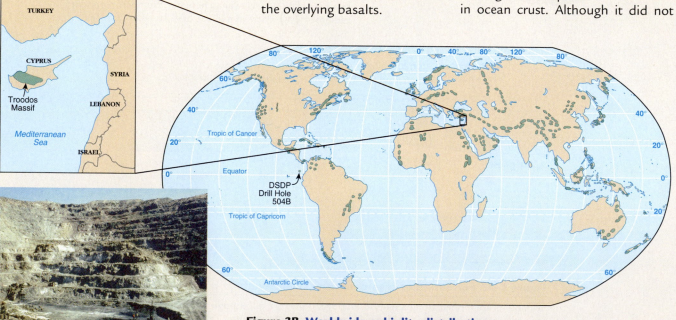

Figure 3B   Worldwide ophiolite distribution.

Ophiolites found near the margins of continents where subduction occurs are generally younger (less than 200 million years old) than those embedded deep within continents, some of which are as much as 1.2 billion years old. The location of DSDP Drill Hole No. 504B—the deepest hole drilled through ocean crust—is also shown. Inset shows location of the Troodos Massif ophiolite on the island of Cyprus in the eastern Mediterranean Sea. Photo shows mining of the Troodos Massif sulfide ore, which is the dark material in the bottom of the pit.

penetrate through the complete ophiolite sequence, the rocks it did penetrate show how well this sequence resembles those observed in ophiolites studied on land (Figure 3C, *part a, right*).

Many ophiolites contain rich metallic ores with patterns of enrichment similar to those occurring in the sediments and oceanic crust near deep-sea hydrothermal vents on the sea floor (Figure 3C, *part b*). However, there is increasing evidence that many ophiolites may have formed at back-arc spreading centers associated with subduction zones (which will be discussed in Chapter 4) rather than at oceanic ridges and rises.

One of the best-studied ophiolite sequences is the Troodos Massif on the island of Cyprus in the eastern Mediterranean Sea (Figure 3B, *inset*). Troodos Massif is also one of the oldest known ore deposits, having been mined for copper[11] since the time of the Phoenicians. In fact, this copper deposit was largely responsible for helping civilization advance from the Stone Age to the Bronze Age. Archeological evidence indicates in about 2760 B.C., people living on Cyprus discovered that copper could be hardened with the addition of tin during

Continued...

---

[11]"Cyprus" means copper; even today, it is unclear if the island is named for the metal, or vice versa.

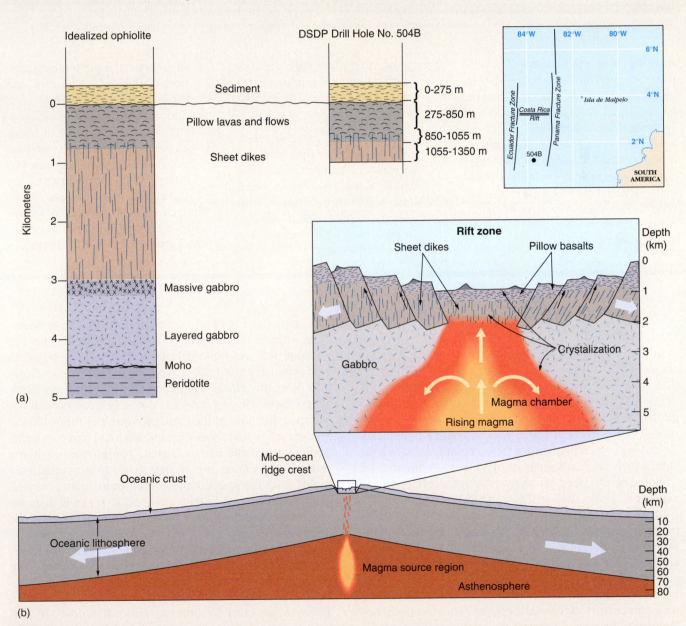

Figure 3C Ophiolite rock types.

(a) The column at left shows an idealized sequence of rock types found in ophiolite complexes such as the Troodos Massif of Cyprus. On the right is the sequence of rocks encountered in Deep Sea Drilling Project (DSDP) Drill Hole No. 504B, which closely resembles those in the upper idealized ophiolite complex. (b) Diagram showing the environment of formation for pillow basalts, sheet dikes, and gabbros of an ophiolite complex.

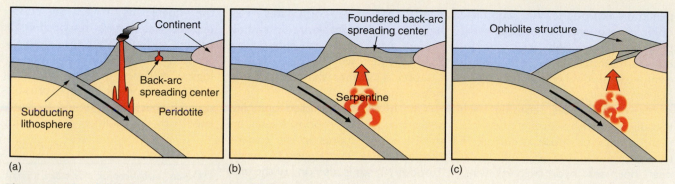

Figure 3D The obduction of ophiolites onto a continent.

(a) Subducting lithosphere contains water-bearing minerals. (b) Water-bearing minerals are heated and hot water is released into the peridotite, which is altered to lower-density serpentine. The expansion of the serpentine lifts the overlying mantle and oceanic crust. (c) Continued compressional forces break the uplifted segments of oceanic crust and wedge them onto the margin of a continent where they remain as ophiolites.

smelting. The smelted bronze could then be fashioned into various weapons and tools that, at the time, had no equal. If the copper deposits of Troodos Massif were not so readily available, humankind might have remained in the Stone Age much longer!

Still, it is not clear how ophiolites—which have high density—become obducted onto continents. Figure 3D shows how this might be accomplished:

a. An oceanic plate subducts beneath a plate containing a continent and a section of oceanic lithosphere.

b. The subducting oceanic plate contains water-bearing minerals such as zeolites and amphiboles, which release their water when heated. This heated water converts the mantle rock, peridotite, to serpentine, which is much less dense than the surrounding peridotite. The serpentine rises and pushes up the overlying crust and mantle above sea level.

c. Continued compressional forces cause the oceanic crust and associated serpentine to break into wedges that are obducted onto the margin of a continent.

Further study is needed to answer questions that remain about the origin and emplacement of ophiolites.

Coral Reef Development

On his voyage aboard the HMS *Beagle*, the famous naturalist **Charles Darwin** noticed a progression of stages in **coral reef** development. He hypothesized that the origin of coral reefs depended on the subsidence (sinking) of volcanic islands (Figure 3–34) and published the concept in *The Structure and Distribution of Coral Reefs* in 1842. What Darwin's hypothesis lacked was a mechanism for how volcanic islands subside. Much later, advances in plate tectonic theory and samples of the deep structure of coral reefs provided evidence to help support Darwin's hypothesis.

Reef-building corals are colonial animals that live in shallow, warm, tropical seawater and produce a hard skeleton of limestone. Once corals are established in an area that has the conditions necessary for their growth, they continue to grow upward layer by layer with each new generation attached to the skeletons of its predecessors. Over millions of years, a thick sequence of coral reef deposits may develop if the conditions remain favorable.

The three stages of development in coral reefs are called fringing, barrier, and atoll. **Fringing reefs** (Figure 3–34a) initially develop along the margin of a landmass (an island or a continent) where the temperature, salinity, and turbidity (cloudiness) of the water are suitable for reef-building corals. Often, fringing reefs are associated with active volcanoes whose lava flows run down the flanks of the volcano and kill the coral. Thus, these fringing reefs are not very thick or well developed. Because of the close proximity of the landmass to the reef, runoff from the landmass can carry so much sediment that the reef is buried. The amount of living coral in a fringing reef at any given time is relatively small, with the greatest concentration in areas protected from sediment and salinity changes. If sea level does not rise or the land does not subside, the process stops at the fringing reef stage.

The **barrier reef** stage follows the fringing reef stage. Barrier reefs are linear or circular reefs separated from the landmass by a well-developed lagoon (Figure 3–34b). As the landmass subsides, the reef maintains its position close to sea level by growing upward. Studies of reef growth rates indicate most have grown 3 to 5 meters (10 to 16 feet) per 1000 years during the recent geologic past. Evidence suggests that some fast-growing reefs in the Caribbean have grown more than 10 meters (33 feet) per 1000 years. Note that if the landmass subsides at a rate

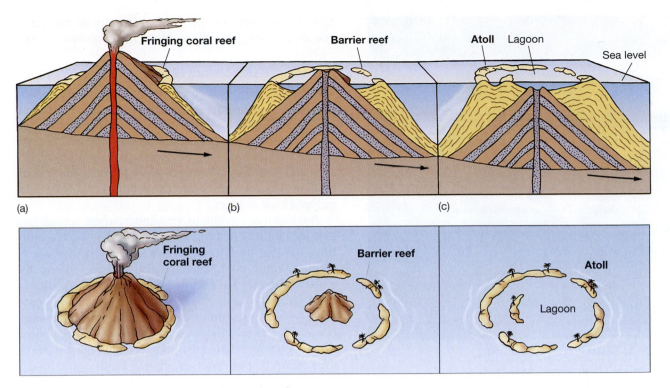

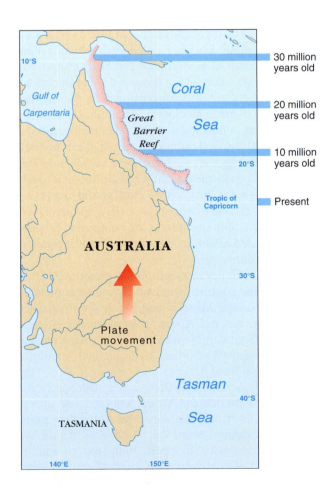

Figure 3–34 Stages of development in coral reefs.

Cross-sectional view (*above*) and map view (*below*) of **(a)** fringing reef, **(b)** barrier reef, and **(c)** atoll. With the right conditions for coral growth, coral reefs change through time from fringing reef to barrier reef to atoll.

Figure 3–35 Australia's Great Barrier Reef records plate movement.

About 30 million years ago, the Great Barrier Reef began to develop as northern Australia moved into tropical waters that allowed coral growth.

faster than coral can grow upward, the coral reef will be submerged in water too deep for it to live.

The largest reef system in the world is Australia's **Great Barrier Reef**, a series of over 3000 individual reefs collectively in the barrier reef stage of development, home to hundreds of coral species and thousands of other reef-dwelling organisms. The Great Barrier Reef lies 40 kilometers (25 miles) or more offshore, averages 150 kilometers (90 miles) in width, and extends for more than 2000 kilometers (1200 miles) along Australia's shallow northeastern coast. The effects of the Indian-Australian plate moving north toward the Equator from colder Antarctic waters are clearly visible in the age and structure of the Great Barrier Reef (Figure 3–35). It is oldest (around 25 million years old) and thickest at its northern end because the northern part of Australia reached water warm enough to grow coral before the southern parts did. In other areas of the Pacific, Indian, and Atlantic Oceans, smaller barrier reefs are found around the tall volcanic peaks that form tropical islands.

Figure 3–36 Barrier reefs and atolls.

A portion of the Society Islands in the Pacific Ocean as photographed from the space shuttle. From lower right, the islands of Raiatea, Tahaa, and Bora-Bora are in the barrier reef stage of development, while Motu Iti is an atoll.

The **atoll** (*atar* = to be crowded together) stage (Figure 3–34c) comes after the barrier reef stage. As a barrier reef around a volcano continues to subside, coral builds up toward the surface. After millions of years, the volcano becomes completely submerged, but the coral reef continues to grow. If the rate of subsidence is slow enough for the coral to keep up, a circular reef called an atoll is formed. The atoll encloses a lagoon usually not more than 30 to 50 meters (100 to 165 feet) deep. The reef generally has many channels that allow circulation between the lagoon and the open ocean. Buildups of crushed-coral debris often form narrow islands that encircle the central lagoon (Figure 3–36) and are large enough to allow human habitation.

Detecting Plate Motion with Satellites

Since the late 1970s, orbiting satellites allow the accurate positioning of locations on Earth (this technique is also used for navigation by ships at sea; see Box 1–1). If the plates are moving, satellite positioning should show this movement over time. The map in Figure 3–37 shows locations that have been measured in this manner over a 20-year period. It demonstrates that locations on Earth are moving in good agreement with the direction and rate of motion predicted by plate tectonics. Successful predic-

tion that locations on Earth are moving with respect to one another very strongly supports plate tectonic theory.

The Past: Paleoceanography

The study of historical changes of continental shapes and positions is called **paleogeography** (*paleo* = ancient, *geo* = earth, *graphy* = the name of a descriptive science). **Paleoceanography** (*paleo* = ancient, *ocean* = the marine environment, *graphy* = the name of a descriptive science) is the study of changes in the physical shape, composition, and character of the oceans brought about by paleogeographical changes.

An important point to remember when considering past locations of features is that nothing is permanently fixed in place on Earth's surface. During the passage of geologic time, continents have moved all around the globe and ocean basins open and close with regularity. Even the mid-ocean ridge moves relative to the continents, which sometimes causes segments of the mid-ocean ridge to be subducted! In addition, the mid-ocean ridge is moving relative to a fixed location outside Earth. This means that an observer orbiting Earth would notice, after only a few million years, that most continental and sea floor features—even plate boundaries—are moving. Hotspots, however, are the exception. They seem to be relatively stationary and can be used to determine the relative motions of other features.

Changes on Earth through Time   Figure 3–38 is a series of world maps showing paleogeographical reconstructions of Earth at 60-million-year intervals. At 540 million years ago, many of the present-day continents are barely recognizable. North America was on the Equator and rotated 90 degrees clockwise. Antarctica was on the Equator and was connected to many other continents.

Between 540 and 300 million years ago, the continents began to come together to form Pangaea. Notice that Alaska had not yet formed. Continents are thought to add material through the process of **continental accretion** (*ad* = toward, *crescere* = to grow). Like adding layers onto a snowball, bits and pieces of continents, islands, and volcanoes are added to the edges of continents and create larger landmasses.

From 180 million years ago to the present, Pangaea separated and the continents moved toward their present-day positions. North America and South America rifted away from Europe and Africa to produce the Atlantic Ocean. In the Southern Hemisphere, South America and a continent composed of India, Australia, and Antarctica begin to separate from Africa.

By 120 million years ago, there was a clear separation between South America and Africa, and India had moved northward, away from the Australia-Antarctica mass, which began moving toward the South Pole. As the Atlantic Ocean continued to open, India moved rapidly northward and collided with Asia about 45 million years

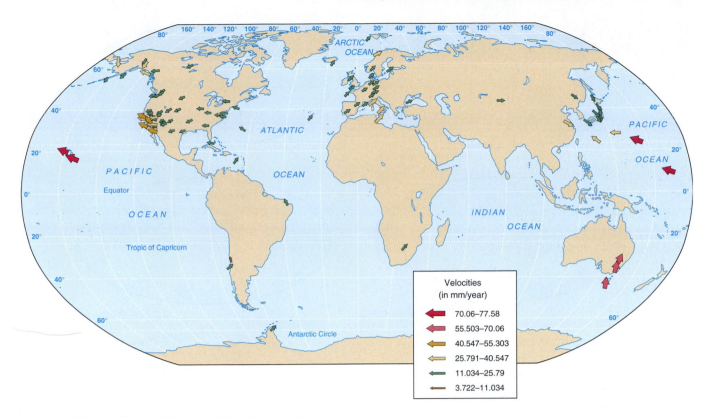

Figure 3–37 Satellite positioning of locations on Earth.

Arrows show direction of motion based on satellite measurement of positions on Earth during the period 1979–1997. Rate of plate motion in millimeters per year is indicated with different colored arrows (*see legend*).

ago. Australia had also begun a rapid journey to the north since separating from Antarctica.

Studies of plate motion suggest that rates of spreading have historically been higher at lower latitudes. This is apparent in the pattern of formation of the North Atlantic Ocean: Europe and North America have separated at a much lower rate than have Africa and North America. Further, North America and South America were not fully connected by the Isthmus of Panama until about 5 million years ago, which had a marked effect on ocean circulation patterns and the distribution of marine life.

One major result of global plate tectonic events over the past 180 million years has been the creation of the Atlantic Ocean, which continues to grow as sea floor spreads along the Mid-Atlantic Ridge. At the same time, the Pacific Ocean continues to shrink due to subduction along the many trenches that surround it and continental plates that bear in from both the east and west.

Laurasia and Gondwanaland Figure 3–39 is a series of detailed paleogeographic maps showing Earth before the time of Pangaea to the present, including the distribution of coral and other fossils. The present-day pattern of these fossils can be explained using paleogeographic re-

constructions. For example, Figure 3–39a (*red stripes*) shows that the exact same species of fossilized coral are found in a band of 350 million-year-old rocks in western Europe and eastern North America, as well as throughout the Alps and Himalayas. This pattern implies that these areas must have been in geographic proximity at that time, even though they are widely separated today (Figure 3–39c).

Additionally, several lines of evidence suggest that from 350 to 250 million years ago the shallow Tethys Sea separated the supercontinent **Laurasia** (composed of what are now North America, Europe, and Asia) from the supercontinent **Gondwanaland** (composed of what are now South America, Africa, India, Australia, and Antarctica). In time, these two supercontinents bridged parts of the Tethys Sea and combined to form the giant landmass of Pangaea about 200 million years ago (Figure 3–39b).

Fossils from sediments that were laid down on land aid in determining the latitudinal positions of the two supercontinents. From about 350 to 280 million years ago, there were two distinct floral (*flora* = flower) assemblages, one on each supercontinent. The *Laurasian floral assemblage* (Figure 3–39, *green shading*) in-

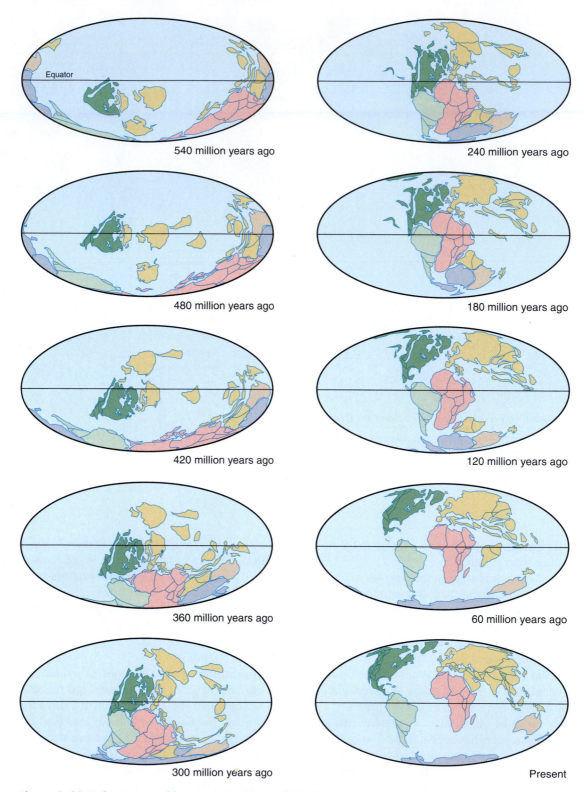

540 million years ago

480 million years ago

420 million years ago

360 million years ago

300 million years ago

240 million years ago

180 million years ago

120 million years ago

60 million years ago

Present

Figure 3–38 Paleogeographic reconstructions of Earth.

The positions of the continents at 60-million-year intervals.

cluded many species of tropical plants that were incorporated into the sediment that formed the extensive coal beds mined throughout the eastern United States and Europe. These tropical plants indicate that Laurasia occupied low latitudes during that time. The *Gondwanaland floral assemblage* (Figure 3–39, *green*

dots) is represented by a few species of plants thought to have grown in cold climates, presumably at a high southern latitude. Supporting evidence for this includes the occurrence of glacial deposits in South America, Africa, India, and Australia, all of which at that time must have been very near the southern polar

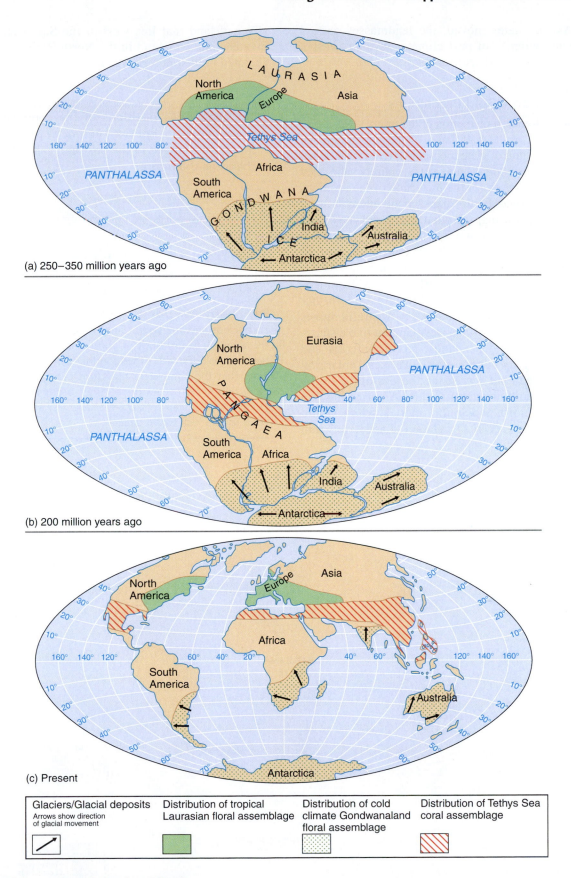

(a) 250–350 million years ago

(b) 200 million years ago

(c) Present

Glaciers/Glacial deposits	Distribution of tropical Laurasian floral assemblage	Distribution of cold climate Gondwanaland floral assemblage	Distribution of Tethys Sea coral assemblage
Arrows show direction of glacial movement			

Figure 3–39 Laurasia and Gondwanaland.

Paleogeographic maps showing the distribution of the tropical Laurasian floral assemblage (*green shading*), the cold-climate Gondwanaland floral assemblage (*green dots*), and the shallow-water Tethys Sea coral assemblage (*red stripes*). Arrows show the direction of ancient glacial flow from the south polar region. **(a)** The distribution of continents 250–350 million years ago. **(b)** By 200 million years ago, Laurasia and Gondwanaland combined to produce the single large continent Pangaea. **(c)** The present-day configuration of the continents.

region. As the plates moved, the landmasses carried this ancient evidence of past climates to their present geographic positions (Figure 3–39c).

The Future: Some Bold Predictions

Using plate tectonics, a prediction of the future positions of features on Earth can be made based on the assumption that the rate and direction of plate motion will remain the same. Although these assumptions may not be entirely valid, they do provide a framework for the prediction of the positions of continents and other Earth features in the future.

Figure 3–40 is a map of what the world may look like 50 million years from now, showing many notable differences as compared to today. For instance, the east African rift valleys may enlarge to form a new linear sea and the Red Sea may be greatly enlarged from rifting there. India may continue to plow into Asia, further uplifting the Himalaya Mountains. As Australia moves north toward Asia, it may use New Guinea like a snowplow to accrete various islands. North America and South America may continue to move west, enlarging the Atlantic Ocean but decreasing the size of the Pacific Ocean. The land bridge of Central America may no longer connect North and South America, which would dramatically alter ocean circulation. Lastly, the thin sliver of land that lies west of the San Andreas Fault may become an island in the North Pacific, soon to be accreted onto southern Alaska.

> The geographic positions of the continents and ocean basins are not fixed in time or place. Rather, they have changed in the past and will continue to change in the future.

? STUDENTS SOMETIMES ASK...

Will the continents come back together and form a single landmass anytime soon?

Yes, it is very likely that the continents will come back together, but not anytime soon. Since all of the continents are on the same planetary body, a continent can travel only so far before it collides with other continents. Recent research suggests that the continents may form a supercontinent once every 500 million years or so. Since it has been 200 million years since Pangaea split up, we have only about 300 million years to establish world peace!

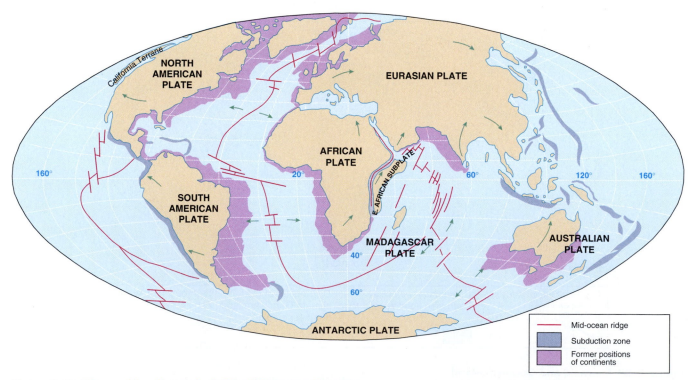

Figure 3–40 **The world as it may look 50 million years from now.**

Dark blue shadows indicate the present-day positions of continents and tan shading indicates positions of the continents in 50 million years. Arrows indicate direction of plate motion.

Plate Tectonics ... To Be Continued

Since its inception by Alfred Wegener nearly 100 years ago, plate tectonics has been supported by a wealth of scientific evidence (some of which has been presented in this chapter). Although there are still details to be worked out (such as the exact driving mechanism), it has been universally accepted by Earth scientists today because it helps explain so many observations about our planet. Further, it has led to predictive models that have been used to successfully understand Earth behavior. One such example is the **Wilson cycle** (Figure 3–41), named in honor of *John Tuzo Wilson* for his contribution to the early ideas of plate tectonics. The Wilson cycle uses plate tectonic processes to show the distinctive life cycle of ocean basins during their formation, growth, and destruction.

Not only is plate tectonic activity primarily responsible for the creation of landforms, it also plays a prominent role in the development of ocean floor features—which is the topic of the next chapter. Armed with the knowledge of plate tectonic processes you've gained from this chapter, understanding the history and development of ocean floor features in various marine provinces will be a much simpler task.

STAGE		MOTION	PHYSIOGRAPHY	EXAMPLE
EMBRYONIC 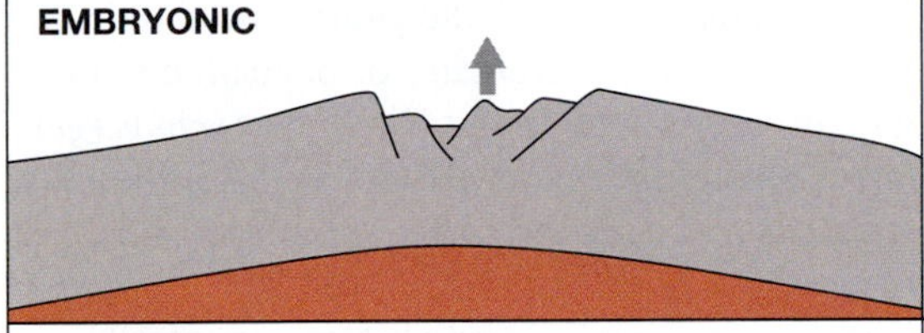		Uplift	Complex system of linear rift valleys on continent	East African rift valleys
JUVENILE 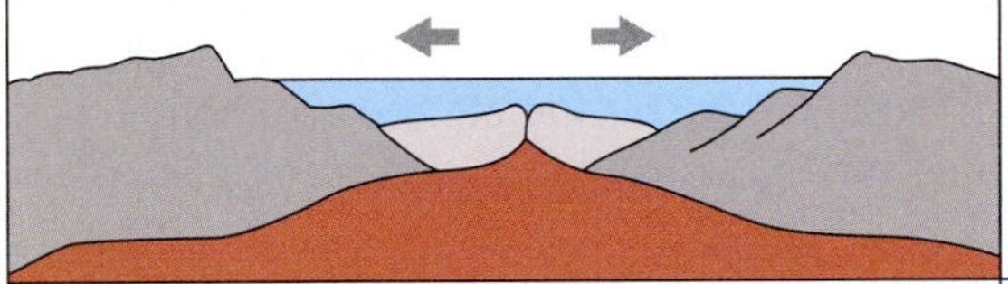		Divergence (spreading)	Narrow seas with matching coasts	Red Sea
MATURE 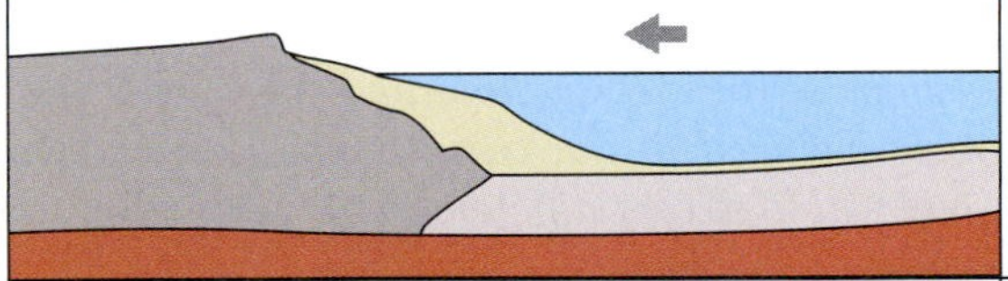		Divergence (spreading)	Ocean basin with continental margins	Atlantic and Arctic Oceans
DECLINING 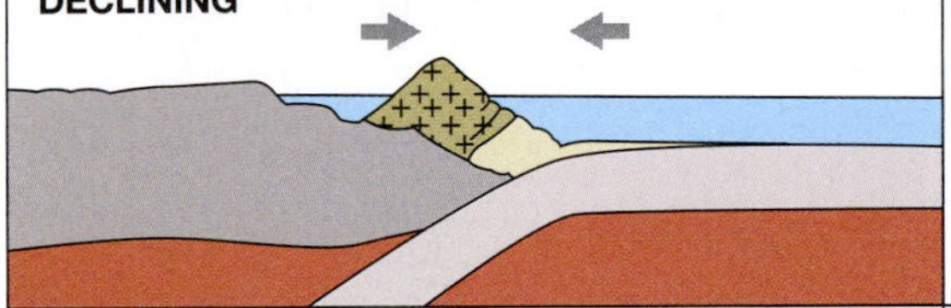		Convergence (subduction)	Island arcs and trenches around basin edge	Pacific Ocean
TERMINAL 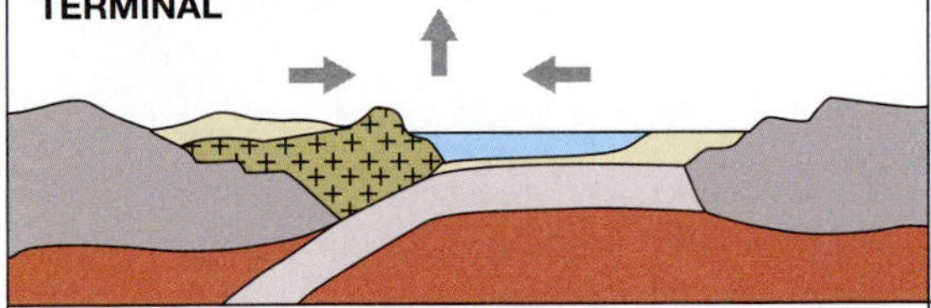		Convergence (collision) and uplift	Narrow, irregular seas with young mountains	Mediterranean Sea
SUTURING 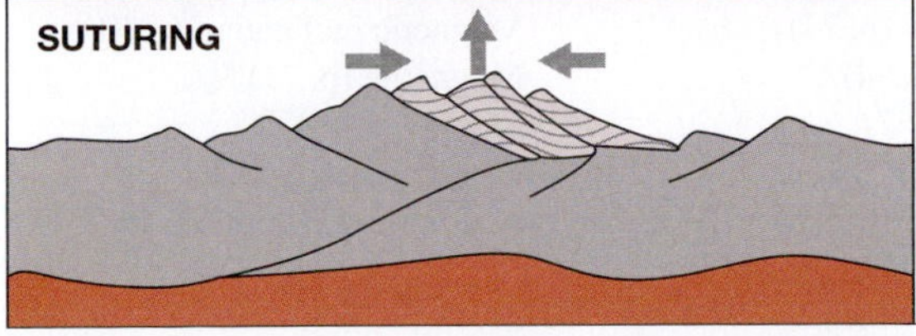		Convergence and uplift	Young to mature mountain belts	Himalaya Mountains

Figure 3–41 The Wilson cycle of ocean basin evolution.

The Wilson cycle depicts the stages of ocean basin development, from the initial embryonic stage of formation to the destruction of the basin as continental masses collide and undergo suturing.

Chapter in Review

- According to the theory of plate tectonics, *the outer-most portion of Earth is composed of a patchwork of thin, rigid lithospheric plates that move horizontally* with respect to one another. The idea began as a hypothesis called continental drift proposed by *Alfred Wegener* at the start of the 20th century. He suggested that *about 200 million years ago, all the continents were combined* into one large continent (*Pangaea*) surrounded by a single large ocean (*Panthalassa*).

- *Many lines of evidence were used to support the idea of continental drift*, including the similar shape of nearby continents, matching sequences of rocks and mountain chains, glacial ages and other climate evidence, and the distribution of fossil and present-day organisms. Although this evidence suggested that continents have drifted, other incorrect assumptions about the mechanism involved caused many geologists and geophysicists to discount this hypothesis throughout the first half of the twentieth century.

- *More convincing evidence for drifting continents was introduced in the 1960s when paleomagnetism—the study of Earth's ancient magnetic field—was developed* and the significance of features of the ocean floor became better known. The paleomagnetism of the ocean floor is permanently recorded in oceanic crust and reveals stripes of normal and reverse magnetic polarity in a symmetric pattern relative to the mid-ocean ridge.

- *Harry Hess advanced the idea of sea floor spreading.* New *sea floor is created at the crest of the mid-ocean ridge* and moves apart in opposite directions and is *eventually destroyed by subduction into an ocean trench.* This helps explain the pattern of magnetic stripes on the sea floor and why sea floor rocks increase linearly in age in either direction from the axis of the mid-ocean ridge. Other supporting evidence for plate tectonics includes oceanic heat flow measurements and the pattern of worldwide earthquakes. The combination of evidence *convinced geologists* of Earth's dynamic nature and helped advance the idea of continental drift into the more encompassing plate tectonic theory.

- *Studies of Earth structure indicate that Earth is a layered sphere*, with the brittle plates of the *lithosphere* riding on a plastic, high-viscosity *asthenosphere*. Near the surface, the lithosphere is composed of continental and oceanic crust. Continental crust consists mostly of granite, but oceanic crust consists mostly of basalt. Continental crust is lower in density, lighter in color, and thicker than oceanic crust. Both types float isostatically on the denser mantle below.

- *As new crust is added to the lithosphere at the mid-ocean ridge* (*divergent boundaries* where plates move apart), the opposite ends of the *plates are subducted into the mantle at ocean trenches* or beneath continental mountain ranges such as the Himalayas (*convergent boundaries* where plates come together). Additionally, oceanic ridges and rises are offset and plates slide past one another along transform faults (*transform boundaries* where plates slowly grind past one another).

- *Tests of the plate tectonic model indicate that many features and phenomena provide support for shifting plates.* These include mantle plumes and their associated hotspots that record the motion of plates past them, the origin of flat-topped tablemounts, stages of coral reef development, and the detection of plate motion by accurate positioning of locations on Earth using satellites.

- *The positions of various sea floor and continental features have changed in the past, continue to change today, and will look very different in the future.* Before the formation of Pangaea, two large continents appear to have existed, *Laurasia* in the north and *Gondwanaland* to the south.

Key Terms

Andesite (p. 91)	Coral reef (p. 100)	Hawaiian Islands–Emperor Seamount Chain (p. 95)	Magnetic anomaly (p. 73)
Asthenosphere (p. 79)	Core (p. 79)	Heat flow (p. 77)	Magnetic dip (p. 71)
Atoll (p. 102)	Crust (p. 79)	Hess, Harry (p. 74)	Magnetic field (p. 70)
Barrier reef (p. 100)	Curie point (p. 71)	Hot spot (p. 94)	Magnetic inclination (p. 71)
Basalt (p. 81)	Darwin, Charles (p. 100)	Ice age (p. 67)	Magnetite (p. 71)
Continental accretion (p. 102)	Divergent boundary (p. 84)	Inner core (p. 79)	Mantle (p. 79)
Continental arc (p. 91)	East Pacific Rise (p. 88)	Isostatic adjustment (p. 83)	Mantle plume (p. 94)
Continental crust (p. 82)	Fringing reef (p. 100)	Isostatic rebound (p. 84)	Matthews, Drummond (p. 74)
Continental drift (p. 65)	Gondwanaland (p. 103)	Laurasia (p. 103)	Mesosaurus (p. 67)
Continental transform fault (p. 94)	Granite (p. 82)	Lithosphere (p. 65)	Mesosphere (p. 79)
Convection cell (p. 74)	Great Barrier Reef (p. 101)	Loihi (p. 96)	Mid-Atlantic Ridge (p. 88)
Convergent boundary (p. 84)	Guyot (p. 96)		Mid-ocean ridge (p. 74)

Nematath (p. 96)
Ocean trench (p. 74)
Oceanic crust (p. 81)
Oceanic ridge (p. 88)
Oceanic rise (p. 88)
Oceanic transform fault (p. 94)
Outer core (p. 79)
Paleoceanography (p. 102)
Paleogeography (p. 102)

Paleomagnetism (p. 71)
Pangaea (p. 65)
Panthalassa (p. 65)
Plate tectonics (p. 65)
Polar wandering curve (p. 71)
Polarity (p. 73)
Rift valley (p. 84)
Rifting (p. 87)
San Andreas Fault (p. 94)

Sea floor spreading (p. 74)
Seamount (p. 96)
Seismic moment magnitude
  (M_w) (p. 89)
Spreading center (p. 74)
Subduction (p. 74)
Subduction zone (p. 74)
Tablemount (p. 96)
Tethys Sea (p. 65)

Transform boundary (p. 84)
Transform faulting (p. 94)
Transform fault (p. 94)
Vine, Frederick (p. 74)
Viscosity (p. 83)
Volcanic arc (p. 90)
Wegener, Alfred (p. 65)
Wilson cycle (p. 107)

Questions and Exercises

1. Cite the lines of evidence Alfred Wagener used to support his idea of continental drift. Why did scientists doubt that continents drifted?

2. If you could travel back in time with three figures (illustrations) from this chapter to help Alfred Wegener convince the scientists of his day that continental drift does exist, what would they be and why?

3. How does the dip of magnetic particles found in igneous rocks tell us at what latitude they were formed?

4. Why was the pattern of alternating reversals of Earth's magnetic field as recorded in the sea-floor rocks such an important piece of evidence for advancing plate tectonics?

5. Describe sea floor spreading and why it was an important piece of evidence in support of plate tectonics.

6. Describe the general relationships that exist among distance from the spreading centers, heat flow, age of the ocean crustal rock, and ocean depth.

7. Why does a map of worldwide earthquakes closely match the locations of worldwide plate boundaries?

8. Discuss how the chemical composition of Earth's interior differs from its physical properties. Include specific examples.

9. What are differences between the lithosphere and the asthenosphere?

10. Describe differences between granite and basalt. Which property is responsible for making the granitic crust "float" higher in the mantle?

11. List and describe the three types of plate boundaries. Include in your discussion any sea floor features that are related to these plate boundaries, and include a real-world example of each. Construct a map view and cross section showing each of the three boundary types and direction of plate movement.

12. Most lithospheric plates contain both oceanic- and continental-type crust. Use plate boundaries to explain why this is true.

13. Describe the differences between oceanic ridges and oceanic rises. Include in your answer why these differences exist.

14. Convergent boundaries can be divided into three types based on the type of crust contained on the two colliding plates. Compare and contrast the different types of convergent boundaries that result from these collisions.

15. Describe the difference in earthquake magnitudes that occur between the three types of plate boundaries, and include why these differences occur.

16. How can plate tectonics be used to help explain the difference between a seamount and a tablemount?

17. How is the age distribution pattern of the Hawaiian Islands–Emperor Seamount Chain explained by the position of the Hawaiian hotspot? What could have caused the curious bend in the chain?

18. Describe the differences in origin between the Aleutian Islands and the Hawaiian Islands. Provide evidence to support your explanation.

19. What are differences between a mid-ocean ridge and a hotspot?

20. Using the paleogeographic reconstructions shown in Figure 3–38, determine when the following events first appear in the geologic record:
    a. North America lies on the Equator.
    b. The continents come together as Pangaea.
    c. The North Atlantic Ocean opens.
    d. India separates from Antarctica.

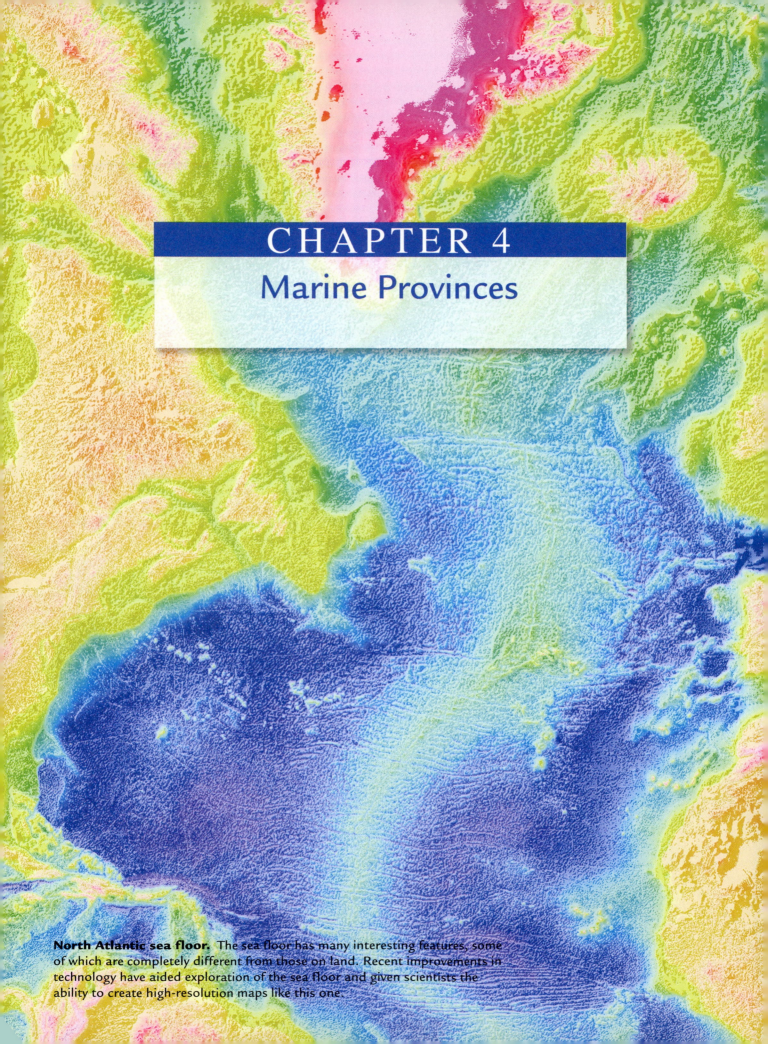

CHAPTER 4
Marine Provinces

North Atlantic sea floor. The sea floor has many interesting features, some of which are completely different from those on land. Recent improvements in technology have aided exploration of the sea floor and given scientists the ability to create high-resolution maps like this one.

Key Questions

- How do scientists collect information about the depth and shape of the sea floor?
- What is the difference between a passive and an active continental margin?
- How are submarine canyons formed?
- Where are the deepest areas of the ocean?
- What kinds of features are found at the mid-ocean ridge?
- What are differences between transform faults and fracture zones?

The answers to these questions (and much more) can be found in the highlighted concept statements within this chapter.

"Could the waters of the Atlantic be drawn off so as to expose to view this great sea-gash which separates the continents, and extends from the Arctic to the Antarctic, it would present a scene most rugged, grand, and imposing."

—Matthew Fontaine Maury (1854), commenting about the Mid-Atlantic Ridge

What does the ocean floor look like? Over a century and a half ago, most scientists believed that the ocean floor was completely flat and carpeted with a thick layer of muddy sediment containing little of scientific interest. Further, it was believed that the deepest parts were somewhere in the middle of the ocean basins. However, as more and more vessels crisscrossed the seas to map the ocean floor and to lay transoceanic cables, scientists found the terrain of the sea floor was highly varied and included deep troughs, ancient volcanoes, submarine canyons, and great mountain chains. It was unlike anything on land and, as it turns out, some of the deepest parts of the oceans are actually close to land!

As marine geologists and oceanographers began to analyze the features of the ocean floor, they realized that certain features had profound implications not only for the history of the ocean floor, but also for the history of Earth. How could all these remarkable features have formed, and how can their origin be explained? Over long periods of time, the shape of the ocean basins has changed as continents have ponderously migrated across Earth's surface in response to forces within Earth's interior. The ocean basins as they

presently exist reflect the processes of plate tectonics (the topic of the previous chapter), which help explain the origin of sea floor features.

Bathymetry

Bathymetry (*bathos* = depth, *metry* = measurement) is the measurement of ocean depths and the charting of the shape or topography (*topos* = place, *graphy* = discription of) the ocean floor. Determining bathymetry involves measuring the vertical distance from the ocean surface down to the mountains, valleys, and plains of the sea floor.

Bathymetric Techniques

The first recorded attempt to measure the ocean's depth was conducted in the Mediterranean Sea in about 85 B.C. by a Greek named Posidonius. His mission was to answer an age-old question: How deep is the ocean? Posidonius's crew made a **sounding** by letting out nearly 2 kilometers (1.2 miles) of line before the heavy weight on the end of the line touched bottom. Sounding lines were used for the next 2000 years by voyagers who used them to probe the ocean's depths. The standard unit of ocean depth is the **fathom** (*fathme* = outstretched arms[1]) and is equal to 1.8 meters (6 feet).

The first systematic bathymetric measurements of the oceans were made in 1872 aboard the HMS *Challenger* during its historic three-and-a-half-year voyage. Every so often, *Challenger's* crew stopped and measured the depth, along with many other ocean properties. These measurements indicated that the deep ocean floor was not flat but had significant *relief* (variations in elevation), just as dry land does. However, determining bathymetry by making occasional soundings rarely gives a complete picture of the ocean floor. For instance, imagine trying to determine what the surface features on land look like while flying in a blimp at an altitude of several kilometers on a foggy night using only a long weighted rope to determine your height above the surface. This is analogous to how bathymetric measurements were collected from ships using sounding lines.

As described in Chapter 1, it wasn't until the early 1900s that ships began to use **echo sounders** to more clearly delineate ocean floor features. An echo sounder sends a sound signal (called a *ping*) into the ocean to produce echoes when the sound bounces off any density difference, such as marine organisms or the ocean floor (Figure 4–1). Water is a good transmitter of sound, so the time it takes for the echoes to return[2] is used to determine the depth and corresponding shape of the ocean floor. Recall from Chapter 1, for example, that the German vessel *Meteor* used echo soundings in 1925 to identify a mountain range running through the center of the South Atlantic Ocean.

---

[1]This term is derived from how depth sounding lines were brought back on board a vessel by hand. While hauling in the line, the workers counted the number of arm-lengths collected. By measuring the length of the person's outstretched arms, the amount of line taken in could be calculated. Much later, the distance of 1 fathom was standardized to equal exactly 6 feet.)

---

[2]This technique uses the speed of sound in seawater, which is 1507 meters (4945 feet) per second.

Figure 4–1 An echo sounder record.

An echo sounder record from the east coast U.S. shows the provinces of the sea floor. Vertical exaggeration (the amount of expansion of the vertical scale) is 12 times. The scattering layer probably represents a concentration of marine organisms.

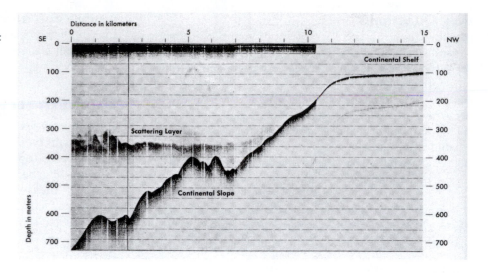

Echo sounding, however, lacks detail and often gives an inaccurate view of the relief of the sea floor. For instance, the sound beam emitted from a ship 4000 meters (13,100 feet) above the ocean floor widens to a diameter of about 4600 meters (15,000 feet) at the bottom. Consequently, the first echoes to return from the bottom are usually from the closest (highest) peak within this broad area. Nonetheless, most of our knowledge of ocean bathymetry has been provided by the echo sounder.

During and after World War II, there was great improvement in sonar technology. For example, the **precision depth recorder (PDR)**, which was developed in the 1950s, uses a focused high-frequency sound beam to measure depths to a resolution of about 1 meter (3.3 feet). Throughout the 1960s, PDRs were used extensively and provided a reasonably good representation of the ocean floor. From thousands of research vessel tracks, the first reliable global maps of sea floor bathymetry were produced. These maps helped confirm the ideas of sea floor spreading and plate tectonics.

Today, *multibeam echo sounders* (which use multiple frequencies of sound simultaneously) and side-scan **sonar** (an acronym for *so*und *na*vigation *a*nd *r*anging) give oceanographers a more precise picture of the ocean floor. **SeaBeam**—the first multibeam echo sounder—made it possible for a survey ship to map the features of the ocean floor along a strip up to 60 kilometers (37 miles) wide. The system uses sound emitters directed away from both sides of the ship, with receivers permanently mounted on the hull of the ship. Side-scan sonar systems such as **Sea MARC** (*Sea Mapping a*nd *R*emote *C*haracterization) and **GLORIA** (*G*eological *L*ong-*R*ange *I*nclined *A*coustical instrument) can be towed behind a survey ship to produce a strip map of ocean floor bathymetry (Figure 4–2).

To make a more detailed picture of the ocean floor, a side-scan instrument can be towed behind a ship on a cable so that it "flies" just above the ocean floor. One newly developed deep-tow system combines a side-scan sonar instrument with a sub-bottom imaging package (Figure 4–3). This allows a simultaneous view of the surface of the ocean floor and a cross-section of the sediment below at water depths to 6500 meters (21,325 feet).

STUDENTS SOMETIMES ASK...

All this sea floor mapping is interesting, but haven't we found everything on the sea floor already?

Certainly not! The world's ocean floor—equal in area to almost two moons plus two Mars-sized planets—is one of the most poorly mapped surfaces in the solar system. Recent satellite missions to other planets and their moons have produced stunning high-resolution images of these worlds. In contrast, the ocean floor image produced from sea surface height data shown in Figure 4C is still an order of magnitude lower in resolution that the optic images obtained from a satellite flyby of Mercury in the early 1970s!

Despite satellites that monitor ocean surface conditions and recent advances in mapping technology, only 5% of the ocean floor has been mapped as precisely as the surface of the Moon. In fact, there are areas of the ocean floor as large as the state of Kansas where no ship soundings have been made, and even well-surveyed areas are based on widely separated ship tracks. The great depth of the oceans along with seawater's opaque character has hindered mapping efforts, leaving room for major discoveries in the future (such as new sea floor features, shipwrecks, and mineral deposits).

Although multibeam and side-scan sonar produce fairly detailed bathymetric maps, mapping the sea floor by ship is a time-consuming process. A research vessel must tediously travel back and forth throughout an area (called "mowing the lawn") to produce an accurate map of bathymetric features. A satellite, on the other hand, can observe large areas of the ocean at one time. Consequently, satellites are increasingly

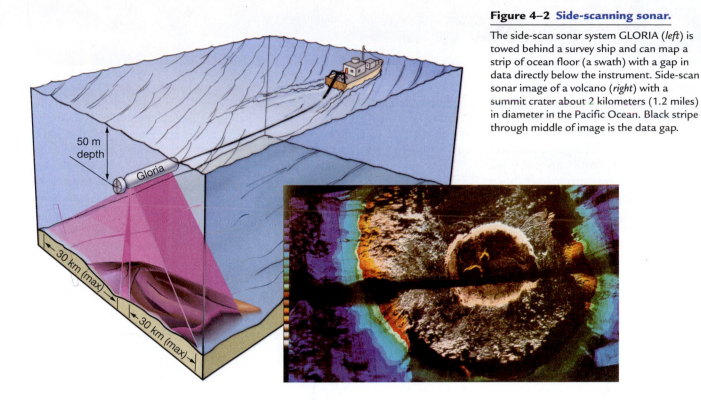

Figure 4–2 Side-scanning sonar.
The side-scan sonar system GLORIA (*left*) is towed behind a survey ship and can map a strip of ocean floor (a swath) with a gap in data directly below the instrument. Side-scan sonar image of a volcano (*right*) with a summit crater about 2 kilometers (1.2 miles) in diameter in the Pacific Ocean. Black stripe through middle of image is the data gap.

50 m depth

Gloria

30 km (max)

30 km (max)

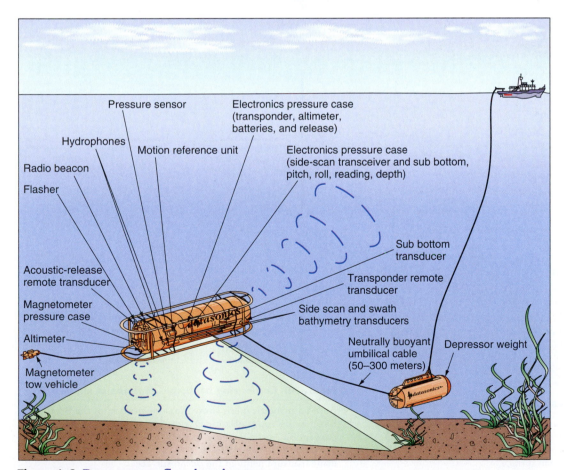

Pressure sensor

Electronics pressure case (transponder, altimeter, batteries, and release)

Hydrophones

Motion reference unit

Electronics pressure case (side-scan transceiver and sub bottom, pitch, roll, reading, depth)

Radio beacon

Flasher

Sub bottom transducer

Transponder remote transducer

Acoustic-release remote transducer

Side scan and swath bathymetry transducers

Magnetometer pressure case

Neutrally buoyant umbilical cable (50–300 meters)

Depressor weight

Altimeter

Datasonics

Magnetometer tow vehicle

Figure 4–3 Deep-tow sea floor imaging system.

Deep-tow side-scan sonar systems are towed close to the ocean floor and provide detailed sonar maps of the ocean floor as well as a profile view of the sediment below.

Recently, satellite measurements of the ocean surface have been used to make maps of the sea floor. How does a satellite—which orbits at a great distance above the planet and can only view the ocean's *surface*—obtain a picture of the sea *floor*?

The answer lies in the fact that sea floor features directly influence Earth's gravitational field. Deep areas such as trenches correspond to a lower gravitational attraction, and large undersea objects such as seamounts exert an extra gravitational pull. These differences affect sea level, causing the ocean surface to bulge outward and sink inward mimicking the relief of the ocean floor. A 2000-meter (6500-foot)-high seamount, for example, exerts a small but measurable gravitational pull on the water around it, creating a bulge 2 meters (7 feet) high on the ocean surface. These irregularities are easily detectable by satellites, which use microwave beams to measure sea level to within 4 centimeters (1.5 inches) accuracy. After corrections are made for waves, tides, currents, and atmospheric effects, the resulting pattern of lumps and bulges at the ocean surface can be used to indirectly reveal ocean floor bathymetry (Figure 4A). For example, Figure 4B compares two different maps of the same area: one based on bathymetric data from ships (*top*) and the other based on satellite measurements (*bottom*), which shows much higher resolution of sea floor features.

Data from the European Space Agency's ERS-1 satellite and from Geosat, a U.S. Navy satellite, were

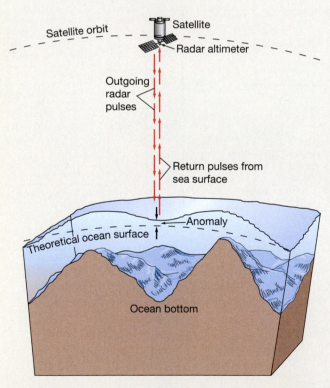

Figure 4A **Satellite measurements of the ocean surface.**

A satellite measures the variation of ocean surface elevation, which is caused by gravitational attraction and mimics the shape of the sea floor. The sea surface *anomaly* is the difference between the measured and theoretical ocean surface.

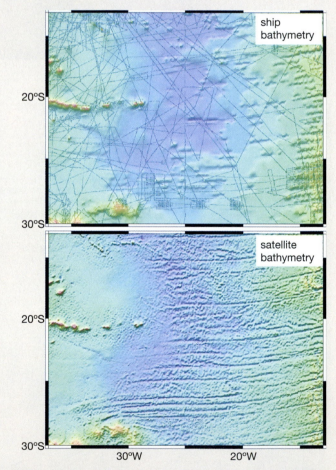

Figure 4B **Comparing bathymetric maps of the sea floor.**

Both bathymetric maps show the same portion of the Brazil Basin in the South Atlantic Ocean. *Top:* A map made using conventional echo sounder records from ships (ship tracks shown by thin lines). *Bottom:* A map from satellite data made using measurements of the ocean surface.

114

collected during the 1980s. When this information was recently declassified, Walter Smith of the National Oceanic and Atmospheric Administration and David Sandwell of Scripps Institution of Oceanography began producing sea floor maps based on the shape of the sea surface. What is unique about these researchers' maps is that they provide a view of Earth similar to being able to drain the oceans and view the ocean floor directly. Their map of ocean surface gravity (Figure 4C) uses depth soundings to calibrate the gravity measurements. Although gravity is not exactly bathymetry, this new map of the ocean floor clearly delineates many ocean floor features, such as the mid-ocean ridge, trenches, seamounts, and nemataths (island chains). In addition, this new mapping technique has revealed sea floor bathymetry in areas where research vessels have not conducted surveys.

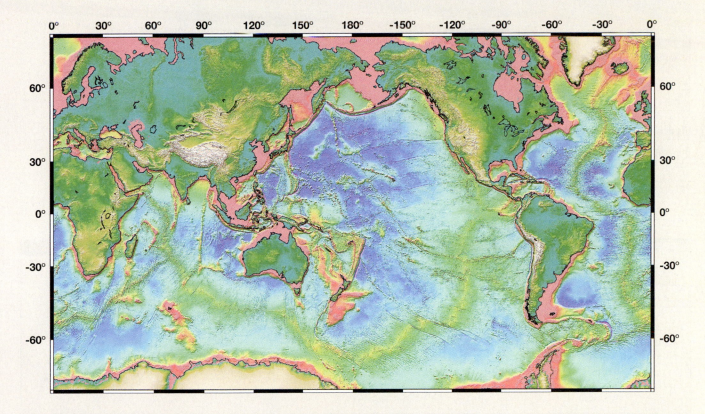

Figure 4C **Global sea surface elevation map from satellite data.**

Map showing the satellite-derived global gravity field, which, when adjusted using measured depths, closely corresponds to ocean depth. Purple indicates deep water; the mid-ocean ridge (intermediate water depths) is mostly light green and yellow; brown indicates shallowest water. Map also shows land surface elevations, with dark green color indicating low elevations and white color indicating high elevations.

used to determine ocean properties (see Chapter 1 and Table 1–1). Remarkably, satellite measurements allow the ocean floor to be mapped from space (Box 4–1).

Oceanographers who want to know about ocean structure beneath the sea floor use strong low-frequency sounds produced by explosions or air guns, as shown in Figure 4–4. These sounds penetrate beneath the sea floor and reflect off the boundaries between different rock or sediment layers, producing **seismic reflection profiles**, which have applications in mineral and petroleum exploration.

> Sending pings of sound into the ocean (echo sounding) is commonly used to determine ocean bathymetry. More recently, satellites have also been used to map sea floor features.

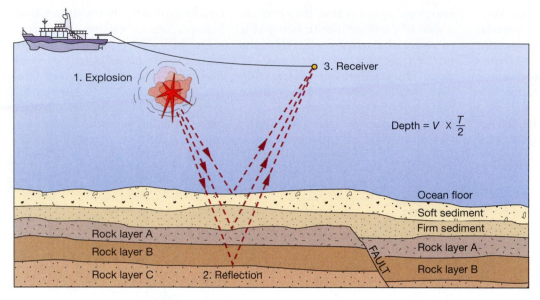

Figure 4–4 Seismic profiling.

An air-gun explosion emits low frequency sounds (1) that can penetrate bottom sediments and rock layers. The sound reflects off the boundaries between these layers (2) and returns to the receiver (3).

EXAMPLE 4–1

Using the equation shown in Figure 4-4, how deep is the water in an area where it takes a ping of sound 10 seconds to reach the bottom and return?

If we know the time it takes for sound to be transmitted, we can use the equation in Figure 4-4 to determine the water depth (the speed of sound in seawater is a constant value, given in footnote 2). Remember, the reason we need to divide the time by 2 is to account for the sound traveling to the bottom *and back to the surface*! The general equation is

$$\text{Depth} = \text{Velocity} \times \frac{\text{Time}}{2}$$

Since the velocity of sound in seawater is 1507 meters per second and the time is 10 seconds, this gives

$$\text{Depth} = 1507 \text{ meters per second} \times \frac{10 \text{ seconds}}{2}$$

So, the water depth is 7535 meters (24,700 feet), which is quite deep!

The Hypsographic Curve

Figure 4–5 illustrates Earth's **hypsographic** (*hypsos* = height, *graphic* = drawn) **curve**, which shows the relationship between the height of the land and the depth of the oceans. The bar graph (Figure 4–5, *left*) gives the percentage of Earth's surface area at various ranges of elevation and depth. The cumulative curve (Figure 4–5, *right*) gives the percentage of surface area from the highest peaks to the deepest depths of the oceans. Together, they show that 70.8% of Earth's surface is covered by oceans and that the average depth of the ocean is 3729 meters (12,234 feet) while the average height of the land is only 840 meters (2756 feet). The difference, recalling from our study of isostasy in Chapter 3, results from the greater density and lesser thickness of oceanic crust as compared to continental crust.

The cumulative hypsographic curve (Figure 4–5, *right*) shows five differently sloped segments. On land, the first steep segment of the curve represents tall mountains while the gentle slope represents low coastal plains (and continues just offshore, representing the shallow parts of the continental margin). The first slope below sea level represents steep areas of the continental margins and also includes the mountainous mid-ocean ridge. Further offshore, the longest, flattest part of the whole curve represents the deep-ocean basins, followed by the last steep part, which represents ocean trenches.

The shape of the hypsographic curve can be used to support the existence of plate tectonics on Earth. Specifically, the two flat areas and three sloped areas of the curve show that there is a very uneven distribution of area at different depths and elevations. If there were no active mechanism involved in creating such features on

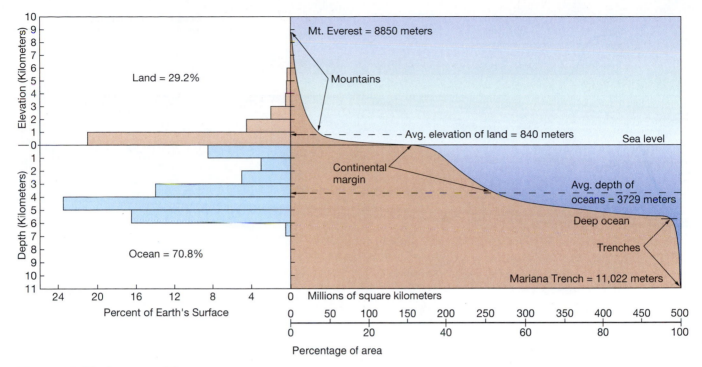

Figure 4–5 The hypsographic curve.

The bar graph (*left*) gives the percentage of Earth's surface area at various ranges of elevation and depth. The cumulative hypsographic curve (*right*) gives the percentage of surface area from the highest peaks to the deepest depths of the oceans. Also shown is the average ocean depth and land elevation.

Earth, the bar graph portions would all be about the same length and the cumulative curve would be a straight line. Instead, the variations in the curve suggest that plate tectonics is actively working to modify Earth's surface. The flat portions of the curve represent various intraplate elevations both on land and underwater while the slopes of the curve represent mountains, continental slopes, the mid-ocean ridge, and deep-ocean trenches, all of which are created by plate tectonic processes. Interestingly, hypsographic curves constructed for other planets and moons using satellite data have been used to determine if plate tectonics is actively modifying the surface of these worlds.

Provinces of the Ocean Floor

The ocean floor can be divided into three major provinces (Figure 4–6): (1) **continental margins**, which are shallow-water areas close to continents; (2) **deep-ocean basins**, which are deep-water areas farther from land; and (3) the **mid-ocean ridge**, which is comprised of shallower areas near the middle of an ocean. Plate tectonic processes (which were discussed in the previous chapter) are integral to the formation of these provinces: Through the process of sea floor spreading, mid-ocean ridges and deep-ocean basins are created; elsewhere, as a continent is split apart, new continental margins are formed.

Features of Continental Margins

Passive and Active Continental Margins Continental margins can be classified as either passive or active depending on their proximity to plate boundaries. **Passive margins** (Figure 4–7, *left*) are imbedded within the interior of lithospheric plates and are therefore not in close proximity to any plate boundary. Thus, passive margins usually lack major tectonic activity (large earthquakes, eruptive volcanoes, and mountain building).

The east coast of the United States, where there is no plate boundary, is an example of a passive continental margin. Passive margins are usually produced by rifting of continental landmasses and continued sea floor spreading over geologic time. Features of passive continental margins include the continental shelf, the continental slope, and the continental rise that extends toward the deep-ocean basins (Figures 4–7 and 4–8).

Active margins (Figure 4–7, *right*) are associated with lithospheric plate boundaries and are marked by a high degree of tectonic activity. Two types of active margins exist. **Convergent active margins** are associated with oceanic–continental convergent plate boundaries. Features include a continental arc onshore, a narrow shelf, a steep slope, and an offshore trench that delineates the plate boundary. Western South America, where the Nazca Plate is being subducted beneath the South American Plate, is an example of a convergent active margin. **Transform active margins** are less

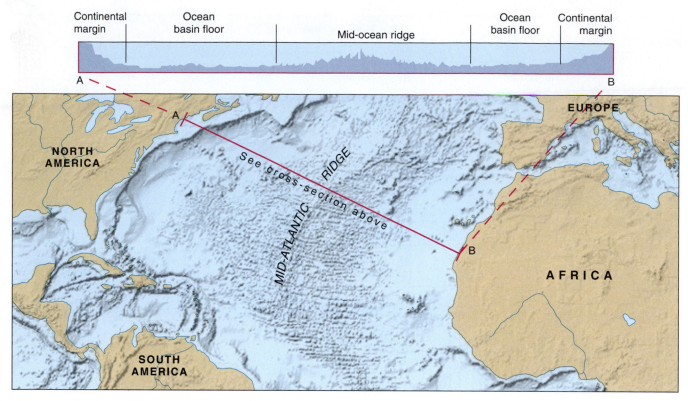

Figure 4–6 **Major regions of the North Atlantic Ocean floor.**

Map view below and profile view above, showing that the ocean floor can be divided into three major provinces: continental margins, deep ocean basins, and mid-ocean ridge.

common and are associated with transform plate boundaries. At these locations, faults that parallel the transform plate boundary create linear islands, banks (shallowly submerged areas), and deep basins close to shore. Coastal California along the San Andreas Fault is an example of a transform active margin.

> Passive continental margins lack a plate boundary and have different features than active continental margins, which include a plate boundary (either convergent or transform).

Continental Shelf The **continental shelf** is defined as a generally flat zone extending from the shore beneath the ocean surface to a point at which a marked increase in slope angle occurs, called the **shelf break** (Figure 4–8). It is usually flat and relatively featureless because of marine sediment deposits but can contain coastal islands, reefs, and raised banks. The underlying rock is granitic continental crust, so the continental shelf is geologically part of the continent. The general bathymetry of the continental shelf can usually be predicted by examining the topography of the adjacent coastal region. With few exceptions, the coastal topography extends beyond the shore and onto the continental shelf.

The average width of the continental shelf is about 70 kilometers (43 miles), but it varies from a few tens

of meters to 1500 kilometers (930 miles). The broadest shelves occur off the northern coasts of Siberia and North America in the Arctic Ocean. The average depth at which the shelf break occurs is about 135 meters (443 feet). Around the continent of Antarctica, however, the shelf break occurs at 350 meters (2200 feet). The average slope of the continental shelf is only about a tenth of a degree, similar to the slope given to a large parking lot for drainage purposes.

Sea level has fluctuated over the history of Earth, causing the shoreline to migrate back and forth across the continental shelf. When colder climates prevailed during the Ice Age, for example, more of Earth's water was frozen as glaciers on land, so sea level was lower than it is today. During this time, more of the continental shelf was exposed.

The type of continental margin will determine the shape and features associated with the continental shelf. For example, the east coast of South America has a broader continental shelf than its west coast. The east coast is a passive margin, which typically has a wider shelf. In contrast, the convergent active margin present along the west coast of South America is characterized by a narrow continental shelf and a shelf break close to shore. For transform active margins such as along California, the presence of offshore faults produces a continental shelf that is not flat. Rather, it is marked by a high degree of relief (islands, shallow banks, and deep basins) called a **continental borderland.**

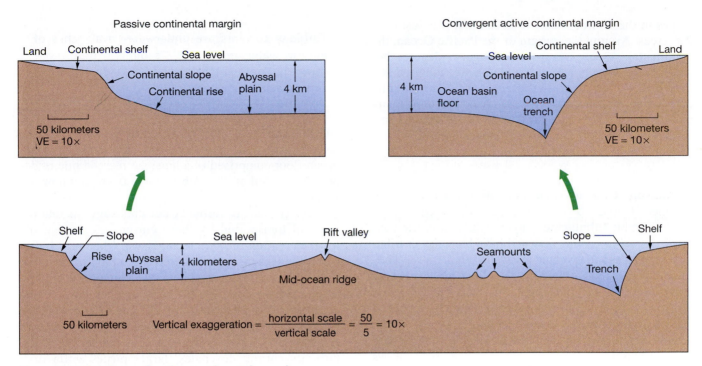

Figure 4–7 Passive and active continental margins.

Cross-sectional view (*below*) of typical features across an ocean basin, including a passive continental margin (*left enlargement*) and a convergent active continental margin (*right enlargement*). Vertical exaggeration is 50 times.

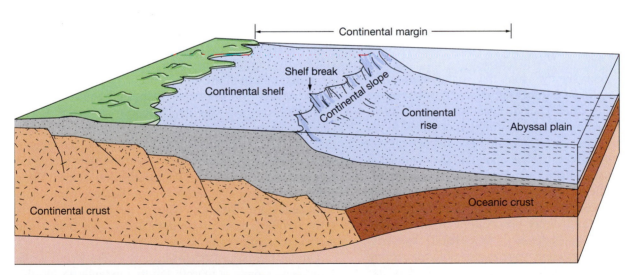

Figure 4–8 Features of a passive continental margin.

Schematic view showing the main features of a passive continental margin.

Continental Slope The **continental slope**, which lies beyond the shelf break, is where the deep-ocean basins begin. Total relief in this region is similar to that found in mountain ranges on the continents. The break at the top of the slope may be from 1 to 5 kilometers (0.6 to 3 miles) above the deep-ocean basin at its base. Along convergent active margins where the slope descends into submarine trenches, even greater vertical relief is measured. Off the west coast of South America, for in-

stance, the total relief from the top of the Andes Mountains to the bottom of the Peru-Chile Trench is about 15 kilometers (9.3 miles).

Worldwide, the slope of the continental slopes averages about 4 degrees, but varies from 1 to 25 degrees.[3] A study that compared the average of five different continental

---

[3]For comparison, the windshield of an aerodynamically designed car has a slope of about 25 degrees.

slopes in the United States revealed that it is just over 2 degrees. Around the margin of the Pacific Ocean, the continental slopes average more than 5 degrees because of the presence of convergent active margins that drop directly into deep offshore trenches. The Atlantic and Indian Oceans, on the other hand, contain many passive margins, which lack plate boundaries. Thus, the amount of relief is lower and slopes in these oceans average about 3 degrees.

Submarine Canyons and Turbidity Currents The continental slope—and, to a lesser extent, the continental shelf—exhibit **submarine canyons**. Submarine canyons are V-shaped in profile view and have branches or tributaries with steep to overhanging walls (Figure 4–9). They resemble canyons formed on land that are carved by rivers and can be quite large. In fact, the Monterey Canyon off California is comparable in depth, steepness, and length to Arizona's Grand Canyon.

How are submarine canyons formed? Initially it was thought submarine canyons were ancient river valleys created by the erosive power of rivers when sea level was lower and the continental shelf was exposed. Although some canyons are directly offshore from where rivers enter the sea, the majority of them are not. Many, in fact, are confined exclusively to the continental slope. Additionally, submarine canyons continue to the base of the continental slope, which averages some 3500 meters (11,500 feet) below sea level. There is no evidence, however, that sea level has ever been lowered by that much.

Side-scan sonar surveys along the Atlantic coast indicate that the continental slope is dominated by submarine canyons from Hudson Canyon near New York City to Baltimore Canyon in Maryland. Canyons confined to the continental slope are straighter and have steeper canyon floor gradients than those that extend into the continental shelf. These characteristics suggest the canyons are created on the continental slope by some marine process and enlarge into the continental shelf through time.

Both indirect and direct observation of the erosive power of **turbidity** (*turbidus* = disordered) **currents** (Box 4–2) has suggested they are responsible for carving submarine canyons. Turbidity currents are underwater avalanches of muddy water mixed with rocks and other debris. When sediment moves across the continental shelf into the head of the canyon and accumulates there, turbidity currents may result from shaking by an earthquake, the oversteepening of sediment that accumulates on the shelf, hurricanes passing over the area, or the rapid input of sediment from flood waters. The mass moves down the slope under the force of gravity when set in motion, carving the canyon as it goes, analogous to a flash flood on land. Turbidity currents are strong enough to carry large rocks down submarine canyons and do a considerable amount of erosion over time.

> Turbidity currents are underwater avalanches of muddy water mixed with sediment that move down the continental slope and are responsible for carving submarine canyons.

Continental Rise The **continental rise** is a transition zone between the continental margin and the deep-ocean floor comprised of a huge submerged pile of debris. Where did all this debris come from, and how did it get there?

The existence of turbidity currents suggests that the material transported by these currents is responsible for the creation of continental rises. When a turbidity current moves through and erodes a submarine canyon, it exits through the mouth of the canyon. The slope angle decreases and the turbidity current slows, causing suspended material to settle out in a distinctive type of layering called **graded bedding** that *grades in size upward* (Figure 4–9a, *inset*). As the energy of the turbidity current dissipates, larger pieces settle first, then progressively smaller pieces settle, and eventually even very fine pieces settle out, which may occur weeks or months later.

An individual turbidity current deposits one graded bedding sequence. The next turbidity current may partially erode the previous deposit and then deposit another graded bedding sequence on top of the previous one. After some time, a thick sequence of graded bedding deposits can develop one on top of another. These stacks of graded bedding are called **turbidite deposits**, of which the continental rise is composed.

As viewed from above, the deposits at the mouths of submarine canyons are fan-, lobate-, or apron-shaped (Figures 4–9a and 4–9c). Consequently, these deposits are called **deep-sea fans** or **submarine fans**. These deep-sea fans create the continental rise when they merge together along the base of the continental slope. Along convergent active margins, however, the steep continental slope leads directly into a deep-ocean trench. Sediment from turbidity currents accumulates in the trench and there is no continental rise.

One of the largest deep-sea fans in the world is the Indus Fan, a passive margin fan that extends 1800 kilometers (1100 miles) south of Pakistan (Figure 4–9c). The Indus River carries extensive amounts of sediment from the Himalaya Mountains to the coast. This sediment eventually makes its way down the submarine canyon and builds the fan, which, in some areas, has sediment that is more than 10 kilometers (6.2 miles) thick. The Indus Fan has a main submarine canyon channel extending seaward onto the fan but soon divides into several branching distributary channels. These distributary channels are similar to those found on deltas, which form at the mouths of streams. On the lower fan, the surface has a very low slope and the flow is no

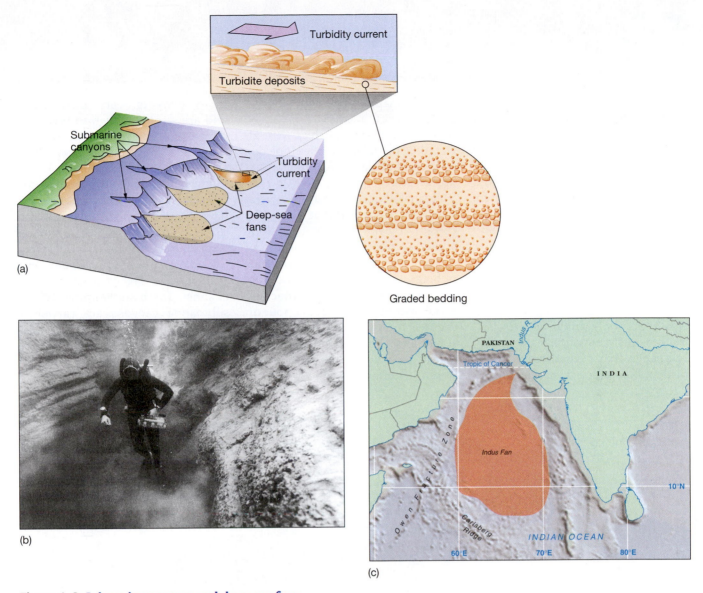

Turbidity current

Turbidite deposits

Submarine canyons

Turbidity current

Deep-sea fans

(a)

Graded bedding

(b)

PAKISTAN

Indus R.

Tropic of Cancer

INDIA

Owen Fracture Zone

Indus Fan

Carlsberg Ridge

INDIAN OCEAN

10°N

60°E   70°E   80°E

(c)

Figure 4–9 Submarine canyons and deep-sea fans.

(a) Turbidity currents move downslope, eroding the continental margin to enlarge submarine canyons. Deep-sea fans are created by turbidite deposits, which consist of sequences of graded bedding (*inset*). **(b)** A diver descends into La Jolla Submarine Canyon, offshore California. **(c)** Map of the Indus Fan, a large but otherwise typical example of a passive margin fan.

longer confined to channels, so it spreads out and forms layers of fine sediment across the fan surface. The Indus Fan has so much sediment, in fact, that it partially buries an active mid-ocean ridge, the Carlsberg Ridge!

Features of the Deep-Ocean Basin

The deep-ocean floor lies beyond the continental margin province (the shelf, slope, and the rise).

Abyssal Plains   Extending from the base of the continental rise into the deep-ocean basins are flat depositional surfaces with slopes that average a small fraction of a degree and cover extensive portions of the deep-ocean basins.

These **abyssal** (*a* = without, *byssus* = bottom) **plains** average between 4500 meters (15,000 feet) and 6000 meters (20,000 feet) deep. They are not literally bottomless, but they are some of the deepest (and flattest) regions on Earth.

Abyssal plains are formed by fine particles of sediment slowly drifting onto the deep-ocean floor. Over millions of years, a thick blanket of sediment is produced by **suspension settling**, where fine particles (analogous to "marine dust") accumulate on the ocean floor. With enough time, these deposits cover most irregularities of the deep ocean, as shown in Figure 4–10. In addition, sediment traveling in turbidity currents from land adds to the sediment load.

BOX 4–2   Research Methods in Oceanography

A GRAND "BREAK": EVIDENCE FOR TURBIDITY CURRENTS

How do earthquakes and telephone cables help explain how turbidity currents move across the ocean floor and carve submarine canyons? In 1929, the Grand Banks earthquake in the North Atlantic Ocean severed some of the trans-Atlantic telephone and telegraph cables that lay across the sea floor south of Newfoundland near the earthquake epicenter (Figure 4D). At first, it was assumed that sea floor movement caused all these breaks. However, analysis of the data revealed that the cables closest to the earthquake broke simultaneously with the earthquake, but cables that crossed progressively further downslope from the epicen-

ter were snapped one after another like a string of firecrackers. It seemed unusual that certain cables were affected by the failure of the slope due to ground shaking, but others were broken several minutes later.

Reanalysis of the pattern several years later suggested that a turbidity current moving down the slope could account for the pattern of cable breaks. Based on the sequence of breaks, the turbidity current must have reached speeds approaching 80 kilometers (50 miles) per hour on the steep portions of the continental slope, and about 24 kilometers (15 miles) per hour on the more gently sloping continental rise. Thus, tur-

bidity currents reach high speeds and are strong enough to break underwater cables, suggesting that they must be powerful enough to erode submarine canyons.

Further evidence of turbidity currents comes from several studies that have documented turbidity currents using sonar. For instance, a study of Rupert Inlet in British Columbia, Canada, monitored turbidity currents moving through an underwater channel. These studies indicate that submarine canyons are carved by turbidity currents over long periods of time, just as canyons on land are carved by running water.

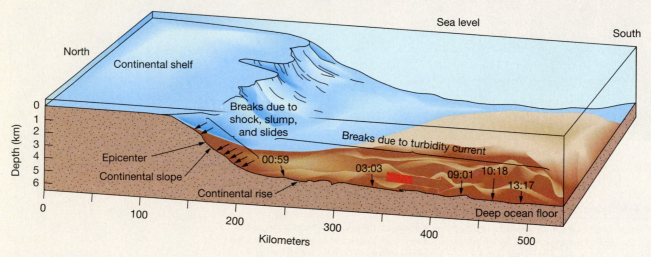

Figure 4D Grand Banks earthquake.

Diagrammatic view of the sea floor showing the sequence of events for the 1929 Grand Banks Earthquake. The epicenter is the point on Earth's surface directly above the earthquake.

The type of continental margin determines the distribution of abyssal plains. For instance, few abyssal plains are located in the Pacific Ocean; instead, most occur in the Atlantic and Indian Oceans. The deep-ocean trenches found on the convergent active margins of the Pacific Ocean prevent sediment from moving past the continental slope. In essence, the trenches act like a gutter that traps sediment transported off the land by turbidity currents. On the passive margins of the Atlantic and Indian Oceans, however, turbidity currents travel directly down the continental margin and deposit sediment on the abyssal plains. In addition, the great distance from the

continental margin to the floor of the deep-ocean basins in the Pacific Ocean is so great that most of the suspended sediment settles out before it reaches these distant regions. Conversely, the smaller size of the Atlantic and Indian Oceans does not prevent suspended sediment from reaching their deep-ocean basins.

Volcanic Peaks of the Abyssal Plains   Poking through the sediment cover of the abyssal plains are a variety of volcanic peaks, which extend to various elevations above the ocean floor (see Figure 4–2). Some even extend above sea level to form islands. Those that are below sea level

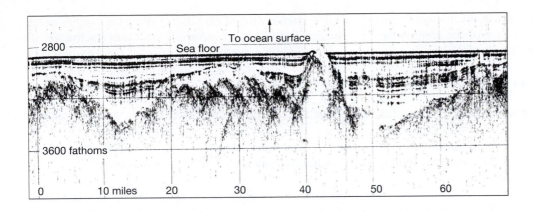

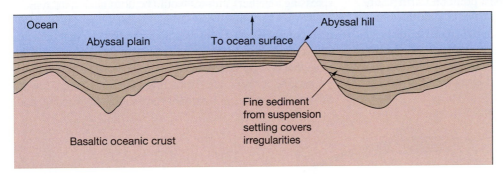

Figure 4–10 Abyssal plain formed by suspension settling.

Seismic cross-section (*above*) and matching drawing (*below*) across part of the deep Madeira abyssal plain in the eastern Atlantic Ocean showing irregular volcanic terrain buried by sediments.

but rise more than 1 kilometer (0.6 mile) above the deep-ocean floor are called *seamounts*. Analysis of satellite bathymetry data suggests there may be as many as 100,000 seamounts on the ocean floor. If seamounts have flattened tops, they are called *tablemounts*, or *guyots*. The origin of seamounts and tablemounts was discussed as a piece of supporting evidence for plate tectonics in Chapter 3 (see Figure 3–33).

Volcanic features on the ocean floor that are less than 1000 meters (0.6 mile) tall—the minimum height of a seamount—are called **abyssal hills** or **seaknolls**. Abyssal hills are one of the most abundant features on the planet (several hundred thousand have been identified) and cover a large percentage of the entire ocean basin floor. Many are gently rounded in shape and they have an average height of about 200 meters (650 feet). Many abyssal hills are found buried beneath the sediments of the abyssal plains of the Atlantic and Indian Oceans. In the Pacific Ocean, the abundance of active margins means the rate of sediment deposition is lower. Consequently, extensive regions dominated by abyssal hills have resulted, which are called **abyssal hill provinces**. The evidence of volcanic activity on the bottom of the Pacific Ocean is particularly widespread—more than 20,000 volcanic peaks exist there.

Flood Basalts Some marine volcanic activity produces widespread, generally flat surfaces where large volumes of lava flowed out in broad sheets and solidified. Such features commonly surround volcanic is-

lands but sometimes are large enough to form elevated plateaus called *large igneous provinces*.

When found on the continents, such volcanic accumulations are called **continental flood basalts**. They are formed where a mantle plume reaches the bottom of the lithosphere and forms a broad rounded head that may range from 1000 to 2000 kilometers (620 to 1240 miles) in diameter. The temperature within this head is hotter than the surrounding mantle and produces a wide uplifted dome at Earth's surface from thermal expansion of the rock below. If the continental crust above thins either from erosion of the uplifted dome or from the early stages of continental rifting, hot magma of basaltic composition escapes and floods out at the surface. The Colombia River and Snake River flood basalts in Washington, Oregon, and Idaho were formed in this manner.

The volume of lava released during the formation of flood basalts is impressive. Continental flood basalts, for example, typically contain up to 2 million cubic kilometers (480,000 cubic miles) of lava and erupt at rates of about 1 cubic kilometer (0.24 cubic mile) per year over a relatively short interval of about 1 or 2 million years. However, the 120-million-year-old Ontong Java Plateau in the equatorial western Pacific contains about 65 million cubic kilometers (15.6 million cubic miles) of lava—more than 30 times the volume of most continental flood basalts. Remarkably, this large volume was produced over a period of no more than 3 million years, which must have required an average eruption rate of about 22 cubic kilometers (5.3 cubic miles) per year.

Ocean Trenches Along passive margins, the continental rise commonly occurs at the base of the continental slope and merges smoothly into the abyssal plain. In convergent active margins, however, the slope descends into a long, narrow, steep-sided **ocean trench**. Ocean trenches are deep linear scars in the ocean floor, caused by the collision of two plates along convergent plate margins (as discussed in Chapter 3). The landward side of the trench rises as a **volcanic arc** that may produce islands (such as the islands of Japan, an **island arc**) or a volcanic mountain range along the margin of a continent (such as the Andes Mountains, a **continental arc**).

The deepest portions of the world's oceans are found in these trenches. In fact, the deepest point on Earth's surface—11,022 meters (36,161 feet)—is found in the Challenger Deep area of the Mariana Trench. The majority of ocean trenches are found along the margins of the Pacific Ocean (Figure 4–11) while only a few exist in the Atlantic and Indian Oceans. Table 4–1 compares the dimensions of selected trenches.

The **Pacific Ring of Fire** occurs along the margins of the Pacific Ocean. It has the majority of Earth's active volcanoes and large earthquakes because of the prevalence of convergent plate boundaries along the Pacific Rim. A part of the Pacific Ring of Fire is South America's western coast, including the Andes Mountains and the associated Peru-Chile Trench. Figure 4–12 shows a cross-sectional view of this area, illustrating the tremendous amount of relief at convergent plate boundaries where deep-ocean trenches are associated with tall volcanic arcs.

Back-Arc Spreading Centers When plates collide at subduction zones, you might suspect that compressional (pressing together) stresses would be dominant in the region. However, tensional (pull-apart) stresses are much more commonly observed. This results from the fact that trenches are not stationary; rather, they move in the direction of the plate that will be subducted in a process called *seaward migration*. As a result, the overriding plate gets pulled toward the trench, producing tensional forces, crustal thinning, and,

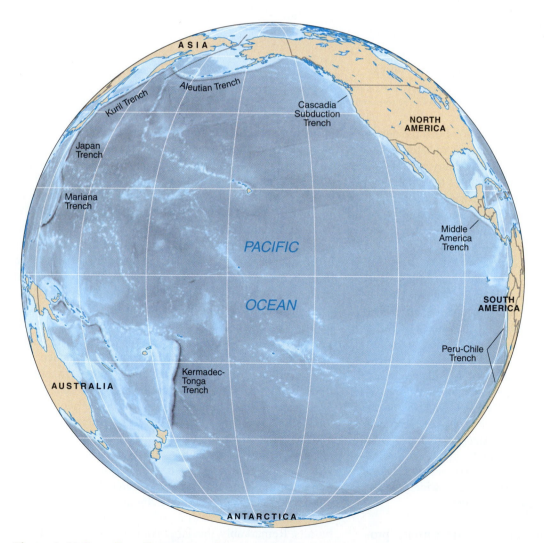

Figure 4–11 *Location of ocean trenches.*

The majority of ocean trenches are along the margins of the Pacific Ocean where plates are being subducted. Most of the world's large earthquakes (due to subduction) and active volcanoes (as volcanic arcs) occur around the Pacific Rim, which is why the area is also called the Pacific Ring of Fire.

TABLE 4–1 **Dimensions of selected trenches.**

Trench	Ocean	Depth (kilometers)	Average width (kilometers)	Length (kilometers)
Middle America	Pacific	6.7	40	2800
Java	Indian	7.5	80	4500
Aleutian	Pacific	7.7	50	3700
Peru-Chile	Pacific	8.0	100	5900
South Sandwich	Atlantic	8.4	90	1450
Japan	Pacific	8.4	100	800
Puerto Rico	Atlantic	8.4	120	1550
Kermadec-Tonga	Pacific	10.0	50	2900
Philippine	Pacific	10.5	60	1400
Kuril	Pacific	10.5	120	2200
Mariana	Pacific	11.0	70	2550

in some cases, a spreading center behind (landward of) the volcanic arc called a **back-arc spreading center**. A back-arc spreading center has all the features of a full mid-ocean ridge, which will be discussed in the next section.

Figure 4–13 shows a cross-sectional view of a well-developed back-arc spreading center created by the Mariana Trench/Island Arc subduction system. The Pacific Plate is being subducted beneath the Philippine Plate, creating tensional stresses and the back-arc spreading center. An ancient *remnant arc* called the West Mariana Ridge exists about 200 kilometers (124 miles) to the west of the back-arc spreading center and indicates a former location of back-arc spreading.

Unusual seamounts composed of serpentine[4] are sometimes associated with a subduction zone's *fore-arc* region (the portion of the overriding plate that lies seaward of the volcanic arc) (Figure 4–13). It is thought

that the serpentine formed when seawater either entered the mantle through fractures in the fore-arc crust or was available from dewatering of subducted sediments. Because of its low density, the serpentine flowed up to the surface of the fore-arc, carrying with it blocks of peridotite (mantle rock) and basalt (possibly from subducting ocean crust) to produce volcano-like seamounts near the seaward front of the fore-arc structure.

> Deep-ocean trenches and volcanic arcs are a result of the collision of two plates at convergent plate boundaries and mostly occur along the margins of the Pacific Ocean (Pacific Ring of Fire).

Features of the Mid-Ocean Ridge

The global mid-ocean ridge is a continuous, fractured-looking mountain ridge that extends through all the ocean basins. The portion of the mid-ocean ridge found in the North Atlantic Ocean is called the Mid-Atlantic

---

[4]Recall from Box 3–2 that serpentine is a mineral that forms when water reacts with *peridotite*, which is rock from Earth's mantle.

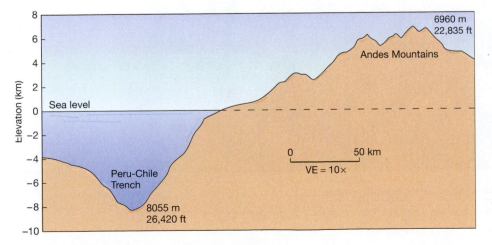

Figure 4–12 Profile across the Peru-Chile Trench and the Andes Mountains.

Over a distance of 200 kilometers (125 miles), there is a change in elevation of more than 14,900 meters (49,000 feet) from the Peru-Chile Trench to the Andes Mountains. This dramatic relief is a result of plate interactions at a convergent active margin, producing a deep-ocean trench and associated continental volcanic arc. Vertical scale is exaggerated 10 times.

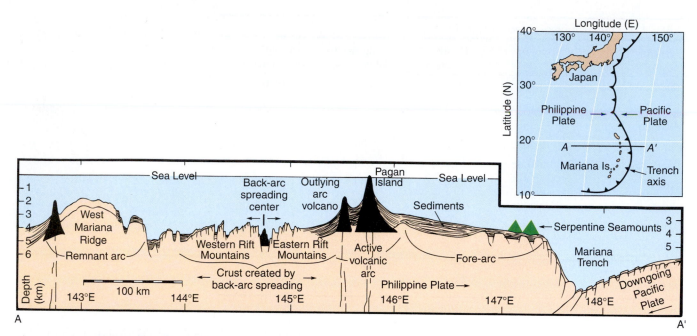

Figure 4–13 Mariana Trench and back-arc spreading.

Cross-sectional view of features created by the Mariana island arc subduction system. From left to right: the West Mariana Ridge is an ancient remnant arc; back-arc spreading occurs as the trench migrates seaward, creating tensional stresses; Pagan Island is part of the active volcanic arc; rare serpentine seamounts occur on the leading edge of the fore-arc; the Mariana Trench is a result of subduction of the Pacific Plate beneath the Philippine Plate. Vertical exaggeration is 14 times.

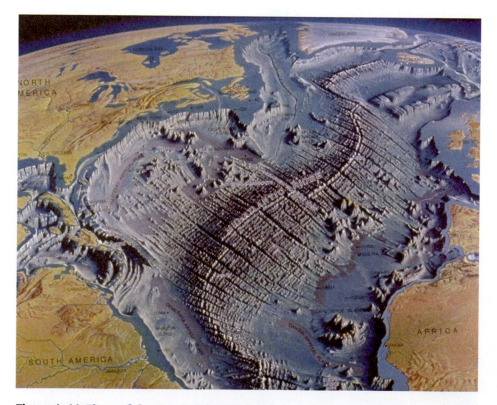

Figure 4–14 Floor of the North Atlantic Ocean.

The global mid-ocean ridge cuts through the center of the Atlantic Ocean, where it is called the Mid-Atlantic Ridge.

Ridge and is shown in Figure 4–14. As discussed in Chapter 3, the mid-ocean ridge results from sea floor spreading along divergent plate boundaries. The mid-ocean ridge forms Earth's longest mountain chain, extending across some 75,000 kilometers (46,600 miles) of the deep-ocean basin. The width of the mid-ocean ridge varies along its length but averages about 1000 kilometers (620 miles). The mid-ocean ridge is a topographically high feature, extending an average of 2.5 kilometers (1.5 miles) above the surrounding sea floor. In some areas, such as in Iceland, the mid-ocean ridge even extends above sea level. Remarkably, the mid-ocean ridge covers 23% of Earth's surface.

The mid-ocean ridge is entirely volcanic and is composed of basaltic lavas characteristic of the oceanic crust. Along its crest is a central downdropped **rift valley** (Figure 4–15) created by sea floor spreading (rifting) where two plates diverge. Cracks called *fissures* (*fissus* = split) and faults are commonly observed in the central rift valley. Swarms of small earthquakes occur along the central rift valley caused by underground movement of magma or rifting along faults.

Volcanic features include volcanoes (seamounts[5]; Figure 4–16a) and recent underwater lava flows. When hot basaltic lava spills onto the sea floor, it is exposed to cold seawater that chills the margins of the lava. This creates **pillow lavas** or **pillow basalts**, which are smooth, rounded lobes of rock that resemble a stack of bed pillows (Figure 4–16b). Although most people are not usually aware of it, frequent volcanic activity is common along the mid-ocean ridge. In fact, every year about 12 cubic kilometers (3 cubic miles) of molten rock erupts underwater—enough to fill 20 Olympic-sized swimming pools every minute. Bathymetric studies along the Juan de Fuca Ridge off Washington and Oregon, for example, revealed that 50 million cubic meters (1800 million cubic feet) of new lava was released sometime between 1981 and 1987. Subsequent surveys of the area indicated many changes along the mid-ocean ridge, including new volcanic features, recent lava flows, and depth changes of up to 37 meters (121 feet).

---

[5]In a number of cases, researchers have discovered seamounts that initially formed along the crest of the mid-ocean ridge and have been split in two as the plates spread apart.

Figure 4–15 Rift valley fissures.

A large fissure in the rift valley of Iceland (*above*). A smaller fissure in the rift valley of the Mid-Atlantic Ridge (*below*), which was photographed by researchers in the submersible *Alvin*.

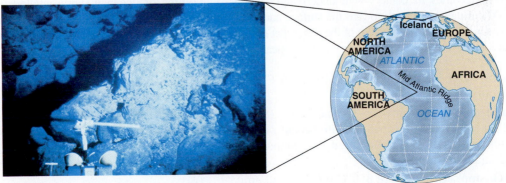

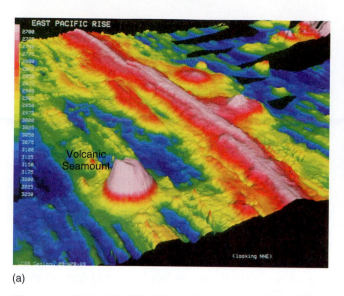

(a)

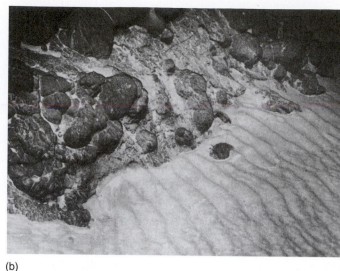

(b)

Figure 4–16 East Pacific Rise volcanoes and pillow lava.

(a) False-color perspective view based on sonar mapping of a portion of the East Pacific Rise (*center*) showing volcanic seamount (*left*). Depth in meters indicated by the color scale along the left margin; vertical exaggeration is six times. **(b)** Recently formed pillow lava along the East Pacific Rise. Photo shows an area of the sea floor about 3 meters (10 feet) across that also displays ripple marks from deep ocean currents.

S STUDENTS SOMETIMES ASK ...

Has anyone ever seen pillow lava forming?

Amazingly, yes! In the 1960s, an underwater film crew ventured to Hawaii during an eruption of the volcano Kilauea where lava spilled into the sea. They braved high water temperatures and risked being burned on the red-hot lava, but filmed some incredible footage. Underwater, the formation of pillow lava occurs where a tube emits molten lava directly into the ocean. When hot lava comes into contact with cold seawater, it forms the characteristic smooth and rounded margins of pillow basalt. The divers also experimented with a hammer on newly formed pillows and were able to initiate new lava outpourings.

Other features in the central rift valley include hot springs called **hydrothermal** (*hydro* = water, *thermo* = heat) **vents**. Seawater seeps along fractures in the ocean crust and is heated when it comes in contact with underlying magma (Figure 4–17). It then rises back toward the surface and exits through the sea floor. The temperature of the water that rushes out of a particular hydrothermal vent determines its appearance:

- **Warm-water vents** are below 30°C (86°F) and generally emit water clear in color.
- **White smokers** are between 30° and 350°C (86° to 662°F) and emit water that is white because of the presence of various light-colored compounds, including barium sulfide.
- **Black smokers** are above 350°C (662°F) and emit water that is black because of the presence of dark-

colored **metal sulfides**, including lead, iron, nickel, copper, zinc, and chromium. Black smokers were named for their resemblance to factory smokestacks belching clouds of smoke.

STUDENTS SOMETIMES ASK ...

If black smokers are so hot, why isn't there steam coming out of them instead of hot water?

Indeed, black smokers emit water that can be up to three and a half times the boiling point of water at the surface. However, the depth where black smokers are found results in much higher pressure than at the surface. At these higher pressures, water has a higher boiling point. Thus, water from hydrothermal vents remains in the liquid state instead of turning into water vapor (steam).

Chemical studies of seawater indicate that the entire volume of ocean water is cycled through hydrothermal circulation systems about every 3 million years. As a result, chemical exchange between ocean water and basaltic crust has a significant influence on the chemical composition of seawater.[6]

Many black smokers spew out of chimneylike structures (Figure 4–17b) that can be up to 60 meters (200 feet) high. The dissolved metal particles often come out of solution, or **precipitates**[7] when the hot water mixes with cold seawater,

---

[6]For more details on this process, see Chapter 6, "Water and Seawater."

[7]A chemical precipitate is formed whenever dissolved materials change from existing in the dissolved state to existing in the solid state.

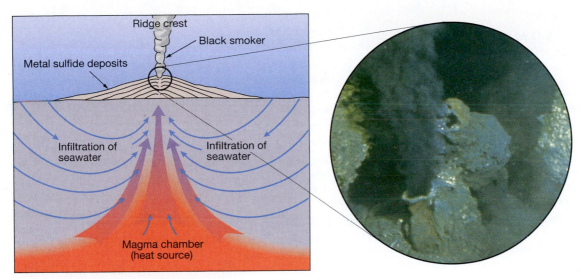

(a)

(b)

Figure 4–17 Hydrothermal vents.

(a) Diagram showing hydrothermal circulation along the mid-ocean ridge and the creation of black smokers. Photo (*inset*) shows a close-up view of a black smoker along the East Pacific Rise. **(b)** Black smoker chimney and fissure at Susu north active site, Manus Basin, western Pacific Ocean. Chimney is about 3 meters (10 feet) tall.

creating coatings of mineral deposits on nearby rocks. Chemical analyses of these deposits reveal that they are composed of various metal sulfides and sometimes even silver and gold. Mining these modern sea floor deposits is neither economically nor politically feasible at present but has often been considered. However, many deposits of these metals that are mined on land today probably originated as hydrothermal deposits at ancient mid-ocean ridges. One such example is the copper deposits that have been mined since antiquity in Cyprus (see Box 3–2).

In addition, most hydrothermal vents support unusual biological communities, including large clams, mussels, tubeworms, and many other organisms—most of which were new to science when they were first encountered. These organisms are able to survive in the absence of sunlight because the vents discharge hydrogen sulfide gas. Archaeons[8] and bacteria oxidize the hydrogen sulfide gas to provide a food source for other organisms in the community. The interesting associations of these organisms are discussed in Chapter 16, "Animals of the Benthic Environment."

Segments of the mid-ocean ridge called **oceanic ridges** have a prominent rift valley and steep, rugged slopes while **oceanic rises** have slopes that are gentler and less rugged. As explained in Chapter 3, the differences in overall shape are caused by the fact that oceanic ridges (such as the Mid-Atlantic Ridge) spread more slowly than oceanic rises (such as the East Pacific Rise).

---

[8]Archaeons are microscopic bacteria-like organisms—a newly discovered domain of life.

What effect does all this volcanic activity along the mid-ocean ridge have at the ocean's surface?

Sometimes the underwater volcanic eruption is large enough to create what is called a *"megaplume"* of warm, mineral-rich water that is lower in density than the surrounding seawater and thus rises to the surface. Remarkably, a few research vessels have reported experiencing the effects of a megaplume at the surface while directly above an erupting sea floor volcano! Researchers on board describe bubbles of gas and steam at the surface, a marked increase in water temperature, and the presence of enough volcanic material to turn the water cloudy. In terms of warming the ocean, the heat released into the ocean at mid-ocean ridges is probably not very significant, mostly because the ocean is so good at absorbing and redistributing heat.

> The mid-ocean ridge is created by plate divergence and typically includes a central rift valley, faults and fissures, seamounts, pillow basalts, hydrothermal vents, and metal sulfide deposits.

Fracture Zones and Transform Faults The mid-ocean ridge is cut by a number of **transform faults**, which offset the spreading zones. Oriented perpendicular to the spreading zones, transform faults give the mid-ocean ridge the zigzag appearance seen in Figure 4–14. As described in Chapter 3, transform faults occur to accommodate spreading of a linear ridge system on a spherical Earth and because different segments of the mid-ocean ridge spread apart at different rates.

In the Pacific Ocean, where scars are less rapidly covered by sediment than in other ocean basins, transform faults can be seen to extend for thousands of kilometers away from the mid-ocean ridge and have widths of up to 200 kilometers (120 miles). These ex-

tensions, however, are not transform faults. Instead, they are **fracture zones**.

What is the difference between a transform fault and a fracture zone? Figure 4–18 shows that both run along the same long linear zone of weakness in Earth's crust. In fact, by following the same zone of weakness from one end to the other, it changes from a fracture zone to a transform fault and back again to a fracture zone. A transform fault is a seismically active area that offsets the axis of a mid-ocean ridge. A fracture zone, on the other hand, is a seismically inactive area that shows evidence of past transform fault activity. A helpful way to visualize the difference is that transform faults occur *between* offset segments of the mid-ocean ridge, while fracture zones occur *beyond* the offset segments of the mid-ocean ridge.

The relative direction of plate motion across transform faults and fracture zones further differentiates these two features. Across a transform fault, two lithospheric plates are moving in opposite directions. Across a fracture zone (which occurs entirely within a plate), there is no relative motion because the parts of the lithospheric plate cut by a fracture zone are moving in the same direction (Figure 4–18). Transform faults are actual plate boundaries, whereas fracture zones are not. Rather, fracture zones are ancient, inactive fault scars embedded in a plate.

In addition, earthquake activity is different in transform faults and fracture zones. Earthquakes shallower than 10 kilometers (6 miles) are common when plates move in opposite directions along transform faults. Along fracture zones, where plate motion is in the same direction, seismic activity is almost completely absent. Table 4–2 summarizes the differences between transform faults and fracture zones.

Many fracture zones exhibit dramatic relief. For example, the Mendocino Fracture Zone in the North Pacific Ocean (see map inside the front cover) is more than 1000 meters (3300 feet) deeper on the south side than on the north side, creating the *Mendocino Escarpment*. The Owen Fracture Zone in the Indian Ocean (Figure 4–19)

Figure 4–18 Transform faults and fracture zones.

Transform faults are active transform plate boundaries that occur *between* the segments of the mid-ocean ridge. Fracture zones are inactive intraplate features that occur *beyond* the segments of the mid-ocean ridge.

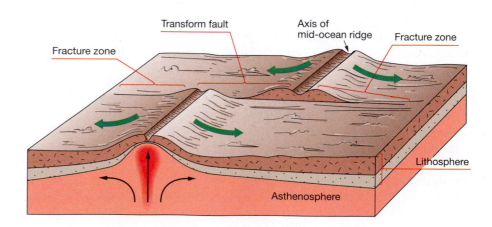

TABLE 4–2 **Comparison between transform faults and fracture zones.**

	Transform faults	Fracture zones
Plate boundary?	Yes—a transform plate boundary	No—an intraplate feature
Relative movement across feature	Movement in opposite directions	Movement in the same direction
	←	←
	——	——
	→	←
Earthquakes?	Many	Few
Relationship to mid-ocean ridge	Occur *between* offset mid-ocean ridge segments	Occur *beyond* offset mid-ocean ridge segments
Geographic examples	San Andreas Fault, Alpine Fault, Dead Sea Fault	Mendocino Fracture Zone, Molokai Fracture Zone

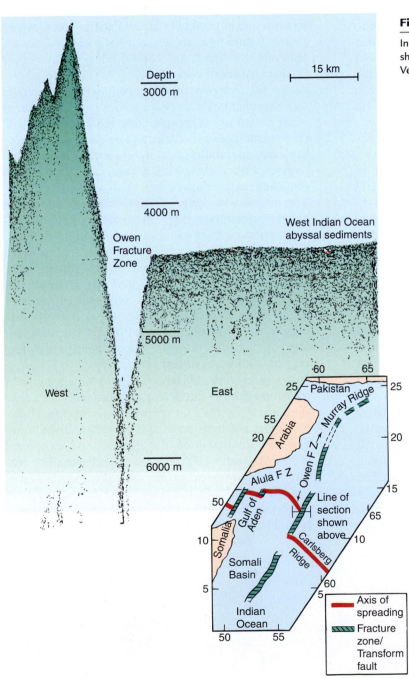

Figure 4–19 Owen Fracture Zone.

Interpreted seismic profile of the Owen Fracture Zone showing high relief, which is typical of many fracture zones. Vertical exaggeration is 25 times.

has relief of over 2000 meters (6600 feet). Much of this difference in elevation is attributable to the vast age difference of the ocean floor on either side of a fracture zone: The older ocean floor undergoes much more thermal contraction and subsequent deepening than the younger ocean floor adjacent to it.

Transform faults are plate boundaries that occur *between* offset segments of the mid-ocean ridge while fracture zones are intraplate features that occur *beyond* the offset segments of the mid-ocean ridge.

Chapter in Review

- *Bathymetry is the measurement of ocean depths and the charting of ocean floor topography.* The varied bathymetry of the ocean floor was first determined using *soundings* to measure water depth. Later, the development of the *echo sounder* gave ocean scientists a more detailed representation of the sea floor. Today, much of our knowledge of the ocean floor has been obtained using various *multibeam echo sounders* or *side-scan sonar instruments* (to make detailed bathymetric maps of a small area of the ocean floor), *satellite measurement* of the ocean surface (to produce maps of the world ocean floor), and *seismic reflection profiles* (to examine ocean structure beneath the sea floor).

- *Earth's hypsographic curve* shows the amount of Earth's surface area at different elevations and depths. The *distribution of area is uneven* with respect to height above or below sea level. The shape of the curve also *reflects the existence of plate tectonic processes.*

- *Continental margins can be either passive* (not associated with any plate boundaries) *or active* (associated with convergent or transform plate boundaries). Extending from the shoreline is the generally shallow, low relief, and gently sloping *continental shelf* that can contain various features such as coastal islands, reefs, and banks. The boundary between the continental slope and the *continental shelf* is marked by an increase in slope that occurs at the *shelf break.* Cutting deep into the slopes are *submarine canyons,* which resemble canyons on land but are created by erosive turbidity currents. *Turbidity currents* deposit their sediment load at the base of the continental slope, creating deep-sea fans that merge to produce a gently sloping continental rise. The deposits from turbidity currents (called *turbidite deposits*) have characteristic sequences of graded bedding. Active margins have similar features, although they are modified by their associated plate boundary.

- The *continental rises* gradually become flat, extensive, deep-ocean *abyssal plains,* which form by *suspension settling* of fine sediment. Poking through the sediment cover of the abyssal plains are numerous *volcanic peaks,* including volcanic islands, seamounts, tablemounts, and abyssal hills. In the Pacific Ocean, where sedimentation rates are low, abyssal plains are not extensively developed, and abyssal hill provinces cover broad expanses of ocean floor.

- *Along the margins of many continents*—especially those around the *Pacific Ring of Fire*—are *deep linear scars called ocean trenches* that are associated with convergent plate boundaries and volcanic arcs. Due to seaward migration of a trench creating tensional stresses, a *back-arc spreading center* can form behind (landward of) a volcanic arc.

- The *mid-ocean ridge is a continuous mountain range* that winds through all ocean basins and is entirely volcanic in origin. Common features associated with the mid-ocean ridge include a *central rift valley, faults* and *fissures, seamounts, pillow basalts, hydrothermal vents, deposits of metal sulfides,* and *unusual life forms.* Segments of the mid-ocean ridge are either *oceanic ridges* if steep with rugged slopes (indicative of slow sea floor spreading) or *oceanic rises* if sloped gently and less rugged (indicative of fast spreading).

- *Long linear zones of weakness—fracture zones and transform faults*—cut across vast distances of ocean floor and *offset the axes of the mid-ocean ridge.* Fracture zones and transform faults are differentiated from one another based on the direction of movement across the feature. *Fracture zones* (an intraplate feature) *have movement in the same direction,* whereas *transform faults* (a transform plate boundary) *have movement in opposite directions.* Many fracture zones have dramatic relief.

Key Terms

Abyssal hill (p. 123)
Abyssal hill province (p. 123)
Abyssal plain (p. 121)
Active margin (p. 117
Back-arc spreading center (p. 125)
Bathymetry (p. 111)
Black smoker (p. 128)
Continental arc (p. 124)
Continental borderland (p. 118)
Continental flood basalt (p. 123)
Continental margin (p. 117)
Continental rise (p. 120
Continental shelf (p. 118)
Continental slope (p. 119)

Convergent active margin (p. 117)
Deep-ocean basin (p. 117)
Deep-sea fan (p. 120)
Echo sounder (p. 111)
Fathom (p. 111)
Fracture zone (p. 130)
GLORIA (p. 112)
Graded bedding (p. 120)
Hydrothermal vent (p. 128)
Hypsographic curve (p. 116)
Island arc (p. 124)
Metal sulfides (p. 128)
Mid-ocean ridge (p. 117)
Ocean trench (p. 124)

Oceanic ridge (p. 129)
Oceanic rise (p. 129)
Pacific Ring of Fire (p. 124)
Passive margin (p. 117)
Pillow basalt (p. 127)
Pillow lava (p. 127)
Precipitate (p. 128)
Precision depth recorder (PDR) (p. 112)
Rift valley (p. 127)
SeaBeam (p. 112)
Seaknoll (p. 123)
Sea MARC (p. 112)
Seismic reflection profile (p. 115)

Shelf break (p. 118)
Sonar (p. 112)
Sounding (p. 111)
Submarine canyon (p. 120)
Submarine fan (p. 120)
Suspension settling (p. 121)
Transform active margin (p. 117)
Transform fault (p. 130)
Turbidite deposit (p. 120)
Turbidity current (p. 120)
Volcanic arc (p. 124)
Warm-water vent (p. 128)
White smoker (p. 128)

Questions and Exercises

1. What is bathymetry?

2. Discuss the development of bathymetric techniques, indicating significant advancements in technology.

3. Describe what is shown by a hypsographic curve and explain why its shape reflects the presence of active tectonic processes on Earth.

4. Describe the differences between passive and active continental margins. Be sure to include how these features relate to plate tectonics, and include an example of each type of margin.

5. Describe the major features of a passive continental margin: continental shelf, continental slope, continental rise, submarine canyon, and deep-sea fans.

6. Explain how submarine canyons are created.

7. What are differences between a submarine canyon and an ocean trench?

8. Explain what graded bedding is and how it forms.

9. Describe the process by which abyssal plains are created.

10. Discuss the origin of the various volcanic peaks of the abyssal plains: seamounts, tablemounts, and abyssal hills.

11. In which ocean basin are most ocean trenches found? Use plate tectonic processes to help explain why.

12. Describe characteristics and features of the mid-ocean ridge, including the difference between oceanic ridges and oceanic rises.

13. List and describe the different types of hydrothermal vents.

14. What kinds of unusual life can be found associated with hydrothermal vents? How do these organisms survive?

15. Use pictures and words to describe differences between fracture zones and transform faults.

CHAPTER 5
Marine Sediments

Arranged diatoms. The objects in this photomicrograph are diatoms, which are microscopic sea creatures that exist in incredible abundance in the ocean. This image was made by carefully arranging various diatoms under a microscope.

Key Questions

- How are marine sediments collected?
- What do marine sediments indicate about past environmental conditions on Earth?
- How are the four main types of marine sediment formed?
- Where can each of the four main types of marine sediment be found?
- Which types of marine sediment comprise coastal and deep-sea deposits?

The answers to these questions (and much more) can be found in the highlighted concept statements within this chapter.

"When I think of the floor of the deep sea, the single, overwhelming fact that possesses my imagination is the accumulation of sediments. I see always the steady, unremitting, downward drift of materials from above, flake upon flake, layer upon layer. . . . For the sediments are the materials of the most stupendous snowfall the Earth has ever seen."

—Rachel Carson, *The Sea Around Us* (1956)

Why are **sediments** (*sedimentum* = settling) interesting to oceanographers? Although ocean sediments are little more than eroded particles and fragments of dirt, dust, and other debris that have settled out of the water and accumulated on the ocean floor (Figure 5–1), they reveal much about Earth's history. For example, sediments provide clues to past climates, movements of the ocean floor, ocean circulation patterns, and nutrient supplies for marine organisms. By examining cores of sediment retrieved from ocean drilling (Figure 5–2) and interpreting them, oceanographers can ascertain the timing of major extinctions, global climate change, and the movement of plates. In fact, most of what is known of Earth's past geology, climate, and biology has been learned through studying ancient marine sediments.

Over time, sediments can become *lithified* (*lithos* = stone, *fic* = making)—turned to rock—and form *sedimentary rock*. More than half of the rocks exposed on the continents are sedimentary rocks deposited in ancient ocean environments and uplifted onto land by plate tectonic processes. Even in the tallest mountains on the continents, far from any ocean, telltale marine fossils indicate that these rocks originated on the ocean floor in the geologic past.

Particles of sediment come from worn pieces of rocks, as well as living organisms, minerals dissolved in water, and outer space. Table 5–1 show a classification of marine sediments according to type, composition, sources, and main locations found. Clues to sediment origin are found in its mineral composition and its **texture** (the size and shape of its particles).

> Marine sediments accumulate on the ocean floor and contain a record of recent Earth history including past environmental conditions. Most sediment cores are obtained by rotary drilling.

Lithogenous Sediment

Lithogenous (*lithos* = stone, *generare* = to produce) **sediment** is derived from preexisting rock material. Because most lithogenous sediment comes from the landmasses, it is also called **terrigenous** (*terra* = land, *generare* = to produce) **sediment**. Volcanic islands in the open ocean are also important sources of lithogenous sediment.

Figure 5–1 Oceanic sediment.

View of the deep-ocean floor from a submersible. Most of the deep-ocean floor is covered with particles of material that have settled out through the water.

BOX 5–1   Research Methods in Oceanography

COLLECTING THE HISTORICAL RECORD OF THE DEEP-OCEAN FLOOR

During early exploration of the oceans, a bucketlike device called a *dredge* was used to scoop up sediment from the deep-ocean floor for analysis. This technique, however, was limited to gathering samples from the *surface* of the ocean floor. Later, the *gravity corer*—a hollow steel tube with a heavy weight on top—was thrust into the sea floor to collect the first *cores* (cylinders of sediment and rock). Although the gravity corer could sample below the surface, its depth of penetration was limited.

In 1963, the National Science Foundation of the United States funded a program that borrowed drilling technology from the offshore oil industry to obtain long sections of core from deep below the surface of the ocean floor. The program united four leading oceanographic institutions (Scripps Institution of Oceanography in California; Rosenstiel School of Atmospheric and Oceanic Studies at the University of Miami, Florida; Lamont-Doherty Earth Observatory of Columbia University in New York; and the Woods Hole Oceanographic Institution in Massachusetts) to form the *Joint Oceanographic Institutions for Deep Earth Sampling (JOIDES)*. JOIDES was later joined by the oceanography departments of several other leading universities.

The first phase of the **Deep Sea Drilling Project (DSDP)** was initiated in 1968 when the specially designed drill ship *Glomar Challenger* was launched. It had a tall drilling rig resembling a steel tower. Cores could be collected by drilling into the ocean floor in water up to 6000 meters (3.7 miles) deep. From the initial cores collected, scientists confirmed the existence of sea floor spreading by documenting that (1) the age of the ocean floor increased progressively with distance from the mid-ocean ridge; (2) sediment thickness increased progressively with distance from the mid-ocean ridge; and (3) Earth's magnetic field polarity reversals were recorded in ocean floor rocks.

Although the oceanographic research program was initially financed by the U.S. government, it became international in 1975 when West Germany, France, Japan, the United Kingdom, and the Soviet Union also provided financial and scientific support. In 1983, the Deep Sea Drilling Project became the **Ocean Drilling Program (ODP)** with 20 participating countries under the supervision of Texas A&M University and a broader objective of drilling the thick sediment layers near the continental margins.

In 1985, the *Glomar Challenger* was decommissioned and replaced by the drill ship **JOIDES Resolution** (Figure 5A). The new ship also has a tall metal drilling rig to conduct **rotary drilling**. The drill pipe is in individual sections of 9.5 meters (31 feet) and sections can be screwed together to make a single string of pipe up to 8200 meters (27,000 feet) long (Figure 5B). The drill bit, located at the end of the pipe string, rotates as it is pressed against the ocean bottom and can drill up to 2100 meters (6900 feet) below the sea floor. Like twirling a soda straw into a layer cake, the drilling operation crushes the rock around the outside and retains a cylinder of rock (a core sample) on the inside of the hollow pipe, which can then be raised on board the ship. Cores are retrieved from inside the pipe and analyzed with state-of-the-art laboratory facilities on board the *Resolution*. Worldwide, more than 2000 holes have been drilled into the sea floor using this method, allowing the collection of cores that provide scientists with valuable information about Earth history as recorded in sea floor sediments.

The ODP was replaced in 2003 by the **Integrated Ocean Drilling Program (IODP)**, which is led by the United States and Japan. The program features two new drill ships: one for drilling shallow high-resolution cores, and one with advanced technology that can drill up to 7000 meters (23,000 feet) below the sea floor. The primary objective of the new program is to collect cores that will allow scientists to better understand Earth history

Origin

Lithogenous sediment begins as rocks on continents or islands. Over time, **weathering** agents such as water, temperature extremes, and chemical effects break rocks into smaller pieces, as shown in Figure 5–3. When rocks are in smaller pieces, they can be more easily **eroded** (picked up) and transported. This eroded material is the basic component of which all lithogenous sediment is composed.

Eroded material from the continents is carried to the oceans by streams, wind, glaciers, and gravity (Figure 5–4). Each year, streamflow alone carries about 20 billion metric tons (22 billion short tons) of sediment to Earth's continental margins; almost 80% is provided by runoff from Asia.

Figure 5–5, is a map of the oceans' major sources of river-, wind-, and glacial-borne sediments. The figure shows that the source of stream-transported sediment occurs in regions on land with high rainfall and resulting large runoff. In particular, several major rivers carry large volumes of sediment from the Asian continent. The figure also shows that windblown sediment is largely derived from the world's arid regions and that the source of glacial sediment is in high latitude regions associated with large continental ice sheets.

Figure 5A The drill ship *JOIDES Resolution*.

and Earth system processes, including the properties of the deep crust, climate change patterns, earthquake mechanisms, and the microbiology of the deep ocean floor.

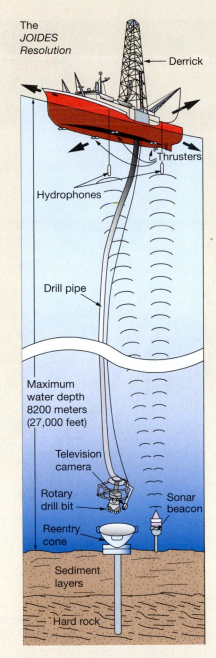

Figure 5B Rotary drilling from the *JOIDES Resolution*.

Transported sediment can be deposited in many environments, including bays or lagoons near the ocean, as deltas at the mouths of rivers, along beaches at the shoreline, or further offshore across the continental margin. It can also be carried beyond the continental margin to the deep-ocean basin by turbidity currents, as discussed in Chapter 4.

The greatest quantity of lithogenous material by far is found around the margins of the continents, where it is constantly moved by high-energy currents along the shoreline and in deeper turbidity currents. Lower-energy currents distribute finer components that settle out onto the deep-ocean basins. Microscopic particles from wind-blown dust or volcanic eruptions can even be carried far out over the open ocean by prevailing winds. These particles either settle into fine layers as the velocity of the wind decreases or disperse into the ocean when they serve as nuclei around which raindrops and snowflakes form.

Figure 5–2 Examining deep-ocean sediment cores.
Long cylinders of sediment and rock called cores are cut in half and examined, revealing interesting aspects of Earth history.

Composition

The composition of lithogenous sediment reflects the material from which it was derived. As discussed in Chapter 2, rocks are composed of discrete crystals of naturally occurring compounds called *minerals*. One of the most abundant, chemically stable, and durable minerals in Earth's crust is **quartz** (SiO_2), which is composed of silicon and oxygen tetrahedran and has the same composition as ordinary glass. Quartz is the major component of nearly all rocks. Because quartz is resistant to abrasion, it can be transported long distances and deposited far from its source area. The majority of lithogenous deposits—such as beach sands—are composed primarily of quartz (Figure 5–6).

A large percentage of lithogenous particles that find their way into deep-ocean sediments far from continents are transported by prevailing winds that remove small particles from the continents' subtropical desert regions. The map in Figure 5–7, shows a close relationship between the location of microscopic fragments of lithogenous quartz in the surface sediments of the

ocean floor and the strong prevailing winds in the desert regions of Africa, Asia, and Australia. Satellite observations of dust storms (Figure 5–7, *inset*) confirm this relationship.

? STUDENTS SOMETIMES ASK ...
How effective is wind as a transporting agent?

Any material that gets into the atmosphere—including dust from dust storms, soot from forest fires, specks of pollution, and ash from volcanic eruptions—is transported by wind and can be found as deposits on the ocean floor. Surprisingly, wind can transport huge volumes of silt- and clay-sized particles great distances. For example, a recent study indicated that each year an average of 11.5 million metric tons (12.6 million short tons) of dust from Africa's Sahara Desert is carried downwind up to 6500 kilometers (4000 miles) across the Atlantic Ocean (see Figure 5–7). Much of this dust falls in the Atlantic. That's why ships traveling downwind from the Sahara Desert often arrive at their destinations quite dusty. Some of it falls in the Caribbean (where it has been linked to stress and disease among coral reefs), the Amazon (where it fertilizes nutrient-poor soil), and across the southern U.S. as far west as New Mexico.

Sediment Texture

One of the most important properties of lithogenous sediment is its texture, including its **grain**[1] **size**. The **Wentworth scale of grain size** (Table 5–2) indicates that particles can be classified as boulders (largest), cobbles, pebbles, granules, sand, silt, or clay (smallest). Grain size is proportional to the energy needed to lay down a deposit.

Figure 5–8 is a graph showing the relationship between grain size and horizontal current velocities, which results in erosion, transportation, or deposition of sediment (colored fields on the graph).[2] The deposition curve (the line between the tan and blue fields on Figure 5–8) shows that at high current velocities, only larger particles settle out (the smaller particles are carried along in the current). Thus, deposits laid down where current action is strong (areas of high energy) are composed primarily of larger particles. The deposition curve also shows that small particles can be transported until current velocities are quite low. As a result, fine-grained particles are deposited where the energy level is low and the current speed is minimal.

---

[1]Sediment grains are also known as particles, fragments, or clasts.

[2]Note that both the horizontal and vertical scales on the graph are logarithmic—meaning that they increase by powers of ten—so the graph is said to have a *log-log scale*. For more about logarithmic relationships, see Appendix I, "Metric and English Units Compared."

TABLE 5–1 **Classification of marine sediments.**

Type	Composition			Sources		Main locations found
Lithogenous	Continental Margin	Rock fragments		Rivers; coastal erosion; landslides		Continental shelf
		Quartz sand		Glaciers		Continental shelf in high latitudes
		Quartz silt		Turbidity currents		Continental slope and rise; ocean basin margins
		Clay				
	Oceanic	Quartz silt		Wind-blown dust; rivers		Deep-ocean basins
		Clay				
		Volcanic ash		Volcanic eruptions		
Biogenous	Calcium carbonate ($CaCO_3$)	Calcareous ooze (microscopic)	Warm surface water	Coccolithophores (algae); Foraminifers (protozoans)		Low-latitude regions; sea floor above CCD; along mid-ocean ridges & the tops of volcanic peaks
		Shell/coral fragments (macroscopic)		Macroscopic shell-producing organisms		Continental shelf; beaches
				Coral reefs		Shallow low-latitude regions
	Silica ($SiO_2 \cdot nH_2O$)	Siliceous ooze	Cold surface water	Diatoms (algae); Radiolarians (protozoans)		High-latitude regions; sea floor below CCD; surface current divergence near the Equator
Hydrogenous		Manganese nodules (manganese, iron, copper, nickel, cobalt)		Precipitation of dissolved materials directly from seawater due to chemical reactions		Abyssal plain
		Phosphorite (phosphorous)				Continental shelf
		Oolites ($CaCO_3$)				Shallow shelf in low-latitude regions
		Metal sulfides (iron, nickel, copper, zinc, silver)				Hydrothermal vents at mid-ocean ridges
		Evaporites (gypsum, halite, other salts)				Shallow restricted basins where evaporation is high in low-latitude regions
Cosmogenous		Iron-nickel spherules Tektites (silica glass)		Space dust		In very small proportions mixed with all types of sediment and in all marine environments
		Iron-nickel meteorites Silicate chondrites		Meteors		Localized near meteor impact structures

Figure 5–8 also shows the energy needed to erode various sizes of sediment. The erosion curve (the diffuse line between the blue and purple fields on Figure 5–8) shows that the velocity required to erode sediment is much higher than for deposition. Generally, the larger the grain size, the higher the velocity is required to erode it. Surprisingly, erosion of finer clay-sized particles requires higher velocities than larger sand-sized particles. This is because clay-sized particles—many of which are flat—tend to stick together by cohesive forces. Consequently, higher-energy conditions than what would be expected based on grain size alone are required to erode and transport clays.

The texture of lithogenous sediment also depends on its **sorting**. Sorting is a measure of the uniformity of

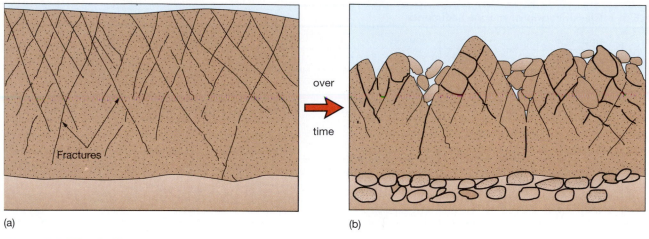

(a)

over

time

(b)

Figure 5–3 Weathering.

Weathering often occurs along fractures in rock, breaking the rocks into smaller fragments over time.

grain sizes and indicates the selectivity of the transportation process. For example, sediments composed of particles that are primarily the same size are well sorted—such as in coastal sand dunes, where winds can only pick up a certain size particle. Poorly sorted deposits, on the other hand, contain a variety of different sized particles and indicate a transportation process capable of picking up clay- to boulder-sized particles. An example of poorly sorted sediment is that which is carried by a glacier and left behind when the glacier melts.

The texture of lithogenous sediment also depends on its **maturity**. Sediment maturity increases as (1) clay content decreases; (2) sorting increases; (3) non-quartz minerals decrease; and (4) grains within the deposit become more rounded. Particles increase in maturity as they are carried from the source to their point of deposition because more time is available during transportation to (1) remove clays (which are carried in suspension and washed out to sea); (2) sort the sediment; (3) eliminate non-quartz minerals (which lack durability); and (4) round particles through abrasion.

A poorly sorted glacial deposit, which contains relatively large quantities of clay-sized particles and poorly rounded larger particles, is immature. Well-sorted beach sand, on the other hand, which contains well-rounded particles and very little clay, is a mature sedimentary deposit. Figure 5–9, illustrates the difference between mature and immature sediments.

Distribution

Marine sedimentary deposits can be categorized as neritic or pelagic. **Neritic** (*neritos* = of the coast) **deposits** are found along continental margins and near islands, and **pelagic** (*pelagios* = of the sea) **deposits** are found in the deep-ocean basins. Lithogenous sediment in the ocean is ubiquitous: At least a small percentage of lithogenous sediment is found nearly everywhere on the ocean floor.

Neritic Deposits Lithogenous sediment dominates most neritic deposits. Lithogenous sediment is derived from rocks on nearby landmasses, consists of coarse-grained deposits, and accumulates rapidly on the continental shelf, slope, and rise. Examples of lithogenous neritic deposits include beach deposits, continental shelf deposits, turbidite deposits, and glacial deposits.

Beach Deposits Beaches are made of whatever materials are locally available. Beach materials are composed mostly of quartz-rich sand that is washed down to the coast by rivers but can also be composed of a wide variety of sizes and compositions. This material is transported by waves that crash against the shoreline, especially during storms.

Continental Shelf Deposits At the end of the last Ice Age (about 18,000 years ago), glaciers melted and sea level rose. As a result, many rivers of the world today deposit their sediment in drowned river mouths rather than carry it onto the continental shelf as they did during the geologic past. In many areas, the sediments that cover the continental shelf—called *relict* (*relict* = left behind) *sediments*—were deposited from 3000 to 7000 years ago and have not yet been covered by more recent deposits. These sediments presently cover about 70% of the world's continental shelves. In other areas, deposits of sand ridges on the continental shelves appear to have been formed more recently than the Ice Age and at present water depths.

Turbidite Deposits As discussed in Chapter 4, **turbidity currents** are underwater avalanches that periodically move down the continental slopes and carve submarine canyons. Turbidity currents also carry vast

(a)

(b)

(c)

(d)

Figure 5–4 Sediment transporting media.

Sediment transporting media include: **(a)** Streams. **(b)** Wind. **(c)** Glaciers. **(d)** Gravity, which creates landslides.

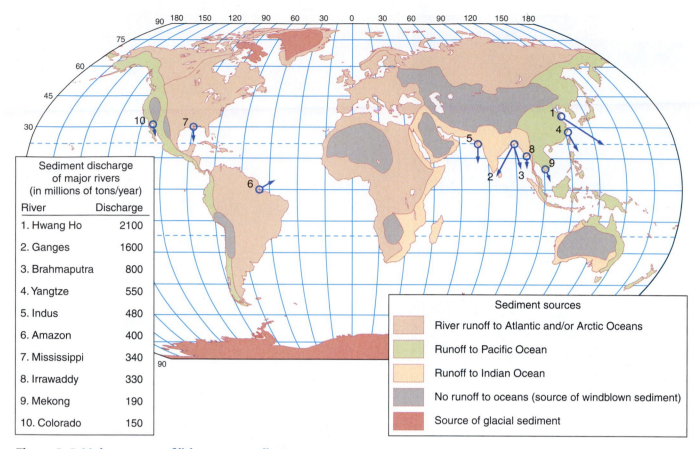

Sediment discharge of major rivers (in millions of tons/year)	
River	Discharge
1. Hwang Ho	2100
2. Ganges	1600
3. Brahmaputra	800
4. Yangtze	550
5. Indus	480
6. Amazon	400
7. Mississippi	340
8. Irrawaddy	330
9. Mekong	190
10. Colorado	150

Sediment sources

River runoff to Atlantic and/or Arctic Oceans

Runoff to Pacific Ocean

Runoff to Indian Ocean

No runoff to oceans (source of windblown sediment)

Source of glacial sediment

Figure 5–5 Major sources of lithogenous sediment.

Map showing the global distribution of major sources of lithogenous sediment. Blue circles with arrows indicate mouths of selected rivers; length of arrow is proportional to the average yearly amount of sediment discharged.

Figure 5–6 Lithogenous beach sand.

Lithogenous beach sand is composed mostly of particles of white quartz, plus small amounts of other minerals. This sand is from North Beach, Hampton, New Hampshire and is magnified approximately 23 times.

amounts of neritic material. This material spreads out as deep-sea fans, comprises the continental rise, and gradually thins toward the abyssal plains. These deposits are called **turbidite deposits** and are composed of characteristic layering called *graded bedding* (see Figure 4–9).

Glacial Deposits Poorly sorted deposits containing particles ranging from boulders to clays may be found in the high-latitude[3] portions of the continental shelf. These **glacial deposits** were laid down after the Ice Age when glaciers that covered the continental shelf eventually melted. Glacial deposits are currently forming around the continent of Antarctica and around Greenland by **ice rafting**. In this process, rock particles trapped in glacial ice are carried out to sea by icebergs that break away from coastal glaciers. As the icebergs melt, lithogenous particles of many sizes are released and settle onto the ocean floor.

---

[3]High-latitude regions are those far from the Equator (either north or south); low latitudes are areas close to the Equator.

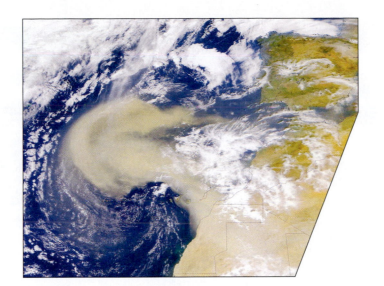

Figure 5–7 Lithogenous quartz in surface sediments of the world's oceans.

High concentrations of microscopic lithogenous quartz in deep-sea sediment match prevailing winds from land (*arrows*). SeaStar SeaWiFS satellite photo (*inset*) on February 26, 2000, shows a Sahara dust storm off the northwest coast of Africa that has spread out for 1000 miles (1600 kilometers) across the Atlantic Ocean.

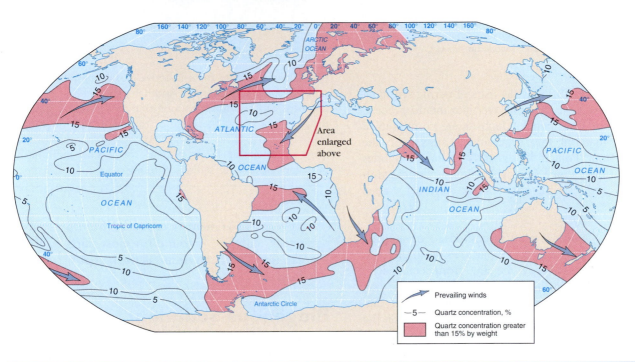

TABLE 5–2 **Wentworth scale of grain size for sediments**

Size range (millimeters)	Particle name	Grain size	Example	Energy conditions
Above 256	Boulder	Coarse-grained	Coarse material found in	High energy
64 to 256	Cobble		stream beds near the source	
4 to 64	Pebble		areas of rivers	
2 to 4	Granule			
$1/16$ to 2	Sand		Beach sand	
$1/256$ to $1/16$	Silt		Feels gritty in teeth	
$1/4096$ to $1/256$	Clay	Fine-grained	Microscopic; feels sticky	Low energy

(Gravel: Boulder–Granule)

0   10   20   30   40   50   60
Scale in millimeters

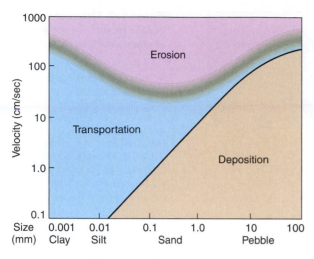

Figure 5–8 Sediment erosion, transportation, and deposition.

Relationship between horizontal current velocity and sediment grain size, showing conditions necessary for sediment erosion (*purple*), transportation (*blue*) and deposition (*brown*).

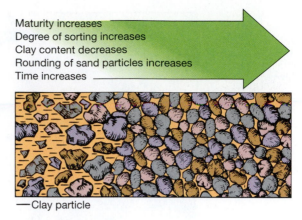

Maturity increases
Degree of sorting increases
Clay content decreases
Rounding of sand particles increases
Time increases

—Clay particle

Figure 5–9 Sediment maturity.

As sediment maturity increases with time (*left to right*), the degree of sorting and rounding of particles increases, whereas clay content decreases.

Pelagic Deposits Turbidite deposits of neritic sediment on the continental rise can spill over into the deep-ocean basin. However, most pelagic deposits are composed of fine-grained material that accumulates slowly on the deep-ocean floor. Pelagic lithogenous sediment includes particles that have come from volcanic eruptions, wind-blown dust, and fine material that is carried by deep ocean currents.

Abyssal Clay **Abyssal clay** is composed of at least 70% (by weight) fine clay-sized particles from the continents. Even though they are far from land, deep abyssal plains contain thick sequences of abyssal clay deposits composed of particles transported great distances by winds or ocean currents and deposited on the deep ocean floor. Because abyssal clays contain oxidized iron, they are commonly red-brown or buff in color and are sometimes referred to as **red clays**. The predominance of abyssal clay on abyssal plains is caused not by an abundance of clay settling on the ocean floor but by the absence of other material that would otherwise dilute it.

STUDENTS SOMETIMES ASK...
Are there any areas of the ocean floor where no sediment is being deposited?

Various types of sediment accumulate on nearly all areas of the ocean floor in the same way dust accumulates in all parts of your home (which is why marine sediment is often referred to as "marine dust"). Even the deep-ocean floor far from land receives small amounts of wind-blown material, microscopic biogenous particles, and space dust.

There are a few places in the ocean, however, where very little sediment accumulates. One such place is along the con-tinental slope, where there is active erosion by turbidity and other deep-ocean currents. Another place where very little sediment can be found is along the mid-ocean ridge. Here, the sea floor along the crest of the mid-ocean ridge is so young (because of sea floor spreading) and the rates of sediment accumulation far from land are so slow there hasn't been enough time for sediments to form.

> Lithogenous sediment is produced from preexisting rock material, is found on most parts of the ocean floor, and can occur as thick deposits close to land.

Biogenous Sediment

Biogenous (*bio* = life, *generare* = to produce) **sediment** is derived from the remains of hard parts of once-living organisms.

Origin

Biogenous sediment begins as the hard parts (shells, bones, and teeth) of living organisms ranging from minute algae and protozoans to fish and whales. When organisms that produce hard parts die, their remains settle onto the ocean floor and can accumulate as biogenous sediment.

Biogenous sediment can be classified as either macroscopic or microscopic. **Macroscopic biogenous sediment** is large enough to be seen without the aid of a microscope and includes shells, bones, and teeth of large organisms. Except in certain tropical beach localities where shells and coral fragments are numerous, this type of sediment is relatively rare in the marine environment, especially in deep water where fewer organisms live. Much more abundant is **microscopic biogenous sediment**, which contains particles so small they can only be seen well through a microscope.

Microscopic organisms produce tiny shells called **tests** (*testa* = shell) that begin to sink after the organisms die and continually rain down in great numbers onto the ocean floor. These microscopic tests can accumulate on the deep-ocean floor and form deposits called **ooze**(*wose* = juice). As its name implies, ooze resembles very fine-grained mushy material.[4] Technically, biogenous ooze must contain at least 30% biogenous test material by weight. What comprises the other part—up to 70% of an ooze? Commonly, it is fine-grained lithogenous clay that is deposited along with biogenous tests in the deep ocean. By volume, much more microscopic ooze than macroscopic biogenous sediment exists on the ocean floor.

The organisms that contribute to biogenous sediment are chiefly **algae** (*alga* = seaweed) and **protozoans** (*proto* = first, *zoa* = animal). Algae are primarily aquatic, eukaryotic,[5] photosynthetic organisms, ranging in size from microscopic single cells to large organisms like giant kelp. Protozoans are any of a large group of single-celled, eukaryotic, usually microscopic organisms that are generally not photosynthetic.

Composition

The two most common chemical compounds in biogenous sediment are **calcium carbonate** ($CaCO_3$, which forms the mineral **calcite**) and **silica** (SiO_2). Often, the silica is chemically combined with water to produce $SiO_2 \cdot nH_2O$, the hydrated form of silica, which is called *opal*.

Silica Most of the silica in biogenous ooze comes from microscopic algae called **diatoms** (*diatoma* = cut in half) and protozoans called **radiolarians** (*radio* = a spoke or ray).

Because diatoms photosynthesize, they need strong sunlight and are found only within the upper sunlit surface waters of the ocean. Most diatoms are free-floating or **planktonic** (*planktos* = wandering). The living organism builds a glass greenhouse out of silica as a protective covering and lives inside. Most species have two parts to their test that fit together like a petri dish or pillbox (Figure 5–10a). The tiny tests are perforated with small holes in intricate patterns to allow nutrients to pass in and waste products to pass out. Where diatoms are abundant at the ocean surface, thick deposits of diatom-rich ooze can accumulate below on the ocean floor. When this ooze lithifies, it becomes **diatomaceous earth**,[6] a lightweight white rock composed of diatom tests and clay that has many uses (Box 5–2).

Radiolarians are microscopic single-celled protozoans, most of which are also planktonic. As their name implies, they often have long spikes or rays of silica protruding from their siliceous shell (Figure 5–10b). They do not photosynthesize but rely on external food sources such as bacteria and other plankton. Radiolarians typically display well-developed symmetry, which is why they have been described as living snowflakes of the sea.

The accumulation of siliceous tests of diatoms, radiolarians, and other silica-secreting organisms produces **siliceous ooze** (Figure 5–10c).

Calcium Carbonate Two significant sources of calcium carbonate biogenous ooze are the **foraminifers** (*foramen* = an opening)—close relatives of radiolarians—and microscopic algae called **coccolithophores** (*coccus* = berry; *lithos* = stone; *phorid* = carrying).

Coccolithophores are single-celled algae, most of which are planktonic. Coccolithophores produce thin plates or shields made of calcium carbonate, 20 or 30 of which overlap to produce a spherical test (Figure 5–11a). Like diatoms, coccolithophores photosynthesize, so they need sunlight to live. Coccolithophores are about 10 to 100 times smaller than most diatoms (Figure 5–11b), which is why coccolithophores are often called **nannoplankton** (*nanno* = dwarf, *planktos* = wandering).

When the organism dies, the individual plates (called **coccoliths**) disaggregate and can accumulate on the ocean floor as coccolith-rich ooze. When this ooze lithifies over time, it forms a white deposit called **chalk**, which is used for a variety of purposes (including writing on chalkboards). The White Cliffs of southern England are composed of hardened coccolith-rich calcium carbonate ooze, which was deposited on the ocean floor and has been uplifted onto land (Figure 5–12). Deposits of chalk the same age as the White Cliffs are so common throughout Europe, North America, Australia, and the Middle East that the geologic period in which these deposits formed is named the Cretaceous (*creta* = chalk) Period.

Foraminifers are single-celled protozoans, many of which are planktonic, ranging in size from microscopic to macroscopic. They do not photosynthesize, so they must ingest other organisms for food. Foraminifers produce a hard calcium carbonate test in which the organism lives (Figure 5–11c,). Most foraminifers produce a segmented or chambered test, and all tests have a prominent opening in one end. Although very small in size, the tests of foraminifers resemble the large shells that one might find at a beach.

Deposits comprised primarily of tests of foraminifers, coccoliths, and other calcareous-secreting organisms are called **calcareous ooze** (Figure 5–11d).

Distribution

Biogenous sediment is commonly found in pelagic deposits but only rarely found as neritic deposits. The distribution of biogenous sediment on the ocean floor depends on three fundamental processes: productivity, destruction, and dilution.

---

[4]Ooze has the consistency of toothpaste mixed about half and half with water. To help you remember this term, imagine walking barefoot across the deep-ocean floor and how the sediment would *ooze* between your toes.

[5]As discussed in Chapter 2, eukaryotic (*eu* = good, *karyo* = the nucleus) cells contain a distinct membrane-bound nucleus.

[6]Diatomaceous earth is also called diatomite, tripolite, or kieselguhr.

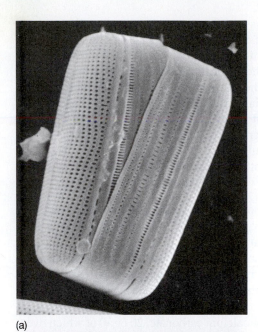

(a)

Figure 5–10 Microscopic siliceous tests.

Scanning electron micrographs: **(a)** Diatom (length = 30 micrometers, equal to 30 millionths of a meter), showing how the two parts of the diatom's test fit together. **(b)** Radiolarian (length = 100 micrometers). **(c)** Siliceous ooze, showing mostly fragments of diatom tests (magnified 250 times).

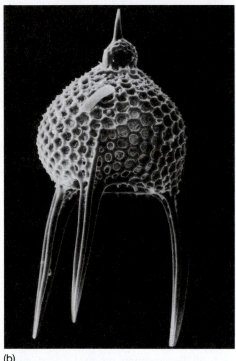

(b)

(c)

Productivity is the number of organisms present in the surface water above the ocean floor. Surface waters with high biologic productivity contain many living and reproducing organisms—conditions that are likely to produce biogenous sediments. Conversely, surface waters with low biologic productivity contain too few organisms to produce biogenous oozes on the ocean floor. *Destruction* occurs when skeletal remains (tests) dissolve in seawater at depth.

Dilution occurs when other sediments are prevalent enough to keep the amount of biogenous test material below 30%. Dilution occurs most often because of the abundance of coarse-grained lithogenous material in ner-

itic environments, so biogenous oozes are uncommon along continental margins.

Neritic Deposits Although neritic deposits are dominated by lithogenous sediment, both microscopic and macroscopic biogenous material may be incorporated into lithogenous sediment in neritic deposits. In addition, biogenous carbonate deposits are common in some areas.

Carbonate Deposits **Carbonate** minerals are those that contains CO_3 in its chemical formula—such as calcium carbonate, $CaCO_3$. Rocks from the marine environment

BOX 5–2   People and Ocean Environment

DIATOMS: THE MOST IMPORTANT THINGS YOU HAVE (PROBABLY) NEVER HEARD OF

"Few objects are more beautiful than the minute siliceous cases of the diatomaceae: were these created that they might be examined and admired under the higher powers of the microscope?"

—Charles Darwin (1872)

Although most people are scarcely aware of it, diatoms are incredibly important to life on Earth. They are also used to produce a variety of common products. What exactly are diatoms?

Diatoms are microscopic single-celled photosynthetic organisms. Each one lives inside a protective silica test, most of which contain two halves that fit together like a shoebox and its lid. First described with the aid of a microscope in 1702, their tests are exquisitely ornamented with holes, ribs, and radiating spines unique to individual species. The fossil record indicates that diatoms have been on Earth since the Jurassic Period (180 million years ago) and at least 70,000 species of diatoms have been identified.

Diatoms live for a few days to as much as a week, can reproduce sexually or asexually, and occur individually or linked together into long communities. They are found in great abundance floating in the ocean and in certain freshwater lakes but can also be found in many diverse environments, such as on the undersides of polar ice, on the skins of whales, in soil, in thermal springs, and even on brick walls.

Continued...

Figure 5C **Products containing or produced using diatomaceous earth (diatom *Thalassiosira eccentrica*, inset).**

When marine diatoms die, their tests rain down and accumulate on the sea floor as siliceous ooze. Hardened deposits of siliceous ooze, called diatomaceous earth, can be as much as 900 meters (3000 feet) thick. Diatomaceous earth consists of billions of minute silica tests and has many unusual properties: It is lightweight, has an inert chemical composition, is resistant to high temperatures, and has excellent filtering properties. Diatomaceous earth is used to produce a variety of common products (Figure 5C). The main uses of diatomaceous earth include

- filters (for refining sugar, separating impurities from wine, straining yeast from beer, and filtering swimming pool water)

- mild abrasives (in toothpaste, facial scrubs, matches, and household cleaning and polishing compounds)

- absorbents (for chemical spills and pest control)

- chemical carriers (in pharmaceuticals, paint, and dynamite)

Other products from diatomaceous earth include optical-quality glass (because of the pure silica content of diatoms), space shuttle tiles (because they are lightweight and provide good insulation), an additive in concrete, a filler in tires, an anti-caking agent, and even building stone for constructing houses.

Further, the vast majority of oxygen that all animals breathe is a byproduct of photosynthesis by diatoms. In addition, each living diatom contains a tiny droplet of oil. When diatoms die, their tests containing droplets of oil accumulate on the sea floor and are the beginnings of petroleum deposits, such as those found offshore California.

Given their many practical applications, it is difficult to imagine how different our lives would be without diatoms!

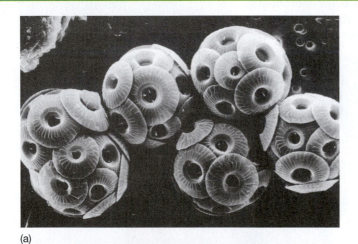

(a)

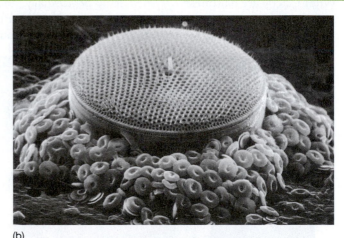

(b)

(c)

Radiolarians

Foraminifers

(d)

Figure 5–11 Microscopic calcareous tests.

Scanning electron micrographs: **(a)** Coccolithophores (diameter of individual coccolithophores = 20 micrometers, equal to 20 millionths of a meter). **(b)** Diatom (siliceous) and coccoliths (diameter of diatom = 70 micrometers). **(c)** Foraminifers (most species 400 micrometers in diameter). **(d)** Calcareous ooze, which also includes some siliceous radiolarian tests (magnified 160 times).

Figure 5–12 The White Cliffs of southern England.

The White Cliffs near Dover in southern England are composed of hardened coccolith-rich calcareous ooze (chalk).

composed primarily of calcium carbonate are called **limestones**. Most limestones contain fossil marine shells, suggesting a biogenous origin, while others appear to have formed directly from seawater without the help of any marine organism. Modern environments where calcium carbonate is currently forming (such as in the Bahama Banks, Australia's Great Barrier Reef, and the Persian Gulf) suggest that these carbonate deposits occurred in shallow, warm-water shelves and around tropical islands as coral reefs and beaches.

Ancient marine carbonate deposits constitute 2% of Earth's crust and 25% of all sedimentary rocks on Earth. These limestone deposits form the bedrock underlying Florida and many Midwestern states from Kentucky to Michigan and from Pennsylvania to Colorado. Percolation of groundwater through these deposits has dissolved the limestone to produce sinkholes and spectacular caverns.

Stromatolites (*stromat* = covering, *ite* = stone) are lobate structures consisting of fine layers of carbonate that form in specific warm, shallow-water environments such as the high salinity tidal pools in Shark Bay, western Australia, and the shifting carbonate sand shoals on Eleuthera Bank in the Bahamas (Figure 5–13). Cyanobacteria[7] produce these deposits by trapping fine sediment in mucous mats. Other types of algae produce long filaments that bind the particles together. As layer upon layer of these algae colonize the surface, a bulbous structure is formed. In the geologic past—particularly from about 1 to 3 billion years ago—conditions were ideal for stromatolites, so stromatolitic structures hundreds of meters high are common in rocks from these ages.

Pelagic Deposits Microscopic biogenous sediment (ooze) is common on the deep-ocean floor because there is so little lithogenous sediment deposited at great distances from the continents that could dilute the biogenous material.

Siliceous Ooze Siliceous ooze contains at least 30% (by weight) of the hard remains of silica-secreting organisms. When the siliceous ooze consists mostly of diatoms, it is called *diatomaceous ooze*. When it consists mostly of radiolarians, it is called *radiolarian ooze*. When it consists mostly of single-celled silicoflagellates—another type of alga—it is called *silicoflagellate ooze*.

The ocean is undersaturated with silica at all depths, so seawater slowly but continually dissolves silica. In addition, bacterial decomposition of the protective protein coating on diatom tests also aids the dissolution process. How can siliceous ooze accumulate on the ocean floor if it is being dissolved? One way is to accumulate the siliceous tests faster than seawater can dissolve them. For instance, many tests sinking at the same time will create a deposit of siliceous ooze on the sea floor below (Figure 5–14).[8] Once buried beneath other siliceous tests, they are no longer exposed to the dissolving effects of seawater. Thus, siliceous ooze is commonly found in areas below surface waters with high biologic productivity of silica-secreting organisms, such as in equatorial and high-latitude regions (Figure 5–15).

Calcareous ooze Calcareous ooze contains at least 30% (by weight) of the hard remains of calcareous-secreting organisms. When it consists mostly of coccolithophores, it is called *coccolith ooze*. When it consists mostly of foraminifers, it is called *foraminifer ooze*. One of the most common types of foraminifer ooze is *Globigerina ooze*,

---

[7]Cyanobacteria (*kuanos* = dark blue) are descendants of the first photosynthetic organisms, as discussed in Chapter 2.

[8]An analogy to this is trying to get a layer of sugar to form on the bottom of a cup of hot coffee. If a few grains of sugar are slowly dropped into the cup, a layer of sugar won't accumulate. However, if a whole bowl full of sugar is dumped into the coffee, a thick layer of sugar will form on the bottom of the cup.

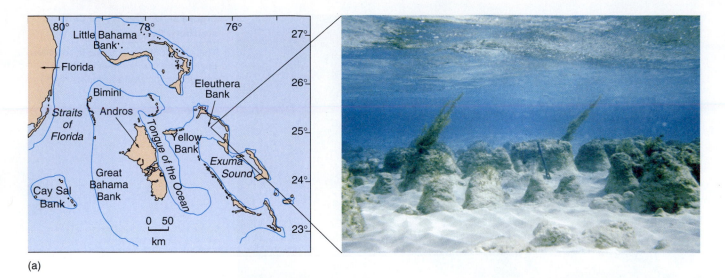

(a)

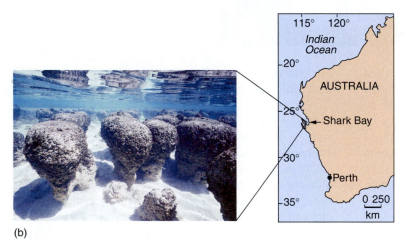

(b)

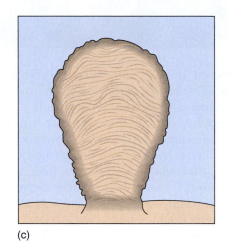

(c)

Figure 5–13 Stromatolites.

(a) Subtidal oolitic stromatolites on the crest of an oolitic tidal bar on Eleuthera Bank, Bahamas. Most reach a maximum height of about 1 meter (3.3 feet). **(b)** Shark Bay stromatolites, which form in high salinity tidal pools and also reach a maximum height of about 1 meter (3.3 feet). **(c)** Cross-section through a stromatolite, showing internal fine layering.

Figure 5–14 Accumulation of siliceous ooze.

Siliceous ooze accumulates on the ocean floor beneath areas of high productivity, where the rate of accumulation of siliceous tests is greater than the rate at which silica is being dissolved.

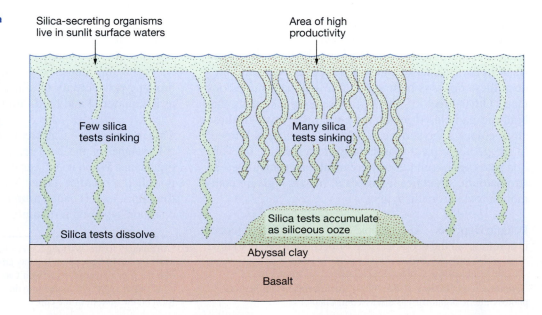

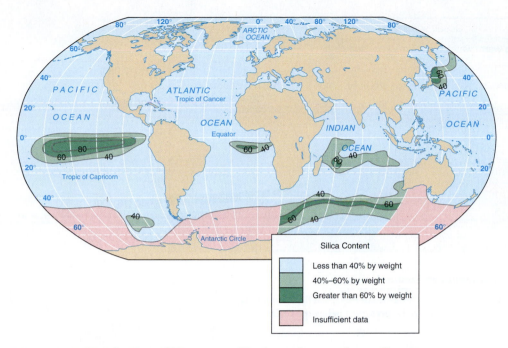

Figure 5–15 **Distribution of biogenous silica in modern surface sediments.**

The distribution of biogenous silica ($SiO_2 \cdot nH_2O$, which is opal) in modern surface sediments shows maximum concentrations associated with areas of highest biological productivity. In equatorial regions, high silica content in sediment is produced predominantly by radiolarians in surface waters above; in high latitudes, high silica content in sediment is the result of high surface water concentrations of diatoms.

named for a foraminifer that is especially widespread in the Atlantic and South Pacific Oceans. Other calcareous oozes include *pteropod oozes* and *ostracod oozes*.

The destruction (solubility) of calcium carbonate (calcite) varies with depth. At the warmer surface and in the shallow parts of the ocean, seawater is generally saturated with calcium carbonate, so calcite does not dissolve. In the deep ocean, however, the colder water contains greater amounts of carbon dioxide, which forms carbonic acid and causes calcareous material to dissolve. The higher pressure at depth also helps speed the dissolution of calcium carbonate.

The depth in the ocean at which the pressure is high enough and the amount of carbon dioxide in deep-ocean waters is great enough to begin dissolving calcium carbonate is called the **lysocline** (*lusis* = a loosening, *cline* = slope). Below the lysocline, calcium carbonate dissolves at an increasing rate with increasing depth until the **calcite compensation depth (CCD)**[9] is reached. At the CCD and greater depths, sediment does not usually contain much calcite because it readily dissolves—even the thick tests of foraminifers dissolve within a day or two. The CCD, on average, is 4500 meters (15,000 feet) below sea level, but, depending on the chemistry of the deep ocean, may be as deep as 6000 meters (20,000 feet)

in portions of the Atlantic Ocean, or as shallow as 3500 meters (11,500 feet) in the Pacific Ocean. The depth of the lysocline also varies from ocean to ocean but averages about 4000 meters (13,100 feet).

Because of the CCD, modern carbonate oozes are generally rare below 5000 meters (16,400 feet). Still, buried deposits of ancient calcareous ooze are found beneath the CCD. How can calcareous ooze exist below the CCD? The necessary conditions are shown in Figure 5–16. The mid-ocean ridge is a topographically high feature that rises above the sea floor. It often pokes up above the CCD, even though the surrounding deep-ocean floor is below the CCD. Thus, calcareous ooze deposited on top of the mid-ocean ridge will not be dissolved. Sea floor spreading causes the newly created sea floor and the calcareous sediment on top of it to move into deeper water away from the ridge, eventually being transported below the CCD. This calcareous sediment will dissolve below the CCD, unless it is covered by a deposit that is unaffected by the CCD (such as siliceous ooze or abyssal clay).

The map in Figure 5–17 shows the percentage (weight) of calcium carbonate in the modern surface ments of the ocean basins. High concentrations careous particles (sometimes exceeding 80%) along segments of the mid-ocean ridge, but in deep-ocean basins below the CCD. For northern Pacific Ocean—one of the d world ocean—there is very little calciu.

[9]Because the mineral calcite is composed of calcium carbonate, the calcite compensation depth is also known as the calcium carbonate compensation depth, or the carbonate compensation depth. All go by the handy abbreviation of CCD.

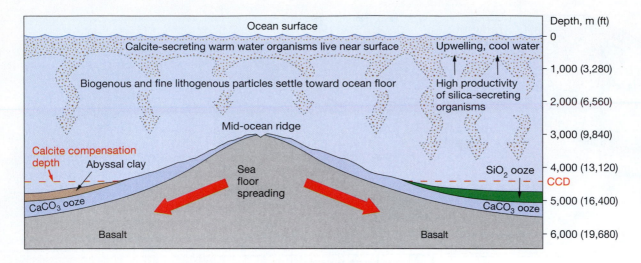

Figure 5–16 Sea floor spreading and sediment accumulation.

Relationships among carbonate compensation depth, the mid-ocean ridge, sea floor spreading, productivity, and destruction that allow calcareous ooze to be preserved below the CCD.

Figure 5–17 Distribution of calcium carbonate in modern surface sediments.

The distribution of calcium carbonate ($CaCO_3$) in modern surface sediments shows that high percentages of calcareous ooze closely follow the mid-ocean ridge, which is above the CCD.

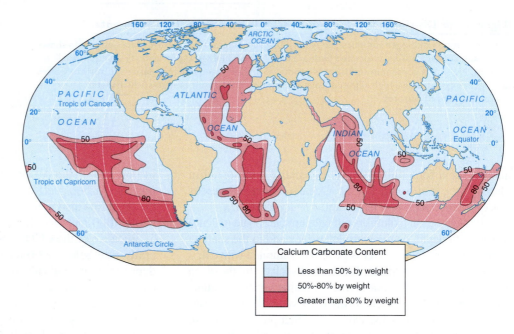

TABLE 5–3 **Comparison of environments interpreted from deposits of siliceous and calcareous ooze in surface sediments.**

	Siliceous ooze	Calcareous ooze
Surface water temperature above sea floor deposits	Cool	Warm
Main location found	Sea floor beneath cool surface water in high latitudes	Sea floor beneath warm surface water in low latitudes
Other factors	Upwelling brings deep, cold, nutrient-rich water to the surface	Calcareous ooze dissolves below the CCD
Other locations found	Sea floor beneath areas of upwelling, including along the Equator	Sea floor beneath warm surface water in low latitudes along the mid-ocean ridge

sediment. Calcium carbonate is also rare in sediments accumulating beneath cold, high-latitude waters where calcareous-secreting organisms are relatively uncommon.

Table 5–3 compares the environmental differences that can be inferred from siliceous and calcareous oozes. It shows that siliceous ooze typically forms below cool surface water regions, including areas of **upwelling** where deep ocean water comes to the surface and supplies nutrients that stimulate high rates of biological productivity. Calcareous ooze, on the other hand, is found on the shallower areas of the ocean floor beneath warmer surface water.

> Biogenous sediment is produced from the hard remains of once-living organisms. Microscopic biogenous sediment is especially widespread and forms deposits of ooze on the ocean floor.

Hydrogenous Sediment

Hydrogenous (*hydro* = water, *generare* = to produce) **sediment** is derived from the dissolved material in water.

Origin

Seawater contains many dissolved materials. Chemical reactions within seawater cause certain minerals to come out of solution or **precipitate** (change from the dissolved to the solid state). Precipitation usually occurs when there is a *change in conditions*, such as a change in temperature or pressure or the addition of chemically active fluids. To make rock candy, for instance, a pan of water is heated and sugar is added. When the water is hot and the sugar dissolved, the pan is removed from the heat and the sugar water is allowed to cool. The *change in temperature* causes the sugar to become oversaturated, which causes it to precipitate. As the water cools, the sugar precipitates on anything that is put in the pan, such as pieces of string or kitchen utensils.

Composition and Distribution

Although hydrogenous sediments represent a relatively small portion of the overall sediment in the ocean, they have many different compositions and are distributed in diverse environments of deposition. Several types of hydrogenous deposits have economic potential, which will be discussed in Chapter 17, "Marine Resources."

Manganese Nodules **Manganese nodules** are rounded, hard lumps of manganese, iron, and other metals typically 5 centimeters (2 inches) in diameter up to a maximum of about 20 centimeters (8 inches). When cut in half, they often reveal a layered structure formed by precipitation around a central nucleation object (Figure 5–18a). The nucleation

Figure 5–18 Manganese nodules.

(a) Manganese nodules cut in half, revealing their central nucleation object and layered internal structure. **(b)** A portion of the South Pacific Ocean floor about 4 meters (13 feet) across showing an abundance of manganese nodules.

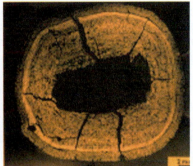

(a)

(b)

object may be a piece of lithogenous sediment, coral, volcanic rock, a fish bone, or a shark's tooth. Manganese nodules are found on the deep-ocean floor at concentrations of about 100 nodules per square meter (square yard). In some areas, they occur in even greater abundance (Figure 5–18b), resembling a scattered field of baseball-sized nodules. The formation of manganese nodules requires extremely low rates of lithogenous or biogenous input so that these sediments do not bury them.

The major components of these nodules are manganese dioxide (around 30% by weight) and iron oxide (around 20%). The element manganese is important for making high-strength steel alloys. Other accessory metals present in manganese nodules include copper (used in electrical wiring, pipe, and in making brass and bronze), nickel (used to make stainless steel), and cobalt (used as an alloy with iron to make strong magnets and steel tools). Although the concentration of these accessory metals is usually less than 1%, they can exceed 2% by weight, which may make them attractive exploration targets in the future.

The origin of manganese nodules has puzzled oceanographers since manganese nodules were first discovered in 1872 during the voyage of HMS *Challenger*.[10] If manganese nodules are truly hydrogenous and precipitate from seawater, then how can they have such high concentrations of manganese (which occurs in seawater at concentrations often too small to measure accurately)? Furthermore, why are the nodules on *top* of ocean floor sediment and not buried by the constant rain of sedimentary particles?

Unfortunately, nobody has definitive answers to these questions. Perhaps the creation of manganese nodules is the result of one of the slowest chemical reactions known—on average, they grow at a rate of about 5 millimeters (0.2 inch) per *million years*. Recent research suggests the formation of manganese nodules may be aided by bacteria and an as-yet-unidentified marine organism that intermittently lifts and rotates them. Other studies reveal that the nodules don't form continuously over time but in spurts that are related to specific conditions such as a low sedimentation rate of lithogenous clay and strong deep-water currents. Interestingly, the larger the nodules are, the faster they grow. The origin of manganese nodules is widely considered the most interesting unresolved problem in marine chemistry.

Phosphates Phosphorus-bearing compounds (**phosphates**) occur abundantly as coatings on rocks and as nodules on the continental shelf and on banks at depths shallower than 1000 meters (3300 feet). Concentrations of phosphates in such deposits commonly reach 30% by weight and indicate abundant biological activity in surface water above where they accumulate. Phosphates are valuable as fertilizers, and ancient marine deposits on land are extensively mined to supply agricultural needs.

Carbonates The two most important carbonate minerals in marine sediment are **aragonite** and calcite. Both are composed of calcium carbonate ($CaCO_3$), but aragonite has a different crystalline structure that is less stable and changes into calcite over time.

As previously discussed, most carbonate deposits are of biogenous origin. However, hydrogenous carbonate deposits can precipitate directly from seawater in tropical climates to form aragonite crystals less than 2 millimeters (0.08 inch) long. Additionally, **oolites** (*oo* = egg, *ite* = stone) are small calcite spheres 2 millimeters (0.08 inch) in diameter or less that have layers like an onion and form in some shallow tropical waters where concentrations of $CaCO_3$ are high. Oolites are thought to precipitate around a nucleus and grow larger as they roll back and forth on beaches by wave action, but some evidence suggests that a type of algae may aid their formation.

Metal Sulfides Deposits of **metal sulfides** are associated with hydrothermal vents and black smokers along the mid-ocean ridge. These deposits contain iron, nickel, copper, zinc, silver, and other metals in varying proportions. Transported away from the mid-ocean ridge by sea floor spreading, these deposits can be found throughout the ocean floor and can even be uplifted onto continents (see Box 3–2).

Evaporites **Evaporite minerals** form where there is restricted open ocean circulation and where evaporation rates are high (see Box 12–1). As water evaporates from these areas, the remaining seawater becomes saturated with dissolved minerals, which then begin to precipitate. Heavier than seawater, they sink to the bottom or form a white crust of evaporite minerals around the edges of these areas (Figure 5–19). Collectively termed "salts," some evaporite minerals taste salty, such as *halite* (common table salt, NaCl), and some do not, such as the calcium sulfate minerals *anhydrite* ($CaSO_4$) and *gypsum* ($CaSO_4 \cdot 2H_2O$).

? STUDENTS SOMETIMES ASK ...

I've been to Hawaii and seen a black sand beach. Because it forms by lava flowing into the ocean that is broken up by waves, is it hydrogenous sediment?

No. Many active volcanoes in the world have black sand beaches that are created when waves break apart dark-colored volcanic rock. The material that produces the black sand is derived from a continent or an island, so it is considered lithogenous sediment. Even though molten lava sometimes flows into the ocean, the resulting black sand could never be considered hydrogenous sediment because the lava was never *dissolved* in water.

---

[10]For more information about the accomplishments of the *Challenger* Expedition, see Chapter 1.

Figure 5–19 Evaporite salts.

Due to a high evaporation rate, salts (white material) precipitate onto the floor of Death Valley, California.

> Hydrogenous sediment is produced when dissolved materials precipitate out of solution and includes a variety of materials found in local concentrations on the ocean floor.

Cosmogenous Sediment

Cosmogenous (*cosmos* = universe, *generare* = to produce) **sediment** is derived from extraterrestrial sources.

Origin, Composition, and Distribution

Forming an insignificant portion of the overall sediment on the ocean floor, cosmogenous sediment consists of two main types: microscopic **spherules**, and macroscopic **meteor** debris.

Microscopic spherules are small globular masses. Some spherules are composed of silicate rock material and show evidence of being formed by extraterrestrial impact events on Earth or other planets that eject small molten pieces of crust into space. These **tektites** (*tektos* = molten) then rain down on Earth and can form *tektite fields*. Other spherules are composed mostly of iron and nickel (Figure 5–20) that form in the asteroid belt between the orbits of Mars and Jupiter and are produced when asteroids collide. This material constantly rains down on Earth as a general component of *space dust* or *micrometeorites* that float harmlessly through the atmosphere. Although about 90% of micrometeorites are destroyed by frictional heating as they enter the atmosphere, it has been estimated that as much as 300,000 metric tons (331,000 short tons) of space dust reach Earth's surface each year. The iron-rich space dust that lands in the oceans often dissolves in seawater. Glassy tektites, however, do not dissolve as easily and sometimes comprise minute proportions of various marine sediments.

Macroscopic meteor debris is rare on Earth but can be found associated with meteor impact sites. Evidence suggests that throughout time, meteors have collided with Earth at great speeds and that some larger ones have released energy equivalent to the explosion of multiple large nuclear bombs (Box 5–3). The debris from meteors—called **meteorite** material—settles out around the impact site and is either composed of silicate rock material (called *chondrites*) or iron and nickel (called *irons*).

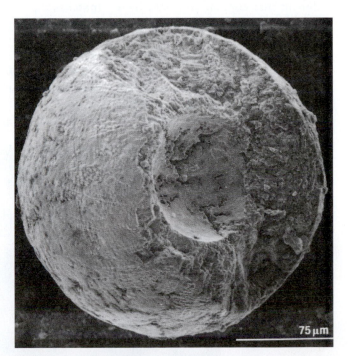

Figure 5–20 Microscopic cosmogenous spherule.

Scanning electron micrograph of an iron-rich spherule of cosmic dust. Bar scale of 75 micrometers is equal to 75 millionths of a meter.

BOX 5–3   Research Methods in Oceanography

WHEN THE DINOSAURS DIED: THE CRETACEOUS–TERTIARY (K–T) EVENT

The extinction of the dinosaurs—and two-thirds of all plant and animal species on Earth (including many marine species)—occurred 65 million years ago. This extinction marks the boundary between the Cretaceous (K) and Tertiary (T) Periods of geologic time and is known as the "**K–T event**." Did slow climate change lead to the extinction of these organisms, or was it a catastrophic event? Was their demise related to disease, diet, predation, or volcanic activity? Earth scientists have long sought clues to this mystery.

In 1980, geologist Walter Alvarez, his father, Nobel physics laureate Luis Alvarez, and two nuclear chemists, Frank Asaro and Helen Michel, reported that deposits collected in northern Italy from the K–T boundary contained a clay layer with high proportions of the metallic element iridium (Ir). Iridium is rare in rocks from Earth but occurs in greater concentration in meteorites. Therefore, layers of sediment that contain unusually high concentrations of iridium suggest that the material may be of extraterrestrial origin. Additionally, the clay layer contained shocked quartz grains, which indicate that an event occurred with enough force to fracture and partially melt pieces of

quartz. Other deposits from the K–T boundary revealed similar features, supporting the idea that Earth experienced an extraterrestrial impact at the same time that the dinosaurs died.

One problem with the impact hypothesis, however, was that volcanic eruptions on Earth could create similar clay deposits enriched in iridium and containing shocked quartz. In fact, large outpourings of basaltic volcanic rock in India (called the Deccan Traps) and other locations had occurred at about the same time as the dinosaur extinction. Also, if there was a catastrophic meteor impact, where was the crater?

In the early 1990s the *Chicxulub* (pronounced "SCHICK-sue-lube") *Crater* off the Yucatán coast in the Gulf of Mexico was identified as a likely candidate because of its structure, age, and size. Its structure is comparable to other impact craters in the solar system, and its age matches the K–T event. At 190 kilometers (120 miles) in diameter, it is the largest impact crater on Earth. To create a crater this large, a 10-kilometer (6-mile)-wide meteoroid composed of rock and/or ice traveling at speeds up to 72,000 kilometers (45,000 miles) per hour must have

slammed into Earth (Figure 5D). The impact probably bared the sea floor in the area and created huge waves—estimated to be up to 914 meters (3000 feet) high—that traveled throughout the oceans. This impact is thought to have kicked up so much dust that it blocked sunlight, chilled Earth's surface, and brought about the extinction of dinosaurs and other species. In addition, acid rains and global fires may have added to the environmental disaster.

Supporting evidence for the meteor impact hypothesis was provided in 1997 by the Ocean Drilling Program (ODP). Previous drilling close to the impact site did not reveal any K–T deposits. Evidently, the impact and resulting huge waves had stripped the ocean floor of its sediment. However, at 1600 kilometers (1000 miles) from the impact site, some of the telltale sediments were preserved on the sea floor. Drilling into the continental margin off Florida into an underwater peninsula called the Blake Nose, the ODP scientific party recovered cores from the K–T boundary that contain a complete record of the impact (Figure 5E).

The cores reveal that before the impact, the layers of Cretaceous age

Figure 5D The K–T meteorite impact event.

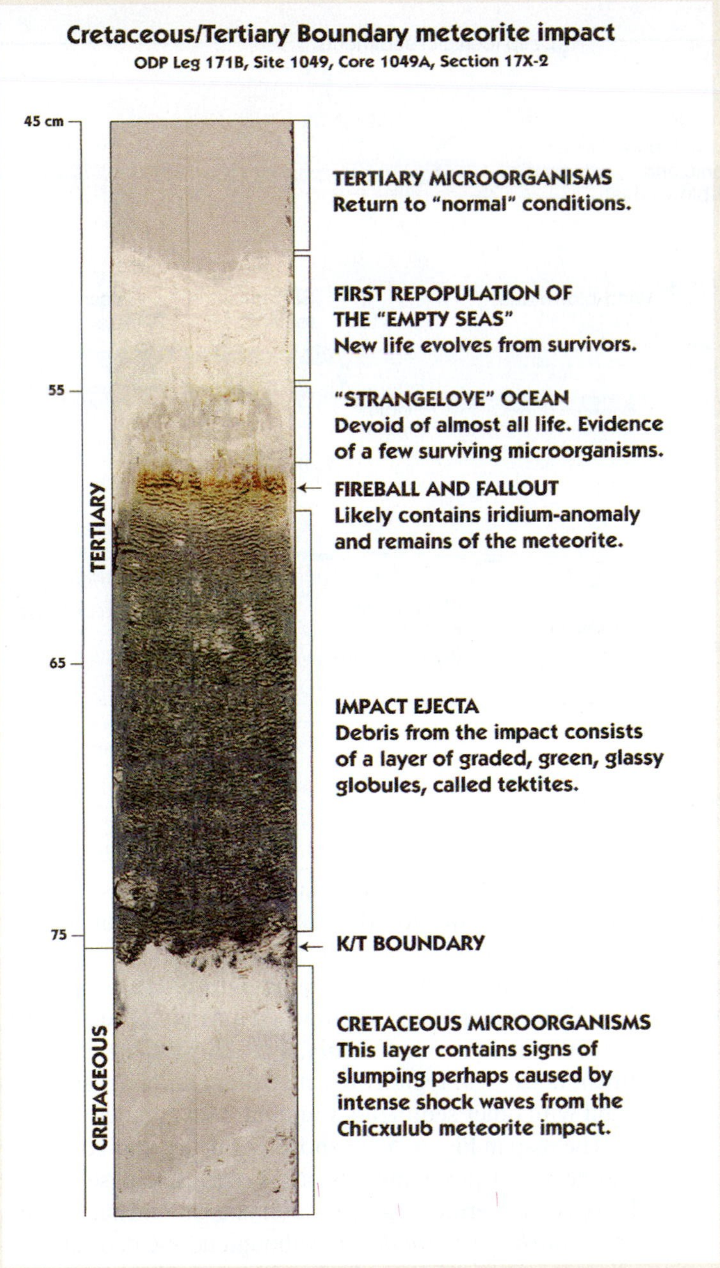

Cretaceous/Tertiary Boundary meteorite impact
ODP Leg 171B, Site 1049, Core 1049A, Section 17X-2

TERTIARY MICROORGANISMS
Return to "normal" conditions.

FIRST REPOPULATION OF THE "EMPTY SEAS"
New life evolves from survivors.

"STRANGELOVE" OCEAN
Devoid of almost all life. Evidence of a few surviving microorganisms.

← **FIREBALL AND FALLOUT**
Likely contains iridium-anomaly and remains of the meteorite.

IMPACT EJECTA
Debris from the impact consists of a layer of graded, green, glassy globules, called tektites.

← **K/T BOUNDARY**

CRETACEOUS MICROORGANISMS
This layer contains signs of slumping perhaps caused by intense shock waves from the Chicxulub meteorite impact.

Figure 5E K–T boundary meteorite impact core.

sediment are filled with abundant fossils of calcareous coccoliths and foraminifers and show signs of underwater landslide activity—perhaps the effect of an impact-triggered earthquake. Above this calcareous ooze is a 20-centimeter (8-inch) thick layer of rubble containing evidence of an impact: spherules, tektites, shocked quartz from hard-hit terrestrial rock—even a 2-centimeter (1-inch) piece of reef rock from the Yucatán peninsula! This layer is also rich in iridium, just like other K–T boundary sequences. Atop this layer is a thick gray clay deposit containing meteor debris and severely reduced numbers of coccoliths and foraminifers. Life in the ocean apparently recovered slowly, taking at least 5000 years before sediment teeming with new, Tertiary-age microorganisms began to be deposited.

Convincing evidence of the K–T impact from this and other cores along with the observation of Comet Shoemaker-Levy's 1994 spectacular collision with Jupiter suggests that Earth has experienced many such extraterrestrial impacts over geologic time. In fact, nearly 200 impact craters have been identified on Earth so far. Statistics show that an impact the size of the K–T event should occur on Earth about once every 100 million years. Each impact would severely affect life on Earth as it did for the dinosaurs. Nevertheless, their extinction made it possible for mammals to eventually rise to the position of dominance they hold on Earth today.

> Cosmogenous sediment is produced from materials originating in outer space and includes microscopic space dust and macroscopic meteor debris.

Mixtures

Lithogenous and biogenous sediment rarely occur as an absolutely pure deposit that does not contain other types of sediment. For instance,

- Most calcareous oozes contain some siliceous material, and vice versa (see, for example, Figure 5–11d).
- The abundance of clay-sized lithogenous particles throughout the world and the ease with which they are transported by winds and currents means that these particles are incorporated into every sediment type.
- The composition of biogenous ooze includes up to 70% fine-grained lithogenous clays.
- Most lithogenous sediment contains small percentages of biogenous particles.
- Other types of sediment can be incorporated into hydrogenous sediment.
- For many types of hydrogenous sediment, evidence suggests they may be formed with the aid of certain marine organisms instead of by purely chemical processes.
- Tiny amounts of cosmogenous sediment are mixed in with all other sediment types.

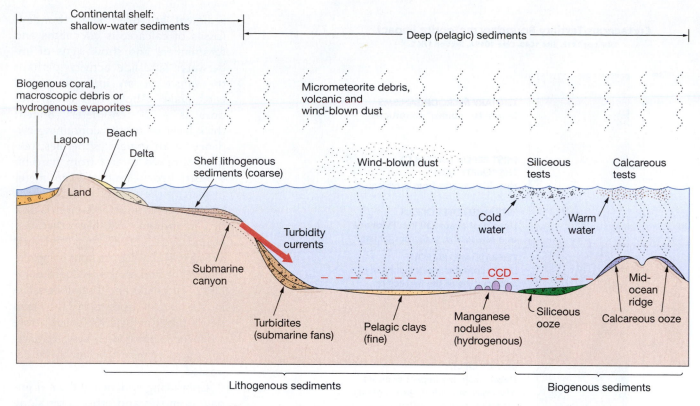

Figure 5–21 **Distribution of sediment across a passive continental margin.**

Each deposit is a mixture of different sediment types. Figure 5–21 shows the distribution of sediment across a passive continental margin and illustrates how mixtures can occur. Typically, however, one type of sediment dominates, which allows the deposit to be classified as primarily lithogenous, biogenous, hydrogenous, or cosmogenous.

Distribution of Neritic and Pelagic Deposits: A Summary

Neritic (nearshore) deposits cover about one-quarter of the ocean floor while pelagic (deep-ocean basin) deposits cover the other three-quarters.

Neritic Deposits

Neritic deposits are composed of materials that are strongly influenced by latitude. Figure 5–22 shows the distribution of neritic sediment types at various latitudes. The figure indicates that silt and clay are most abundant in equatorial and tropical latitudes, accounting for about 50% of the total sediment at the Equator, about 20% in the tropics, and about 15% at higher latitudes. Sand particles are most abundant at mid-latitudes, accounting for about 60% of the sediment, decreasing to about 30% at the Equator and to about 35% at high latitudes. Coarse, poorly sorted deposits (rock and gravel) are found primarily at high latitudes where they are deposited by glaciers and icebergs. Remarkably, about 6 to 7% of neritic sediment at all latitudes is composed of shell fragments.

Figure 5–22 also shows that coral reef debris is significant only at low latitudes, which is a result of the fact that corals cannot survive in the cooler waters of higher latitudes. Although some tropical locations near reefs contain sediment that is composed entirely of reef debris, coral debris accounts for only about 20% of the total sediment in tropical regions because these amounts are averaged with other sediments found at low latitudes.

The map in Figure 5–23 shows the distribution of neritic and pelagic deposits in the world's oceans. Coarse-grained lithogenous neritic deposits dominate continental margin areas (*dark brown shading*). Although neritic deposits usually contain biogenous, hydrogenous, and cosmogenous particles, these constitute only a minor percentage of the total sediment mass.

Pelagic Deposits

Figure 5–23 shows that pelagic deposits are dominated by biogenous calcareous oozes (*blue shading*), which are found on the relatively shallow deep-ocean areas along the mid-ocean ridge. Biogenous siliceous oozes are found beneath areas of unusually high biologic productivity such as the North Pacific, Antarctic (*light green shading*, where diatomaceous ooze occurs), and the equatorial Pacific (*dark green shading*, where radiolarian ooze occurs). Fine lithogenous pelagic deposits of abyssal clays (*light brown shading*) are common in deeper areas of the ocean basins. Hydrogenous and cosmogenous sediment comprise only a small proportion of pelagic deposits in the ocean.

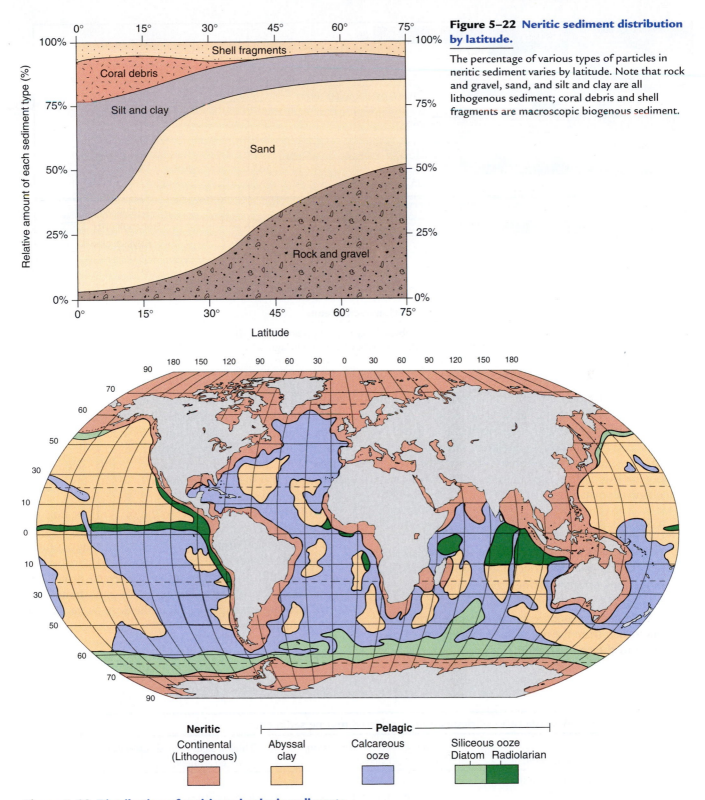

Figure 5–22 Neritic sediment distribution by latitude.

The percentage of various types of particles in neritic sediment varies by latitude. Note that rock and gravel, sand, and silt and clay are all lithogenous sediment; coral debris and shell fragments are macroscopic biogenous sediment.

Neritic
Continental (Lithogenous)

Pelagic
Abyssal clay

Calcareous ooze

Siliceous ooze
Diatom Radiolarian

Figure 5–23 Distribution of neritic and pelagic sediments.

The bar graph in Figure 5–24 shows the proportion of each ocean floor that is covered by pelagic calcareous ooze, siliceous ooze, or abyssal clay. Calcareous oozes predominate, covering almost 48% of the world's deep-ocean floor. Abyssal clay covers 38% and siliceous oozes 14% of the world ocean floor area. The graph also shows that the amount of ocean basin floor covered by calcareous ooze decreases in deeper basins because they generally lie beneath the CCD. The dominant oceanic sediment in the deepest basin—the Pacific—is abyssal clay (see also Figure 5–23). Calcareous ooze is the most widely deposited sediment in the shallower Atlantic and Indian

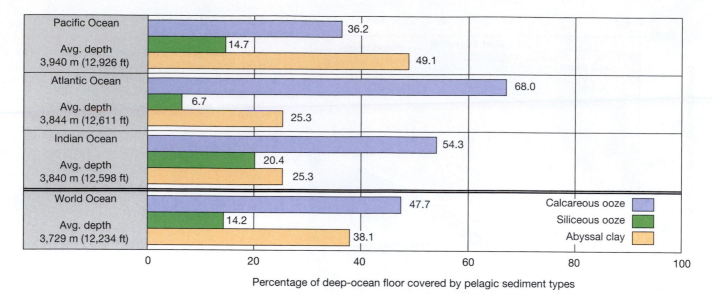

Figure 5–24 Percentage of pelagic sediment types within each ocean.

Distribution of pelagic calcareous ooze, siliceous ooze, and abyssal clay in each ocean, and for the world ocean. The average depths shown exclude shallow adjacent seas, where little pelagic sediment accumulates.

Oceans. Siliceous oozes cover a smaller percentage of the ocean bottom in all the oceans because regions of high productivity of siliceous-secreting organisms are generally restricted to the equatorial region (for radiolarians) and high latitudes (for diatoms). Table 5–4 shows the average rates of deposition of selected marine sediments in neritic and pelagic deposits.

> Neritic deposits occur close to shore and are dominated by coarse lithogenous material. Pelagic deposits occur in the deep ocean and are dominated by biogenous oozes and fine lithogenous clay.

Microscopic biogenous tests should take from 10 to 50 years to sink from the ocean surface where the organisms lived to the abyssal depths where biogenous ooze

accumulates. During this time, even a sluggish horizontal ocean current of only 0.05 kilometer (0.03 mile) per hour could carry tests as much as 22,000 kilometers (13,700 miles) before they settled onto the deep-ocean floor. Why, then, do biogenous tests on the deep-ocean floor closely reflect the population of organisms living in the surface water directly above? Remarkably, about 99% of the particles that fall to the ocean floor do so as part of *fecal pellets*, which are produced by tiny animals that eat algae and protozoans living in the water column, digest their tissues, and excrete their hard parts. These pellets are full of the remains of algae and protozoans from the surface waters (Figure 5–25) and, though still small, are large enough to sink to the deep ocean floor in only 10 to 15 days.

TABLE 5–4 **Average rates of deposition of selected marine sediments.**

Type of sediment/deposit	Average rate of deposition (per 1000 years)	Thickness of deposit after 1000 years equivalent to ...
Coarse lithogenous sediment, neritic deposit	1 meter (3.3 feet)	A meter stick
Biogenous ooze, pelagic deposit	1 centimeter (0.4 inch)	The diameter of a dime
Abyssal clay, pelagic deposit	1 millimeter (0.04 inch)	The thickness of a dime
Manganese nodule, pelagic deposit	0.001 millimeter (0.00004 inch)	A microscopic dust particle

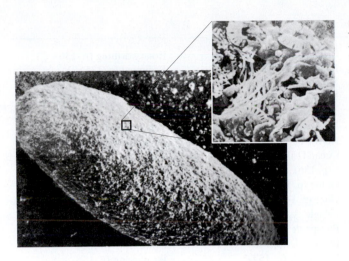

Figure 5–25 **Fecal pellet.**
A 200-micrometer (0.008 inch)-long fecal pellet, which is large enough to sink rapidly from the surface to the ocean floor. Close-up of the surface of a fecal pellet (*inset*) shows the remains of coccoliths and other debris.

Chapter in Review

- The existence of *sea floor spreading was confirmed when the* Glomar Challenger *began the Deep Sea Drilling Project* to sample ocean sediments and the underlying crust, which was continued by the *Ocean Drilling Program's* JOIDES Resolution. Today, the *Integrated Ocean Drilling Program* continues the important work of retrieving sediments from the ocean floor. Analysis and interpretation of marine sediments reveal that *Earth has had an interesting and complex history* including mass extinctions, global climate change, and movement of plates.

- Sediments that accumulate on the ocean floor are *classified by origin as lithogenous* (derived from rock), *biogenous* (derived from organisms), *hydrogenous* (derived from water), or *cosmogenous* (derived from outer space).

- *Lithogenous sediments reflect the composition of the rock from which they were derived.* Sediment *texture*—determined in part by the size, sorting, and rounding of particles—is affected greatly by how the particles were transported (by water, wind, ice, or gravity) and the energy conditions under which they were deposited. Coarse lithogenous material dominates neritic deposits that accumulate rapidly along the margins of continents while fine abyssal clays are found in pelagic deposits.

- *Biogenous sediment consists of the hard remains* (shells, bones, and teeth) *of organisms.* These are composed of either *silica* (SiO_2) from diatoms and radiolarians or *calcium carbonate* ($CaCO_3$) from foraminifers and coccolithophores. *Accumulations of microscopic shells* (tests) of organisms must comprise at least 30% of the deposit for it to be classified as *biogenic ooze.* Biogenous oozes are the most common type of pelagic deposits. The rate of biological productivity relative to the rates of destruction and dilution of biogenous sediment determines whether abyssal clay or oozes will form on the ocean floor. *Siliceous ooze* will only form below areas of high biologic productivity of silica-secreting organisms at the surface. *Calcareous ooze* will only form above the *calcite compensation depth (CCD)*—the depth where seawater dissolves calcium carbonate—although it can be covered and transported into deeper water through sea floor spreading.

- *Hydrogenous sediment* includes manganese nodules, phosphates, carbonates, metal sulfides, and evaporites that *precipitate directly from water* or are formed by the interaction of substances dissolved in water with materials on the ocean floor. Hydrogenous sediments represent a relatively small proportion of marine sediment and are distributed in many diverse environments.

- *Cosmogenous sediment is composed of either macroscopic meteor debris* (such as that produced during the K–T impact event) *or microscopic iron-nickel and silicate spherules* that result from asteroid collisions or extraterrestrial impacts. Minute amounts of cosmogenous sediment are mixed into most other types of ocean sediment.

- Although *most ocean sediment is a mixture of various sediment types*, it is usually dominated by lithogenous, biogenous, hydrogenous, or cosmogenous material.

- *The distribution of neritic and pelagic sediment is influenced by many factors*, including proximity to sources of lithogenous sediment, productivity of microscopic marine organisms, depth of the ocean floor, and the distribution of various sea floor features. *Fecal pellets* rapidly transport biogenous particles to the deep-ocean floor and cause the composition of sea floor deposits to match the organisms living in surface waters immediately above them.

Key Terms

Abyssal clay (p. 144)
Algae (p. 145)
Aragonite (p. 154)
Biogenous sediment (p. 144)
Calcareous ooze (p. 145)
Calcite (p. 145)
Calcite compensation depth (CCD) (p. 151)
Calcium carbonate (p. 145)
Carbonate (p. 146)
Chalk (p. 145)
Coccolith (p. 145)
Coccolithophore (p. 145)
Core (p. 136)
Cosmogenous sediment (p. 155)
Deep Sea Drilling Project (DSDP) (p. 136)
Diatom (p. 145)
Diatomaceous earth (p. 145)

Eroded (p. 136)
Evaporite mineral (p. 154)
Foraminifer (p. 145)
Glacial deposit (p. 142)
Glomar Challenger (p. 136)
Grain size (p. 138)
Hydrogenous sediment (p. 153)
Ice rafting (p. 142)
Integrated Ocean Drilling Program (IODP) (p. 136)
JOIDES Resolution (p. 136)
K–T event (p. 156)
Limestone (p. 149)
Lithogenous sediment (p. 135)
Lysocline (p. 151)
Macroscopic biogenous sediment (p. 144)
Manganese nodule (p. 153)
Maturity (p. 140)

Metal sulfide (p. 154)
Meteor (p. 155)
Meteorite (p. 155)
Microscopic biogenous sediment (p. 144)
Nannoplankton (p. 145)
Neritic deposit (p. 140)
Ocean Drilling Program (ODP) (p. 136)
Oolite (p. 154)
Ooze (p. 145)
Pelagic deposit (p. 140)
Phosphate (p. 154)
Planktonic (p. 145)
Precipitate (p. 153)
Protozoan (p. 145)
Quartz (p. 136)
Radiolarian (p. 145)
Red clay (p. 144)

Rotary drilling (p. 136)
Sediment (p. 135)
Silica (p. 145)
Siliceous ooze (p. 145)
Sorting (p. 139)
Spherule (p. 155)
Stromatolite (p. 149)
Tektite (p. 155)
Terrigenous sediment (p. 135)
Test (p. 145)
Texture (p. 135)
Turbidite deposit (p. 142)
Turbidity current (p. 140)
Upwelling (p. 153)
Weathering (p. 136)
Wentworth scale of grain size (p. 138)

Questions and Exercises

1. Describe the process of how a drilling ship like the *JOIDES Resolution* obtains core samples from the deep-ocean floor.

2. What kind of information can be obtained by examining and analyzing core samples?

3. List and describe the characteristics of the four basic types of marine sediment.

4. How does lithogenous sediment originate?

5. Why is most lithogenous sediment composed of quartz grains? What is the chemical composition of quartz?

6. If a deposit has a coarse grain size, what does this indicate about the energy of the transporting medium? Give several examples of various transporting media that would produce such a deposit.

7. What characteristics of marine sediment indicate increasing maturity? Give an example of a mature and immature sediment.

8. List the two major chemical compounds of which most biogenous sediment is composed and the organisms that produce them. Sketch these organisms.

9. What are several reasons why diatoms are so remarkable? List products that contain or are produced using diatomaceous earth.

10. If siliceous ooze is slowly but constantly dissolving in seawater, how can deposits of siliceous ooze accumulate on the ocean floor?

11. How do oozes differ from abyssal clay? Discuss how productivity, destruction, and dilution combine to determine whether an ooze or abyssal clay will form on the deep-ocean floor.

12. Describe the environmental conditions (e.g., surface water temperature, productivity, dissolution, etc.) that influence the distribution of siliceous and calcareous ooze.

13. Explain the stages of progression that result in calcareous ooze existing below the CCD.

14. Describe manganese nodules, including what is currently known about how they form.

15. Describe the most common types of cosmogenous sediment and give the probable source of these particles.

16. Describe the K–T event, including evidence for it and its effect on the environment.

17. Why is lithogenous sediment the most common neritic deposit? Why are biogenous oozes the most common pelagic deposits?

18. How do fecal pellets help explain why the particles found in the ocean surface waters are closely reflected in the particle composition of the sediment directly beneath? Why would one not expect this?

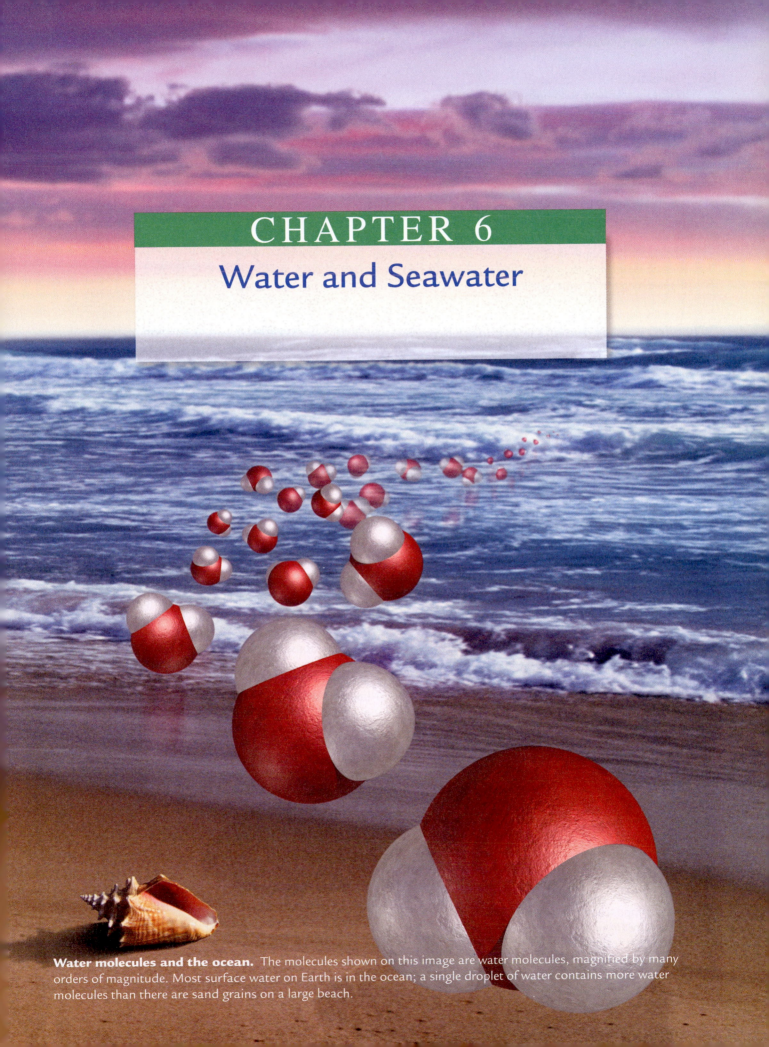

CHAPTER 6
Water and Seawater

Water molecules and the ocean. The molecules shown on this image are water molecules, magnified by many orders of magnitude. Most surface water on Earth is in the ocean; a single droplet of water contains more water molecules than there are sand grains on a large beach.

Key Questions

■ Why does water have such unusual chemical properties?

■ How have water's thermal properties been important for all life on Earth?

■ How salty is the ocean?

■ What is the pH of seawater and how does ocean buffering work?

■ What processes affect seawater salinity?

■ What factors affect seawater density?

■ What are the halocline, pycnocline, and thermocline?

The answers to these questions (and much more) can be found in the highlighted concept statements within this chapter.

"Chemistry . . . is one of the broadest branches of science, if for no other reason that, when we think about it, everything is chemistry."

—Luciano Caglioti, *The Two Faces of Chemistry* (1985)

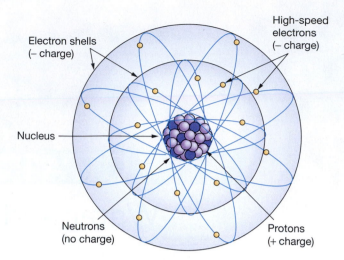

Figure 6–1 Simplified model of an atom.

An atom consists of a central nucleus composed of protons and neutrons that is encircled by electrons.

W hy are temperature extremes found at places far from the ocean, while those areas close to the ocean are moderated? The mild climates found in coastal regions are made possible by the unique thermal properties of water. These and other properties of water, which stem from the arrangement of its atoms and how its molecules stick together, give water the ability to store vast quantities of heat, to stick together, and to dissolve almost everything. Water is so common we often take it for granted, yet it is one of the most remarkable substances on Earth.

The chemical properties of water are also essential for sustaining all forms of life. In fact, the primary component of all living organisms is water. The water content of organisms, for instance, ranges from about 65% (humans) to 95% (most plants). Water is the ideal medium to have within our bodies because it facilitates chemical reactions. Our blood—so important for transporting nutrients and removing wastes within our bodies—is 83% water. The very presence of water on our planet makes life possible, and the unusual properties of water make our planet livable. To understand these unusual properties of water, let's examine water's chemical structure.

Atomic Structure

Atoms (*a* = not, *tomos* = cut) are the basic building blocks of all matter. Every physical substance in our world—chairs, tables, books, people, the air we breathe—is composed of atoms. An atom resembles a microscopic sphere (Figure 6–1) and was originally thought to be the smallest form of matter. Additional study has revealed that atoms are composed of even smaller particles, called subatomic particles.[1] As shown in Figure 6–1, the **nucleus** (*nucleos* = a little nut) of an atom is composed

of **protons** (*protos* = first) and **neutrons** (*neutr* = neutral), which are bound together by strong forces. Protons have a positive electrical charge, whereas neutrons have no electrical charge. The masses of protons and neutrons are about the same, but both are extremely small. Surrounding the nucleus are particles called **electrons** (*electro* = electricity), which have about $\frac{1}{2000}$ the mass of either protons or neutrons. Electrical attraction between positively charged protons and negatively charged electrons holds electrons in layers or shells around the nucleus.

The overall electrical charge of atoms is balanced because each atom contains an equal number of protons and electrons. An oxygen atom, for example, has eight protons and eight electrons. Most oxygen atoms also have eight neutrons, which do not affect the overall electrical charge because neutrons are electrically neutral. The number of protons is what distinguishes atoms of the 115 known chemical elements from one another. For example, an oxygen atom (and only an oxygen atom) has eight protons. Similarly, a hydrogen atom (and only a hydrogen atom) has one proton, a helium atom has two protons, and so on. In some cases, an atom will lose or gain one or more electrons and thus have an overall electrical charge. These atoms are called **ions** (*ienai* = to go).

The Water Molecule

A **molecule** (*molecula* = a mass) is a group of two or more atoms held together by mutually shared electrons. It is the smallest form of a substance that can exist yet still retain the original properties of that substance. When atoms combine with other atoms to form molecules, they share or trade electrons and establish chemical bonds. For instance,

---

[1]It has recently been discovered that subatomic particles themselves are composed of even smaller particles, called quarks.

the chemical formula for water—H_2O—indicates that a water molecule is composed of two hydrogen atoms chemically bonded to one oxygen atom.

Geometry

Atoms can be represented as spheres of various sizes, with the more electrons the atom contains, the larger the sphere. It turns out that an oxygen atom (with eight electrons) is about twice the size of a hydrogen atom (with one electron). A water molecule consists of a central oxygen atom covalently bonded to the two hydrogen atoms, which are separated by an angle of about 105 degrees (Figure 6–2a). The **covalent** (*co* = with, *valere* = to be strong) **bonds** in water are due to the sharing of electrons between oxygen and each hydrogen atom. They are relatively strong chemical bonds, so a lot of energy is needed to break them. Figure 6–2b shows a water molecule in a more compact representation, and in Figure 6–2c, letter symbols are used to represent the atoms in water (*O* for oxygen, *H* for hydrogen). Instead of all atoms being in a straight line, *both hydrogen atoms are on the same side of the oxygen atom.* This curious bend in the geometry of the water molecule is the underlying cause of most of the unique and unusual properties of water.

Polarity

The bent geometry of the water molecule gives a slight overall negative charge to the side of the oxygen atom and a slight overall positive charge to the side of the hydrogen atoms (Figure 6–2a). This slight separation of charges gives the entire molecule an electrical **polarity** (*polus* = pole, *ity* = having the quality of), so water molecules are **dipolar** (*di* = two, *polus* = pole). Other common dipolar objects are flashlight batteries, car batteries, and bar magnets. Although the electrical charges are weak, water molecules behave as if they contain a tiny bar magnet.

Interconnections of Molecules

If you've ever experimented with bar magnets, you know they have polarity and orient themselves relative to one another such that the positive end of one bar magnet is attracted to the negative end of another. Water molecules have polarity, too, so they orient themselves relative to one another. In water, the positively charged hydrogen area of one water molecule interacts with the negatively charged oxygen end of an adjacent water molecule, forming a **hydrogen bond** (Figure 6–3). The hydrogen bonds between water molecules are much weaker than the covalent bonds that hold individual water molecules together. In essence, weaker hydrogen bonds form *between* adjacent water molecules and stronger covalent bonds occur *within* water molecules.

Even though hydrogen bonds are weaker than covalent bonds, they are strong enough to cause water molecules to stick to one another and exhibit **cohesion** (*cohaesus* = to cling together). The cohesive properties of water cause it to "bead up" on a waxed surface, such as a freshly waxed car.

Cohesion also gives water its **surface tension** (Figure 6–4a). Water's surface has a thin "skin" that allows a glass to be filled just above the brim without spilling any

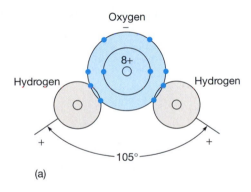

(a)

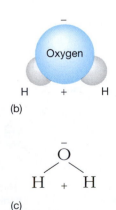

(b)

(c)

Figure 6–2 The water molecule.

(a) Geometry of a water molecule. The oxygen end of the molecule is negatively charged, and the hydrogen regions exhibit a positive charge. Covalent bonds occur between the oxygen and the two hydrogen atoms. **(b)** A three-dimensional representation of the water molecule. **(c)** The water molecule represented by letters (*H* = hydrogen, *O* = oxygen).

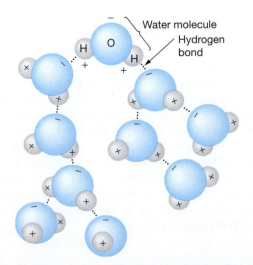

Figure 6–3 Hydrogen bonding in water.

Dashed lines indicate locations of hydrogen bonds, which occur between water molecules.

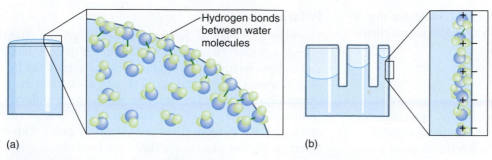

(a)   (b)

Figure 6–4 Surface tension and capillarity.
(a) The high surface tension of water causes it to pile up above the top of a container and is created by hydrogen bonds holding water molecules together against the force of gravity (*inset*). **(b)** Capillarity causes water to climb up the side of a container because of the attraction between the positive charge of water molecules and the negatively charged surface of the glass (*inset*).

of the water. Surface tension results from the formation of hydrogen bonds between the outermost layer of water molecules and the underlying molecules, allowing the water to pile up above a container's rim (Figure 6–4a). In fact, water's ability to form hydrogen bonds causes it to have the highest surface tension of any liquid (except for the element mercury, which is the only metal that is a liquid at normal surface temperatures).

Many organisms take advantage of the high surface tension of water. Insects such as the water strider can often be seen walking across the surface of a quiet pool of water as though it were solid. Other creatures are adapted to hang down in batlike fashion into the water from the underside of the surface, counteracting gravitational forces that pull them downward. It is also possible to float small or thin objects that are much denser than water on water surfaces. For example, a razor blade, a metal paper clip, or a steel needle—all of which are about five times denser than water—will float if laid carefully on a quiet water surface.

Water also clings to the surfaces of many substances, which is referred to as *adhesion* or *wetting*. For example, when water is poured into a glass container, it strongly adheres to the sides, and the force of attraction between the water molecules and the glass molecules even cause the water to climb a little way up the sides (Figure 6–4b). On a molecular scale, the positively charged portions of the water molecules (the hydrogen ends) are attracted to the negatively charged electrons in the oxygen atoms of the glass (Figure 6–4b, *inset*). In very thin tubes, the area of contact between the water and the glass is maximized, and the water's free surface (where hydrogen bonding forces are dominant) is minimized. Such thin tubes are called *capillary tubes*, and this property of liquids is called **capillarity** (*capillaris* = hair). Many organisms make use of the capillary properties of water; for example, plants use capillarity to raise water to the tops of plants and even tall trees.

Water: The Universal Solvent

Water molecules stick not only to other water molecules, but also to other polar chemical compounds. In doing so,

water molecules can reduce the attraction between ions of opposite charges by as much as 80 times. For instance, ordinary table salt—sodium[2] chloride, NaCl—consists of an alternating array of positively charged sodium ions and negatively charged chloride ions (Figure 6–5a). The **electrostatic** (*electro* = electricity, *stasis* = standing) **attraction** between oppositely charged ions produces an **ionic** (*ienai* = to go) **bond**. When solid NaCl is placed in water, the electrostatic attraction (ionic bonding) between the sodium and chloride ions is reduced by 80 times. This, in turn, makes it much easier for the

---

[2]Sodium is represented by the letters Na because the Latin term for sodium is *natrium*.

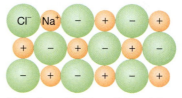

(a) Sodium chloride, solid crystal structure

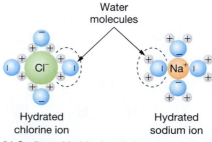

Water molecules

Hydrated chlorine ion   Hydrated sodium ion

(b) Sodium chloride, in solution

Figure 6–5 Water as a solvent.
(a) Table salt, composed of sodium chloride (Na^+ = sodium ion, Cl^- = chlorine ion). **(b)** As sodium chloride is dissolved, the positively charged ends of water molecules are attracted to the negatively charged Cl^- ion, while the negatively charged ends are attracted to the positively charged Na^+ ion.

sodium ions and chloride ions to separate. When the ions separate, the positively charged sodium ions become attracted to the negative ends of the water molecules, the negatively charged chloride ions become attracted to the positive ends of the water molecules (Figure 6–5b), and the salt is dissolved in water. The process by which water molecules completely surround ions is called *hydration* (*hydra* = water, *ation* = action or process).

Because water molecules interact with other water molecules and other polar molecules, water is able to dissolve nearly everything.[3] Given enough time, water can dissolve more substances and in greater quantity than any other known substance. This is why water is called "the universal solvent." It is also why the ocean contains so much dissolved material—an estimated 50 quadrillion tons (50 million billion tons) of salt—which makes seawater taste "salty."

A water molecule has a bend in its geometry, with the two hydrogen atoms on the same side of the oxygen atom, which gives water its polarity and ability to form hydrogen bonds.

Water's Thermal Properties

Water exists on Earth as a solid, liquid, and a gas and has the capacity to store and release great amounts of heat. Water's thermal properties influence the world's heat budget, moderate coastal temperatures, and are in part responsible for the development of tropical cyclones, worldwide wind belts, and ocean surface currents.

Heat, Temperature, and Changes of State

Matter can exist in three states: solid, liquid, and gas.[4] What must happen to change the state of a compound? The attractive forces between molecules or ions in the substance must be overcome if the state of the substance is to be changed from solid to liquid or from liquid to gas. These attractive forces include hydrogen bonds and van der Waals forces. The **van der Waals forces**—named for Dutch physicist Johannes Diderik van der Waals (1837–1923)—are relatively weak interactions that become significant only when molecules are very close together, as in the solid and liquid states (but not the gaseous state). Energy must be added to the molecules or ions so that they can move fast enough to overcome these attractions.

What form of energy changes the state of matter? Very simply, adding or removing heat is what causes a substance to change its state of matter. For instance, adding heat to ice cubes causes them to melt and removing heat from water causes ice to form. Before proceeding, let's clarify the difference between heat and temperature:

- **Heat** is the *energy of moving molecules*. It is proportional to the energy level of molecules, and thus is the total **kinetic** (*kinetos* = moving) **energy** of a substance. Water is a solid, liquid, or gas depending on the amount of heat added. Heat may be generated by combustion (a chemical reaction commonly called "burning"), through other chemical reactions, by friction, or from radioactivity. A **calorie** (*calor* = heat) is the amount of heat required to raise the temperature of 1 gram of water[5] by 1 degree centigrade. The calories used to measure heat are 1000 times smaller than the familiar calories used to measure the energy content of food.

- **Temperature** is the *direct measure of the average kinetic energy of the molecules that make up a substance*. The greater the temperature, the greater the kinetic energy of the substance. Temperature changes when heat energy is added to or removed from a substance. Temperature is usually measured in degrees centigrade (°C) or degrees Fahrenheit (°F).

Figure 6–6 shows water molecules in the solid, liquid, and gaseous states. In the *solid state* (ice), water has a rigid structure and does not flow. Intermolecular bonds are constantly being broken and reformed, but the molecules remain firmly attached. That is, the molecules vibrate with energy but remain in relatively fixed positions. As a result, solids do not conform to the shape of their container.

In the *liquid state* (water), water molecules still interact with each other, but they have enough kinetic energy to flow past each other and take the shape of their container. Intermolecular bonds are being formed and broken at a much greater rate than in the solid state.

In the *gaseous state* (water **vapor**), water molecules no longer interact with one another except during random collisions. Water vapor molecules flow very freely, filling the volume of whatever container they are placed in.

Water's Freezing and Boiling Points

If enough heat energy is added to a solid, it melts to a liquid. The temperature at which melting occurs is the substance's **melting point**. If enough heat energy is removed from a liquid, it freezes to a solid. The temperature at which freezing occurs is the substance's **freezing point**, which is the same temperature as the melting point (Figure 6–6). For pure water, melting and freezing occur at 0°C (32°F).[6]

[3]If water is such a good solvent, why doesn't oil dissolve in water? As you might have guessed, the chemical structure of oil is remarkably nonpolar. With no positive or negative ends to attract the polar water molecule, oil will not dissolve in water.

[4]*Plasma*, an electrically neutral, highly ionized gas composed of ions, electrons, and neutral particles, is a phase of matter distinct from solids, liquids, and normal gases.

[5]One gram of water is equal to about 10 drops.

[6]All melting/freezing/boiling points discussed in this chapter assume a standard sea level pressure of 1 atmosphere (14.7 pounds per square inch).

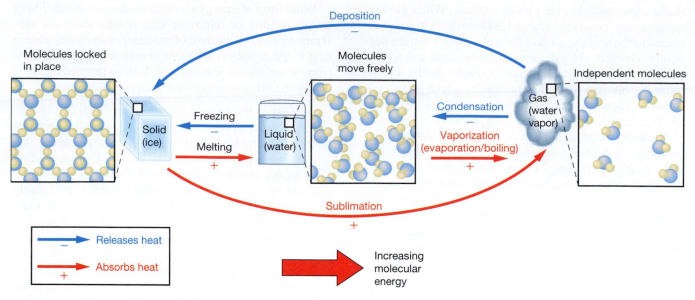

Figure 6–6 Water in the three states of matter.

Blue arrows (−) indicate heat released by water (which warms the environment) as it changes state; red arrows (+) indicate heat absorbed by water (which cools the environment).

If enough heat energy is added to a liquid, it converts to a gas. The temperature at which boiling occurs is the substance's **boiling point**. If enough heat energy is removed from a gas, it **condenses** to a liquid. The temperature at which condensation occurs is the substance's **condensation point**, which is at the same as the boiling point (Figure 6–6). For water, boiling and condensation occur at 100°C (212°F).

Based on the pattern of similar chemical compounds, a molecule as small and lightweight as water should melt at −90°C (−130°F) and boil at −68°C (−90°F). If that were the case, all water on Earth would be in the gaseous state. Instead, water melts and boils at the relatively high temperatures of 0°C (32°F) and 100°C (212°F),[7] respectively, because additional heat energy is required to overcome its hydrogen bonds and van der Waals forces. Thus, if not for the unusual geometry and resulting polarity of the water molecule, life as we know it would not exist on Earth.

Water's Heat Capacity

Heat capacity is *the amount of heat required to raise the temperature of 1 gram of any substance by 1 degree centigrade.*[8] A substance with a high heat capacity can absorb (or lose) large quantities of heat with only a small change in temperature. A substance that changes temperature rapidly when heat is applied has a low heat capacity (for instance, oil and all metals have low heat capacity). Water has a high

heat capacity that is exactly 1 calorie per gram, whereas most other substances have a much lower heat capacity.

Water has the highest heat capacity of all common substances. It takes more energy to increase the kinetic energy of hydrogen-bonded water molecules than it does for substances in which the dominant intermolecular interaction is the much weaker van der Waals force. As a result, water gains or loses much more heat than other common substances while undergoing an equal temperature change. Additionally, water resists any change in temperature, as you may have observed when heating a large pot of water. When heat is applied to the pot, which is made of metal and has a low heat capacity, the pot heats up quickly. The water *inside* the pot, however, takes a long time to heat up (hence, the tale that a watched pot never boils but an unwatched pot boils over!). Making the water boil takes even more heat because all the hydrogen bonds must be broken. The exceptional capacity of water to absorb large quantities of heat helps explain why water is used in home heating, industrial and automobile cooling systems, and home cooking applications.

STUDENTS SOMETIMES ASK ...

What is the strategy behind adding salt to a pot of water when making pasta? Does it make the water boil faster?

Adding salt to water will not make the water boil faster. It will, however, make the water boil at a slightly higher temperature because dissolved substances *raise* its boiling point (and *lower* its freezing point; see Table 6–7). Thus, the pasta will cook in less time. Additionally, the salt adds flavoring, so the pasta may taste better, too. Be sure to add the salt *after* the water has come to a boil, though, or it will take longer to reach a boil. This is a wonderful use of chemical principles—helping you to cook better!

---

[7]Note that the temperature scale *centigrade* (*centi* = a hundred, *grad* = step) is based on 100 even divisions between the melting and boiling points of pure water.

[8]Note that the heat capacity of water is used as the unit of heat quantity, the calorie. Thus, water is the standard against which the heat capacities of other substances are compared.

Water's Latent Heats

When water undergoes a change of state—that is, when ice melts or water freezes, or when water boils or water vapor condenses—a large amount of heat is absorbed or released. The amount of heat absorbed or released is due to water's high latent (*latent* = hidden) heats and is closely related to water's unusually high heat capacity. As water evaporates from your skin, it cools your body by absorbing heat (this is why sweating cools your body). Conversely, if you ever have been scalded by water vapor—steam—you know that steam releases an enormous amount of latent heat when it condenses to a liquid.

Latent Heat of Melting The graph in Figure 6–7 shows how latent heat affects the amount of energy needed to increase water temperature and change its state. Beginning with 1 gram of ice (*lower left*), the addition of 20 calories of heat raises its temperature by 40 degrees, from –40°C to 0°C (point *a* on the graph). The temperature remains at 0°C (32°F) even though more heat is being added, as shown by the plateau on the graph between points *a* and *b*. The temperature of the water does not change until 80 more calories of heat energy have been added. The **latent heat of melting** is the energy needed to break the intermolecular bonds that hold water molecules rigidly in place in ice crystals. The temperature remains unchanged until most of the bonds are broken and the mixture of ice and water has changed completely to 1 gram of water.

After the change from ice to liquid water has occurred at 0°C, additional heat raises the water temperature between points *b* and *c* in Figure 6–7. As it does, it takes 1 calorie of heat to raise the temperature of water 1°C (or 1.8°F). Therefore, another 100 calories must be added before the gram of water reaches the boiling point of 100°C (212°F). So far, a total of 200 calories have been added to reach point *c*.

Latent Heat of Vaporization The graph in Figure 6–7 flattens out again at 100°C, between points *c* and *d*. This plateau represents the **latent heat of vaporization**, which is 540 calories for water. This is the amount of heat that must be added to 1 gram of a substance at its boiling point to break the intermolecular bonds and complete the change of state from liquid to vapor (gas).

The drawings in Figure 6–8 which show the structure of water molecules in the solid, liquid, and gaseous states, help explain why the latent heat of vaporization is so much greater than the latent heat of melting. To go from a solid to a liquid, just enough hydrogen bonds must be broken to allow water molecules to slide past one another. To go from a liquid to a gas, however, all of the hydrogen bonds must be completely broken so that individual water molecules can move about freely.

Latent Heat of Evaporation Sea-surface temperatures average 20°C (68°F) or less. How, then, does liquid water convert to vapor at the surface of the ocean? The conversion of a liquid to a gas below the boiling point is called **evaporation**. At ocean surface temperatures, individual molecules converted from the liquid to the gaseous state have less energy than do water molecules at 100°C. To gain the additional energy necessary to break free of the surrounding ocean water molecules, an individual molecule must capture heat energy from its neighbors. In other words, the molecules left behind have lost heat energy to those that evaporate, which explains the cooling effect of evaporation.

It takes more than 540 calories of heat to produce 1 gram of water vapor from the ocean surface at temperatures less than 100°C. At 20°C (68°F), for instance, the **latent heat of evaporation** is 585 calories per gram. More heat is required because more hydrogen bonds must be broken. At higher temperatures, liquid water has fewer hydrogen bonds because the molecules are vibrating and jostling about more.

Latent Heat of Condensation When water vapor is cooled sufficiently, it condenses to a liquid and releases its **latent heat of condensation** into the surrounding air. On a small scale, the heat released is enough to cook food, which is how a "steamer" works. On a large scale, the heat

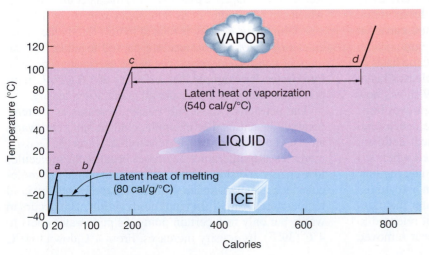

Figure 6–7 Latent heats and changes of state of water.

The latent heat of melting (80 calories) is much less than the latent heat of vaporization (540 calories). See text for description of points *a, b, c,* and *d*.

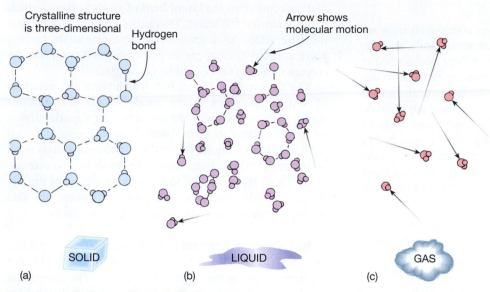

Crystalline structure
is three-dimensional

Hydrogen
bond

Arrow shows
molecular motion

SOLID

LIQUID

GAS

(a)

(b)

(c)

Figure 6–8 Hydrogen bonds in H₂O.

(a) In the solid state, water exists as ice, in which there are hydrogen bonds between all water molecules. **(b)** In the liquid state, there are some hydrogen bonds. **(c)** In the gaseous state, there are no hydrogen bonds and the water molecules are moving rapidly and independently.

released is sufficient to power large thunderstorms and hurricanes (see Chapter 7, "Air–Sea Interaction").

Latent Heat of Freezing Heat is also released when water freezes. The amount of heat released when water freezes is the same amount that was absorbed when the water was melted in the first place. Thus, the **latent heat of freezing** is identical to the latent heat of melting. Similarly, the latent heats of vaporization and condensation are identical.

Global Thermostatic Effects

The **thermostatic** (*thermos* = heat, *stasis* = standing) **effects** of water are those properties that act to moderate changes in temperature, which in turn affect Earth's climate. For example, the huge amount of heat energy exchanged in the evaporation–condensation cycle helps make life possible on Earth. The Sun radiates energy to Earth, where some is stored in the oceans. Evaporation removes this heat energy from the oceans and carries it high into the atmosphere. In the cooler upper atmosphere, water vapor condenses into clouds, which are the basis of **precipitation** (mostly rain and snow) that releases latent heat of condensation. The map in Figure 6–9 shows how this cycle of evaporation and condensation removes huge amounts of heat energy from the low latitude oceans and adds huge amounts of heat energy to the heat-deficient higher latitudes. In addition, the heat released when sea ice forms further moderates Earth's high-latitude regions.

The exchange of latent heat between ocean and atmosphere is very efficient. For every gram of water that condenses in cooler latitudes, the amount of heat released to warm these regions equals the amount of heat removed

from the tropical ocean when that gram of water was evaporated initially. The end result is that the thermal properties of water have prevented wide variations in Earth's temperature, moderating Earth's climate. Because rapid change is the enemy of all life, Earth's moderated climate is one of the main reasons life exists on Earth.

> Water's unique thermal properties include water's latent heats and high heat capacity, which redistribute heat on Earth and have moderated Earth's climate.

Water Density

Recall from Chapter 2 that density is mass per unit volume, which can be thought of as *how heavy something is for its size*. Ultimately, density is related to how tightly the molecules or ions of a substance are packed together. Typical units of density are grams per cubic centimeter (g/cm^3). Pure water, for example, has a density of 1.0 g/cm^3. Temperature, salinity, and pressure all affect water density.

The density of most substances increases as its temperature decreases. For example, cold air sinks and warm air rises because cold air is denser than warm air. Density increases as temperature decreases because the molecules lose energy and slow down, so the same number of molecules occupies less space. This shrinkage caused by cold temperatures, called **thermal contraction**, also occurs in water, but only to a certain point. As pure water cools to 4°C (39°F), its density increases. From 4°C down to 0°C

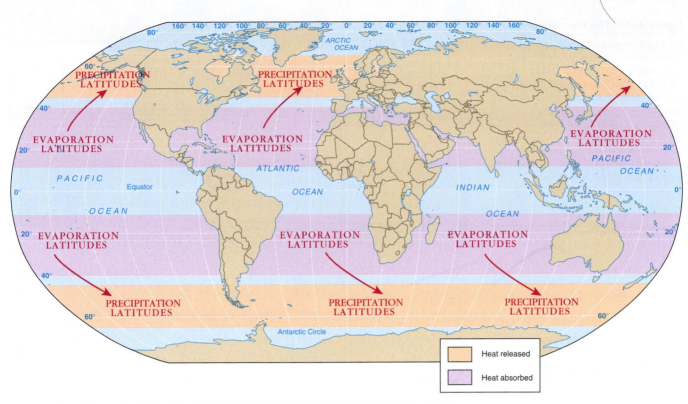

Figure 6–9 Atmospheric transport of surplus heat from low latitudes into heat-deficient high latitudes.

The heat removed from the tropical ocean (*evaporation latitudes*) is carried toward the poles and is released at higher latitudes through precipitation (*precipitation latitudes*), thus moderating Earth's climate.

(32°F), however, its density *decreases*. In other words, water stops contracting and actually expands, which is highly unusual among Earth's many substances. The result is that ice is less dense than liquid water, so ice floats on water. For most other substances, the solid state is denser than the liquid state, so the solid sinks.

Why is ice less dense than water? Figure 6–10a shows how molecular packing changes as water approaches its freezing point. From points *a* to *c* in the figure, the temperature decreases from 20°C (68°F) to 4°C (39°F) and the density increases from 0.9982 g/cm^3 to 1.0000 g/cm^3. Density increases because the amount of thermal motion decreases, so the water molecules occupy less volume. As a result, the window at point *c* contains more water molecules than the windows at points *a* or *b*. When the temperature is lowered below 4°C (39°F), the overall volume increases again because ice crystals become more abundant. Ice crystals are bulky, open, six-sided structures in which water molecules are widely spaced. Their characteristic hexagonal shape (Figure 6–11) mimics the hexagonal molecular structure resulting from hydrogen bonding between water molecules (see Figure 6–8a). By the time water fully freezes (point *e*), the density of the ice is much less than that of water at 4°C (39°F), the temperature at which water achieves its maximum density.

When water freezes, its volume increases by about 9%. Anyone who has put a beverage in a freezer for "just a few minutes" to cool it down and inadvertently forgot-

ten about it has experienced the volume increase associated with water's expansion as it freezes—usually resulting in a burst beverage container. The force exerted when ice expands is powerful enough to break apart rocks, split pavement on roads and sidewalks, and crack water pipes.

Increasing the pressure or adding dissolved substances decreases the temperature of maximum density for fresh water because the formation of bulky ice crystals is inhibited. Increasing pressure increases the number of water molecules in a given volume and inhibits the number of ice crystals that can be created. Increasing amounts of dissolved substances inhibits the formation of hydrogen bonds, which further restricts the number of ice crystals that can form. To produce ice crystals equal in volume to those that could be produced at 4°C (39°F) in fresh water, more energy must be removed, causing a reduction in the temperature of maximum density. Figure 6–10b shows the effect of salinity on the temperature of maximum density and the freezing temperature of water.

Because increased amounts of dissolved solids reduce the freezing point of water, most seawater never freezes, except near Earth's frigid poles (and even then, only at the surface). It's also the reason why salt is spread on roads and sidewalks during the winter in cold climates. The salt lowers the freezing point of water, allowing ice-free roads and sidewalks at temperatures that are several degrees below freezing.

Table 6–1 summarizes the physical and biological significance of the unusual properties of water.

Figure 6–10 The formation of ice.

(a) The formation of ice clusters in fresh water, showing how water reaches its maximum density at 4°C. Below 4°C, water becomes less dense as ice begins to form. At 0°C, ice forms, the crystal structure expands, and density decreases, causing ice to float.
(b) Affect of salinity on the freezing point of water. Note that the 0‰ salinity line on this diagram corresponds to the graph shown in part (a).

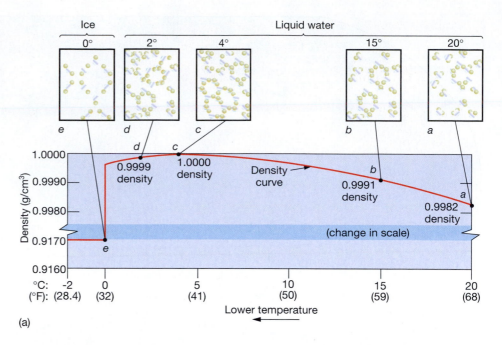

(a)

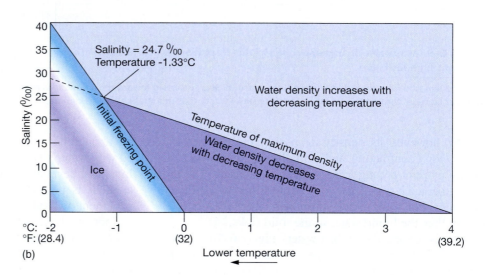

(b)

Figure 6–11 Snowflakes.

Hexagonal snowflakes indicate the internal structure of water molecules held together by hydrogen bonds.

TABLE 6-1 **Summary of the properties of water and their significance.**

Property	Comparison with other substances	Physical and/or biological significance
Physical states: gas (*water vapor*), liquid (*water*), solid (*ice*).	Water is the only substance that occurs as a gas, liquid, and solid within the range of surface temperatures on Earth.	*Water vapor* is an important component of the atmosphere, because water vapor transfers great quantities of heat from warm, low latitudes to cold, high latitudes.
		Liquid water runs across land, dissolving minerals and carrying them to the oceans. Most organisms are composed primarily of water. Serves as a transport medium and facilitates chemical reactions.
		Ice formation at the surface protects the water and life below it from freezing.
Solvent property: the ability to dissolve substances.	Water can dissolve more substances than any other liquid; hence it is often called "the universal solvent."	*Wide-ranging implications* in both physical and biological phenomena. Helps explain why seawater is "salty." In addition, seawater carries dissolved within it the nutrients required by marine algae and the oxygen needed by animals.
Surface tension: cohesive attraction of hydrogen bonds causes a molecule-thick "skin" to form on water surfaces.	Water has the greatest surface tension of all common liquids, which makes capillarity possible.	*Causes water to form drops,* to "bead up," and to overfill a glass without spilling. Water striders use this "skin" as a walking surface, while other organisms hang from its under-surface. Capillarity assists in raising water to the tops of tall plants such as trees.
Heat capacity: the quantity of heat required to change the temperature of 1 g of a substance by 1°C. The heat capacity of water is used as the unit of heat quantity, the *calorie (cal)*.	Water has the highest heat capacity of all common liquids and gains or loses much more heat than other common substances while undergoing an equal temperature change.	*A major factor* in moderating Earth's climate, it helps explain the narrow range of temperature change occurring near the oceans as compared to interior regions.
Latent heat of melting: the quantity of heat gained or lost per gram by a substance changing from a solid to a liquid, or from a liquid to a solid, without a temperature change.	For water, it is 80 cal at 0°C (32°F), the highest of any common substance.	*Heat energy lost when ice forms* is mostly absorbed by the heat-deficient atmosphere at high latitudes.
		Heat energy gained by water when ice melts is manifested as molecular energy of the liquid water. This prevents the high-latitude ocean from becoming much warmer or colder than the freezing temperature of ocean water.
Latent heat of vaporization: the quantity of heat gained or lost per gram by a substance changing from a liquid to a gas, or from a gas to a liquid, without a temperature change.	It is greater for water, 540 cal at 100°C (212°F), than for any other common substance. At ocean surface temperatures of 20°C (68°F), it is 585 cal.	*Extremely important* in global heat and water transfer in the atmosphere. Evaporation from the low-latitude ocean removes a great amount of excess heat energy that is released through precipitation at heat-deficient higher latitudes. This greatly moderates temperatures at the poles and Equator, which otherwise would be far more extreme.
Density: mass per unit volume (g/cm^3).	Water's density increases as water cools, but it *decreases* below 4°C (39°F), which is highly unusual. Water density also increases as salinity and pressure increase.	*Plankton* that stay near the surface through buoyancy and frictional resistance to sinking are greatly influenced by the effect of temperature on density. In low-density warm water, plankton must be smaller or more ornate to obtain the increased ratio of surface area to body mass necessary to remain afloat. Also produces layering of ocean water.
Thermal expansion: the expansion of a substance when it is heated, and the contraction of a substance when it is cooled.	As water is cooled, it contracts. Below 4°C (39°F), it expands, which is highly unusual for a substance.	*Water expands* by 9% when frozen, causing ice to float. In the ocean at high latitudes where sea ice forms, ice provides a layer below which seawater does not freeze, protecting marine life.

Seawater

What is the difference between pure water and seawater? One of the most obvious differences is that seawater contains dissolved substances that give it a distinctly salty taste. These dissolved substances are not simply sodium chloride (table salt)—they include various other salts, metals, and dissolved gases. The oceans contain enough salt to cover the entire planet with a layer more than 150 meters (500 feet) thick (about the height of a 50-story skyscraper). Unfortunately, the salt content of seawater makes it unsuitable for drinking or irrigating most crops and causes it to be highly corrosive to many materials.

Salinity

Salinity (*salinus* = salt) is the total amount of solid material dissolved in water, including dissolved gases, because even gases become solids at low enough temperatures. Salinity does *not* include fine particles being held in suspension (turbidity) or solid material in contact with water because these materials are not dissolved. Salinity is the ratio of the mass of dissolved substances to the mass of the water sample.

The salinity of seawater is typically about 3.5%, about 220 times saltier than fresh water. Seawater with a salinity of 3.5% indicates that it also contains 96.5% pure water, as shown in Figure 6–12. Because seawater is mostly pure water, its physical properties are very similar to those of pure water, with only slight variations.

Figure 6–12 and Table 6–2 show that the elements chlorine, sodium, sulfur (as the sulfate ion), magnesium,

calcium, and potassium account for over 99% of the dissolved solids in seawater. Over 80 other chemical elements have been identified in seawater—most in extremely small amounts—and probably all of Earth's naturally occurring elements exist in the sea. Remarkably, some of the trace amounts of dissolved components in seawater are vital for human survival (Box 6–1).

Salinity is often expressed in **parts per thousand** (‰). Just as 1% is one part in 100, 1‰ is one part in 1000. When converting from percent to parts per thousand, the decimal is simply moved over one place to the right. Typical seawater salinity, for instance, averages 3.5% or 35‰. Advantages of expressing salinity in parts per thousand are that decimals are often avoided and values convert directly to grams of salt per kilogram of seawater. For example, 35‰ seawater has 35 grams of salt in every 1000 grams of seawater.

Salinity Variations

Salinity varies within the oceans. In the open ocean far from land, it varies between about 33 and 38‰. In coastal areas, salinity variations can be extreme. In the Baltic Sea, for example, salinity averages only 10‰ because physical conditions create **brackish** (*brak* = salt, *ish* = somewhat) water. Brackish water is produced in areas where fresh water (from rivers and high rainfall) and seawater mix. In the Red Sea, on the other hand, salinity averages 42‰ because physical conditions produce **hypersaline** (*hyper* = excessive, *salinus* = salt) water. Hypersaline water is typical of seas and inland bodies of water that experience high evaporation rates and limited open-ocean circulation.

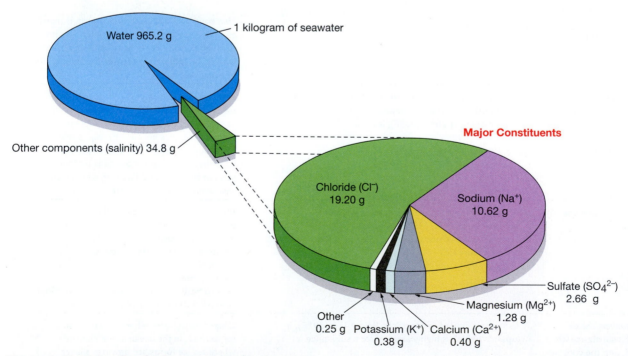

Figure 6–12 Major dissolved components in seawater.

Diagrammatic representation of the most abundant components in a kilogram of 35‰ salinity seawater. Constituents are listed in grams per kilogram, which is equivalent to parts per thousand (‰).

TABLE 6-2 **Selected dissolved materials in 35‰ seawater.**

Major constituents (in parts per thousand, ‰)		
Constituent	Concentration (‰)	Ratio of constituent/total salts (%)
Chloride (Cl^-)	19.2	55.04
Sodium (Na^+)	10.6	30.61
Sulfate (SO_4^{2-})	2.7	7.68
Magnesium (Mg^{2+})	1.3	3.69
Calcium (Ca^{2+})	0.40	1.16
Potassium (K^+)	0.38	1.10
Total	34.58	99.28%

Minor constituents (in parts per million, ppm [a])					
Gases		Nutrients		Others	
Constituent	Concentration (ppm)	Constituent	Concentration (ppm)	Constituent	Concentration (ppm)
Carbon dioxide (CO_2)	90	Silicon (Si)	3.0	Bromide (Br^-)	65.0
Nitrogen (N_2)	14	Nitrogen (N)	0.5	Carbon (C)	28.0
Oxygen (O_2)	6	Phosphorous (P)	0.07	Strontium (Sr)	8.0
		Iron (Fe)	0.002	Boron (B)	4.6

Trace constituents (in parts per billion, ppb [b])					
Constituent	Concentration (ppb)	Constituent	Concentration (ppb)	Constituent	Concentration (ppb)
Lithium (Li)	185	Zinc (Zn)	10	Lead (Pb)	0.03
Rubidium (Rb)	120	Aluminum (Al)	2	Mercury (Hg)	0.03
Iodine (I)	60	Manganese (Mn)	2	Gold (Au)	0.005

[a]Note that 1000 ppm = 1‰.
[b]Note that 1000 ppb = 1 ppm.

STUDENTS SOMETIMES ASK...
What would happen to a person if he or she drank seawater?

It depends on the quantity. The salinity of seawater is about four times greater than that of your body fluids. In your body, seawater causes your internal membranes to lose water through *osmosis* (*osmos* = to push), which transports water molecules from higher concentrations (the normal body chemistry of your internal fluids) to areas of lower concentrations (your digestive tract containing seawater). Thus, your natural body fluids would move into your digestive tract and eventually be expelled, causing dehydration.

Don't worry too much if you've inadvertently swallowed some seawater. As a nutritional drink, seawater provides seven important nutrients and contains no fat, cholesterol, or calories. Some people even claim that drinking a small amount of seawater daily gives them good health! However, beware of microscopic contaminants that are often present in seawater.

Some of the most hypersaline water in the world is found in inland lakes, which are often called seas because they are so salty. The Great Salt Lake in Utah, for example, has a salinity of 280‰, and the Dead Sea on the border of Israel and Jordan has a salinity of 330‰. The water in the Dead Sea, therefore, contains 33% dissolved solids and is almost *10 times saltier than seawater*. As a result, hypersaline waters are so dense that one can easily float—with arms and legs sticking up above water level! Hypersaline waters also taste much saltier than seawater.

Salinity of seawater in coastal areas also varies seasonally. For example, the salinity of seawater off Miami Beach, Florida, varies from about 34.8‰ in October to 36.4‰ in May and June when evaporation is high. Offshore of Astoria, Oregon, seawater salinity is always relatively low because of the vast fresh water input from the Columbia River. Here, seawater salinity varies from 0.3‰ in April and May (when the Columbia River is at its maximum flow rate) to 2.6‰ in October (the dry season).

Other types of water have much lower salinity. Tap water, for instance, has salinity somewhere below 0.8‰,

BOX 6–1 People and Ocean Environment
HOW TO AVOID GOITERS

The nutritional label on containers of salt usually proclaims "this product contains iodine, a necessary nutrient." Why is iodine necessary in our diets? It turns out that if a person's diet contains an insufficient amount of iodine, a potentially life-threatening disease called *goiters* (*guttur* = throat) may result (Figure 6A).

Iodine is used by the thyroid gland, which is a butterfly-shaped organ located in the neck in front of and on either side of the trachea (windpipe). The thyroid gland manufactures hormones that regulate cellular metabolism essential for mental development and physical growth. If people lack iodine in their diet, their thyroid glands cannot function properly. Often, this results in the enlargement or swelling of the thyroid gland. Severe symptoms include dry skin, loss of hair, puffy face, weakness of muscles, weight increase, diminished vigor, mental sluggishness, and a large nodular growth on the neck called a goiter. If proper steps are not taken to correct this disease, it can lead to cancer. Iodine ingested regularly often begins to reverse the effects. In advanced stages, surgery to remove the goiter or exposure to radioactivity is the only course of action.

How can you avoid goiters? Fortunately, goiters are preventable by a diet with just *trace amounts* of iodine. Where can you get iodine in your diet? All products from the sea contain trace amounts of iodine because iodine is one of the many elements dissolved in seawater. Sea salt, seafood, seaweed, and other sea products contain plenty of iodine to help prevent goiters. Although goiters are rarely a problem in developed nations like the United States, goiters pose a serious health hazard in many underdeveloped nations, especially those far from the sea. In the U.S., however, many people get too much iodine in their diet, leading to the overproduction of hormones by the thyroid gland. That's why most stores that sell iodized salt also carry noniodized salt for those people who have a *hyperthyroid* (*hyper* = excessive, *thyroid* = the thyroid gland) *condition* and must restrict their intake of iodine.

Figure 6A A woman with goiters.

and good-tasting tap water is usually below 0.6‰. Salinity of premium bottled water is on the order of 0.3‰, with the salinity often displayed prominently on its label, usually as total dissolved solids (TDS) in units of parts per million (ppm), where 1000 ppm equals 1‰.

> Average seawater salinity is 35‰ but varies widely from brackish (low salinity) to hypersaline (high salinity).

Determining Salinity

Early methods of determining seawater salinity involved evaporating a carefully weighed amount of seawater and weighing the salts that precipitated from it. For example, if you weighed out 1000 grams of typical 35‰ seawater and set it aside to evaporate, you would find that the remaining salts would weigh about 35 grams. However, the accuracy of this time-consuming method is limited because some water can remain bonded to salts that precipitate and some substances can evaporate along with the water.

Another way to measure salinity is to use the *principle of constant proportions*, which was firmly established by chemist William Dittmar when he analyzed the water samples collected during the *Challenger* Expedition.[9] The **principle of constant proportions** states that the major dissolved constituents responsible for the salinity of seawater occur nearly everywhere in the ocean in the exact same proportions, independent of salinity. The ocean, therefore, is well mixed. When salinity changes, moreover, the salts don't leave (or enter) the ocean but water molecules do. Seawater has *constancy of composition*, so the concentration of a single major constituent can be measured to determine the total salinity of a given water sample. The constituent that occurs in the greatest abundance and is the easiest to measure accurately is the chloride ion, Cl^-. The weight of this ion in a water sample is its **chlorinity**.

In any sample of ocean water worldwide, the chloride ion accounts for 55.04% of the total proportion of dissolved solids (Figure 6–12 and Table 6–2). Therefore, by measuring only the chloride ion concentration, the total salinity of a seawater sample can be determined using the following relationship:

$$\text{Salinity (‰)} = 1.80655 \times \text{chlorinity (‰)}^* \quad (6\text{-}1)$$

For example, the average chlorinity of the ocean is 19.2‰ so the average salinity is 1.80655 × 19.2‰,

which rounds to 34.7‰. In other words, on average there are 34.7 parts of dissolved material in every 1000 parts of seawater.

Standard seawater consists of ocean water analyzed for chloride ion content to the nearest ten-thousandth of a part per thousand by the Institute of Oceanographic Services in Wormly, England. It is then sealed in small glass vials called *ampules* and sent to laboratories throughout the world for use as a reference standard in calibrating analytical equipment.

Seawater salinity can be measured very accurately with advanced oceanographic instruments such as a **salinometer** (*salinus* = salt, *meter* = measure). A salinometer measures seawater's *electrical conductivity* (the ability of a substance to transmit electric current), which increases as more dissolved substances are dissolved in water (Figure 6–13). Salinometers can determine salinity to resolutions of better than 0.003‰.

Dissolved Components Added and Removed from Seawater

Why does seawater salinity vary from place to place? Interestingly, dissolved substances do not remain in the ocean forever. Instead, they are cycled into and out of seawater by the processes shown in Figure 6–14. These processes include stream *runoff* (river discharge) in which streams dissolve ions from continental rocks and carry them to the sea, and volcanic eruptions, both on the land and on the sea floor. Other sources include the atmosphere (which contributes gases) and biologic interactions.

Stream runoff is the primary method by which dissolved substances are added to the oceans. Table 6–3 compares the major components dissolved in stream water

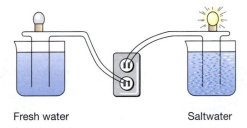

Figure 6–13 Salinity affects water conductivity.

Increasing the amount of dissolved substances increases the conductivity of the water. A light bulb with bare electrodes shows that the higher the salinity, the more electricity is transmitted, and the brighter the bulb will be lit.

---

[9]For more information about the accomplishments of the *Challenger* Expedition (which include the beginnings of the study of ocean chemistry), see Chapter 1.

---

*The number 1.80655 comes from dividing 1 by 0.5504 (the chloride ion's proportion in seawater of 55.04%). However, if you actually divide this, you will get 1.81686, which is different from the original value by 0.57%. Empirically, oceanographers found that seawater's constancy of composition is an approximation and have agreed to use 1.080655 because it more accurately represents the total salinity of seawater.

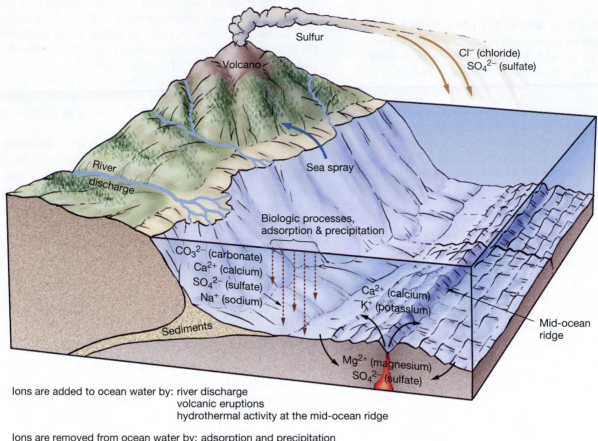

Figure 6–14 The cycling of dissolved components in seawater.

Dissolved components are added to seawater primarily by river discharge and volcanic eruptions, while they are removed by adsorption, precipitation, ion entrapment in sea spray, and marine organisms that produce hard parts. Chemical reactions at the mid-ocean ridge add and remove various dissolved components.

TABLE 6–3 **Comparison of major dissolved components in streams with those in seawater.**

Constituent	Concentration in streams (parts per million)	Concentration in seawater (parts per million)
Carbonate ion (HCO_3^-)	58.4	trace
Calcium ion (Ca^{2+})	15.0	400
Silicate (SiO_2)	13.1	3
Sulfate ion (SO_4^{2-})	11.2	2700
Chloride ion (Cl^-)	7.8	19,200
Sodium ion (Na^+)	6.3	10,600
Magnesium ion (Mg^{2+})	4.1	1300
Potassium ion (K^+)	2.3	380
Total (parts per million)	119.2	34,793
Total (‰)	0.1192‰	34.8‰

with those in seawater. It shows that streams have far lower salinity and a vastly different composition of dissolved substances than seawater. For example, carbonate ion (HCO_3^-) is the most abundant dissolved constituent in stream water yet is found in only trace amounts in seawater. Conversely, the most abundant dissolved component in seawater is the chloride ion (Cl^-), which exists in very small concentrations in streams.

If stream water is the main source of dissolved substances in seawater, why don't the components of the two match each other more closely? One of the reasons is that some dissolved substances stay in the ocean and accumulate over time.

Residence Time

Residence time is the average length of time that a substance resides in the ocean. The formula for residence time is

$$\text{Residence time} = \frac{\begin{array}{c}\text{Amount of substance} \\ \text{in the oceans}\end{array}}{\begin{array}{c}\text{Rate at which substance is added to} \\ \text{or removed from the oceans}\end{array}} \quad (6\text{-}2)$$

The residence time of a substance in the ocean depends on how reactive it is with the marine environment: Compounds that are less reactive have a longer residence time. Long residence times, in turn, generally lead to higher concentrations of the dissolved substance (Table 6–4). The ion sodium (Na^+), for instance, has a residence time of 260 million years, and, as a result, has a high concentration in the ocean. Others elements such as aluminum (a highly reactive substance) have a residence time of only 100 years and occur in seawater in much lower concentrations.

Water is continually cycled through the oceans (as will be discussed later in this chapter in the section titled The Hydrologic Cycle), so it has a specific residence time in the ocean, too. Calculations suggest that shallow surface water resides in the ocean for periods that average hundreds of years, while deep water averages between 1000 and 2000 years.

Ocean Salinity Change through Time

Since new dissolved components are being added to the oceans all the time and because various salts have long residence times in the ocean, are the oceans getting saltier through time? Analysis of ancient marine organisms and sea floor sediments suggest that the oceans have not increased in salinity over time. This must be because the rate at which an element is added to the ocean equals the rate at which it is removed again, so the average *amounts* of various elements remain constant (which is called a *steady-state* condition).

Materials added to the oceans are counteracted by several processes that cycle dissolved substances out of seawater. When waves break at sea, for example, salt spray releases tiny salt particles into the atmosphere where they may be blown over land before being washed back to Earth. The amount of material leaving the ocean in this way is enormous: According to a recent study, as much as 3.3 billion metric tons (3.6 billion short tons) of salt spray enter the atmosphere each year. Another example is the infiltration of seawater along mid-ocean ridges near hydrothermal vents (see Figure 4–17), which incorporates magnesium and sulfate ions into sea floor mineral deposits. In fact, chemical studies of seawater indicate that the *entire volume of ocean water* is recycled through this hydrothermal circulation system at the mid-ocean ridge approximately every 3 million years. As a result, the chemical exchange between ocean water and the basaltic crust has a major influence on the composition of ocean water.

Dissolved substances are also removed from seawater in other ways. Calcium, carbonate, sulfate, sodium, and silicon are deposited in ocean sediments within the shells of dead microscopic organisms and animal feces. Vast amounts of dissolved substances can be removed when inland arms of seas dry up, leaving salt deposits called *evaporites* (such as those beneath the Mediterranean Sea; see Box 12–1). In addition, ions dissolved in ocean water are removed by adsorption (physical attachment) to the surfaces of sinking clay and biologic particles.

TABLE 6–4 **Average amount and residence time of selected elements in the ocean.**

Element	Amount in the oceans (trillion kilograms)	Residence time (years)
Sodium (Na)	14,700,000	260,000,000
Potassium (K)	530,000	11,000,000
Calcium (Ca)	560,000	8,000,000
Silicon (Si)	5200	10,000
Manganese (Mn)	14	7,000
Iron (Fe)	14	140
Aluminum (Al)	14	100

Dissolved Gases in Seawater

The presence of dissolved gases in the ocean is extremely important for marine organisms. For example, most animals living in the ocean rely on minute quantities of dissolved oxygen in seawater to supply their oxygen needs. Although most atmospheric gases readily dissolve in seawater, the amount of dissolved gases in seawater depends on several factors.

Time to Grab a Soda

As an aid to understanding the behavior of gases in water, try a few simple experiments with a bottle of carbonated soda, which is a handy example of slightly impure water containing dissolved gases. Carbonated sodas are made by injecting the gas CO_2 into the beverage under high pressure. If you have a clear glass or plastic bottle of soda, note how few bubbles there are in the liquid prior to opening it.

Begin by opening the soda, which releases the excess pressure in the container and causes the gas to bubble out of the liquid—the reason for the fizz. After a short time, the initial rush of bubbles abates. Notice how bubbles continue to escape but at a slower rate. Now shake your soda a bit, and you'll produce a new rush of bubbles as they come out of solution. As the soda warms, it will also increase the amount of gas that escapes. Eventually, most of the gas will be gone, and your soda will have gone "flat."

From these experiments, you have seen that the gas content of the soda depends on three factors: pressure, temperature, and its ability to escape to the atmosphere. You decreased the gas content first by decreasing the pressure when you opened the container. As the soda warmed, it caused more gas to escape. By stirring, you also increased the atmospheric escape. The final gas content of your soda is the equilibrium volume that can dissolve in water at room temperature and atmospheric pressure given the CO_2 content of the atmosphere.

Gas Exchange between the Atmosphere and Ocean

At the surface of the ocean, where seawater is in contact with the atmosphere, a small amount of each atmospheric gas will dissolve into the water. Wave action enhances this air–water contact in the same way that you did when you stirred your soda. As wind velocities increase and waves grow in size, there is an increase in turbulence and the depth to which it extends. Thus, the amount of mixing of atmospheric gas into water is proportional to wind velocity.

The total amount of gas that can be dissolved eventually reaches an equilibrium concentration in the same way your soda did as it went "flat." The equilibrium concentration for any gas is proportional to the atmospheric concentration of that gas and its solubility (the amount of a substance that can be dissolved under a given pressure and temperature). Note that the solubility of a gas increases as pressure increases (that's why an unopened soda can have high gas content: It's under high pressure). Also, solubility of a gas increases as temperature decreases (cold water holds more dissolved gas, which is why a warm soda goes "flat" faster than a cold one).

The most abundant atmospheric gases are shown in Table 6–5. The table also shows the concentration of these gases in the ocean, which is largely due to their solubility in water but also varies due to chemical reactions that occur in seawater. For example, some nitrogen gas (N_2) is converted to nitrate ions (NO_3^-) by bacteria.

The constituents of seawater that occur in constant proportions or change only very slowly through time are called **conservative constituents**. Examples of conservative constituents are the major dissolved components of seawater—which also have long residence times in the oceans (see Table 6–4). Other examples of conservative constituents are argon and the other noble gases (helium, neon, krypton, and radon), which are chemically unreactive. Although they are present in very small quantities, noble gases serve as important tracers of ocean physical processes.

On the other hand, **nonconservative constituents** are altered appreciably by many biological and chemical processes in the sea. As such, they can be used to trace biological and chemical processes. They also have short residence times. Examples of nonconservative constituents are the gases oxygen and carbon dioxide.

Oxygen Figure 6–15a shows the concentration of oxygen in the North Pacific and North Atlantic Oceans and how the concentration changes with depth. At the surface, oxygen concentration is relatively high, largely because of

TABLE 6–5 **Major gases in the atmosphere and ocean**

Gas	Percent of gas in atmosphere (by volume)	Solubility of gas in seawater (ml/l at 20°C and 1 atm pressure)	Concentration in seawater in parts per million (by mass)
Nitrogen (N_2)	78.1%	10	10–18 ppm
Oxygen (O_2)	20.9%	22	0–13 ppm
Argon (Ar)	0.9%	26	1–1.5 ppm
Carbon dioxide (CO_2)	0.037%	800	64–107 ppm

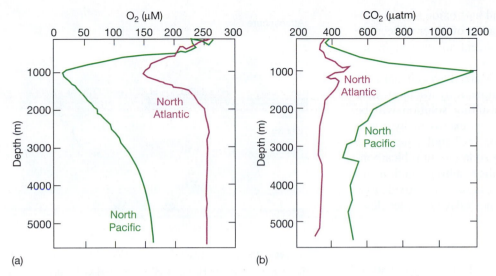

Figure 6–15 **Seawater dissolved oxygen and carbon dioxide variation with depth.**

Depth profiles for the North Pacific and North Atlantic Oceans showing concentration of **(a)** oxygen and **(b)** carbon dioxide.

photosynthesis by marine algae. As algae photosynthe-size, they release oxygen to seawater as a byproduct (see the photosynthesis equation in Figure 2–10). Additional oxygen is supplied by exchange with the atmosphere and by mechanisms that create a rough sea surface (such as waves and even rain falling on the ocean).

Oxygen content decreases rapidly with depth in the ocean because algae live only in sunlit surface waters. Further, oxygen is used up by animals living in deeper depths (see the respiration equation in Figure 2–10). Cu-riously, deep ocean oxygen content is typically higher than that at intermediate depths. This is because deep ocean water originates at the surface in polar regions, where water temperatures are low and the gas content is high (recall that cold water holds more dissolved gas). As a result, this cold water holds more dissolved oxygen, which it retains as it travels slowly along the sea floor.[10]

Carbon Dioxide Although carbon dioxide is present in the atmosphere in very small amounts, it is one of the most abundant gases dissolved in seawater. In fact, seawa-ter can hold perhaps a thousand times more carbon diox-ide than either nitrogen or oxygen at saturation. This is because carbon dioxide has one of the highest gas solubil-ities (Table 6–5). However, carbon dioxide is quickly used during marine photosynthesis, so dissolved quantities of carbon dioxide are usually much les than the theoretical maximum. Still, there is about 60 times more carbon diox-ide dissolved in the ocean as in the atmosphere.

Figure 6–15b shows the concentration of carbon diox-ide in the North Pacific and North Atlantic Oceans and how the concentration changes with depth. Surface ocean

carbon dioxide concentrations are in equilibrium with the atmosphere. Since oxygen and carbon dioxide are related by the processes of photosynthesis and respiration (see Figure 2–10), the curves are mirror images of each other. Production of carbon dioxide by respiration causes the amount of carbon dioxide to increase rapidly over the same depth interval at which oxygen is rapidly depleted. In the deep ocean, carbon dioxide values decrease again for the same reason that the oxygen values increase—the concen-trations are those the waters acquired at the surface in polar regions. Carbon dioxide levels are also affected by volcanic activity and sediment–water interactions on the sea floor.

Acidity and Alkalinity of Seawater

An **acid** is a compound that releases hydrogen ions (H^+) when dissolved in water. The resulting solution is said to be *acidic*. A strong acid readily and completely releases hydrogen ions when dissolved in water. An **alkaline** or **base** is a compound that releases hydroxide ion (OH^-) when dissolved in water. The resulting solution is said to be *alkaline* or *basic*. A strong base readily and complete-ly releases hydroxide ions when dissolved in water.

Both hydrogen ions and hydroxide ions are present in extremely small amounts at all times in water because water molecules dissociate and reform. Chemically, this is represented as

$$
\underset{\text{reform}}{\overset{\text{dissociate}}{\rightleftharpoons}}
$$

$$
H_2O \rightleftharpoons H^+ + OH^- \tag{6-3}
$$

reform

[10]For more details on deep currents, see Chapter 8, "Ocean Circulation."

Note that if the hydrogen ions and hydroxide ions in a solution are due only to the dissociation of water molecules, they are always found in equal concentrations, and the solution is consequently neutral.

When substances dissociate in water, they can make the solution acidic or basic. For example, if hydrochloric acid (HCl) is added to water, the resulting solution will be acidic because there will be a large excess of hydrogen ions from the dissociation of the HCl molecules. Conversely, if a base such as baking soda (sodium bicarbonate, $NaHCO_3$) is added to water, the resulting solution will be basic because there will be an excess of hydroxide ions (OH^-) from the dissociation of the $NaHCO_3$ molecules.

The pH Scale

Figure 6–16 shows the **pH** (potential of hydrogen) **scale**, which is a measure of the acidity or alkalinity of a solution. Values for pH range from 0 (strongly acidic) to 14 (strongly alkaline or basic). The pH of a **neutral** solution such as pure water[11] is 7.0, whereas the pH of an acidic solution is less than 7.0 and the pH of an alkaline solution is greater than 7.0.

Water in the ocean combines with carbon dioxide to form a weak acid, called carbonic acid (H_2CO_3), which dissociates and releases hydrogen ions (H^+):

$$H_2O + CO_2 \rightarrow H_2CO_3 \rightarrow H^+ + HCO_3^- \qquad (6\text{-}4)$$

This reaction would seem to make the ocean slightly acidic. Carbonic acid, however, keeps the ocean slightly alkaline through the process of *buffering*.

The Carbonate Buffering System

The chemical reactions shown in Figure 6–17 [and Equation (6–4)] show that carbon dioxide (CO_2) combines

---

[11]Water's neutral pH might seem surprising in light of water's tremendous ability to dissolve substances; it seems intuitive that water should be acidic and thus have a low pH. However, pH measures the activity of the hydrogen ion in solution, not the ability of a substance to dissolve by forming hydrogen bonds (as water does).

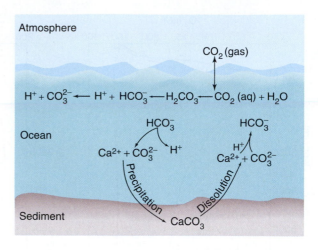

Seawater too basic: $H_2CO_3 \longrightarrow HCO_3^- + H^+$ pH drops
Seawater too acidic: $HCO_3^- + H^+ \longrightarrow H_2CO_3$ pH rises

Figure 6–17 The carbonate buffering system.
Atmospheric carbon dioxide (CO_2) enters the ocean and undergoes chemical reactions. If seawater is too basic, chemical reactions occur that release H^+ into seawater and lowers pH. If seawater is too acidic, chemical reactions occur that remove H^+ from seawater and cause pH to rise. Thus, buffering keeps the pH of seawater constant.

with water (H_2O) to form carbonic acid (H_2CO_3). Carbonic acid can then lose a hydrogen ion (H^+) to form the negatively charged bicarbonate ion (HCO_3^-). The bicarbonate ion can lose its hydrogen ion, too, though it does so less readily than carbonic acid. When the bicarbonate ion loses its hydrogen ion, it forms the double-charged negative carbonate ion (CO_3^{2-}), some of which combines with calcium ions to form calcium carbonate ($CaCO_3$). Some of the calcium carbonate is precipitated and deposited on the ocean floor, which can cycle back into the ocean by dissolving at depth.

The equations below Figure 6–17 show how these chemical reactions involving carbonate minimize changes in the pH of the ocean, in a process called **buffering**. Buffering protects the ocean from getting too acidic or too basic, similar to how buffered aspirin protects sensitive stomachs. For example, if the pH of the ocean increases (becomes too basic), it causes H_2CO_3 to release H^+ and pH drops. Conversely, if the pH of the ocean decreases (becomes too acidic), HCO_3^- combines with H^+ to remove it, causing pH to rise. In this way, buffering maintains the pH of the ocean at an average of 8.1 with little variation through time.

Deep-ocean water contains more carbon dioxide than surface water because deep water is cold and has the ability to dissolve more gases. Also, the higher pressures of the deep ocean further aid the dissolution of gases in seawater. Because carbon dioxide combines with water to form carbonic acid, why isn't the cold water of the deep ocean highly acidic? When microscopic marine organisms that make their shells out of calcium carbonate (calcite) die and sink into the deep ocean, they neutralize the

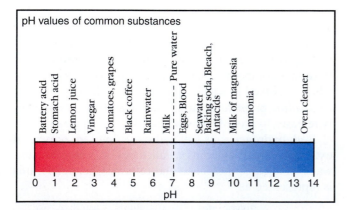

Figure 6–16 The pH scale.
The pH scale ranges from 0 (highly acidic) to 14 (highly alkaline). A pH of 7 is neutral; the pH values of common substances are also shown.

acid through buffering. In essence, these organisms act as an "antacid" for the deep ocean analogous to the way commercial antacids use calcium carbonate to neutralize excess stomach acid. As explained in Chapter 5, these shells are readily dissolved below the calcite (calcium carbonate) compensation depth (CCD).

The availability of carbonate sediments gives the oceans a large buffering capacity for carbon dioxide, which has important implications for climate change because carbon dioxide is one of the main greenhouse gases.[12] Remarkably, the oceans appear to have removed a large amount of the atmospheric CO_2 added by fossil fuel burning during the industrial age without a significant change in seawater pH.

Figure 6–18 shows how the pH of seawater varies with depth in the North Atlantic and North Pacific Oceans. Notice how the pH profile is the inverse of the carbon dioxide profile (see Figure 6–15). This is because an increase in carbon dioxide causes pH to decrease, which is why the pH minimum in the middle depths coincides with the carbon dioxide maximum (and the oxygen minimum). Only in deeper water where carbonate buffering becomes important does pH start to rise again.

> Reactions involving carbonate chemicals serve to buffer the ocean and keep its pH at an average value of 8.1 (slightly alkaline).

---

[12]For more about the atmosphere's greenhouse effect, see Chapter 7, "Air–Sea Interaction."

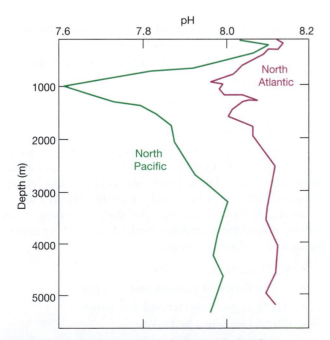

Figure 6–18 Seawater pH variation with depth.

Depth profile for the North Pacific and North Atlantic Oceans showing variation of pH with depth.

Processes Affecting Seawater Salinity

Processes affecting seawater salinity change either the amount of dissolved substances within the water or the amount of water. Adding more water, for instance, dilutes the dissolved components, and lowers the salinity of the sample. Conversely, removing water increases salinity. Changing the salinity in these ways does not affect the *amount* or the *composition* of the dissolved components, which remain in constant proportions.

Processes That Decrease Seawater Salinity

Table 6–6 summarizes the processes affecting seawater salinity. Precipitation, runoff, melting icebergs, and melting sea ice *decrease* seawater salinity by adding more fresh water to the ocean. Precipitation is the way atmospheric water returns to Earth as rain, snow, sleet, and hail. Worldwide, about three-quarters of all precipitation falls directly back into the ocean and one-quarter falls onto land. Precipitation falling directly into the oceans adds fresh water, reducing seawater salinity.

Most of the precipitation that falls on land returns to the oceans indirectly as stream runoff. Even though this water dissolves minerals on land, the runoff is relatively pure water, as shown in Table 6–3. Runoff, therefore, adds mostly water to the ocean, causing seawater salinity to decrease.

Icebergs are chunks of ice that have broken free (*calved*) from a glacier when it flowed into an ocean or marginal sea and began to melt. Glacial ice originated as snowfall in high mountain areas, so icebergs are composed of fresh water. When icebergs melt in the ocean, they add fresh water, which is another way in which seawater salinity is reduced.

Sea ice forms when ocean water freezes in high-latitude regions and is composed primarily of fresh water. When warmer temperatures return to high-latitude regions in the summer, sea ice melts in the ocean, adding mostly fresh water with a small amount of salt to the ocean. Seawater salinity, therefore, is decreased.

Processes That Increase Seawater Salinity

The formation of sea ice and evaporation increase seawater salinity by removing water from the ocean (Table 6–6). Sea ice forms when seawater freezes. Depending on the salinity of seawater and the rate of ice formation, about 30% of the dissolved components in seawater are retained in sea ice. This means that 35‰ seawater creates sea ice with about 10‰ salinity (30% of 35‰ is 10‰). Consequently, the formation of sea ice removes mostly fresh water from seawater, increasing the salinity of the remaining unfrozen water. High-salinity water also has a high density, so it sinks below the surface.

TABLE 6-6 **Processes affecting seawater salinity**

Process	How accomplished	Adds or removes	Effect on salt in seawater	Effect on H₂O in seawater	Salinity increase or decrease?	Source of fresh water from the sea?
Precipitation	Rain, sleet, hail, or snow falls directly on the ocean	Adds very fresh water	None	More H_2O	Decrease	—
Runoff	Streams carry water to the ocean	Adds mostly fresh water	Negligible addition of salt	More H_2O	Decrease	—
Icebergs melting	Glacial ice calves into the ocean and melts	Adds very fresh water	None	More H_2O	Decrease	Yes, icebergs from Antarctic have been towed to South America
Sea ice melting	Sea ice melts in the ocean	Adds mostly fresh water and some salt	Adds a small amount of salt	More H_2O	Decrease	Yes, sea ice can be melted and is better than drinking seawater
Sea ice forming	Seawater freezes in cold ocean areas	Removes mostly fresh water	30% of salts in seawater are retained in ice	Less H_2O	Increase	Yes, through multiple freezings, called *freeze separation*
Evaporation	Seawater evaporates in hot climates	Removes very pure water	None (essentially all salts are left behind)	Less H_2O	Increase	Yes, through evaporation of seawater and condensation of water vapor, called *distillation*

STUDENTS SOMETIMES ASK ...

You mentioned that when seawater freezes, it produces ice with about 10‰ salinity. Once that ice melts, can a person drink it with no ill effects?

Early Arctic explorers found out the answer to your question by necessity. Some of these explorers who traveled by ship in high-latitude regions became inadvertently or purposely entrapped by sea ice (see, for example, the voyage of the *Fram*, which is described in Chapter 1). Lacking other water sources, they used melted sea ice. Although newly formed sea ice contains little salt, it does trap a significant amount of brine (drops of salty water). Depending on the rate of freezing, newly formed ice may have a total salinity from 4 to 15‰. The more rapidly it forms, the more brine it captures and the higher the salinity. Melted sea ice with salinity this high doesn't taste very good, and it still causes dehydration, but not as quickly as drinking 35‰ seawater does. Over time, however, the brine will trickle down through the coarse structure of the sea ice, so its salinity decreases. By the time it is a year old, sea ice normally becomes relatively pure. Drinking melted sea ice enabled these early explorers to survive.

Recall that evaporation is the conversion of water molecules from the liquid state to the vapor state at temperatures below the boiling point. Evaporation removes water from the ocean, leaving its dissolved substances behind. Evaporation, therefore, increases seawater salinity.

Various surface processes either decrease seawater salinity (precipitation, runoff, icebergs melting, or sea ice melting) or increase salinity (sea ice forming and evaporation).

The Hydrologic Cycle

Figure 6–19 shows how the **hydrologic** (*hydro* = water, *logus* = study of) **cycle** relates the processes that affect seawater salinity. These processes recycle water among the ocean, the atmosphere, and the continents, so water is in continual motion between the different components (*reservoirs*) of the hydrologic cycle. Earth's water supply exists in the following proportions:

- 97.2% in the world ocean
- 2.15% frozen in glaciers and ice caps
- 0.62% in groundwater and soil moisture
- 0.02% in streams and lakes
- 0.001% as water vapor in the atmosphere

Additionally, Figure 6–19 shows the average yearly amounts of transfer or *flux* of water between various reservoirs.

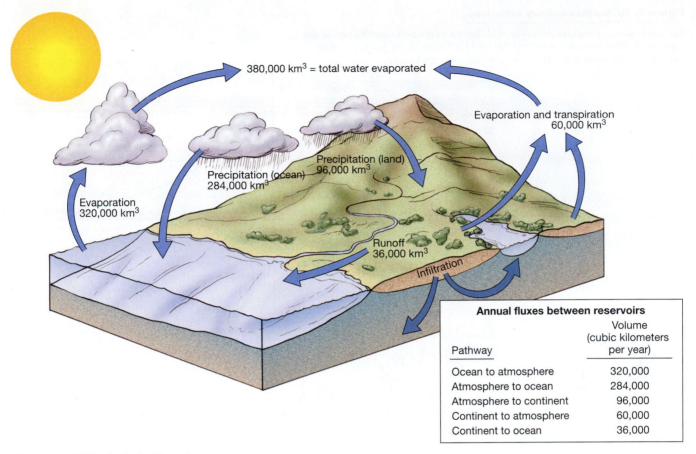

Figure 6–19 The hydrologic cycle.

All water is in continual motion between the various components (reservoirs) of the hydrologic cycle. Volumes are Earth's average yearly amounts in cubic kilometers; table shows average yearly flux between reservoirs; ice not shown.

Labels in figure:

380,000 km³ = total water evaporated

Evaporation and transpiration 60,000 km³

Precipitation (land) 96,000 km³

Precipitation (ocean) 284,000 km³

Evaporation 320,000 km³

Runoff 36,000 km³

Infiltration

Annual fluxes between reservoirs

Pathway	Volume (cubic kilometers per year)
Ocean to atmosphere	320,000
Atmosphere to ocean	284,000
Atmosphere to continent	96,000
Continent to atmosphere	60,000
Continent to ocean	36,000

Surface and Depth Salinity Variation

Average seawater salinity is 35‰, but it varies significantly from place to place at the surface and also with depth.

Surface Salinity Variation

Figure 6–20 shows how salinity varies at the surface with latitude. The red curve shows temperature, which decreases at high latitudes and increases near the Equator. The green curve shows salinity, which is lowest at high latitudes, highest at the Tropics of Cancer and Capricorn, and dips near the Equator. Why does surface salinity vary this way? At high latitudes, abundant precipitation and runoff, and the melting of freshwater icebergs all decrease salinity. In addition, cool temperatures limit the amount of evaporation that takes place (which would increase salinity). The formation and melting of sea ice balance each other out in the course of a year and are not a factor in changes in salinity.

The pattern of Earth's atmospheric circulation (see Chapter 7, "Air–Sea Interaction") causes warm dry air to descend at lower latitudes near the Tropics of Cancer and Capricorn, so evaporation rates are high and salinity increases. Additionally, little precipitation and runoff occur to decrease salinity. As a result, the regions near the Tropics of Cancer and Capricorn are the continental *and* maritime deserts of the world.

Temperatures are warm near the Equator, so evaporation rates are high enough to increase salinity. Increased precipitation and runoff partially offsets the high salinity, though. For example, daily rain showers are common along the Equator, adding water to the ocean and lowering its salinity.

The map in Figure 6–21 shows how ocean surface salinity varies worldwide. Notice how the overall pattern matches the graph in Figure 6–20.

Depth Salinity Variation

Figure 6–22 shows how seawater salinity varies with depth. The graph shows one curve for high-latitude regions and one for low-latitude regions.

For low-latitude regions, the curve begins at the surface with relatively high salinity (as was discussed in the preceding section). Even at the Equator, surface salinity is still relatively high. With increasing depth, the curve swings toward an intermediate salinity value.

Figure 6–20 Surface salinity variation.

Surface seawater temperature (*red curve*) is lowest at the poles and highest at the Equator. Surface seawater salinity (*green curve*) is lowest at the poles, peaks at the Tropics of Cancer and Capricorn, and dips near the Equator.

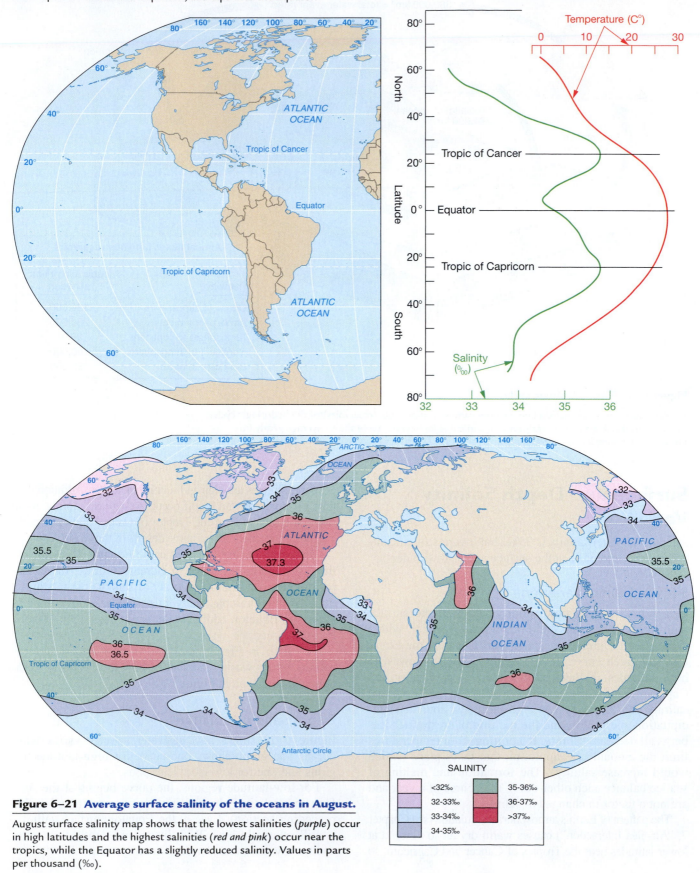

Figure 6–21 Average surface salinity of the oceans in August.

August surface salinity map shows that the lowest salinities (*purple*) occur in high latitudes and the highest salinities (*red and pink*) occur near the tropics, while the Equator has a slightly reduced salinity. Values in parts per thousand (‰).

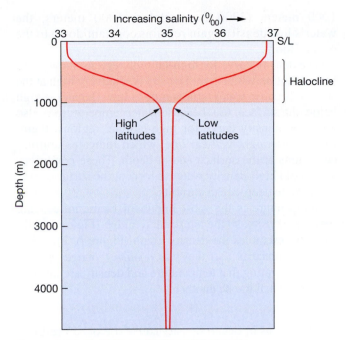

Figure 6–22 Salinity variation with depth.

Vertical profile showing high- and low-latitude salinity variation (horizontal scale in ‰) with depth (vertical scale in meters with sea level at the top). The zone of rapidly changing salinity with depth is the halocline.

For high-latitude regions, the curve begins at the surface with relatively low salinity (again, see the discussion in the preceding section). With increasing depth, the curve also swings toward an intermediate salinity value that approaches the value of the low-latitude salinity curve at the same depth.

These two curves, which together resemble the outline of a martini glass, show that salinity varies widely at the surface, but very little in the deep ocean. Why is this so? It occurs because all the processes that affect seawater salinity (precipitation, runoff, melting icebergs, melting sea ice, sea ice forming, and evaporation) occur at the *surface* and thus have no effect on the deep water below.

Halocline Both curves in Figure 6–22 show a rapid change in salinity between the depths of about 300 meters (980 feet) and 1000 meters (3300 feet). For the low-latitude curve, the change is a *decrease* in salinity. For the high-latitude curve, the change is an *increase* in salinity. In both cases, this layer of rapidly changing salinity with depth is called a **halocline** (*halo* = salt, *cline* = slope). Haloclines separate layers of different salinity in the ocean.

Seawater Density

The density of pure water is 1.000 gram per cubic centimeter (g/cm^3) at 4°C (39°F). This value serves as a standard against which the density of all other substances can be measured. Seawater contains various dissolved substances that increase its density. In the

open ocean, seawater density averages between 1.022 and 1.030 g/cm^3 (depending on its salinity). Thus, the density of seawater is 2 to 3% greater than pure water. Unlike fresh water, seawater continues to increase in density until it freezes at a temperature of −1.9°C (28.6°F) (recall that below 4°C, the density of fresh water actually *decreases*; see Figure 6–10a). At its freezing point, however, seawater behaves similar to fresh water: Its density decreases dramatically, which is why sea ice floats, too.

Density is an important property of ocean water because density differences determine the vertical position of ocean water and cause water masses to float or sink, thereby creating deep ocean currents. For example, if seawater with a density of 1.030 g/cm^3 were added to fresh water with a density of 1.000 g/cm^3, the denser seawater would sink below the fresh water, initiating a deep current.

The ocean, like Earth's interior, is layered according to density. Low-density water exists near the surface and higher-density water occurs below. Except for some shallow inland seas with a high rate of evaporation, the highest-density water is found at the deepest ocean depths. Temperature, salinity, and pressure influence seawater density in the following ways:

- As temperature increases, density decreases[13] (due to thermal expansion).
- As salinity increases, density increases (due to the addition of more dissolved material).
- As pressure increases, density increases (due to the compressive effects of pressure).

Of these three factors, only temperature and salinity influence the density of surface water. Pressure influences seawater density only when very high pressures are encountered, such as in deep ocean trenches. Still, the density of seawater in the deep ocean is only about 5% greater than at the ocean surface, showing that despite tons of pressure per square centimeter, water is nearly incompressible. Unlike air, which can be compressed and put in a tank for scuba diving, the molecules in liquid water are already close together and cannot be compressed much more. Therefore, pressure has the least effect on influencing the density of surface water and can largely be ignored.

Temperature, on the other hand, has the greatest influence on surface seawater density because the range of surface seawater temperature is greater than that of salinity. In fact, only in the extreme polar areas of the ocean, where temperatures are low and remain relatively constant, does salinity significantly affect density. Cold water that also has high salinity is some of the highest-density water in the world. The density of seawater—the result of its salinity and temperature—influences currents

---

[13]A relationship where one variable *decreases* as a result of another variable's *increase* is known as an inverse relationship, in which the two variables are *inversely proportional*.

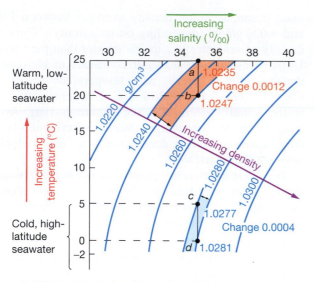

Figure 6–23 Seawater density varies with temperature and salinity.

Blue curves show density, which increases with increasing salinity and decreases with increasing temperature. In high-latitude areas of cold water, temperature has less effect on density than in high-temperature, low-latitude areas.

in the deep ocean because high-density water sinks below less-dense water.

The effects of temperature and salinity changes on density are shown in Figure 6–23. Temperature change affects density (*blue curving lines*) much more in warm, low-latitude regions than in high-latitude regions. Points *a*, *b*, *c*, and *d* mark different densities of water at a constant salinity of 35‰ Line *a–b* shows that the density change across a temperature range of 20 to 25°C (68 to 77°F) is 0.0012 g/cm³. Line *c–d* shows that the density change across a temperature range of 0 to 5°C (32 to 41°F) is only 0.0004 g/cm³. This difference illustrates that a change in temperature of warm, low-latitude water has about three times the effect on density as an equal change in temperature of colder, high-latitude water.

The graph in Figure 6–24a shows how seawater density varies with depth, while the graph in Figure 6–24b shows how seawater temperature varies with depth. Each of these two graphs shows one curve for low-latitude regions (*colored line*) and one for high-latitude regions (*dashed line*).

The density curve for low-latitude regions in Figure 6–24a shows that density is relatively low at the surface. Density is low because temperature is high. (Remember that temperature has the greatest influence on density and temperature is inversely proportional to density.) Below the surface, density remains constant until a depth of about 300 meters (980 feet) because of good surface mixing mechanisms such as surface currents, waves, and tides. Below 300 meters (980 feet), the density increases rapidly until a depth of about

1000 meters (3300 feet). Below 1000 meters, the water's low density again remains constant down to the ocean floor.

The density curve for high-latitude regions shows very little variation with depth. Density is relatively high at the surface because water temperature is low. Density is high below the surface, too, because water temperature is also low. The density curve for high-latitude regions, therefore, is a straight vertical line, which indicates uniform conditions at the surface and at depth. These conditions allow cold high-density water to form at the surface, sink, and initiate deep-ocean currents.

Temperature is the most important factor influencing seawater density, so the temperature graph (Figure 6–24b) strongly resembles the density graph (Figure 6–24a). The only difference is that they are *a mirror image of each other*, illustrating that temperature and density are inversely proportional to one another.

> Differences in ocean density cause the ocean to be layered. Seawater density increases with decreased temperature, increased salinity, and increased pressure.

Pycnocline and Thermocline

Analogous to the halocline (the layer of rapidly changing salinity in Figure 6–22), the low-latitude curve in Figure 6–24a shows a layer of rapidly changing density called a **pycnocline** (*pycno* = density, *cline* = slope), and the low-latitude graph in Figure 6–24b shows a layer of rapidly changing temperature called a **thermocline** (*thermo* = heat, *cline* = slope). Like the halocline, the pycnocline and thermocline typically occur between about 300 meters (980 feet) and 1000 meters (3300 feet) below the surface.

When a pycnocline is established in an area, it presents an incredible barrier to mixing between low-density water above and high-density water below. A pycnocline has a high gravitational stability and thus physically isolates adjacent layers of water.[14] The pycnocline results from the combined effect of the thermocline and the halocline, because temperature and salinity influence density. The interrelation of these three layers determines the degree of separation between the upper-water and deep-water masses. Figure 6–24c shows that the **mixed surface layer** occurs above a strong permanent thermocline (and corresponding pycnocline). The water is uniform because it is well mixed by surface currents, waves, and tides. The thermocline and pycnocline occur in a relatively low-density layer called the **upper water**, which is well developed throughout the low and middle latitudes. Denser and

---

[14]This is similar to a temperature inversion in the atmosphere, which traps cold (high-density) air underneath warm (low-density) air.

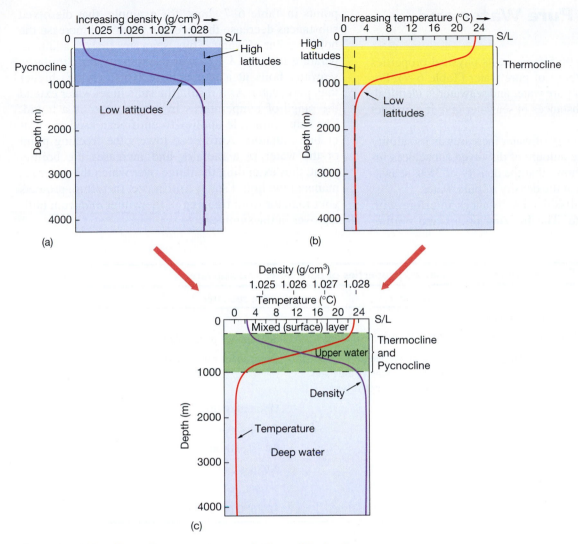

Figure 6–24 Density and temperature variations with depth.

(a) Density variation with depth for high- and low-latitude regions. The zone of rapidly changing density with depth in low latitudes is the pycnocline. **(b)** Temperature variation with depth for high- and low-latitude regions. The zone of rapidly changing temperature with depth in low latitudes is the thermocline. **(c)** A typical ocean profile showing the variation of temperature and density with depth, showing the inverse relationship between temperature and density.

colder **deep water** extends from below the thermocline/pycnocline to the deep-ocean floor.

At depths above the main thermocline, divers often experience lesser thermoclines (and corresponding pycnoclines) when descending into the ocean. Thermoclines can develop in swimming pools, ponds, and lakes, too. During the spring and fall, when nights are cool but days can be quite warm, the Sun heats the surface water of the pool yet the water below the surface can be quite cold. If the pool has not been mixed, a thermocline isolates the warm surface layer from the deeper cold water. The cold water below the thermocline can be quite a surprise for anyone who dives into the pool!

In high-latitude regions, the temperature of the surface water remains cold year round, so there is very little difference between the temperature at the surface and in deep water below. Thus, a thermocline and corresponding pycnocline rarely develop in high-latitude regions. Only during the short summer when the days are long does the Sun begin to heat the surface water. Even then, the water does not heat up very much. Nearly all year, then, the water column in high latitudes is **isothermal** (*iso* = same, *thermo* = heat) and **isopycnal** (*iso* = same, *pycno* = density), allowing good vertical mixing between surface and deeper water.

A halocline is a layer of rapidly changing salinity, a pycnocline is a layer of rapidly changing density, and a thermocline is a layer of rapidly changing temperature.

Comparing Pure Water and Seawater

Because seawater is 96.5% water, its physical properties are very similar to those of pure water (Table 6–7). For instance, the color of pure water and seawater is identical, yet the dissolved substances in seawater give it a distinct odor and taste.

Recall that the density of water increases as its salinity increases because the amount of dissolved substances increases. Table 6–7 shows that the density of 35‰ seawater is 2.8% higher than the density of pure water.

Recall also that dissolved substances interfere with water changing state. The freezing points and boiling points in Table 6–7 show, for example, that dissolved substances decrease the freezing point and increase the boiling point of water. Thus, seawater freezes at a temperature 1.9°C (3.4°F) lower than pure water. Similarly, seawater boils at a temperature 0.6°C (1.1°F) higher than pure water. As a result, the salts in seawater extend the range of temperatures in which water is a liquid. The same principle applies to antifreeze used in automobile radiators. Antifreeze lowers the freezing point of the water in a radiator and increases the boiling point, thus extending the range over which the water remains in the liquid state. Antifreeze, therefore, protects your radiator from freezing in the winter *and* from boiling over in the summer.

TABLE 6–7 Comparison of selected properties of pure water and seawater.

Property	Pure water	35‰ seawater
Color (light transmission)		
· Small quantities of water	Clear (high transparency)	Same as for pure water
· Large quantities of water	Blue-green because water molecules scatter blue and green wavelengths best	Same as for pure water
Odor	Odorless	Distinctly marine
Taste	Tasteless	Distinctly salty
pH	7.0 (neutral)	8.1 (slightly alkaline)
Density at 4°C (39°F)	1.000 g/cm³	1.028 g/cm³
Freezing point	0°C (32°F)	−1.9°C (28.6°F)
Boiling point	100°C (212°F)	100.6°C (213°F)

Chapter in Review

- *Water's remarkable properties help make life as we know it possible on Earth.* These properties include the arrangement of its atoms, how its molecules stick together, its ability to dissolve almost everything, and its heat storage capacity.

- *The water molecule is composed of one atom of oxygen and two atoms of hydrogen (H_2O).* The two hydrogen atoms, which are covalently bonded to the oxygen atom, are *attached to the same side of the oxygen atom* and produce a bend in the geometry of a water molecule. This geometry makes water molecules *polar*, which allows them to form hydrogen bonds with other water molecules or other substances and gives water its remarkable properties. Water, for example, is *"the universal solvent"* because it can hydrate charged particles (ions), thereby dissolving them.

- *Water is one of the few substances that exists naturally on Earth in all three states of matter* (solid, liquid, gas). Hydrogen bonding gives water *unusual thermal properties*, such a high freezing point (0°C or 32°F) and boiling point (100°C or 212°F), a high heat capac-

ity (1 calorie), a high latent heat of melting (80 calories per gram), and a high latent heat of vaporization (540 calories per gram). Water's high heat capacity and latent heats have important implications in regulating global thermostatic effects.

- *The density of water increases as temperature decreases*, similar to most substances, and reaches a maximum density at 4°C (39°F). *Below 4°C, however, water density decreases with temperature*, due to the formation of bulky ice crystals. *As water freezes, it expands by about 9% in volume*, so ice floats on water.

- *Salinity is the amount of dissolved solids in ocean water.* It averages about 35 grams of dissolved solids per kilogram of ocean water [*35 parts per thousand (‰)*] but ranges from brackish to hypersaline. Six ions—chloride, sodium, sulfate, magnesium, calcium, and potassium—account for over 99% of the dissolved solids in ocean water. These ions always occur in a *constant proportion* in any seawater sample, so salinity can be determined by measuring the concentration of only one—typically, the chloride ion. The residence time of various elements indicates how long they stay in the ocean and imply that *ocean salinity has remained constant through time.*

- The abundance of *dissolved gases in seawater* is regulated by the solubility of the gas in water and by various processes, including photosynthesis and respiration. A *natural buffering system* exists in the ocean, based on the chemical reaction of carbon dioxide in water. This buffering system soaks up excess atmospheric carbon dioxide and *prohibits large changes in pH*, which creates a stable ocean environment.

- *Dissolved components in seawater are added and removed by a variety of processes.* Precipitation, runoff, and the melting of icebergs and sea ice add fresh water to seawater and decrease its salinity. The formation of sea ice and evaporation remove fresh water from seawater and increase its salinity. The hydrologic cycle includes all the reservoirs of water on Earth, including the oceans, which contain 97% of Earth's water.

- *The salinity of surface water varies considerably due to surface processes*, with the maximum salinity found near the Tropics of Cancer and Capricorn and the minimum salinity found in high-latitude regions. Salinity also varies with depth down to about 1000 meters but below that, the salinity of deep water is very consistent. A *halocline* is a layer of rapidly changing salinity.

- *Seawater density increases as temperature decreases and salinity increases*, though temperature influences surface seawater density more strongly than salinity (the influence of pressure is negligible). Density and temperature vary considerably with depth in low-latitude regions, creating a *pycnocline* and corresponding *thermocline*, which are generally absent in high latitudes.

- *The physical properties of pure water and seawater are remarkably similar*, with a few notable exceptions. Compared to pure water, seawater has a higher pH, density, and boiling point (but a lower freezing point).

Key Terms

Acid (p. 181)
Alkaline (p. 181)
Atom (p. 164)
Base (p. 181)
Boiling point (p. 168)
Brackish (p. 174)
Buffering (p. 182)
Calorie (p. 167)
Capillarity (p. 166)
Chlorinity (p. 177)
Cohesion (p. 165)
Condensation point (p. 168)
Condense (p. 168)
Conservative constituent (p. 180)
Covalent bond (p. 165)
Deep water (p. 189)
Dipolar (p. 165)

Electron (p. 164)
Electrostatic attraction (p. 166)
Evaporation (p. 169)
Freezing point (p. 167)
Halocline (p. 187)
Heat (p. 167)
Heat capacity (p. 168)
Hydrogen bond (p. 165)
Hydrologic cycle (p. 184)
Hypersaline (p. 174)
Ionic bond (p. 166)
Ion (p. 164)
Isopycnal (p. 189)
Isothermal (p. 189)
Kinetic energy (p. 167)
Latent heat of condensation
(p. 169)

Latent heat of evaporation
(p. 169)
Latent heat of freezing (p. 170)
Latent heat of melting (p. 169)
Latent heat of vaporization
(p. 169)
Melting point (p. 167)
Mixed surface layer (p. 188)
Molecule (p. 164)
Neutral (p. 182)
Neutron (p. 164)
Nonconservative constituent
(p. 180)
Nucleus (p. 164)
Parts per thousand (‰) (p. 174)
pH scale (p. 182)
Polarity (p. 165)

Precipitation (p. 170)
Principle of constant
proportions (p. 177)
Proton (p. 164)
Pycnocline (p. 188)
Residence time (p. 179)
Salinity (p. 174)
Salinometer (p. 177)
Surface tension (p. 165)
Temperature (p. 167)
Thermal contraction (p. 170)
Thermocline (p. 188)
Thermostatic effect (p. 170)
Upper water (p. 188)
van der Waals force (p. 167)
Vapor (p. 167)

Questions and Exercises

1. Sketch a model of an atom, showing the positions of the sub-atomic particles protons, neutrons, and electrons.

2. Describe what condition exists in water molecules to make them dipolar.

3. Sketch several water molecules, showing all covalent and hydrogen bonds. Be sure to indicate the polarity of each water molecule.

4. How does hydrogen bonding produce the surface tension phenomenon of water?

5. Discuss how the dipolar nature of the water molecule makes it such an effective solvent of ionic compounds.

6. Describe the differences between the three states of matter, using the arrangement of molecules in your explanation.

7. Why are the freezing and boiling points of water higher than would be expected for a compound of its molecular makeup?

8. How does the heat capacity of water compare with that of other substances? Describe the effect this has on climate.

9. The heat energy added as latent heat of melting and latent heat of vaporization does not increase water temperature. Explain why this occurs and where the energy is used.

10. Why is the latent heat of vaporization so much greater than the latent heat of melting?

11. Describe how excess heat energy absorbed by Earth's low-latitude regions is transferred to heat-deficient higher latitudes through a process that uses water's latent heat of evaporation.

12. As water cools, two distinct changes take place in the behavior of molecules: Their slower movement tends to increase density, whereas the formation of bulky ice crystals decreases density. Describe how the relative rates of their occurrence cause pure water to have a temperature of maximum density at 4°C (39.2°F) and make ice less dense than liquid water.

13. What is your state sales tax, in parts per thousand?

14. What are goiters? How can they be avoided?

15. What physical conditions create brackish water in the Baltic Sea and hypersaline water in the Red Sea?

16. What condition of salinity makes it possible to determine the total salinity of ocean water by measuring the concentration of only one constituent, the chloride ion?

17. Describe the factors that control the amount of gas that dissolves in water.

18. Explain the difference between an acid and an alkali (base) substance. How does the ocean's buffering system work?

19. Describe the ways that dissolved components are added and removed from seawater.

20. List the components (reservoirs) of the hydrologic cycle that hold water on Earth and the percentage of Earth's water in each one. Describe the processes by which water moves among these reservoirs.

21. Explain why there is such a wide variation of surface salinity, but such a narrow range of salinity at depth.

22. Why is there such a close association between (a) the curve showing seawater density variation with ocean depth and (b) the curve showing seawater temperature variation with ocean depth?

Iceberg above and below water. Full view of an iceberg, showing that 90% of an iceberg's mass is below water. Interactions between sea ice, the ocean, and the atmosphere help regulate Earth's climate.

Key Questions

■ What causes Earth's seasons?

■ How does the Coriolis effect influence moving objects?

■ How and why does the atmosphere move?

■ What are the characteristics of Earth's major wind belts and boundaries?

■ How does a hurricane work?

■ What is the difference between sea ice and icebergs?

■ What causes the greenhouse effect?

The answers to these questions (and much more) can be found in the highlighted concept statements within this chapter.

"When the still sea conspires an armor

And her sullen and aborted

Currents breed tiny monsters,

True sailing is dead.

Awkward instant

And the first animal is jettisoned . . . "

—The Doors, *Horse Latitudes* (1972)

The atmosphere and the ocean act as one interdependent system. What happens in one causes changes in the other, and the two are linked by complex feedback loops. Surface currents in the oceans, for instance, are a direct result of Earth's atmospheric wind belts. Conversely, certain atmospheric weather phenomena are manifested in the oceans. If either the oceans or the atmosphere is to be understood, then their interactions and relationships must be studied.

Atmospheric winds create most of the surface currents and waves in the ocean. Solar energy, in turn, which heats Earth's surface, creates the winds. Radiant energy from the Sun, therefore, is responsible for motion in the atmosphere and the ocean. Recall from Chapter 6 that the atmosphere and ocean use the high heat capacity of water to constantly exchange this energy, shaping Earth's global weather patterns in the process.

Periodic extremes of atmospheric weather, such as droughts and profuse precipitation, are related to periodic changes in oceanic conditions. For instance, it was recognized as far back as the 1920s that El Niño—an ocean event—was tied to catastrophic weather events worldwide. What is as yet unclear, however, is whether changes in the ocean produce changes in the atmosphere that lead to the El Niño phenomenon—or vice versa. El Niño–Southern Oscillation events are discussed in Chapter 8, "Ocean Circulation."

Air–sea interactions influence the greenhouse effect, too. In the 1980s, studies confirmed that Earth's average temperature had risen over the past century. Global warming may be due to human-caused increases in carbon dioxide and other gases that absorb heat. However, atmospheric carbon dioxide has increased only half as much as predicted from human activities, so where did the rest of the carbon dioxide go? If the oceans are removing carbon dioxide from the atmosphere, how does carbon dioxide enter the oceans and where does it go?

These questions are answered slowly as new research is completed and more sophisticated computer models are developed. In this chapter, we examine the redistribution of solar heat by the atmosphere and its influence on oceanic conditions. First, large-scale phenomena that influence air–sea interactions are studied, and then smaller-scale phenomena are examined.

Uneven Solar Heating on Earth

The side of Earth facing the Sun (the daytime side) receives a tremendous dose of intense solar energy. This energy drives the global ocean–atmosphere engine, creating pressure and density differences that stir currents and waves in both the atmosphere and the ocean.

Distribution of Solar Energy

If Earth were a flat plate in space, with its flat side directly facing the Sun, then sunlight would fall equally on all parts of Earth. Earth is spherical, however, so the amount and intensity of solar radiation received at higher latitudes is much less than at lower latitudes. The following factors influence the amount of radiation received at low and high latitudes:

- Sunlight strikes low latitudes at a high angle, so the radiation is concentrated in a relatively small area (*area A* in Figure 7-1). Sunlight strikes high latitudes at a low angle, so the same amount of radiation is spread over a larger area (*area B* in Figure 7-1).

- Earth's atmosphere absorbs some radiation, so less radiation strikes Earth at high latitudes than at low latitudes, because sunlight passes through more of the atmosphere at high latitudes.

- The *albedo* (*albus* = white) of various Earth materials is the percentage of incident radiation that is reflected back to space. The average albedo of Earth's surface is about 30%. More radiation is reflected back to space at high latitudes because ice has a much higher albedo than soil or vegetation.

- The angle at which sunlight strikes the ocean surface determines how much is absorbed and how much is reflected. If the Sun shines down on a smooth sea from directly overhead, only 2% of the radiation is reflected, but if the Sun is only 5 degrees above the horizon, 40% is reflected back into the atmosphere (Table 7-1). Thus, the ocean reflects more radiation at high latitudes than at low latitudes.

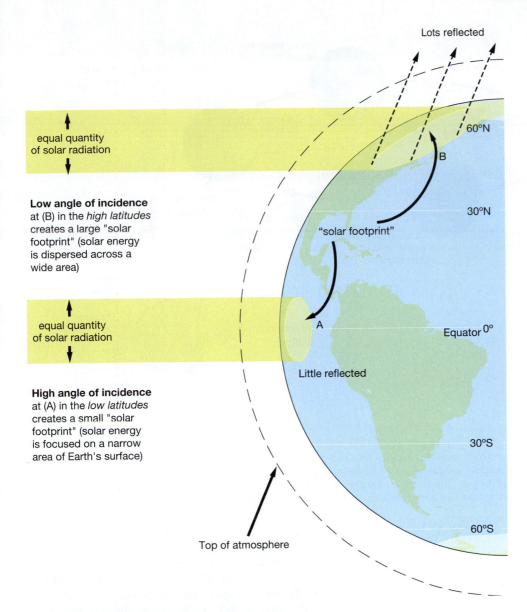

Lots reflected

equal quantity
of solar radiation

60°N

B

Low angle of incidence
at (B) in the *high latitudes*
creates a large "solar
footprint" (solar energy
is dispersed across a
wide area)

30°N

"solar footprint"

equal quantity
of solar radiation

A

Equator 0°

Little reflected

High angle of incidence
at (A) in the *low latitudes*
creates a small "solar
footprint" (solar energy
is focused on a narrow
area of Earth's surface)

30°S

60°S

Top of atmosphere

Figure 7–1 Solar radiation received on Earth.
Two identical beams of sunlight strike Earth. At *A*, the light beam is focused on a narrow area of
Earth's surface and produces a smaller "solar footprint"; at *B*, the light beam is dispersed across a
wide area and produces a larger "solar footprint." Additionally, more light is reflected at *B*. Thus,
the amount of solar energy received at higher latitudes is much less than that at lower latitudes.

Because of all these reasons, the intensity of radiation at
high latitudes is greatly decreased compared with that
falling in the equatorial region.

In addition, the amount of radiation received at Earth's
surface varies *daily* because Earth rotates on its axis so
the surface experiences daylight and darkness each day.
The amount of radiation also varies *annually* due to
Earth's seasons.

Earth's Seasons

Earth revolves around the Sun along an elliptical path
(Figure 7-2). The plane traced by Earth's orbit is called
the *ecliptic*. Earth's axis of rotation is not perpendicular
("upright") on the ecliptic; rather, it tilts at an angle of
23.5 degrees. Figure 7-2 shows that throughout the yearly
cycle, Earth's axis *always points in the same direction*,
which is toward Polaris, the North Star.

TABLE 7-1 **Reflection and absorption of solar energy relative to the angle
of incidence on a flat sea.**

Elevation of the Sun above the horizon	90°	60°	30°	15°	5°
Reflected radiation (%)	2	3	6	20	40
Absorbed radiation (%)	98	97	94	80	60

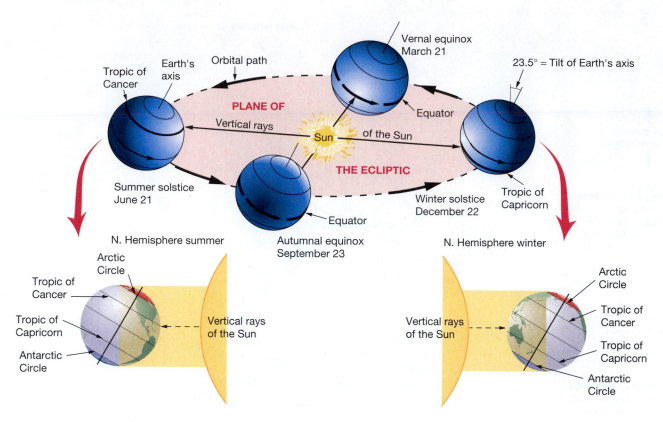

Figure 7–2 Earth's seasons.

As Earth orbits the Sun during one year, its axis of rotation constantly tilts 23.5 degrees from perpendicular (relative to the plane of the ecliptic) and always points in the same direction, causing different areas to experience vertical rays of the Sun. This tilt causes Earth to have seasons, such as when the Northern Hemisphere is tilted toward the Sun during the summer (*left inset*), and away from the Sun during the winter (*right inset*).

The tilt of Earth's rotational axis (and not its elliptical path) causes Earth to have seasons. Spring, summer, fall, and winter occur as follows:

- At the **vernal equinox** (*vernus* = spring; *equi* = equal, *noct* = night), which occurs on or about March 21, the Sun is directly overhead along the Equator. During this time, all places in the world experience equal lengths of night and day (hence the name *equinox*). In the Northern Hemisphere, the vernal equinox is also known as the spring equinox.

- At the **summer solstice** (*sol* = the Sun, *stitium* = a stoppage), which occurs on or about June 21, the Sun reaches its most northerly point in the sky, directly overhead along the **Tropic of Cancer**, at 23.5 degrees north latitude (Figure 7-2, *left inset*). To an observer on Earth, the noonday Sun reaches its northernmost or southernmost position in the sky at this time and appears to pause—hence the term *solstice*—before beginning its next 6-month cycle.

- At the **autumnal** (*autumnus* = fall) **equinox**, which occurs on or about September 23, the Sun is directly overhead along the Equator again. In the Northern Hemisphere, the autumnal equinox is also known as the fall equinox.

- At the **winter solstice**, which occurs on or about December 22, the Sun is directly overhead along the **Tropic of Capricorn**,[1] at 23.5 degrees south latitude (Figure 7-2, *right inset*). In the Southern Hemisphere, the seasons are reversed. Thus, the winter solstice is the time when the Southern Hemisphere is most directly facing the Sun, which is the beginning of the Southern Hemisphere summer.

Because Earth's rotational axis is tilted 23.5 degrees, the Sun's *declination* (angular distance from the equatorial plane) varies between 23.5 degrees north and 23.5 degrees south of the Equator on a yearly cycle. As a result, the region between these two latitudes (called the **tropics**) receives much greater annual radiation than polar areas.

Seasonal changes in the angle of the Sun and the length of day profoundly influence Earth's climate. In the Northern Hemisphere, for example, the longest day occurs on the summer solstice and the shortest day on the winter solstice.

Daily heating of Earth also influences climate in most locations. Exceptions to this pattern occur north of the

---

[1]The tropics are named for their constellations: Directly over the Tropic of Cancer is the constellation Cancer (the crab) and over the Tropic of Capricorn is Capricorn (the goat).

Arctic Circle (66.5 degrees north latitude) and south of the **Antarctic Circle** (66.5 degrees south latitude), which at certain times of the year do not experience daily cycles of daylight and darkness. For instance, during the Northern Hemisphere winter, the area north of the Arctic Circle receives no direct solar radiation at all and experiences up to six months of darkness. At the same time, the area south of the Antarctic Circle receives continuous radiation ("midnight Sun"), so it experiences up to six months of light. Half a year later, during the Northern Hemisphere summer (the Southern Hemisphere winter), the situation is reversed.

> Earth's axis is tilted at an angle of 23.5 degrees, causing the Northern and Southern Hemispheres to take turns "leaning toward" the Sun every six months, resulting in the change of seasons.

Oceanic Heat Flow

Close to the poles, much incoming solar radiation strikes Earth's surface at low angles. Combined with the high albedo of ice, more energy is reflected back into space than absorbed. In contrast, between 35 degrees north latitude and 40 degrees south latitude,[2] sunlight strikes Earth at much higher angles and more energy is absorbed than reflected back into space. The graph in Figure 7-3 shows how incoming sunlight and outgoing heat combine on a

---

[2]The reason this latitudinal range extends farther into the Southern Hemisphere is because the Southern Hemisphere has more ocean surface area in the mid-latitudes than the Northern Hemisphere does.

daily basis for a net heat gain in low-latitude oceans and a net heat loss in high-latitude oceans.

Based on Figure 7-3, you might expect the equatorial zone to grow progressively warmer and the polar regions to grow progressively cooler. The polar regions are always considerably colder than the equatorial zone, but the temperature *difference* remains the same because excess heat is transferred from the equatorial zone to the poles. How is this accomplished? Circulation in both the oceans and the atmosphere transfers the heat.

The Atmosphere: Physical Properties

The atmosphere transfers heat and water vapor from place to place on Earth. Within the atmosphere, complex relationships exist among air composition, temperature, density, water vapor content, and pressure. Before we apply these relationships, let's examine the atmosphere's composition and some of its physical properties.

Composition   Table 7-2 shows that the atmosphere consists almost entirely of nitrogen and oxygen gases. Other

TABLE 7-2 **Composition of dry air.**

Gas	Concentration (%)
Nitrogen (N_2)	78.1
Oxygen (O_2)	20.9
Argon (Ar)	0.9
Carbon dioxide (CO_2)	0.037
All others	trace

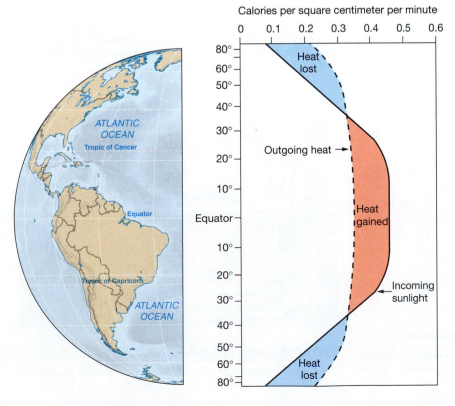

Calories per square centimeter per minute

Figure 7–3 Heat gain and heat loss from the oceans at various latitudes.

Heat gained by the oceans in equatorial latitudes (*red portion of graph*) equals heat lost in polar latitudes (*blue portion of graph*). On average, the two balance each other and the excess heat from equatorial latitudes is transferred to heat-deficient polar latitudes by both oceanic and atmospheric circulation.

gases include argon (an inert gas), carbon dioxide, and others in trace amounts. Although these gases are present in very small amounts, they can trap significant amounts of heat within the atmosphere.

Temperature   Figure 7-4 shows the temperature profile of the atmosphere. The lowermost portion of the atmosphere, which extends from the surface to about 12 kilometers (7 miles), is called the **troposphere** (*tropo* = turn, *sphere* = a ball) and is where all weather is produced. The abundance of atmospheric mixing in the troposphere gives this layer its name. Within the troposphere, temperature gets cooler with altitude. At high altitudes, the air temperature is well below freezing. If you have ever flown in a jet airplane, for instance, you may have noticed that any water on the wings or inside your window freezes during a high altitude flight.

Density   It may seem surprising that air has density, but since air is composed of molecules, it certainly does. Temperature has a dramatic affect on the density of air. At higher temperatures, for example, air molecules move more quickly, take up more space, and density is decreased. Thus, the general relationship between density and temperature is as follows:

- Warm air is less dense, so it rises, which is commonly expressed as "heat rises."
- Cool air is more dense, so it sinks.

Figure 7-5 shows how a radiator (heater) uses convection to heat a room. The heater warms the nearby air and causes it to expand. This expansion makes the air less dense, causing it to rise. Conversely, a cold window cools the nearby air and causes it to contract, thereby becoming more dense, which causes it to sink. A **convection** (*con* = with, *vect* = carried) **cell** forms, composed of the rising and sinking air moving in a circular fashion, similar to the convection in Earth's mantle discussed in Chapter 3.

Water Vapor Content   The amount of water vapor in air depends in part on the air's temperature. Warm air, for instance, can hold more water vapor than cold air because the air molecules are moving more quickly and come into contact with more water vapor. Thus, warm air is typically moist, and, conversely, cool air is typically dry. As a result, a warm, breezy day speeds evaporation when you hang your laundry outside to dry.

Water vapor influences the density of air. The addition of water vapor decreases the density of air because water vapor has a lower density than air. Thus, humid air is less dense than dry air.

Pressure   Atmospheric pressure is 1.0 atmosphere[3] (14.7 pounds per square inch) at sea level and decreases with increasing altitude. Atmospheric pressure depends on the weight of the column of air above. For instance, a thick column of air produces higher atmospheric pressure than a thin column of air. An analogy to this is water pressure in a swimming pool: The thicker the column of water above, the higher the water pressure. Thus, the highest pressure in a pool is at the bottom of the deep end.

---

[3]The *atmosphere* is a unit of pressure; 1.0 atmosphere is the average pressure exerted by the overlying atmosphere at sea level and is equivalent to 760 millimeters of mercury.

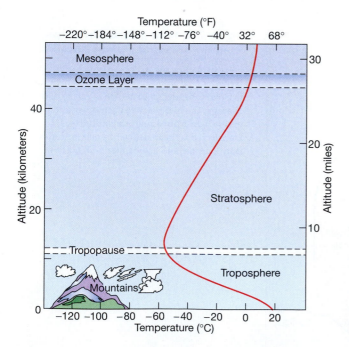

Figure 7–4   Temperature profile of the atmosphere.

Within the troposphere, the atmosphere gets cooler with increasing altitude and above the troposphere, the atmosphere generally warms.

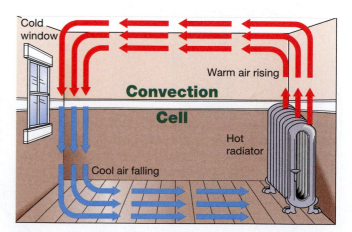

Figure 7–5   Convection in a room.

A circular-moving loop of air (a *convection cell*) is caused by warm air rising and cool air sinking.

Similarly, the thicker column of air at sea level means air pressure is high at sea level and decreases with increasing elevation. When sealed bags of potato chips or pretzels are taken to a high elevation, the pressure is much lower than where they were sealed, sometimes causing the bags to burst! You may have also experienced this change in pressure when your ears "popped" during the takeoff or landing of an airplane or while driving on steep mountain roads.

Changes in atmospheric pressure cause air movement as a result of changes in the molecular density of the air. The general relationship is as follows (Figure 7-6):

- A column of cool, dense air causes high pressure at the surface, which will lead to sinking air (movement *toward* the surface and compression).
- A column of warm, less dense air causes low pressure at the surface, which will lead to rising air (movement *away from* the surface and expansion).

In addition, sinking air tends to warm because of its compression, while rising air tends to cool due to expansion. Note that there are complex relationships among air composition, temperature, density, water vapor content, and pressure.

Movement Air *always* moves from high-pressure regions toward low-pressure regions. This moving air is called *wind*. If a balloon is inflated and let go, what happens to the air inside the balloon? It rapidly escapes, moving from a high-pressure region inside the balloon (caused by the balloon pushing on the air inside) to the lower-pressure region outside the balloon.

> The atmosphere's changing temperature, density, water vapor content, and pressure cause atmospheric movement, initiating wind.

An Example: A Nonspinning Earth

Imagine for a moment that Earth is not spinning on its axis but that the Sun rotates around Earth, with the Sun directly above Earth's Equator at all times (Figure 7-7). Because more solar radiation is received along the Equator than at the poles, the air at the Equator in contact with Earth's surface is warmed. This warm, moist air rises, creating low pressure at the surface. This rising air cools (see Figure 7-4) and releases its moisture as rain. Thus, a zone of low pressure and much precipitation occurs along the Equator.

As the air along the Equator rises, it reaches the top of the troposphere and begins to move toward the poles. Because the temperature is much lower at high altitudes, the air cools and its density increases. This cool, dense air sinks at the poles, creating high pressure at the surface. The sinking air is quite dry because cool air cannot hold much water vapor. Thus, there is high pressure and clear, dry weather at the poles.

Which way will surface winds blow? Air always moves from high pressure to low pressure, so air travels from the high pressure at the poles toward the low pressure at the Equator. Thus, there are strong northerly winds in the Northern Hemisphere and strong southerly winds in the Southern Hemisphere.[4] The air warms as it

---

[4]Notice that winds are named based on the direction *from which they are moving*.

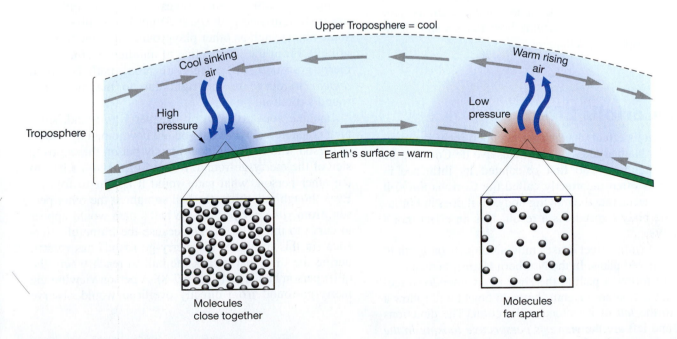

Upper Troposphere = cool

Cool sinking air

Warm rising air

High pressure

Low pressure

Troposphere

Earth's surface = warm

Molecules close together

Molecules far apart

Figure 7–6 High and low atmospheric pressure zones.

A column of cool, dense air causes high pressure at the surface (*left*), which will lead to sinking air. A column of warm, less dense air causes low pressure at the surface (*right*), which will lead to rising air.

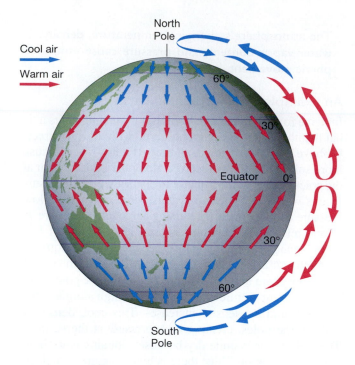

Cool air

Warm air

North Pole

60°

30°

Equator 0°

30°

60°

South Pole

Figure 7–7 Atmospheric circulation on a nonspinning Earth.

A fictional nonspinning Earth with the Sun rotating around Earth directly above Earth's Equator at all times. Arrows show the pattern of winds that would develop due to uneven solar heating on Earth.

makes its way back to the Equator, completing the loop (called a *convection* or *circulation cell*; see Figure 7-5).

Is this fictional case of a nonspinning Earth a good analogy to what is really happening on Earth? Actually, it is not, even though the *principles* that drive the physical movement of air remain the same whether Earth is spinning or not. Let's now examine how Earth's spin influences atmospheric circulation.

The Coriolis Effect

The **Coriolis effect** changes the intended path of a moving body. Named after Gaspard Gustave de Coriolis, the French engineer who first calculated its influence in 1835, it is often incorrectly called the Coriolis *force*. It does not accelerate the moving body, so it does not influence the body's speed. As a result, it is an effect, not a true force.

The Coriolis effect causes moving objects on Earth to follow curved paths. In the Northern Hemisphere, an object will follow a path to the *right* of its intended direction; in the Southern Hemisphere, an object will follow a path to the *left* of its intended direction. The directions right and left are the *viewer's perspective looking in the direction in which the object is traveling*. For example, the Coriolis effect very slightly influences the movement of a ball thrown between two people. In the Northern

Hemisphere, the ball will veer slightly to its right *from the thrower's perspective*.

The Coriolis effect acts on all moving objects. However, it is much more pronounced on objects traveling long distances, especially north or south. This is why the Coriolis effect has a dramatic effect on atmospheric circulation and the movement of ocean currents.

The Coriolis effect is a result of Earth's rotation toward the east. More specifically, the *difference* in the speed of Earth's rotation at different latitudes causes the Coriolis effect. In reality, objects travel along straight-line paths,[5] but Earth rotates underneath them, making the objects appear to curve. Let's look at two examples to help clarify this.

Example 1: Perspectives and Frames of Reference on a Merry-Go-Round

A merry-go-round is a useful experimental apparatus with which to test some of the concepts of the Coriolis effect. A merry-go-round is a large circular wheel that rotates around its center. It has bars that people hang onto while the merry-go-round spins, as shown in Figure 7-8.

Imagine that you are on a merry-go-round that is spinning counterclockwise as viewed from above (Figure 7-8). As you are spinning, what will happen to you if you let go of the bar? If you guessed that you would fly off along a straight-line path perpendicular to the merry-go-round (Figure 7-8, *path A*), that's not quite right. Your angular momentum would propel you in a straight line *tangent* to your circular path on the merry-go-round at the point where you let go (Figure 7-8, *path B*). The law of inertia states that a moving object will follow a straight-line path until it is compelled to change that path by other forces. Thus, you would follow a straight-line path (*path B*) until you collide with some object such as other playground equipment or the ground. From the perspective of another person on the merry-go-round, your departure along path B would *appear* to curve to the right due to the merry-go-round's rotation.

Imagine you are again on the merry-go-round, spinning counterclockwise, but you are now joined by another person who is facing you directly but on the opposite side of the merry-go-round. If you were to toss a ball to the other person, what path would it appear to follow? Even though you threw the ball straight at the other person, from *your perspective* the ball's path would appear to curve to the right. That's because the frame of reference (in this example, the merry-go-round) has rotated during the time that it took the ball to reach where the other person had been (Figure 7-8). A person viewing the merry-go-round from directly overhead would observe

---

[5]Newton's first law of motion (the law of inertia) states that every body persists in its state of rest or of uniform motion in a straight line unless it is compelled to change that state by forces imposed on it.

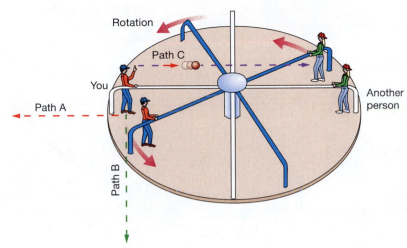

that the ball did indeed travel along a straight-line path (Figure 7–8, *path C*), just as your path was straight when you let go of the merry-go-round bar. Similarly, the perspective of being on the rotating Earth causes objects to appear to travel along curved paths. This is the Coriolis effect. The merry-go-round spinning in a counterclockwise direction is analogous to the Northern Hemisphere because, as viewed from above the North Pole, Earth is spinning counterclockwise. Thus, moving objects appear to follow curved paths to the *right* of their intended direction in the Northern Hemisphere.

If the other person on the merry-go-round had thrown a ball toward you, it would also appear to have curved. From the perspective of the other person, the ball would appear to curve to its right, just as the ball you threw curved. From your perspective, however, the ball thrown toward you would appear to curve to its left. The perspective to keep in mind when considering the Coriolis effect is the one *looking in the same direction that the object is moving.*

To simulate the Southern Hemisphere, the merry-go-round would need to rotate in a *clockwise* direction, which is analogous to Earth when viewed from above the South Pole. Thus, moving objects appear to follow curved paths to the *left* of their intended direction in the Southern Hemisphere.

STUDENTS SOMETIMES ASK...

Is it true that the Coriolis effect causes water to drain one way in the Northern Hemisphere and the other way in the Southern?

In most cases, no. If all other effects are nullified, however, the Coriolis effect comes into play and makes draining water spiral counterclockwise north of the Equator and the other way in the Southern Hemisphere (the same direction that hurricanes spin). But the Coriolis effect is extremely weak on small systems like a basin of water. The shape and irregularities of the basin, local slopes, or any external movement can easily outweigh the Coriolis effect in determining the direction water drains.

Example 2: A Tale of Two Missiles

The distance that a point on Earth has to travel in a day is shorter with increasing latitude. A location near the pole, for example, travels in a circle not nearly as far in a day as will an area near the Equator. Since both areas travel their respective distances in one day, the velocity of the two areas must not be the same. Figure 7-9a, shows that as Earth rotates on its axis, the velocity decreases with latitude, ranging from over 1600 kilometers (1000 miles) per hour at the Equator to 0 kilometers per hour at the poles. *This change in velocity with latitude is the true cause of the Coriolis effect.* The following example illustrates how velocity changes with latitude.

Imagine we have two missiles that fly in straight lines toward their destinations. For simplicity, assume that the flight of each missile takes one hour regardless of the distance flown. The first missile is launched from the North Pole toward New Orleans, Louisiana, which is at 30 degrees north latitude (Figure 7-9b). Does the missile land in New Orleans? Actually, no. Earth rotates eastward at 1400 kilometers (870 miles) per hour along the 30 degrees latitude line (Figure 7-9a), so the missile lands somewhere near El Paso, Texas, 1400 kilometers west of its target. From your perspective at the North Pole, the path of the missile appears to curve *to its right* in accordance with the Coriolis effect. In reality, New Orleans has moved out of the line of fire due to Earth's rotation.

The second missile is launched toward New Orleans from the Galápagos Islands, which are directly south of New Orleans along the Equator (Figure 7-9b). From their position on the Equator, the Galápagos Islands are moving east at 1600 kilometers (1000 miles) per hour, 200 kilometers (124 miles) per hour faster than New Orleans (Figure 7-9a). At takeoff, therefore, the missile is also moving toward the east 200 kilometers per hour faster than New Orleans. Thus, when the missile returns to Earth one hour later at the latitude of New Orleans, it will land offshore of Alabama, 200 kilometers east of New Orleans. Again, from your perspective on the Galápagos Islands, the missile appears to curve *to its right*. Keep in mind that both of these missile examples ignore

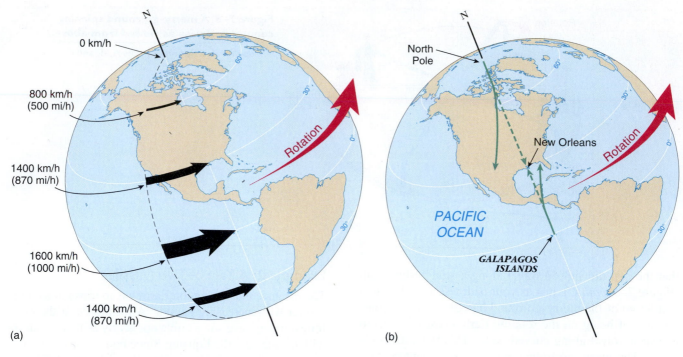

Figure 7–9 The Coriolis effect and missile paths.

(a) The velocity of any point on Earth varies with latitude from about 1600 kilometers (1000 miles) per hour at the Equator to 0 kilometers per hour at either pole. **(b)** The paths of missiles shot toward New Orleans from the North Pole and from the Galápagos Islands on the Equator. Dashed lines indicate intended paths; solid lines indicate paths that the missiles would travel as viewed from Earth's surface.

friction, which would greatly reduce the amount the missiles deflect to the right of their intended courses.

STUDENTS SOMETIMES ASK...

If Earth is spinning so fast, why don't we feel it?

In spite of Earth's constant rotation, we have the illusion that Earth is still. The reason that we don't feel the motion is because Earth rotates smoothly and quietly, with no bumps, no accelerations or decelerations and the atmosphere moves along with us. Thus, all sensations we receive tell us there is no motion and the ground is comfortably at rest—even though most of the United States is continually moving at speeds greater than 800 kilometers (500 miles) per hour!

Changes in the Coriolis Effect with Latitude

The first missile (shot from the North Pole) missed the target by 1600 kilometers (1000 miles), while the second missile (shot from the Galápagos Islands) missed its target by only 200 kilometers (124 miles). What was responsible for the difference? Not only does the rotational velocity of points on Earth range from 0 kilometers per hour at the poles to more than 1600 kilometers (1000 miles) per hour at the Equator, but the *rate of change* of the rotational velocity (per degree of latitude) increases as the pole is approached from the Equator.

For example, the rotational velocity differs by 200 kilometers (124 miles) per hour between the Equator (0 degrees) and 30 degrees north latitude. From 30 degrees north latitude to 60 degrees north latitude, however, the rotational velocity differs by 600 kilometers (372 miles) per hour. Finally, from 60 degrees north latitude to the North Pole (where the rotational velocity is zero), the rotational velocity differs by over 800 kilometers (500 miles) per hour.

Thus, the maximum Coriolis effect is at the poles, and there is no Coriolis effect at the Equator. The magnitude of the Coriolis effect depends much more, however, on the length of time the object (such as an air mass or ocean current) is in motion. Even at low latitudes, where the Coriolis effect is small, a large Coriolis deflection is possible if an object is in motion for a long time. In addition, because the Coriolis effect is caused by the *difference* in velocity of different latitudes on Earth, there is no Coriolis effect for those objects moving due east or due west along the Equator.

A summary of the Coriolis effect is presented in Table 7-3.

> The Coriolis effect causes moving objects to curve to the right in the Northern Hemisphere and to the left in the Southern Hemisphere. It is maximized at the poles and is zero at the Equator.

TABLE 7-3 **The Coriolis effect: A summary.**

- The Coriolis effect is caused by Earth's rotation and the resulting decrease in velocity with increasing latitude.
- The Coriolis effect influences all moving objects, especially those that move over large distances.
- The Coriolis effect changes only the direction of a moving object, never its speed.
- Coriolis deflection is to the right in the Northern Hemisphere and to the left in the Southern Hemisphere.
- The Coriolis effect is zero at the Equator and increases with increasing latitude, reaching its maximum strength at the poles.

Atmospheric Circulation Cells on a Spinning Earth

Figure 7-10 shows atmospheric circulation and the corresponding wind belts on a spinning Earth, which presents a more complex pattern than that of the fictional nonspinning Earth (Figure 7-7).

Circulation Cells

The greater heating of the atmosphere over the Equator causes the air to expand, to decrease in density, and to rise. As the air rises, it cools by expansion because the pressure is lower, and the water vapor it contains condenses and falls as rain in the equatorial zone. The resulting dry air mass travels north or south of the Equator. Around 30 degrees north and south latitude, the air cools off enough to become denser than the surrounding air, so it begins to descend, completing the loop (Figure 7-10). These circulation cells are called **Hadley cells** after noted English meteorologist George Hadley (1685–1768).

In addition to Hadley cells, each hemisphere has a **Ferrel cell** between 30 and 60 degrees latitude, and a **polar cell** between 60 and 90 degrees latitude. The Ferrel cell—named after American meteorologist William Ferrel (1817–1891), who invented the three-cell per hemisphere model for atmospheric circulation—is not driven solely by differences in solar heating; if it were, air within it would circulate in the opposite direction. Similar to the movement of interlocking gears, the Ferrel cell moves in the direction that coincides with the movement of the two adjoining circulation cells.

Pressure

A column of cool, dense air moves *toward* the surface and creates high pressure. The descending air at about 30 degrees north and south latitude creates high-pressure zones called the **subtropical highs**. Similarly, descending air at the poles creates high-pressure regions called the **polar highs**.

What kind of weather is experienced in these high-pressure areas? Descending air is quite dry and it tends to warm under its own weight, so these areas typically experience dry, clear, fair conditions. The conditions are not necessarily warm (such as at the poles)—just dry and associated with clear skies.

A column of warm, low density air rises *away* from the surface and creates low pressure. Thus, rising air creates a band of low pressure at the Equator—the **equatorial low**—and at about 60 degrees north and south latitude—the **subpolar low**. The weather in areas of low pressure is dominated by cloudy conditions with lots of precipitation, because rising air cools and cannot hold its water vapor.

Wind Belts

The lowermost portion of the circulation cells—that is, the part that is closest to the surface—generates the major wind belts of the world. The masses of air that move across Earth's surface from the subtropical high-pressure belts toward the equatorial low-pressure belt constitute the **trade winds**. These steady winds are named from the term *to blow trade*, which means to blow in a regular course. If Earth did not rotate, these winds would blow in a north–south direction. In the Northern Hemisphere, however, the **northeast trade winds** curve to the right due to the Coriolis effect and blow from *northeast* to *southwest*. In the Southern Hemisphere, on the other hand, the **southeast trade winds** curve to the left due to the Coriolis effect and blow from *southeast* to *northwest*.

Some of the air that descends in the subtropical regions moves along Earth's surface to higher latitudes as the **prevailing westerly wind belts**. Because of the Coriolis effect, the prevailing westerlies blow from southwest to northeast in the Northern Hemisphere and from northwest to southeast in the Southern Hemisphere.

Air moves away from the high pressure at the poles, too, producing the **polar easterly wind belts**. The Coriolis effect is maximized at high latitudes, so these winds are deflected strongly. The polar easterlies blow from the northeast in the Northern Hemisphere, and from the southeast in the Southern Hemisphere. When the polar easterlies come into contact with the prevailing westerlies near the subpolar low pressure belts (at 60 degrees north and south latitude), the warmer, less dense air of the prevailing westerlies rises above the colder, more dense air of the polar easterlies.

Boundaries

The boundary between the two trade wind belts along the Equator is known as the **doldrums** (*doldrum* = dull) because, long ago, sailing ships were becalmed there by the

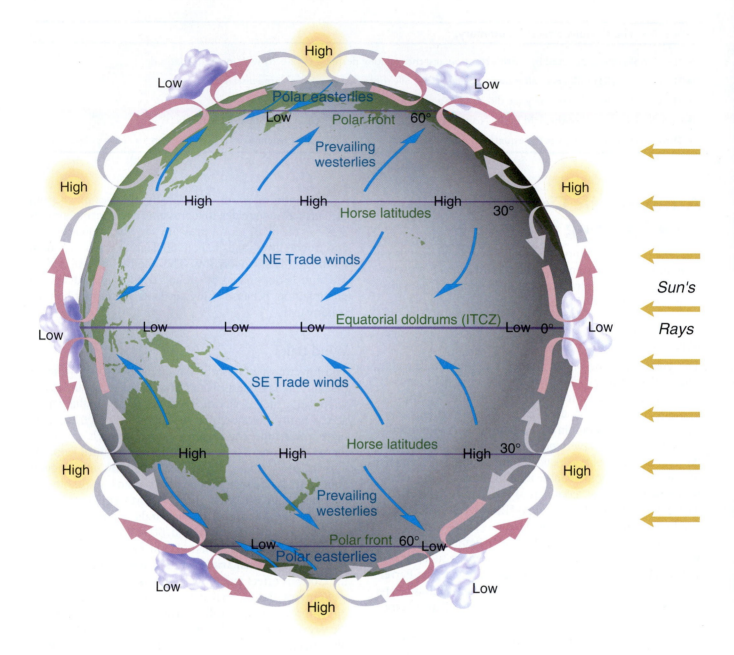

Figure 7–10 Atmospheric circulation and wind belts of the world.

The three-cell model of atmospheric circulation creates the major wind belts of the world. Boundaries between wind belts and surface atmospheric pressures are also shown. The general pattern of wind belts is modified by seasonal changes and the distribution of continents.

lack of winds. Sometimes stranded for days or weeks, the situation was unfortunate but not life-threatening: Daily rainshowers supplied sailors with plenty of fresh water. Today, meteorologists refer to this region as the **Intertropical Convergence Zone (ITCZ)**, because it is the region between the tropics where the trade winds converge (Figure 7-10).

The boundary between the trade winds and the prevailing westerlies (centered at 30 degrees north or south latitude) is known as the **horse latitudes**. High pressure and clear, dry, fair conditions dominate these regions, where the air is sinking and surface winds are light and variable. For sailors on a ship powered only by sails, becoming stuck in

the horse latitudes was a potentially serious problem because they could run out of fresh water. Becalmed here for weeks, crews would cast their horses overboard to conserve drinking water, hence the name "horse latitudes."

The boundary between the prevailing westerlies and the polar easterlies at 60 degrees north or south latitude is known as the **polar front**. This is a battleground for different air masses, so cloudy conditions and lots of precipitation are common here.

Clear, dry, fair conditions are associated with the high pressure at the poles, so precipitation is minimal. The poles are often classified as cold deserts because the annual precipitation is so low.

Circulation Cells: Idealized or Real?

The three-cell model of atmospheric circulation first proposed by Ferrel provides a simplified model of the general circulation pattern on Earth. This circulation model is idealized and does not always match the complexities observed in nature, particularly for the location and direction of motion of the Ferrel and polar cells. Nonetheless, it generally matches the pattern of major wind belts of the world and provides a general framework for understanding why they exist.

Further, the following factors significantly alter the idealized wind, pressure, and atmospheric circulation patterns illustrated in Figure 7-10:

1. The tilt of Earth's rotation axis, which produces seasons.

2. The lower heat capacity of continental rock compared to seawater,[6] which makes the air over continents colder in winter and warmer in summer than the air over adjacent oceans.

3. The uneven distribution of land and ocean over Earth's surface, which particularly affects patterns in the Northern Hemisphere.

During winter, therefore, the continents usually develop atmospheric high-pressure cells from the weight of cold air centered over them and, during the summer, they usually

develop low-pressure cells (Figure 7-11). In fact, such seasonal shifts in atmospheric pressure over Asia cause *monsoon winds,* which have a dramatic effect on Indian Ocean currents and will be discussed in Chapter 8, "Ocean Circulation." In general, however, the pattern of atmospheric high- and low-pressure zones shown in Figure 7-11 corresponds closely to those shown in Figure 7-10.

Table 7-4, summarizes the characteristics of global wind belts and boundaries. The world's wind belts closely match the pattern of ocean surface currents, which are discussed in Chapter 8, "Ocean Circulation."

> The major wind belts in each hemisphere are the trade winds, the prevailing westerlies, and the polar easterlies. The boundaries between these wind belts include the doldrums, the horse latitudes, the polar front, and the polar high.

The Oceans, Weather, and Climate

Weather describes the conditions of the atmosphere at a given time and place. **Climate** is the long-term average of weather. If we observe the weather conditions in an area over a long period, we can begin to draw some conclusions about its climate. For instance, if the weather in an area is dry over many years, we can say that the area has an arid climate.

---

[6]An object that has low heat capacity heats up quickly when heat energy is applied. Recall from Chapter 6 that water has high heat capacity.

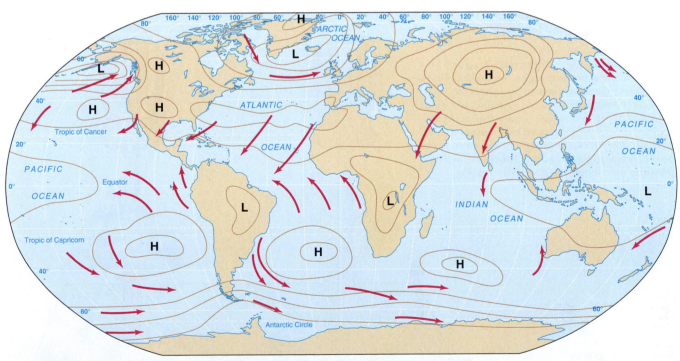

Figure 7–11 January sea-level atmospheric pressures and winds.

Average atmospheric pressure pattern for January. High (*H*) and low (*L*) atmospheric pressure zones correspond closely to those shown in Figure 7-10 but are modified by the change of seasons and the distribution of continents. Arrows show direction of winds, which move from high-to-low pressure regions.

TABLE 7-4 **Characteristics of wind belts and boundaries.**

Region (north or south latitude)	Wind belt or boundary name	Atmospheric pressure	Characteristics
Equatorial (0–5 degrees)	Doldrums	Low	Light, variable winds. Abundant cloudiness and much precipitation. Breeding ground for hurricanes.
5–30 degrees	Trade winds	—	Strong, steady winds generally from the east.
30 degrees	Horse latitudes	High	Light, variable winds. Dry, clear, fair weather with little precipitation. Major deserts of the world.
30–60 degrees	Prevailing westerlies	—	Winds generally from the west. Brings storms that influence weather across U.S.
60 degrees	Polar front	Low	Variable winds. Stormy, cloudy weather year-round.
60–90 degrees	Polar easterlies	—	Cold, dry winds generally from the east.
Poles (90 degrees)	Polar high pressure	High	Variable winds. Clear, dry, fair conditions, cold temperatures, and minimal precipitation. Cold deserts.

Winds

The movement of air is called *wind*. Recall that air always moves from high pressure toward low pressure. However, as air moves away from high-pressure regions and toward low-pressure regions, the Coriolis effect modifies its direction. In the Northern Hemisphere, for example, air moving from high to low pressure curves to the right and results in a counterclockwise[7] flow of air around low-pressure cells [called **cyclonic** (*kyklon* = moving in a circle) **flow**]. Similarly, as the air leaves the high-pressure region

---

[7]These directions are reversed in the Southern Hemisphere.

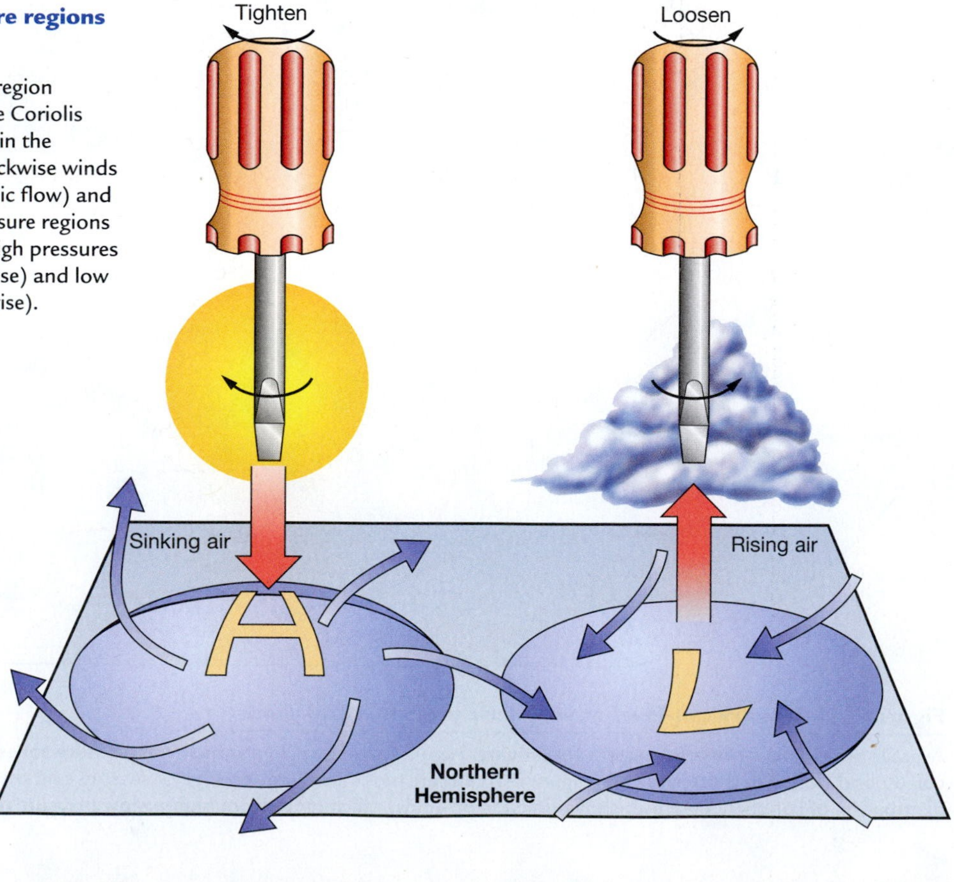

and curves to the right, it establishes a clockwise flow of air around high-pressure cells (called **anticyclonic flow**). Figure 7-12 shows how a screwdriver can help you remember how air moves around high and low pressure regions. High pressures are similar to a high screw that needs tightening, so a screwdriver would be turned clockwise; low pressures are similar to a tightened screw that needs loosening, so a screwdriver would be turned counterclockwise. Because winter high-pressure cells are replaced by summer low-pressure cells over the continents, wind patterns associated with continents often reverse themselves seasonally.

Figure 7–12 High- and low-pressure regions and air flow.

As air moves away from a high-pressure region (*H*) toward a low-pressure region (*L*), the Coriolis effect causes the air to curve to the right in the Northern Hemisphere. This results in clockwise winds around high-pressure regions (anticyclonic flow) and counterclockwise winds around low-pressure regions (cyclonic flow). It can be thought of as high pressures need tightening by a screwdriver (clockwise) and low pressures need loosening (counterclockwise).

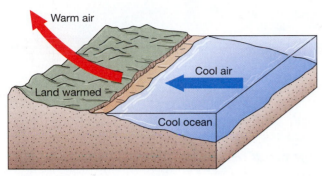

(a) Sea breeze

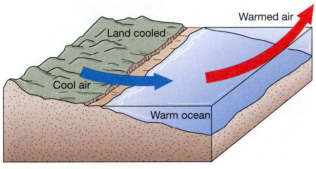

(b) Land breeze

Figure 7–13 Sea and land breezes.

(a) Sea breezes occur when air warmed by the land rises and is replaced by cool air from the ocean. **(b)** Land breezes occur when the land has cooled, causing dense air to sink and flow toward the warmer ocean.

Other factors that influence regional winds, especially in coastal areas, are **sea breezes** and **land breezes** (Figure 7-13). When an equal amount of solar energy is applied to both land and ocean, the land heats up about five times more due to its lower heat capacity. The land heats the air around it and, during the afternoon, the warm, low-density air over the land rises. Rising air creates a low-pressure region over the land, pulling the cooler air over the ocean toward land, creating what is known as a sea breeze. At night, the land surface cools about five times more rapidly than the ocean and cools the air around it. This cool, high-density air sinks, creating a high-pressure region that causes the wind to blow from the land. This is known as a land breeze, and is most prominent in the late evening and early morning hours.

Storms

At high and low latitudes, there is little daily and minor seasonal change in weather.[8] Equatorial regions are usually warm, damp, and typically calm, because the dominant direction of air movement in the doldrums is upward. Midday rains are common, even during the supposedly "dry" season. It is within the mid-latitudes where storms are common.

Storms are atmospheric disturbances characterized by strong winds, precipitation, and often thunder and lightning. Due to the seasonal change of pressure systems over continents, air masses from the high and low latitudes may move into the mid-latitudes, meet, and produce severe storms. **Air masses** are large volumes of air that have a definite area of origin and distinctive characteristics. Several air masses influence the United States, including polar air masses and tropical air masses (Figure 7-14). Some air masses originate over land (c = continental) and are therefore dryer, but most originate over the sea (m = maritime) and are moist. Some are colder (P = polar; A = Arctic) and some are warm (T = tropical). Typically, the United States is influenced more by polar air masses during the winter and more by tropical air masses during the summer.

As polar and tropical air masses move into the mid-latitudes, they also move gradually in an easterly direction. A **warm front** is the contact between a warm air mass moving into an area occupied by cold air. A **cold front** is the contact between a cold air mass moving into an area occupied by warm air (Figure 7-15).

These confrontations are brought about by the movement of the **jet stream**, which is a narrow, fast moving, easterly flowing air mass. It exists above the mid-latitudes just below the top of the troposphere, centered at an altitude of about 10 kilometers (6 miles). It usually follows a wavy path and may cause unusual weather by

---

[8]In fact, in equatorial Indonesia, the vocabulary of Indonesians doesn't include the word *seasons*.

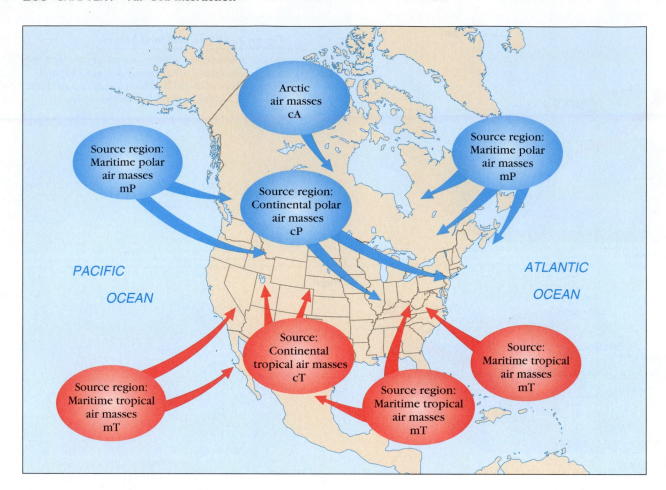

Figure 7–14 Air masses that affect U.S. weather.

Polar air masses are shown in blue, and tropical air masses are shown in red. Air masses are classified based on their source region: The designation continental (c) or maritime (m) indicates moisture content, whereas polar (P), Arctic (A), and tropical (T) indicate temperature conditions.

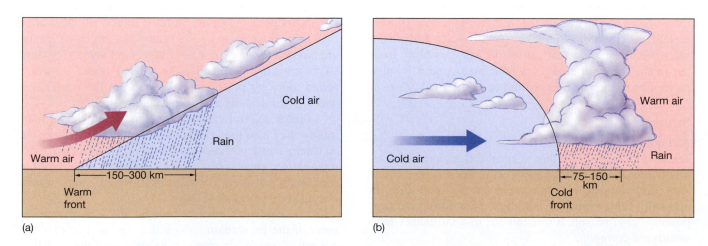

Figure 7–15 Warm and cold fronts.

Cross-sections through a gradually rising warm front **(a)** and a steeper cold front **(b)**. With both fronts, warm air rises and precipitation is produced.

steering a polar air mass far to the south or a tropical air mass far to the north.

Regardless of whether a warm front or cold front is produced, the warmer, less-dense air always rises above the denser cold air. The warm air cools as it rises, so its water vapor condenses as precipitation. A cold front is usually steeper, and the temperature difference across it is greater than a warm front. Therefore, rainfall along a cold front is usually heavier and briefer than rainfall along a warm front.

Tropical Cyclones (Hurricanes)

Tropical cyclones (*kyklon* = moving in a circle) are huge rotating masses of low pressure characterized by strong winds and torrential rain. They are the largest storm systems on Earth, though they are not associated with any fronts. In North and South America, tropical cyclones are called **hurricanes** (*Huracan* = Taino god of wind); in the western North Pacific Ocean, they are called **typhoons** (*tai-fung* = great wind); and in the Indian Ocean, they are called **cyclones**. No matter what they are called, tropical cyclones can be highly destructive. In fact, the energy contained in a *single* hurricane is greater than that generated by all energy sources over the last 20 years in the United States.

Origin Tropical cyclones begin as low-pressure cells that break away from the equatorial low-pressure belt and grow as they pick up heat energy from the warm ocean. Surface winds feed moisture (in the form of water vapor) into the storm. When water evaporates, it stores tremendous amounts of latent heat of evaporation,[9] which is released as latent heat of condensation when the water

---

[9]For a discussion of water's latent heat of evaporation, see Chapter 6.

---

vapor condenses to form liquid water (rain). The release of vast amounts of water's latent heat of condensation powers tropical cyclones.

Tropical storms are classified according to their maximum sustained wind speed:

- If winds are less than 61 kilometers (38 miles) per hour, the storm is classified as a *tropical depression*.
- If winds are between 61 and 120 kilometers (38 and 74 miles) per hour, the storm is called a *tropical storm*.
- If winds exceed 120 kilometers (74 miles) per hour, the storm is a *tropical cyclone*.

The **Saffir–Simpson Scale** of hurricane intensity further divides tropical cyclones into categories based on wind speed and damage (Table 7-5). In some cases, the wind in tropical cyclones attains speeds as high as 400 kilometers (250 miles) per hour!

Worldwide, about 100 storms grow to hurricane status each year. The conditions needed to create a hurricane are as follows:

- Ocean water with a temperature greater than 25°C (77°F), which provides an abundance of water vapor to the atmosphere through evaporation.
- Warm moist air, which supplies vast amounts of latent heat as the water vapor in the air condenses and fuels the storm.
- The Coriolis effect, which causes the hurricane to spin counterclockwise in the Northern Hemisphere and clockwise in the Southern Hemisphere. No hurricanes occur directly on the Equator because the Coriolis effect is negligible there.

These conditions are found during the late summer and early fall, when the tropical and subtropical oceans are at their maximum temperature. Hurricane season in the Atlantic Basin is officially from June 1 to

---

TABLE 7-5 **The Saffir–Simpson Scale of hurricane intensity.**

Category	Wind speed (kilometers per hour)	(miles per hour)	Typical storm surge (meters)	(feet)	Damage
1	120–153	74–95	1.2–1.5	4–5	Minimal: Minor damage to buildings
2	154–177	96–110	1.8–2.4	6–8	Moderate: Some roofing material, door, and window damage; some trees blown down
3	178–209	111–130	2.7–3.7	9–12	Extensive: Some structural damage and wall failures; foliage blown off trees and large trees blown down
4	210–249	131–155	4.0–5.5	13–18	Extreme: More extensive structural damage and wall failures; most shrubs, trees, and signs blown down
5	>250	>155	>5.8	>19	Catastrophic: Complete roof failures and entire building failures common; all shrubs, trees, and signs blown down; flooding of lower floors of coastal structures

November 30, but hurricanes can form outside of this time frame, although only rarely.

Movement When hurricanes are initiated, they typically remain in the tropics. On rare occasions, they pass through the tropics and affect the mid-latitudes (Figure 7-16). If a hurricane moves over land, its energy source is cut off and it eventually dissipates. The trade winds drive hurricanes, so they move from east to west across the oceans and typically last five to ten days.

The diameter of a hurricane can exceed 800 kilometers (500 miles); more typically, though, the diameter is less than 200 kilometers (124 miles). As air moves across the ocean surface toward the low-pressure center, it is drawn up around the **eye of the hurricane** (Figure 7-17). The air in the vicinity of the eye spirals upward, so horizontal wind speeds may be less than 15 kilometers (25 miles) per hour. The eye of the hurricane, therefore, is usually calm. Hurricanes are composed of spiral rain bands where intense rainfall caused by severe thunderstorms can produce tens of centimeters (several inches) of rainfall in an hour.

Types of Destruction Destruction from hurricanes is caused by high winds and flooding from intense rainfall. **Storm surge**, however, causes the majority of a hurricane's coastal destruction. In fact, storm surge is responsible for 90% of the deaths associated with hurricanes.

When a hurricane develops over the ocean, its low-pressure center produces a low "hill" of water (Figure 7-18). As the hurricane migrates across the open ocean, the hill moves with it. As the hurricane approaches shallow water near shore, the portion of the hill over which the wind is blowing shoreward produces a mass of elevated, wind-driven water. This mass of water—the storm surge—can

be as high as 12 meters (40 feet), resulting in a dramatic increase in sea level at the shore, large storm waves, and tremendous destruction to low-lying coastal areas (particularly if it occurs at high tide). In addition, the area of the coast that is hit with the leading quadrant of the storm surge—where onshore winds further pile up water—experiences the most severe storm surge (Figure 7-18). Table 7-5 shows typical storm surge heights associated with Saffir–Simpson hurricane intensities.

Historic Destruction Periodic destruction from hurricanes occurs along the East Coast and the Gulf Coast regions of the United States. The most deadly natural disaster in U.S. history was caused by a hurricane that hit Galveston, Texas, in September 1900 (Box 7-1). Flooding caused by the hurricane's storm surge combined with intense rainfall and winds of 135 kilometers (84 miles) per hour killed at least 4000 people in Galveston, 1000 people in other parts of low-lying Galveston Island, and another 1000 people on the mainland.

Category 4 hurricanes like the one that hit Galveston have been surpassed by Category 5 hurricanes only three times in the United States. In 1935, on Labor Day, one flattened the Florida Keys, and in 1969, Hurricane Camille roared through Mississippi. In August 1992, Hurricane Andrew came ashore in southern Florida and did more than $20 billion in damage before crossing the Gulf of Mexico and inflicting another $2 billion in damage along the Gulf Coast (Figure 7-19). Winds reached 258 kilometers (160 miles) per hour as Andrew crossed the Everglades, ripping down every tree in its path. More property was damaged by this hurricane than in any other natural disaster in U.S. history, and over 250,000 people were left homeless. Fortunately, many people heeded the warnings to evacuate, so only 54 were killed.

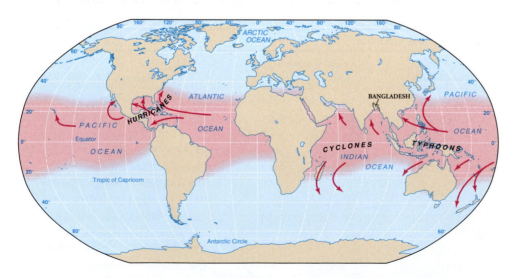

Figure 7–16 Paths of major tropical cyclones.

Cyclones originate in tropical areas where ocean surface temperatures are warm (*red shaded areas*) and migrate across the tropics (*red arrows*), sometimes reaching the mid-latitudes. Depending on the area, cyclones may also be called *hurricanes* or *typhoons*.

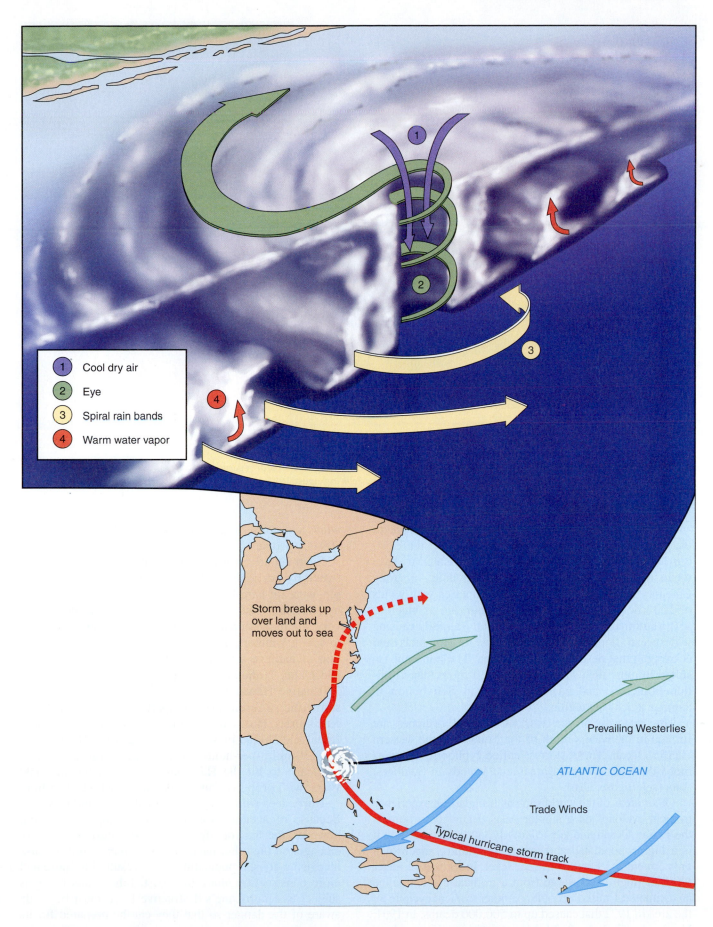

Figure 7–17 **Typical North Atlantic hurricane storm track and internal structure** *(inset).*

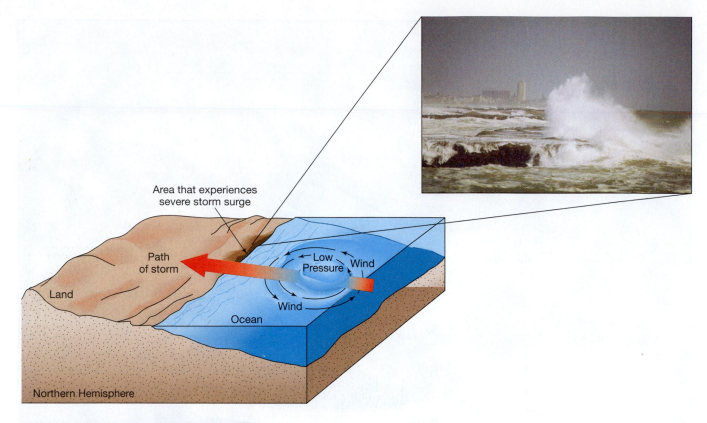

Figure 7–18 A storm surge hits the coast.

As a tropical cyclone in the Northern Hemisphere moves ashore, the low-pressure center around which the storm winds blow combined with strong onshore winds produce a high-water storm surge that floods and batters the coast. A storm surge caused by Hurricane Felix in August 1995 along New Jersey (*inset*).

In October 1998, Hurricane Mitch proved to be one of the most devastating tropical cyclones to affect the Western Hemisphere. At its peak, it was estimated to have winds of 290 kilometers (180 miles) per hour—a strong Category 5 hurricane. It hit Central America with winds of 160 kilometers (100 miles) per hour and as much as 130 centimeters (51 inches) of total rainfall, causing widespread flooding and mudslides in Honduras and Nicaragua that destroyed entire towns. The hurricane resulted in more than 11,000 deaths, left more than 2 million homeless, and caused more than $10 billion in damage across the region.

The majority of the world's tropical cyclones are formed in the waters north of the Equator in the western Pacific Ocean. These storms, called typhoons, do enormous damage to coastal areas and islands in Southeast Asia (see Figure 7-16).

Other areas of the world such as Bangladesh experience tropical cyclones on a regular basis. Bangladesh borders the Indian Ocean and is particularly vulnerable because it is a highly populated and low-lying country, much of it only 3 meters (10 feet) above sea level. In 1970, a 12-meter (40-foot)-high storm surge from a tropical cyclone killed an estimated 1 million people. Another tropical cyclone hit the area in 1972 that caused up to 500,000 deaths. In 1991, Hurricane Gorky's 233-kilometer (145-mile) per-hour winds and storm surge caused extensive damage and killed 200,000 people.

Even islands near the centers of ocean basins can be struck by hurricanes. The Hawaiian Islands, for example, were hit hard by Hurricane Dot in August 1959 and by Hurricane Iwa in November 1982. Hurricane Iwa hit very late in the hurricane season and produced winds up to 130 kilometers (81 miles) per hour. Over $100 million of damage was sustained on the islands of Kauai and Oahu. Niihau, a small island that is inhabited by 230 native Hawaiians, was directly in the path of the storm and suffered severe property damage but no serious injuries. Hurricane Iniki roared across the islands of Kauai and Niihau in September 1992, with 210-kilometer (130-mile)-per-hour winds. It was the most powerful hurricane to hit the Hawaiian Islands in the last 100 years, with property damage that approached $1 billion.

Hurricanes will always be a threat to life and property. Because of accurate forecasts and prompt evacuation, however, the loss of life has been decreasing. Property damage, on the other hand, has been *increasing* because increasing coastal populations have resulted in more and more construction along the coast. Inhabitants of areas subject to a hurricane's destructive force must be made aware of the danger so that they can be prepared for its eventuality.

BOX 7–1 Historical Features
THE STORM OF THE CENTURY: GALVESTON, TEXAS (1900)

"If we had known then what we know now . . . we would have known earlier the terrors of the storm which these swells . . . told us in unerring language was coming."

—Report from Isaac Cline, Local Forecaster for the U.S. Weather Bureau, after the Galveston disaster

Among the natural disasters in U.S. history, many have resulted in large numbers of deaths. The Chicago fire of 1871, for instance, killed 250 people; the San Francisco earthquake of 1906 killed 503; a 1928 hurricane in Florida killed 1836; and the 1889 Johnstown flood killed 2209. The single deadliest U.S. natural disaster, however, was a hurricane that struck Galveston Island, Texas, on September 8, 1900, resulting in over 6000 deaths.

Galveston Island is a thin strip of sand called a barrier island located in the Gulf of Mexico off Texas (Figure 7A, *inset*). In 1900, it averaged only 1.5 meters (5 feet) above sea level and its highest point was 2.7 meters (9 feet) above sea level. During construction of the city of Galveston, protective sand dunes up to 4.6 meters (15 feet) high had been removed to improve beach access. Galveston became a popular beach resort that boasted a milder climate than mainland areas, so it drew many summer visitors. The beach sloped off into deep water so gradually that most believed any destructive waves from the Gulf would break and dissipate before reaching shore.

The loss of life resulted not from lack of warning but from a failure to take the threat seriously. Two days earlier, a strong storm was reported moving westward into the Gulf of Mexico off Cuba, and ships returning from sea reported encountering the storm offshore the day before it made landfall. Warnings of the danger seem to have been largely unheeded, in part because most U.S. meteorologists erroneously believed it was virtually impossible for a storm in the Caribbean to move across the Gulf. Furthermore, the Local Forecaster for the U.S. Weather Bureau in Galveston, Isaac Cline, did not recognize indications of how severe the storm would become and so downplayed its danger.

Continued...

Figure 7A **Destruction from the 1900 hurricane at Galveston and location map of Galveston, Texas** *(inset).*

On the morning the storm hit, curious residents flocked to the beach to observe the storm's powerful waves crashing against the shore. In spite of the rain, a holiday atmosphere prevailed in Galveston, with most people believing that the flooding already apparent was the result of another minor storm. According to Cline's account, he realized how powerful the storm might be shortly before the storm hit and rode along the beach, urging residents to evacuate or seek shelter. Other accounts indicated that there were few clear warnings of the severity of the storm until it was too late.

By the time the hurricane neared Galveston and people realized they were in eminent danger, the escape routes to the mainland had been severed and the people were trapped on the island. Only a few hours later, the hurricane's 6-meter (20-foot)-high storm surge came ashore and completely submerged the island. The high water combined with intense wind—measured at 160 kilometers (100 miles) per hour until the weather station's anemometer blew away—and large waves destroyed many structures on the island (Figure 7A). Even Cline's house, which was built to withstand a hurricane, collapsed with 50 people inside. The hurricane claimed the lives of at least 6000 people, including Cline's wife. Most of the people who survived—including Cline, his three daughters, and his brother—rode out the storm by clinging to floating debris.

The horror did not end when the hurricane passed. Thousands of dead bodies were found in the wreckage of the town and more washed ashore. Unable to bury them in the saturated ground, some were taken out to sea on barges and unceremoniously dumped, only to wash back to shore. Ultimately, most of the bodies were cremated where they were found.

Today, Galveston has been shored up and a 5.2-meter (17-foot)-high protective seawall has been built to prevent similar hurricane disasters. In addition, satellite tracking techniques have led to improved forecasts of approaching hurricanes and more timely warnings. The grim catastrophe of Galveston, however, remains as stark testimony to the destructive power of hurricanes.

(a)

(b)

Figure 7–19 Hurricane Andrew, the most destructive hurricane in U.S. history.

(a) A satellite view of Hurricane Andrew, which had a diameter of about 200 kilometers (124 miles); its counterclockwise direction of flow and prominent central eye are also visible. **(b)** The hurricane struck Florida in 1992, causing extensive damage.

> Hurricanes are intense and sometimes destructive tropical storms that form where water temperatures are high, where there is an abundance of warm moist air, and where they can spin.

The Ocean's Climate Patterns

Just as land areas have climate patterns, so do regions of the oceans. The open ocean is divided into climatic regions that run generally east–west (parallel to lines of latitude) and have relatively stable boundaries that are somewhat modified by ocean surface currents (Figure 7-20).

The **equatorial** region spans the Equator. The major air movement there is upward because heated air rises. Surface winds, therefore, are weak and variable, which is why this region is called the doldrums. Surface waters are warm and the air is saturated with water vapor. Daily rainshowers are common, which keeps surface salinity relatively low. The equatorial regions just north or south of the Equator are also the breeding grounds for tropical cyclones.

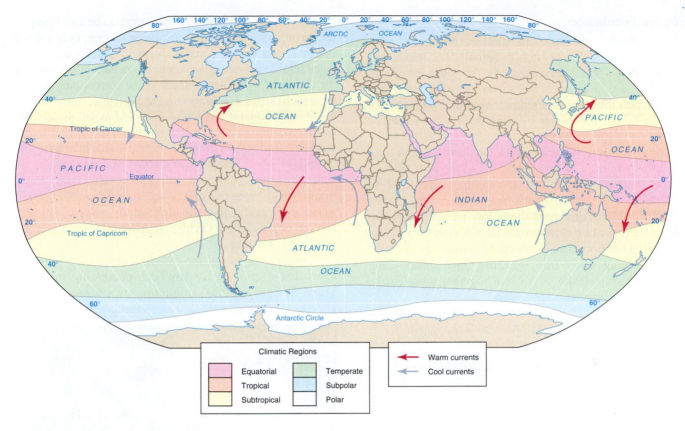

Figure 7–20 The ocean's climatic regions.

The ocean's climatic regions are defined primarily by latitude but modified by ocean currents and wind belts. Red arrows indicate warm surface currents; blue arrows indicate cool surface currents.

Tropical regions extend north or south of the equatorial region up to the Tropic of Cancer and the Tropic of Capricorn, respectively. They are characterized by strong trade winds, which blow from the northeast in the Northern Hemisphere and from the southeast in the Southern Hemisphere. These winds push the equatorial currents and create moderately rough seas. Relatively little precipitation falls at higher latitudes within tropical regions, but precipitation increases toward the Equator. Once tropical cyclones form, they gain energy here as large quantities of heat are transferred from the ocean to the atmosphere.

Beyond the tropics are the **subtropical** regions. Belts of high pressure are centered there, so the dry, descending air produces little precipitation and a high rate of evaporation, resulting in the highest surface salinities in the open ocean (see Figure 6-21). Winds are weak and currents are sluggish, typical of the horse latitudes. However, strong boundary currents (along the boundaries of continents) flow north and south, particularly along the western margins of the subtropical oceans.

The **temperate** regions (also called the *mid-latitudes*) are characterized by strong westerly winds (the prevailing westerlies) blowing from the southwest in the Northern Hemisphere and from the northwest in the Southern Hemisphere (see Figure 7-10). Severe storms are common, especially during winter, and precipitation is heavy. In fact, the North Atlantic is noted for fierce storms, which have claimed many ships and numerous lives over the centuries.

The **subpolar** region experiences extensive precipitation due to the subpolar low. Sea ice covers the subpolar ocean in winter, but it melts away, for the most part, in summer. Icebergs are common, and the surface temperature seldom exceeds 5°C (41°F) in the summer months.

Surface temperatures remain at or near freezing in the **polar** regions, which are covered with ice throughout most of the year. The polar high pressure dominates the area, which includes the Arctic Ocean and the ocean adjacent to Antarctica. There is no sunlight during the winter and constant daylight during the summer.

Sea Ice

Low temperatures in high latitude regions cause a permanent or nearly permanent ice cover on the sea surface. The term **sea ice** is used to distinguish such masses of frozen seawater from **icebergs**, which are also found at sea, but originate by breaking off (*calving*) from glaciers that originate on land. Sea ice is found throughout the year around the margin of Antarctica, within the Arctic Ocean, and in the extreme high-latitude region of the North Atlantic Ocean.

Sea Ice Formation

Sea ice begins as small, needlelike, hexagonal (six-sided) crystals. These crystals eventually become so numerous that a *slush* develops. As the slush begins to form into a thin sheet, it is broken by wind stress and wave action into disk-shaped pieces called **pancake ice** (Figure 7-21a). As further freezing occurs, the pancakes coalesce to form **ice floes**.

The rate at which sea ice forms is closely tied to temperature conditions. Large quantities of ice form in relatively short periods when the temperature falls to extremely low levels [such as temperatures below $-30°C$ $(-22°F)$]. Even at these low temperatures, the rate of ice formation slows as sea ice thickens because the ice (which has poor heat conduction) effectively insulates the underlying water from freezing. In addition, calm water enables pancake ice to join together more easily, which aids the formation of sea ice.

The process of sea ice formation tends to be a self-perpetuating process. As sea ice forms at the surface, only a small percentage of the dissolved components can be accommodated into the crystalline structure of ice. As a result, most of the dissolved substances remain in the surrounding seawater, which causes its salinity to increase. Recall from Chapter 6 that increasing the amount of dissolved materials decreases the freezing point of water, which doesn't appear to enhance ice formation. However, also recall that increasing the salinity of water increases its density and its tendency to sink. As it sinks below the surface, it is replaced by lower salinity (and lower density) water from below, which will freeze more readily than the high salinity water it replaced, thereby establishing a circulation pattern that enhances the formation of sea ice.

Satellite analysis of the extent of Arctic sea ice indicates that Arctic sea ice has been decreasing in recent years. In fact, the 2002 minimum sea ice record in the

(a)

(b)

(c) Rafted ice · Ridged ice · Weathered ridge ice · Hummocked ice

Figure 7–21 Sea ice.

(a) Pancake ice, which is freezing slush that is broken by wind stress and wave action into disk-shaped pieces. **(b)** Aerial view of the Larsen ice shelf on the Antarctic Peninsula, where ribbons of sea ice (*top*) remain seaward of the shelf ice during September (the beginning of the spring season in the Southern Hemisphere). **(c)** Ice structures associated with rafted ice, which is created as ice floes expand and raft onto one another.

Arctic is the lowest since the early 1950s and represents a 17% decrease in overall ice extent. In addition, the interior ice is unusually thin and spread out, which has resulted in wide patches of ice-free ocean during the summer—even at the North Pole. The accelerated melting appears to be linked to shifts in Northern Hemisphere atmospheric circulation patterns that have caused the region to experience anomalous warming.

Sea Ice in the Arctic Ocean

Sea ice in the Arctic Ocean, which occupies high latitudes in the Northern Hemisphere, can be classified into one of three types: pack ice, polar ice, or fast ice.

Pack Ice   Floating ice that has been driven together into a single mass is called **pack ice**, which commonly forms around the margins of the Arctic Ocean. The ice extends through the Bering Strait into the Bering Sea on the Pacific side, and as far south as Newfoundland and Nova Scotia in the North Atlantic. Pack ice reaches its maximum extent during May and its minimum extent when it breaks up in September. This ice, which can be penetrated by ships with reinforced hulls called *icebreakers*, reaches a maximum thickness of about 2 meters (6.6 feet) in winter.

Pack ice is driven primarily by wind, although it also responds to surface currents, which produce stresses that continually break and reform its structure. Pack ice formation is achieved by the expansion of ice floes that begin to raft onto one another as they expand (Figure 7-21c).

Polar Ice   Thicker ice that forms near the North Pole is called **polar ice**, which covers the greatest portion of the Arctic Ocean and attains a maximum thickness in excess of 50 meters (164 feet). During the brief summer season, melting may produce areas of open water surrounded by sea ice called *polynyas* (*polynya* = opening), although the polar ice never totally disappears. Its average thickness during summer is over 2 meters (6.6 feet). Polar ice is constantly being exchanged, because floes from the adjoining pack ice are carried into the polar region during winter and floes break out of the polar ice and reenter the pack ice during summer. Circling in a clockwise direction around the Arctic Ocean, about one-third of the pack and polar ice is carried into the North Atlantic by the East Greenland Current each year.

Fast Ice   Fast ice develops in the winter from the shore out to the pack ice and it completely melts during the summer. This fast ice (so-called because it is "held fast" or firmly attached to the shore) attains a winter thickness exceeding 2 meters (6.6 feet).

Sea Ice near Antarctica

In the Southern Hemisphere, the polar region is occupied by the continent of Antarctica rather than by an ocean. As a result, sea ice can only form around the margins of Antarctica and is limited to pack ice and fast ice that have a rather temporary existence. In fact, pack ice rarely extends north of 55 degrees south latitude and breaks up almost completely during the Southern Hemisphere summer from October to January (Figure 7-21b). In addition, strong winds are typical in this region and so prevent the formation of pack ice around the Antarctic continent.

Icebergs

Icebergs are bodies of floating ice broken away from a glacier (Figure 7-22) and so are quite distinct from sea ice. Icebergs are formed by vast ice sheets on land, which grow from the accumulation of snow and slowly flow outward to the sea. Once at the sea, the ice either breaks up and produces icebergs there, or, because it is less dense than water, it floats on top of the water, often extending a great distance away from shore before breaking up under the stress of current, wind, and wave action. Most calving occurs during the summer months, when temperatures are highest.

In the Arctic, icebergs originate primarily from glaciers that follow narrow valleys into the sea along the western coast of Greenland (Figure 7-22b). Icebergs are also produced along the eastern coasts of Greenland, Ellesmere Island, and other Arctic islands. The East Greenland Current and the West Greenland Current (Figure 7-22b, *arrows*) carry the icebergs into North Atlantic shipping lanes, where they become navigational hazards. The large size of some of these icebergs takes them several years to melt, and, in that time, they may be carried as far south as 40 degrees north latitude, the same latitude as Philadelphia.

The RMS *Titanic* Disaster

The luxury liner **RMS *Titanic*** (see Figure 1-26), at 269 meters (882 feet) long (not quite the length of three football fields), was at the time of its manufacture the world's largest passenger ship and the world's largest movable object. She was designed to be unsinkable, with multiple compartments that would keep her afloat even if some of the compartments filled with water. As a result, the *Titanic* had only enough lifeboats for about half of its passengers. Operated by the White Star Line, she departed in April 1912 on her maiden voyage from England to the United States across the North Atlantic Ocean.

On the evening of April 14, the *Titanic* was traveling under a clear sky and calm seas near the Grand Banks south of Newfoundland (see Figure 7-22b). She was running ahead of schedule and the officials of the White Star Line were hoping to make a good impression by arriving in New York early. Despite repeated warnings of icebergs in the area, the crew maintained an excessive speed of 41 kilometers (25.5 miles) per hour. About half an hour before midnight, she sideswiped a huge iceberg and sustained a 100-meter (330-foot) gash in her hull. As the

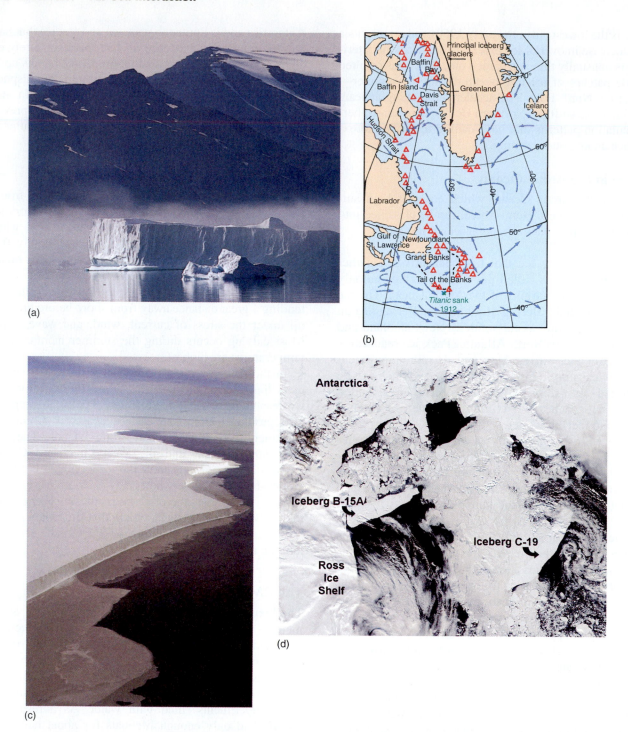

(a)

(b)

(c)

(d)

Figure 7–22 Icebergs.

(a) Icebergs, such as this small North Atlantic berg, are formed when pieces of ice calve from glaciers that reach the sea. **(b)** Map showing North Atlantic currents (*blue arrows*), typical iceberg distribution (*red triangles*), and site of the 1912 *Titanic* disaster (*green*). **(c)** Part of a large tabular Antarctic iceberg. **(d)** Satellite view of iceberg C-19, which broke off from Antarctica's Ross Ice Shelf in May 2002. Also shown is iceberg B-15A, which is part of the larger B-15 iceberg that was the size of Connecticut when it calved in March 2000.

Titanic's front compartments began to fill with water, the ship tipped and allowed more compartments to fill with water. Within $2\frac{1}{2}$ hours after striking the iceberg, the ship tipped so far that it broke in half and sank. Because of cold water temperatures, nighttime conditions, lack of aid from other ships in the area, and a shortage of lifeboats, 1513 of the 2224 passengers perished.

The tragedy brought about more stringent safety rules for ships and the formation of an iceberg patrol. The ice patrol regularly surveys an area larger than the state of Pennsylvania to prevent further disasters of this kind. Originated by the U.S. Navy, the ice patrol became international in 1914. Today, the International Ice Patrol is maintained by the U.S. Coast Guard.

? STUDENTS SOMETIMES ASK ...

How effective has the ice patrol been in preventing accidents like the Titanic *disaster?*

Since the International Ice Patrol was established in 1914, not one ship has been lost to ice within the patrolled area (except during wartime). However, accidents still do occur outside the patrolled area. For example, in 1989 the Soviet cruise ship *Maxim Gorky* accidentally rammed an iceberg in the Arctic Ocean well north of the Arctic Circle between Greenland and Spitzbergen. Fortunately, quick action by the Norwegian Coast Guard prevented any loss of life during evacuation of the more than 1300 passengers and crew on board.

Shelf Ice

In Antarctica, where glaciers nearly cover the entire continent, the edges of glaciers form thick floating sheets of ice called **shelf ice** that break off and produce vast plate-like icebergs (Figures 7-22c and d). In March 2000, for example, a Connecticut-sized iceberg (11,000 square kilometers or 4250 square miles) known as B-15 and nicknamed "Godzilla" broke loose from the Ross Ice Shelf into the Ross Sea. By comparison, the largest iceberg ever recorded in Antarctic waters measured an incredible 335 by 97 kilometers (208 by 60 miles)—about the same size as Connecticut and Massachusetts combined.

The icebergs have flat tops that may stand as much as 200 meters (650 feet) above the ocean surface, although most rise less than 100 meters (330 feet) above sea level, and as much as 90% of their mass is below waterline. Once created, ocean currents driven by strong winds carry the icebergs north, where they eventually melt. Because this region is not a major shipping route, the icebergs pose little serious navigation hazard except for supply ships traveling to Antarctica. Officers aboard ships sighting these gigantic bergs have, in some cases, mistaken them for land!

The rate at which Antarctica is producing icebergs—especially large icebergs—has increased in recent years. For example, the Antarctic Peninsula's Larsen B ice shelf has decreased by 40% since 1997, including a huge release of 3250 square kilometers (1255 square miles) of ice during two months in early 2002. Scientists attribute the retreat to regional climatic warming, which has been present since at least the late 1940s. The rate of warming is approximately 0.5°C (0.9°F) per decade and may be related to global warming.

Sea ice is created when seawater freezes; icebergs form when chunks of ice break off from glaciers on land that reach the sea.

The Atmosphere's Greenhouse Effect

The worldwide average temperature of Earth and the troposphere is about 15°C (59°F). If the atmosphere contained no water vapor, carbon dioxide, methane, or other trace gases, however, the average worldwide temperature would be −18°C(−4°F). At this temperature, all of the water on Earth would be frozen and it would be too cold to support the present distribution of life. Instead, the greenhouse effect of gases in the atmosphere helps create and sustain the more pleasant temperatures to which life has grown accustomed.

The **greenhouse effect** gets its name because it keeps Earth's surface and lower atmosphere warm the same way a greenhouse keeps plants warm enough to grow regardless of outside conditions (Figure 7-23). Energy radiated by the Sun covers the full electromagnetic spectrum, but most of the energy that reaches Earth's surface is short wavelengths, in and near the visible portion of the spectrum. In a greenhouse, short-wave sunlight passes through the glass or plastic covering, where it strikes the plants, the floor, and other objects inside, and is converted into longer-wavelength infrared radiation (heat). Some of this heat energy escapes from the greenhouse and some is trapped for a while by the glass or plastic covering, which keeps the greenhouse nice and snug—much

Figure 7–23 How a greenhouse works.

The glass of a greenhouse allows incoming sunlight to pass through but traps heat. Similarly, gases like water, carbon dioxide, and methane in Earth's atmosphere act just like the glass of a greenhouse by allowing sunlight to pass through but trapping heat.

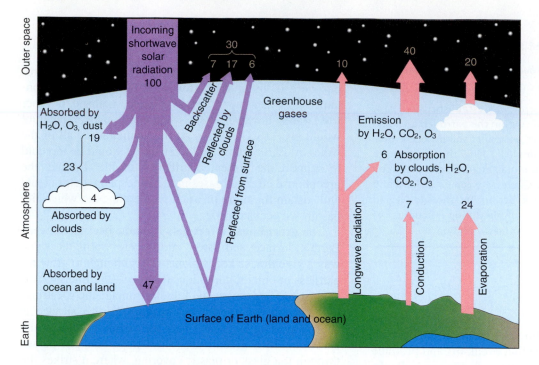

Figure 7–24 Earth's heat budget.

One hundred units of short-wave solar radiation from the Sun (mostly visible light) is reflected, scattered, and absorbed by various components of the Earth-atmosphere system. The absorbed energy is radiated back into space from Earth as long-wave infrared radiation (heat). If this infrared radiation does not leave Earth, global warming will occur.

like what happens in Earth's atmosphere. In addition, the lack of circulation inside the greenhouse also adds to the warming effect. The greenhouse effect is the same process that causes the temperature inside a car parked in the Sun to increase so dramatically.

Figure 7-24 diagrams the roles of the various components of Earth's **heat budget**. In the upper atmosphere, most solar radiation within the visible spectrum penetrates the atmosphere to Earth's surface, like sunlight coming through greenhouse glass. After scattering by atmospheric molecules and reflection off clouds, about 47% of the solar radiation that is directed towards Earth is absorbed by the oceans and continents. About 23% is absorbed by the atmosphere and clouds, and about 20% is reflected into space by atmospheric backscatter, clouds, and Earth's surface.

Figure 7-25 shows that most of the energy coming to Earth from the Sun is within the visible spectrum and peaks at a wavelength of 0.48 micrometers[10] (0.0002 inch). The atmosphere is transparent to much of this radiation, but it is absorbed by materials such as water and rocks at Earth's surface. These materials reradiate this energy back toward space as longer-wavelength infrared (heat) radiation, with a peak at a wavelength of 10 micrometers (0.004 inch). Earth has maintained a constant average temperature over long periods of time, so the rates of energy absorption and reradiation back into space must be equal.

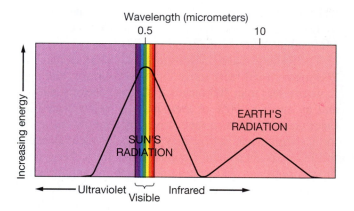

Figure 7–25 Energy radiated by the Sun and Earth.

The intensity of energy radiated by the Sun peaks at a wavelength of 0.48 micrometer (0.0002 inch), which is in the visible part of the spectrum. Some of this energy is absorbed or reflected while some re-radiates from Earth in the infrared (heat) range at a wavelength of 10 micrometers (0.004 inch).

Figure 7-26 shows that most of the solar radiation that is not reflected back to space passes through the atmosphere and is absorbed at Earth's surface. Earth's surface, in turn, emits longer wavelength infrared radiation (heat). A portion of this energy is absorbed by certain heat-trapping gases in the atmosphere such as water vapor, carbon dioxide, and other gases. These gases then reradiate some of the infrared energy, producing the greenhouse effect. Thus, *the change of wavelengths from visible to infrared is the key to how the greenhouse effect works.*

[10]A micrometer (μm) is one-millionth of a meter.

Figure 7–26 The heating of Earth's atmosphere.

Most of the solar radiation that is not reflected back to space passes through the atmosphere and is absorbed at Earth's surface (*1*). Earth's surface, in turn, emits longer wavelength infrared (heat) radiation (*2*). A portion of this energy is absorbed by certain gases in the atmosphere and reradiated back to Earth, thus trapping heat and warming Earth (*3*).

Some of the infrared energy absorbed in the atmosphere becomes reabsorbed by Earth to continue the process (the rest is lost to space). The solar radiation received at the surface, therefore, is retained for a time within our atmosphere, where it moderates temperature fluctuations between night and day and also between seasons.

Which Gases Contribute to the Greenhouse Effect?

Water vapor contributes more to the greenhouse effect than any other gas. It occurs naturally in the atmosphere,

however, and human activities are not thought to affect the amount of water vapor in the atmosphere. Table 7-6 shows the greenhouse gases that have been increasing as a result of human activities. Some of these gases also occur naturally and have been in the atmosphere prior to human activities (such as carbon dioxide) whereas others are clearly human induced (such as chlorofluorocarbons).

Of the gases increased as a result of human activities, carbon dioxide makes the greatest relative contribution to the greenhouse effect (Table 7-6). As a result of human activities, the atmospheric concentration of carbon dioxide

TABLE 7-6 **Greenhouse gases and their contribution to increasing the greenhouse effect.**

Gas	Concentration (ppbv[a])	Rate of increase (% per year)	Relative contribution to increasing the greenhouse effect (%)	Infrared radiation absorption per molecule (number of times greater than CO_2)
Carbon dioxide (CO_2)	370,000	0.5	60	1
Methane (CH_4)	1,700	1.0	15	25
Nitrous oxide (N_2O)	31	0.2	5	200
Tropospheric ozone (O_3)	10–50	0.5	8	2,000
Chlorofluorocarbon (CFC-11)	0.28	4.0	4	12,000
Chlorofluorocarbon (CFC-12)	0.48	4.0	8	15,000
Total	—	—	100	—

[a]ppbv = parts per billion by volume (not by weight).

has increased 30% over the past 200 years (Figure 7-27). Currently 370 parts per million, the concentration increases by 1.2 parts per million each year.

The other trace gases—methane, nitrous oxide, tropospheric ozone, and chlorofluorocarbons—are present in far lower concentrations. They are important, however, because they absorb many times more infrared radiation per molecule than carbon dioxide (Table 7-6, *last column*). Still, these gases have a smaller overall contribution to increasing the greenhouse effect because their concentrations are so low. Nevertheless, all of these gases must be taken into account when considering the total amount of greenhouse warming.

> The greenhouse effect is caused by gases such as water vapor and carbon dioxide that allow sunlight to pass through but trap heat energy before it is reradiated back to space.

What Changes Will Occur as a Result of Increased Global Warming?

Earth's average surface temperature has risen at least 0.6°C (1.1°F) in the last 130 years (Figure 7-28). Is this warming attributable to the increase in atmospheric carbon dioxide and other greenhouse gases, or is it related to a long-term natural climate cycle? Unfortunately, there is no clear proof to answer this question and many scientists debate the relative merits of information in support of human-in-

duced greenhouse warming. In fact, some scientists who develop sophisticated computer models of climate have evidence that additional greenhouse warming may evaporate more seawater, using up much of the excess heat and generating more cloud cover, which will block the Sun's rays and significantly reduce the warming effect.

If global temperatures continue to increase as a result of global warming, researchers believe there will be many changes, but not all agree as to what those changes will be or how severe they might become. Some of the predicted changes include the following:

- Higher than normal sea surface temperatures, which cause more frequent and more intense tropical storms. In addition, high water temperatures affect temperature-sensitive marine organisms (such as corals) and may alter the ocean's deep-water circulation pattern, profoundly affecting Earth's climate.

- More severe droughts in certain areas and more precipitation in other areas, leading to increased chances of flooding.

- Water contamination issues that lead to larger outbreaks of water-borne diseases such as malaria, yellow fever, and dengue fever.

- Longer and more intense heat waves.

- Shifts in the distribution of plant and animal communities.

- The melting of polar ice caps, resulting in a rise in sea level that could flood low-lying coastal areas.

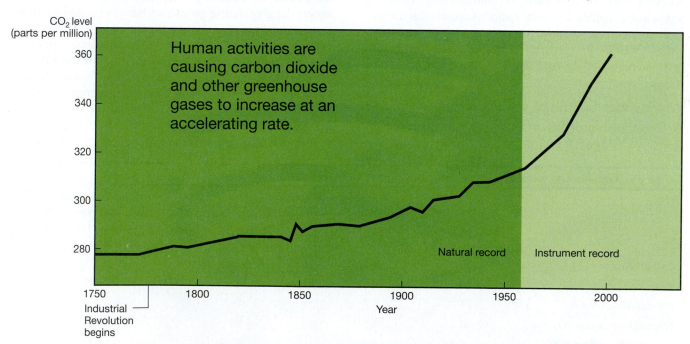

Figure 7–27 Amount of carbon dioxide in the atmosphere since 1750.

There has been a dramatic increase of average worldwide atmospheric carbon dioxide since the Industrial Revolution began in the late 1700s. Values for 1958–present are from instrumental measurement of CO_2 at Mauna Loa Observatory in Hawaii; natural record values prior to 1958 are estimated from air bubbles in polar ice cores.

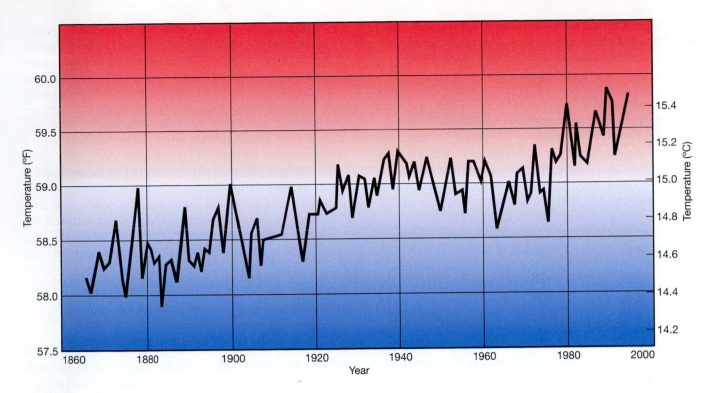

Figure 7–28 Instrumental temperature record since 1865.

The record of global average surface air temperature from thermometer readings indicates a global warming of at least 0.6°C (1.1°F) over the past 130 years. The peaks and troughs indicate natural year-to-year variability of climate.

STUDENTS SOMETIMES ASK...

Since scientists don't know for sure that the greenhouse effect is actually increasing global temperature, why should we do anything about it?

The kinds of measures that prevent the greenhouse effect—being more fuel efficient, protecting plant communities, eliminating chlorofluorocarbons, reducing carbon dioxide emissions—are all sound practices for preserving the environment regardless of reducing the greenhouse effect. Besides, if we find out in the next few decades that there really has been increased global warming due to the addition of human-caused gases, we'll wish that we had started making changes much earlier.

If global temperatures increase, it is also possible that the ice caps may actually *enlarge*. How can global warming cause ice caps to enlarge? If the atmosphere is warmer, then evaporation will occur more quickly (which, incidentally, will probably make tropical storms more intense). Increased evaporation will cause more water vapor to be present in the atmosphere. Much of this water vapor will fall as precipitation on land, thus potentially increasing the amount of snowfall for the formation of ice caps. An increase in the amount of water stored in ice caps (instead of in the ocean) will lower sea level.

In addition, not all predicted changes have negative consequences. For instance, increased warming will provide a longer growing season for some crops and increased atmospheric carbon dioxide helps promote productivity in plants. However, a number of uncertainties remain in understanding regional effects of climate change, and various systems may respond to these changes in unanticipated ways.

The global climate system is complex and contains many poorly understood feedback loops, including the role of clouds and atmospheric water. The successful modeling of Earth's climate, therefore, is one of the biggest scientific challenges today, even with some of the world's most powerful computers.

The IPCC and the Kyoto Protocol

In 1988, the United Nations Environment Programme and the World Meteorological Organization sponsored the **Intergovernmental Panel on Climate Change (IPCC)**, a group of over 200 scientists worldwide, to study human effects on global warming. The group's report, published in 1995, states that "the balance of evidence suggests a discernable human influence on global climate" and that global warming "is unlikely to be entirely due to natural causes." Since the Industrial Revolution in the 1700s, humans have been burning large quantities of fossil fuels—coal, oil, and gas—that had been buried in sedimentary rocks. Mechanization in agriculture has led to the removal of thousands of acres of forests. Both of these are natural storage places for Earth's organic carbon. Deforestation and fossil fuel burning releases previously stored carbon, increasing the amount of CO_2 in the atmosphere (see Figure 7-27).

Mounting scientific evidence that global warming and some of its side effects are occurring has led to international efforts to address the human contribution to the greenhouse effect. A number of international conferences have resulted in an agreement amongst 60 nations to voluntarily limit greenhouse gas emissions. This agreement, which is called the **Kyoto Protocol** (because it was created at an international conference held in 1997 in Kyoto, Japan) sets target reductions for each country. For example, the U.S. agreed to reduce all greenhouse gas emissions to 7% below 1992 emissions by 2007. More recently, however, the U.S. government has withdrawn its support for the protocol, citing potential harm to the economy. The protocol also establishes processes by which technology may be transferred to developing nations to enable them to industrialize without becoming producers of large quantities of greenhouse gases.

In 2001, a second IPCC group published its report, which was prepared by 426 authors and unanimously accepted by more than 160 delegates from 100 countries. The report states that recent regional climate changes already have affected many physical and biological systems on Earth and that projected climate change—as well as changes in climate extremes—could have major consequences. The report also increased the estimate of the world's temperature rise between 1990 and 2100 from 1.0 to 3.5°C (1.8 to 6.3°F) to 1.4 to 5.8°C (2.5 to 10.4°F) according to new climate models.

The Ocean's Role in Reducing the Greenhouse Effect

The ocean is extremely important for reducing the amount of carbon dioxide in the atmosphere, thus reducing the greenhouse effect. In fact, the vast majority of carbon dioxide in the ocean–atmosphere system is found in the ocean because carbon dioxide is approximately 30 times more soluble in water than are other common gases.

What happens to the carbon dioxide that enters the ocean? Most of it is incorporated into organisms through photosynthesis and through their secretion of carbonate shells. Over geologic time, more than 99% of the carbon dioxide added to the atmosphere by volcanic activity has been removed by the ocean and deposited in marine sediments as biogenous calcium carbonate and fossil fuels (oil and natural gas). Thus, the ocean acts as a *repository* (or *sink*) for carbon dioxide, soaking it up and removing it from the environment as sea floor deposits.

In light of the historic Kyoto Protocol, the ocean's ability to remove excess carbon dioxide has been investigated by many countries. Experiments have been conducted that capture emissions before they are released into the atmosphere and pumping the gas into the deep ocean or underground reservoirs (thereby removing carbon dioxide and reducing global warming). However, there may be unwanted side effects when vast amounts of carbon dioxide are added to the deep ocean.

In addition, the thermal properties of the ocean make it ideal for minimizing changes in temperature that may be brought about by global warming. The oceans act as a "thermal sponge," absorbing heat but not increasing much in temperature. To determine if the oceans are warming, the global monitoring of ocean temperature has been initiated (Box 7-2).

What Should We Do About the Increasing Greenhouse Gases?

Stimulating productivity in the ocean has been proven to remove carbon dioxide from the atmosphere. Through photosynthesis, microscopic marine algae convert carbon dioxide dissolved in the ocean to carbohydrate and oxygen gas. By removing more carbon dioxide from the ocean, the ocean can, in turn, absorb more carbon dioxide from the atmosphere.

Areas of the ocean that have relatively low productivity, such as the tropical oceans, are a good place to stimulate productivity and thus increase the amount of carbon dioxide removed from the atmosphere. In 1987, oceanographer John Martin determined that the absence of iron limited productivity in tropical oceans, so he proposed fertilizing the ocean with iron to increase its productivity (the "**iron hypothesis**"). In 1993, Martin's associates added finely ground iron to a test area of the ocean near the Galápagos Islands in the Pacific Ocean. Their results and the results of other "Iron Ex" open-ocean experiments in 1995 and 1999 showed that adding iron to the ocean increased productivity up to 30 times.

Although these results are promising, there were problems grinding the iron fine enough, dispersing the iron, and keeping it in suspension for long periods of time. Moreover, the long-term global environmental effects of adding additional iron and large amounts of carbon dioxide to the ocean are as yet unknown.

What is very clear, however, is that our agricultural and industrial activities are changing the environment. Although we cannot completely eliminate the changes, we can at least reduce our impact on Earth. For instance, we can reduce the rate of increasing combustion of fossil fuels and the widespread deforestation that accompanies agricultural development. In order to halt further damage, we must also preserve Earth's plant communities and replace much of what has been removed. Additionally, we must modify activities that affect the environment most severely, while continuing to pursue research that improves our understanding of how Earth's systems work.

Worldwide, there is a layer in the ocean at a depth of about 1000 meters (3300 feet) caused by temperature and pressure conditions that causes sound originating above and below it to become refracted, or bent, into the layer (Figure 7B, *lower inset*). Once in this layer, called the *SOFAR channel* (an acronym for *SOund Fixing And Ranging*), sound is efficiently trapped and transmitted long distances. For instance, certain whales may use the SOFAR channel to send sounds across entire ocean basins.

Walter Munk of the Scripps Institution of Oceanography (Figure 7B, *upper inset*) came up with the idea of sending sounds through the SOFAR channel in an effort to detect the amount of ocean warming as a result of the greenhouse effect. His *Acoustic Thermometry of Ocean Climate (ATOC)* experiment is designed to accurately measure the travel time of similar low-frequency sound signals through the SOFAR channel now and in the future. The speed of sound in seawater increases as temperature increases, so

sound should take less time to travel the same distance in the future if, in fact, the oceans are warming.

In 1991, Munk's group successfully tested ATOC at Heard Island in the southern Indian Ocean, from which sound can reach many different
Continued…

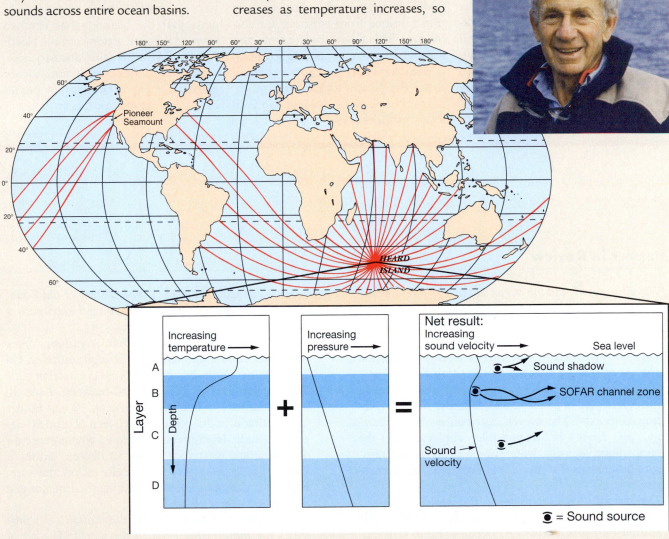

Figure 7B   The ATOC experiment.

Map showing Heard Island sound travel paths, oceanographer Walter Munk (*upper inset*), and the SOFAR channel (*lower inset*). Temperature and pressure conditions in the ocean combine to produce the SOFAR channel (*layer B*) that traps and transmits sound energy. Sound waves created in layers A and C are bent into the SOFAR channel and remain within it, allowing the sound to be transmitted across entire ocean basins.

receiving sites along straight-line paths (Figure 7B). The researchers used an underwater array similar to a series of loudspeakers deployed from a ship to send acoustical (akouein = to hear) signals for six days that were refracted into the SOFAR channel and transmitted throughout the oceans. These signals were received at shipboard recording stations up to three-and-a-half hours later after traveling as far as 19,000 kilometers (11,800 miles), where the precise time of their arrival was noted.

The success of ship-based testing at Heard Island led to the establishment of a fixed sound source near Pioneer Seamount off California and an array of fixed receivers throughout the Pacific Ocean. In 1995, ATOC sound signals were sent again. Even though many precautions were taken to avoid any unwanted effects on marine mammals, three humpback whales were found dead in the area a few days after the sound transmissions had begun. The sounds were halted while the U.S. National Marine Fisheries Service (NMFS) conducted research to determine whether the sounds affected the hearing of nearby whales and contributed to their deaths. After extensive scientific study, their results indicated that marine mammals are unaffected by the transmissions and that the dead whales were an unfortunate coincidence. The long-term effect of ATOC signals on marine mammals—including marine mammal communication through the SOFAR channel—is poorly understood but is also likely to be minimal. The NMFS approved the project and transmissions from offshore California and Hawaii have been conducted without incident, producing a wealth of data that indicates a warming trend in the Pacific and establishing an important baseline for comparison with future measurements.

A similar but smaller-scale operation, the *Transarctic Acoustic Propagation Experiment (TAP)*, was successfully undertaken in 1994. Scientists sent sound signals through Arctic waters from near Spitzbergen in the North Atlantic Ocean to receivers off the north coast of Baffin Island and in the Beaufort Sea. Because heat exchange mechanisms are so important in high latitudes, it is thought that any global change in temperature would be experienced first in high-latitude waters. When compared with 10-year-old temperature measurements, it was discovered that the sound traveled faster than predicted, indicating that Arctic waters have warmed. Further research suggests that the warming has resulted from an influx of water from the North Atlantic. What remains unclear, however, is if this influx occurred because of global warming or if it is part of a natural cycle in the Atlantic Ocean.

Chapter in Review

- *The atmosphere and the ocean act as one interdependent system*, linked by complex feedback loops. There is a close association between many atmospheric and oceanic phenomena.
- *The Sun heats Earth's surface unevenly due to the change of seasons (caused by the tilt of Earth's rotational axis*, which is 23.5 degrees from vertical) and the daily cycle of sunlight and darkness (Earth's rotation on its axis). The *uneven distribution of solar energy on Earth* is responsible for creating the temperature, density, water vapor content, and pressure differences that produce atmospheric and oceanic movement.
- *The Coriolis effect influences the paths of objects* moving in a north or south direction on Earth and is *caused by Earth's rotation*. Because Earth's surface rotates at different velocities at different latitudes, *objects in motion tend to veer to the right in the Northern Hemisphere and to the left in the Southern Hemi-sphere*. The Coriolis effect is *nonexistent at the Equator but increases with latitude*, reaching a maximum at the poles.
- *More solar energy is received than is radiated back into space at low latitudes than at high latitudes*. On the spinning Earth, this creates *three circulation cells in each hemisphere*: a Hadley cell between 0 and 30 degrees latitude, a Ferrel cell between 30 and 60 degrees latitude, and a polar cell between 60 and 90 degrees latitude. High-pressure regions, where dense air descends, are located at about 30 degrees north or south latitude and at the poles. Belts of low pressure, where air rises, are generally found at the Equator and at about 60 degrees latitude.
- *The movement of air within the circulation cells produces the major wind belts of the world*. The air at Earth's surface that is moving away from the subtropical highs produces *trade winds* moving toward the Equator and *prevailing westerlies* moving toward

higher latitudes. The air moving along Earth's surface from the polar high to the subpolar low creates the *polar easterlies.*

- *Calm winds characterize the boundaries between the major wind belts of the world.* The boundary between the two trade wind belts is called the *doldrums,* which coincides with the Intertropical Convergence Zone (ITCZ). The boundary between the trade winds and the prevailing westerlies is called the *horse latitudes.* The boundary between the prevailing westerlies and the polar easterlies is called the *polar front.*

- The *tilt of Earth's axis of rotation,* the *lower heat capacity of rock material* compared to seawater, and the *distribution of continents modify the wind and pressure belts of the idealized three-cell model.* However, the three-cell model closely matches the pattern of the major wind belts of the world.

- *Weather describes the conditions of the atmosphere at a given place and time, while climate is the long-term average of weather.* Atmospheric motion (*wind*) *always moves from high- to low-pressure regions.* In the Northern Hemisphere, therefore, there is a *counterclockwise cyclonic movement* of air around low-pressure cells and a *clockwise anticyclonic movement* around high-pressure cells. Coastal regions commonly experience *sea and land breezes,* due to the daily cycle of heating and cooling.

- *Many storms are due to the movement of air masses.* In the mid-latitudes, cold air masses from higher latitudes meet warm air masses from lower latitudes and create *cold and warm fronts* that move from west to east across Earth's surface. *Tropical cyclones (hurricanes) are large, powerful storms that mostly affect tropical regions of the world.* Destruction caused by hurricanes is caused by storm surge, high winds, and intense rainfall.

- *The ocean's climate patterns are closely related to the distribution of solar energy and the wind belts of the world.* Ocean surface currents somewhat modify oceanic climate patterns.

- *In high latitudes, low temperatures freeze seawater and produce sea ice,* which forms as a slush and breaks into pancakes that ultimately grow into ice floes. Types of sea ice include *pack ice,* which forms each winter and melts almost entirely each summer; *polar ice,* which is permanent in polar regions of the Arctic Ocean; and *fast ice,* which forms frozen to the shore during the winter. *Icebergs form when chunks of ice break off glaciers* that form on Antarctica, Greenland, and some Arctic islands. *Ice sheets* in Antarctica produce the largest icebergs.

- *Energy reaching Earth from the Sun is mostly in the ultraviolet and visible regions* of the electromagnetic spectrum, whereas *energy radiated back to space from Earth is primarily in the infrared (heat) region.* Water vapor, carbon dioxide, and other trace gases absorb infrared radiation and heat the atmosphere, creating *the greenhouse effect. Earth's average surface temperature has warmed over the last century* and there is concern that the human-caused increases of certain heat-trapping gases have enhanced the greenhouse effect.

- *A low-velocity SOFAR sound channel can transmit sound over great distances in the oceans.* This sound channel is being used in an experiment to document global warming in the ocean.

Key Terms

Air mass (p. 207)
Antarctic Circle (p. 197)
Anticyclonic flow (p. 206)
Arctic Circle (p. 197)
Autumnal equinox (p. 196)
Climate (p. 205)
Cold front (p. 207)
Convection cell (p. 198)
Coriolis effect (p. 200)
Cyclone (p. 209)
Cyclonic flow (p. 206)
Doldrums (p. 203)
Equatorial (p. 214)
Equatorial low (p. 203)
Eye of the hurricane (p. 210)
Fast ice (p. 217)
Ferrel cell (p. 203)
Greenhouse effect (p. 219)

Hadley cell (p. 203)
Heat budget (p. 220)
Horse latitudes (p. 204)
Hurricane (p. 209)
Ice floe (p. 216)
Iceberg (p. 215)
Intergovernmental Panel on
 Climate Change
 (IPCC) (p. 223)
Intertropical Convergence Zone
 (ITCZ) (p. 204)
Iron hypothesis (p. 224)
Jet stream (p. 207)
Kyoto Protocol (p. 224)
Land breeze (p. 207)
Northeast trade winds (p. 203)
Pack ice (p. 217)
Pancake ice (p. 216)

Polar (p. 215)
Polar cell (p. 203)
Polar easterly wind belt (p. 203)
Polar front (p. 204)
Polar high (p. 203)
Polar ice (p. 217)
Prevailing westerly wind belt
 (p. 203)
Saffir–Simpson Scale (p. 209)
Sea breeze (p. 207)
Sea ice (p. 215)
Shelf ice (p. 219)
Southeast trade winds (p. 203)
Storm surge (p. 210)
Storm (p. 207)
Subpolar (p. 215)
Subpolar low (p. 203)
Subtropical (p. 215)

Subtropical high (p. 203)
Summer solstice (p. 196)
Temperate (p. 215)
Titanic, RMS (p. 217)
Trade winds (p. 203)
Tropic of Cancer (p. 196)
Tropic of Capricorn (p. 196)
Tropical (p. 215)
Tropical cyclone (p. 209)
Tropics (p. 196)
Troposphere (p. 198)
Typhoon (p. 209)
Vernal equinox (p. 196)
Warm front (p. 207)
Weather (p. 205)
Winter solstice (p. 196)

Questions and Exercises

1. Describe the effect on Earth as a result of Earth's axis of rotation being angled 23.5 degrees from perpendicular relative to the ecliptic. What would happen if Earth were not tilted on its axis?

2. Along the Arctic Circle, how would the Sun appear during the summer solstice? During the winter solstice?

3. Since there is a net annual heat loss at high latitudes and a net annual heat gain at low latitudes, why does the temperature difference between these regions not increase?

4. Describe the physical properties of the atmosphere, including its composition, temperature, density, water vapor content, pressure, and movement.

5. Describe the Coriolis effect in the Northern and Southern Hemispheres and include a discussion of why the effect increases with increased latitude.

6. Sketch the pattern of surface wind belts on Earth, showing atmospheric circulation cells, zones of high and low pressure, the names of the wind belts, and the names of the boundaries between wind belts.

7. Discuss why the idealized belts of high and low atmospheric pressure shown in Figure 7-10 are modified (see Figure 7-11).

8. What is the difference between weather and climate? If it rains in a particular area during a day, does that mean that the area has a wet climate? Explain.

9. Describe the difference between cyclonic and anticyclonic flow, and show how the Coriolis effect is important in producing both a clockwise and a counterclockwise flow pattern.

10. How do sea breezes and land breezes form? During a hot summer day, which one would be most common and why?

11. Name the polar and tropical air masses that affect U.S. weather. Describe the pattern of movement across the continent and patterns of precipitation associated with warm and cold fronts.

12. What are the conditions needed for the formation of a tropical cyclone? Why do most mid-latitude areas only rarely experience a hurricane? Why are there no hurricanes at the Equator?

13. Describe the types of destruction caused by hurricanes. Of those, which one causes the majority of fatalities and destruction?

14. How are the ocean's climate belts (Figure 7-20) related to the broad patterns of air circulation described in Figure 7-10? What are some areas where the two are not closely related?

15. What caused the sinking of RMS *Titanic*? What international organization was formed after the sinking of the *Titanic* to prevent other such disasters?

16. How is sea ice different from icebergs? Describe how each is formed.

17. Describe the fundamental difference between solar radiation absorbed at Earth's surface and the radiation that is primarily responsible for heating Earth's atmosphere.

18. Discuss the greenhouse gases in terms of their relative concentrations and relative contributions to any increased greenhouse effect.

19. Describe the iron hypothesis, and discuss the relative merits and dangers of undertaking a project that could cause dramatic changes in the global environment.

20. What physical conditions produce a SOFAR, or sound channel, below the ocean's surface? How is the SOFAR channel being used to determine if global warming has occurred in the oceans?

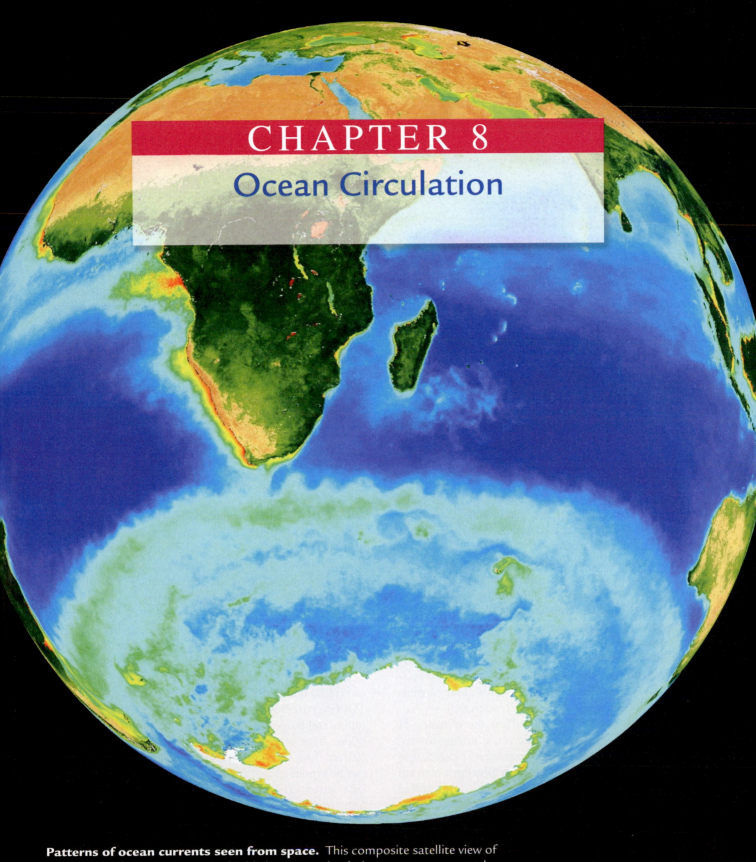

CHAPTER 8
Ocean Circulation

Patterns of ocean currents seen from space. This composite satellite view of Earth during the austral summer highlights ocean circulation patterns near southern Africa, where the Agulhas Retroflection occurs. As the Agulhas Current flows south past the east coast of Africa, it meets the strong eastward-flowing Antarctic Circumpolar Current, which makes it turn abruptly and creates the wavy current pattern between Africa and Antarctica.

Key Questions

- How are ocean currents measured?
- How are surface currents organized in each ocean basin?
- What is western intensification?
- Why is upwelling associated with abundant marine life?
- What environmental effects do El Niños and La Niñas produce?
- What is thermohaline circulation?

The answers to these questions (and much more) can be found in the highlighted concept statements within this chapter.

"The general circulation of the world's oceans is a matter of great interest not only from various practical points of view—climate, fishing, dumping of radioactive wastes and so forth—but primarily from the standpoint of understanding the dynamics and history of the planet on which we live."

—Physical Oceanographer Henry Stommel (1955)

Ocean currents are masses of ocean water that flow from one place to another. The amount of water can be large or small, currents can be at the surface or deep below, and the phenomena that create them can be simple or quite complex. Simply put, currents are *water masses in motion*.

Huge current systems dominate the surfaces of the major oceans. These currents transfer heat from warmer to cooler areas on Earth, just as the major wind belts of the world do. Wind belts transfer about two-thirds of the total amount of heat from the tropics to the poles; ocean surface currents transfer the other third. Ultimately, energy from the Sun drives surface currents and they closely follow the pattern of the world's major wind belts. Harnessing the vast renewable energy of winds and currents is discussed in Chapter 17, "Marine Resources."

More locally, surface currents affect the climates of coastal continental regions. Cold currents flowing toward the Equator on the western sides of continents produce arid conditions. Conversely, warm currents flowing poleward on the eastern sides of continents produce warm, humid conditions. Additionally, ocean currents contribute to the mild climate of northern Europe and Iceland, whereas conditions at similar latitudes along the Atlantic coast of North America (such as Labrador) are much colder.

Currents profoundly affect ocean life, especially those organisms in the deep sea, where currents provide a continuing supply of oxygen. This oxygen is carried there by cold, dense water that sinks in polar regions and spreads across the deep-ocean floor. Ocean currents influence the abundance of life in surface waters by affecting the growth of microscopic algae, which is the basis of most oceanic food chains. Currents have also aided the travel of prehistoric peoples from Europe and Africa to the New World and throughout the Pacific Ocean islands.

Measuring Ocean Currents

Ocean currents are either *wind driven* or *density driven*. Moving air masses—particularly the major wind belts of the world—set wind-driven currents in motion. This motion is parallel to the surface (horizontal) and occurs primarily in the ocean's surface waters, so these currents are called **surface currents**. Density-driven circulation, on the other hand, moves vertically and accounts for the thorough mixing of the deep masses of ocean water. Temperature and salinity conditions at the surface that produce high-density water initiate density-driven circulation. The dense water sinks and spreads slowly beneath the surface, so these currents are called **deep currents**.

Surface currents rarely flow in the same direction and at the same rate for very long, so measuring average flow rates can be difficult. Some consistency, however, exists in the *overall* surface current pattern worldwide. Surface currents can be measured directly or indirectly.

Two main methods are used to *directly* measure currents. In one, a floating device is released into the current and tracked through time. Typically, radio-transmitting float bottles or other devices are used (Figure 8-1a), but other accidentally released items also make good drift meters (Box 8-1). The other method is done from a fixed position (such as a pier) where a current-measuring device, such as the propeller flow meter shown in Figure 8-1b, is lowered into the water. Propeller devices can also be towed behind ships, and the ship's speed is then subtracted to determine a current's true flow rate.

Three different methods can be used to *indirectly* measure surface currents. Water flows parallel to a pressure gradient, so one method is to determine the internal distribution of density and the corresponding pressure gradient across an area of the ocean. A second method uses radar altimeters, such as the one launched aboard the TOPEX/Poseidon satellite in 1992, to determine the lumps and bulges at the ocean surface, which are a result of the shape of the underlying sea floor (see Box 4-1) and current flow. From these data, *dynamic topography* maps can be produced that show the speed and direction of surface currents (Figure 8-2). A third method uses a *Doppler flow meter* to transmit low-frequency sound signals through the water. The flow meter measures the shift in frequency between the sound waves emitted and those backscattered by particles in the water to determine current movement.

The location of deep currents far below the surface makes them even more difficult to measure. Often, they are mapped using devices that are carried with the current or by tracking telltale chemical tracers. Some tracers are

(a)

Figure 8–1 Current-measuring devices.

(a) Drift current meter. Depth of metal vanes is 1 meter (3.3 feet).
(b) Propeller-type flow meter. Length of instrument is 0.6 meter (2 feet).

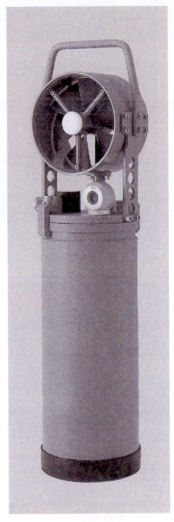

(b)

BOX 8–1   Research Methods in Oceanography
RUNNING SHOES AS DRIFT METERS: JUST DO IT

Any floating object can serve as a makeshift drift meter, as long as it is known where the object entered the ocean and where it was retrieved. The path of the object can then be inferred, providing information about the movement of surface currents. If the time of release and retrieval are known, the speed of currents can also be determined. Oceanographers have long used *drift bottles* (a floating "message in a bottle" or a radio-transmitting device set adrift in the ocean) to track the movement of currents.

Many objects have inadvertently become drift meters when ships have lost some (or all) of their cargo at sea. In this way, Nike athletic shoes

and colorful floating bathtub toys (Figure 8A, *inset*) have advanced the understanding of current movement in the North Pacific Ocean.

In May 1990, the container vessel *Hansa Carrier* was en route from Korea to Seattle, Washington, when it encountered a severe North Pacific storm. The ship was transporting 12.2-meter (40-foot)-long rectangular metal shipping containers, many of which were lashed to the ship's deck for the voyage. During the storm, the ship lost 21 deck containers overboard, including five that held Nike athletic shoes. The shoes floated, so those that were released from their containers were carried east by the North Pacific Current. Within six

months, thousands of the shoes began to wash up along the beaches of Alaska, Canada, Washington, and Oregon (Figure 8A), over 2400 kilometers (1500 miles) from the site of the spill. A few shoes were found on beaches in northern California, and over two years later shoes from the spill were even recovered from the north end of the Big Island of Hawaii!

Even though the shoes had spent considerable time drifting in the ocean, they were in good shape and wearable (after barnacles and oil were removed). Because the shoes were not tied together, many beachcombers found individual shoes or pairs that did not match. Many of the shoes retailed for around $100,

Continued...

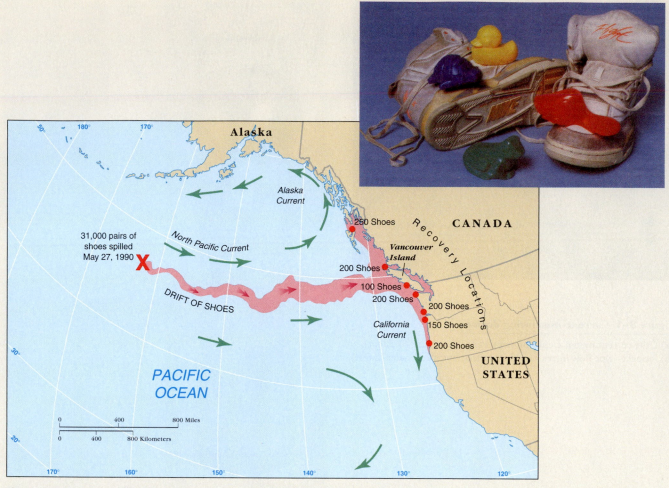

Figure 8A Path of drifting shoes and recovery locations from the 1990 spill; recovered shoes and plastic bathtub toys (*inset*).

so people interested in finding matching pairs placed ads in newspapers or attended local swapmeets.

With help from the beachcombing public (as well as lighthouse operators), information on the location and number of shoes collected was compiled during the months following the spill. Serial numbers inside the shoes were traced to individual containers, and they indicated that only four of the five containers had released their shoes; evidently, one entire container sank without opening. Thus, a maximum of 30,910 pairs of shoes (61,820 individual shoes) were released. The almost instantaneous release of such a large number of drift items helped oceanographers refine computer models of North Pacific circulation. Before the shoe spill, the largest number of drift bottles purposefully released at one time by

oceanographers was about 30,000. Although only 2.6% of the shoes were recovered, this compares favorably with the 2.4% recovery rate of drift bottles released by oceanographers conducting research.

In January 1992, another cargo ship lost 12 containers during a storm to the north of where the shoes had previously spilled. One of these containers held 29,000 packages of small, floatable, colorful plastic bathtub toys in the shapes of blue turtles, yellow ducks, red beavers, and green frogs (Figure 8A, *inset*). Even though the toys were housed in plastic packaging glued to a cardboard backing, studies showed that after 24 hours in seawater, the glue deteriorated and over 100,000 of the toys were released.

The floating bathtub toys began to come ashore in southeast Alaska

10 months later, verifying the computer models. The models indicate that many of the bathtub toys will continue to be carried by the Alaska Current, eventually dispersing throughout the North Pacific Ocean. Some may find their way into the Arctic Ocean, where they could spend time within the Arctic Ocean ice pack. From there, the toys may drift into the North Atlantic, eventually washing up on beaches in northern Europe, thousands of kilometers from where they were accidentally released into the ocean.

Since 1992, oceanographers have continued to study ocean currents by tracking other floating items spilled from cargo ships, including 34,000 hockey gloves, 5 million plastic Lego pieces, at least 3000 computer monitors, and an unidentified number of small plastic doll parts.

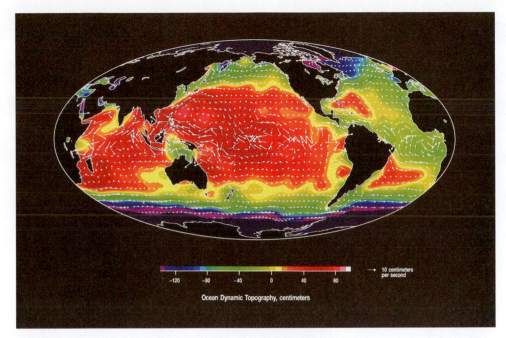

Figure 8–2 Satellite view of ocean dynamic topography.

Map showing TOPEX/Poseidon radar altimeter data in centimeters from September 1992 to September 1993. Red colors are areas that have higher than normal sea level; blue colors are areas that are lower than normal. White arrows indicate the flow direction of currents, with longer arrows indicating faster flow rates.

naturally absorbed into seawater, while others are intentionally added. Some useful tracers that have inadvertently been added to seawater include tritium (a radioactive isotope of hydrogen produced by nuclear bomb tests in the 1950s and early 1960s) and chlorofluorocarbons (freons and other gases now thought to be depleting the ozone layer). Other techniques used to identify deep currents include measuring the distinctive temperature and salinity characteristics of a deep-water mass.

Wind-induced surface currents are measured with floating objects, by satellites, or by other techniques. Density-induced deep currents are measured using tracers or other devices.

Surface Currents

Surface currents develop from friction between the ocean and the wind that blows across its surface. Only about 2% of the wind's energy is transferred to the ocean surface, so a 50-knot[1] wind will create a 1-knot current. You can simulate this on a tiny scale simply by blowing gently and steadily across a cup of coffee.

If there were no continents on Earth, the surface currents would simply follow the major wind belts of the world. In each hemisphere, therefore, a current would

flow between 0 and 30 degrees latitude due to the trade winds, a second would flow between 30 and 60 degrees latitude due to the prevailing westerlies, and a third would flow between 60 and 90 degrees latitude due to the polar easterlies.

However, the distribution of continents on Earth influences the nature and the direction of flow of surface currents. For example, Figure 8-3 shows how the trade winds and prevailing westerlies create large circular-moving loops of water in the Atlantic Ocean. Other ocean basins show a similar pattern, and surface currents are also influenced by gravity, friction, and the Coriolis effect.

Surface currents occur within and above the *pycnocline* (layer of rapidly changing density) to a depth of about 1 kilometer (0.6 mile) and affect only about 10% of the world's ocean water.

Equatorial Currents, Boundary Currents, and Gyres

The trade winds, which blow from the southeast in the Southern Hemisphere and from the northeast in the Northern Hemisphere, set in motion the water masses between the tropics. The resulting currents are called **equatorial currents**, which travel westward along the Equator (Figure 8-4). They are called north or south equatorial currents, depending on their position relative to the Equator.

When equatorial currents reach the western portion of an ocean basin, they must turn because they cannot cross land. The Coriolis effect deflects these currents away from the Equator as **western boundary currents**. The

[1]A *knot* is one nautical mile per hour. A nautical mile is defined as the distance of one minute of latitude and is equivalent to 1.15 statute (land) miles or 1.85 kilometers.

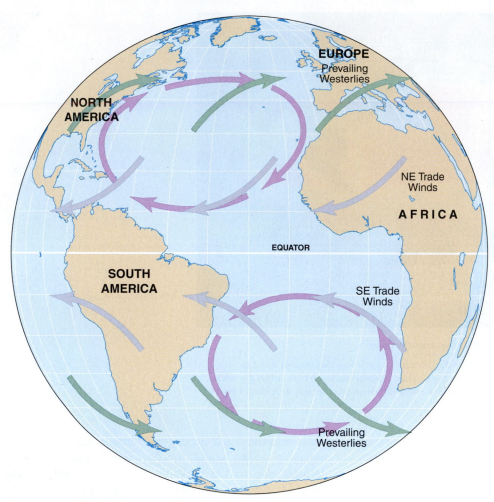

Figure 8–3 **Atlantic Ocean surface circulation pattern.**

The trade winds (*blue arrows*) in conjunction with the prevailing westerlies (*green arrows*) create circular-moving loops of water (*underlying purple arrows*) at the surface in both parts of the Atlantic Ocean basin. If there were no continents, the ocean's surface circulation pattern would closely match the major wind belts of the world.

name means they are currents traveling along the western boundary of their ocean basin.[2] The Gulf Stream and the Brazil Current, which are shown in Figure 8-4, are western boundary currents. They come from equatorial regions, where water temperatures are warm, so they carry warm water to higher latitudes. Figure 8-4 shows warm currents as red arrows.

Between 30 and 60 degrees latitude, the prevailing westerlies blow from the northwest in the Southern Hemisphere and from the southwest in the Northern Hemisphere. These winds direct ocean surface water in an easterly direction across the ocean basin, as shown in Figure 8-4 by the North Atlantic Current and the West Wind Drift.

When currents flow back across the ocean basin, the Coriolis effect and continental barriers turn them toward the Equator, creating **eastern boundary currents** along the eastern boundary of the ocean basins. The Canary Current and the Benguela Current, which are shown in Figure 8-4, are eastern boundary currents.[3] They come from high-latitude regions where water temperatures are cool, so they carry cool water to lower latitudes. Figure 8-4 shows cold currents as blue arrows.

The equatorial, western boundary, prevailing westerly, and eastern boundary currents combine to create a circular flow within an ocean basin called a **gyre** (*gyros* = a circle). Figure 8-4 shows the world's five **subtropical gyres**: (1) the *North Atlantic Gyre*, (2) the *South Atlantic Gyre*, (3) the *North Pacific Gyre*, (4) the *South Pacific Gyre*, and (5) the *Indian Ocean Gyre* (which is mostly within the Southern Hemisphere). The center of each subtropical gyre coincides with the subtropics at 30 degrees north or south latitude. As shown in

---

[2]Notice that the western boundary currents are off the *eastern* coasts of continents. This sounds confusing but is a result of the fact that we have a land-based perspective. From an oceanic perspective, the western side of the ocean basin is where the western boundary current resides.

[3]Currents are sometimes named for a prominent geographic location near where they pass. For instance, the Canary Current passes the Canary Islands; the Benguela Current is named for the Benguela Province in Angola, Africa.

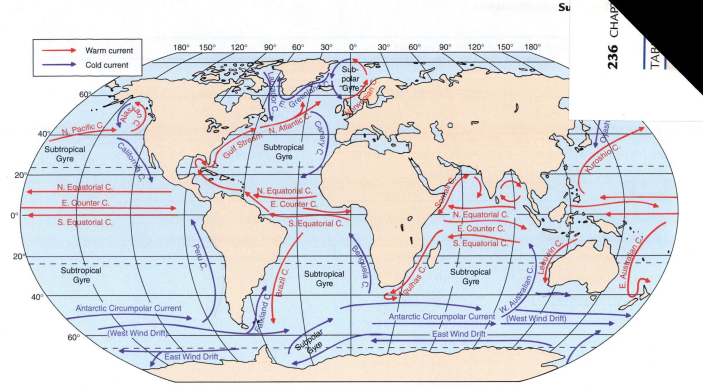

Figure 8–4 Wind-driven surface currents.

Major wind-driven surface currents of the world's oceans during February–March. The five major subtropical gyres are the North and South Pacific Ocean Gyres, the North and South Atlantic Ocean Gyres, and the Indian Ocean Gyre. The smaller subpolar gyres rotate in the reverse direction of the adjacent subtropical gyres.

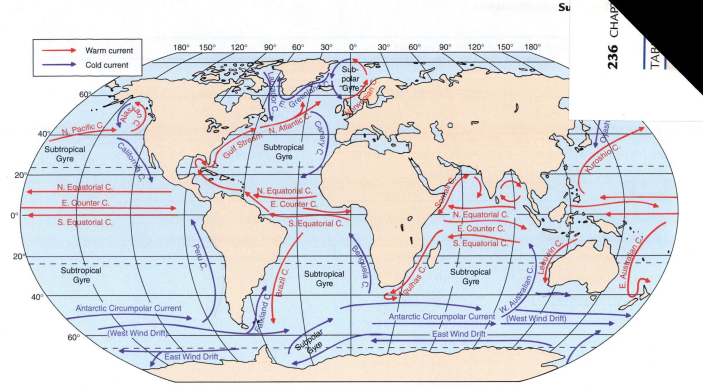

Figures 8-3 and 8-4, subtropical gyres rotate clockwise in the Northern Hemisphere and counterclockwise in the Southern Hemisphere.

Generally, each subtropical gyre is composed of four main currents that flow progressively into one another (Table 8-1). The North Atlantic Gyre, for instance, is composed of the North Equatorial Current, the Gulf Stream, the North Atlantic Current, and the Canary Current (Figure 8-4).

Surface currents moving eastward as a result of the prevailing westerlies approach subpolar latitudes (about 60 degrees north or south latitude). Here, they are driven in a westerly direction by the polar easterlies, producing **subpolar gyres** that rotate opposite the adjacent subtropical gyres. Subpolar gyres are smaller and fewer than subtropical gyres and are best developed in the Atlantic Ocean between Greenland and Europe and the Weddell Sea off Antarctica (Figure 8-4).

> The principal ocean surface current pattern consists of subtropical and subpolar gyres that are large circular-moving loops of water powered by the major wind belts of the world.

Ekman Spiral and Ekman Transport

During the voyage of the *Fram* (see Chapter 1), Norwegian explorer Fridtjof Nansen observed that Arctic Ocean ice moved 20 to 40 degrees to the *right* of the wind blowing across its surface (Figure 8-5). Surface water in the

Northern Hemisphere behaves similarly and, in the Southern Hemisphere, surface currents move to the left of the wind direction. Why does surface water move in a direction different than the wind? *V. Walfrid Ekman* (1874–1954), a Swedish physicist, developed a circulation model called the **Ekman spiral** (Figure 8-6) that explains Nansen's observations in accordance with the Coriolis effect.

The Ekman spiral describes the speed and direction of flow of surface waters at various depths. It is caused by wind blowing across the surface and is modified by the Coriolis effect. Ekman's model assumes that a uniform column of water is set in motion by wind blowing across its surface. Because of the Coriolis effect, the immediate surface water moves in a direction 45 degrees to the right of the wind (in the Northern Hemisphere). The surface water moves as a thin "layer" on top of deeper layers of water. As the surface layer moves, it also sets in motion other layers beneath it, thus passing the energy of the wind down through the water column.

Current speed decreases with increasing depth, however, and the Coriolis effect increases curvature to the right (like a spiral). Thus, each successive layer of water is set in motion at a progressively slower velocity, and in a direction progressively to the right of the one above it. At some depth, a layer of water may move in a direction *exactly opposite to the wind direction that initiated it*! If the water is deep enough, friction will consume the energy imparted by the wind and no motion will occur below that depth. Although it depends on wind speed

LE 8–1   **Subtropical gyres and surface currents.**

Pacific Ocean	Atlantic Ocean	Indian Ocean
North Pacific Gyre	**North Atlantic Gyre**	**Indian Ocean Gyre**
North Pacific Current	North Atlantic Current	South Equatorial Current
California Current[a]	Canary Current[a]	Agulhas Current[b]
North Equatorial Current	North Equatorial Current	West Wind Drift
Kuroshio (Japan) Current[b]	Gulf Stream[b]	West Australian Current[a]
South Pacific Gyre	**South Atlantic Gyre**	**Other Major Currents**
South Equatorial Current	South Equatorial Current	Equatorial Countercurrent
East Australian Current[b]	Brazil Current[b]	North Equatorial Current
West Wind Drift	West Wind Drift	Leeuwin Current
Peru (Humboldt) Current[a]	Benguela Current[a]	Somali Current
Other Major Currents	**Other Major Currents**	
Equatorial Countercurrent	Equatorial Countercurrent	
Alaskan Current	Florida Current	
Oyashio Current	East Greenland Current	
	Labrador Current	
	Falkland Current	

[a]Denotes an eastern boundary current of a gyre, which is relatively *slow, wide,* and *shallow* (and is also a *cold-water* current).
[b]Denotes a western boundary current of a gyre, which is relatively *fast, narrow,* and *deep* (and is also a *warm-water* current).

Figure 8–5   Transport of floating objects.

Fridtjof Nansen first noticed that floating objects, such as icebergs and ships, were carried to the right of the wind direction in the Northern Hemisphere.

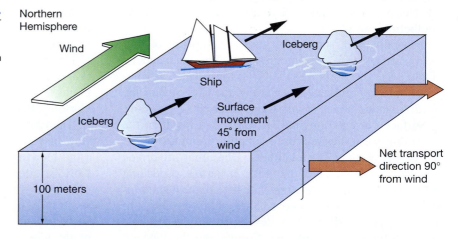

and latitude, this stillness normally occurs at a depth of about 100 meters (330 feet).

Figure 8-6 shows the spiral nature of this movement with increasing depth from the ocean's surface. The length of each arrow in Figure 8-6 is proportional to the velocity of the individual layer, and the direction of each arrow indicates the direction it moves.[4] Under ideal conditions, therefore, the surface layer should move at an angle of 45 degrees from the direction of the wind. All the layers combine, however, to create a net water movement that is 90 degrees from the direction of the wind. This average movement, called **Ekman transport**, is 90 degrees to the *right* in the Northern Hemisphere and 90 degrees to the *left* in the Southern Hemisphere.

"Ideal" conditions rarely exist in the ocean, so the actual movement of surface currents deviates slightly from the angles shown in Figure 8-6. Generally, surface currents move at an angle somewhat less than 45 degrees from the direction of the wind and Ekman transport in the open ocean is typically about 70 degrees from the wind direction. In shallow coastal waters, Ekman transport may be very nearly the same direction as the wind.

---

[4]The name Ekman *spiral* refers to the spiral observed by connecting the tips of the arrows shown in Figure 8-6.

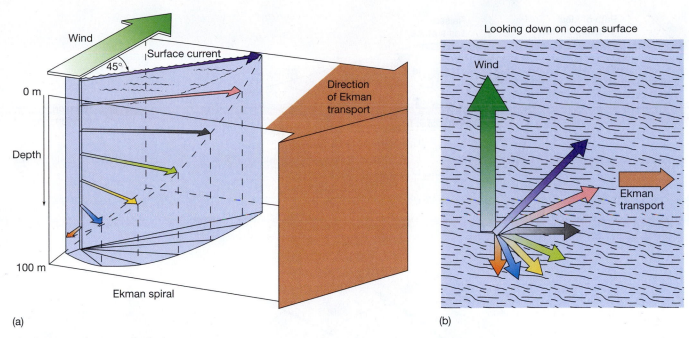

(a)

(b)

Figure 8–6 Ekman spiral.

Perspective view **(a)** and top view **(b)** of Ekman spiral and Ekman transport. Wind drives surface water in a direction 45 degrees to the right of the wind in the Northern Hemisphere. Deeper water continues to deflect to the right and moves at a slower speed with increased depth, causing the Ekman spiral. Ekman transport, which is the net water movement, is at a right angle (90 degrees) to the wind direction.

STUDENTS SOMETIMES ASK...

What does an Ekman spiral look like at the surface? Is it strong enough to disturb ships?

The Ekman spiral creates different layers of surface water that move in slightly different directions at slightly different speeds. It is too weak to create eddies or whirlpools (vortexes) at the surface and so presents no danger to ships. In fact, the Ekman spiral is unnoticeable at the surface. It can be observed, however, by lowering oceanographic equipment over the side of a vessel. At various depths, the equipment can be observed to drift at various angles from the wind direction according to the Ekman spiral.

Geostrophic Currents

Ekman transport deflects surface water to the right in the Northern Hemisphere, so a clockwise rotation develops within an ocean basin and produces a **subtropical convergence** of water in the middle of the gyre, causing water literally to pile up in the center of the subtropical gyre. Thus, there is a hill of water within all subtropical gyres that is as much as 2 meters (6.6 feet) high.

Surface water in a subtropical convergence tends to flow downhill in response to gravity. The Coriolis effect opposes gravity, however, deflecting the water to the right in a curved path (Figure 8-7a) into the hill again. When these two factors balance, the net effect is a **geostrophic** (*geo* = earth, *strophio* = turn) **current** that moves in a circular path around the hill.[5] In Figure 8-7a it is labeled as

the *path of ideal geostrophic flow*. Friction between water molecules, however, causes the water to move gradually down the slope of the hill as it flows around it. This is the *path of actual geostrophic flow* labeled in Figure 8-7a.

Western Intensification

Figure 8-7a shows that the apex (top) of the hill formed within a rotating gyre is closer to the western boundary than the center of the gyre. As a result, the western boundary currents of the subtropical gyres are faster, narrower, and deeper than their eastern boundary current counterparts. For example, the Kuroshio Current (a western boundary current) of the North Pacific Subtropical Gyre is up to 15 times faster, 20 times narrower, and five times as deep as the California Current (an eastern boundary current). This phenomenon is called **western intensification**, and currents affected by this phenomenon are said to be western intensified. *The western boundary currents of all subtropical gyres are western intensified, even in the Southern Hemisphere.*

A number of factors cause western intensification, including the Coriolis effect. The Coriolis effect increases toward the poles, so eastward-flowing high-latitude water turns toward the Equator more strongly than westward-flowing equatorial water turns toward higher latitudes. This causes a wide, slow, and shallow

---

[5]The term *geostrophic* for these currents is appropriate, since the currents behave as they do because of Earth's rotation.

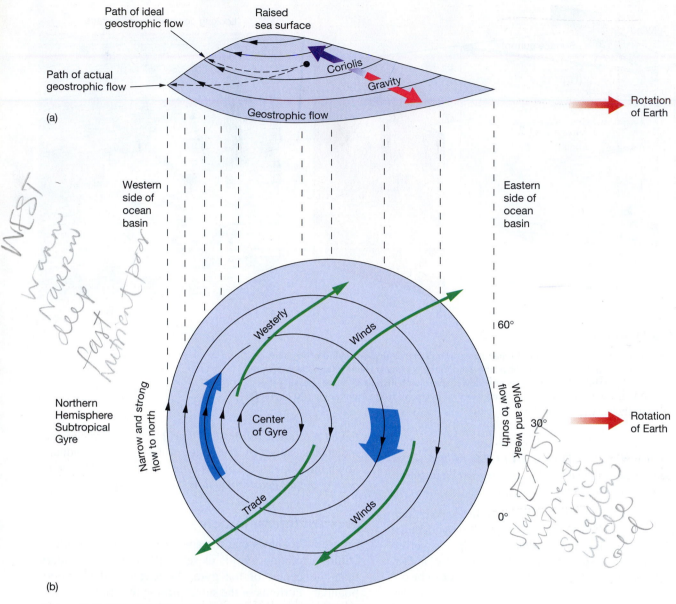

Figure 8–7 Geostrophic current and western intensification.

(a) A cross-sectional view of a subtropical gyre showing how water literally piles up in the center, forming a hill up to 2 meters (6.6 feet) high. Gravity and the Coriolis effect balance to create an ideal geostrophic current that flows in equilibrium around the hill. However, friction makes the current gradually run downslope (*actual geostrophic flow*). **(b)** A map view of the same subtropical gyre, showing that the flow pattern is restricted (lines are closer together) on the western side of the gyre, resulting in western intensification.

flow of water toward the Equator across most of each subtropical gyre, leaving only a narrow band through which the poleward flow can occur along the western margin of the ocean basin. If a constant volume of water rotates around the apex of the hill in Figure 8-7b, then the velocity of the water along the western margin will be much greater than the velocity around the eastern side.[6] In Figure 8-7b, the lines are close together along the western margin, indicating the faster flow. The end result is a high-speed western boundary current that flows along the hill's steeper westward slope and a slow drift of water toward the Equator along the more gradual eastern slope. Table 8-2 summarizes the

differences between western and eastern boundary currents of subtropical gyres.

> Western intensification is a result of Earth's rotation and causes the western boundary current of all subtropical gyres to be fast, narrow, and deep.

[6]A good analogy for this phenomenon is a funnel: In the narrow end of a funnel, the flow rates are speeded up (such as in western intensified currents); in the wide end, the flow rates are sluggish (such as in eastern boundary currents).

TABLE 8-2 **Characteristics of western and eastern boundary currents of subtropical gyres.**

Current type (examples)	Width	Depth	Speed	Transport volume (millions of cubic meters per second [a])	Comments
Western boundary currents (Gulf Stream, Brazil Current, Kuroshio Current)	*Narrow*: usually less than 100 kilometers (60 miles)	*Deep*: to depths of 2 kilometers (1.2 miles)	*Fast*: hundreds of kilometers per day	*Large*: as much as 100 Sv[a]	Waters derived from low latitudes and are warm; little or no upwelling
Eastern boundary currents (Canary Current, Benguela Current, California Current)	*Wide*: up to 1000 kilometers (600 miles)	*Shallow*: to depths of 0.5 kilometer (0.3 mile)	*Slow*: tens of kilometers per day	*Small*: typically 10 to 15 Sv[a]	Waters derived from mid-latitudes and are cool; coastal upwelling common

[a]One million cubic meters per second is a flow rate equal to one Sverdrup (Sv).

Equatorial Countercurrents

A large volume of water is driven westward due to the north and south equatorial currents. The Coriolis effect is minimal near the Equator, so much of the water is not turned toward higher latitudes. Instead, it piles up at the western margins of the ocean basins, which causes average sea level on the western side of the basin to be as much as 2 meters (6.6 feet) higher than on the eastern side. The water on the western margins then flows downhill under the influence of gravity, creating narrow **equatorial countercurrents** that flow to the east *counter to* and *between* the adjoining equatorial currents.

Figure 8-4 shows that an equatorial countercurrent is particularly apparent in the western Pacific Ocean, where a dome of equatorial water is trapped in the island-filled embayment between Australia and Asia. This dome of water with very weak current flow has the highest year-round ocean surface temperatures found anywhere in the world ocean, as shown in Figure 8-8. Continual influx of water from equatorial currents builds the dome and creates an eastward countercurrent that stretches across the Pacific toward South America.

If you reexamine the satellite image of sea surface elevation in Figure 8-2, you'll see that the hills of water within the subtropical gyres of the Atlantic Ocean are clearly visible. The hill in the North Pacific is visible as well, but the elevation of the equatorial Pacific is not as low as expected because the map shows conditions during a moderate El Niño event,[7] so there is a well-developed warm and anomalously high equatorial countercurrent. Figure 8-2 also shows very little distinction between the North and South Pacific gyres. Moreover, the South Pacific subtropical gyre is less intense than other gyres, mostly because it covers such a large area, it lacks confinement by continental barriers along its western margin, and it has numerous islands (really the tops of tall sea floor mountains). The South Indian Ocean hill is rather well developed in the figure, although its northeastern boundary stands high because of the influx of warm Pacific Ocean water through the East Indies islands.

Ocean Currents and Climate

Ocean surface currents directly influence the climate of adjoining landmasses. For instance, warm ocean currents warm the nearby air. This warm air can hold more water vapor, which puts more moisture (high humidity) in the atmosphere. When this warm, moist air travels over a continent, it releases its water vapor in the form of precipitation. Continental margins that have warm ocean currents offshore (Figure 8-8, *red arrows*) typically have a humid climate. The presence of a warm current off the east coast of the United States helps explain why the area experiences such high humidity, especially in the summer.

Conversely, cold ocean currents cool the nearby air, which cannot hold as much water vapor. When the cool, dry air travels over a continent, it results in very little precipitation. Continental margins that have cool ocean currents offshore (Figure 8-8, *blue arrows*) typically have a dry climate. The presence of a cold current off California is part of the reason why it has such an arid climate.

Upwelling and Downwelling

Upwelling is the vertical movement of cold, deep, nutrient-rich water to the surface; **downwelling** is the vertical movement of surface water to deeper parts of the ocean. Upwelling hoists chilled water to the surface. This cold water, rich in nutrients, creates high **productivity** (an abundance of microscopic algae), which establishes the base of the food web and, in turn, supports incredible numbers of

---

[7]El Niño events are discussed later in this chapter under "Pacific Ocean Circulation."

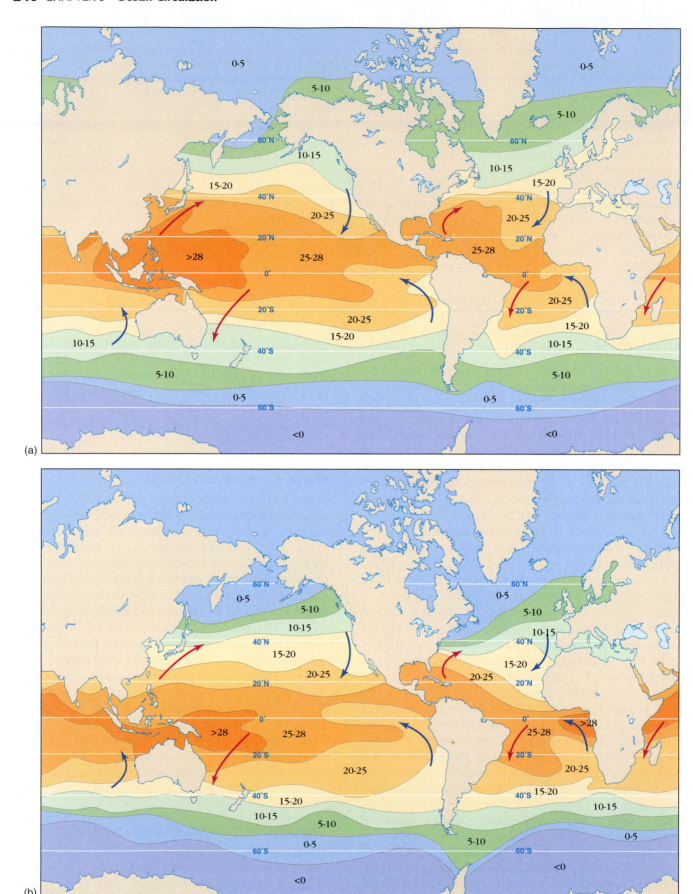

(a)

(b)

Figure 8–8 Surface temperature of the world ocean.

Average sea surface temperature distribution in degrees centigrade for August **(a)** and for February **(b)**. Note that temperatures migrate north–south with the seasons. Red arrows indicate warm surface currents; blue arrows indicate cool surface currents.

larger marine life like fish and whales. Downwelling, on the other hand, is associated with much lower amounts of surface productivity but carries necessary dissolved oxygen to those organisms living on the deep-sea floor.

Upwelling and downwelling provide important mixing mechanisms between surface and deep waters and are accomplished by a variety of methods.

Diverging Surface Water

Current divergence occurs when surface waters move *away from* an area on the ocean's surface, such as along the Equator. As shown in Figure 8-9, the South Equatorial Current occupies the area along the *geographical Equator* (most notably in the Pacific Ocean; see Figure 8-4), while the *meteorological Equator* (where the doldrums exist) typically occurs a few degrees of latitude to the north. As the southeast trade winds blow across this region, Ekman transport causes surface water north of the Equator to veer to the right (northward) and water south of the Equator to veer to the left (southward). The net result is a divergence of surface currents along the geographical Equator, which causes upwelling of cold, nutrient-rich water. Since this type of upwelling is common along the Equator—especially in the Pacific—it is called **equatorial upwelling** and creates areas of high productivity that are some of the most prolific fishing grounds in the world.

Converging Surface Water

Current convergence occurs when surface waters move *toward* each other. In the North Atlantic Ocean, for instance, the Gulf Stream, the Labrador Current, and the East Greenland Current all come together in the same vicinity. When currents converge, water stacks up and has no place to go but downward. The surface water slowly sinks in a process called downwelling (Figure 8-10). Unlike upwelling, areas of downwelling are not associated with prolific marine life because the necessary nutrients are not continuously replenished from the cold, deep, nutrient-rich water below the surface. Consequently, downwelling areas have low productivity.

Coastal Upwelling and Downwelling

Coastal winds can cause upwelling or downwelling due to Ekman transport. Figure 8-11 shows a coastal region along the west coast of a continent in the Southern

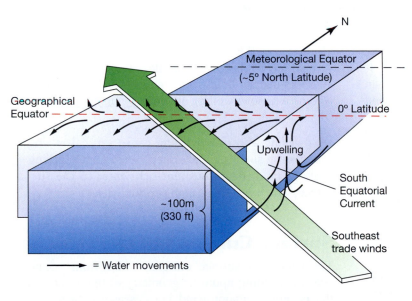

Figure 8–9 Equatorial upwelling.

As the southeast trade winds pass over the geographical Equator to the meteorological Equator, they cause water within the South Equatorial Current north of the Equator to veer to the right (northward) and water south of the Equator to veer to the left (southward). Thus, surface water diverges, which causes equatorial upwelling.

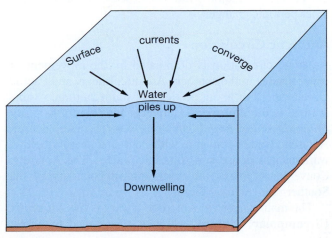

Figure 8–10 Downwelling caused by convergence of surface currents.

Where surface currents converge, water piles up and slowly sinks downward, creating downwelling.

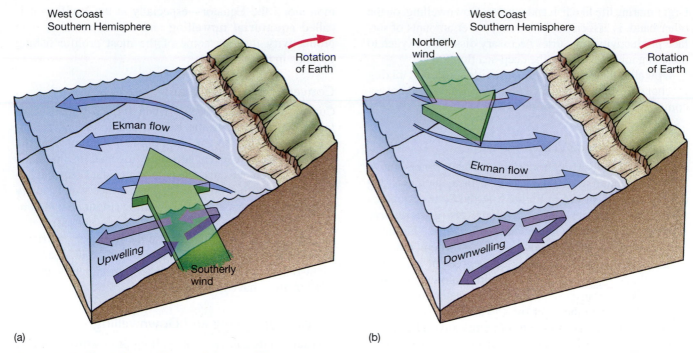

(a)   (b)

Figure 8–11 Coastal upwelling and downwelling.

(a) Where southerly coastal winds blow parallel to a west coast in the Southern Hemisphere, Ekman transport carries surface water away from the continent. Upwelling of deeper water occurs to replace the surface water that has moved away from the coast. **(b)** A reversal of the direction of the winds that cause upwelling causes water to pile up against the shore and causes downwelling.

Hemisphere with winds moving parallel to the coast. If the winds are from the south (Figure 8-11a), Ekman transport moves the coastal water to the left of the wind direction, causing the water to flow *away from* the shoreline. Water rises from below to replace the water moving away from shore in a process called **coastal upwelling**. Areas where coastal upwelling occurs, such as the West Coast of the United States, are characterized by high concentrations of nutrients, resulting in high biological productivity and rich marine life. This coastal upwelling also creates low water temperatures in areas such as San Francisco that provide a natural form of air conditioning (and much cool weather and fog) in the summer.

If the winds are from the north, Figure 8-11b shows that Ekman transport still moves the coastal water to the left of the wind direction but, in this case, the water flows *toward* the shoreline. This causes the water to stack up along the shoreline, where it has nowhere to go but down, in a process called **coastal downwelling**. Areas where coastal downwelling occurs have low productivity. Coastal downwelling can occur in areas that typically experience coastal upwelling when the winds reverse.

Other Upwelling

Figure 8-12 shows how upwelling can be created by offshore winds, sea floor obstructions, or a sharp bend in a coastline. Upwelling also occurs in high-latitude regions, where there is no pycnocline (a layer of rapidly changing density). The absence of a pycnocline allows significant

vertical mixing between high-density cold surface water and high-density cold deep water below. Thus, both upwelling and downwelling are common in high latitudes.

> Upwelling and downwelling cause vertical mixing between surface and deep waters. Upwelling brings cold, deep, nutrient-rich water to the surface, which results in high productivity.

Surface Currents of the Oceans

The pattern of surface currents varies from ocean to ocean depending upon the geometry of the ocean basin, the pattern of major wind belts, seasonal factors, and other periodic changes.

Antarctic Circulation

Antarctic circulation is dominated by the movement of water masses in the southern Atlantic, Indian, and Pacific Oceans south of about 50 degrees south latitude. At this latitude is the **Antarctic Convergence** (Figure 8-13) or *Antarctic Polar Front*, which is where colder, denser, Antarctic waters converge with (and sink sharply below) warmer, less dense sub-Antarctic waters. The Antarctic Convergence marks the northernmost boundary of the Southern or Antarctic Ocean.

The main current in Antarctic waters is the **Antarctic Circumpolar Current**, which is also called the **West**

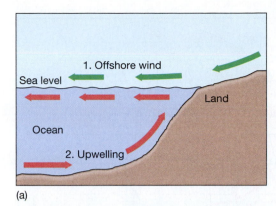

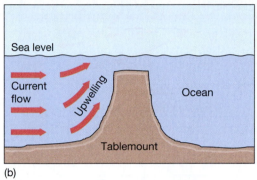

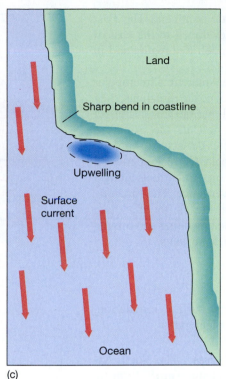

(a)

(b)

(c)

Figure 8–12 Other types of upwelling.

Upwelling can be caused by (a) offshore winds; (b) a sea floor obstruction: in this case, a tablemount; (c) a sharp bend in coastal geometry.

Wind Drift. It encircles Antarctica and flows from west to east at approximately 50 degrees south latitude but varies between 40 and 65 degrees south latitude. At about 40 degrees south latitude is the Subtropical Convergence (Figure 8-13), which forms the northernmost boundary of the Antarctic Circumpolar Current. The Antarctic Circumpolar Current is driven by the powerful prevailing westerly wind belt, which creates winds so strong that these Southern Hemisphere latitudes have been called the "Roaring Forties," "Furious Fifties," and "Screaming Sixties."

The Antarctic Circumpolar Current is the only current that completely circumscribes Earth and is allowed to do so because of the lack of land at high southern latitudes. It meets its greatest restriction as it passes through the Drake Passage (named for explorer Sir Francis Drake) between the Antarctic Peninsula and the southern islands of South America, which is about 1000 kilometers (600 miles) wide. Although the current is not speedy [its maximum surface velocity is about 2.75 kilometers (1.65 miles) per hour], it transports more water (an average of about 130 million cubic meters per second[8] than any other surface current.

The **East Wind Drift**, a surface current propelled by the polar easterlies, moves from an easterly direction

---

[8]One million cubic meters per second is a useful flow rate for describing ocean currents, so it has become a standard unit, named the **Sverdrup (Sv)** after Norwegian explorer Otto Sverdrup.

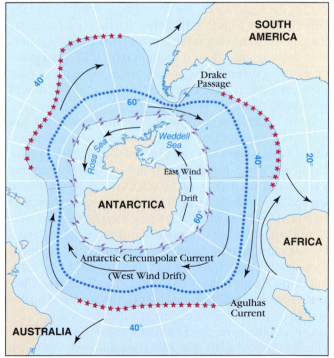

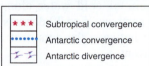

✶✶✶	Subtropical convergence
••••••	Antarctic convergence
⌇⌇⌇	Antarctic divergence

Figure 8–13 Antarctic surface circulation.

The East Wind Drift is driven by the polar easterlies and flows around Antarctica from the east. The Antarctic Circumpolar Current (West Wind Drift) flows around Antarctica from the west but is further from the continent and is a result of the strong prevailing westerlies. Antarctic Convergence and Divergence is caused by interactions at the boundaries of these two currents.

around the margin of the Antarctic continent. The East Wind Drift is most extensively developed to the east of the Antarctic Peninsula in the Weddell Sea region and in the area of the Ross Sea (Figure 8-13).

As the East Wind Drift and the Antarctic Circumpolar Current flow around Antarctica in opposite directions, they create a surface divergence. Recall that the Coriolis effect deflects moving masses to the left in the Southern Hemisphere, so the East Wind Drift is deflected toward the continent and the Antarctic Circumpolar Current is deflected away from it. This creates a divergence of currents along a boundary called the **Antarctic Divergence**. The Antarctic Divergence has abundant marine life in the Southern Hemisphere summer because of the upwelling and mixing of these two currents, which supplies nutrients.

Atlantic Ocean Circulation

Figure 8-14 shows Atlantic Ocean surface circulation, which consists of two large subtropical gyres.

The North and South Atlantic Gyres The **North Atlantic Gyre** rotates clockwise and the **South Atlantic Gyre** rotates counterclockwise, due to the combined effects of the trade winds, the prevailing westerlies, and the Coriolis effect. Figure 8-14 shows that each gyre consists of a poleward-moving warm current (*red*) and an equatorward-moving cold "return" current (*blue*). The two gyres are partially offset by the shapes of the surrounding continents and the *Atlantic Equatorial Countercurrent* moves in between them.

In the South Atlantic Gyre, the **South Equatorial Current** reaches its greatest strength just below the Equator, where it encounters the coast of Brazil and splits in two. Part of the South Equatorial Current moves off along the northeastern coast of South America toward the Caribbean Sea and the North Atlantic. The rest is turned southward as the Brazil Current, which ultimately merges with the West Wind Drift and moves eastward across the South Atlantic. The **Brazil Current** is much smaller than its Northern Hemisphere counterpart, the Gulf Stream, due to the splitting of the South Equatorial Current. The **Benguela Current**, slow-moving and cold, flows toward the Equator along Africa's western coast, completing the gyre.

Outside the gyre, the *Falkland Current* (Figure 8-14), which is also called the *Malvinas Current*, moves a significant amount of cold water along the coast of Argentina as far north as 25 to 30 degrees south latitude, wedging its way between the continent and the southbound Brazil Current.

The Gulf Stream The **Gulf Stream** is the best studied of all ocean currents. It moves northward along the East Coast of the United States, warming coastal states and moderating winters in these and northern European regions.

Figure 8-15 shows the network of currents in the North Atlantic Ocean that contribute to the flow of the Gulf

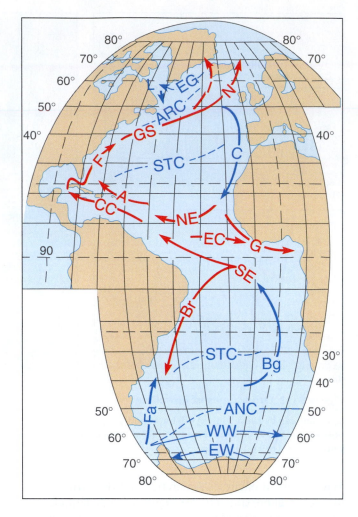

TEMPERATURE
cold →
warm →

CONVERGENCES
ARC=Arctic
STC=Subtropical
ANC=Antarctic

CURRENTS

A=Antilles	G=Guinea
Bg=Benguela	GS=Gulf Stream
Br=Brazil	I=Irminger
C=Canary	L=Labrador
CC=Caribbean	NE=North Equatorial
EG=East Greenland	N=Norwegian
EW=East Wind Drift	SE=South Equatorial
EC=Equatorial Counter	WW=West Wind Drift
Fa=Falkland	(Antarctic Circumpolar
F=Florida	Current)

Figure 8–14 **Atlantic Ocean surface currents.**

Atlantic Ocean surface circulation is composed primarily of two subtropical gyres.

Stream. The **North Equatorial Current** moves parallel to the Equator in the Northern Hemisphere, where it is joined by the portion of the South Equatorial Current that turned northward along the South American coast. This flow then splits into the *Antilles Current*, which passes along the Atlantic side of the West Indies, and the *Caribbean Current*,

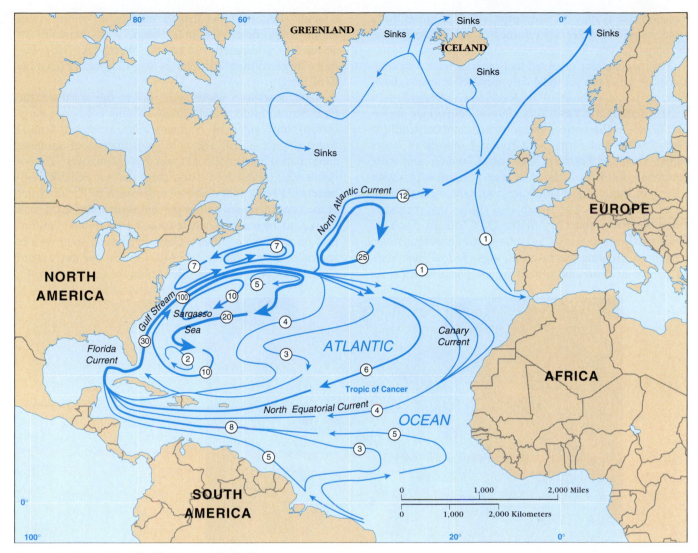

Figure 8–15 North Atlantic Ocean circulation.

The North Atlantic Gyre, showing average flow rates in Sverdrups (1 Sverdrup = 1 million cubic meters per second). The four major currents include the western intensified Gulf Stream, the North Atlantic Current, the Canary Current, and the North Equatorial Current. Some water splits off in the North Atlantic, where it becomes cold and dense, so it sinks. The Sargasso Sea occupies the stagnant eddy in the middle of the subtropical gyre.

which passes through the Yucatán Channel into the Gulf of Mexico. These masses reconverge as the *Florida Current.*

The Florida Current flows close to shore over the continental shelf at a rate that at times exceeds 35 Sverdrups. As it moves off North Carolina's Cape Hatteras and flows across the deep ocean in a northeasterly direction, it is called the Gulf Stream. The Gulf Stream is a western boundary current, so it is subject to western intensification. Thus, it is only 50 to 75 kilometers (31 to 47 miles) wide, but it reaches depths of 1.5 kilometers (1 mile) and speeds from 3 to 10 kilometers (2 to 6 miles) per hour, making it the fastest current in the world ocean.

The western boundary of the Gulf Stream is usually abrupt, but it periodically migrates closer to and farther away from the shore. Its eastern boundary is very difficult to identify because it is usually masked by meandering water masses that change their position continuously.

The Gulf Stream gradually merges eastward with the water of the **Sargasso Sea**. The Sargasso Sea is the water that circulates around the rotation center of the North Atlantic gyre. The Sargasso Sea, therefore, is the stagnant eddy of the North Atlantic Gyre. Its name is derived from a type of floating marine alga called *Sargassum* (*sargassum* = grapes) that abounds on its surface.

The transport rate of the Gulf Stream off Chesapeake Bay is about 100 Sverdrups,[9] which suggests that a large volume of water from the Sargasso Sea has combined with the Florida Current to produce the Gulf Stream. By the time the Gulf Stream nears Newoundland, however, the

---

[9]The Gulf Stream's flow of 100 Sverdrups equates to a volume of about 100 major league football stadiums passing by the southeast U.S. coast *each second* and is more than 100 times greater than the combined flow of *all* the world's rivers!

transport rate is only 40 Sverdrups, which suggests that a large volume of water has returned to the diffuse flow of the Sargasso Sea.

The mechanisms involved that produce such a dramatic loss of water are yet to be determined. Meanders, however, may cause much of it. *Meanders* (*Menderes* = a river in Turkey that has a very sinuous course) are snakelike bends in the current that often disconnect from the Gulf Stream and form large rotating masses of water called *vortexes* (*vertere* = to turn), which are more commonly known as *eddies* or *rings*. Figure 8-16 shows several of these rings, which are noticeable near the center of each image. The figure also shows that meanders along the north boundary of the Gulf Stream pinch off and trap warm Sargasso Sea water in eddies that rotate clockwise, creating **warm-core rings** (*yellow*) surrounded by cooler (*blue* and *green*) water. These warm rings

contain shallow, bowl-shaped masses of warm water about 1 kilometer (0.6 mile) deep, with diameters of about 100 kilometers (60 miles). Warm-core rings remove large volumes of water as they disconnect from the Gulf Stream.

Cold nearshore water spins off to the south of the Gulf Stream as counterclockwise-rotating **cold-core rings** (*green*) surrounded by warmer (*yellow* and *red-orange*) water (Figure 8-16). The cold rings consist of spinning cone-shaped masses of cold water that extend over 3.5 kilometers (2.2 miles) deep. These rings may exceed 500 kilometers (310 miles) in diameter at the surface. The diameter of the cone increases with depth and sometimes reaches all the way to the sea floor, where cones have a tremendous impact on sea floor sediment. Cold rings move southwest at speeds of 3 to 7 kilometers (2 to 4 miles) per day toward Cape Hatteras, where they often rejoin the Gulf Stream.

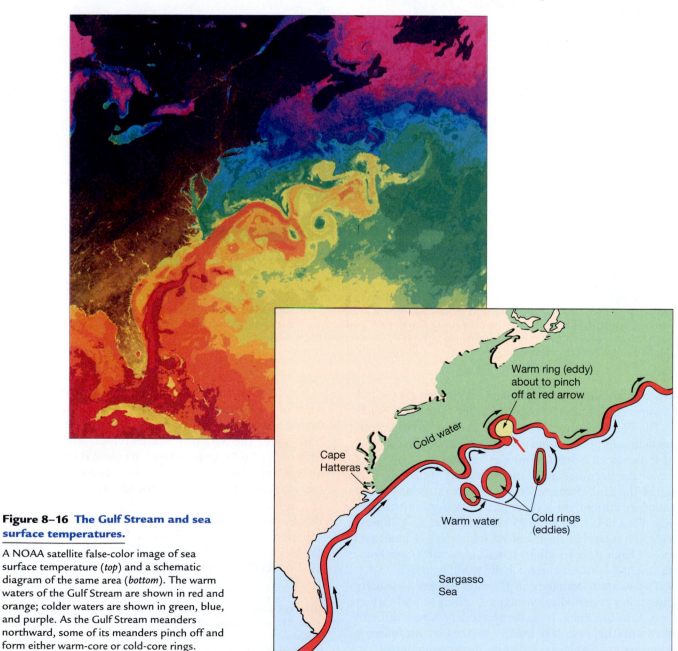

Figure 8–16 The Gulf Stream and sea surface temperatures.

A NOAA satellite false-color image of sea surface temperature (*top*) and a schematic diagram of the same area (*bottom*). The warm waters of the Gulf Stream are shown in red and orange; colder waters are shown in green, blue, and purple. As the Gulf Stream meanders northward, some of its meanders pinch off and form either warm-core or cold-core rings.

Both warm- and cold-core rings maintain not only unique temperature characteristics but also unique biological populations. Studies of rings have found they are isolated habitats for either warm-water organisms in a cold ocean or, conversely, cold-water organisms in a warmer ocean. The organisms can survive as long as the ring does; in some cases, rings have been documented to last as long as two years.

STUDENTS SOMETIMES ASK ...
Is the Gulf Stream rich in life?

The Gulf Stream *itself* isn't, but its *boundaries* often are. The oceanic areas that have abundant marine life are typically associated with cool water—either in high-latitude regions, or in any region where upwelling occurs. These areas are constantly resupplied with oxygen- and nutrient-rich water, which results in high productivity. Warm-water areas develop a prominent thermocline that isolates the surface water from colder, nutrient-rich water below. Nutrients used up in warm waters tend not to be resupplied. The Gulf Stream, therefore, which is a western intensified, warm-water current, is associated with low productivity and an absence of marine life. The reason New England fishers knew about the Gulf Stream (see the section about Benjamin Franklin in Chapter 1) was because they sought their catch along the *sides* of the current, where mixing and upwelling occur.

Actually, all western intensified currents are warm and are associated with low productivities. The Kuroshio Current in the North Pacific Ocean, for example, is named for its conspicuous absence of marine life. In Japanese, *Kuroshio* means "black current," in reference to its clear, lifeless waters.

Southeast of Newfoundland, the Gulf Stream continues in an easterly direction across the North Atlantic (Figure 8-15). Here the Gulf Stream breaks into numerous branches, many of which become cold and dense enough to sink beneath the surface. As shown in Figure 8-14, one major branch combines the cold water of the *Labrador Current* with the warm Gulf Stream, producing abundant fog in the North Atlantic. This branch eventually breaks into the *Irminger Current*, which flows along Iceland's west coast, and the *Norwegian Current*, which moves northward along Norway's coast. The other major branch crosses the North Atlantic as the **North Atlantic Current** (also called the *North Atlantic Drift*, emphasizing its sluggish nature) and turns southward to become the cool Canary Current. The **Canary Current** is a broad, diffuse southward flow that eventually joins the North Equatorial Current, thus completing the gyre.

Climatic Effects of North Atlantic Currents The warming effects of the Gulf Stream are far ranging. It not only moderates temperatures along the East Coast of the United States, but also in Northern Europe (in conjunction with heat transferred by the atmosphere). Thus, the temperatures across the Atlantic at different latitudes are much higher in Europe than in North America because of the effects of heat transfer from the Gulf Stream to Europe. For example, Spain and Portugal have warm climates yet they are at the same latitude as the New England states, which are known for severe winters. The warming that Northern Europe experiences because of the Gulf Stream is as much as 9°C (20°F), which is enough to keep high-latitude Baltic ports ice free throughout the year.

The warming effects of western boundary currents in the North Atlantic Ocean can be seen on the average sea surface temperature map for February shown in Figure 8-8b. Off the east coast of North America from latitudes 20 degrees north (the latitude of Cuba) to 40 degrees north (the latitude of Philadelphia), for example, there is a 20°C (36°F) difference in sea surface temperatures. On the eastern side of the North Atlantic, on the other hand, there is only a 5°C (9°F) difference in temperature between the same latitudes, indicating the moderating effect of the Gulf Stream.

The average sea surface temperature map for August (see Figure 8-8a) also shows how the North Atlantic and Norwegian Currents (branches of the Gulf Stream) warm northwestern Europe compared with the same latitudes along the North American coast. On the western side of the North Atlantic, the southward-flowing Labrador Current—which is cold and often contains icebergs from western Greenland—keeps Canadian coastal waters much cooler. During the Northern Hemisphere winter (Figure 8-8b), North Africa's coastal waters are cooled by the southward-flowing Canary Current and are much cooler than waters near Florida and the Gulf of Mexico.

Pacific Ocean Circulation

Two large subtropical gyres dominate the circulation pattern in the Pacific Ocean, resulting in surface water movement and climatic effects similar to those found in the Atlantic. However, the Equatorial Countercurrent is much better developed in the Pacific Ocean than in the Atlantic (Figure 8-17), because the Pacific Ocean basin is larger and more unobstructed than the Atlantic Ocean basin.

Normal Conditions Figure 8-17 shows the **North Pacific Gyre**, which consists in part of the North Equatorial Current, which flows westward into the western intensified **Kuroshio Current**[10] near Asia. Because of its proximity to Japan, the Kuroshio Current is also called the *Japan Current*. Its warm waters make Japan's climate warmer than would be expected for its latitude. The

---

[10]Kuroshio is pronounced "kuhr-ROH-shee-oh."

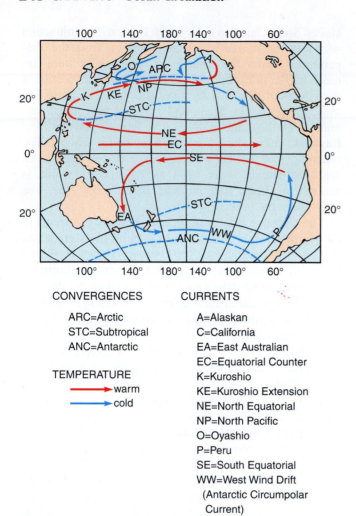

CONVERGENCES

ARC=Arctic
STC=Subtropical
ANC=Antarctic

TEMPERATURE

→ warm
→ cold

CURRENTS

A=Alaskan
C=California
EA=East Australian
EC=Equatorial Counter
K=Kuroshio
KE=Kuroshio Extension
NE=North Equatorial
NP=North Pacific
O=Oyashio
P=Peru
SE=South Equatorial
WW=West Wind Drift
(Antarctic Circumpolar
Current)

Figure 8–17 Pacific Ocean surface currents.

Similar to the Atlantic Ocean, the Pacific contains two large subtropical gyres. However, the equatorial countercurrent is more strongly developed here than in smaller ocean basins.

northern part of the Kuroshio Current is called the *Kuroshio Extension*, which flows to about 170 degrees east longitude. This current flows into the **North Pacific Current**, which connects to the cool-water **California Current**. The California Current flows south along the coast of California to complete the loop. Some North Pacific Current water also flows to the north and merges into the **Alaskan Current** in the Gulf of Alaska.

Figure 8-17 also shows the **South Pacific Gyre**, which consists in part of the South Equatorial Current, which flows westward into the western intensified **East Australian Current**.[11] From there, it joins the Antarctic Circumpolar Current (West Wind Drift) and completes the gyre as the **Peru Current** (also called the *Humboldt Current*, after German naturalist Friedrich Heinrich Alexander von Humboldt). Flowing underneath the South Equatorial Current of this gyre is a thin, ribbonlike **Equatorial Un-**

---

[11]Note that the western intensified East Australian Current was named because it lies off the *East Coast* of Australia, even though it occupies a position along the *western* margin of the Pacific Ocean basin.

dercurrent that flows in an easterly direction along the Equator and extends for more than 6000 kilometers (3700 miles) across the Pacific at depths of up to 200 meters (656 feet). Although it is only 0.2 kilometer (0.12 mile) deep and about 300 kilometers (186 miles) wide, it has a volume transport of approximately 40 Sverdrups.

The cool water of the Peru Current has historically been one of Earth's richest fishing grounds. What conditions produce such an abundance of fish? Figure 8-18a shows that along the west coast of South America, coastal winds create Ekman transport that moves water away from shore, causing upwelling of cool, nutrient-rich water. This upwelling increases productivity and results in an abundance of marine life, including small silver-colored fish called *anchovetas* (anchovies) that become particularly plentiful near Peru and Ecuador. Anchovies provide a food source for many larger marine organisms and also supply Peru's commercial fishing industry, which was established in the 1950s. Anchovies are so abundant in the waters off South America that by 1970, Peru was the largest producer of fish from the sea in the world, with a peak production of 12.3 million metric tons (13.5 million short tons), accounting for about one-quarter of *all* fish from the sea worldwide.

STUDENTS SOMETIMES ASK ...

The amount of anchovies produced by Peru is impressive! Besides a topping for pizza, what are some other uses of anchovies?

Anchovies are an ingredient in certain dishes, hors d'oeuvres, sauces, and salad dressing, and they are also used as bait by fishers. Historically, however, most of the *anchoveta* caught in Peruvian waters were exported and used as fishmeal (consisting of ground anchovies). The fishmeal, in turn, was used largely in pet food and as a high-protein chicken feed. As unbelievable as it may seem, El Niños affected the price of eggs! Prior to the collapse of the Peruvian *anchoveta* fishing industry in 1972–1973, El Niño events significantly reduced the availability of *anchoveta*. This drastically cut the export of anchovies from Peru, causing U.S. farmers to pursue more expensive options for chicken feed. Thus, egg prices typically increased.

The collapse of the *anchoveta* fishing industry in Peru was triggered by the 1972–1973 El Niño event but was caused by chronic overfishing in prior years (see Box 17-1). Interestingly, the shortage of fishmeal after 1972–1973 led to an increased demand for soyameal, an alternative source of high-quality protein. Increased demand for soyameal increased the price of soy commodities, thereby encouraging U.S. farmers to plant soybeans instead of wheat. Reduced production of wheat, in turn, caused a major global food crisis—all triggered by an El Niño event.

Figure 8-18a shows that high pressure and sinking air dominates the coastal region of South America, resulting in clear, fair, and dry weather. On the other side of the Pacific, a low-pressure region and rising air creates cloudy

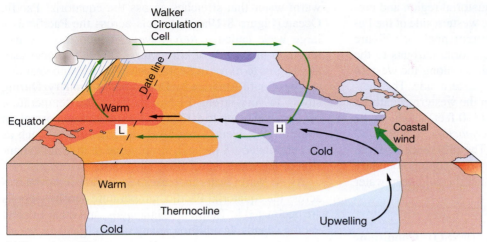

(a) Normal conditions

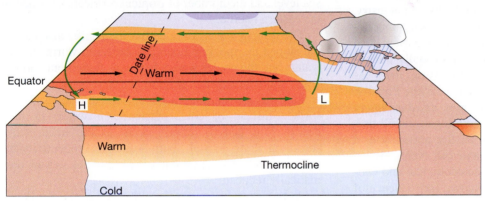

(b) El Niño conditions

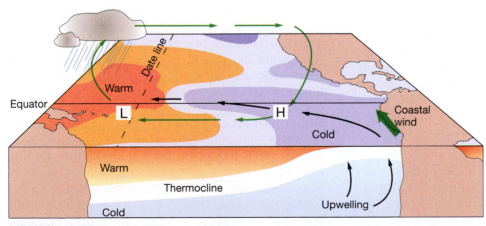

(c) La Niña conditions

Figure 8–18 Normal, El Niño, and La Niña conditions.

(a) Normal oceanic and atmospheric conditions in the equatorial Pacific. **(b)** El Niño (ENSO warm phase) conditions. **(c)** La Niña (ENSO cool phase) conditions.

conditions with plentiful precipitation in Indonesia, New Guinea, and northern Australia. This pressure difference causes the strong southeast trade winds to blow across the equatorial South Pacific. The resulting atmospheric circulation cell in the equatorial South Pacific Ocean is named

the **Walker Circulation Cell** (*green arrows*) after Sir Gilbert T. Walker, the British meteorologist who first described the effect in the 1920s.

The southeast trade winds set ocean water in motion, which also moves across the Pacific toward the west. The water warms as it flows in the equatorial region and creates a wedge of warm water on the western side of the Pacific Ocean, called the **Pacific warm pool** (see Figure 8-8). Due to the movement of equatorial currents to the west, the Pacific warm pool is thicker along the western side of the Pacific than along the eastern side. The thermocline beneath the warm pool in the western equatorial Pacific occurs below 100 meters (330 feet) depth. In the eastern Pacific, however, the *thermocline* is within 30 meters (100 feet) of the surface. The difference in depth of the thermocline can be seen by the sloping boundary between the warm surface water and the cold deep water in Figure 8-18a.

El Niño–Southern Oscillation (ENSO) Conditions

Historically, Peru's residents knew that every few years, a current of warm water reduced the population of anchovies in coastal waters. The decrease in anchovies caused a dramatic decline not only in the fishing industry, but also in marine life such as sea birds and seals that depended on anchovies for food. The warm current also brought about changes in the weather—usually intense rainfall—and even brought such interesting items as floating coconuts from tropical islands near the Equator. At first, these events were called *años de abundancia* (years of abundance) because the additional rainfall dramatically increased plant growth on the normally arid land. What was once thought of as a joyous event, however, soon became associated with the ecological and economic disaster that is now a well-known consequence of the phenomenon.

This warm-water current usually occurred around Christmas and thus was given the name **El Niño**, Spanish for "the child," in reference to baby Jesus. In the 1920s, Walker was the first to recognize that an east–west atmospheric pressure seesaw accompanied the warm current and called the phenomenon the **Southern Oscillation**. Today, the combined oceanic and atmospheric effects are called **El Niño–Southern Oscillation (ENSO)**, which periodically alternate between warm and cold phases and cause dramatic environmental changes.

ENSO Warm Phase (El Niño)

Figure 8-18b shows the atmospheric and oceanic conditions during an ENSO warm phase, which is known as El Niño. The high pressure along the coast of South America weakens, reducing the difference between the high- and low-pressure regions of the Walker Circulation Cell. This, in turn, causes the southeast trade winds to diminish. In very strong El Niño events, the trade winds actually blow in the *reverse* direction.

Without the trade winds, the Pacific warm pool that has built up on the western side of the Pacific begins to flow back across the ocean toward South America. Aided by an increase in the flow of the Equatorial Countercurrent, the Pacific warm pool creates a band of warm water that stretches across the equatorial Pacific Ocean (Figure 8-19a). It travels across the Pacific as a large wave called a *Kelvin wave*, which has a wavelength of thousands of kilometers. The warm water usually begins to move in September of an El Niño year and reaches South America by December or January. During strong to very strong El Niños, the water temperature off Peru can be up to 10°C (18°F) higher than normal. In addition, the average sea level can increase as much as 20 centimeters (8 inches), simply due to thermal expansion of the warm water along the coast.

As the warm water increases sea surface temperatures across the equatorial Pacific, temperature-sensitive corals are decimated in Tahiti, the Galápagos, and other tropical Pacific islands. In addition, many other organisms are affected by the warm water (Box 8-2). Once the warm water reaches South America, it moves north and south along the west coast of the Americas, increasing average sea level and the number of tropical hurricanes formed in the eastern Pacific.

The flow of warm water across the Pacific also causes the sloped thermocline boundary between warm surface waters and the cooler waters below to flatten out and become more horizontal (Figure 8-19b). Near Peru, upwelling brings warmer, nutrient-depleted water to the surface instead of cold, nutrient-rich water. In fact, *downwelling* can sometimes occur as the warm water stacks up along coastal South America. Productivity diminishes and most types of marine life in the area are dramatically reduced.

As the warm water moves to the east across the Pacific, the low-pressure zone also migrates. In a strong to very strong El Niño event, the low pressure can move entirely across the Pacific and remain over South America. The low pressure substantially increases precipitation along coastal South America. Conversely, high pressure replaces the Indonesian low, bringing dry conditions or, in strong to very strong El Niño events, drought conditions to Indonesia and northern Australia.

ENSO Cool Phase (La Niña)

In some instances, conditions opposite of El Niño prevail in the equatorial South Pacific, which is known as ENSO cool phase or **La Niña** (Spanish for "the female child"). Figure 8-19c shows La Niña conditions, which are similar to normal conditions but more intensified because there is a larger pressure difference across the Pacific Ocean. This larger pressure difference creates stronger Walker Circulation and stronger trade winds, which in turn causes more upwelling, a shallower thermocline in the eastern Pacific, and a band of cooler than normal water that stretches across the equatorial South Pacific (Figure 8-19b).

La Niña conditions commonly occur following an El Niño. For instance, the 1997–1998 El Niño was followed by several years of persistent La Niña conditions. The alternating pattern of El Niño–La Niña conditions since 1950 is shown by the multivariate **ENSO index** (Figure 8-20),

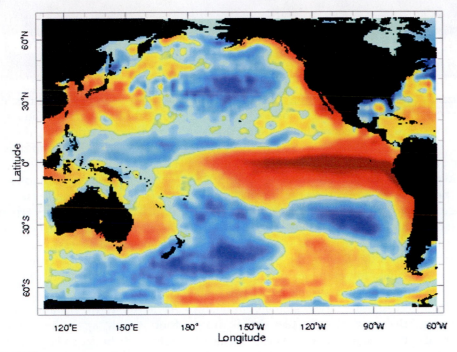

(a) Jan 1998

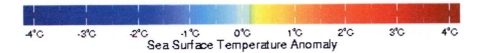

(b) Jan 2000

Figure 8–19 **Sea surface temperature anomaly maps.**

Maps showing sea surface temperature anomalies, which represent departures from normal conditions. Red colors indicate water warmer than normal and blue colors represent water cooler than normal. **(a)** Map of the Pacific Ocean in January 1998, showing the anomalous warming during the 1997–1998 El Niño. **(b)** Map of the same area in January 2000, showing cooling in the equatorial Pacific related to La Niña.

BOX 8–2   Research Methods in Oceanography

EL NIÑO AND THE INCREDIBLE SHRINKING MARINE IGUANAS OF THE GALÁPAGOS ISLANDS

The world's only marine lizards are the marine iguanas (*Amblyrhynchus cristatus*) of Ecuador's Galápagos Islands (Figure 8B). The iguanas are vegetarians that eat only marine algae and have adaptations that enable them to spend long periods foraging for food in the ocean. The islands they inhabit are severely affected by El Niño events, which cause periodic food shortages for the iguanas.

When El Niño conditions occur, warm water from the Pacific warm pool moves to the east along the Equator. This change, accompanied by a decrease in upwelling along the eastern Pacific, causes ocean surface temperatures to rise by as much as 10°C (18°F). In the Galápagos Islands, these conditions cause severe hardship for species that like cooler water temperatures, such as green and red algae that are the iguana's preferred food. Instead, these types of algae are replaced by brown algae, which are harder for the iguanas to digest. Unlike many marine animals that can migrate to other areas where food supplies are plentiful, the iguanas are confined to the islands where food supplies become limited. During a severe El Niño, up to 90% of the iguana population can die of starvation.

Two studies covering 8 and 18 years reveal that the iguana's main adaptation for this lack of food is to shrink in size, allowing iguanas to utilize meager food supplies more efficiently. During the 1997–1998 El Niño, for example, Galápagos marine iguanas shrank by as much as 20%, with larger animals shrinking the most. Also, females shrank more than similar-sized males, probably because of the additional energy females expended producing eggs the previous year.

About half of the shrinkage can be attributed to decreases in the mass of cartilage and connective tissue, while bone absorption may account for the remainder. In addition, the iguanas don't forage for food and thus get little exercise during El Niño events, which may result in additional shrinkage (similar to the decrease in weight associated with inactivity experienced by astronauts that spend long periods in weightlessness).

Following El Niño conditions, upwelling is reestablished, which creates cooler surface water temperatures. As a result, the iguana's preferred food supply again becomes abundant and the iguanas grow back to normal size. Remarkably, the marine iguanas of the Galápagos Islands repeatedly shrink and regrow in response to El Niño-induced changes in the environment.

Figure 8B Marine iguana and location maps of the Galápagos Islands.

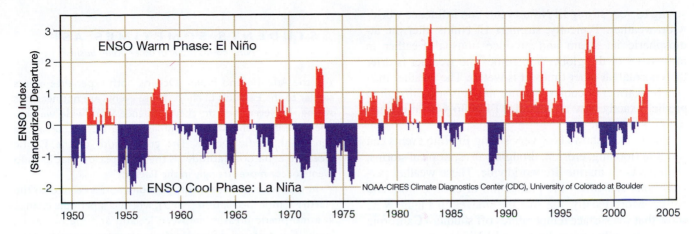

Figure 8-20 Multivariate ENSO index 1950–present.

The multivariate ENSO index is calculated using a variety of atmospheric and oceanic factors. ENSO index values greater than zero (*red areas*) indicate El Niño conditions while ENSO index values less than zero (*blue areas*) indicate La Niña conditions. The greater the value is from zero, the stronger the corresponding El Niño or La Niña.

which is calculated using a weighted average of atmospheric and oceanic factors including atmospheric pressure, winds, and sea surface temperatures. Positive ENSO index numbers indicate El Niño conditions whereas negative numbers reflect La Niña conditions. Normal conditions are indicated by a value near zero, and the greater the index value differs from zero (either negative or positive), the stronger the respective condition.

How Often Do El Niño Events Occur? Records of sea surface temperatures over the past 100 years reveal that throughout the twentieth century, El Niño conditions occur on average about every 2 to 10 years, but in a highly irregular pattern. In some decades, for instance, there has been an El Niño event every few years, while in others there may have been only one. Figure 8-20 shows the pattern since 1950, revealing that the equatorial Pacific fluctuates between El Niño and La Niña conditions, with only a few years that could be considered "normal" conditions (represented by an ENSO index value close to zero). Typically, El Niño events last for 12 to 18 months and are followed

by La Niña conditions that exist for a similar length of time. However, some El Niño or La Niña conditions can last for several years.

El Niño events—especially severe ones—may occur more frequently as a result of increased global warming. For instance, the two most severe El Niño events in the 20th century occurred in 1982–1983 and 1997–1998. Presumably, increased ocean temperatures could trigger more frequent and more severe El Niños. However, this pattern could also be a part of a long-term natural climate cycle. Recently, oceanographers have recognized a phenomenon called the **Pacific Decadal Oscillation (PDO)**, which lasts 20 to 30 years and appears to influence Pacific sea surface temperatures. Analysis of TOPEX/Poseidon satellite data suggests that the Pacific Ocean has been in the warm phase of the PDO from 1977 to 1999 and that it has just entered the cool phase, which may suppress the initiation of El Niño events.

Effects of El Niños and La Niñas Mild El Niño events influence only the equatorial South Pacific Ocean while

---

TABLE 8-3 El Niño–Southern Oscillation: A summary.

- El Niño (ENSO warm phase) is the name of a warm water current that occurs periodically around Christmastime in the equatorial Pacific Ocean; Southern Oscillation describes the switching of atmospheric pressure that accompanies El Niño.

- El Niño events are characterized by warmer than normal water and a deepened thermocline in the east Pacific, a rise in sea level along the Equator, a decrease or reversal of the southeast trade winds, and reduced upwelling and less abundant marine life in waters near Peru and Ecuador.

- El Niños of various strengths occur every 2 to 10 years, but in a highly irregular pattern.

- Strong El Niño events produce unusual weather worldwide.

- La Niña conditions (ENSO cool phase) are characterized by conditions opposite of El Niño; the Pacific Ocean experiences alternating El Niño and La Niña events.

strong to very strong El Niño events can influence world-wide weather. Typically, stronger El Niños alter the atmospheric jet steam and produce unusual weather in most parts of the globe. Sometimes the weather is drier than normal; at other times, it is wetter. The weather may also be warmer or cooler than normal. It is still difficult to predict exactly how a particular El Niño will affect any region's weather.

Figure 8-21 shows how very strong El Niño events can result in flooding, erosion, droughts, fires, tropical storms, and effects on marine life worldwide. These weather perturbations also affect the production of corn, cotton, and coffee. More locally, the satellite images in Figure 8-22 show that sea surface temperatures off southern California are significantly higher during an El Niño year.

Even though severe El Niños are typically associated with vast amounts of destruction, they can be beneficial in some areas. Tropical hurricane formation, for instance, is generally suppressed in the Atlantic Ocean, some desert regions receive much-needed rain, and organisms adapted to warm-water conditions thrive in the Pacific.

La Niña events are associated with sea surface temperatures and weather phenomena opposite to those of El Niño. Indian Ocean monsoons, for instance, are typically drier than usual in El Niño years but wetter than usual in La Niña years.

STUDENTS SOMETIMES ASK ...
Do El Niño events occur in other ocean basins?

Yes, the Atlantic and Indian Oceans both experience events similar to the Pacific's El Niño. These events are not nearly as strong, however; nor do they influence worldwide weather phenomena to the same extent as those that occur in the equatorial Pacific Ocean. The great width of the Pacific Ocean in equatorial latitudes is the main reason that El Niño events occur more strongly in the Pacific.

In the Atlantic Ocean, this phenomenon is related to the North Atlantic Oscillation (NAO), which is a periodic change in atmospheric pressure between Iceland and the Azores Islands. This pressure difference determines the strength of the prevailing westerlies in the North Atlantic, which in turn affects ocean surface currents there. The Atlantic Ocean periodically experiences NAO events, which sometimes cause intense cold in the northeast U.S., unusual weather in Europe, and heavy rainfall along the normally arid coast of southwest Africa.

Examples from Recent El Niños   Recent El Niños provide an indication of the variability of the effects of El Niño events. For instance, in the winter of 1976, a moderate

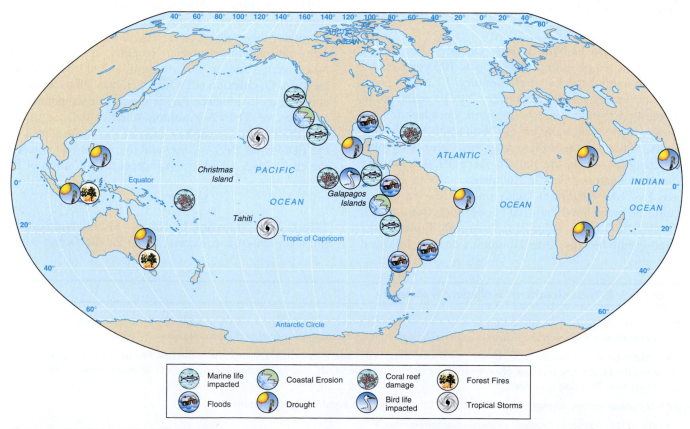

Figure 8–21 Effects of severe El Niños.

Flooding, erosion, droughts, fires, tropical storms, and effects on marine life are all associated with severe El Niño events.

(a) Normal   (b) El Niño

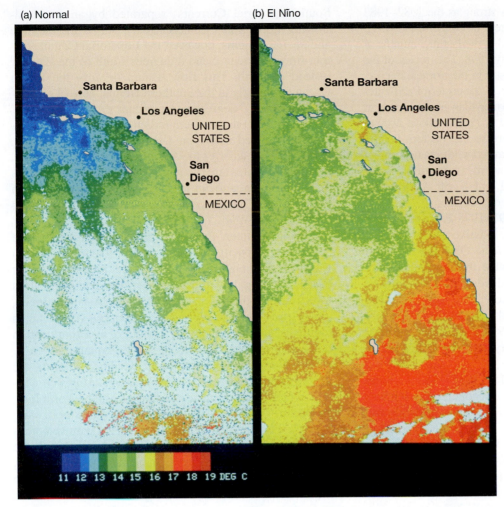

11 12 13 14 15 16 17 18 19 DEG C

Figure 8–22 Sea surface temperatures off southern California.

Sea surface temperature maps (in °C) for southern California from data collected by the satellite-mounted Advanced Very High Resolution Radiometer. Blue is cold water; red is warm water. **(a)** Water temperatures in January 1982, a non–El Niño year. **(b)** Water temperatures a year later, during an El Niño event.

El Niño event coincided with northern California's worst drought of the last century, showing that El Niño events don't always bring torrential rains to the western United States. During that same winter, the eastern United States experienced record cold temperatures.

The 1982–1983 El Niño is the strongest ever recorded, causing far-ranging effects across the globe. Not only was there anomalous warming in the tropical Pacific, but the warm water spread along the coast of North America, influencing sea surface temperatures from California (Figure 8-22) to Alaska. Sea level was higher than normal (due to thermal expansion of the water), which, when high surf was experienced, caused damage to coastal structures and increased coastal erosion. In addition, the jet stream swung much farther south than normal across the United States, bringing a series of powerful storms that resulted in three times normal rainfall across the southwestern United States. The increased rainfall caused severe flooding and landslides as well as higher than normal snowfall in the Rocky Mountains. Alaska and western Canada had a relatively warm winter, and the eastern United States had its mildest winter in 25 years.

The full strength of El Niño was experienced in western South America. Normally arid Peru was drenched with more than 3 meters (10 feet) of rain, causing extreme flooding and landslides. Sea surface temperatures were so high for so long that temperature-sensitive coral reefs across the equatorial Pacific were decimated. Marine mammals and sea birds, which depend on the food normally available in the highly productive waters along the west coast of South America, went elsewhere or died. In the Galápagos Islands, for example, over half of the island's seals and sea lions died of starvation during the 1982–1983 El Niño.

French Polynesia had not experienced a hurricane in 75 years; in 1983, it endured six. The Hawaiian Island of Kauai also experienced a rare hurricane. Meanwhile, in Europe, severe cold weather prevailed. Worldwide, droughts occurred in Australia, Indonesia, China, India, Africa, and Central America. In all, over 2000 deaths and at least $10 billion in property damage ($2.5 billion in the United States) were attributed to the 1982–1983 El Niño event.

The 1997–1998 El Niño event began several months earlier than normal and peaked in January 1998. The amount of Southern Oscillation and sea surface warming in the equatorial Pacific was initially as strong as the 1982–1983 El Niño, which caused a great deal of concern. However, the 1997–1998 El Niño weakened in the last few months of 1997 before reintensifying in early 1998. The impact of the 1997–1998 El Niño was felt mostly in the tropical Pacific, where surface water temperatures in the eastern Pacific averaged more than 4°C (7°F) warmer than normal, and, in some locations, reached up to 9°C (16°F) above normal (see Figure 8-19a). High pressure in the western Pacific brought drought conditions that caused wildfires to burn out of control in Indonesia. Also, the warmer than normal water along the west coast of Central and North America increased the number of hurricanes off Mexico.

In the United States, the 1997–1998 El Niño caused killer tornadoes in the southeast, massive blizzards in the upper Midwest, and flooding in the Ohio River Valley. Most of California received twice the normal rainfall, which caused flooding and landslides in many parts of the state. The lower Midwest, the Pacific Northwest, and the eastern seaboard, on the other hand, had relatively mild weather. In all, the 1997–1998 El Niño caused 2100 deaths and $33 billion in property damage worldwide.

Predicting El Niño Events The 1982–1983 El Niño event was not predicted; nor was it recognized until it was near its peak. Because it affected weather worldwide and caused such extensive damage, the **Tropical Ocean–Global Atmosphere (TOGA)** program was initiated in 1985 to study how El Niño events developed. The goal of the TOGA program was to monitor the equatorial South Pacific Ocean during El Niño events to enable scientists to model and predict future El Niño events. The 10-year program studied the ocean from research vessels, analyzed surface and subsurface data from radio-transmitting sensor buoys, monitored oceanic phenomena by satellite, and developed computer models.

These models made it possible to predict El Niño events since 1987 as much as one year in advance. After the completion of TOGA, the **Tropical Atmosphere and Ocean (TAO)** project (sponsored by the United States, Canada, Australia, and Japan) has continued to monitor the equatorial Pacific Ocean with a series of 70 moored buoys, providing real-time information about the conditions of the tropical Pacific that is available on the Internet. Although monitoring has improved, the causes of El Niño events are still not fully understood.

> El Niño is a combined oceanic–atmospheric phenomenon that occurs periodically in the tropical Pacific Ocean, bringing warm water to the east. La Niña describes conditions opposite of El Niño.

Indian Ocean Circulation

From November to March, equatorial circulation in the Indian Ocean is similar to that in the other oceans, with two westward-flowing equatorial currents (North and South Equatorial Currents) separated by an eastward-flowing Equatorial Countercurrent. Unlike the Atlantic and Pacific systems, however, the Equatorial Countercurrent flows between 2 and 8 degrees south of the Equator instead of north. This flow occurs because the Indian Ocean lies mostly in the Southern Hemisphere (it extends only to about 20 degrees north latitude).

The winds of the northern Indian Ocean have a seasonal pattern called **monsoon** (*mausim* = season) winds. During winter, air over the Asian mainland rapidly cools, creating high pressure, which forces atmospheric masses off the continent and out over the ocean (*green arrows* in Figure 8-23a). These northeast trade winds are called the *northeast monsoon*.

During summer, the winds reverse. Because of the lower heat capacity of rocks and soil compared with water, the Asian mainland warms faster than the adjacent ocean, creating low pressure over the continent. This forces air over the Indian Ocean onto the Asian landmass, giving rise to the *southwest monsoon* (*green arrows* in Figure 8-23b), which may be thought of as a continuation of the southeast trade winds across the Equator.

The North Equatorial Current disappears during the summer and is replaced by the *Southwest Monsoon Current*, which flows from west to east across the North Indian Ocean. The **Somali Current**, which flows northward from the Equator along the coast of Africa with velocities approaching 4 kilometers (2.5 miles) per hour, feeds the Southwest Monsoon Current. In September or October, the northeast trade winds are reestablished, and the North Equatorial Current reappears (Figure 8-23a).

Surface circulation in the southern Indian Ocean (the **Indian Ocean Gyre**) is similar to subtropical gyres observed in other southern oceans. When the northeast trade winds blow, the South Equatorial Current provides water for the Equatorial Countercurrent and the **Agulhas Current**, which flows southward along Africa's east coast and joins the Antarctic Circumpolar Current (West Wind Drift). The *Agulhas Retroflection* is created when the Agulhas Current makes an abrupt turn as it meets the strong Antarctic Circumpolar Current. Turning northward out of the Antarctic Circumpolar Current is the **West Australian Current**, an eastern boundary current that merges with the South Equatorial Current, completing the gyre.

Eastern boundary currents in other subtropical gyres are cold drifts toward the Equator that produce arid coastal climates [that is, they receive less than 25 centimeters (10 inches) of rain per year]. In the southern Indian Ocean, however, the *Leeuwin Current* displaces the West Australian Current offshore. The Leeuwin Current is driven southward along the Australian coast from the warm-water dome piled up in the East Indies by the Pacific equatorial currents.

The Leeuwin Current produces a mild climate in southwestern Australia, which receives about 125 centimeters (50 inches) of rain per year. During El Niño events, however, the Leeuwin Current weakens, so the cold Western Australian Current brings drought instead.

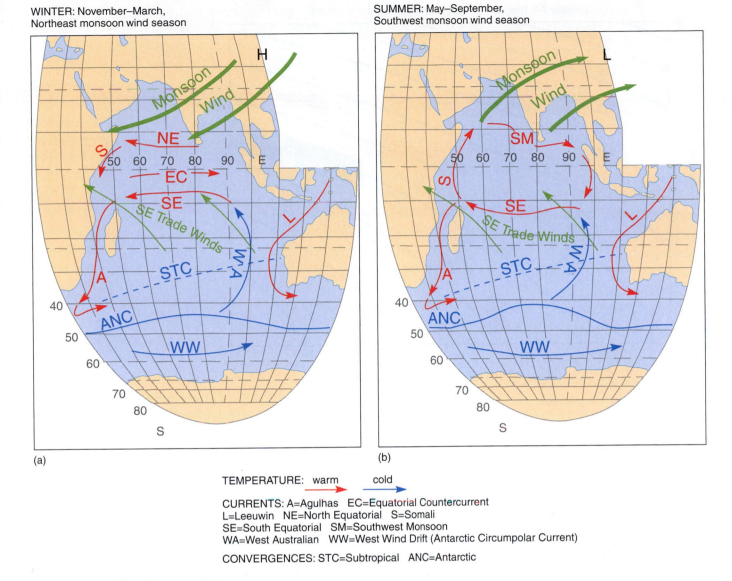

WINTER: November–March,
Northeast monsoon wind season

SUMMER: May–September,
Southwest monsoon wind season

(a)

(b)

TEMPERATURE: warm   cold

CURRENTS: A=Agulhas   EC=Equatorial Countercurrent
L=Leeuwin   NE=North Equatorial   S=Somali
SE=South Equatorial   SM=Southwest Monsoon
WA=West Australian   WW=West Wind Drift (Antarctic Circumpolar Current)

CONVERGENCES: STC=Subtropical   ANC=Antarctic

Figure 8–23 Indian Ocean surface currents.

Surface currents in the Indian Ocean are influenced by the seasonal monsoons. (a) Northeast monsoon, which occurs during winter. (b) Southwest monsoon, which occurs during summer.

Deep Currents

Deep currents occur in the deep zone below the pycnocline, so they influence about 90% of all ocean water. Density differences create deep currents. Although these density differences are usually small, they are large enough to cause denser waters to sink. Because the density variations that cause deep ocean circulation are caused by differences in temperature and salinity, deep ocean circulation is also referred to as **thermohaline** (*thermo* = heat, *haline* = salt) **circulation**.

Origin of Thermohaline Circulation

Recall from Chapter 6 that an increase in seawater density can be caused by a *decrease* in temperature or an *increase* in salinity. Temperature, though, has the greater influence on density. Density changes due to salinity are

important only in very high latitudes, where water temperature remains low and relatively constant.

Most water involved in deep-ocean currents (thermohaline circulation) begins in high latitudes *at the surface*. In these regions, surface water becomes cold and its salinity increases as sea ice forms. When this surface water becomes dense enough, it sinks, initiating deep-ocean currents. Once this water sinks, it is removed from the physical processes that increased its density in the first place, and so its temperature and salinity don't change very much during the time it spends in the deep ocean. Thus, a **temperature–salinity (T–S) diagram** can be used to identify deep-water masses based on their characteristic temperature, salinity, and resulting density. Figure 8-24 shows a T–S diagram for the North Atlantic Ocean.

As these surface-water masses become dense and are sinking (downwelling) in high-latitude areas, deep-water

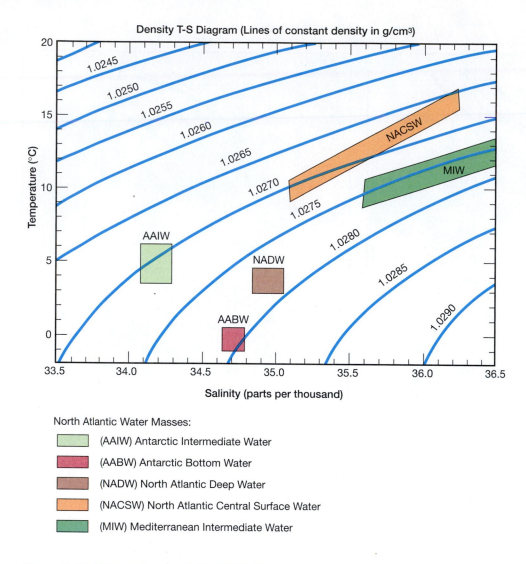

Figure 8–24 Temperature–salinity (T–S) diagram.

A density T–S diagram for the North Atlantic Ocean. Lines of constant density are in grams per cubic centimeter (g/cm^3). After various deep-water masses sink below the surface, they can be identified based on their characteristic temperature, salinity, and resulting density.

masses are also rising to the surface (upwelling). Because the water temperature in high latitude regions is the same at the surface as it is down below, the water column is isothermal, there is no thermocline or associated pycnocline (see Chapter 6), and upwelling and downwelling can easily be accomplished.

Deep-water currents move larger volumes of water and are much slower than surface currents. Typical speeds of deep currents range from 10 to 20 kilometers (6 to 12 miles) per year. Thus, it takes a deep current *an entire year* to travel the same distance that a western intensified surface current can move in *one hour*.

Sources of Deep Water

In southern subpolar latitudes, huge masses of deep water form beneath sea ice along the margins of the Antarctic continent. Here, rapid winter freezing produces very cold, high-density water that sinks down the continental slope of Antarctica and becomes **Antarctic Bottom Water**, the densest water in the open ocean (Figure 8-25). Antarctic Bottom Water slowly sinks beneath the surface and spreads into all the world's ocean basins, eventually returning to the surface perhaps 1000 years later.

In the northern subpolar latitudes, large masses of deep water form in the Norwegian Sea. From there, it flows as a subsurface current into the North Atlantic, where it becomes part of the **North Atlantic Deep Water**. North Atlantic Deep Water also comes from the margins of the Irminger Sea off southeastern Greenland, the Labrador Sea, and the dense, salty Mediterranean Sea. Like Antarctic Bottom Water, North Atlantic Deep Water spreads throughout the ocean basins. It is less dense, however, so it layers on top of the Antarctic Bottom Water (Figure 8-25).

Surface-water masses converge within the subtropical gyres and in the Arctic and Antarctic. Subtropical convergences do not produce deep water, however, because the

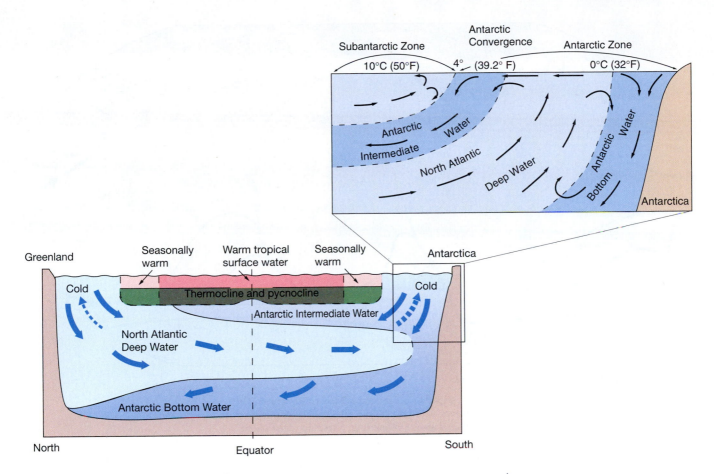

Figure 8–25 Atlantic Ocean subsurface water masses.

Schematic diagram of the various water masses in the Atlantic Ocean. Similar but less distinct layering based on density occurs in the Pacific and Indian Oceans as well. Upwelling and downwelling occurs in the North Atlantic and near Antarctica (*inset*), creating deep-water masses.

density of warm surface waters is too low for them to sink. Major sinking does occur, however, along the **Arctic Convergence** and Antarctic Convergence (Figure 8-25, *inset*). The deep-water mass formed from sinking at the Antarctic convergence is called the **Antarctic Intermediate Water** mass (Figure 8-25). Scientists have not yet thoroughly studied Antarctic Intermediate Water, so it remains a true frontier of knowledge, awaiting further exploration.

Figure 8-25 also shows that the highest-density water is found along the ocean bottom, with less-dense water above. In low-latitude regions, the boundary between the warm surface water and the deeper cold water is marked by a prominent thermocline and corresponding pycnocline that prevent vertical mixing. There is no pycnocline in high-latitude regions, so substantial vertical mixing (upwelling and downwelling) occurs.

This same general pattern of layering based on density occurs in the Pacific and Indian Oceans as well. They have no source of Northern Hemisphere deep water, however, so they lack a deep-water mass. In the northern Pacific Ocean, the low salinity of surface waters prevents it from sinking into the deep ocean. In the northern Indian Ocean, surface waters are too warm to sink. **Oceanic Common Water,** which is created when Antarctic Bottom Water and North Atlantic Deep Water mix, lines the bottoms of these basins.

Worldwide Deep-Water Circulation

For every liter of water that sinks from the surface into the deep ocean, a liter of deep water must return to the surface somewhere else. It is difficult to identify specifically where this vertical flow to the surface is occurring. It is generally believed that it occurs as a gradual, uniform upwelling throughout the ocean basins. It may be somewhat greater in low-latitude regions, where surface temperatures are higher.

Figure 8-26 shows a general deep-water circulation model of the world ocean. The most intense deep-water flow in each ocean basin is along the western side, due to the Coriolis effect and bathymetric features along the sea floor (such as the mid-ocean ridge).

Conveyer-Belt Circulation An integrated model combining deep thermohaline circulation and surface currents is shown in Figure 8-27. Because the overall circulation pattern resembles a large conveyer belt, the model is called **conveyer-belt circulation**. Beginning in the North Atlantic, surface water carries heat to high latitudes via the Gulf Stream. During the cold winter months, this heat is transferred to the overlying atmosphere, warming northern Europe.

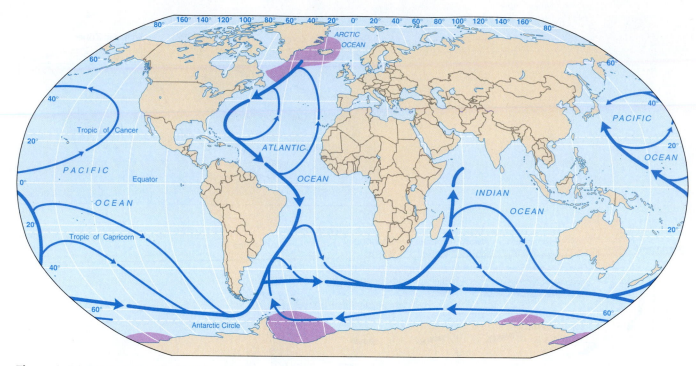

Figure 8–26 Deep-water circulation model.

Schematic model of deep-water circulation first developed by oceanographer Henry Stommel in 1958. Heavy lines mark the major western boundary currents, which result from the same forces that produce western intensified surface currents. The purple shaded areas indicate source areas for deep water.

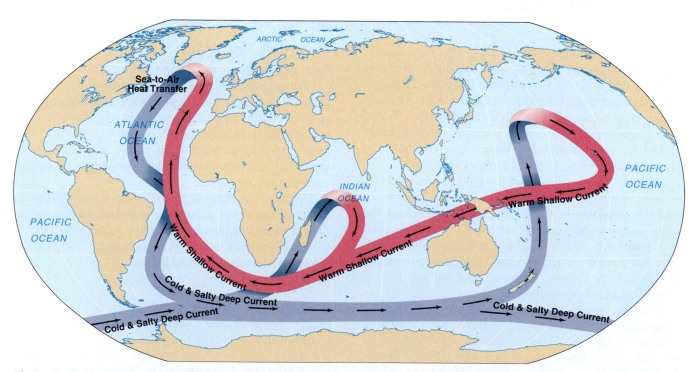

Figure 8–27 Conveyer-belt circulation.

Conveyer-belt circulation is initiated in the North Atlantic Ocean, where warm water cools and sinks below the surface. This water moves southward as a subsurface flow and joins water near Antarctica. This deep water spreads into the Indian and Pacific Oceans, where it slowly rises and completes the conveyer as it travels along the surface into the North Atlantic Ocean.

Cooling in the North Atlantic increases the density of this surface water to the point where it sinks to the bottom and flows southward, initiating the lower limb of the "conveyor." Here, seawater flows downward at a rate equal to 100 Amazon Rivers and begins its long journey into the deep basins of all of the world's oceans. This limb extends all the way to the southern tip of Africa, where it joins the deep water that encircles Antarctica. The deep water that encircles Antarctica includes deep water that descends along the margins of the Antarctic continent. This mixture of deep waters flows northward into the deep Pacific and Indian Ocean basins, where it eventually surfaces and completes the conveyer belt by flowing west and then north again into the North Atlantic Ocean.

Dissolved Oxygen in Deep Water   Cold water can dissolve more oxygen than warm water. Thus, deep-water circulation brings dense, cold, oxygen-enriched water from the surface to the deep ocean. During its time in the deep ocean, deep water becomes enriched in nutrients, as well, due to decomposition of dead organisms and the lack of organisms using nutrients there.

At various times in the geologic past, warmer water probably constituted a larger proportion of deep oceanic waters. As a result, the oceans had a lower oxygen concentration than today because warm water cannot hold as much oxygen. In essence, the oxygen content of the oceans has probably fluctuated widely throughout time.

If high-latitude surface waters did not sink and eventually return from the deep sea to the surface, the distribution of life in the sea would be considerably different. There would be no life in the deep ocean, for instance, because there would be no oxygen for organisms to breathe. Life in surface waters would be significantly reduced and confined to the extreme margins of the oceans, where the only source of oxygen and nutrients would be runoff from streams.

Thermohaline Circulation and Climate Change
Evidence from deep-sea sediments and recently developed computer models indicate that changes in the global deep-water circulation pattern can dramatically and abruptly change climate. If surface waters stopped sinking, for instance, the oceans would absorb and redistribute heat from solar radiation less efficiently. This might cause much warmer surface water temperatures and much higher land temperatures than we have now.

Alternatively, the buildup of greenhouse gases in the atmosphere may change ocean circulation. Warmer temperatures, for example, may increase the rate at which glaciers in Greenland melt, forming a pool of fresh, low-density surface water in the North Atlantic Ocean. This fresh water could inhibit the downwelling that generates North Atlantic Deep Water, altering global deep-water circulation patterns and causing a corresponding change in climate. Because changes such as these can occur rapidly, it would be difficult for plant and animal life to adapt successfully to the new conditions on the planet.

Based on historical oceanic observations during the past 50 years, global surface water temperatures have been increasing. In fact, the top 3000 meters (9840 feet) of ocean water has warmed by an average of 0.06°C (0.11°F). Although this doesn't seem like much, it represents an enormous amount of energy absorbed by the oceans because of water's tremendous ability to absorb large quantities of heat. What is the cause of this warming? Although it is tempting to point to global warming as the cause, it could just as easily be due to natural variations or—more likely—a combination of the two. Only continued monitoring of ocean properties will help determine its cause as well as its potential effect on ocean circulation.

> Thermohaline circulation describes the movement of deep currents, which are created at the surface in high latitudes where they become cold and dense, so they sink.

Chapter in Review

- *Ocean currents are masses of water that flow* from one place to another and can be divided into *surface currents that are wind driven* or *deep currents that are density driven.* Currents can be measured directly or indirectly.
- *Surface currents occur within and above the pycnocline.* They consist of circular-moving loops of water called *gyres,* set in motion by the major wind belts of the world. They are modified by the positions of the continents, the Coriolis effect, and other factors. There are *five major subtropical gyres* in the world, which rotate *clockwise in the Northern Hemisphere* and *counterclockwise in the Southern Hemisphere.* Water is pushed toward the center of the gyres, forming low "hills" of water.

- The *Ekman spiral* influences shallow surface water and is *caused by winds and the Coriolis effect.* The average net flow of water affected by the Ekman spiral causes the water to move at *90-degree angles to the wind direction.* At the center of a gyre, the Coriolis effect deflects the water so that it tends to move into the hill, whereas gravity moves the water down the hill. When gravity and the Coriolis effect balance, a *geostrophic current* flowing parallel to the contours of the hill is established.
- *The apex (top) of the hill is located to the west of the geographical center of the gyre* due to Earth's rotation. A phenomenon called *western intensification* occurs in which western boundary currents of subtropical gyres are faster, narrower, and deeper than their eastern boundary counterparts.

- *Upwelling and downwelling help vertically mix deep and surface waters.* Upwelling—the movement of cold, deep, nutrient-rich water to the surface—stimulates biologic productivity and creates large amounts of marine life. Upwelling and downwelling can occur in a variety of ways.

- *Antarctic circulation is dominated by a single large current, the Antarctic Circumpolar Current* (West Wind Drift), which flows in a clockwise direction around Antarctica and is driven by the Southern Hemisphere's prevailing westerly winds. Between the Antarctic Circumpolar Current and the Antarctic continent is a current called the East Wind Drift, which is powered by the polar easterly winds. The two currents flow in opposite directions, so the Coriolis effect deflects them away from each other, creating the Antarctic Divergence, an area of abundant marine life due to upwelling and current mixing.

- *The North Atlantic Gyre and the South Atlantic Gyre dominate circulation in the Atlantic Ocean.* A poorly developed equatorial countercurrent separates these two subtropical gyres. The highest-velocity and best-studied ocean current is the Gulf Stream, which carries warm water along the southeastern U.S. Atlantic coast. Meanders of the Gulf Stream produce warm- and cold-core rings. The warming effects of the Gulf Stream extend along its route and reach as far away as Northern Europe.

- *Circulation in the Pacific Ocean consists of two subtropical gyres: the North Pacific Gyre and the South Pacific Gyre*, which are separated by a well-developed equatorial countercurrent.

- *A periodic disruption of normal sea surface and atmospheric circulation patterns in the Pacific Ocean is called El Niño–Southern Oscillation (ENSO).* The *warm phase of ENSO (El Niño)* is associated with the eastward movement of the Pacific warm pool, halting or reversal of the trade winds, a rise in sea level along the Equator, a decrease in productivity along the west coast of South America, and, in very strong El Niños, worldwide changes in weather. El Niños fluctuate with the *cool phase of ENSO (La Niña conditions)*, which are associated with cooler than normal water in the eastern tropical Pacific.

- *The Indian Ocean consists of one gyre, the Indian Ocean Gyre*, which exists mostly in the Southern Hemisphere. The *monsoon wind system*, which changes direction with the seasons, dominates circulation in the Indian Ocean. The monsoons blow from the northeast in the winter and from the southwest in the summer.

- *Deep currents occur below the pycnocline.* They affect much larger amounts of ocean water and move much more slowly than surface currents. Changes in temperature and/or salinity at the surface create slight increases in density, which set deep currents in motion. Deep currents, therefore, are called *thermohaline circulation*.

- *The deep ocean is layered based on density.* Antarctic Bottom Water, the densest deep-water mass in the oceans, forms near Antarctica and sinks along the continental shelf into the South Atlantic Ocean. Farther north, at the Antarctic Convergence, the low-salinity *Antarctic Intermediate Water* sinks to an intermediate depth dictated by its density. Sandwiched between these two masses is the *North Atlantic Deep Water*, rich in algal nutrients after hundreds of years in the deep ocean. Layering in the Pacific and Indian oceans is similar, except there is no source of Northern Hemisphere deep water.

- *Worldwide circulation models that include both surface and deep currents resemble a conveyer belt.* Deep currents carry oxygen into the deep ocean, which is extremely important for life on the planet. Recent investigations indicate that *worldwide deep-water circulation is closely linked to global climate change.*

Key Terms

Agulhas Current (p. 256)
Alaskan Current (p. 248)
Antarctic Bottom Water (p. 258)
Antarctic Circumpolar Current (p. 242)
Antarctic Convergence (p. 242)
Antarctic Divergence (p. 244)
Antarctic Intermediate Water (p. 259)
Arctic Convergence (p. 259)
Benguela Current (p. 244)
Brazil Current (p. 244)
California Current (p. 248)
Canary Current (p. 247)
Coastal downwelling (p. 242)
Coastal upwelling (p. 242)

Cold-core ring (p. 246)
Conveyer-belt circulation (p. 259)
Deep current (p. 230)
Downwelling (p. 239)
East Australian Current (p. 248)
East Wind Drift (p. 243)
Eastern boundary current (p. 234)
Ekman spiral (p. 235)
Ekman transport (p. 236)
El Niño (p. 250)
El Niño–Southern Oscillation (ENSO) (p. 250)
ENSO index (p. 250)
Equatorial countercurrent (p. 239)

Equatorial current (p. 233)
Equatorial Undercurrent (p. 248)
Equatorial upwelling (p. 241)
Geostrophic current (p. 237)
Gulf Stream (p. 244)
Gyre (p. 234)
Indian Ocean Gyre (p. 256)
Kuroshio Current (p. 247)
La Niña (p. 250)
Monsoon (p. 256)
North Atlantic Current (p. 247)
North Atlantic Deep Water (p. 258)
North Atlantic Gyre (p. 244)
North Equatorial Current (p. 244)

North Pacific Current (p. 248)
North Pacific Gyre (p. 247)
Ocean current (p. 230)
Oceanic Common Water (p. 259)
Pacific Decadal Oscillation (PDO) (p. 253)
Pacific warm pool (p. 250)
Peru Current (p. 248)
Productivity (p. 239)
Sargasso Sea (p. 245)
Somali Current (p. 256)
South Atlantic Gyre (p. 244)
South Equatorial Current (p. 244)
South Pacific Gyre (p. 248)
Southern Oscillation (p. 250)
Subpolar gyre (p. 235)

Subtropical convergence
(p. 237)
Subtropical gyre (p. 234)
Surface current (p. 230)
Sverdrup (Sv) (p. 243)

Temperature–salinity (T–S)
diagram (p. 257)
Thermohaline circulation (p. 257)
Tropical Atmosphere and Ocean
(TAO) (p. 256)

Tropical Ocean–Global
Atmosphere (TOGA) (p. 256)
Upwelling (p. 239)
Warm-core ring (p. 246)
Walker Circulation Cell (p. 249)

West Australian Current (p. 256)
West Wind Drift (p. 242)
Western boundary current
(p. 233)
Western intensification (p. 237)

Questions and Exercises

1. Compare the forces that are directly responsible for creating horizontal and deep vertical circulation in the oceans. What is the ultimate source of energy that drives both circulation systems?

2. Describe the different ways in which currents are measured.

3. What would the pattern of ocean surface currents look like if there were no continents on Earth?

4. On a base map of the world, plot and label the major currents involved in the surface circulation gyres of the oceans. Use colors to represent warm versus cool currents and indicate which currents are western intensified. On an overlay, super-impose the major wind belts of the world on the gyres and de-scribe the relationship between wind belts and currents.

5. What atmospheric pressure is associated with the centers of subtropical gyres? With subpolar gyres? Explain why the sub-tropical gyres in the Northern Hemisphere move in a clock-wise fashion while the subpolar gyres rotate in a counterclockwise pattern.

6. Diagram and discuss how Ekman transport produces the "hill" of water within subtropical gyres that causes geostroph-ic current flow. As a starting place on the diagram, use the wind belts (the trade winds and the prevailing westerlies).

7. What causes the apex of the geostrophic "hills" to be offset to the west of the center of the ocean gyre systems?

8. Draw or describe several different oceanographic conditions that produce upwelling.

9. During flood stage, the largest river in the world—the mighty Amazon River—dumps 200,000 cubic meters of water into the Atlantic Ocean each second. Compare its flow rate with the volume of water transported by the West Wind Drift and the Gulf Stream. How many times larger than the Amazon is each of these two ocean currents?

10. Observing the flow of Atlantic Ocean currents in Figure 8-14, offer an explanation as to why the Brazil Current has a much lower velocity and volume transport than the Gulf Stream.

11. Explain why Gulf Stream eddies that develop northeast of the Gulf Stream rotate clockwise and have warm-water cores, whereas those that develop to the southwest rotate counter-clockwise and have cold-water cores.

12. Describe changes in oceanographic phenomena, including Walker Circulation, the Pacific warm pool, trade winds, equa-torial countercurrent flow, upwelling/downwelling, and abun-dance of marine life, that occur during an El Niño event. What are some global effects of El Niño?

13. How often do El Niño events occur? Using Figure 8-20, deter-mine how many years since 1950 have been El Niño years. Has the pattern of El Niño events occurred at regular intervals?

14. How is La Niña different from El Niño? Describe the pattern of La Niña events in relation to El Niños since 1950 (see Figure 8-20).

15. Describe the relationship between atmospheric pressure, winds, and surface currents during the monsoons of the Indian Ocean.

16. Discuss the origin of thermohaline vertical circulation. Why do deep currents form only in high-latitude regions?

17. Name the two major deep-water masses and give the loca-tions of their formation at the ocean's surface.

18. The Antarctic Intermediate Water can be identified throughout much of the South Atlantic based on its temper-ature, salinity, and dissolved oxygen content. Why is it colder and less salty—and contain more oxygen—than the surface-water mass above it and the North Atlantic Deep Water below it?

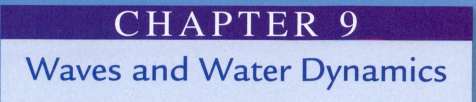

CHAPTER 9
Waves and Water Dynamics

Wipeout at Maverick's. On December 19, 1994, 16-year old Jay Moriarity, one of surfing's bright young stars, caught this wave at Maverick's in central California. At the beginning of his takeoff, he slipped off his board and fell down the 60-foot (18-meter) face of the wave, creating what would be called the most spectacular wipeout ever caught on film. Fortunately, Moriarity survived the tumble. A combination of oceanographic factors produces waves so large here that only the most accomplished surfers even attempt to wade into the surf zone. In fact, professional surfer Mark Foo was killed here only four days after Moriarity's dramatic wipeout.(©*Bob Barbour*)

Key Questions
- What causes waves?
- How do waves move?
- What are differences between deep- and shallow-water waves?
- How do constructive and destructive interference affect waves?
- What physical changes occur in waves as they approach shore and break?
- How is wave refraction different from wave reflection?
- How are tsunami formed and what kinds of destruction do they cause?

The answers to these questions (and much more) can be found in the highlighted concept statements within this chapter.

"There's no time to put on survival suits or grab a life vest; the boat's moving through the most extreme motion of her life and there isn't even time to shout. The refrigerator comes out of the wall and crashes across the galley. Dirty dishes cascade out of the sink. The TV, the washing machine, the VCR tapes, the men, all go flying. And, seconds later, the water moves in."

—Sebastian Junger, *The Perfect Storm* (1997)

What combination of oceanographic factors cause waves to reach extreme heights at places such as Maverick's? This site, which is rated as the world's premier big wave surf spot, is located 0.5 kilometer (0.3 mile) offshore of Pillar Point in Half Moon Bay along the central California coast. One factor is that Maverick's is located offshore of a prominent point of land, which, as will be explained in this chapter, tends to concentrate wave energy due to *wave refraction*. Another factor is that the point juts directly into the North Pacific Ocean, which is known for its wintertime storms and giant waves. Still another factor is that the shoreline abruptly rises from deep depths to a shallowly submerged rock reef, which causes waves to build up to extreme heights in a very short distance. These factors combined with cold ocean temperatures, unforgiving boulders just below the surface, and the presence of large sharks make the site challenging to even the most skilled surfers. Still, surfing competitions are held every year at Maverick's for those brave enough to catch some of the world's most extreme waves.

Most waves are driven by the wind and are relatively small, so release their energy gently, although ocean storms can build up waves to extreme heights. When these waves come ashore, they often produce devastating effects—or, in the case at Maverick's, a wild ride. Waves are *moving energy* traveling along the interface between ocean and atmosphere, often transferring energy from a storm far out at sea over distances of several thousand kilometers. That's why even on calm days, the ocean is in continual motion as waves travel across its surface.

What Causes Waves?

All waves begin as *disturbances*; the energy that causes ocean waves to form is called a **disturbing force**. A rock thrown into a still pond creates waves that radiate out in all directions. Releases of energy, similar to the rock hitting the water, are the cause of all waves.

Wind blowing across the surface of the ocean generates most ocean waves. The waves radiate out in all directions, just as when the rock is thrown into the pond, but on a much larger scale.

The movement of fluids with different densities can also create waves. These waves travel along the interface (boundary) between the two different fluids. Both the air and the ocean are fluids, so waves can be created along interfaces *between* and *within* these fluids as follows:

- Along an *air–water interface*, the movement of air across the ocean surface creates **ocean waves** (simply called *waves*).
- Along an *air–air interface*, the movement of different air masses creates **atmospheric waves**, which are often represented by ripplelike clouds in the sky. Atmospheric waves are especially common when cold fronts (high-density air) invade an area.
- Along a *water–water interface*, the movement of water of different densities creates **internal waves**, as shown in Figure 9-1a. Because these waves travel along the boundary between waters of different density, they are associated with a *pycnocline*.[1] Internal waves can be much larger than surface waves, with heights exceeding 100 meters (330 feet), but they are not as energetic. Tidal movement, turbidity currents, wind stress, or even passing ships at the surface create internal waves, which can sometimes be observed from space (Figure 9-1b). Internal waves can even be a hazard for submarines: If submarines are caught in an internal wave while testing their depth limits, the submarines can inadvertently be carried to depths exceeding their designed pressure strength. At the surface, parallel slicks caused by a film of surface debris may indicate the presence of internal waves below. On a smaller scale, internal waves are prominently featured sloshing back and forth in "desktop oceans," which contain two fluids that do not mix.

---

[1]As discussed in Chapter 6, a *pycnocline* is a layer of rapidly changing density.

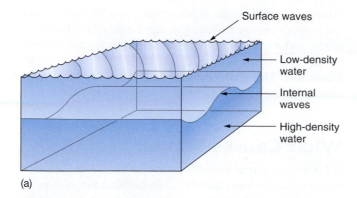

(a)

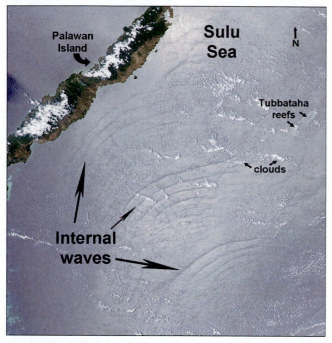

(b) Internal waves, Sulu Sea

Figure 9–1 Internal wave.

(a) An internal wave moving along the density interface (*pycnocline*) below the ocean surface. **(b)** Internal waves in the Sulu Sea between the Philippines and Malaysia. Image taken April 8, 2003 by the Moderate Resolution Imaging Spectroradiometer (MODIS) instrument aboard the Aqua satellite.

STUDENTS SOMETIMES ASK ...
Can internal waves break?

Internal waves do not break in the way that surface waves break in the surf zone because the density difference across an interface at depth is much smaller than that between the atmosphere and the surface. When internal waves approach the edges of continents, however, they do undergo similar physical changes as waves in the surf zone. This causes the waves to build up and expend their energy with much turbulent motion, in essence "breaking" against the continent.

Mass movement into the ocean, such as coastal landslides and calving icebergs, also creates waves. These waves are commonly called *splash waves* (see Box 9-2 for a description of a large splash wave).

Sea floor movement, which changes the shape of the ocean floor and can release large amounts of energy to the entire water column (compared to wind-driven waves, which affect only surface water), can create very large waves. Examples include underwater avalanches (turbidity currents), volcanic eruptions, and fault slippage. The resulting waves are called *seismic sea waves* or *tsunami*. Fortunately, tsunami occur infrequently. When they do, however, they can flood coastal areas and cause large amounts of destruction.

The gravitational pull of the Moon and the Sun tug on every part of Earth's oceans and create vast, low, highly predictable waves called *tides*. Tides are discussed in Chapter 10.

Human activities also cause ocean waves. When ships travel across the ocean, they leave behind a wake, which is a wave. In fact, smaller boats are often carried along in the wake of larger ships, and marine mammals sometimes play there. Also, the detonation of nuclear devices at or near sea level releases huge amounts of energy that creates waves.

In all cases, though, some type of energy release creates waves. Figure 9-2 shows the distribution of energy in waves, indicating that most ocean waves are wind-generated.

> Most ocean waves are caused by wind, but many other types of waves are created by releases of energy in the ocean, including internal waves, splash waves, tsunami, tides, and human-induced waves.

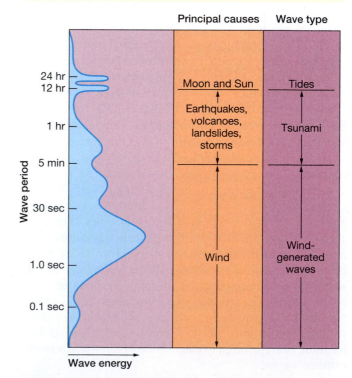

Figure 9–2 Distribution of energy in ocean waves.

Most of the energy possessed by ocean waves exists as wind-generated waves while other peaks of wave energy represent tsunami and ocean tides.

How Waves Move

Waves are energy in motion. Waves transmit energy by means of cyclic movement through matter. The medium itself (solid, liquid, or gas) does not actually travel in the direction of the energy that is passing through it. The particles in the medium simply oscillate, or cycle, back and forth, up and down, or around and around, transmitting energy from one particle to another. If you thump your fist on a table, for example, the energy travels through the table as waves that someone sitting at the other end can feel, but the table itself does not move.

Waves move in different ways. Simple *progressive waves* (Figure 9-3a) are waves that oscillate uniformly and *progress* or travel without breaking. Progressive waves may be *longitudinal, transverse*, or a combination of the two motions, called *orbital*.

In **longitudinal waves** (also known as push-pull waves), the particles that vibrate "push and pull" in the same direction that the energy is traveling, like a spring whose coils are alternately compressed and expanded.[2] The shape of the wave (called a *waveform*) moves through the medium by compressing and decompressing as it goes. Sound, for instance, travels as longitudinal waves. Clapping your hands initiates a percussion that compresses and decompresses the air as the sound moves

through a room. Energy can be transmitted through all states of matter—gaseous, liquid, or solid—by this longitudinal movement of particles.

In **transverse waves** (also known as side-to-side waves), energy travels at right angles to the direction of the vibrating particles.[3] If one end of a rope is tied to a doorknob while the other end is moved up and down (or side to side) by hand, for example, a waveform progresses along the rope and energy is transmitted from the motion of the hand to the doorknob. The waveform moves up and down (or side to side) with the hand, but the motion is at right angles to the direction in which energy is transmitted (from the hand to the doorknob). Generally, transverse waves transmit energy only through solids, because the particles in solids are bound to one another strongly enough to transmit this kind of motion.

Longitudinal and transverse waves are called *body waves* because they transfer energy through a body of matter. Ocean waves are body waves, too, because they transmit energy through the upper part of the ocean near the interface between the atmosphere and the ocean. The

---

[2]Examples of longitudinal (compressional) waves are seismic P waves, which are discussed in Chapter 3 (see Figure 3-16a).

[3]Examples of transverse waves are seismic S waves, which are discussed in Chapter 3 (see Figure 3-16b).

LONGITUDINAL WAVE
Particles (color) move back and forth in direction of energy transmission. These waves transmit energy through all states of matter.

TRANSVERSE WAVE
Particles (color) move back and forth at right angles to direction of energy transmission. These waves transmit energy only through solids.

ORBITAL WAVE
Particles (color) move in orbital path. These waves transmit energy along interface between two fluids of different density (liquids and/or gases).

(a) Types of progressive waves

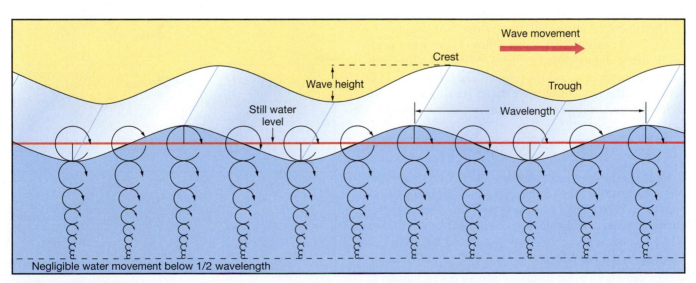

(b) Wave characteristics

Figure 9–3 Types and characteristics of progressive waves.

(a) Types of progressive waves. (b) A diagrammatic view of an idealized ocean wave showing its characteristics.

movement of particles in ocean waves involves components of *both* longitudinal and transverse waves, so particles move in circular orbits. Thus, waves at the ocean surface are **orbital waves** (also called *interface waves*).

Wave Characteristics

Figure 9-3b shows the characteristics of an idealized ocean wave. The simple, uniform, moving waveform transmits energy from a single source and travels along the ocean–atmosphere interface. These waves are also called *sine waves* because their uniform shape resembles the oscillating pattern expressed by a sine curve. Even though idealized waveforms do not exist in nature (actual waves have sharper crests and elongated troughs), they help us understand wave characteristics.

As the idealized wave passes a permanent marker, such as a pier piling, a succession of high parts of the waves, called **crests**, alternate with low parts, called **troughs**. Halfway between the crests and the troughs is the **still water level**, or *zero energy level*. This is the level of the water if there were no waves. The **wave height**, designated by the symbol H, is the vertical distance between a crest and a trough.

The horizontal distance between any two corresponding points on successive waveforms, such as from crest to crest or from trough to trough, is the **wavelength**, L. **Wave steepness** is the ratio of wave height to wavelength:

$$\text{Wave steepness} = \frac{\text{wave height }(H)}{\text{wavelength }(L)} \quad (9\text{-}1)$$

If the wave steepness exceeds $\frac{1}{7}$, the wave *breaks* (spills forward) because the wave is too steep to support itself. A wave can break anytime the 1:7 ratio is exceeded, either along the shoreline or out at sea. This ratio also dictates the maximum height of a wave. For example, a wave 7 meters long can only be 1 meter high or it will break.

The time it takes one full wave—one wavelength—to pass a fixed position (like a pier piling) is the **wave period**, T, often simply called the *period*. Typical wave periods range between 6 and 16 seconds. The **frequency** (f) is defined as the number of wave crests passing a fixed location per unit of time and is the inverse of the period:

$$\text{Frequency }(f) = \frac{1}{\text{period }(T)} \quad (9\text{-}2)$$

Because the speed and wavelength of ocean waves are such that less than one wavelength passes a point per second, the preferred unit of time is period (rather than frequency), when calculating the speed of ocean waves.

Circular Orbital Motion

Waves can travel great distances across ocean basins. In one study, waves generated near Antarctica were tracked as they traveled through the Pacific Ocean basin. After more than 10,000 kilometers (over 6000 miles), the

waves finally expended their energy a week later along the shoreline of the Aleutian Islands of Alaska. The water itself doesn't travel the entire distance, but the waveform does. As the wave travels, the water passes the energy along by moving in a circle. This movement is called **circular orbital motion**.

Observation of an object floating in the waves reveals that it moves not only up and down, but also slightly forward and backward with each successive wave. Figure 9-4 shows that a floating object moves up and backward as the crest approaches, up and forward as the crest passes, down

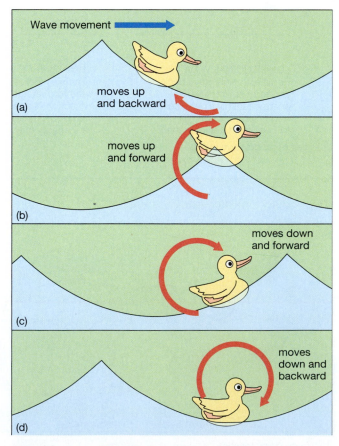

(a) moves up and backward

(b) moves up and forward

(c) moves down and forward

(d) moves down and backward

Wave movement

Figure 9–4 A rubber duck in water.

As waves pass, the motion of a floating rubber duck resembles that of a circular orbit.

and forward after the crest, down and backward as the trough approaches, and rises and moves backward again as the next crest advances. When the movement of the rubber duck shown in Figure 9-4 is traced as a wave passes, it can be seen that the duck moves in a circle and returns close to its original position.[4] This motion allows a waveform (the wave's shape) to move forward through the water while the individual water particles that transmit the wave move around in a circle and return to essentially the same place. Wind moving across a field of wheat causes a similar phenomenon: The wheat itself doesn't travel across the field, but the waves do.

The circular orbits of an object floating at the surface have a diameter equal to the wave height (Figure 9-3b). Figure 9-5 shows that circular orbital motion dies out quickly below the surface. At some depth below the surface, the circular orbits become so small that movement is negligible. This depth is called the **wave base**, and it is equal to one-half the wavelength ($L/2$) *measured from still water level*. Thus, only wavelength controls the depth of the wave base such that the longer the wave, the deeper the wave base.

The decrease of orbital motion with depth has many practical applications. For instance, submarines can avoid large ocean waves simply by submerging below the wave base. Even the largest storm waves will go unnoticed if a submarine submerges to only 150 meters (500 feet). Floating bridges and floating oil rigs are constructed so that most of their mass is below wave base, so that they will be unaffected by wave motion. In fact, offshore floating airport runways have been designed using similar principles. Additionally, seasick scuba divers find relief when they submerge into the calm, motionless water below wave base. Finally, as you walk from the beach into the ocean, you reach a point where it is easier to dive under an incoming wave than to jump over it. That is, it is easier to swim through the smaller orbital motion below the surface than to fight the large waves at the surface.

---

[4]Actually, the circular orbit does not quite return the floating object to its original position because the half of the orbit accomplished in the trough is slower than the crest half of the orbit. This slight forward movement (net mass transport) is called *wave drift*.

The ocean transmits wave energy by circular orbital motion, where the water particles move in circular orbits and return to approximately the same location.

Deep-Water Waves

If the water depth (d) is greater than the wave base ($L/2$), the waves are called **deep-water waves** (Figure 9-6a). Deep-water waves have no interference with the ocean bottom, so they include all wind-generated waves in the open ocean, where water depths far exceed wave base.

Wave speed (S) is defined as

$$\text{Wave speed } (S) = \frac{\text{wavelength } (L)}{\text{period } (T)} \quad (9\text{-}3)$$

Wave speed is more correctly known as *celerity* (C). Celerity is different from the traditional concept of speed in that it is used only in relation to waves where no mass is in motion, just the wave form.

EXAMPLE 9–2

If a wave has a wavelength of 156 meters and a period of 10 seconds, what is its speed?

Since we have wavelength and period, we can use Equation (9-3):

$$\text{Wave speed } (S) = \frac{\text{wavelength } (L)}{\text{period } (T)} = \frac{156 \text{ meters}}{10 \text{ seconds}}$$

$$= 15.6 \text{ meters per second}$$

According to the equations that govern the movement of progressive waves, the speed of deep-water waves is dependent upon (1) wavelength and (2) several other variables (such as gravitational attraction) that remain constant on Earth. So, by filling in the constants with numbers, the equation for wave speed of deep-water waves varies only with wavelength and becomes (in meters per second)

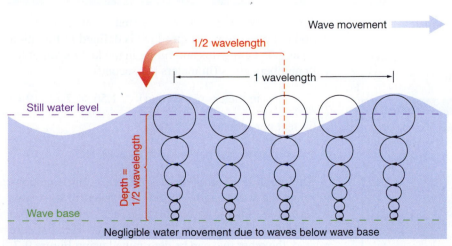

Wave movement

Figure 9–5 Orbital motion in waves.

The orbital motion of water particles in waves extends to a depth of one-half the wavelength, measured from still water level, which is the wave base.

1/2 wavelength

1 wavelength

Still water level

Depth = 1/2 wavelength

Wave base

Negligible water movement due to waves below wave base

Figure 9–6 Characteristics of deep-water, shallow-water, and transitional waves.

(a) Deep-water waves, showing the diminishing size of the circular orbits with increasing depth. (b) Shallow-water waves, where the ocean floor interferes with circular orbital motion, causing the orbits to become more flattened. (c) Transitional waves, which are intermediate between deep-water and shallow-water waves. All diagrams are not to scale.

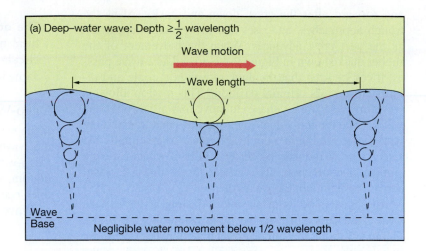

(a) Deep–water wave: Depth $\geq \frac{1}{2}$ wavelength

Wave motion

Wave length

Wave Base

Negligible water movement below 1/2 wavelength

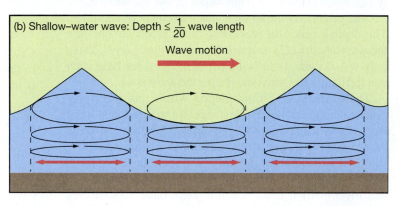

(b) Shallow–water wave: Depth $\leq \frac{1}{20}$ wave length

Wave motion

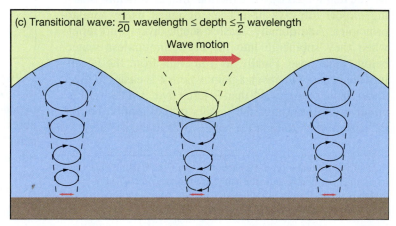

(c) Transitional wave: $\frac{1}{20}$ wavelength $\leq$ depth $\leq \frac{1}{2}$ wavelength

Wave motion

$$S \text{ (in meters per second)} = 1.25\sqrt{L \text{ (in meters)}} \quad (9\text{-}4)$$

or, in feet per second,

$$S \text{ (in feet per second)} = 2.26\sqrt{L \text{ (in feet)}} \quad (9\text{-}5)$$

EXAMPLE 9–3

If a wave has a wavelength of 156 meters, what is its speed?
Since wave speed varies as a function of wavelength and we have wavelength known in meters, we can use Equation (9-4):

Wave speed $(S) = 1.25\sqrt{L \text{ (in meters)}}$
$1.25\sqrt{156 \text{ meters}} = 15.6$ meters per second

Note that this is the same answer as in Example 9-2.

We can also determine wave speed knowing only the period (T) because wave speed (S) is defined in Equation (9-3) as L/T. Doing this and filling in the known variables with numbers gives (in meters per second)

$$S \text{ (in meters per second)} = 1.56 \times T \quad (9\text{-}6)$$

or, in feet per second,

$$S \text{ (in feet per second)} = 5.12 \times T \quad (9\text{-}7)$$

The graph in Figure 9-7 uses the preceding equations to relate the wavelength, period, and speed of de   0ep-water waves. Of the three variables, the wave period is usually easiest to measure. Since all three variables are related, the other two can be determined using Figure 9-7.

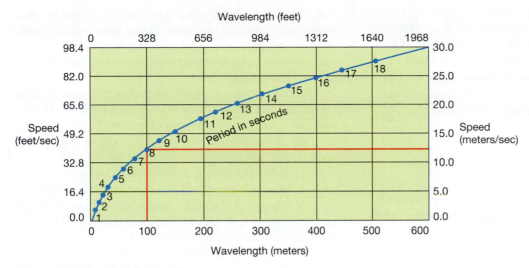

Figure 9–7 Speed of deep-water waves.

Ideal relations among wavelength, period (*blue line*), and wave speed for deep-water waves. Red lines show an example wave with a wavelength of 100 meters, a period of 8 seconds, and a speed of 12.5 meters per second.

EXAMPLE 9–4

Using Figure 9-7, what is the wave speed for a wave with a wavelength of 100 meters and a period of 8 seconds?

The vertical red line on Figure 9-7 shows a wave with a wavelength of 100 meters, which has a period of 8 seconds (where the vertical red line intersects the blue line). Thus, the speed of the wave is shown by the horizontal red line on the graph, which is 12.5 meters per second.

Note that we can also use Equation (9-3) to calculate the answer:

$$\text{Speed }(S) = \frac{L}{T} = \frac{100 \text{ meters}}{8 \text{ seconds}}$$

$$= 12.5 \text{ meters per second}$$

The general relationship shown by Equations (9-3) through (9-7) (and shown in Figure 9-7) for deep-water waves is *the longer the wavelength, the faster the wave travels*. A fast wave does not necessarily have a large wave height, however, because wave speed depends *only* on wavelength.

Shallow-Water Waves

Waves in which depth (*d*) is less than $\frac{1}{20}$ of the wavelength (*L*/20) are called **shallow-water waves**, or *long waves* (Figure 9-6b). Shallow-water waves are said to *touch bottom* or *feel bottom* because the ocean floor interferes with their orbital motion.

The speed of shallow-water waves is influenced only by gravitational acceleration and the water depth (*d*). Since gravitational acceleration remains constant on

Earth, the equation for wave speed becomes (in meters per second)

$$S \text{ (in meters per second)} = 3.13 \sqrt{d \text{ (in meters)}} \quad (9\text{-}8)$$

or, in feet per second,

$$S \text{ (in feet per second)} = 5.67 \sqrt{d \text{ (in feet)}} \quad (9\text{-}9)$$

These equations show that wave speed in shallow-water waves is determined *only* by water depth, where *the deeper the water, the faster the wave travels*.

Shallow-water waves include wind-generated waves that have moved into shallow nearshore areas; *tsunami* (seismic sea waves), generated by earthquakes in the ocean floor; and the *tides*, which are a type of wave generated by the gravitational attraction of the Moon and the Sun. Tsunami and tides are very long-wavelength waves, which far exceed even the deepest ocean water depths.

Particle motion in shallow-water waves is in a very flat elliptical orbit that approaches horizontal (back-and-forth) oscillation. The vertical component of particle motion decreases with increasing depth, causing the orbits to become even more flattened.

Transitional Waves

Waves that have some characteristics of shallow-water waves and some of deep-water waves are called **transitional waves** or *intermediate waves*. The wavelengths of transitional waves are between two times and 20 times the water depth (Figure 9-6c). Recall that the wave speed of shallow-water waves is a function of water depth; for deep-water waves, wave speed is a function of wavelength. Because transitional waves are intermediate between the two, their wave speed depends partially on water depth and partially on wavelength.

> Deep-water waves exist in water that is deeper than wave base and move at speeds controlled by wavelength; shallow-water waves exist in water in which depth is less than $\frac{1}{20}$ the wavelength and at speeds controlled by water depth; transitional waves are intermediate between the two.

Wind-Generated Waves

The life history of a wind-generated wave includes its origin in a windy region of the ocean, its movement across great expanses of open water without subsequent aid of wind, and its termination when it breaks and releases its energy, either in the open ocean or against the shore.

"Sea"

As the wind blows over the ocean surface, it creates pressure and stress. These factors deform the ocean surface into small, rounded waves with V-shaped troughs and wavelengths less than 1.74 centimeters (0.7 inch). Commonly called *ripples*, oceanographers call them **capillary** (*cappilaris* = hair) **waves** (Figure 9-8, *left*). The name comes from *capillarity*, a property that results from the surface tension of water. Capillarity is the dominant **restoring force** that works to destroy these tiny waves, restoring the smooth ocean surface once again.

As capillary wave development increases, the sea surface takes on a rougher appearance. The water "catches" more of the wind, allowing the wind and ocean surface to interact more efficiently. As more energy is transferred to the ocean, **gravity waves** develop, which are symmetric waves that have wavelengths exceeding 1.74 centimeters (0.7 inch) (Figure 9-8, *middle*). Because they reach greater height at this stage, gravity replaces capillarity as the dominant restoring force, giving these waves their name.

The length of gravity waves is generally 15 to 35 times their height. As additional energy is gained, wave height increases more rapidly than wavelength. The crests become pointed and the troughs are rounded, resulting in a *trochoidal* (*trokhos* = wheel) waveform (Figure 9-8, *right*).

Energy imparted by the wind increases the height, length, and speed of the wave. When wave speed equals wind speed, neither wave height nor length can change because there is no net energy exchange and the wave has reached its maximum size.

The area where wind-driven waves are generated is called **"sea"** or the *sea area*. It is characterized by choppiness and waves moving in many directions. The waves have a variety of periods and wavelengths (most of them short) due to frequently changing wind speed and direction.

Factors that determine the amount of energy in waves are (1) the *wind speed*, (2) the *duration*—the length of time during which the wind blows in one direction, and (3) the *fetch*—the distance over which the wind blows in one direction, as shown in Figure 9-9.

Wave height is directly related to the energy in a wave. Wave heights in a sea area are usually less than 2 meters (6.6 feet), but waves with heights of 10 meters (33 feet) and periods of 12 seconds are not uncommon. As "sea" waves gain energy, their steepness increases. When steepness reaches a critical value of $\frac{1}{7}$, open ocean breakers—called *whitecaps*—form. The appearance of a sea surface as it changes from calm to the condition that results from hurricane-force winds is described in Table 9-1.

Figure 9-10 is a map based on satellite data of average wave heights during October 3–12, 1992. The waves in the Southern Hemisphere are particularly large because the prevailing westerlies between 40 and 60 degrees south latitude reach the highest average wind speeds on Earth, creating the latitudes called the "Roaring Forties," "Furious Fifties," and "Screaming Sixties."

How high can waves be? According to a U.S. Navy Hydrographic Office bulletin published in the early 1900s, the theoretical maximum height of wind-generated waves should be no higher than 18.3 meters (60 feet), which became known as the "60-foot rule." Although there were some isolated eyewitness accounts of larger waves, the U.S. Navy considered any sightings of waves over 60 feet to be exaggerations. Certainly, embellishment of reported wave height under conditions of extremely heavy seas would be understandable. For many years, the "60-foot rule" was accepted as fact.

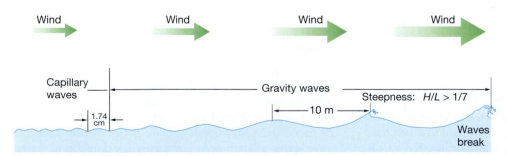

Figure 9–8 Wind creates capillary and gravity waves.

As wind increases (*left to right*), the height and wavelength of waves increases, beginning as capillary waves and progressing to gravity waves. When the wave steepness (*H/L*) exceeds a 1:7 ratio, the waves become unstable and break. Not to scale.

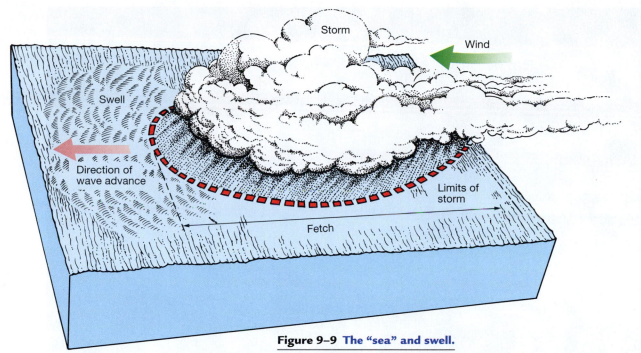

Figure 9–9 **The "sea" and swell.**
As wind blows across the "sea" (*red dash*), wave size increases with increasing wind speed, duration, and fetch. As waves advance beyond their area of origination, they advance across the ocean surface and become sorted into uniform, symmetric swell.

TABLE 9-1 **Beaufort Wind Scale and the state of the sea.**

Beaufort number	Descriptive term	Wind speed (km/h)	(mi/h)	Appearance of the sea
0	Calm	<1	<1	Like a mirror
1	Light air	1–5	1–3	Ripples with the appearance of scales, no foam crests
2	Light breeze	6–11	4–7	Small wavelets; crests of glassy appearance, no breaking
3	Gentle breeze	12–19	8–12	Large wavelets; crests begin to break, scattered whitecaps
4	Moderate breeze	20–28	13–18	Small waves, becoming longer; numerous whitecaps
5	Fresh breeze	29–38	19–24	Moderate waves, taking longer form; many whitecaps, some spray
6	Strong breeze	39–49	25–31	Large waves begin to form, whitecaps everywhere, more spray
7	Near gale	50–61	32–38	Sea heaps up and white foam from breaking waves begins to be blown in streaks
8	Gale	62–74	39–46	Moderately high waves of greater length, edges of crests begin to break into spindrift, foam is blown in well-marked streaks
9	Strong gale	75–88	47–54	High waves, dense streaks of foam and sea begins to roll, spray may affect visibility
10	Storm	89–102	55–63	Very high waves with overhanging crests; foam is blown in dense white streaks, causing the sea to appear white; the rolling of the sea becomes heavy; visibility reduced
11	Violent storm	103–117	64–72	Exceptionally high waves (small and medium-sized ships might for a time be lost from view behind the waves), the sea is covered with white patches of foam, everywhere the edges of the wave crests are blown into froth, visibility further reduced
12	Hurricane	118+	73+	The air is filled with foam and spray, sea completely white with driving spray, visibility greatly reduced

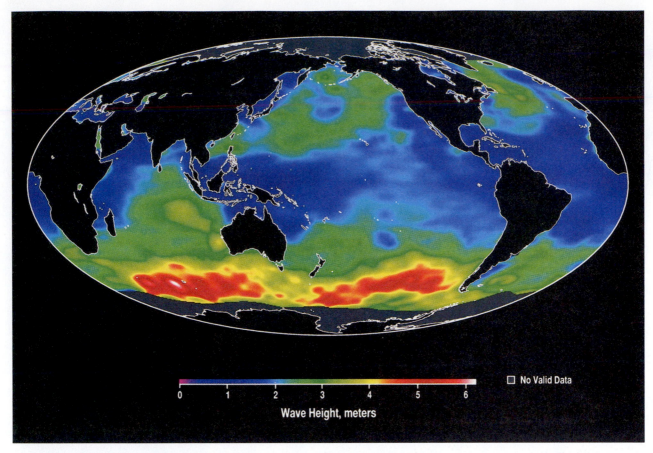

Figure 9–10 TOPEX/Poseidon wave height, October 3–12, 1992.

The TOPEX/Poseidon satellite receives a return of stronger signals from calm seas and weaker signals from seas with large waves. Based on these data, a map of wave height can be produced. The largest average wave heights (*red areas*) are in the prevailing westerly wind belt in the Southern Hemisphere. Scale in meters.

However, careful observations made aboard the 152-meter (500-foot)-long U.S. Navy tanker USS *Ramapo* in 1935 proved otherwise. The ship was caught in a typhoon in the western Pacific Ocean and encountered 108-kilometer (67-mile)-per-hour winds en route from the Philippines to San Diego. The resulting waves were symmetrical, uniform,

and had a period of 14.8 seconds. Because the *Ramapo* was traveling with the waves, the vessel's officers were able to measure the waves accurately. The officers used the dimensions of the ship, including the *eye height* of an observer on the ship's bridge (Figure 9-11). Geometric relationships revealed that the waves were 34 meters (112 feet) high, which

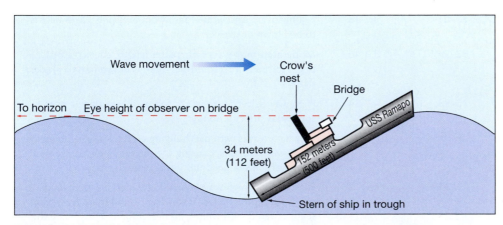

Figure 9–11 USS *Ramapo* in heavy seas.

Bridge officers aboard the USS *Ramapo* in 1935 measured the largest authentically recorded wave by sighting from the bridge across the crow's nest to the horizon while the ship's stern was directly in the trough of a large wave. A wave height of 34 meters (112 feet) was calculated based on geometric relationships of the vessel and the waves.

made them taller than an 11-story building! These waves proved to be a record that still stands today for the largest authentically recorded wind-generated waves, shattering the "60-foot rule." Although the *Ramapo* was largely undamaged, other ships traveling in heavy seas aren't always so lucky (Figure 9-12).

For a given wind speed, Table 9-2 lists the minimum fetch and duration of wind beyond which the waves cannot grow. Waves cannot grow because an equilibrium condition, called a **fully developed sea**, has been achieved. Waves can grow no further in a fully developed sea because they lose as much energy breaking as whitecaps under the force of gravity as they receive from the wind. Table 9-2 also lists the average characteristics of waves resulting from a fully developed sea, including the height of the highest 10% of the waves.

EXAMPLE 9–5

Based on the period of the giant waves the USS Ramapo experienced, what were the waves' speed and wavelength?

Since the waves the USS *Ramapo* experienced were in the open ocean, they must have been deep-water waves. Knowing that the period (T) = 14.8 seconds, we can use Equation (9-6) to determine wave speed (S):

$$S = 1.56T = 1.56 \times 14.8 \text{ seconds}$$
$$= 23.1 \text{ meters per second}$$

Note that 23.1 meters per second = 83.1 kilometers (51.6 miles) per hour.

Now that wave speed is known, wavelength (L) can be determined using Equation (9-3):

$$S = \frac{L}{T}, \text{ so } L = S \times T$$

$$L = 23.1 \text{ meters per second} \times 14.8 \text{ seconds}$$
$$= 342 \text{ meters (1121 feet)}$$

Figure 9–12 Wave damage on the aircraft carrier *Bennington*.

The *Bennington* returns from heavy seas encountered in a typhoon off Okinawa in 1945 with part of its reinforced steel flight deck bent down over the bow. Damage to the flight deck, which is 16.5 meters (54 feet) above still water level, was caused by large waves.

TABLE 9-2 **Conditions necessary to produce a fully developed sea at various wind speeds and the characteristics of the resulting waves.**

Wind speed in km/h (mi/h)	Fetch in km (mi)	Duration in hours	Average height in m (ft)	Average wavelength in m (ft)	Average period in seconds	Highest 10% of waves in m (ft)
20 (12)	24 (15)	2.8	0.3 (1.0)	10.6 (34.8)	3.2	0.8 (2.5)
30 (19)	77 (48)	7.0	0.9 (2.9)	22.2 (72.8)	4.6	2.1 (6.9)
40 (25)	176 (109)	11.5	1.8 (5.9)	39.7 (130.2)	6.2	3.9 (12.8)
50 (31)	380 (236)	18.5	3.2 (10.5)	61.8 (202.7)	7.7	6.8 (22.3)
60 (37)	660 (409)	27.5	5.1 (16.7)	89.2 (292.6)	9.1	10.5 (34.4)
70 (43)	1093 (678)	37.5	7.4 (24.3)	121.4 (398.2)	10.8	15.3 (50.2)
80 (50)	1682 (1043)	50.0	10.3 (33.8)	158.6 (520.2)	12.4	21.4 (70.2)
90 (56)	2446 (1517)	65.2	13.9 (45.6)	201.6 (661.2)	13.9	28.4 (93.2)

Swell

As waves generated in a sea area move toward its margins, wind speeds diminish and the waves eventually move faster than the wind. When this occurs, wave steepness decreases, and waves become long-crested waves called **swells** (*swellan* = swollen). Swells are uniform, symmetrical waves that have traveled out of the area where they originated. Swells move with little loss of energy over large stretches of the ocean surface, transporting energy away from one sea area and depositing it in another. Thus, there can be waves at distant shorelines where there is no wind.

STUDENTS SOMETIMES ASK...
I know that swell is what surfers hope for. Is swell always big?

Not necessarily. Swell is defined as waves that have moved out of their area of origination, so these waves do not have to be a certain wave height to be classified as swell. It is true, however, that the uniform and symmetrical shape of most swell delights surfers.

Waves with longer wavelengths travel faster, and thus leave the sea area first. They are followed by slower, shorter **wave trains**, or groups of waves. The progression from long, fast waves to short, slow waves illustrates the principle of **wave dispersion** (*dis* = apart, *spargere* = to scatter)—the sorting of waves by their wavelength.

Waves of many wavelengths are present in the generating area. Wave speed depends on wavelength in deep water (see Figure 9-7), however, so the longer waves "outrun" the shorter ones. The distance over which waves change from a choppy "sea" to uniform swell is called the **decay distance**, which can be up to several hundred kilometers.

As a group of waves leaves a sea area and becomes a swell *wave train*, the leading wave keeps disappearing. However, the same number of waves always remains in the group because as the leading wave disappears, a new wave replaces it at the back of the group (Figure 9-13). For example, if four waves are generated, the lead wave keeps dying out as the wave train travels, but one is created in the back, so the wave train stays four waves. Because of the progressive dying out and creation of new waves, the group moves across the ocean surface at only *half* the velocity of an individual wave in the group.

Interference Patterns When swells from different storms run together, the waves clash, or interfere with one another, giving rise to **interference patterns**. An interference pattern produced when two or more wave systems collide is the sum of the disturbance that each wave would have produced individually. Figure 9-14 shows that the result may be a larger or smaller trough or crest, depending on conditions.

When swells from two storm areas collide, the interference pattern may be constructive or destructive, but it is more likely to be mixed. **Constructive interference**

Figure 9–13
Movement of a wave train.

As energy in the leading waves (**a**, *waves 1 and 2*) is transferred into circular orbital motion, the waves in front die out and are replaced by new waves from behind (**b**). Even though new waves take up the lead (**c** and **d**), the length of the wave train and the total number of waves remain the same. This causes the group speed to be one-half that of the individual wave.

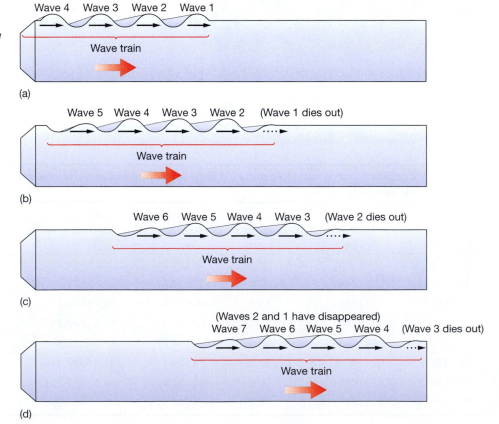

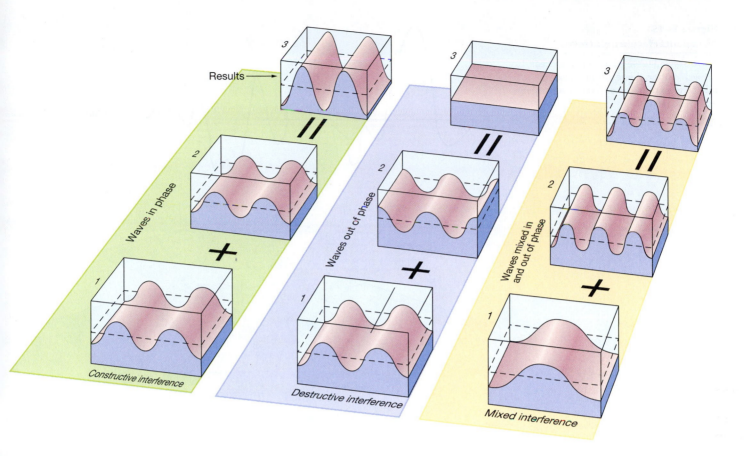

Results —→

Waves in phase

Constructive interference

Waves out of phase

Destructive interference

Waves mixed in and out of phase

Mixed interference

Figure 9–14 Constructive, destructive, and mixed interference patterns.

Constructive interference (*left*) occurs when waves of the same wavelength come together in phase (crest to crest and trough to trough), producing waves of greater height. Destructive interference (*center*) occurs when overlapping waves have identical characteristics but come together out of phase, resulting in a canceling effect. More commonly, waves of different lengths and heights encounter one another and produce a complex, mixed interference pattern (*right*).

occurs when wave trains having the same wavelength come together *in phase*, meaning crest to crest and trough to trough. If the displacements from each wave are added together, the interference pattern results in a wave with the same wavelength as the two overlapping wave systems, but with a wave height equal to the sum of the individual wave heights (Figure 9-14, *left*).

Destructive interference occurs when wave trains having the same wavelength come together *out of phase*, meaning the crest from one wave coincides with the trough from a second wave. If the waves have identical heights, the sum of the crest of one and the trough of another is zero, so the energies of these waves cancel each other (Figure 9-14, *center*).

It is more likely, however, that the two swells consist of waves of various heights and lengths that come together with a mixture of constructive and destructive interference. Thus, a more complex **mixed interference** pattern develops (Figure 9-14, *right*), which explains the varied sequence of high and lower waves (called *surf beat*) and other irregular wave patterns that occur when swell approaches the shore. In the open ocean,

several swell systems often interact, creating complex wave patterns (Figure 9-15).

Constructive interference results from in phase overlapping of waves and creates larger waves, while destructive interference results from waves overlapping out of phase, reducing wave height.

Free and Forced Waves   Swell is an example of a **free wave**, which is a wave moving with the momentum and energy imparted to it in the sea area but it is not experiencing a maintaining force that keeps it in motion. A **forced wave** is one that is maintained by a force that has a periodicity coinciding with the period of the wave. For most ocean waves, this force is the wind. Because of the high variability of wind in a storm, many wave systems in the sea area alternate between forced and free waves. Another example of a forced wave is the tides, which are always maintained by the gravitational attraction of the Moon and the Sun.

Figure 9–15
Mixed interference pattern.

The observed wave pattern in the ocean (*above*) is often the result of mixed interference of many different overlapping wave sets (*below*).

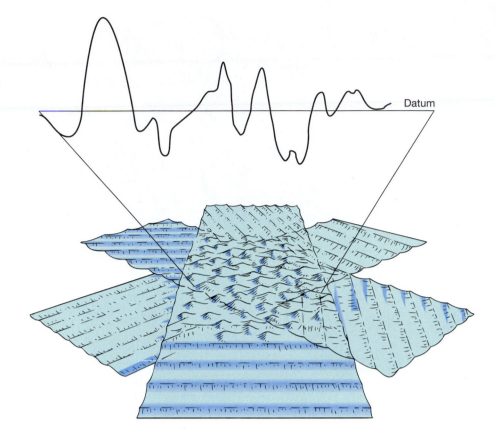

Datum

Rogue Waves

Rogue waves are massive, solitary waves that can reach enormous height and often occur at times when normal ocean waves are not unusually large. In a sea of 2-meter (6.5-foot) waves, for example, a 20-meter (65-foot) rogue wave may suddenly appear. *Rogue* means "unusual" and, in this case, the waves are unusually large. Rogue waves—sometimes called *superwaves*—can be quite destructive and have been popularized in literature and movies such as *The Perfect Storm*.

In the open ocean, one wave in 23 will be over twice the height of the wave average, one in 1175 will be three times as high, and one in 300,000 will be four times as high. The chances of a truly monstrous wave, therefore, are only one in several billion. Nevertheless, rogue waves do occur, though no one knows specifically when or where they will arise. For instance, the 17-meter (56-foot) NOAA research vessel R/V *Ballena* was flipped and sunk in November 2000 by a 15-foot (4.6-meter) rogue wave off the California coast while conducting a survey in shallow, calm water. Fortunately, the three people on board survived the incident.

Even with satellites that can measure average wave size and forecast storms, about 10 large ships each year are reported missing without a trace. Worldwide, the total number of vessels lost of all sizes may reach 1000 per year, some of which are the victims of rogue waves. Recent satellites designed to observe the ocean (see Table 1-1) have provided a wealth of data about ocean waves but still don't allow the prediction of rogue waves.

The main cause of rogue waves is theorized to be an extraordinary case of constructive wave interference where multiple waves overlap in phase to produce an extremely large wave (Box 9-1). Rogue waves also tend to occur more frequently downwind from islands or shoals. In addition, rogue waves can occur when storm-driven waves move against strong ocean currents, causing the waves to steepen, shorten, and become larger. These conditions exist along the "Wild Coast" off the southeast coast of Africa, where the Agulhas Current flows directly against large Antarctic storm waves, creating rogue waves that can crash onto the bow of a ship, overcome its structural capacity, and cause the ship to sink (Figure 9-16). This stretch of water is probably responsible for sinking more ships than any other place on Earth.

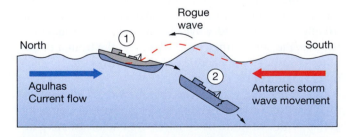

Figure 9–16 Rogue waves along Africa's "Wild Coast."

Rogue waves can be formed where the Algulhas Current flows directly against large Antarctic Storm waves along Africa's "Wild Coast." These large and steep rogue waves can crash onto the bow of a ship, causing it to sink.

BOX 9–1   People and Ocean Environment

YACHTING IN MONSTER SEAS: THE SYDNEY TO HOBART AND FASTNET RACING DISASTERS

"The 1998 Sydney to Hobart took us beyond sport, beyond the drive to compete. We found ourselves at the edge of life-and-death survival."

—Ed Psaltsis, winning skipper of the 1998 Sydney to Hobart Yacht Race

The world championship of yacht racing is determined each year by standings in the five Admiral's Cup Series races. Two of these races are the Sydney to Hobart Race off Australia and the Fastnet Race off England. Recently, both of these challenging yachting events proved fatal because of high winds (created by low pressure) and rogue waves (created by overlapping seas).

On December 26, 1998, 115 boats sailed out of Sydney Harbor at the start of the 1180-kilometer (735-mile)-long Sydney to Hobart Race. Ranked as one of the most treacherous races in the world, its course takes sailors along the east coast of Australia, across Bass Strait, and down the length of Tasmania (Figure 9A). Prior to the race, the Australian Bureau of Meteorology issued a warning for winds of 90 kilometers (56 miles) per hour in Bass Strait, later upping the severity of the storm and describing the conditions as "atrocious," which proved to be an understatement.

The first storm hit the fleet around midnight, causing many boats to turn and sail for home. By midafternoon on December 27, three massive low-pressure weather systems collided in Bass Strait, producing a killer sea. Over the next 24 hours, mammoth waves of 20 meters (66 feet) and winds of 161 kilometers (100 miles) per hour battered the remaining boats. Eyewitnesses reported waves tall enough to send 25-ton yachts "spearing into midair," then "plunging down into the trough... like repeatedly launching a truck off a 30-foot ramp and awaiting the

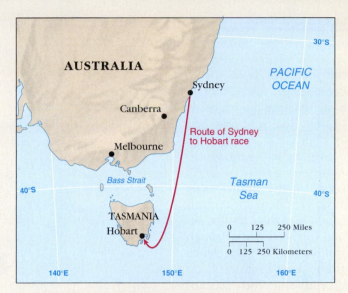

Figure 9A **Route of the Sydney to Hobart Race off Australia.**

Figure 9B **Rescue in extreme conditions at sea.**

A helicopter rescues the crew of a stricken yacht in Bass Strait during the 1998 Sydney to Hobart Race.

crash." Survivors' testimony indicates that large rogue waves dismasted, capsized, rolled and sank several vessels (Figure 9B). By the end of the day on December 29, five boats had been sunk, 24 boats were abandoned, 55 sailors had been rescued under near-impossible circumstances, and six lives were lost.

Even more people lost their lives in the 1979 Fastnet Race. The 1000-kilometer (620-mile) race starts from the Isle of Wight off the south coast

of England and requires rounding Fastnet Rock off the southern tip of Ireland with a return to Plymouth, England (Figure 9C). On August 12, the second day of that year's Fastnet race, a low-pressure system approached the British Isles, but the storm's winds were not excessive and it did not appear to pose a threat to the race at the time.

By August 13, however, the storm intensified and its pressure dropped rapidly. As atmospheric pressure

Continued...

Figure 9C **Route of the Fastnet Race off England.**

In the Fastnet Race, yachts sail from the Isle of Wight, around Fastnet Rock just off Ireland, and back to Plymouth, England. The track of the 1979 storm is also shown.

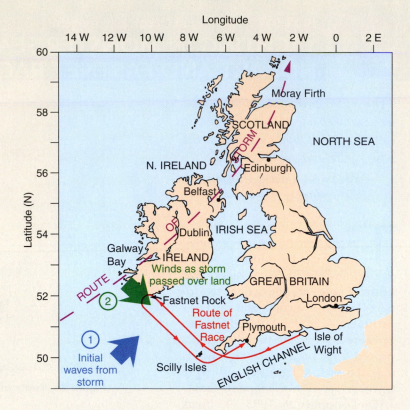

drops, winds increase because the lower the pressure, the greater the pressure gradient force, which drives winds. With increased wind speed, waves become larger, too. Although the yachts in the lead had already rounded Fastnet Rock and now had the wind at their backs, most of the fleet was struggling into west-south-west winds and rising seas.

On August 14, the storm center arrived in Galway Bay in central-western Ireland and the pressure decreased even further. As the trough of the storm hit the fleet, the winds shifted from west-southwest to northwest, producing wind gusts approaching 145 kilometers (90 miles) per hour and waves as high as 15 meters (49 feet). Soon thereafter, the storm center moved to Moray Firth in northeastern Scotland, and the seas subsided dramatically.

Analysis of the oceanographic conditions present during the race revealed that the right-angle change in the winds as the storm crossed over the British Isles was a major factor in the tragedy. As the crests of incoming waves from the northwest merged with the crests of waves created by west-southwest winds, *constructive interference* produced very short, steep rogue waves (Figure 9D). These compact waves caused many of the yachts to be rolled on their sides so far that their masts tipped into the water—an extremely vulnerable position. As the next wave struck the ship's keel, the yacht would have a tendency to roll upside down in the water and even continue through for a complete roll! Many of the yachts rolled in this way, and some yachts reported being rolled a number of times. Eyewitnesses recalled how some of the yachts crashed into others as they were tossed and rolled in the large waves.

During the 1979 Fastnet Race, only 85 out of the 303 yachts made it back to the finish line. In all, 23 yachts were sunk or abandoned, 114 people were rescued, and 15 died, making the 1979 Fastnet Race the worst disaster in the history of yachting. Although weather forecasters were criticized for not giving sufficient warning to allow the yachts to seek safety, they did not have adequate advance knowledge that conditions would become so severe. Even with adequate warnings, it is unlikely that many yacht captains would have chosen to withdraw from the race considering the competitive nature of yachting enthusiasts. In fact, Allan Green of the Royal Ocean Racing Club, which organized the event, said, "The lessons of the Fastnet [disaster] should be studied carefully and applied sensibly but in the knowledge that they can never expel the danger from yachting … it will be a sad and bad day when the seafaring people declines the challenge of the ocean."

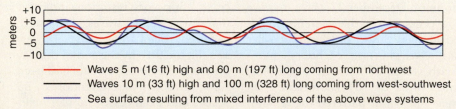

Figure 9D **Wave interference creates monster waves.**

The large waves created during the 1979 Fastnet Race were caused by mixed interference between west-southwest waves (*black line*) and northwest waves (*red line*), both of which were produced by a storm that swept through the area. The maximum amount of constructive interference produced 15-meter (49.2-foot) waves (*blue line*).

Surf

Most waves generated in the sea area by storm winds move across the ocean as swell. These waves then release their energy along the margins of continents in the **surf zone**, which is the zone of breaking waves. Breaking waves exemplify power and persistence, and sometimes they have the ability to move objects weighing several tons. In doing so, energy from a distant storm can travel thousands of kilometers until it is finally expended along a distant shoreline in a few wild moments.

As deep-water waves of swell move toward continental margins over gradually **shoaling** (*shold* = shallow) water, they eventually encounter water depths that are less than one-half of their wavelength (Figure 9-17) and become transitional waves. Actually, any shallowly submerged obstacle (such as a coral reef, sunken wreck, or sand bar) will cause waves to release some energy. Navigators have long known that breaking waves indicate dangerously shallow water.

Many physical changes occur to a wave as it encounters shallow water, becomes a shallow-water wave, and breaks. The shoaling depths interfere with water particle movement at the base of the wave, so the *wave speed decreases*. As one wave slows, the following waveform, which is still moving at its original speed, moves closer to the wave that is being slowed, causing a *decrease in wavelength*. Although some wave energy is lost due to friction, the wave energy that remains must go somewhere, so *wave height increases*. This increase in wave height combined with the decrease in wavelength causes an *increase in wave steepness (H/L)*. When the wave steepness reaches the 1:7 ratio, the waves break as surf (Figure 9-17).

If the surf is swell that has traveled from distant storms, breakers will develop relatively near shore in shallow water. The horizontal motion characteristic of shallow-water waves moves water alternately toward and away from the shore as an oscillation. The surf will be characterized by parallel lines of relatively uniform breakers.

If the surf consists of waves generated by local winds, the waves may not have been sorted into swell. The surf may be mostly unstable, deep-water, high-energy waves with steepness already near the 1:7 ratio. In this case, the waves will break shortly after feeling bottom some distance from shore, and the surf will be rough, choppy, and irregular.

When the water depth is about one and one-third times the wave height, the crest of the wave breaks, producing surf.[5] When the water depth becomes less than $\frac{1}{20}$ the wavelength, waves in the surf zone begin to behave as shallow-water waves (see Figure 9-6). Particle motion is greatly impeded by the bottom, and a significant transport of water toward the shoreline occurs (Figure 9-17).

Waves break in the surf zone because particle motion near the bottom of the wave is severely restricted, slowing the waveform. At the surface, however, individual orbiting water particles have not yet been slowed because they have no contact with the bottom. In addition, the wave height increases in shallow water. The difference in speed between the top and bottom parts of the wave cause the top part of the wave to overrun the lower part, which results in the wave toppling over and breaking. Breaking waves are analogous to a person who leans too far forward. If you don't catch yourself, you may also "break" something when you fall.

---

[5]This is a handy way of estimating water depth in the surf zone: The depth of the water where waves are breaking is one and one-third times the breaker height.

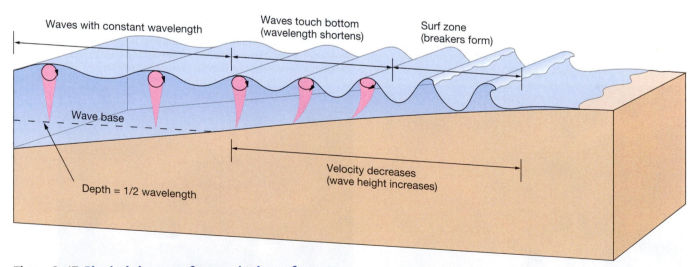

Figure 9–17 Physical changes of a wave in the surf zone.

As waves approach the shore and encounter water depths of less than one-half wavelength, the waves "feel bottom." The *wave speed decreases* and waves stack up against the shore, causing the *wavelength to decrease*. This results in an *increase in wave height* to the point where the *wave steepness is increased* beyond the 1:7 ratio, causing the wave to pitch forward and break in the surf zone.

As waves come into shallow water and feel bottom, their speed and wavelength decrease while their wave height and wave steepness increase, causing the wave to break.

Breakers and Surfing Figure 9-18a shows a **spilling breaker**. Spilling breakers result from a gently sloped ocean bottom, which extracts energy from the wave more gradually, producing a turbulent mass of air and water that runs down the front slope of the wave instead of producing a spectacular cresting curl. Spilling breakers have a longer life span and give surfers a long—but somewhat less exciting—ride than other breakers.

Figure 9-18b shows a **plunging breaker**, which has a curling crest that moves over an air pocket. The curling crest occurs because the particles in the crest literally outrun the wave, and there is nothing beneath them to support their motion. Plunging breakers form on moderately steep beach slopes, and are the best waves for surfing.

(a)

(b)

STUDENTS SOMETIMES ASK ...
Why is surfing so much better along the west coast of the United States than along the east coast?

There are three main reasons why the west coast has better surfing conditions:

• The waves are generally bigger in the Pacific. The Pacific is larger than the Atlantic, so the fetch is larger, allowing bigger waves to develop in the Pacific.

• The beach slopes are generally steeper along the west coast. Along the east coast, the gentle slopes often create spilling breakers, which are not as favorable for surfing. The steeper beach slopes along the west coast cause plunging breakers, which are better for surfing.

• The wind is more favorable. Most of the United States is influenced by the prevailing westerlies, which blow toward shore and enhance waves along the west coast. Along the east coast, the wind blows away from shore.

When the ocean bottom has an abrupt slope, the wave energy is compressed into a shorter distance and the wave will surge forward, creating a **surging breaker** (Figure 9-18c). These waves build up and break right at the shoreline, so board surfers tend to avoid them. For body surfers, however, these waves present the greatest challenge.

Surfing is analogous to riding a gravity-operated water sled by balancing the forces of gravity and buoyancy. The particle motion of ocean waves (see Figure 9-3b) shows that water particles move up into the front of the crest. This force, along with the buoyancy of the surfboard, helps maintain a surfer's position in front of a

Figure 9–18 Types of breakers.

(a) Spilling breaker, resulting from a gradual beach slope. (b) Plunging breaker at Oahu, Hawaii, resulting from a steep beach slope. (c) Surging breaker, resulting from an abrupt beach slope.

(c)

breaking wave. The trick is to perfectly balance the force of gravity (directed downward) with the buoyant force (directed perpendicular to the wave face) to enable a surfer to be propelled forward by the wave's energy. A skillful surfer, by positioning the board properly on the wave front, can regulate the degree to which the propelling gravitational forces exceed the buoyancy forces, and speeds up to 40 kilometers (25 miles) per hour can be obtained while moving along the face of a breaking wave. When the wave passes over water that is too shallow to allow the upward movement of water particles to continue, the ride is over.

Wave Refraction

Waves seldom approach a shore at a perfect right angle (90 degrees). Therefore, as a wave approaches at a slight angle to the shore, some segment of the wave will "feel bottom" first and will slow before the rest of the wave. This results in **wave refraction** (*refringere* = to break up) or the *bending* of each wave crest (also called a wave front) as the waves approach the shore.

Figure 9-19a shows how waves coming toward a straight shoreline are refracted and tend to align themselves *nearly* parallel to the shore. This explains why all

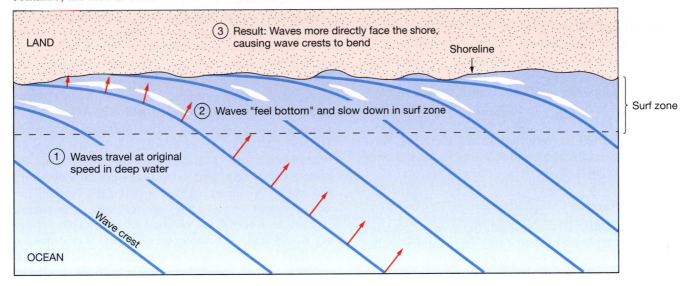

(a)

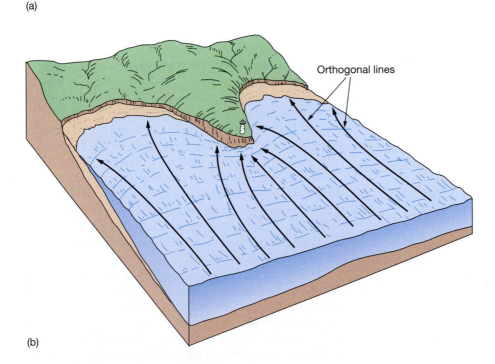

(b)

Figure 9–19 Wave refraction.

(a) Wave refraction along a straight shoreline. Waves approaching the shore at an angle first "feel bottom" close to shore. This causes the segment of the wave in shallow water to slow, causing the crest of the wave to refract or bend so that the waves arrive at the shore nearly parallel to the shoreline. Red arrows represent direction and speed of the wave. (b) Wave refraction along an irregular shoreline. As waves first "feel bottom" in the shallows off the headlands, they are slowed, causing the waves to refract and align nearly parallel to the shoreline. Evenly spaced orthogonal lines (*black arrows*) show that wave energy is concentrated on headlands (causing erosion) and dispersed in bays (resulting in deposition).

waves come almost straight in toward a beach, no matter what their original orientation was.

Figure 9-19b shows how waves coming toward an irregular shoreline refract so that they, too, nearly align with the shore. However, the refraction of waves along an irregular shoreline distributes wave energy unevenly along the shore.

The long black arrows in Figure 9-19b are called **orthogonal** (*ortho* = straight, *gonia* = angle) **lines** or *wave rays*. Orthogonal lines are drawn perpendicular to the wave fronts (so they indicate the direction that waves travel) and are spaced so that the energy between lines is equal at all times. They help show how energy is distributed along the shoreline by breaking waves.

The orthogonals in Figure 9-19b are equally spaced far from shore. As they approach the shore, however, the orthogonals *converge* on headlands that jut into the ocean, and *diverge* in bays. This means that wave energy is concentrated against the headlands, but dispersed in bays. The result is heavy erosion of headlands and deposition of sediment in bays. The greater energy of waves breaking on headlands is reflected in an increased wave height.[6] Conversely, the smaller waves in bays provide areas for good boat anchorages.

Wave Diffraction

Wave diffraction (*dis* = apart, *frangere* = to break) results from wave energy being transferred around or behind barriers that impede the wave's forward motion. Diffraction occurs because any point on a wave front is a source from which energy can propagate in all directions. For example, as waves move past a barrier at the entrance to a harbor, wave diffraction causes some wave energy to move laterally along the wave crest, thereby producing diffracted waves that move into the harbor (Figure 9-20).

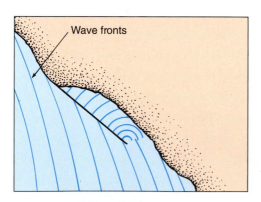

Figure 9–20 Wave diffraction in a harbor.

As waves move past a barrier such as a harbor breakwater, wave diffraction causes wave energy to be transferred behind the barrier. By diffraction, wave energy can spread to even the most protected areas within a harbor.

---

[6]Sailors have long known that "the points draw the waves." Surfers also know how wave refraction causes good "point breaks."

In this way, at least some wave energy can spread to even the most protected areas within a harbor.

Wave Reflection

Not all wave energy is expended as waves rush onto the shore. A vertical barrier, such as a seawall or a rock ledge, can reflect waves back into the ocean with little loss of energy—a process called **wave reflection** (*reflecten* = to bend back), which is similar to how a mirror reflects (bounces) back light. If the incoming wave strikes the barrier at a right (90-degree) angle, the wave energy is reflected back parallel to the incoming wave, often interfering with the next incoming wave and creating unusual waveforms. More commonly, waves approach the shore at an angle, causing wave energy to be reflected at an angle equal to the angle at which the wave approached the barrier.

An outstanding example of wave reflection occurs in an area called "The Wedge," which develops west of the jetty that protects the harbor entrance at Newport Harbor, California (Figure 9-21). The jetty is a solid human-made object that extends into the ocean 400 meters (1300 feet) and has a near-vertical side facing the waves. As incoming waves strike the vertical side of the jetty at an angle, they are reflected at an equivalent angle. Because the original waves and the reflected waves have the same wavelength, a constructive interference pattern develops, creating plunging breakers that may exceed 8 meters (26 feet) in height (Figure 9-21, *inset*). Too dangerous for board surfers, these waves present a fierce challenge to the most experienced body surfers. The Wedge has crippled or even killed many who have come to try it.

Standing waves (or *stationary waves*) can be produced when waves are reflected at right angles to a barrier. Standing waves are the sum of two waves with the same wavelength moving in opposite directions, resulting in no net movement. Although the water particles continue to move vertically and horizontally, there is none of the circular motion that is characteristic of a progressive wave.

Figure 9-22 shows the movement of water during the wave cycle of a standing wave. Lines along which there is no vertical movement are called *nodes* (*node* = knot), or nodal lines. *Antinodes*, crests that alternately become troughs, are the points of greatest vertical movement within a standing wave.

There is no particle motion when an antinode is at its greatest vertical displacement, and the maximum particle movement occurs when the water surface is level. At this time, the maximum movement of the water is in a horizontal direction directly beneath the nodal lines. The movement of water particles beneath the antinodes is entirely vertical.

We consider standing waves further when we discuss tidal phenomena in Chapter 10, "Tides." Under certain conditions, the development of standing waves significantly affects the tidal character of coastal regions.

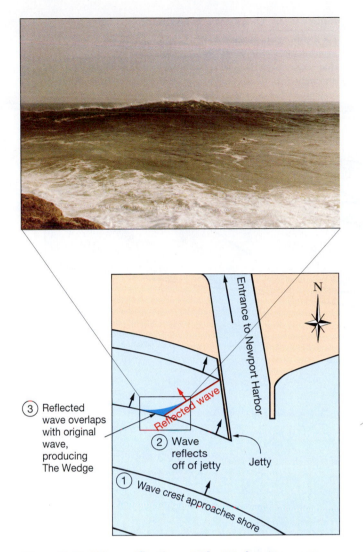

Figure 9–21 Wave reflection at The Wedge, Newport Harbor, California.

As waves approach the shore (1), some of the wave energy is reflected off the long jetty at the entrance to the harbor (2). The reflected wave overlaps and constructively interferes with the original wave (3), resulting in a wedge-shaped wave (*dark blue triangle*) that may reach heights exceeding 8 meters (26 feet). Photo of The Wedge (*inset*) shows three dots in front of the wave that are the heads of body surfers.

Tsunami

The Japanese term for the large, sometimes destructive waves that occasionally roll into their harbors is **tsunami** (*tsu* = harbor, *nami* = wave). Tsunami originate from sudden changes in the topography of the sea floor caused by slippage along underwater faults, underwater avalanches, or underwater volcanic eruptions. Many people mistakenly call them "tidal waves," but tsunami are unrelated to the tides. The mechanisms that trigger tsunami are typically seismic events, so tsunami are *seismic sea waves*.

The majority of tsunami are caused by *fault movement*. Underwater fault movement displaces Earth's crust, generates earthquakes, and, if it ruptures the sea floor, produces a sudden change in water level at the ocean surface (Figure 9-23a). Faults that produce *vertical* displacements (the uplift or downdropping of ocean floor) change the volume of the ocean basin, which affects the entire water column and generates tsunami. Conversely, faults that produce *horizontal* displacements (such as the lateral movement associated with transform faulting) generally do not generate tsunami because the side-to-side movement of these faults does not change the volume of the ocean basin. Much less common events, such as underwater avalanches triggered by shaking, meteorite impacts, or underwater volcanic eruptions—which create the largest waves—also produce tsunami.

The wavelength of a typical tsunami exceeds 200 kilometers (125 miles). In the open ocean, tsunami move at well over 700 kilometers (435 miles) per hour—they could easily keep pace with a jet airplane—and have heights of only about 0.5 meter (1.6 feet). Even though they are fast, tsunami are small in the open ocean and pass unnoticed in deep water until they reach shore, where they slow in the shallow water, and the water begins to pile up.

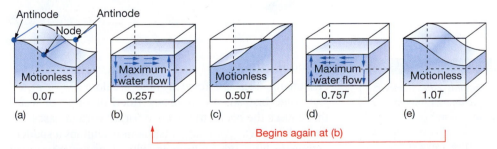

Figure 9–22 Sequence of motion in a standing wave.

In a standing wave, water is motionless when antinodes reach maximum displacement (*a, c,* and *e; a* and *e* are identical). Water movement is at a maximum (*blue arrows*) when the water is horizontal (*b* and *d*). Movement is vertical beneath the antinodes, and maximum horizontal movement occurs beneath the node. After *e,* cycle begins again at *b.*

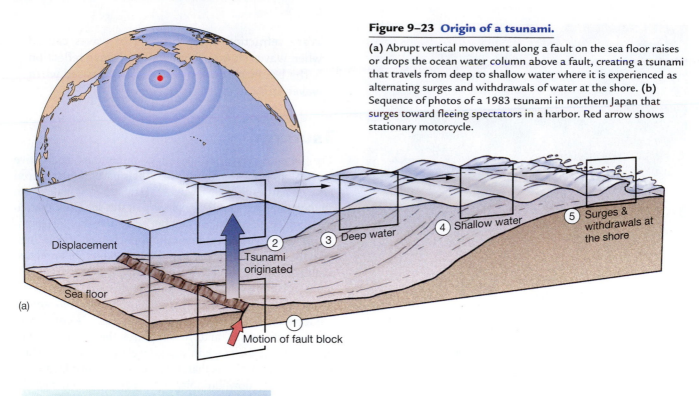

Figure 9–23 Origin of a tsunami.

(a) Abrupt vertical movement along a fault on the sea floor raises or drops the ocean water column above a fault, creating a tsunami that travels from deep to shallow water where it is experienced as alternating surges and withdrawals of water at the shore. **(b)** Sequence of photos of a 1983 tsunami in northern Japan that surges toward fleeing spectators in a harbor. Red arrow shows stationary motorcycle.

EXAMPLE 9–6

Given the typical wavelength of a tsunami is 200 kilometers (125 miles) and the deepest depth of the oceans is 11 kilometers (7 miles), can a tsunami ever be a deep-water wave?

Recall that the depth of the wave base is one-half a wave's wavelength. For a tsunami with a wavelength of 200 kilometers, its wave base would be 200 ÷ 2 = 100 kilometers (62 miles), which is deeper than even the deepest ocean trenches. Remarkably, tsunami are shallow-water waves everywhere in the ocean! Because they are shallow-water dwaves, remember that a tsunami's speed is determined *only* by water depth.

Coastal Effects

A tsunami does not form a huge breaking wave at the shoreline. Instead, it is a strong flood or surge of water that causes the ocean to advance (or, in certain cases, retreat) dramatically. In fact, a tsunami resembles a sudden, *extremely* high tide, which is why they are misnamed "tidal waves." It takes several minutes for the tsunami to express itself fully, during which time sea level can rise up to 40 meters (131 feet) above normal, with normal waves superimposed on top of the higher sea level. The strong surge of water can rush into low-lying areas with destructive results (Figure 9-23b).

As the trough of the tsunami arrives at the shore, the water will rapidly drain off the land. In coastal areas, it will look like a sudden and *extremely* low tide, where sea level is many meters lower than even the lowest low tide. Because tsunami are typically a series of waves, there are often an alternating series of dramatic surges and withdrawals of water, separated by only a few minutes. The first surge may not always be the largest; the third, fourth, or even seventh surge may be, instead.

In some cases, the trough of a tsunami arrives at the coast first, exposing parts of the lowermost shoreline that are rarely seen. For people at the shoreline, the temptation is to explore these newly exposed areas and catch stranded fish. Within a few minutes, however, a strong surge of water (the crest of the tsunami) is due to arrive.

The alternating surges and retreats of water by tsunami can severely damage coastal structures. Tsunami can be deadly as well. The speed of the advance—up to 4 meters (13 feet) per second—is faster than a person can run. Those who are trapped by tsunami are often drowned or crushed by floating debris (Figure 9-24).

Historic Tsunami

Many small tsunami are created each year, and go largely unnoticed. On average, 57 tsunami occur every decade, with a large tsunami occurring somewhere in the world every two to three years and an extremely large and damaging one occurring every 15 to 20 years. About 86% of all great waves are generated in the Pacific Ocean because large-magnitude earthquakes occur along the series of trenches that ring its ocean basin where oceanic plates are subducted along convergent plate boundaries. Volcanic activity is also common along the Pacific "Ring of Fire," and the large earthquakes that occur along its margin are capable of producing extremely large tsunami.

One of the most destructive tsunami ever generated came from the eruption of the volcanic island of Krakatau[7] on August 27, 1883. Approximately the size of a small Hawaiian Island in what is now Indonesia, Krakatau exploded with the greatest release of energy from Earth's interior observed in historic times. The island was nearly obliterated and the sound of the explosion was heard up to 4800 kilometers (2980 miles) away. Dust from the explosion ascended into the atmosphere and circled Earth on high-altitude winds, producing unusual and beautiful sunsets for nearly a year.

Not many were killed by the outright explosion of the volcano because the island was uninhabited. However, the displacement of water from the energy released during the explosion was enormous, creating a tsunami that exceeded 35 meters (116 feet)—as high as a 12-story building. It devastated the coastal region of the Sunda Strait between the nearby islands of

[7]The volcanic island Krakatau (which is *west* of Java) is also called Krakatoa.

Figure 9–24 Tsunami damage in Hilo, Hawaii.
Flattened parking meters in Hilo, Hawaii, caused by the 1946 tsunami that resulted in more than $25 million in damage and 159 deaths.

Sumatra and Java, drowning over 1000 villages and taking more than 36,000 lives. The energy carried by this wave reached every ocean basin and was detected by tide recording stations as far away as London and San Francisco.

Several ships were along the coast of Java during the eruption of Krakatau, containing eyewitnesses to the tsunami and its destruction. N. van Sandick, an engineer aboard the Dutch vessel *Loudon*, gave the following account:

> Suddenly we saw a gigantic wave of prodigious height advancing from the sea-shore with considerable speed. Immediately the crew set to under considerable pressure and managed after a fashion to set sail in face of the imminent danger; the ship had just enough time to meet with the wave from the front. After a moment, full of anguish, we were lifted up with a dizzy rapidity. The ship made a formidable leap, and immediately afterwards we felt as though we had plunged into the abyss. But the ship's blade went higher and we were safe. Like a high mountain, the monstrous wave precipitated its journey towards the land. Immediately afterwards another three waves of colossal size appeared. And before our eyes this terrifying upheaval of the sea, in a sweeping transit, consumed in one instant the ruin of the town; the lighthouse fell in one piece, and all the houses of the town were swept away in one blow like a castle of cards. All was finished. There, where a few moments ago lived the town of Telok Betong, was nothing but the open sea.

Another strong tsunami was experienced in the port of Hilo, Hawaii, on April 1, 1946. The tsunami was from a magnitude $M_w = 7.3$ earthquake in the Aleutian Trench off the island of Unimak, Alaska, over 3000 kilometers (1850 miles) away. The bathymetry in horseshoe-shaped Hilo Bay tends to focus a tsunami's energy directly toward town, building up waves to tremendous heights. In this case, the tsunami expressed

BOX 9–2 Historical Feature

THE BIGGEST WAVE IN RECORDED HISTORY: LITUYA BAY, ALASKA (1958)

Lituya Bay is located in southeast Alaska about 200 kilometers (125 miles) west of Juneau, Alaska's capital. It is a deep, T-shaped, 11-kilometer (7-mile)-long bay with a sand bar named La Chaussee Spit that separates it from the Pacific Ocean (Figure 9E). The largest wave ever authentically recorded occurred in Lituya Bay. Remarkably, the wave was witnessed by six people on board three small fishing boats that were near the bay's entrance (Figure 9E, *top*).

Figure 9E Lituya Bay, Alaska, with aerial view before the 1958 splash wave (*top*) and after (*bottom*).

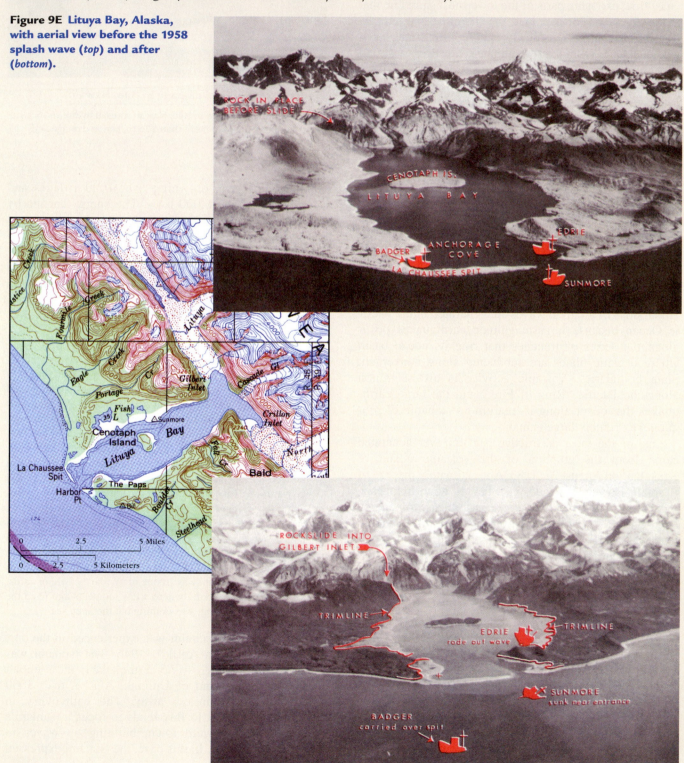

At about 10:00 P.M. on July 9, 1958, an earthquake of magnitude $M_w = 7.9$* occurred along the Fairweather Fault, which runs along the top of the "T" portion of the bay. The earthquake didn't produce the wave directly, but it triggered an enormous rockslide that dumped at least 90 million tons of rock—some of it from as high as 914 meters (3000 feet) above sea level—into the upper part of the bay. The rockslide created a huge **splash wave** (a long-wavelength wave produced when an object splashes into water) that swept over the ridge facing the rockslide area and uprooted, debarked, or snapped off trees up to 530 meters (1740 feet) above the water level of the bay—a full 87 meters (285 feet) *higher* than the world's tallest building, the Sears Tower in Chicago. As the giant wave raced down the bay toward the boats at a speed of over 160 kilometers (100 miles) per hour, it continued to snap off trees and completely overtopped the island in the middle of the bay.

During the summer in Alaska, it was still light enough at 10:00 P.M. for the people on board the boats

*The symbol M_w indicates the *moment magnitude* of an earthquake, as discussed in Chapter 3.

to see the rockslide occur—and the giant wave bearing down on them. The *Badger*, a 13.4-meter (44-foot) fishing vessel, had its anchor chain snapped and was lifted up bow-first into the oncoming wave. Amazingly, the vessel surfed the wave over the sand bar! The two people on board reported looking down from a height of 24 meters (80 feet) above the tops of the trees on the sand bar, in an area where trees reach heights of 30 meters (100 feet). The *Badger* plunged into the Pacific Ocean on the other side of the sand bar stern-first, where it foundered and eventually sank. The people on board were able to launch a small skiff before the *Badger* sank and were rescued a few hours later, shaken but alive.

The *Edrie* was at anchor in the bay when the wave arrived. Its anchor chain snapped, and the vessel (including two people on board) was washed onto land. After the wave passed, the withdrawal of water washed it back into the bay, leaving the vessel largely undamaged. The two people on board the *Sunmore* were not nearly so lucky. The wave hit their vessel broadside, which capsized and sank the *Sunmore*, killing both people on board. The wave spread out into the Pacific and was even detected over six hours later at

a tide-recording station in Hawaii, where the wave was only 10 centimeters (4 inches) tall.

The most noticeable damage to the shoreline of the bay included a trimline of trees that extended around the bay and across the island (Figure 9E, *bottom*). The wave also knocked down all the trees on the sand bar and killed most of the shellfish living near the water's edge. Additionally, floating logs from the destruction filled Lituya Bay for many years.

Older knocked-down trees suggest that Lituya Bay periodically experiences rockslides that generate giant splash waves. For instance, there is evidence of a 120-meter (395-foot) wave in 1853, a 61-meter (200-foot) wave in 1899, and a 150-meter (490-foot) wave on October 27, 1936. Even though other events may have produced larger waves (such as the 914-meter (3000-foot) wave created by a meteorite impact in the Gulf of Mexico about 65 million years ago†), the 1958 splash wave in Lituya Bay stands as the largest wave in recorded history.

†See Box 5-3 in Chapter 5 for a description of the Cretaceous–Tertiary (K–T) impact event.

itself as a strong recession followed by a surge of water nearly 17 meters (55 feet) above normal high tide, causing more than $25 million in damage and killing 159 people. Remarkably, it stands as Hawaii's worst natural disaster (Figure 9-24).

Closer to the source of the earthquake, the tsunami was considerably larger. The tsunami struck Scotch Cap, Alaska, on Unimak Island, where a two-story reinforced concrete lighthouse stood 14 meters (46 feet) above sea level at its base. The lighthouse was destroyed by a wave that is estimated to have reached 36 meters (118 feet), killing all five people inside the lighthouse at the time. Vehicles on a nearby mesa 31 meters (103 feet) above water level were also moved by the onrush of water.

STUDENTS SOMETIMES ASK...

What is the record height of a tsunami?

Japan holds the record because Japan's proximity to subduction zones causes it to endure more tsunami than any other place on Earth (followed by Chile and Hawaii). The largest documented tsunami occurred in the Ryukyu Islands of southern Japan in 1971, when one raised normal sea level by 85 meters (278 feet). In low-lying coastal areas, such an enormous vertical rise can send water many kilometers inland, causing flooding and widespread damage. The most deadly tsunami was probably the one that hit Aura, Japan, in 1703 and was responsible for an estimated 100,000 deaths.

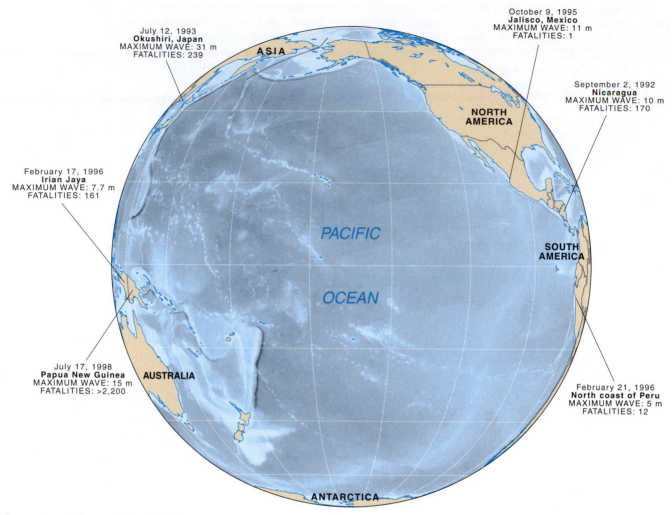

Figure 9–25 Tsunami since 1990.

Ten destructive tsunami have claimed more than 4000 lives since 1990. These killer waves are most often generated by earthquakes along colliding tectonic plates of the Pacific Rim, although the deadly 1998 Papua New Guinea tsunami that killed more than 2200 was generated by an underwater landslide.

Figure 9-25 shows that since 1990, ten destructive tsunami along the Pacific Ring of Fire have claimed more than 4000 lives. Of these tsunami, the one that caused the greatest number of casualties occurred in Papua New Guinea in July 1998. An offshore magnitude $M_w = 7.1$ earthquake was followed shortly thereafter by a 15-meter (49-foot) tsunami, which was up to five times larger than expected for a quake that size. The tsunami completely overtopped a heavily populated low-lying sand bar, destroying three entire villages and resulting in at least 2200 deaths. Researchers who mapped the sea floor after the tsunami discovered the remains of a huge underwater landslide, which was apparently triggered by the shaking and generated the deadliest tsunami in 65 years.

Tsunami Warning System

In response to the tsunami that struck Hawaii in 1946, a tsunami warning system was established throughout the Pacific Ocean. It led to what is now the **Pacific Tsunami Warning Center (PTWC)**, which coordinates information from 25 Pacific Rim countries and is headquartered in Ewa Beach (near Honolulu), Hawaii. In the open ocean, tsunami have small wave heights and are difficult to detect, so the tsunami warning system uses seismic waves—some of which travel through Earth at speeds 15 times faster than tsunami—to forecast destructive tsunami.[8] When a seismic disturbance occurs beneath the ocean surface that is large enough to be tsunamigenic (capable of producing a tsunami), a *tsunami watch* is issued. At this point, a tsunami may or may not have been generated, but the potential for one exists.

The PTWC is linked to over 50 tide-measuring stations throughout the Pacific, so the recording station

---

[8]A new method of tracking tsunami that is currently being tested uses a series of sensitive pressure sensors on the ocean floor that can detect the passage of tsunami in the open ocean.

nearest the earthquake is closely monitored for any indication of unusual wave activity. If unusual wave activity is verified, the tsunami watch is upgraded to a *tsunami warning*. Generally, earthquakes smaller than magnitude $M_w = 6.5$ are not tsunamigenic because they lack the duration of ground shaking necessary to initiate a tsunami. Additionally, transform faults do not usually produce tsunami because lateral movement does not offset the ocean floor and impart energy to the water column in the same way that vertical fault movements do.

? STUDENTS SOMETIMES ASK ...

If there is a tsunami warning issued, what is the best thing to do?

The *smartest* thing to do is to stay out of coastal areas, but people often want to see the tsunami firsthand. For instance, when an earthquake of magnitude $M_w = 7.7$ occurred offshore of Alaska in May 1986, a tsunami warning was issued for the west coast of the United States. In southern California, people flocked to the beach to observe the phenomenon. Fortunately, the tsunami was only a few centimeters high by the time it reached southern California, so it went unnoticed.

If you must go to the beach to observe a tsunami, expect crowds, road closings, and general mayhem. It would be a good idea to stay at least 30 meters (100 feet) above sea level. If you happen to be at a remote beach where the water suddenly withdraws, evacuate immediately to higher ground (Figure 9-26). And, if you happen to be at a beach where an earthquake occurs and shakes the ground so hard that you can't stand up, then *RUN*—don't walk—for high ground as soon as you *can* stand up!

After the first surge of the tsunami, stay out of low-lying coastal areas for several hours because several more surges (and withdrawals) can be expected. There are many documented cases where curious people have been killed when they are trapped by the third or fourth surge of a tsunami.

Once a tsunami is detected, warnings are sent to all the coastal regions that might encounter the destructive wave, along with its estimated time of arrival. This warning, usually just a few hours in advance of the tsunami, makes it possible to evacuate people from low-lying areas and remove ships from harbors before the waves arrive. If the disturbance is nearby, however, there is not enough time to issue a warning because a tsunami travels so rapidly. Unlike hurricanes, whose high winds and waves threaten ships at sea and send them to the protection of a coastal harbor, a tsunami washes ships from their coastal moorings into the open ocean or onto shore. Thus, the best strategy during a tsunami warning is to move ships out of coastal harbors and into deep water, where tsunami are not easily felt.

Since the PTWC was established in 1948, it has effectively prevented loss of life due to tsunami when people have heeded the evacuation warnings. Property damage, however, has increased as more buildings have been constructed close to shore. To combat the damage caused by tsunami, countries that are especially prone to tsunami like Japan have invested in shoreline barriers, seawalls, and other coastal fortifications.

Perhaps one of the best strategies to limit tsunami damage and loss of life is to restrict construction projects in low-lying coastal regions where tsunami have frequently struck in the past. However, the long time interval between large tsunami can lead people to forget past disasters.

> Most tsunami are generated by underwater fault movement, which transfers energy to the entire water column. When these fast and long waves surge ashore, they can do considerable damage.

Figure 9–26 Tsunami warning sign.

This tsunami warning sign in Oregon advises residents to evacuate low-lying areas during a tsunami.

Chapter in Review

- *All ocean waves begin as disturbances caused by releases of energy.* The releases of energy include wind, the movement of fluids of different densities (which create internal waves), mass movement into the ocean, underwater sea floor movements, the gravitational pull of the Moon and the Sun on Earth, and human activities in the ocean.

- Once initiated, *waves transmit energy through matter by oscillatory motion* in the particles that make up the matter. Progressive waves are longitudinal, transverse, or orbital, depending on the pattern of particle oscillation. Particles in ocean waves move primarily in orbital paths.

- *Waves are described according to their wavelength (L), wave height (H), wave steepness (H/L), wave period (T), frequency (f), and wave speed (S).* As a wave travels, the water passes the energy along by moving in a circle, called *circular orbital motion*. This motion advances the waveform, not the water particles themselves. Circular orbital motion decreases with depth, ceasing entirely at wave base, which is equal to one-half the wavelength measured from still water level.

- If water depth is greater than one-half the wavelength, a progressive wave travels as a *deep-water wave with a speed that is directly proportional to wavelength.* If water depth is less than $\frac{1}{20}$ wavelength (L/20), the wave moves as a *shallow-water wave with a speed that is directly proportional to water depth.* Transitional waves have wavelengths between deep- and shallow-water waves, with speeds that depend on both wavelength and water depth.

- As wind-generated waves form in a sea area, *capillary waves* with rounded crests and wavelengths less than 1.74 centimeters (0.7 inch) *form first. As the energy of the waves increases, gravity waves form,* with increased wave speed, wavelength, and wave height. Factors that influence the size of wind-generated waves include wind speed, duration (time), and fetch (distance). An equilibrium condition called a *fully developed sea* is reached when the maximum wave height is achieved for a particular wind speed, duration, and fetch.

- *Energy is transmitted from the sea area across the ocean by uniform, symmetrical waves called swell.*

- Different wave trains of swell can create either constructive, destructive, or mixed interference patterns. Constructive interference produces unusually large waves called rogue waves or superwaves.

- *As waves approach shoaling water near shore, they undergo many physical changes.* Waves release their energy in the surf zone when their steepness exceeds a 1:7 ratio and break. If waves break on a relatively flat surface, they produce spilling breakers. The curling crests of plunging breakers, which are the best for surfing, form on steep slopes and abrupt beach slopes create surging breakers.

- When swell approaches the shore, *segments of the waves that first encounter shallow water are slowed,* whereas other parts unaffected by shallow water move ahead, *causing the wave to refract, or bend.* Refraction concentrates wave energy on headlands, whereas low-energy breakers are characteristically found in bays.

- *Reflection of waves off seawalls or other barriers* can cause an interference pattern called a standing wave. The crests of standing waves do not move laterally as in progressive waves but alternate with troughs at antinodes. Between the antinodes are nodes, where there is no vertical movement of the water.

- *Sudden changes in the elevation of the sea floor, such as from fault movement or volcanic eruptions, generate tsunami, or seismic sea waves.* These waves often have lengths exceeding 200 kilometers (125 miles) and travel across the open ocean with undetectable heights of about 0.5 meter (1.6 feet) at speeds in excess of 700 kilometers (435 miles) per hour. Upon approaching shore, a tsunami produces a series of rapid withdrawals and surges, some of which may increase the height of sea level by 40 meters (131 feet) or more. Most tsunami occur in the Pacific Ocean, where they have caused millions of dollars of coastal damage and taken tens of thousands of lives. The Pacific Tsunami Warning Center (PTWC) has dramatically reduced fatalities by successfully predicting tsunami using real-time seismic information.

Key Terms

Atmospheric wave (p. 265)
Capillary wave (p. 272)
Circular orbital motion (p. 268)
Constructive interference (p. 276)
Crest (p. 268)
Decay distance (p. 276)
Deep-water wave (p. 269)
Destructive interference (p. 277)
Disturbing force (p. 265)
Free wave (p. 277)
Forced wave (p. 277)
Frequency (p. 268)
Fully developed sea (p. 275)

Gravity wave (p. 271)
Interference pattern (p. 276)
Internal wave (p. 265)
Longitudinal wave (p. 267)
Mixed interference (p. 277)
Ocean wave (p. 265)
Orbital wave (p. 268)
Orthogonal line (p. 284)
Pacific Tsunami Warning Center (PTWC) (p. 290)
Plunging breaker (p. 282)
Restoring force (p. 272)
Rogue wave (p. 278)

Sea (p. 272)
Shallow-water wave (p. 271)
Shoaling (p. 281)
Splash wave (p. 289)
Spilling breaker (p. 282)
Standing wave (p. 284)
Still water level (p. 268)
Surf zone (p. 281)
Surging breaker (p. 282)
Swell (p. 276)
Transitional wave (p. 271)
Transverse wave (p. 267)
Trough (p. 268)

Tsunami (p. 285)
Wave base (p. 269)
Wave diffraction (p. 284)
Wave dispersion (p. 276)
Wave height (p. 268)
Wave period (p. 268)
Wave reflection (p. 284)
Wave refraction (p. 283)
Wave speed (p. 269)
Wave steepness (p. 268)
Wave train (p. 276)
Wavelength (p. 268)

Questions and Exercises

1. Discuss several different ways in which waves form. How are most ocean waves generated?

2. Why is the development of internal waves likely within the pycnocline?

3. Discuss longitudinal, transverse, and orbital wave phenomena, including the states of matter in which each can transmit energy.

4. Draw a diagram of a simple progressive wave. From memory, label the crest, trough, wavelength, wave height, and still water level.

5. Can a wave with a wavelength of 14 meters ever be more than 2 meters high? Why or why not?

6. What physical feature of a wave is related to the depth of the wave base? On the diagram that you drew for Question 4, add the wave base. What is the difference between the wave base and still water level?

7. Explain why the following statements for deep-water waves are either true or false:
   a. The longer the wave, the deeper the wave base.
   b. The greater the wave height, the deeper the wave base.
   c. The longer the wave, the faster the wave travels.
   d. The greater the wave height, the faster the wave travels.
   e. The faster the wave, the greater the wave height.

8. Calculate the speed (S) in meters per second for deep-water waves with the following characteristics:
   a. $L = 351$ meters, $T = 15$ seconds
   b. $T = 12$ seconds
   c. $f = 0.125$ wave per second

9. Define *swell*. Does swell necessarily imply a particular wave size? Why or why not?

10. Waves from separate sea areas move away as swell and produce an interference pattern when they come together. If Sea A has wave heights of 1.5 meters (5 feet) and Sea B has wave heights of 3.5 meters (11.5 feet), what would be the height of waves resulting from constructive interference and destructive interference? Illustrate your answer (see Figure 9-14).

11. Describe the physical changes that occur to a wave's wave speed (S), wavelength (L), height (H), and wave steepness (H/L) as a wave moves across shoaling water to break on the shore.

12. Describe the three different types of breakers and indicate the slope of the beach that produces the three types. How is the energy of the wave distributed differently within the surf zone by the three types of breakers?

13. Using examples, explain how wave refraction is different from wave reflection.

14. Using orthogonal lines, illustrate how wave energy is distributed along a shoreline with headlands and bays. Identify areas of high and low energy release.

15. Define the terms *node* and *antinode* as they relate to standing waves.

16. Why is it more likely that a tsunami will be generated by faults beneath the ocean along which vertical rather than horizontal movement has occurred?

17. While shopping in a surf shop, you overhear some surfing enthusiasts mention that they would really like to ride the curling wave of a tidal wave at least once in their life, because it is a single breaking wave of enormous height. What would you say to these surfers?

18. Explain what it would look like at the shoreline if the trough of a tsunami arrives there first. What is the impending danger?

19. What ocean depth would be required for a tsunami with a wavelength of 220 kilometers (136 miles) to travel as a deep-water wave? Is it possible that such a wave could become a deep-water wave any place in the world ocean? Explain.

20. Explain how the tsunami warning system in the Pacific Ocean works. Why must the tsunami be verified at the closest tide recording station?

CHAPTER 10
Tides

Extreme tidal variation. High and low tides at Hall's Harbor in Nova Scotia, Canada, demonstrate the dramatic change of sea level experienced daily in the Bay of Fundy, which has the world's largest tidal range.

Key Questions

- What causes the tides?
- How is a lunar day different from a solar day?
- Which body creates a larger tidal influence on Earth: the Moon or the Sun?
- How do the relative positions of the Earth–Moon–Sun affect the tidal range on Earth?
- What are differences between diurnal, semidiurnal, and mixed tidal patterns?
- Where is the world's largest tidal range?
- What types of tidal currents exist?

The answers to these questions (and much more) can be found in the highlighted concept statements within this chapter

"I derive from the celestial phenomena the forces of gravity with which bodies tend to the sun and several planets. Then from these forces, by other propositions which are also mathematical, I deduce the motions of the planets, the comets, the moon, and the sea."

—Sir Isaac Newton, *Philosophiae Naturalis Principia Mathematica (Philosophy of Natural Mathematical Principles)*
(1686)

Tides are the periodic raising and lowering of average sea level that occurs throughout the oceans of the world. As sea level rises and falls, the edge of the sea slowly shifts landward and seaward daily, often destroying sand castles built during low tide. Knowledge of tides is important in many coastal activities, including tide pooling, shell collecting, surfing, fishing, navigation, and preparing for storms. Tides are so important that accurate records have been kept at nearly every port for several centuries and there are many examples of the term *tide* in everyday vocabulary (for instance, "to tide someone over," "to go against the tide," or to wish someone "good tidings").

People have undoubtedly observed the tides for as long as they have inhabited coastal regions. However, no written record of tides exists before Herodotus' observations of the Mediterranean Sea in about 450 B.C. Even the earliest sailors knew the Moon had some connection with the tides because both followed a similar cyclic pattern. However, it wasn't until **Isaac Newton** (1642–1727) developed the *universal law of gravitation* that the tides could be adequately explained.

Although the study of the tides can be complex, tides are fundamentally very long and regular shallow-water waves. Their wavelengths are measured in thousands of kilometers and their heights range to more than 15 meters (50 feet). The gravitational attraction of the Sun and Moon generate ocean tides, thereby affecting every particle of water from the surface to the deepest ocean basin.

Generating Tides

Fundamentally, tides are generated by forces imposed on Earth that are generated by a combination of *gravity* and *motion* among Earth, the Moon, and the Sun.

Tide-Generating Forces

Newton's work on quantifying the forces involved in the Earth–Moon–Sun system led to the first understanding of why tides behave as they do. It is well known that gravity tethers the Sun, its planets, and their moons together. Most of us are taught that "the Moon orbits Earth," but it is not quite that simple. The two bodies actually rotate around a common center of mass called the **barycenter** (*barus* = heavy, *center* = center), which is located 1600 kilometers (1000 miles) beneath Earth's surface (Figure 10-1a). This can be visualized by imagining Earth and its Moon as ends of a sledgehammer, flung into space, tumbling slowly end over end about its balance point, which is closest to the hammer. The barycenter follows a smooth orbit around the Sun, while Earth and the Moon themselves follow wavy paths (Figure 10-2). Moreover, the Earth–Moon system is involved in a mutual orbit held together by gravity and motion, which prevents the Moon and Earth from colliding. In this way, orbits are established that keep objects at more-or-less fixed distances. Gravity also tugs every particle of water on Earth toward the Moon and the Sun, thus creating tides on Earth.

Gravitational and Centripetal Forces in the Earth–Moon System   To understand how *tide-generating forces* influence the oceans, let's examine how *gravitational forces* and *centripetal forces* affect objects on Earth within the Earth–Moon system. (We'll ignore the influence of the Sun for the moment.)

The **gravitational force** is derived from Newton's law of universal gravitation, which states that *every particle of mass in the universe attracts every other particle of mass*. Mathematically, gravitational force is expressed as

$$\text{Gravitational force} = \frac{Gm_1 m_2}{r^2} \qquad (10\text{-}1)$$

where G is the universal gravitational constant, m_1 and m_2 are two masses, and r is the distance between the two masses. Note that for spherical bodies, all of the mass can be considered to exist at the center of the sphere, and thus r will always be the distance between the centers of bodies being considered.

Equation (10-1) has several implications. For example, it explains why objects with a large mass (such as the Sun) produce a large gravitational force. This is because as mass increases, the gravitational force increases. Also, it shows that gravitational force varies with the square of distance, so even a small *increase* in the distance between two objects significantly *decreases* the gravitational force between them. Thus, the *greater* the mass of the objects

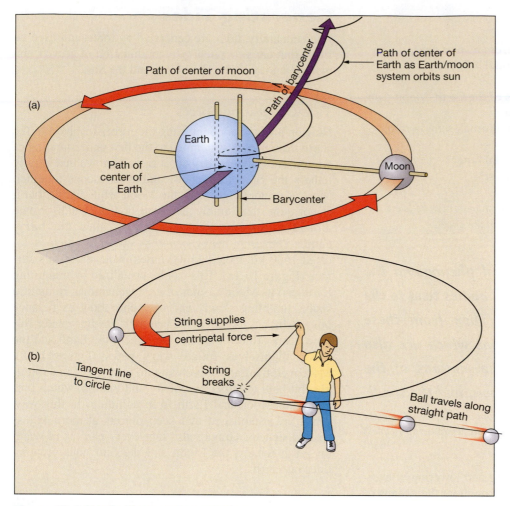

(a)

(b)

Figure 10–1 Earth–Moon system rotation.

(a) The center of mass (barycenter) of the Earth–Moon system moves in a nearly circular orbit around the Sun. **(b)** If a ball with a string attached is swung overhead, it stays in a circular orbit because the string exerts a centripetal (center-seeking) force on the ball. If the string breaks, the ball will fly off along a straight path along a tangent to the circle.

and the *closer* they are together, the greater their gravitational attraction.

Figure 10-3 shows how gravitational forces for points on Earth (caused by the Moon) vary depending on their distances from the Moon. The greatest gravitational attraction (the longest arrow) is at Z, the *zenith* (*zenith* = a path over the head), which is the point closest to the Moon. The gravitational attraction is weakest at N, the *nadir* (*nadir* = opposite the zenith), which is the point farthest from the Moon. The direction of the gravitational attraction between most particles and the center of the Moon is at an angle relative to a line connecting the center of Earth and the Moon (Figure 10-3). This angle causes the force of gravitational attraction between each particle and the Moon to be slightly different.

The **centripetal** (*centri* = the center, *pet* = seeking) **force**[1] required to keep planets in their orbits is provided by the gravitational attraction between each of them and the Sun. Centripetal force "tethers" an orbiting body to its parent, pulling the object *inward* toward the parent, "seek-

ing the center" of its orbit. For example, if you tie a string to a ball and swing the tethered ball around your head (Figure 10-1b), the string pulls the ball toward your hand. The string provides a *centripetal force* on the ball, forcing the ball to *seek the center* of its orbit. If the string should break, the force is gone and the ball can no longer maintain its circular orbit. The ball flies off in a *straight* line,[2] *tangent* (*tangent* = touching) to the circle (Figure 10-1b).

The Earth and Moon are tethered, too, not by strings but by gravity. Gravity provides the centripetal force that holds the Moon and Earth in a mutual orbit. If all gravity in the solar system could be shut off, centripetal force

---

[1]This is not to be confused with the so-called *centrifugal* (*centri* = the center, *fug* = flee) *force*, an apparent force that is oriented outward.

[2]At the moment that the string breaks, the ball will continue along a straight-line path, obeying Newton's first law of motion (the law of inertia), which states that moving objects follow straight-line paths until they are compelled to change that path by other forces.

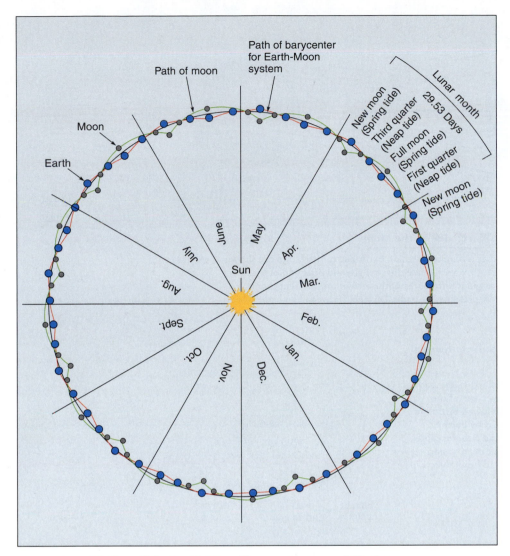

Figure 10–2 Motion of the Earth–Moon barycenter around the Sun.

As the barycenter (center of mass) of the Earth–Moon system orbits the Sun in a nearly circular path (*black line*), both the Moon (*green line*) and Earth (*red line*) follow separate wavy paths as they rotate about the barycenter, which is located 1600 kilometers (1000 miles) beneath Earth's surface. The phases of the Moon and the resulting tidal conditions are explained later in this chapter.

would vanish, and the momentum of the celestial bodies would send them flying off into space along straight-line paths, tangent to their orbits.

As the Earth and Moon rotate around their common barycenter, all particles that make up Earth follow circles of equal radii (Figure 10-4a). If Earth is divided into a great number of particles of equal mass, the centripetal force required to keep each particle of Earth following an identical orbit is the same (Figure 10-4b). The required centripetal force for each particle is supplied by its gravitational attraction to the Moon. In essence, the centripetal force required for all particles is identical and is directed toward the center of each particle's orbit (Figure 10-5).

Resultant and Tide-Generating Forces Gravitational attraction between the particle and the Moon supplies the centripetal force, but the *supplied* force is different than

the *required* force (because gravitational attraction varies with distance from the Moon) except at the center of Earth. This difference creates tiny **resultant forces**, which are the mathematical difference between the two sets of arrows shown in Figures 10-3 and 10-5.

Figure 10-6 combines Figures 10-3 and 10-5 to show that resultant forces are produced by the difference between the required centripetal (C) and supplied gravitational (G) forces. However, do not think that both of these forces are being applied to the points, because (C) is a force that would be required to keep the particles in a perfectly circular path, while (G) is the force actually provided for this purpose by gravitational attraction between the particles and the Moon. The resultant forces (*blue arrows*) are established by constructing an arrow from the tip of the centripetal (*red*) arrow to the tip of the gravity (*black*) arrow and located where the red and black arrows begin.

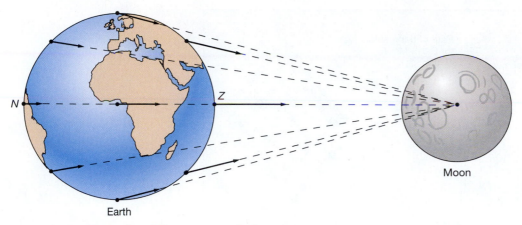

Earth

Figure 10–3 Gravitational forces on Earth due to the Moon.

The gravitational forces on objects located at different places on Earth due to the Moon are shown by arrows. The length and orientation of the arrows indicate the strength and direction of the gravitational force. Notice the length and angular differences of the arrows for different points on Earth. The letter *Z* represents the zenith; *N* represents the nadir. Distance between Earth and Moon not shown to scale.

Figure 10–4 Earth–Moon rotation and centripetal (center-seeking) forces.

(a) The dashed line through the center of Earth is the path of Earth's center as it moves around the barycenter of the Earth–Moon system. The circular paths followed by points *a* and *b* have the same radius as that followed by Earth's center. **(b)** Arrows from points *a, b, c, d,* and *e* to the center of their circular orbits show that the same direction and magnitude of centripetal force is required to hold objects in their orbital paths.

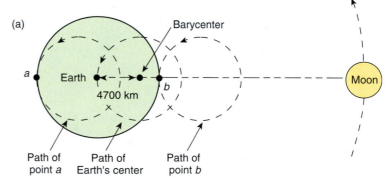

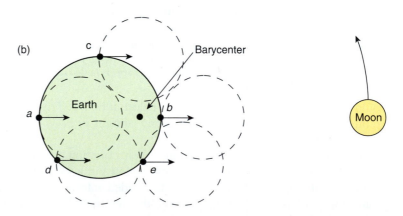

Resultant forces are small, averaging about one-millionth the magnitude of Earth's gravity. If the resultant force is vertical to Earth's surface, as it is at the zenith and nadir (oriented upward) and along an "equator" connecting all points halfway between the zenith and nadir (oriented downward), it has no tide-generating effect (Figure 10-7). However, if the resultant force has a significant *horizontal component*—that is, tangent to Earth's surface—it produces tidal bulges on Earth, creating what are known as the **tide-generating forces**. These tide-gen-

erating forces are quite small but reach their maximum value at points on Earth's surface at a "latitude" of 45 degrees relative to the "equator" between the zenith and nadir (Figure 10-7).

The tide-generating force is the difference between the gravitational force of the tide-generating object acting on a mass at the Earth's surface and at the Earth's center. Although the tide-generating force is derived from the gravitational force [Equation (10-1)], it is not linearly proportional to it. The tide-generating force can be written as

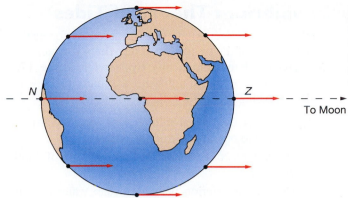

To Moon

Figure 10–5 Required centripetal (center-seeking) forces.

Centripetal forces required to keep identical-sized particles in identical-sized orbits as a result of the rotation of the Earth–Moon system around its barycenter. As in Figure 10-4, notice that the arrows are all the same length, and are oriented in the same direction for all points on Earth. Z = zenith; N = nadir.

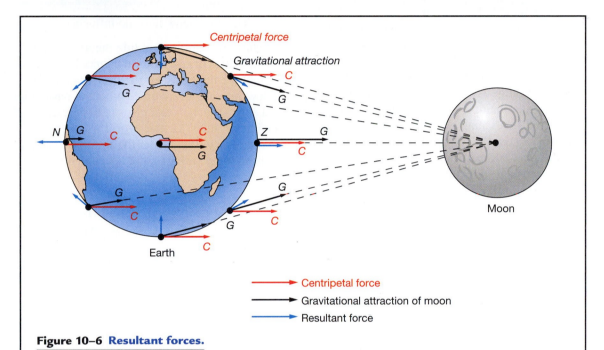

Centripetal force

Gravitational attraction

Moon

Earth

→ Centripetal force

→ Gravitational attraction of moon

→ Resultant force

Figure 10–6 Resultant forces.

Red arrows indicate centripetal forces (C), which are not equal to the black arrows that indicate gravitational attraction (G). The small blue arrows show resultant forces, which are established by constructing an arrow from the tip of the centripetal (*red*) arrow to the tip of the gravity (*black*) arrow and located where the red and black arrows begin. Z = zenith; N = nadir. Distance between Earth and Moon not shown to scale.

Figure 10–7 Tide-generating forces.

Where the resultant force acts vertically relative to Earth's surface, the tide-generating force is zero. This occurs at the zenith (Z) and nadir (N), and along an "equator" connecting all points halfway between the zenith and nadir (*black dots*). However, where the resultant force has a significant *horizontal component*, it produces a tide-generating force on Earth. These tide-generating forces reach their maximum value at points on Earth's surface at a "latitude" of 45 degrees relative to the "equator" mentioned here (*blue arrows*). Distance between Earth and Moon not shown to scale.

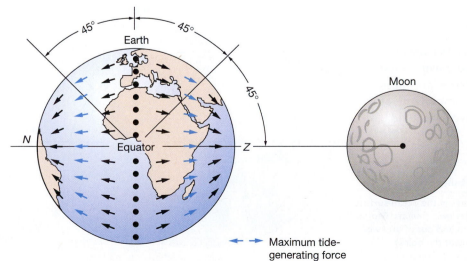

← → Maximum tide-generating force

$$\text{Tide-generating force} \propto \frac{m_1 m_2}{r^3} \qquad (10\text{-}2)$$

where the $\propto$ symbol means "proportional to." Equation (10-2) shows that the tide-generating force varies inversely as the *cube* of the distance from the center of Earth to the center of the tide-generating object (instead of varying inversely as the *square* of the distance as does the gravitational force). In the tide-generating equation, therefore, distance is a more highly weighted variable. So, the greater the distance from Earth the tide-generating body (Moon or Sun) is, the smaller the tide-generating force will be. This is why the Moon influences tides far more than the Sun (even though the Sun is much more massive).

The tide-generating forces push water into two bulges: one on the side of Earth directed *toward* the Moon (the zenith) and the other on the side directed *away from* the Moon (the nadir) (Figure 10-8). On the side directly facing the Moon, the bulge is created because the gravitational force is greater than the required centripetal force. Conversely, on the side facing away from the Moon, the bulge is created because the required centripetal force is greater than the gravitational force. Although the forces are oriented in opposite directions on the two sides of Earth, the resultant forces are equal in magnitude, so the bulges are equal, too.

? STUDENTS SOMETIMES ASK...
Are there also tides in other objects, such as lakes and swimming pools?

The Moon and the Sun act on all objects that have the ability to flow, so there are tides in lakes, wells, and swimming pools. In fact, there are even extremely tiny tidal bulges in a glass of water! However, the tides in the atmosphere and the "solid" Earth have greater significance. Tides in the atmosphere—called *atmospheric tides*—can be miles high. The tides inside Earth's interior—called *Earth tides*—cause a slight but measurable stretching of Earth's crust, typically only a few centimeters high.

> The tides are caused by an imbalance between the required centripetal and the provided gravitational forces acting on Earth. This difference produces residual forces, the horizontal component of which pushes ocean water into two equal tidal bulges on opposite sides of Earth.

Equilibrium Theory of Tides

In the preceding discussion, we considered the forces that form the basis of the **equilibrium tide theory**, which was first developed mathematically by Newton in the 17th century. Some of the simplifying assumptions Newton made to develop this theory include the following:

1. Earth has two equal tidal bulges, one toward the Moon and one away from the Moon.
2. The oceans cover the entire Earth and are of a uniform depth.
3. There is no friction between ocean water and the sea floor.
4. The continents have no influence.

Because the equilibrium tide theory ignores some of the complexities of real tides, it cannot be used to accurately predict the tides at specific locations on Earth. However, it does provide an adequate model of basic tidal phenomena and, as such, it can be used to predict the general behavior of tides in the world's oceans. Later in this chapter, we will consider the *dynamic tide theory*, which addresses the variables not accounted for by the equilibrium tide theory.

Tidal Bulges: The Moon's Effect

In the equilibrium tide theory, the ideal Earth has two tidal bulges, one toward the Moon and one away from the Moon (called the **lunar bulges**) as shown in Figure 10-8. If the Moon is stationary and aligned with the ideal Earth's Equator, the maximum bulge will occur on the Equator on opposite sides of Earth. If you were standing on the Equator, you would experience two high tides each day. The time between high tides, the **tidal period**, would be 12 hours. If you moved to any latitude north or south of the Equator, you would experience the same tidal period, but the high tides would be less high, because you would be at a lower point on the bulge.

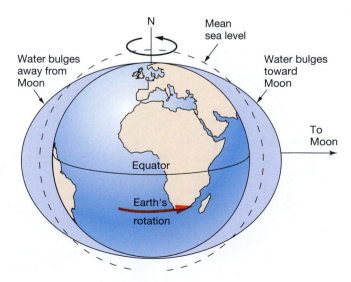

Figure 10–8 Idealized equilibrium tidal bulges.

In an idealized case, the Moon creates two bulges in the ocean surface: One that extends *toward* the Moon and the other *away from* the Moon. As Earth rotates, it carries various locations into and out of the two tidal bulges so that all points on its surface (except the poles) experience two high tides daily.

In most places on Earth, however, high tides occur every 12 hours 25 minutes because tides depend on the lunar day, not the solar day. The **lunar day** (also called a *tidal day*) is measured from the time the Moon is on the meridian of an observer—that is, directly overhead—to the next time the Moon is on that meridian, and is 24 hours 50 minutes.[3] The **solar day** is measured from the time the Sun is on the meridian of an observer to the next time the Sun is on that meridian, and is 24 hours. Why is the lunar day 50 minutes longer than the solar day? During the 24 hours it takes Earth to make a full rotation, the Moon has continued moving another 12.2 degrees to the east in its orbit around Earth (Figure 10-9). Thus, Earth must rotate an additional 50 minutes to "catch up" to the Moon so that the Moon is again on the meridian (directly overhead) of our observer.

The difference between a solar day and a lunar day can be seen in some of the natural phenomena related to the tides. For example, alternating high tides are normally 50

---

[3]A lunar day is exactly 24 hours, 50 minutes, 28 seconds long.

minutes *later* each successive day and the Moon rises 50 minutes *later* each successive night.

A solar day (24 hours) is shorter than a lunar day (24 hours and 50 minutes). The extra 50 minutes is caused by the Moon's movement in its orbit around Earth.

Tidal Bulges: The Sun's Effect

The Sun affects the tides, too. Like the Moon, the Sun produces tidal bulges on opposite sides of Earth, one oriented toward the Sun and one oriented away from the Sun.

Even though the Sun is 27 million times more massive than the Moon, its tide-generating force is not 27 million times greater than the Moon's. This is because the Sun is 390 times farther from Earth than the Moon (Figure 10-10). Recall from Equation (10-2) that tide-generating forces vary inversely as the *cube* of the distance between objects. Thus, the tide-generating force is reduced by the cube of 390, or about 59 million times

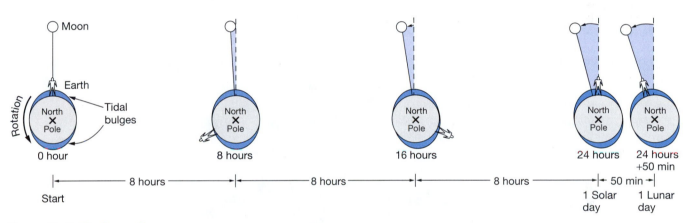

Figure 10–9 The lunar day.

A lunar day is the time that elapses between when the Moon is directly overhead and the next time the Moon is directly overhead. During one complete rotation of Earth (the 24-hour solar day), the Moon moves eastward 12.2 degrees, and Earth must rotate an additional 50 minutes to place the Moon in the exact same position overhead. Thus, a lunar day is 24 hours 50 minutes long.

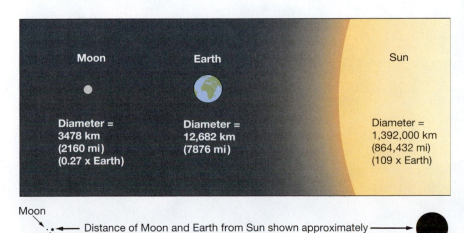

Moon
Earth
Distance of Moon and Earth from Sun shown approximately to scale
Sun

Figure 10–10 Relative sizes and distances of the Moon, Earth, and Sun.

Top: The relative sizes of the Moon, Earth, and Sun, showing the diameter of the Moon is roughly one-fourth that of Earth, while the diameter of the Sun is 109 times the diameter of Earth. *Bottom:* The relative distances of the Moon, Earth, and Sun are shown to scale.

Moon
Diameter = 3478 km (2160 mi) (0.27 x Earth)

Earth
Diameter = 12,682 km (7876 mi)

Sun
Diameter = 1,392,000 km (864,432 mi) (109 x Earth)

compared with that of the Moon. These conditions result in the Sun's tide-generating force being $\frac{27}{59}$ that of the Moon, or 46% (about one-half). As a result, the **solar bulges** are only 46% the size of the lunar bulges.

Earth's Rotation

The tides appear to move water in toward shore (the **flood tide**) and to move water away from shore (the **ebb tide**). However, according to the nature of the idealized tides presented so far, *Earth's rotation carries various locations into and out of the tidal bulges*, which are in fixed positions relative to the Moon and the Sun. In essence, alternating high and low tides are created as Earth constantly rotates inside fluid bulges that are supported by the Moon and the Sun.

Figure 10–11 Earth–Moon–Sun positions and the tides.

Top: When the Moon is in the new or full position, the tidal bulges created by the Sun and Moon are aligned, there is a large tidal range on Earth, and spring tides are experienced. *Bottom:* When the Moon is in the first- or third-quarter position, the tidal bulges produced by the Moon are at right angles to the bulges created by the Sun. Tidal ranges are smaller and neap tides are experienced.

> The lunar bulges are about twice the size of the solar bulges. In an idealized case, the rise and fall of the tides are caused by Earth's rotation carrying various locations into and out of the tidal bulges.

The Monthly Tidal Cycle

The monthly tidal cycle is $29\frac{1}{2}$ days because that's how long it takes the Moon to complete an orbit around Earth.[4] During this time, the phase of the Moon changes dramatically. When the Moon is between Earth and the Sun, it cannot be seen at night, and it is called the **new moon**. When the Moon is on the side of Earth opposite the Sun, its entire disk is brightly visible, and it is called a **full moon**. A **quarter moon**—a moon that is half lit and half dark as viewed from Earth—occurs when the Moon is at right angles to the Sun relative to Earth.

Figure 10-11 shows the positions of the Earth, Moon, and Sun at various points during the $29\frac{1}{2}$-day lunar cycle

---

[4]The $29\frac{1}{2}$-day monthly tidal cycle is also called a *lunar cycle*, a *lunar month*, or a *synodic* (*synodos* = meeting) *month*.

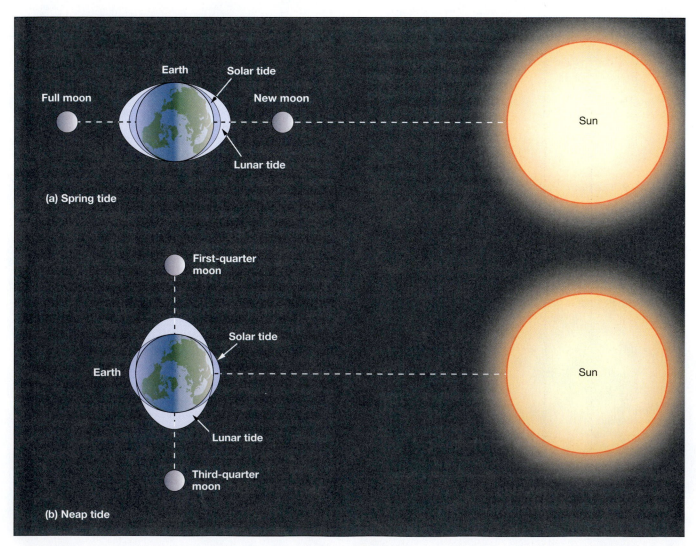

(a) Spring tide

(b) Neap tide

(see also Figure 10-2). When the Sun and Moon are aligned, either with the Moon between Earth and the Sun (new moon; Moon in *conjunction*) or with the Moon on the side opposite the Sun (full moon; Moon in *opposition*), the tide-generating forces of the Sun and Moon combine (Figure 10-11, *top*). At this time, the **tidal range** (the vertical difference between high and low tides) is large (very *high* high tides and quite *low* low tides) because there is *constructive interference*[5] between the lunar and solar tidal bulges (Figure 10-12a). The maximum tidal range is called a **spring** (*springen* = to rise up) **tide**,[6] because the tide is extremely large or "springs forth." When the Earth–Moon–Sun system is aligned, the Moon is said to be in **syzygy** (*syzygia* = union).

When the Moon is in either the first- or third-quarter[7] phase (Figure 10-11, *bottom*), the tide-generating force of the Sun is working at right angles to the tide-generating force of the Moon. The tidal range is small (*lower* high tides and *higher* low tides) because there is destructive interference[8] between the lunar and solar tidal bulges (Figure 10-12b). This is called a **neap** (*nep* = scarcely or barely touching) **tide**,[9] and the Moon is said to be in **quadrature** (*quadra* = four).

The time between successive spring tides (full moon and new moon) or neap tides (first quarter and third quarter) is one-half the monthly lunar cycle, which is about two weeks. The time between a spring tide and a successive neap tide is one-quarter the monthly lunar cycle, which is about one week.

---

[5]As mentioned in Chapter 9, constructive interference occurs when two waves (or, in this case, two tidal bulges) overlap crest to crest and trough to trough.

[6]Spring tides have no connection with the spring season; they occur twice a month during the time when the Earth–Moon–Sun system is aligned.

[7]The third-quarter moon is often called the last-quarter moon, which is not to be confused with certain sports that have a fourth quarter.

[8]Destructive interference occurs when two waves (or, in this case, two tidal bulges) match up crest to trough and trough to crest.

[9]To help you remember a *neap* tide, think of it as one that has been "*nipped* in the bud," indicating a small tidal range.

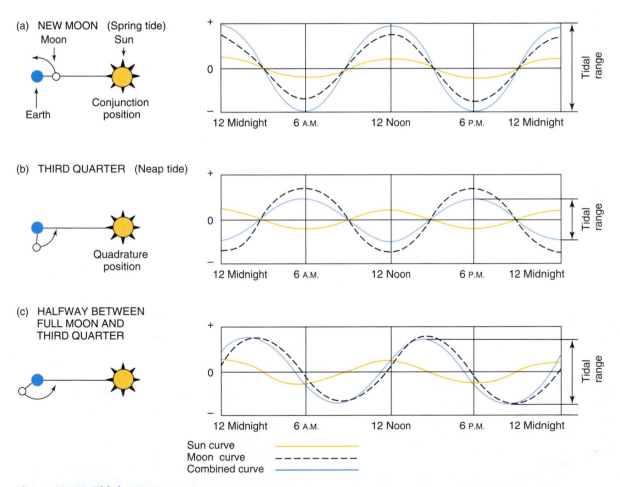

Figure 10–12 Tidal curves.

Tides experienced during various positions of the Earth–Moon–Sun system. Tidal curve graphs show a combined curve (*blue line*), which is produced by constructive and destructive interference between the Sun curve (*yellow line*) and the Moon curve (*black dashed line*). **(a)** New moon (spring tide) shows a maximum tidal range. **(b)** Third-quarter moon (neap tide) shows a minimum tidal range. **(c)** Halfway between full moon and third quarter (waning gibbous phase) shows a tidal range halfway between that of spring and neap tides.

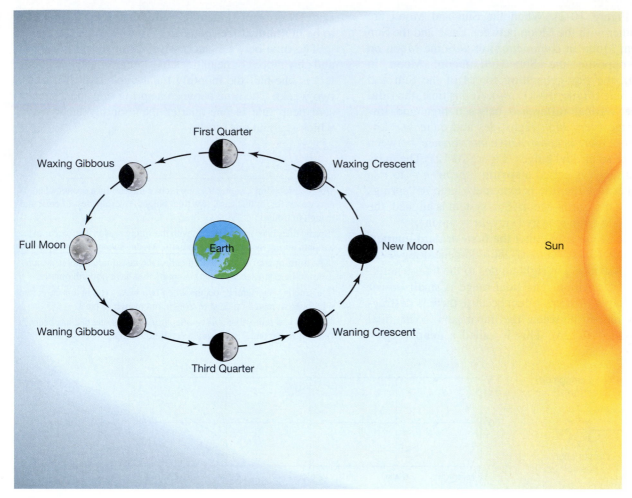

Figure 10–13 Phases of the Moon.

As the Moon moves around Earth during its 29$\frac{1}{2}$-day lunar cycle, its phase changes depending on its position relative to the Sun and Earth. During a new moon, the dark side of the Moon faces Earth while during a full moon, the lit side of the Moon faces Earth. Moon phases are shown diagrammatically as seen from Earth.

Figure 10-13 shows the pattern that the Moon experiences as it moves through its monthly cycle. As the Moon progresses from new moon to first-quarter phase, the Moon is a **waxing crescent** (*waxen* = to increase; *crescere* = to grow). In between the first-quarter and full moon phase, the Moon is a **waxing gibbous** (*gibbus* = hump). Between the Moon's full and third-quarter phase, it is a **waning gibbous** (*wanen* = to decrease). And, in between the third-quarter and new moon phase, the Moon is a **waning crescent**. The Moon has identical periods of rotation on its axis and revolution around Earth (a property called *synchronous rotation*). As a result, the same side of the Moon always faces Earth.

Figure 10-12c shows the tide conditions on Earth during a waning gibbous moon but is also representative of any situation where the Moon is halfway between sygygy and quadrature. The resulting mixed interference pattern shows that the tidal range is less than that of spring tides but greater than that of neap tides.

STUDENTS SOMETIMES ASK ...
I've heard of a blue moon. Is the Moon really blue then?

No. "Once in a blue moon" is just a phrase that has gained popularity and is synonymous with a rather unlikely occurrence. A blue moon is the second full moon of any calendar month, which occurs when the 29$\frac{1}{2}$-day lunar cycle falls entirely within a 30- or 31-day month. Because the divisions between our calendar months were determined arbitrarily, a blue moon has no special significance, other than it occurs only once every 2.72 years (about 33 months). At that rate, it's certainly less common than a month of Sundays!

The origin of the term *blue moon* is not exactly known, but it probably has nothing to do with color (although large forest fires or volcanic eruptions can put enough pollution in the atmosphere to cause the Moon to appear blue). One likely explanation involves the Old English word *belewe*, meaning "to betray." Thus, the Moon is *belewe* because it betrays the usual perception of one full moon per month. Another explanation links the term to the *Farmer's Almanac*, which was first published in color in 1938 and included a calendar designating the first full moon of each month in red color and the second full moon in blue.

Spring tides occur during the full and new moon when the lunar and solar tidal bulges constructively interfere, producing a large tidal range. Neap tides occur during the quarter moon phases when the lunar and solar tidal bulges destructively interfere, producing a small tidal range.

Declination of the Moon and Sun

Up to this point, we have assumed that the Moon and Sun have remained directly overhead at the Equator, but this is not usually the case. Most of the year, in fact, they are either north or south of the Equator. The angular distance of the Sun or Moon above or below Earth's equatorial plane is called **declination** (*declinare* = to turn away).

Earth revolves around the Sun along an invisible ellipse in space. The imaginary plane that contains this ellipse is called the **ecliptic** (*ekleipein* = to fail to appear) (Figure 10-14, *yellow plane*). Recall from Chapter 7 that Earth's axis of rotation is tilted 23.5 degrees with respect to the ecliptic and that this tilt causes Earth's seasons. It also means the maximum declination of the Sun relative to Earth's Equator is 23.5 degrees (Figure 10-14a). Be-

cause of this tilt, the Sun's declination varies between 23.5 degrees north and 23.5 degrees south of the Equator on a yearly cycle.

To complicate matters further, the plane of the Moon's orbit is tilted 5 degrees with respect to the ecliptic (Figure 10-14a). Thus, the maximum declination of the Moon's orbit relative to Earth's Equator is 28.5 degrees (5 degrees plus the 23.5 degrees of Earth's tilt). In addition, the plane of the Moon's orbit also *precesses*, or rotates, while maintaining this 5 degree angle. This **precession** (*praecedere* = to go before) completes a cycle every 18.6 years. Figure 10-14 shows the relationship of the ecliptic, the plane of the Moon's orbit, and the plane of Earth's Equator through one-half a precession (9.3 years).

Meanwhile, the Moon's declination changes from 28.5 degrees south to 28.5 degrees north and back to 28.5 degrees south of the Equator during the multiple lunar cycles within one year. As a result, tidal bulges are rarely aligned with the Equator. Instead, they occur mostly north and south of the Equator. The Moon affects Earth's tides more than the Sun, so tidal bulges follow the Moon, ranging from a maximum of 28.5 degrees north to a maximum of 28.5 degrees south of the Equator (Figure 10-15).

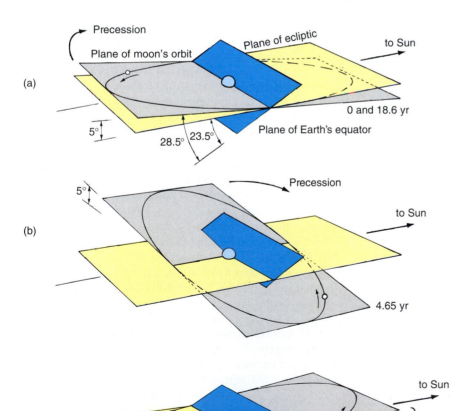

Figure 10–14 The precession of the Moon's declination.

Diagrammatic views of the Moon's orbit (*gray plane*), the plane of the ecliptic (*yellow plane*), and Earth's equatorial plane (*blue plane*). **(a)** The Moon's declination reaches a maximum of 28.5 degrees: 23.5 degrees (the angle of tilt of Earth's equatorial plane relative to the plane of the ecliptic) plus 5 degrees (the angle between the plane of the Moon's orbit and the ecliptic). **(b)** Positions of the planes 4.65 years later when the Moon has achieved one-fourth of its 18.6-year precessional rotation. **(c)** Positions of the planes after 9.3 years or one-half of the Moon's precession. Note that the maximum declination of the Moon relative to Earth's equatorial plane is now 18.5 degrees (23.5 degrees less 5 degrees).

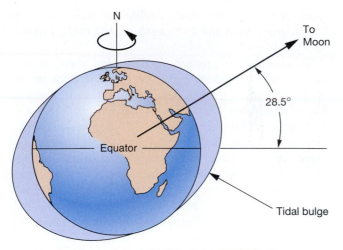

Figure 10–15 Maximum declination of tidal bulges from the Equator.

The center of the tidal bulges may lie at any latitude from the Equator to a maximum of 28.5 degrees on either side of the Equator, depending on the season of the year (solar angle) and the Moon's position.

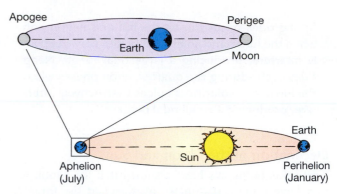

Figure 10–16 Effects of elliptical orbits.

Top: The Moon moves from its most distant point (*apogee*) to its closest point to Earth (*perigee*), which causes greater tidal ranges every 27½ days. *Bottom*: The Earth also moves from its most distant point (*aphelion*) to its closest point (*perihelion*), which causes greater tidal ranges every year in January. Diagram is not to scale.

Effects of Elliptical Orbits

Earth revolves around the Sun in an elliptical orbit (Figure 10-16) such that Earth is 148.5 million kilometers (92.2 million miles) from the Sun during the Northern Hemisphere winter and 152.2 million kilometers (94.5 million miles) from the Sun during summer. Thus, the distance between Earth and the Sun varies by 2.5% over the course of a year. Tidal ranges are largest when Earth is near its closest point, called **perihelion** (*peri* = near, *helios* = Sun) and smallest near its most distant point, called **aphelion** (*apo* = away from, *helios* = Sun). Thus, the greatest tidal ranges typically occur in January each year.

The Moon revolves around Earth in an elliptical orbit, too. The Earth–Moon distance varies by 8% [between 375,000 kilometers (233,000 miles) and 405,800 kilometers (252,000 miles)]. Tidal ranges are largest when the Moon is closest to Earth, called **perigee** (*peri* = near, *geo* = Earth), and smallest when most distant, called **apogee** (*apo* =away from, *geo* = Earth) (Figure 10-16, *top*). The Moon cycles between perigee, apogee, and back to perigee every 27½ days. When spring tides coincide with perigee, the tides—called **proxigean** (*proximus* = nearest, *geo* = Earth) or "closest of the close moon" tides—are especially large, which often result in the flooding of low-lying coastal areas during high tide. If a storm occurs during this time, damage can be extreme. For example, the most damaging winter storm along the U.S. east coast (the Ash Wednesday storm of March 5–8, 1962) occurred during a proxigean tide.

The elliptical orbits of Earth around the Sun and the Moon around Earth change the distances between Earth, the Moon, and the Sun, thus affecting Earth's tides. The net result is that spring tides have greater ranges during the Northern Hemisphere winter than in the summer, and spring tides have greater ranges when they coincide with perigee.

? STUDENTS SOMETIMES ASK ...
How often are conditions right to produce the maximum tide-generating force?

Maximum tides occur when Earth is closest to the Sun (at perihelion), the Moon is closest to Earth (at perigee), and the Earth–Moon–Sun system is aligned (at syzygy) with both the Sun and Moon at zero declination. This rare condition—which creates an absolute *maximum* spring tidal range—occurs once every 1600 years. Fortunately, the next occurrence is predicted for the year 3300.

However, there are other times when conditions produce large tide-generating forces. During early 1983, for example, large, slow-moving low-pressure cells developed in the North Pacific Ocean that caused strong northwest winds. In late January, the winds produced a near fully developed 3-meter (10-foot) swell that affected the west coast from Oregon to Baja California. The large waves would have been trouble enough under normal conditions, but there were also unusually high spring tides of 2.25 meters (7.4 feet) because Earth was near perihelion at the same time that the Moon was at perigee. In addition, a strong El Niño had raised sea level by as much as 20 centimeters (8 inches). When the waves hit the coast during these unusual conditions, they caused over $100 million in damages, including the destruction of 25 homes, damage to 3500 others, the collapse of several commercial and municipal piers, and at least a dozen deaths.

Prediction of Equilibrium Tides

In the equilibrium tide model, the declination of the Moon determines the position of the tidal bulges. The example illustrated in Figure 10-17 shows that the Moon is directly overhead at 28 degrees north latitude when its declination is 28 degrees north of the Equator. If you stand at this latitude when the Moon is directly overhead, it will be high tide (Figure 10-17a). Low tide occurs six lunar hours later (6 hours $12\frac{1}{2}$ minutes solar time) (Figure 10-17b). Another high tide, but one much lower than the first, occurs six lunar hours later (Figure 10-17c). Another low tide occurs six lunar hours later (Figure 10-17d). Six lunar hours later, at the end of a 24-lunar-hour period (24 hours 50 minutes solar time), you will have passed through a complete lunar-day cycle of two high tides and two low tides.

The graphs in Figure 10-17e show the heights of the tides observed during the same lunar day at 28 degrees north latitude, the Equator, and 28 degrees south latitude when the declination of the Moon is 28 degrees north of the Equator. Tide curves for 28 degrees north and 28 degrees south latitude have identically timed highs and lows, but the *higher* high tides and *lower* low tides occur 12 hours later. The reason that they occur out of phase by 12 hours is because the bulges in the two hemispheres are on opposite sides of Earth in relation to the Moon. Table 10-1 summarizes the characteristics of equilibrium tides on the idealized Earth.

STUDENTS SOMETIMES ASK...
What are tropical tides?

Differences between successive high tides and successive low tides occur each lunar day (see, for example, Figure 10-17e). Because these differences occur within a period of one day, they are called diurnal (daily) inequalities. These inequalities are at their greatest when the Moon is at its maximum declination, and such tides are called *tropical tides* because the Moon is over one of Earth's tropics. When the Moon is over the Equator (*equatorial tides*), the difference between successive high tides and low tides is minimal.

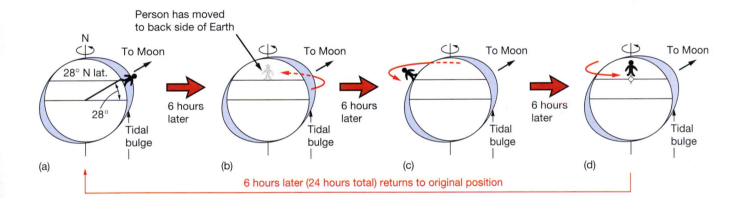

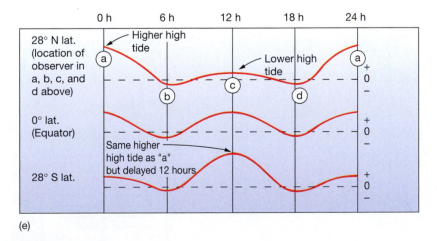

Figure 10–17 Predicted equilibrium tides.
(a)–(d) Sequence showing the tide experienced every 6 lunar hours at 28 degrees north latitude when the declination of the Moon is 28 degrees north. **(e)** Tide curves for 28 degrees north, 0 degrees, and 28 degrees south latitudes during the lunar day shown in the sequence above. The tide curves for 28 degrees north and 28 degrees south latitude show that the higher high tides occur 12 hours later.

TABLE 10–1 **Summary of characteristics of equilibrium tides on the idealized Earth.**

- Any location (except the poles) will have two high tides and two low tides per lunar day.
- Neither the two high tides nor the two low tides are of the same height because of the declination of the Moon and the Sun (except for the rare occasions when the Moon and Sun are simultaneously above the Equator).
- Monthly and yearly cycles of tidal range are related to the changing distances of the Moon and Sun from Earth.
- Each week, there would be alternating spring and neap tides. Thus, in a lunar month, there are two spring tides and two neap tides.

Dynamic Theory of Tides

The equilibrium tide theory uses the model of tidal bulges to explain the tides. Tides in the ocean, however, behave in much more complex ways than predicted by this simplistic model. The **dynamic tide theory** takes into account the factors ignored by the equilibrium tide theory and does a better job of approximating real ocean tides.

For example, if equilibrium tidal bulges are truly wave crests separated by a distance of one-half Earth's circumference—about 20,000 kilometers (12,420 miles)—one would expect the bulges to move across Earth at about 1600 kilometers (1000 miles) per hour. Tides, however, are an extreme example of shallow-water waves, so their speed is proportional to the water depth. For a tide wave to travel at 1600 kilometers (1000 miles) per hour, the ocean would have to be 22 kilometers (13.7 miles) deep! Instead, the average depth of the ocean is only 3.7 kilometers (2.3 miles), so tidal bulges move as *forced waves*, with their speed determined by ocean depth.

Based on the average ocean depth, the average speed at which tide waves can travel across the open ocean is only about 700 kilometers (435 miles) per hour. Thus, the idealized bulges that are oriented toward and away from a tide-generating body cannot exist because they cannot keep up with the rotational speed of Earth. Instead, ocean tides break up into distinct units called *cells*.

Amphidromic Points and Cotidal Lines

In the open ocean, the crests and troughs of the tide wave rotate around an **amphidromic** (*amphi* = around, *dromus* = running) **point** near the center of each cell. There is essentially no tidal range here, but radiating from this point are **cotidal** (*co* = with, *tidal* = tide) **lines**, which connect points where high tide occurs simultaneously. The labels on the cotidal lines in Figure 10-18 indicate the time of high tide in hours after the Moon crosses the Greenwich Meridian.

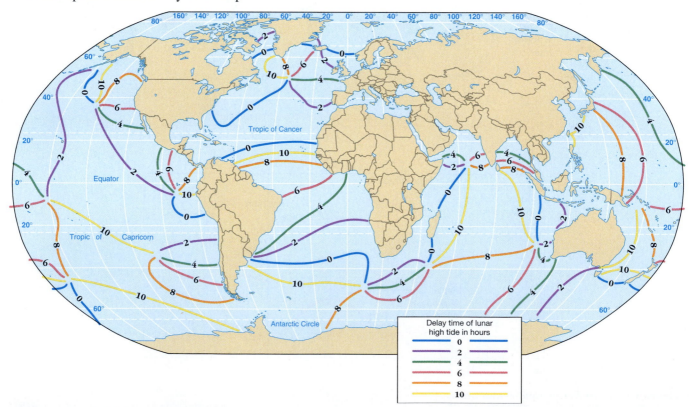

Figure 10–18 Cotidal map of the world.

Cotidal lines indicate times of the main lunar daily high tide in lunar hours after the Moon has crossed the Greenwich Meridian (0 degrees longitude). Tidal ranges generally increase with increasing distance along cotidal lines away from the amphidromic points. Where cotidal lines terminate at both ends in amphidromic points, maximum tidal range will be near the midpoints of the lines.

The times in Figure 10-18 indicate that the tide wave rotates counterclockwise in the Northern Hemisphere and clockwise in the Southern Hemisphere. The wave must complete one rotation during the tidal period (usually 12 lunar hours), so this limits the size of the cells.

Low tide occurs six hours after high tide in an amphidromic cell. If high tide is occurring along the cotidal line labeled "10," for example, then low tide is occurring along the cotidal line labeled "4."

Effect of the Continents

The continents affect tides, too, because they interrupt the free movement of the tidal bulges across the ocean surface. The ocean basins between continents have free standing waves set up within them. The positions and shapes of the continents modify the forced astronomical tide waves that develop within an ocean basin.

Other Considerations

Over 150 different factors affect the tides at a particular coast, which are far more than can be adequately addressed here. One of the results of these factors, however, is that high tide rarely occurs when the Moon is at its highest point in the sky. Instead, the time between the Moon crossing the meridian and a corresponding high tide varies from place to place.

Because of the complexity of the tides, a completely mathematical model of the tides is beyond the limits of marine science. Instead, a combination of mathematical analysis and observation is required to adequately model the tides.

Just as the sea is composed of multiple wave systems, the tides are also composed of multiple tide waves called **partial tides**. A mathematical approach useful in studying tides is called *harmonic analysis*, which takes into account the numerous tide-generating variables that possess a periodicity (cyclic pattern). Moreover, the actual tide observed at any given location is the combined effect of all the partial tides at that point.

Remarkably, a reasonably accurate model of actual tides can be computed considering only the seven major partial tides that affect a coastal area (Figure 10-19a). Combining the periods of each of the partial tides with the amplitudes and phases that can be obtained from observation results in a relatively accurate prediction of the tides (Figure 10-19b). To make the predictions as accurate as possible, the observations must be made throughout a period of at least 18.6 years, which is the period of the precession of the plane of the Moon's orbit through the ecliptic.

Tidal Patterns

In theory, most areas on Earth should experience two high tides and two low tides of unequal heights during a lunar day. In practice, however, the various depths, sizes, and shapes of ocean basins modify tides so they exhibit three different patterns in different parts of the world. The three tidal patterns, which are illustrated in Figure 10-20, are

diurnal (*diurnal* = daily), *semidiurnal* (*semi* = twice, *diurnal* = daily), and *mixed*.[10]

A **diurnal tidal pattern** has a single high and low tide each lunar day. These tides are common in shallow inland seas such as the Gulf of Mexico and along the coast of Southeast Asia. Diurnal tides have a tidal period of 24 hours 50 minutes.

---

[10]Sometimes a *mixed* tidal pattern is referred to as *mixed semidiurnal*.

	Symbol	Period in solar hours	Amplitude $M_2 = 100$	Description
Semidiurnal tides	M_2	12.42	100.00	Main lunar (semidiurnal) constituent
	S_2	12.00	46.6	Main solar (semidiurnal) constituent
	N	12.66	19.1	Lunar constituent due to monthly variation in moon's distance
	K_2	11.97	12.7	Soli-lunar constituent due to changes in declination of sun and moon throughout their orbital cycle
Diurnal tides	K_1	23.93	58.4	Soli-lunar constituent
	O	25.82	41.5	Main lunar (diurnal) constituent
	P	24.07	19.3	Main solar (diurnal) constituent

(a)   The seven most important partial tides

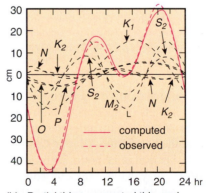

(b)   Partial tides, computed tide, and observed tide at Pula, Yugoslavia (January 6, 1909). Note close fit of computed and observed tides.

Figure 10–19 Partial tides.

(a) Table showing the seven most important partial tides and their characteristics. **(b)** Tidal curves for January 6, 1909 at Pula, Yugoslavia, showing the partial tides described in part *a* along with the computed (*solid red line*) and observed tides (*dashed red line*). Note the close match of the two red lines.

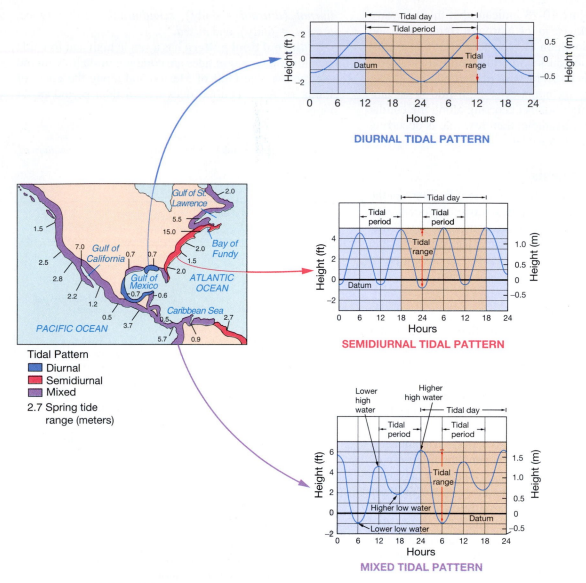

Figure 10–20 Tidal patterns.

Tidal patterns experienced along North and Central American coasts. A diurnal tidal pattern (*top graph*) shows one high and low tide each lunar day. A semidiurnal pattern (*middle graph*) shows two highs and lows of approximately equal heights during each lunar day. A mixed tidal pattern (*bottom graph*) shows two highs and two lows of unequal heights during each lunar day.

A **semidiurnal tidal pattern** has two high and two low tides each lunar day. The heights of successive high tides and successive low tides are approximately the same.[11] Semidiurnal tides are common along the Atlantic Coast of the United States. The tidal period is 12 hours 25 minutes.

A **mixed tidal pattern** may have characteristics of both diurnal and semidiurnal tides. Successive high tides and/or low tides will have significantly different heights, called *diurnal inequalities*. Mixed tides commonly have a tidal period of 12 hours 25 minutes, but they may also have diurnal periods. Mixed tides are the most common type in the world, including along the Pacific Coast of North America.

---

[11]Since tides are always growing higher or lower at any location due to the spring-neap tide sequence, successive high tides and successive low tides can never be *exactly* the same at any location.

STUDENTS SOMETIMES ASK...

Figure 10-20 shows negative tides. How can there ever be a negative tide?

Negative tides occur because the *datum* (starting point or reference point from which tides are measured) is an average of the tides over many years. Along the west coast of the United States, for instance, the datum is Mean Lower Low Water (MLLW), which is the average of the *lower* of the two low tides that occur daily in a mixed tidal pattern. Because the datum is an average, there will be some days when the tide is less than the average (similar to the distribution of exam scores, some of which will be below the average). These lower-than-average tides are given negative values, occur only during spring tides, and are often the best times to visit local tide pool areas.

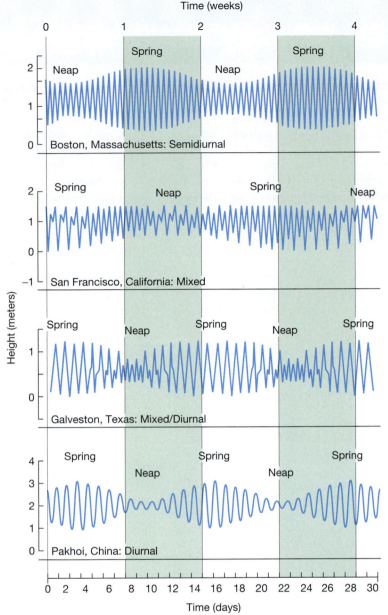

Time (weeks)

Height (meters)

Time (days)

Figure 10–21 **Monthly tidal curves.**

Top: Boston, Massachusetts, showing semidiurnal tidal pattern. *Upper middle*: San Francisco, California, showing mixed tidal pattern. *Lower middle*: Galveston, Texas, showing mixed tidal pattern with strong diurnal tendencies. *Bottom*: Pakhoi, China, showing diurnal tidal pattern.

Figure 10-21 shows examples of monthly tidal curves for various coastal locations. Even though a tide at any particular location follows a single tidal pattern, it still may pass through stages of one or both of the other tidal patterns. Typically, however, the tidal pattern for a location remains the same throughout the year. Also, the tidal curves in Figure 10-21 clearly show the weekly switching of the spring tide–neap tide cycle.

> A diurnal tidal pattern exhibits one high and low tide each lunar day; a semidiurnal tidal pattern exhibits two high and low tides daily of about the same height; a mixed tidal pattern usually has two high and low tides daily of different heights but may also exhibit diurnal qualities.

Tidal Phenomena

Remember that the tides are fundamentally a wave. When tide waves enter coastal waters, they are subject to reflection and amplification similar to what wind-generated waves experience. In certain locations, reflected wave energy causes water to slosh around in a bay, producing *standing waves*.[12] As a result, interesting tidal phenomena are sometimes experienced in coastal waters.

Large lakes and coastal rivers experience tidal phenomena, too. In some low-lying rivers, for instance, a *tidal bore* is produced by an incoming high tide (Box 10-1). Further, the tides profoundly affect the behavior of certain marine organisms (Box 10-2). Using the tides as a source of renewable power is discussed in Chapter 17, "Marine Resources."

---

[12]See Chapter 9 for a discussion of standing waves, including the terms *node* and *antinode*.

BOX 10–1 People and Ocean Environment
TIDAL BORES: BORING WAVES THESE ARE NOT!

A **tidal bore** (*bore*=crest or wave) is a wall of water that moves up certain low-lying rivers due to an incoming tide. Because it is a wave created by the tides, it is a *true* tidal wave. When an incoming tide rushes up a river, it develops a steep forward slope because the flow of the river resists the advance of the tide (Figure 10A). This creates a tidal bore, which may reach heights of 5 meters (16.4 feet) or more and move at speeds up to 22 kilometers (14 miles) per hour.

Tidal bores develop where there is a large tidal range and a low-lying coastal river. Although tidal bores do not attain the size of some waves in the surf zone, tidal bores have been successfully surfed. They can give a surfer a very long ride because the bore travels many kilometers upriver. If you miss the bore, though, you have to wait about half a day before the next one comes along because the incoming high tide occurs only twice a day. In some locations, tidal bore rafting is promoted as a draw for tourists.

The Amazon River probably possesses the longest estuary that is affected by oceanic tides. Tides can be measured as far as 800 kilometers (500 miles) from the river's mouth, although the effects are quite small at this distance. Tidal bores near the mouth of the Amazon River can be up to 5 meters (16.4 feet) high and are locally called *pororocas* (water-falls). Other rivers that have notable tidal bores include the Chientang River in China [which has the largest tidal bores in the world, often reaching 8 meters (26 feet) high]; the Petitcodiac River in New Brunswick, Canada; the River Seine in France; the Trent River in England; and Cook Inlet near Anchorage, Alaska (where the largest tidal bore in the United States can be found). Although the Bay of Fundy has the world's largest tidal range, its tidal bore rarely exceeds 1 meter (3.3 feet), mostly because the bay is so wide.

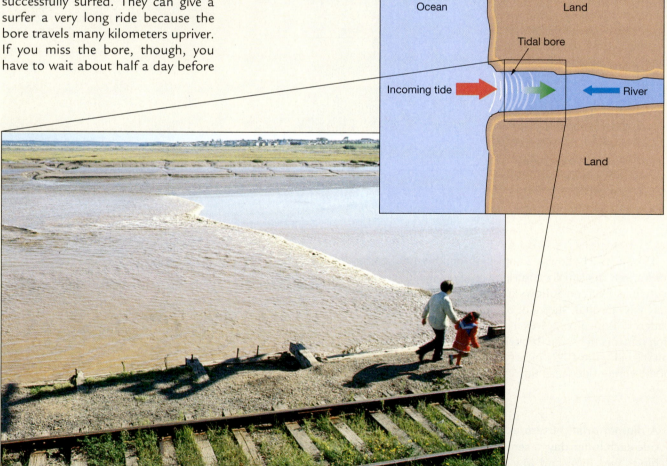

Figure 10A **A tidal bore moving quickly upriver in Chignecto Bay, New Brunswick, Canada.**

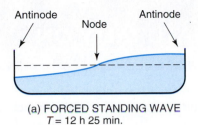

Antinode   Antinode

Node

(a) FORCED STANDING WAVE
T = 12 h 25 min.

Nodal line

(b) FREE STANDING WAVE

$$T = \frac{2L}{\sqrt{gh}}$$

L = length of lake (m)
g = acceleration due to gravity
    (9.8 m/s^2)
h = depth of lake (m)

Figure 10–22 Tides in lakes.

(a) A forced standing wave generated by tide-generating forces, which has a period of 12 hours 25 minutes. **(b)** A single nodal free standing wave (*seiche*), which has a period determined by the size and shape of the basin of the lake (*equations below figure*). If the period of the seiche is approximately equal to (or a multiple of) the forced tide-generated wave, resonance occurs, producing greater displacements at the antinodes.

Tides in Lakes

Although most lake basins are too small to have noticeable tidal effects, tides may be significant in large lakes especially when the long axis of the basin extends in an east-west direction. In any such basin, very small standing waves may be generated with a period equal to that of the tide-generating force, producing a *forced standing wave* (Figure 10-22a).

Of much greater importance is the *free standing wave* (Figure 10-22b) that is initiated by strong winds at the surface (or, less commonly, by a seismic disturbance). The period of a free standing wave is determined by the length and depth of the basin, and is termed the *characteristic period* for the basin. If the characteristic period of the free wave is very near that (or a multiple) of the period of a forced wave resulting from tide-generating forces, the oscillations may reinforce one another and produce a *resonance tide*. Lake Ontario, for example, displays such tides.

Free standing waves of the type described above were first noticed in Lake Geneva in Switzerland and are called **seiches**[13] (*seiche* = exposed lake bottom) because of the way the sloshing water exposed parts of the lake bottom. For seiches that have a single nodal line, the formula for the period in Figure 10-22b gives a close approximation of the period that actually develops. For rectangular basins with two and three nodal lines, the periods would be approximately one-half and one-third those of a single nodal seiche, respectively.

Tides in Narrow Basins Connected to the Ocean

Even under conditions of resonance between free and forced oscillations in lakes, the tides still do not get very large (they only rarely exceed a few centimeters). However, in similarly sized seas, bays, and gulfs that are relatively narrow but open at one end to the ocean, the tides may become much larger. What causes this difference?

To help answer this question, consider a rectangular narrow bay with one end open to the ocean (Figure 10-23). Although one end of the basin is open, the bay still develops free standing waves that are reflected from the open end around the basin. At the open end of the bay, the water must

always be at the same level as the ocean at that location. Therefore, the tidal range is of greater magnitude than it is in closed basins. If the free standing wave (which is caused by wave energy reflected from the closed end of the basin) is in resonance with the forced wave produced by tidal forces, the energy of these two waves produces a standing wave with increased amplitude within the basin.

An Example of Tidal Extremes: The Bay of Fundy

As we have seen, standing waves that have a period near that of the forced tide wave result in constructive interference that produces significant increases in tidal range. Nova Scotia's **Bay of Fundy** is one such place where this occurs, and it is here that the largest tidal range in the world is found. With a length of 258 kilometers (160 miles), the Bay of Fundy has a wide opening into the Atlantic Ocean. At its northern end, however, it splits into two narrow basins, Chignecto Bay and Minas Basin (Figure 10-24).

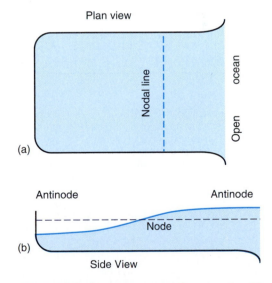

Plan view

Nodal line

Open ocean

Open

(a)

Antinode   Antinode

Node

(b)

Side View

Figure 10–23 Tides in narrow seas, bays, and gulfs.

Top view **(a)** and side view **(b)** of a narrow body of water such as a sea, bay, or gulf. In these bodies, forced standing waves have a greater height than those that develop in lakes because the height of the tide at the open end of the basin must be the same as the open ocean. Therefore, the development of a resonant condition between the free and forced standing waves in such basins produce much greater displacements at the antinodes.

---

[13]Seiche is pronounced "*saysh*."

313

BOX 10–2   Research Methods in Oceanography

GRUNIONS: DOING WHAT COMES NATURALLY ON THE BEACH

From March through September, shortly after the maximum spring tide has occurred, the grunion (*Leuresthes tenuis*) come ashore along the beaches of southern California and Baja California to bury their fertilized eggs in the sand. Grunion—slender, silvery, and 12 to 15 centimeters (4.7 to 6 inches) long—are the only marine fish in the world that come completely out of water to spawn. The name **grunion** comes from the Spanish *gruñón*, which means "grunter" and refers to the faint noise they make during spawning.

A mixed tidal pattern occurs along southern California and Baja California beaches. On most lunar days (24 hours and 50 minutes), there are two high and two low tides. There is usually a significant difference in the heights of the two high tides that

occur each day. During the summer months, the higher high tide occurs at night. The night high tide becomes higher each night as the maximum spring-tide range is approached, causing sand to be eroded from the beach (Figure 10B, *graph*). After the maximum spring tide has occurred, the night high tide diminishes each night. As neap tide is approached, sand is deposited on the beach.

Grunion spawn only after each night's higher high tide has peaked on the three or four nights following the night of the highest spring high tide. This assures that their eggs will be covered deeply in sand deposited by the receding higher high tides each succeeding night. The fertilized eggs buried in the sand are ready to hatch nine days after spawning. By this time, another spring tide is ap-

proaching, so the night high tide is getting progressively higher each night again. The beach sand is eroding again, too, which exposes the eggs to the waves that break ever higher on the beach. The eggs hatch about three minutes after being freed in the water. Tests done in laboratories have shown that the grunion eggs will not hatch until agitated in a manner that simulates that of the eroding waves.

The spawning begins as the grunion come ashore immediately following an appropriate high tide, and it may last from one to three hours. Spawning usually peaks about an hour after it starts and may last an additional 30 minutes to an hour. Thousands of fish may be on the beach at this time. During a run, the females, which are larger than the males, move high on the

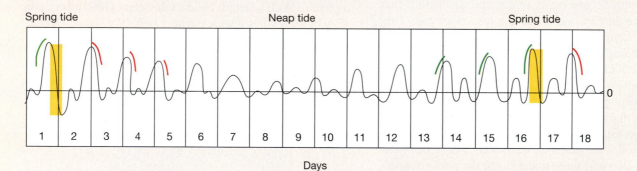

Spring tide   Neap tide   Spring tide

Days

 Grunion deposit eggs in beach sand during early stages of the ebb of higher high tides on the three or four days following maximum spring tidal range.

Flood tides erode sand and free grunion eggs during higher high tide as maximum spring tidal range is approached.

Maximum spring tidal range

Figure 10B **The tidal cycle and spawning grunion.**

beach. If no males are near, a female may return to the water without depositing her eggs. In the presence of males, she drills her tail into the semifluid sand until only her head is visible. The female continues to twist, depositing her eggs 5 to 7 centimeters (2 to 3 inches) below the surface.

The male curls around the female's body and deposits his milt against it (Figure 10B, *picture*). The milt runs down the body of the female to fertilize the eggs. When the spawning is completed, both fish return to the water with the next wave.

Larger females are capable of producing up to 3000 eggs for each series of spawning runs, which are separated by the two-week period between spring tides. As soon as the eggs are deposited, another group of eggs begins to form within the female. They will be deposited during the next spring tide run. Early in the season, only older fish spawn. By May, however, even the one-year-old females are in spawning condition.

Young grunion grow rapidly and are about 12 centimeters (5 inches) long when they are a year old and ready for their first spawning. They usually live two or three years, but four-year-olds have been recovered. The age of a grunion can be determined by the scales. After growing rapidly during the first year, they grow very slowly thereafter. There is no growth at all during the 6-month spawning season, which causes marks to form on each scale that can be used to identify the grunion's age.

It is not known how grunion are able to time their spawning behavior so precisely with the tides. Some investigators believe the grunion are able to sense very small changes in the hydrostatic pressure caused by the changing level of the water associated with rising and falling sea level due to the tides. Certainly, a very dependable detection mechanism keeps the grunion accurately informed of the tidal conditions, because their survival depends on a spawning behavior precisely tuned to tidal motions.

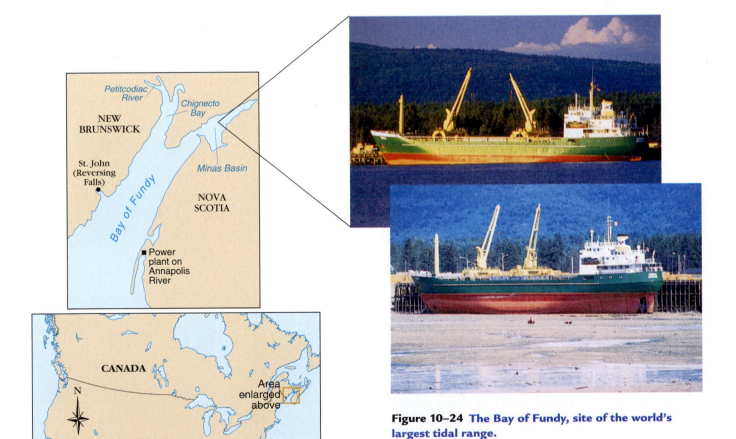

Figure 10–24 The Bay of Fundy, site of the world's largest tidal range.

Even though the maximum spring tidal range at the mouth of the Bay of Fundy is only 2 meters (6.6 feet), amplification of tidal energy causes a maximum tidal range at the northern end of Minas Basin of 17 meters (56 feet), often stranding ships (*insets*).

The period of free oscillation in the bay—the oscillation that occurs when a body is displaced and then released—is very nearly that of the tidal period. The resulting constructive interference—along with the narrowing and shoaling of the bay to the north—causes a buildup of tidal energy in the northern end of the bay. In addition, the bay curves to the right, so the Coriolis effect in the Northern Hemisphere adds to the extreme tidal range.

During maximum spring tide conditions, the tidal range at the mouth of the bay (where it opens to the ocean) is only about 2 meters (6.6 feet). However, the tidal range increases progressively from the mouth of the bay northward. In the northern end of Minas Basin, the maximum spring tidal range is 17 meters (56 feet), which leaves boats high and dry during low tide (Figure 10-24, *insets*).

> The world's largest tides occur in the upper end of the Bay of Fundy, where reflection and amplification produce a maximum spring tidal range of 17 meters (56 feet).

Coastal Tidal Currents

The current that accompanies the slowly turning tide crest in a Northern Hemisphere basin rotates counterclockwise, producing a **rotary current** in the open portion of the basin. Friction increases in nearshore shoaling waters, so the rotary current changes to an alternating or **reversing current** that moves into and out of restricted passages along a coast.

The velocity of rotary currents in the open ocean is usually well below 1 kilometer (0.6 mile) per hour. Reversing currents, however, can reach velocities up to 44 kilometers (28 miles) per hour in restricted channels such as between islands of coastal waters.

Reversing currents also exist in the mouths of bays (and some rivers) due to the daily flow of tides. Figure 10-25 shows that a **flood current** is produced when water rushes into a bay (or river) with an incoming high tide. Conversely, an **ebb current** is produced when water drains out of a bay (or river) because a low tide is approaching. No currents occur for several minutes during either **high slack water** (which occurs at the peak of each high tide) or during **low slack water** (at the peak of each low tide).

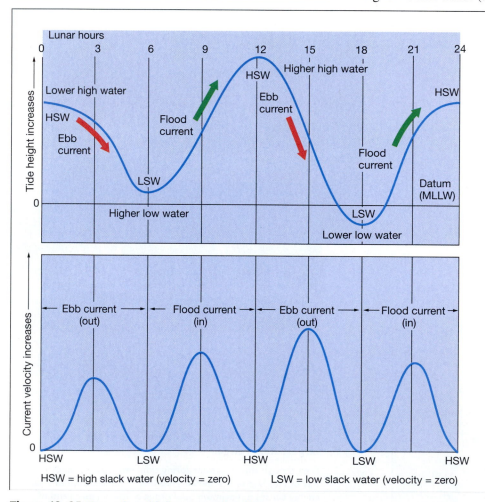

Figure 10–25 Reversing tidal currents in a bay.

Top: Tidal curve for a bay, showing ebb currents are created by an outgoing low tide and flood currents are created by an incoming high tide. No currents occur during either high slack water (*HSW*) or low slack water (*LSW*). The datum of MLLW means mean lower low water, which is the average of the lower of the two low tides that occur daily in a mixed tidal pattern. *Bottom*: Corresponding chart showing velocity of ebb and flood currents.

Reversing currents in bays can sometimes reach speeds of 40 kilometers (25 miles) per hour, creating a navigation hazard for ships. On the other hand, the daily flow of these currents often keeps sediment from closing off the bay and resupplies the bay with new seawater and ocean nutrients.

Tidal currents can be significant even in deep ocean waters. For example, tidal currents were encountered shortly after the discovery of the remains of the *Titanic* at a depth of 3795 meters (12,448 feet) on the continental slope south of Newfoundland's Grand Banks in 1985.

These tidal currents were so strong that they forced researchers to abandon the use of the camera-equipped tethered remotely operated vehicle, *Jason Jr.*

Rotary tidal currents occur in the deep ocean while reversing tidal currents occur close to shore—most notably in bays and rivers—due to the change in the tides.

Chapter in Review

- *Gravitational attraction of the Moon and Sun create Earth's tides*, which are fundamentally long-wavelength waves. According to the simplified *equilibrium theory of tides*, which assumes an ocean of uniform depth and ignores the effects of friction, small horizontal forces (the tide-generating forces, which vary as the cube of distance) tend to push water into *two bulges on opposite sides of Earth*. One bulge is directly facing the tide-generating body (the Moon and the Sun), and the other is directly opposite.

- Despite its vastly smaller size, *the Moon has about twice the tide-generating effect of the Sun* because the Moon is so much closer to Earth. The tidal bulges due to the Moon's gravity (the lunar bulges) dominate, so lunar motions dominate the periods of Earth's tides. However, the changing position of the solar bulges relative to the lunar bulges modifies tides. According to the simplified equilibrium tide theory, *Earth's rotation carries locations on Earth into and out of the various tidal bulges*.

- For most places on Earth, *the time between successive high tides would be 12 hours 25 minutes (half a lunar day). The $29\frac{1}{2}$-day monthly tidal cycle would consist of tides with maximum tidal range (spring tides) and minimum tidal range (neap tides). Spring tides would occur each new moon and full moon, and neap tides would occur each first- and third-quarter phases of the Moon*.

- The *declination of the Moon* varies between 28.5 degrees north or south of the Equator during the lunar month, and the *declination of the Sun* varies between 23.5 degrees north or south of the Equator during the year, so *the location of tidal bulges usually creates two high tides and two low tides of unequal height per lunar day*. Tidal ranges are greatest when Earth is nearest the Sun and Moon.

- *Friction and the true shape of ocean basins are considered in the dynamic theory of tides*, which is more complex but does a better job of approximating real ocean tides. Moreover, the two bulges on opposite sides of Earth cannot exist because they cannot keep up with the rotational speed of Earth. Instead, the bulges are broken up into *several tidal cells that rotate around an amphidromic point*—a point of zero tidal range. Rotation is counterclockwise in the Northern Hemisphere and clockwise in the Southern Hemisphere. *Many other factors influence tides on Earth*, too, such as the placement of the continents and the shapes of the coasts. The seven major partial tides approximate real tidal conditions.

- The *three types of tidal patterns* observed on Earth are *diurnal* (a single high and low tide each lunar day), *semidiurnal* (two high and two low tides each lunar day), and *mixed* (characteristics of both). Mixed tidal patterns usually consist of semidiurnal periods with significant diurnal inequality. Mixed tidal patterns are the most common type in the world.

- *Tidal phenomena include tides in lakes and coastal water bodies*, both of which are influenced by basin size and water depth. All basins have a characteristic *free standing wave*, or *seiche*. It is usually not very large for small basins, but if it is in phase with the *forced standing wave created by tide-generating bodies*, its height can be significant.

- The effects of constructive interference and the shoaling and narrowing of coastal bays creates the *largest tidal range in the world—17 meters (56 feet)—at the northern end of Nova Scotia's Bay of Fundy. Tidal currents* follow a *rotary pattern* in open-ocean basins but are converted to *reversing currents* along continental margins. The maximum velocity of reversing currents occurs during flood and ebb currents when the water is halfway between high and low slack waters. *Tidal bores are true tidal waves* (a wave produced by the tides) that occur in certain rivers and bays due to an incoming high tide.

- *The tides are important to many marine organisms*. For instance, *grunion*—small silvery fish that inhabit waters along the west coast of North America—time their spawning cycle to match the pattern of the tides.

Key Terms

Amphidromic point (p. 308)
Aphelion (p. 306)
Apogee (p. 306)
Barycenter (p. 295)
Bay of Fundy (p. 313)
Centripetal force (p. 296)
Cotidal line (p. 308)
Declination (p. 305)
Diurnal tidal pattern (p. 309)
Dynamic tide theory (p. 308)
Ebb current (p. 316)
Ebb tide (p. 302)
Ecliptic (p. 305)
Equilibrium tide theory (p. 300)

Flood current (p. 316)
Flood tide (p. 302)
Full moon (p. 302)
Gravitational force (p. 295)
Grunion (*Leuresthes tenuis*)
  (p. 314)
High slack water (p. 306)
Low slack water (p. 306)
Lunar bulge (p. 300)
Lunar day (p. 301)
Mixed tidal pattern (p. 310)
Neap tide (p. 303)
New moon (p. 302)
Newton, Isaac (p. 295)

Partial tide (p. 309)
Perigee (p. 306)
Perihelion (p. 306)
Precession (p. 305)
Proxigean (p. 306)
Quadrature (p. 303)
Quarter moon (p. 302)
Resultant force (p. 297)
Reversing current (p. 316)
Rotary current (p. 316)
Seiche (p. 313)
Semidiurnal tidal pattern
  (p. 310)
Solar bulge (p. 302)

Solar day (p. 301)
Spring tide (p. 303)
Syzygy (p. 303)
Tidal bore (p. 312)
Tidal period (p. 300)
Tidal range (p. 303)
Tide-generating force (p. 298)
Tides (p. 303)
Waning crescent (p. 304)
Waning gibbous (p. 304)
Waxing crescent (p. 304)
Waxing gibbous (p. 304)

Questions and Exercises

1. Explain why the Sun's influence on Earth's tides is only 46% that of the Moon's, even though the Sun is so much more massive than the Moon.

2. Why is a lunar day 24 hours 50 minutes long, while a solar day is 24 hours long?

3. Which is more technically correct: The tide comes in and goes out; or Earth rotates into and out of the tidal bulges? Why?

4. From memory, draw the positions of the Earth–Moon–Sun system during a complete monthly tidal cycle. Indicate the tide conditions experienced on Earth, the phases of the Moon, the time between those phases, and syzygy and quadrature.

5. Explain why the maximum tidal range (spring tide) occurs during new and full moon phases and the minimum tidal range (neap tide) at first-quarter and third-quarter moons.

6. If Earth did not have the Moon orbiting it, would there still be tides? Why or why not?

7. Assume that there are two moons in orbit around Earth that are on the same orbital plane but always on opposite sides of Earth and that each moon is the same size and mass of our Moon. How would this affect the tidal range during spring and neap tide conditions?

8. What is declination? Discuss the degree of declination of the Moon and Sun relative to Earth's Equator. What are the effects of declination of the Moon and Sun on the tides?

9. Diagram the Earth–Moon system's orbit about the Sun. Label the positions on the orbit at which the Moon and Sun are clos-

est to and farthest from Earth, stating the terms used to identify them. Discuss the effects of the Moon's and Earth's positions on Earth's tides.

10. Are tides considered deep-water waves anywhere in the ocean? Why or why not?

11. Describe the number of high and low tides in a lunar day, the period, and any inequality of the following tidal patterns: diurnal, semidiurnal, and mixed.

12. What forces produce forced and free standing waves in lakes and narrow ocean embayments?

13. Discuss factors that help produce the world's greatest tidal range in the Bay of Fundy.

14. Discuss the difference between rotary and reversing tidal currents.

15. Of flood current, ebb current, high slack water, and low slack water, when is the best time to enter a bay by boat? When is the best time to navigate in a shallow, rocky harbor? Explain.

16. Describe the spawning cycle of grunion, indicating the relationship between tidal phenomena, where grunion lay their eggs, and the movement of sand on the beach.

17. Observe the Moon from a reference location every night at about the same time for two weeks. Keep track of your observations about the shape (phase) of the Moon and its position in the sky. Then compare these to the reported tides in your area. How do the two compare?

CHAPTER 11
The Coast: Beaches and Shoreline Processes

House falling into the sea at North Carolina's Outer Banks. When coastal structures are built too close to the sea, they could collapse into it, as did this house in Nags Head, North Carolina in 2000. Understanding coastal dynamics and shoreline processes can help prevent damage such as this.

Key Questions

- Which coastal areas are parts of the beach?
- What seasonal changes do beaches experience?
- How are longshore currents created, and what is longshore drift?
- Which coastal features are characteristic of erosional and depositional coasts?
- How has changing sea level affected coastal regions?
- What are differences between various U.S. coastal regions?
- What effects do various forms of hard stabilization have on shorelines?

The answers to these questions (and much more) can be found in the highlighted concept statements within this chapter.

"The waves which dash upon the shore are, one by one, broken, but the ocean conquers nevertheless. It overwhelms the Armada, it wears out the rock."

—Lord Byron (1821)

Humans have always been attracted to the coastal regions of the world for their moderate climate, seafood, transportation, recreational opportunities, and commercial benefits. In the United States, for example, 80% of the population now lives within easy access of the Atlantic, Pacific, and Gulf Coasts, increasing the stress on these important national resources.

The coastal region is constantly changing because waves crash along most shorelines more than 10,000 times a day, releasing their energy from distant storms.

Waves cause erosion in some areas and deposition in others, resulting in changes that occur hourly, daily, weekly, monthly, seasonally, and yearly.

In this chapter, we'll examine the major features of the seacoast and shore and the processes that modify them. We'll also discuss ways people interfere with these processes, creating hazards to themselves and to the environment.

The Coastal Region

The **shore** is a zone that lies between the lowest tide level (low tide) and the highest elevation on land that is affected by storm waves. The **coast** extends inland from the shore as far as ocean-related features can be found (Figure 11-1). The width of the shore varies between a few meters and hundreds of meters. The width of the coast may vary from less than a kilometer (0.6 mile) to many tens of kilometers. The **coastline** marks the boundary between the shore and the coast. It is the landward limit of the effect of the highest storm waves on the shore.

Beach Terminology

The beach profile in Figure 11-1 shows features characteristic of a cliffed shoreline. The shore is divided into the **backshore** and the **foreshore**.[1] The backshore is above the high-tide shoreline and is covered with water only during storms. The foreshore is the portion exposed at low tide and submerged at high tide. The **shoreline** migrates back and forth with the tide and is the water's edge. The **nearshore** extends seaward from

---

[1] The foreshore is often referred to as the *intertidal* or *littoral* (*litoralis* = the shore) zone.

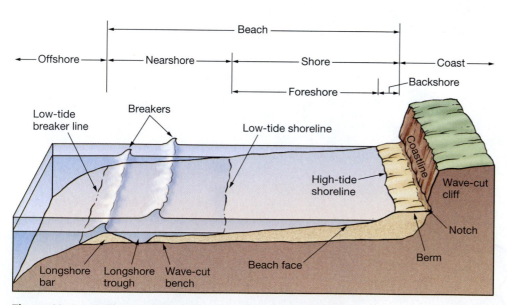

Figure 11–1 Landforms and terminology of coastal regions.

The beach is the entire active area affected by waves that extends from the low tide breaker line to the base of the coastal cliffs.

the low-tide shoreline to the low-tide breaker line. It is never exposed to the atmosphere, but it is affected by waves that touch bottom. Beyond the low-tide breakers is the **offshore** zone, which is deep enough that waves rarely affect the bottom.

A **beach** is a deposit of the shore area. It consists of wave-worked sediment that moves along the **wave-cut bench** (a flat, wave-eroded surface). A beach may continue from the coastline across the nearshore region to the line of breakers. Thus, the beach is the entire active area of a coast that experiences changes due to breaking waves. The area of the beach above the shoreline is often called the *recreational beach*.

The **berm** is the dry, gently sloping region at the foot of the coastal cliffs or dunes. The berm is often composed of sand, making it a favorite place of beachgoers. The **beach face** is the wet, sloping surface that extends from the berm to the shoreline. It is more fully exposed during low tide, and is also known as the *low tide terrace*. The beach face is a favorite place for runners because the sand is wet and hard packed. Offshore beyond the beach face is one or more **longshore bars**—sand bars that parallel the coast. A longshore bar may not always be present throughout the year, but when one is, it may be exposed during extremely low tides. Longshore bars can "trip" waves as they approach shore and cause them to begin breaking. Separating the longshore bar from the beach face is a **longshore trough**.

> The beach is the coastal area affected by breaking waves and includes the berm, beach face, longshore trough, and longshore bar.

Beach Composition and Slope

Beaches are composed of whatever material is locally available. When this material—sediment—comes from the erosion of beach cliffs or nearby coastal mountains, beaches are composed of mineral particles from these rocks and may be relatively coarse in texture. When the sediment comes primarily from rivers that drain lowland areas, beaches are finer in texture. Often, mud flats develop along the shore because only tiny clay-sized and silt-sized particles are emptied into the ocean. Such is the case for muddy coastlines such as along the coast of Suriname in South America and the Kerala coast of southwest India.

Other beaches have a significant biologic component. For example, in low-relief, low-latitude areas such as southern Florida, where there are no mountains or other sources of rock-forming minerals nearby, most beaches are composed of shell fragments and the remains of organisms that live in coastal waters. Many beaches on volcanic islands in the open ocean are composed of black or green fragments of the basaltic lava that comprise the islands, or of coarse debris from coral reefs that develop around islands in low latitudes.

Regardless of the composition, though, the material that comprises the beach does not stay in one place. Instead, the waves that crash along the shoreline are constantly moving it. Thus, beaches can be thought of as *material in transit along the shoreline*.

Measurements of beach slopes reveal that coarser beach materials have steeper beach slopes (Table 11-1). This relationship is caused by waves that wash up onto the beach, which transport sand up the beach, too. If most of the water from a wave percolates into the beach, the sediment it carries remains on the beach and increases the beach slope. As a result, beaches composed of coarse, loosely packed materials have steeper beach slopes. Conversely, a beach composed of fine-grained sand that does not allow water to soak in will have a more gently sloping (and firmer) surface. This is because most of the water from breaking waves runs back toward ocean and has enough energy to carry sand back down the beach face. In this way, an equilibrium is reached and a gentle beach slope is produced.

TABLE 11–1 **Relationship between particle size and beach slope.**

Wentworth particle name	Maximum size (mm)	Average beach slope
Cobble	64	24°
Pebble	4	17°
Granule	2	11°
Very coarse sand	1	9°
Coarse sand	0.5	7°
Medium sand	0.25	5°
Fine sand	0.125	3°
Very fine sand	0.063	1°

Movement of Sand on the Beach

The movement of sand on the beach occurs both perpendicular to the shoreline (*toward* and *away from* shore) and parallel to the shoreline (often referred to as *up-coast* and *down-coast*).

Movement Perpendicular to Shoreline Breaking waves move sand perpendicular to the shoreline. As each wave breaks, water rushes up the beach face toward the berm. Some of this **swash** soaks into the beach and eventually returns to the ocean. However, most of the water drains away from shore as backwash, though usually not before the next wave breaks and sends its swash over the top of the previous wave's **backwash**.

While standing in ankle-deep water at the shoreline, you can see that swash and backwash transports sediment up and down the beach face perpendicular to the shoreline. Whether swash or backwash dominates determines whether sand is deposited or eroded from the berm.

In *light wave activity* (characterized by less energetic waves), much of the swash soaks into the beach, so backwash is reduced. The swash dominates the transport system, therefore, causing a net movement of the sand up the beach face toward the berm, making it wide and well developed.

In *heavy wave activity* (characterized by high-energy waves), the beach is saturated with water from previous waves, so very little of the swash soaks into the beach. Backwash dominates the transport system, therefore, causing a net movement of sand down the beach face, which erodes the berm. When a wave breaks, moreover, the incoming swash comes *on top of* the previous wave's backwash, effectively protecting the beach from the swash and adding to the eroding effect of the backwash.

During heavy wave activity, where does the sand from the berm go? The orbital motion in waves is too shallow to move the sand very far offshore. Thus, the sand accumulates just beyond where the waves break and forms one or more offshore sand bars (the longshore bars).

Light and heavy wave activity alternate seasonally at most beaches, so the characteristics of the beaches change, too (Table 11-2). Light wave activity produces a wide sandy berm and an overall steep beach face—a **summertime beach**—at the expense of the longshore bar (Figure 11-2a). Conversely, heavy wave activity produces a narrow rocky berm and an overall flattened beach face—a **wintertime beach**—and builds prominent longshore bars (Figure 11-2b). A wide berm that takes several months to build can be destroyed in just a few hours by high-energy wintertime storm waves.

> Smaller, low-energy waves move sand up the beach face toward the berm and create a summertime beach while larger, high-energy waves scour sand from the berm and create a wintertime beach.

Movement Parallel to Shoreline   At the same time that movement occurs perpendicular to shore, movement parallel to shoreline also occurs. Recall from Chapter 9 that waves refract (bend) and line up *nearly* parallel to the shore. With each breaking wave, the swash moves up onto the exposed beach at a slight angle, then gravity pulls the backwash straight down the beach face. As a result, water moves in a zigzag fashion along the shore, creating a movement of water within the surf zone called a **longshore current** (Figure 11-3).

Longshore currents have speeds up to 4 kilometers (2.5 miles) per hour. Speeds increase as beach slope increases, as the angle at which breakers arrive at the beach increases, as wave height increases, and as wave frequency increases.

Swimmers can be inadvertently carried by longshore currents and find themselves carried far from where they initially entered the water. This demonstrates that longshore currents are strong enough to move people as well as a vast amount of sand in a zigzag fashion along the shore.

Longshore drift or **longshore transport** is the movement of *sediment* in a zigzag fashion caused by the longshore current (Figure 11-3b). Because a longshore current affects the entire surf zone, the resulting longshore drift works within the entire surf zone as well.

The amount of longshore drift in any coastal region depends on the equilibrium between erosional and depositional forces. Any interference with the movement of sediment along the shore disrupts the equilibrium, forming a new erosional and depositional pattern. Nevertheless, longshore drift moves millions of tons of sediment along coastal regions every year.

Both rivers and coastal zones move water *and* sediment from one area (*upstream*) to another (*downstream*). As a result, the beach has often been referred to as a "river of sand." A longshore current moves in a zigzag fashion, however, and rivers flow mostly in a turbulent, swirling fashion. Additionally, the direction of flow of longshore currents along a shoreline can change, whereas rivers always flow in the same direction (downhill). The longshore current changes direction because the direction with which waves approach the beach changes seasonally. Nevertheless, the longshore current generally flows *southward along both the Atlantic and Pacific shores of the United States.*

? STUDENTS SOMETIMES ASK...

What is the difference between a rip current and a rip tide? Are they the same thing as an undertow?

Like tidal waves (tsunami), rip tides are a misnomer and have nothing to do with the tides. Rip tides are more correctly called rip currents. Perhaps rip currents have incorrectly been called rip tides because they occur suddenly (like an incoming tide).

An undertow, similar to a rip current, is a flow of water away from shore. An undertow is much wider, however, and is usually more concentrated along the ocean floor. An undertow is really a continuation of backwash that flows down the beach face and is strongest during heavy wave activity. Undertows can be strong enough to knock people off their feet, but they are confined to the immediate floor of the ocean and only within the surf zone.

> Longshore currents are produced by waves approaching the beach at an angle and create longshore drift, which transports sand along the coast in a zigzag fashion.

BOX 11–1 People and Ocean Environment
WARNING: RIP CURRENTS … DO YOU KNOW WHAT TO DO?

The backwash from breaking waves usually returns to the open ocean as a flow of water across the ocean bottom, so it is commonly referred to as "sheet flow." Some of this water, however, flows back in surface **rip currents**. Rip currents typically flow perpendicular to the beach and move away from the shore.

Rip currents are between 15 and 45 meters (15 and 150 feet) wide and can attain velocities of 7 to 8 kilometers (4 to 5 miles) per hour—faster than most people can swim for any length of time. In fact, it is useless to swim for long against a current stronger than about 2 kilometers (1.2 miles) per hour. Rip currents can travel hundreds of meters from shore before they break up. If a light-to-moderate swell is breaking, numerous rip currents may develop, which are moderate in size and velocity. A heavy swell usually produces fewer, more concentrated, and stronger rips. They can often be recognized by the way they interfere with incoming waves, by their characteristic brown color caused by suspended sediment, or by their foamy and choppy surface (Figure 11A).

The rip currents that occur during heavy swell are a significant hazard to coastal swimmers. In fact, 80% of rescues at beaches by lifeguards involve people who are trapped in rip currents. Swimmers caught in a rip current can escape by swimming parallel to the shore for a short distance (simply swimming out of the narrow rip current) and then riding the waves in toward the beach. However, even excellent swimmers who panic or try to fight the current by swimming directly into it are eventually overcome by exhaustion and may drown. Even though most beaches have warnings posted and are frequently patrolled by lifeguards, many people lose their lives each year because of rip currents.

Figure 11A Rip currents.

A rip current, which extends outward from shore near the middle of the photo and interferes with incoming waves, and warning sign (*inset*).

TABLE 11–2 **Characteristics of beaches affected by light and heavy wave activity.**

	Light wave activity	Heavy wave activity
Berm/longshore bars	Berm is built at the expense of the longshore bars	Longshore bars are built at the expense of the berm
Wave energy	Low wave energy (non-storm conditions)	High wave energy (storm conditions)
Time span	Long time span (weeks or months)	Short time span (hours or days)
Characteristics	Creates summertime beach: sandy, wide berm, steep beach face	Creates wintertime beach: rocky, narrow berm, flattened beach face

Figure 11–2 Summertime and wintertime beach conditions.

Dramatic differences occur between (**a**) summertime and (**b**) wintertime beach conditions at Boomer Beach in La Jolla, California.

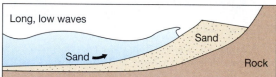

(a) Summertime beach (fair weather)

(b) Wintertime beach (storm)

(a)

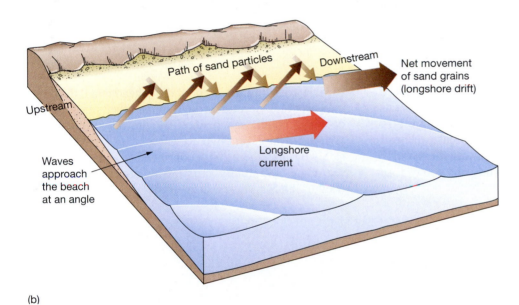

(b)

Figure 11–3 Longshore current and longshore drift.

(a) Waves approaching the beach at a slight angle near Oceanside, California, producing a longshore current moving toward the right of the photo. (b) A longshore current, caused by refracting waves, moves water in a zigzag fashion along the shoreline. This causes a net movement of sand grains (longshore drift) from upstream to downstream ends.

Erosional- and Depositional-Type Shores

Sediment eroded from the beach is transported along the shore and deposited in areas where wave energy is low. Even though all shores experience some degree of both erosion and deposition, shores can often be identified primarily as one type or the other. **Erosional-type shores** typically have well-developed cliffs and are in areas where tectonic uplift of the coast occurs, such as along the U.S. Pacific Coast.

The U.S. southeastern Atlantic Coast and the Gulf Coast, on the other hand, are primarily **depositional-type shores**. Sand deposits and offshore barrier islands are common there because the shore is gradually subsiding. Erosion can still be a major problem on depositional shores, especially when human development interferes with natural coastal processes.

Features of Erosional-Type Shores

Because of wave refraction, wave energy is concentrated on any **headlands** that jut out from the continent, while the

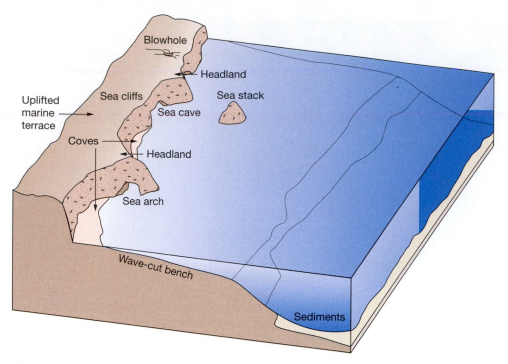

Figure 11–4 Features of erosional coasts.

Diagrammatic view of features characteristic of erosional coasts.

amount of energy reaching the shore in bays is reduced. Headlands, therefore, are eroded and the shoreline retreats. Some of these erosional features are shown in Figure 11-4.

Waves pound relentlessly away at the base of headlands, undermining the upper portions, which eventually collapse to form **wave-cut cliffs**. The waves may form **sea caves** at the base of the cliffs.

As waves continue to pound the headlands, the caves may eventually erode through to the other side, forming openings called **sea arches** (Figure 11-5). Some sea arches are large enough to allow a boat to maneuver safely through them. With continued erosion, the tops of sea arches eventually crumble to produce **sea stacks** (Figure 11-5). Waves also erode the bedrock of the

Figure 11–5 Sea arch and sea stack along the coast of Iceland.

When the roof of a sea arch (*left*) collapses, a sea stack (*right*) is formed.

bench. Uplift of the wave-cut bench creates a gently sloping **marine terrace** above sea level (Figure 11-6).

Rates of coastal erosion are influenced by the degree of exposure to waves, the amount of tidal range, and the composition of the coastal bedrock. Regardless of the erosion rate, all coastal regions follow the same developmental path. As long as there is no change in the elevation of the landmass relative to the ocean surface, the cliffs will continue to erode and retreat until the beaches widen sufficiently to prevent waves from reaching them. The eroded material is carried from high-energy areas and deposited in low-energy areas.

Features of Depositional-Type Shores

Coastal erosion of sea cliffs produces large amounts of sediment. Additional sediment, which is carried to the shore by rivers, comes from the erosion of inland rocks. Waves then distribute all of this sediment along the continental margin.

Figure 11–6 Wave-cut bench and marine terrace.

A wave-cut bench is exposed at low tide along the California coast at Bolinas Point near San Francisco. An elevated wave-cut bench, called a marine terrace, is shown at right.

Figure 11-7 shows some of the features of depositional coasts. These features are primarily deposits of sand moved by longshore drift but are also modified by other coastal processes. Some are partially or wholly separated from the shore.

A **spit** (*spit* = spine) is a linear ridge of sediment that extends in the direction of longshore drift from land into the deeper water near the mouth of a bay. The end of the spit normally curves into the bay due to the movement of currents.

Tidal currents or currents from river runoff are usually strong enough to keep the mouth of the bay open. If not, the spit may eventually extend across the bay and connect to the mainland, forming a **bay barrier** or **bay-mouth bar** (Figure 11-8a), which cuts off the bay from the open ocean. Although bay barriers are a buildup of sand usually less than 1 meter (3.3 feet) above sea level, permanent buildings are often constructed on them.

A **tombolo** (*tombolo* = mound) is a sand ridge that connects an island or sea stack to the mainland (Figure 11-8b). Tombolos can also connect two adjacent islands. Formed in the wave-energy shadow of an island, tombolos are usually perpendicular to the average direction from which waves approach.

Barrier Islands Extremely long offshore deposits of sand lying parallel to the coast are called **barrier islands** (Figure 11-9). They form a first line of defense against storm waves that otherwise would severely assault the shore. Their origin is complex, but many barrier islands seem to have developed during the worldwide rise in sea level that began with the melting of the most recent major glaciers some 18,000 years ago.

At least 280 barrier islands ring the Atlantic and Gulf Coasts of the United States. They are nearly con-

Figure 11–7 Features of depositional coasts.

Diagrammatic view of features characteristic of depositional coasts.

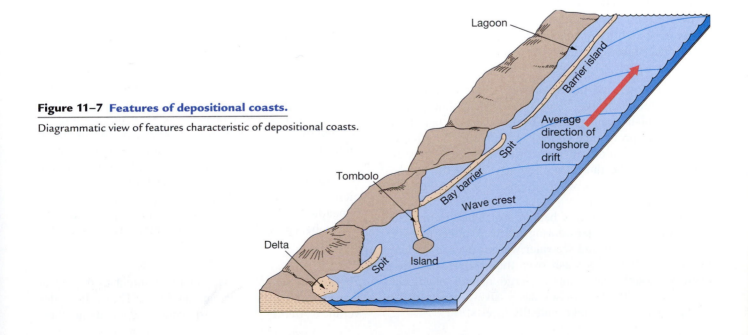

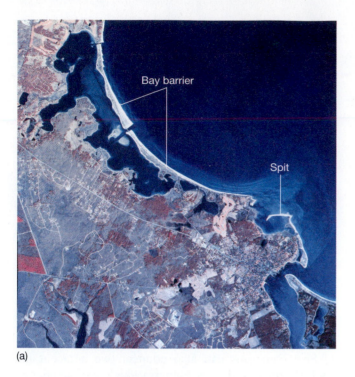

(a)

(b)

Figure 11–8 Coastal depositional features.

(a) Barrier coast, spit, and bay barrier along the coast of Martha's Vineyard, Massachusetts. (b) Tombolo at Goat Rock Beach, California.

tinuous from Massachusetts to Florida and continue through the Gulf of Mexico, where they exist well south of the Mexican border. Barrier islands may exceed 100 kilometers (60 miles) in length, have widths of several kilometers, and are separated from the mainland by a lagoon. Notable barrier islands include Fire Island off the New York coast, North Carolina's Outer Banks, and Padre Island off the coast of Texas.

A typical barrier island has the physiographic features shown in Figure 11-10a. From the ocean landward, they are (1) ocean beach, (2) dunes, (3) barrier flat, (4) high salt marsh, (5) low salt marsh, and (6) lagoon between the barrier island and the mainland.

During the summer, gentle waves carry sand to the *ocean beach*, so it widens and becomes steeper. During the winter, higher-energy waves carry sand offshore and produce a narrow, gently sloping beach.

Winds blow sand inland during dry periods to produce coastal *dunes*, which are stabilized by dune grasses. These plants can withstand salt spray and burial by sand. Dunes protect the lagoon against excessive flooding during storm-driven high tides. Numerous passes exist through the dunes, particularly along the southeastern Atlantic Coast, where dunes are less well developed than to the north.

The *barrier flat* forms behind the dunes from sand driven through the passes during storms. Grasses quickly colonize these flats and seawater washes over them during storms. If storms wash over the barrier flat infrequently enough, the plants undergo natural biological succession, with the grasses successively replaced by thickets, woodlands, and eventually forests.

Salt marshes typically lie inland of the barrier flat. They are divided into the *low marsh*, which extends from about mean sea level to the high neap-tide line, and the *high marsh*, which extends to the highest spring-tide line. The low marsh is by far the most biologically productive part of the salt marsh.

New marshland is formed as overwash carries sediment into the lagoon, filling portions so they become intermittently exposed by the tides. Marshes may be poorly developed on parts of the island that are far from flood-tide inlets. Their development is greatly restricted on barrier islands, where people perform artificial dune enhancement and fill inlets, which are activities that prevent overwashing and flooding.

The gradual sea level rise experienced along the eastern North American coast is causing barrier islands to migrate landward. The movement of the barrier island is similar to a slowly moving tractor tread, with the entire island rolling over itself, impacting structures built on these islands. *Peat deposits*, which are formed by the accumulation of organic matter in marsh environments, provide further evidence of barrier island migration (Figure 11-10b). As the island slowly rolls over itself and migrates toward land, it buries ancient peat deposits. These peat deposits can be found beneath the island and may even be exposed on the ocean beach when the barrier island has moved far enough.

Deltas Some rivers carry more sediment to the ocean than longshore currents can distribute. These rivers develop a **delta** (*delta* = triangular) deposit at their

(a)

(b)

(c)

Figure 11–9 Barrier islands.

(**a**) Barrier islands along North Carolina's Outer Banks. (**b**) Barrier islands along the south Texas coast. (**c**) A portion of a heavily developed barrier island near Tom's River, New Jersey.

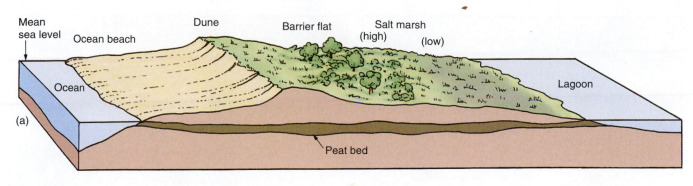

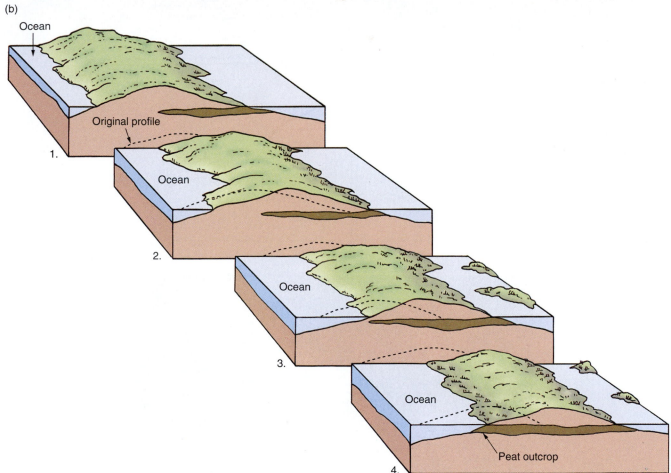

Figure 11–10 Formation of barrier islands.

(a) Diagrammatic view showing the major physiographic zones of a barrier island. The peat bed represents ancient marsh environments. (b) Sequence (*1–4*) showing how a barrier island migrates and exposes peat deposits that have been covered by the island as it migrates toward the mainland in response to rising sea level.

mouths. The Mississippi River, which empties into the Gulf of Mexico (Figure 11-11a), forms one of the largest deltas on Earth. Deltas are fertile, flat, low-lying areas that are subject to periodic flooding.

Delta formation begins when a river has filled its mouth with sediment. The delta then grows through the formation of *distributaries*, which are branching channels that deposit sediment as they radiate out over the delta in fingerlike extensions (Figure 11-11a). When the fingers get too long, they become choked with sediment. At this point, a flood may easily shift the distrib-

utary's course and provide sediment to low-lying areas between the fingers. When depositional processes exceed coastal erosion and transportation processes, a branching "bird's foot" Mississippi-type delta results.

When erosion and transportation processes exceed deposition, on the other hand, a delta shoreline is smoothed to a gentle curve, like that of the Nile River Delta in Egypt (Figure 11-11b). The Nile Delta is presently eroding because sediment is trapped behind the Aswan High Dam. Prior to completion of the dam in 1964, the Nile carried huge volumes of sediment into the Mediterranean Sea.

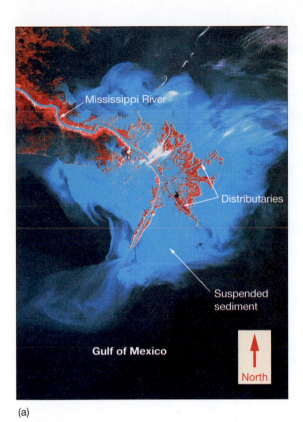

(a)

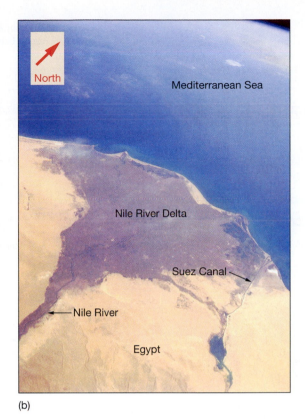

(b)

Figure 11–11 Deltas.

(a) False-color infrared image of the branching "bird's foot" structure of the Mississippi River Delta. Red color is vegetation on land; light blue color is suspended sediment within the water. **(b)** Photograph from the space shuttle of Egypt's Nile River Delta, which has a smooth, curved shoreline as it extends into the Mediterranean Sea.

Beach Compartments **Beach compartments** consist of three components: a series of rivers that supply sand to a beach; the beach itself where sand is moving due to longshore transport; and offshore submarine canyons where sand is drained away from the beach. The map in Figure 11-12 shows that the coast of southern California contains four separate beach compartments.

Primarily rivers, but also coastal erosion, supply sand to the beach within an individual beach compartment (Figure 11-12, *inset*). The sand moves south with the longshore current, so beaches are wider near the southern (*downstream*) end of each beach compartment. Although some sand is washed offshore along the way, most eventually moves near a head of a submarine canyon, where it is diverted away from the beach and onto the ocean floor. When the sand is removed from the coastal environment, it is lost from the beach forever. To the south of this beach compartment, the beaches will be thin and rocky, without much sand. The process begins all over again at the next beach compartment, where rivers add their sediment. Farther downstream, the beach widens and has an abundance of sand until that sand is also moved down a submarine canyon.

STUDENTS SOMETIMES ASK ...
Can submarine canyons fill with sediment?

Yes. In many beach compartments, the submarine canyons that drain sand from the beach empty into deep basins offshore. However, given several million years and tons of sediments per year sliding down the submarine canyons, the offshore basins begin to fill up and can eventually be exposed above sea level. In fact, the Los Angeles basin in California was filled in by sediment derived from local mountains in this manner during the geologic past.

Human activities have altered the natural system of beach compartments. When a dam is built along one of the rivers that feed into the beach compartment, it deprives the beach of sand. Lining rivers with concrete for flood control further reduces the sediment load delivered to coastal regions. Longshore transport continues to sweep the shoreline's sand into the submarine canyons, so the beaches

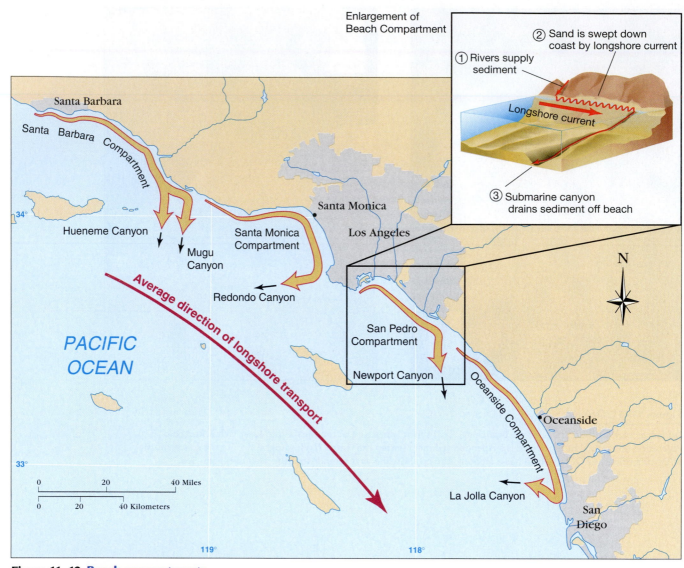

Figure 11–12 Beach compartments.

Southern California has several beach compartments, which include rivers that bring sediment to the beach, the beach that experiences longshore transport, and the submarine canyons that remove sand from the beaches. Average longshore transport is toward the south.

become narrower and experience **beach starvation**. If all the rivers are blocked, the beaches may nearly disappear.

What can be done to prevent beach starvation in beach compartments? One obvious solution is to eliminate the dams, which would allow rivers to supply sand to the beach and return beach compartments to a natural balance. However, most dams are built for flood protection, water storage, and the generation of hydropower, so it is unlikely that many will be removed.

Another option is **beach replenishment** (also called **beach nourishment**), in which sand is added to the beach to replace the sediment held back by dams. Beach replenishment is expensive, however, because huge volumes of sand must be continually supplied to the beach. When dams are built, their effect on beaches far downstream is rarely considered. It's not until beach starvation occurs that the rivers are seen as parts of much larger systems that operate along the coast.

STUDENTS SOMETIMES ASK...
How much does beach replenishment cost?

The cost of beach replenishment depends on the type and quantity of material placed on the beach, how far the material must be transported, and how it is to be distributed on the beach. Most sand used for replenishment comes from offshore areas, but sand that is dredged from nearby rivers, drained dams, harbors, and lagoons is also used.

The average cost of sand used to replenish beaches is between $5 and $10 per 0.76 cubic meter (1 cubic yard). For comparison, a typical top-loading trash dumpster holds about 2.3 cubic meters (3 cubic yards) of material, and a typical dump truck holds about 45 cubic meters (60 cubic yards) of material. The problem with replenishment projects is that a huge volume of sand is needed, and new sand must be supplied continuously. For example, a small beach replenishment project of several hundred cubic meters can cost around

$10,000 per year. Larger projects—several thousand cubic meters of sand—cost several million dollars per year.

Recycled glass that is ground to sand size and spread on the beach has been proposed as a less expensive source for beach replenishment, but the health and safety risks have not been fully explored.

Erosional-type shores are characterized by erosional features such as cliffs, sea arches, sea stacks, and marine terraces. Depositional-type shores are characterized by depositional features such as spits, tombolos, barrier islands, deltas, and beach compartments.

Classification of Coasts

Francis Shepard (1897–1985) was one of the first to study coastal processes and, because of his pioneering work, is considered the "father of marine geology." Among his many accomplishments, he developed a classification of coasts that divided all coasts into one of two types: (1) **primary coasts**, which are younger coasts that have been formed by nonmarine processes; or (2) **secondary coasts**, which have aged to the point where physical and/or biological marine processes dominate the character of the coast. These two main coastal types and various subtypes are shown in Table 11-3.

Primary coasts are controlled by nonmarine processes. For example, *land erosion coasts* include drowned rivers (such as Chesapeake Bay) and drowned glacial-erosion

TABLE 11–3 **Shepard's classification of coasts.**

Coasts Shaped by Nonmarine Processes (Primary Coasts)			Coasts Shaped by Marine Processes or Marine Organisms (Secondary Coasts)		
Land erosion coasts	Drowned rivers		**Wave erosion**	Straightened coasts	
	Drowned glacial-erosion coasts	Fjord (narrow)		Irregular coasts	
		Trough (wide)	**Marine deposition coasts (prograded by waves, currents)**	Barrier coasts	Sand beaches (single ridge)
Subaerial deposition coasts	River deposition coasts	Deltas			Sand islands (multiple ridges, dunes)
		Alluvial plains			Sand spits (connected to mainland)
	Glacial-deposition coasts	Moraines			Bay barriers
		Drumlins		Cuspate forelands (large projecting points)	
	Wind deposition coasts	Dunes		Beach plains	
	Landslide coasts	Sand flats		Mud flats, salt marshes (no breaking waves)	
Volcanic coasts	Lava flow coasts		**Coasts formed by biological activity**	Coral reef, algae (in the tropics)	
	Tephra coasts			Oyster reefs	
	Coasts formed by volcanic collapse or explosion			Mangrove coasts	
Coasts shaped by Earth movements	Faults			Marsh grass	
	Folds			Serpulid reefs (small reefs constructed of serpulid worm tubes)	
	Sedimentary extrusions	Mud lumps			
		Salt domes			
Ice coasts					

Figure 11–13 **Examples of primary and secondary coasts.**

Primary coasts: **(a)** Drowned glacial-erosion coast: fjord. Fjord and Marjorie Glacier, Glacier Bay, Alaska.
(b) Wind deposition coast: dune. Baja California, Mexico. **(c)** Volcanic coast: lava flow. Kilauea, Hawaii.
Secondary coasts: **(d)** Wave erosion coast: straightened. Leucadia, California. **(e)** Barrier coast: sand beach.
Barrier island, North Carolina. **(f)** Coast formed by biological activity: mangrove. Galápagos Islands, Ecuador.

coasts (such as Puget Sound). These coasts were formed by a relative rise in sea level that accompanied the melting of glaciers at the end of the most recent Ice Age. Other types of coasts that have formed recently by nonmarine geologic processes include *subaerial deposition coasts, volcanic coasts, coasts shaped by Earth movements*, and *ice coasts* (Figure 11-13).

With time, exposure of primary coasts to the action of ocean waves or biological processes destroys all evidence of the nonmarine process that produced them, thereby converting them to secondary coasts. For instance, a primary coast initially formed by Earth movements may be eroded sufficiently by wave action to produce a secondary *wave erosion coast*. Further, if the underlying bedrock is of uniform strength, a straightened coast will be produced (Figure 11-13); if the resistance of the bedrock varies along the coast, an irregular coast is produced. In fact, the majority of the Pacific Coast of the United States is classified as a wave erosion coast. Alternatively, most of the Atlantic Coast—especially from Massachusetts south—is classified as a *marine deposition coast*. *Coasts formed by biological activity* include coral reef coasts and mangrove coasts (Figure 11-13), both of which are quite restricted along U.S. shorelines and occur only in low-latitude areas.[2]

Emerging and Submerging Shorelines

Shorelines can also be classified based on their position relative to sea level. *Sea level, however, has changed throughout time.* It can change because the level of the land changes, the level of the sea changes, or a combination of the two. Shorelines that are rising above sea level are called **emerging shorelines** and those sinking below sea level are called **submerging shorelines**.

Marine terraces (Figures 11-6, 11-14, and 11-19) are one feature characteristic of emerging shorelines. Marine terraces are flat platforms backed by cliffs, which form when a wave-cut bench is exposed above sea level. **Stranded beach deposits** and other evidence of marine processes may exist many meters above the present shoreline, indicating that the former shoreline has risen above sea level.

Features characteristic of submerging shorelines include **drowned beaches** (Figure 11-14), **submerged dune topography**, and **drowned river valleys** along the present shoreline.

What causes the changes in sea level that produce submerging and emerging shorelines? One main cause is tectonic and isostatic movements (as discussed in Chapter 3), which raise or lower the land surface relative to sea level. Another is worldwide changes in sea level, which affect the sea itself.

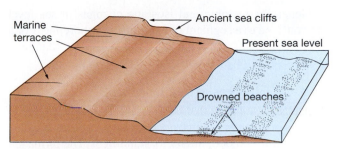

Figure 11–14 Evidence of ancient shorelines.

Marine terraces result from exposure of ancient sea cliffs and wave-cut benches above present sea level. Below sea level, drowned beaches indicate the sea level has risen relative to the land.

Tectonic and Isostatic Movements of Earth's Crust

The most dramatic changes in sea level during the past 3000 years have been caused by *tectonic processes* (movement of the land). These changes include uplift or subsidence of major portions of continents or ocean basins, as well as localized folding, faulting, or tilting of the continental crust.

Earth's crust also undergoes *isostatic adjustment*. It sinks under the accumulation of heavy loads of ice, vast piles of sediment, or outpourings of lava, and it rises when heavy loads are removed.

Most of the U.S. Pacific Coast is an emerging shoreline because continental margins where plate collisions occur are tectonically active, producing earthquakes, volcanoes, and mountain chains paralleling the coast. Most of the U.S. Atlantic Coast, on the other hand, is a submerging shoreline. When a continent moves away from a spreading center (such as the Mid-Atlantic Ridge), its trailing edge subsides because of cooling and the additional weight of accumulating sediment. Passive margins experience only a low level of tectonic deformation, earthquakes, and volcanism, making the Atlantic Coast far more quiet and stable than the Pacific Coast.

At least four major accumulations of glacial ice—and dozens of smaller ones—have occurred in high-latitude regions during the last 2.5 to 3 million years. Although Antarctica is still covered by a very large, thick ice cap, much of the ice that once covered much of northern Asia, Europe, and North America has melted.

The weight of ice sheets as much as 3 kilometers (2 miles) thick caused the crust beneath to sink. Today, these areas are still slowly rebounding, 18,000 years after the ice began to melt. The floor of Hudson Bay, for example, which is now about 150 meters (500 feet) deep, will be close to or above sea level by the time it stops isostatically rebounding. Another example is the Gulf of Bothnia (between Sweden and Finland), which has isostatically rebounded 275 meters (900 feet) during the last 18,000 years.

---

[2]Coral reefs are discussed in Chapter 16, "Animals of the Benthic Environment"; mangroves are discussed in Chapter 12, "Coastal Waters and Marginal Seas."

[3]The term *eustatic* refers to a highly idealized situation in which all of the continents remain static (in *good standing*), while only the sea rises or falls.

Generally, tectonic and isostatic changes in sea level are confined to a segment of a continent's shoreline. For a *worldwide* change in sea level, there must be a change in seawater volume or ocean basin capacity.

Eustatic Changes in Sea Level

A change in sea level that is experienced worldwide due to changes in seawater volume or ocean basin capacity is called **eustatic** (*eu* = good, *stasis* = standing).[3] The formation or destruction of large inland lakes, for example, causes small eustatic changes in sea level. When lakes form, they trap water that would otherwise run off the land into the ocean, so sea level is lowered worldwide. When lakes are drained and release their water back to the ocean, sea level rises.

Changes in sea floor spreading rates can change the capacity of the ocean basin, resulting in eustatic sea level changes. Fast spreading produces larger rises, such as the East Pacific Rise, which displace more water than slow-spreading ridges such as the Mid-Atlantic Ridge. Thus, fast spreading raises sea level, whereas slower spreading lowers sea level worldwide. Significant changes in sea level due to changes in spreading rate typically take hundreds of thousands to millions of years and may have changed sea level by 1000 meters (3300 feet) or more.

Ice ages cause eustatic sea level changes, too. As glaciers form, they tie up vast volumes of water on land, eustatically lowering sea level. An analogy to this effect is a sink of water representing an ocean basin. To simulate an ice age, some of the water from the sink is removed and frozen, causing the water level of the sink to be lower. In a similar fashion, worldwide sea level is lower during an ice age. During interglacial stages (such as the one we are in at present), the glaciers melt and release great volumes of water that drain to the sea, eustatically raising sea level. This would be analogous to putting the frozen chunk of ice on the counter near the sink and letting the ice melt, causing the water to drain into the sink and raise "sink level."

Glaciers during the Pleistocene Epoch[4] advanced and retreated many times on land near the poles, causing sea level to fluctuate considerably. The thermal contraction and expansion of the ocean as its temperature decreased and increased, respectively, affected sea level too. The thermal contraction and expansion of seawater works much like a mercury thermometer: as the mercury inside the thermometer warms, it expands and rises into the thermometer; as it cools, it contracts. Similarly, cooler seawater contracts and occupies less volume, thereby eustatically *lowering* sea level. Warmer seawater expands, eustatically *raising* sea level.

For every 1°C (1.8°F) change in the average temperature of ocean surface waters, sea level changes about 2 meters (6.6 feet). Microfossils in Pleistocene ocean sediments suggest that ocean surface waters may have been as much as 5°C (9°F) lower than at present. Therefore, thermal contraction of the ocean water may have lowered sea level by about 10 meters (33 feet).

---

[4]The Pleistocene Epoch of geologic time (also called the "Ice Age") is from 1.6 million to 10 thousand years ago (see the Geologic Time Scale, Figure 2-15).

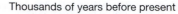
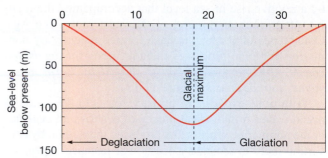

Thousands of years before present

Figure 11–15 Sea level change during the most recent advance and retreat of Pleistocene glaciers.
Sea level dropped worldwide by about 120 meters (400 feet) as the last glacial advance removed water from the oceans and transferred it to continental glaciers. About 18,000 years ago, sea level began to rise as the glaciers melted and water was returned to the oceans.

Although it is difficult to state definitely the range of shoreline fluctuation during the Pleistocene, evidence suggests that it was at least 120 meters (400 feet) below the present shoreline (Figure 11-15). It is also estimated that if *all* the remaining glacial ice on Earth were to melt, sea level would rise another 60 meters (200 feet). Thus, the *minimum* sea level change during the Pleistocene is on the order of 180 meters (600 feet), most of which was due to the capture and release of Earth's water by land-based glaciers. In fact, melting of just the West Antarctic ice sheet, which has lost about two-thirds of its mass over the past 20,000 years, has resulted in an 11-meter (36-foot) increase in worldwide sea level.

The combination of tectonic and eustatic changes in sea level is very complex, so it is difficult to classify coastal regions as purely emergent or submergent. In fact, most coastal areas show evidence of *both* submergence and emergence in the recent past. Evidence suggests, however, that until recently sea level has experienced only minor changes as a result of melting glacial ice during the last 3000 years.

Sea Level and the Greenhouse Effect

As discussed in Chapter 7, carbon dioxide in the atmosphere has increased 30% over the last 200 years and there has been an increase in global temperature of at least 0.6°C (1.1°F) over the last 130 years. Analysis of worldwide tide records indicates that there has also been a eustatic rise in sea level of between 10 and 25 centimeters (4 and 10 inches) over the last 100 years. At certain tide recording stations where data goes back well into the 19th century, there has been an increase in relative sea level of 40 centimeters (16 inches) over the last 150 years (Figure 11-16). In addition, satellite altimeter data since 1993 indicates a global increase in sea level of 2.5 millimeters (0.1 inch) per year.

Clearly, sea level is rising. Is this rise the result of increased global warming because of the greenhouse effect or is it part of a long-term natural cycle? At this point, the answer cannot be easily determined, but evidence suggests that humans are altering the environment on a global scale with emissions that enhance Earth's greenhouse effect. The rise in sea level most likely represents the combined effect of an

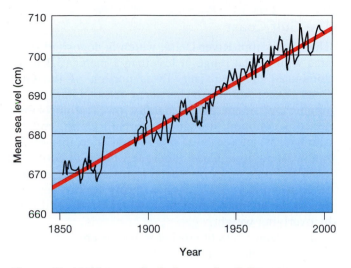

Figure 11–16 Measured relative sea level rise at New York City.

Tide-gauge data from New York City shows an increase in sea level of 40 centimeters (16 inches) since 1850. While some of this rise is due to local effects, the majority is likely caused by thermal expansion of warmer ocean water and the retreat of small ice caps and glaciers.

increase in ocean volume due to thermal expansion and the observed retreat of small ice caps and glaciers that are adding water to the ocean.

According to some estimates, the rate of sea level rise will increase with increased global warming, which may result in as much as a 1-meter (3.3-foot) rise in sea level by 2100. The recent increase in coastal development puts more homes in danger's path and compounds the problem. The great lesson for humankind in all this is that we cannot dominate nature. Rather, we must learn to live within it.

> Sea level is affected by the movement of land and changes in seawater volume or ocean basin capacity. Sea level has changed dramatically in the past because of changes in Earth's climate.

Characteristics of U.S. Coasts

Whether the dominant process along a coast is erosion or deposition depends on the combined effect of many variables, such as composition of coastal bedrock, the degree of exposure to ocean waves, tidal range, tectonic subsidence or emergence, isostatic subsidence or emergence, and eustatic sea level change.

Although many factors contribute to shoreline retreat, sea level rise is the main factor driving worldwide coastal land loss. In fact, more than 70% of the world's sandy beaches are currently eroding, and the percentage increases to nearly 90% for well-studied U.S. sandy coasts. Studies supported by the U.S. Geological Survey produced the rates of shoreline change presented in Figure 11-17, where erosion rates are shown as *negative* values and deposition rates are shown as *positive* values.

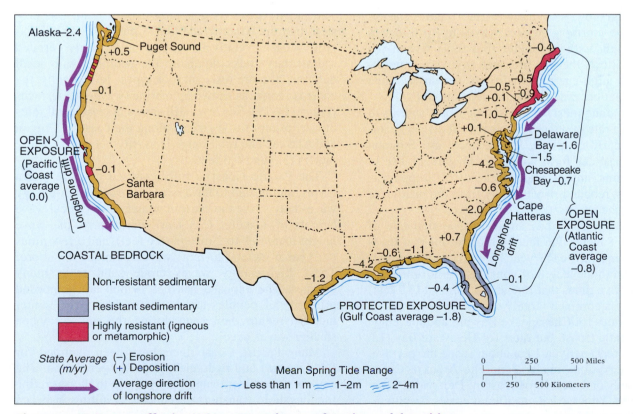

Figure 11–17 Factors affecting U.S. coasts and rates of erosion and deposition.

Map showing United States coastal bedrock type (*red, yellow,* and *blue colors*), the mean spring-tide range (*light blue lines*), degree of exposure, and average direction of longshore drift (*purple arrows*). The map also shows the average rate of erosion (−) or deposition (+) between 1979 and 1983 in meters per year for each coastal state and average rate for each coastal region.

The Atlantic Coast

Figure 11-17 shows that the U.S. Atlantic Coast has a variety of complex coastal conditions:

- Most of the Atlantic Coast is exposed to storm waves from the open ocean. Barrier islands from Massachusetts southward, however, protect the mainland from large storm waves.

- Tidal ranges generally increase from less than 1 meter (3.3 feet) along the Florida coast to more than 2 meters (6.5 feet) in Maine.

- Bedrock for most of Florida is a resistant type of sedimentary rock called *limestone*. Most of the bedrock northward through New Jersey, however, consists of nonresistant sedimentary rocks formed in the recent geologic past. As these rocks rapidly erode, they supply sand to barrier islands and other depositional features common along the coast. The bedrock north of New York consists of very resistant rock types.

- From New York northward, continental glaciers affected the coastal region directly. Many coastal features, including Long Island and Cape Cod, are glacial deposits (called *moraines*) left behind when the glaciers melted.

North of Cape Hatteras in North Carolina, the coast is subject to very high-energy waves during fall and winter when powerful storms called *"nor'easters"* (northeasters) blow in from the North Atlantic. The energy of these storms generates waves up to 6 meters (20 feet) high, with a 1-meter (3.3-foot) rise in sea level that follows the low pressure as it moves northward. Such high-energy conditions seriously erode coastlines that are predominantly depositional.

Sea level along most of the Atlantic Coast appears to be rising at a rate of about 0.3 meter (1 foot) per century. Drowned river valleys, for instance, are common along the coast and form large bays (Figure 11-18). In northern Maine, however, sea level may be dropping as the continent rebounds isostatically from the melting of the Pleistocene ice sheet.

The Atlantic Coast has an average annual rate of erosion of 0.8 meter (2.6 feet), which means that sea is migrating landward each year by a distance approximately equal to the length of your legs! In Virginia, the loss is over five times that rate at 4.2 meters (13.7 feet) per year but is confined largely to barrier islands.

Erosion rates for Chesapeake Bay are about average for the Atlantic Coast, but rates for Delaware Bay [1.6 meters (5.2 feet) per year] are about twice the average. Of the observations made along the Atlantic Coast, 79% showed some degree of erosion. Delaware, Georgia, and New York have depositional coasts despite serious erosion problems in these states as well.

The Gulf Coast

The Mississippi River Delta, which is deposited in an area with a tidal range of less than 1 meter (3.3 feet), dominates the Louisiana–Texas portion of the Gulf Coast. Except during the hurricane season (June to November), wave energy is generally low. Tectonic subsidence is common throughout the Gulf Coast, and the average rate of sea level rise is similar to that of the southeast Atlantic Coast, about 0.3 meter (1 foot) per century. Some areas of coastal Louisiana have experienced a 1-meter (3.3-foot) rise during the last century, due to the compaction of Mississippi River sediments by overlying weight.

The average rate of erosion is 1.8 meters (6 feet) per year in the Gulf Coast. The Mississippi River Delta experiences the greatest rate, averaging 4.2 meters (13.7 feet) per year. Erosion is made worse by barge channels dredged through marshlands, and Louisiana has lost more than 1 million acres of delta since 1900. Louisiana is now losing marshland at a rate exceeding 130 square kilometers (50 square miles) per year.

Although all Gulf states show a net loss of land, and the Gulf Coast has a greater erosion rate than the Atlantic Coast, only 63% of the shore is receding because of erosion. The high average rate of erosion reflects the heavy losses in the Mississippi River Delta.

The Pacific Coast

The Pacific Coast is generally experiencing less erosion than the Atlantic and Gulf Coasts. Along the Pacific Coast, relatively young and easily eroded sedimentary rocks dominate the bedrock, with local outcrops of more resistant rock types. Tectonically, the coast is rising, as shown by marine (wave-cut) terraces (Figure 11-19). Sea level still shows at least small rates of rise, except for segments along the coast of Oregon and Alaska. The tidal range is mostly between 1 and 2 meters (3.3 and 6.6 feet).

The Pacific Coast is fully exposed to large storm waves, and is said to have *open exposure*. High-energy waves may strike the coast in winter, with typical wave heights of 1 meter (3.3 feet). Frequently, the wave height increases to 2 meters (6.6 feet), and a few times per year 6-meter (20-foot) waves hammer the shore! These high-energy waves erode sand from many beaches. The exposed beaches, which are composed primarily of pebbles and boulders during the winter months, regain their sand during the summer when smaller waves occur.

Many Pacific Coast rivers have been dammed for flood control and hydroelectric power generation. The amount of sediment supplied by rivers to the shoreline for longshore transport is reduced, resulting in beach starvation in some areas.

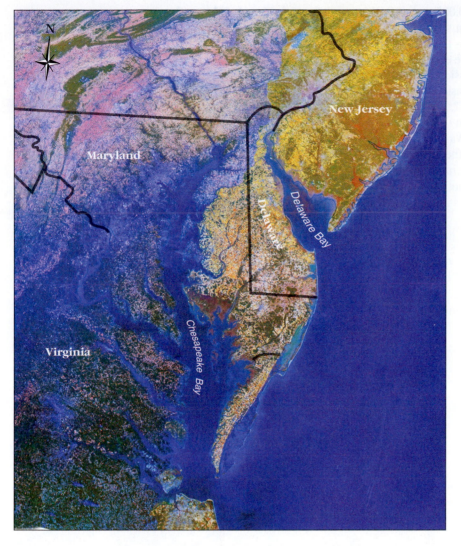

Figure 11–18 Drowned river valleys.

Satellite false-color image of drowned river valleys such as Chesapeake and Delaware Bays along the east coast of the U.S., which were formed by a relative rise in sea level that followed the end of the Pleistocene Ice Age.

Figure 11–19 Marine (wave-cut) terraces.

Each marine terrace on San Clemente Island offshore southern California was created by wave activity at sea level. Subsequently, each terrace has been exposed by tectonic uplift. The highest (oldest) terraces near the top of the photo are now about 400 meters (1320 feet) above sea level.

With an average erosion rate of only 0.005 meter (0.016 foot)[5] per year and only 30% of the coast showing erosion loss, the Pacific Coast is eroded much less than the Atlantic and Gulf Coasts. Nevertheless, high wave energy and relatively soft rocks result in high rates of erosion in some parts of the Pacific Coast. In some parts of Alaska, for example, the average rate of erosion is 2.4 meters (7.9 feet) per year.

Of the Pacific states, only Washington shows a net sediment deposition. The long, protected Washington shoreline within Puget Sound helps skew the Pacific Coast values (Figure 11-17). Although the average erosion rate for California is only 0.1 meter (0.33 foot) per year, over 80% of the California coast is experiencing erosion, with rates up to 0.6 meter (2 feet) per year.

> U.S. coastal regions are affected by many variables, including composition of the coastal bedrock, degree of exposure, and tidal range. Most U.S. coastal regions are experiencing erosion.

[5]0.005 meter is equal to 5 millimeters (0.2 inch).

Hard Stabilization

Coastal residents continually modify coastal sediment erosion/deposition in attempts to improve or preserve their property. Structures built to protect a coast from erosion or to prevent the movement of sand along a beach are known as **hard stabilization**. Hard stabilization can take many forms and often results in predictable yet unwanted outcomes.

Groins and Groin Fields

One type of hard stabilization is a **groin** (*groin* = ground). Groins are built perpendicular to a coastline and are specifically designed to trap sand moving along the coast in longshore transport (Figure 11-20). They are constructed of many types of material, but large blocky material called **rip-rap** is most common. Sometimes, groins are even constructed of sturdy wood pilings (similar to a fence built out into the ocean).

Although a groin traps sand on its *upstream side*, erosion occurs immediately downstream of the groin because the sand that is normally found just downstream

Figure 11–20 Interference of sand movement.

Hard stabilization like the groin shown here interferes with the movement of sand along the beach, causing deposition of sand upstream of the groin and erosion immediately downstream, modifying the shape of the beach.

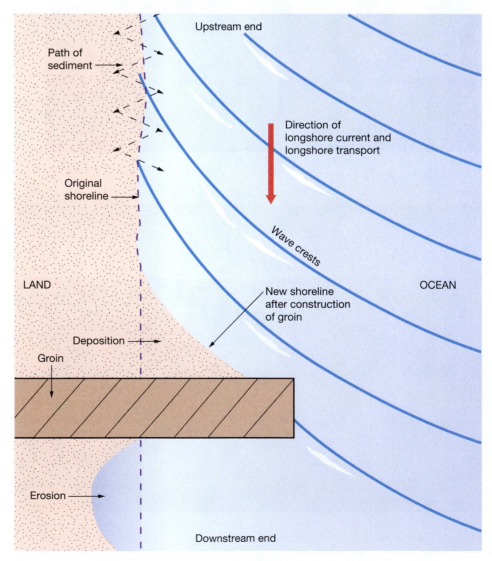

Path of sediment

Upstream end

Direction of longshore current and longshore transport

Original shoreline

Wave crests

LAND

OCEAN

New shoreline after construction of groin

Deposition

Groin

Erosion

Downstream end

of the groin is trapped on the groin's upstream side. To lessen the erosion, another groin can be constructed downstream, which in turn also creates erosion downstream from it. More groins are needed to alleviate the beach erosion, and soon a **groin field** is created (Figure 11-21).

Does a groin (or a groin field) actually retain more sand on the beach? Sand eventually migrates around the end of the groin, so there is no additional sand on the beach; it is only *distributed differently*. With proper engineering, an equilibrium may be reached that allows sufficient sand to move along the coast before excessive erosion occurs downstream from the last groin. However, some serious erosional problems have developed in many areas resulting from attempts to stabilize sand on the beach by the excessive use of groins.

Jetties

Another type of hard stabilization is a **jetty** (*jettee* = to project). A jetty is similar to a groin because it is built perpendicular to the shore and is usually constructed of rip-rap. The purpose of a jetty, however, is to protect harbor entrances from waves and only secondarily does it trap sand (Figure 11-22). Because jetties are usually built in closely spaced pairs and can be quite long, they can cause more pronounced upstream deposition and downstream erosion than groins.

Breakwaters

Figure 11-23 shows a **breakwater**—hard stabilization built parallel to a shoreline—that was constructed to create the harbor at Santa Barbara, California. California's longshore

Figure 11–21 Groin field.

A series of groins has been built along the shoreline north of Ship Bottom, New Jersey, in an attempt to trap sand, altering the distribution of sand on the beach. The view is toward the north, and the primary direction of longshore current is toward the bottom of the photo (toward the south).

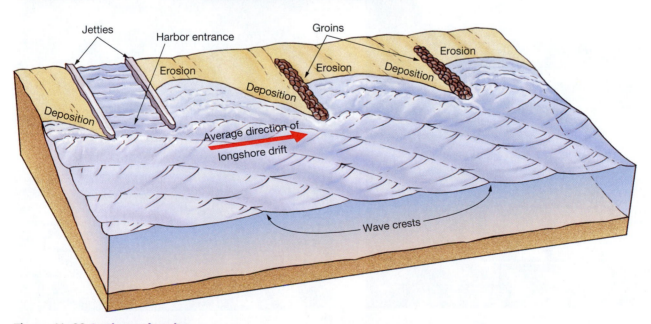

Figure 11–22 Jetties and groins.

Jetties protect a harbor entrance and usually occur in pairs. Groins are built specifically to trap sand moving in the longshore transport system and occur individually or as a groin field. Both structures cause deposits of sand on their upstream sides and an equal amount of erosion downstream.

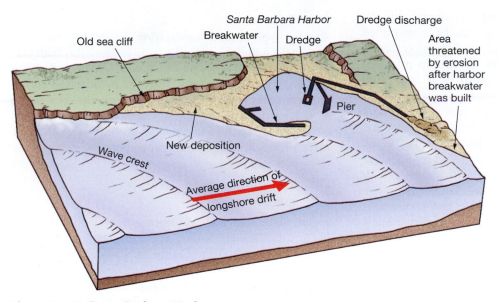

Figure 11–23 Santa Barbara Harbor.

Construction of a breakwater at Santa Barbara Harbor interfered with the longshore drift, creating a broad beach. As the beach extended around the breakwater into the harbor, the harbor was in danger of being closed off by accumulating sand. As a result, dredging operations were initiated to move sand from the harbor downstream, where it helped reduce coastal erosion.

drift is predominantly southward, so the breakwater on the western side of the harbor accumulated sand that had migrated eastward along the coast. The beach to the west of the harbor continued to grow until finally the sand moved around the breakwater and began to fill in the harbor (Figure 11-23).

While abnormal deposition occurred to the west, erosion proceeded at an alarming rate east of the harbor. The waves east of the harbor were no greater than before, but

the sand that had formerly moved down the coast was now trapped behind the breakwater.

A similar situation occurred in Santa Monica, California, where a breakwater was built to provide a boat anchorage. A bulge in the beach soon formed upstream of the breakwater and severe erosion occurred downstream (Figure 11-24). The breakwater interfered with the natural transport of sand by blocking the waves that used to

(a)

(b)

Figure 11–24 Santa Monica breakwater.

(a) The shoreline and pier at Santa Monica as it appeared in 1931. **(b)** The same area in 1949, showing that the construction of a breakwater to create a boat anchorage disrupted the longshore transport of sand and caused a bulge of sand in the beach.

keep the sand moving. If something was not done to put energy back into the system, the breakwater would soon be attached by a tombolo of sand, and further erosion downstream might destroy coastal structures.

In Santa Barbara and Santa Monica, dredging was used to compensate for erosion downstream from the breakwater and to keep the harbor or anchorage from filling with sand. Sand dredged from behind the breakwater is pumped down the coast so it can reenter the longshore drift and replenish the eroded beach.

The dredging operation has stabilized the situation in Santa Barbara, but at a considerable (and ongoing) expense. In Santa Monica, dredging was conducted until the breakwater was largely destroyed during winter storms in 1982–1983. Shortly thereafter, wave energy was able to move sand along the coast again, and the system was restored to near-normal conditions. When people interfere with natural processes in the coastal region, they must provide the energy needed to replace what they have misdirected through modification of the shore environment.

Seawalls

One of the most destructive types of hard stabilization is the **seawall** (Figure 11-25), which is built parallel to the shore along the landward side of the berm. The purpose of a seawall is to armor the coastline and protect landward developments from ocean waves.

Once waves begin breaking against a seawall, however, turbulence generated by the abrupt release of wave energy quickly erodes the sediment on its seaward side, causing it to collapse into the surf (Figure 11-25). Where seawalls have been used to protect property on barrier islands, the seaward slope of the island beach has steepened and the rate of erosion has increased, causing the destruction of the recreational beach.

A well-designed seawall may last for many decades, but the constant pounding of waves eventually takes its toll (Figure 11-26). In the long run, the cost of repairing or replacing seawalls will be more than the property is worth, and the sea will claim more of the coast through the natural processes of erosion. It's just a matter of time for homeowners who live too close to the coast, many of whom are gambling that their houses won't be destroyed in their lifetimes.

Alternatives to Hard Stabilization

Is it better to preserve the houses of a few people who have built too close to the shore at the expense of armoring the coast with hard stabilization and destroying the recreational beach? If you own coastal property, your response would probably be different from the general beachgoing public. Because hard stabilization has been shown to have negative environmental consequences, alternatives have been sought.

Of course, one alternative to the use of hard stabilization is to restrict construction in areas prone to coastal erosion. Unfortunately, this is becoming less and less an option as coastal regions experience population increases and governments increase the risk of damage and injuries

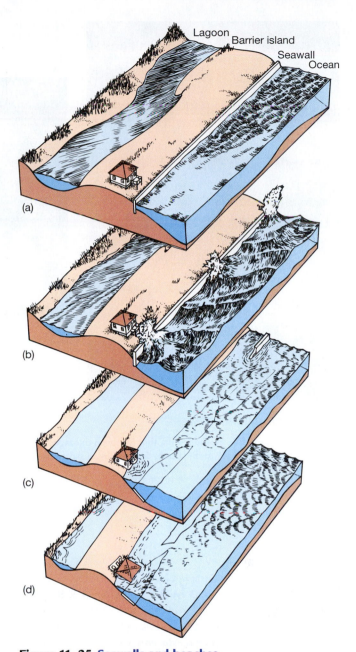

Figure 11–25 Seawalls and beaches.
When a seawall is built along a beach (such as on this barrier island) to protect beachfront property (a), a large storm can remove the beach from the seaward side of the wall and steepen its seaward slope (b). Eventually, the wall is undermined and falls into the sea (c). The property is lost (d) as the oversteepened beach slope advances landward in its effort to reestablish a natural slope angle.

because of programs like the *National Flood Insurance Program* (*NFIP*). Since its inception in 1968, NFIP has paid out billions of dollars in federal subsidy to repair or replace high-risk coastal structures. As a result, NFIP has actually *encouraged* construction in exactly the unsafe locations it was designed to prevent![6] Further, many homeowners spend large amounts of money rebuilding structures and fortifying their property.

---

[6]Recent changes in regulations of the Federal Emergency Management Agency (FEMA), which oversees NFIP, are intended to curb this practice.

Figure 11–26 Seawall damage.

A seawall in Solana Beach, California, that has been damaged by waves and needs repair.
Although seawalls appear to be sturdy, they can be destroyed by the continual pounding of high-
energy storm waves.

? STUDENTS SOMETIMES ASK...

*I have the opportunity to live in a house at the edge of a coastal
cliff where there is an incredible view along the entire coast. Is it
safe from coastal erosion?*

Based on what you've described, most certainly not! Geologists have long known that cliffs are naturally unstable.
Even if the cliffs appear to be stable (or have been stable for
a number of years), one significant storm can seriously
damage the cliff.

The most common cause of coastal erosion is direct
wave attack, which undermines the support and causes the
cliff to fail. You might want to check the base of the cliff and
examine the local bedrock to determine for yourself if you
think it will withstand the pounding of powerful storm
waves that can move rocks weighing several tons. Other
dangers include drainage runoff, weaknesses in the bedrock,
slumps and landslides, seepage of water through the cliff,
and even burrowing animals.

Even though all states enforce a setback from the edge
of the cliff for all new buildings, sometimes that isn't
enough because large sections of "stable" cliffs can fail all
at once. For instance, several city blocks of real estate have
been eroded from the edge of cliffs during the last 100

years in some areas of southern California. Even though
the view sounds outstanding, you may find out the hard
way that the house is built a little too close to the edge of
the cliff!

However, policy has recently shifted from defending
coastal property in high hazard areas to removing structures and letting nature reclaim the beach. This approach is called **relocation**, which involves moving
structures to safer locations as they become threatened
by erosion. One example of the successful use of this
technique is the relocation of the Cape Hatteras Lighthouse in North Carolina (Box 11-2). Relocation, if used
wisely, can allow humans to live in balance with the
natural processes that continually modify beaches.

> Hard stabilization includes groins, jetties, break-
> waters, and seawalls, all of which alter the coastal
> environment, cause erosion and deposition, and
> result in changes in the shape of the beach.

BOX 11–2   People and Ocean Environment

THE MOVE OF THE CENTURY: RELOCATING THE CAPE HATTERAS LIGHTHOUSE

In spite of efforts to protect structures that are too close to the shore, they can still be in danger of being destroyed by receding shorelines and the destructive power of waves. Such was the case for one of the nation's most prominent landmarks, the candy-striped lighthouse at Cape Hatteras, North Carolina, which is 21 stories tall—the nation's tallest lighthouse and the tallest brick lighthouse in the world.

The lighthouse was built in 1870 on the Cape Hatteras barrier island 457 meters (1500 feet) from the shoreline to guide mariners through the dangerous offshore shoals known as the "Graveyard of the Atlantic." As the barrier island began migrating towards land, its beach narrowed. When the waves began to lap just 37 meters (120 feet) from its brick and granite base, there was concern that even a moderate-strength hurricane could trigger beach erosion sufficient to topple the lighthouse.

In 1970, the U.S. Navy built three groins in front of the lighthouse in an effort to protect the lighthouse from further erosion. The groins initially slowed erosion, but disrupted sand flow in the surf zone, which caused the flattening of nearby dunes and the formation of a bay south of the lighthouse. Attempts to increase the width of the beach in front of the lighthouse included beach nourishment and artificial offshore beds of seaweed, both of which failed to widen the beach substantially. In the 1980s, the Army Corps of Engineers proposed building a massive stone seawall around the lighthouse but decided the eroding coast would eventually move out from under the structure, leaving it stranded at sea on its own island. In 1988, the National Academy of Sciences determined that the shoreline in front of the lighthouse would retreat so far as to destroy the lighthouse and recommended its relocation. In 1999, the National Park Service, which owns the lighthouse, finally authorized moving the structure to a safer location.

Moving the lighthouse, which weighs 4395 metric tons (4830 short tons), was accomplished by severing it from its foundation and carefully hoisting it onto a platform of steel beams fitted with roller dollies. Once on the platform, it was slowly rolled along a specially designed steel track using a series of hydraulic jacks. A strip of vegetation was cleared to make a runway along which the lighthouse crept 1.5 meters (5 feet) at a time, with the track picked up from behind and reconstructed in front of the tower as it moved. During June and July 1999, the lighthouse was gingerly transported 884 meters (2900 feet) from its original location, making it one of the largest structures ever successfully moved.

After its $12 million move, the lighthouse now resides in a scrub oak and pine woodland 488 meters (1600 feet) from the shore (Figure 11B). Although it now stands further inland, the light's slightly higher elevation makes it visible just as far out to sea, where it continues to warn mariners of the hazardous shoals. At the current rate of shoreline retreat, the lighthouse should be safe from the threat of waves for at least another century.

Figure 11B Relocation of the Cape Hatteras Lighthouse, North Carolina.

Chapter in Review

- *The coastal region changes continuously.* The *shore* is the region of contact between the oceans and the continents, lying between the lowest low tides and the highest elevation on the continents affected by storm waves. The *coast* extends inland from the shore as far as marine-related features can be found. The *coastline* marks the boundary between the shore and the coast. The shore is divided into the *foreshore*, extending from low tide to high tide, and the *backshore*, extending beyond the high tide line to the coastline. Seaward of the low tide shoreline are the *nearshore* zone, extending to the breaker line, and the *offshore* zone beyond.

- *A beach is a deposit of the shore area*, consisting of wave-worked sediment that moves along a wave-cut bench. It includes the *recreational beach*, *berm*, *beach face*, *low tide terrace*, one or more *longshore bars*, and *longshore trough*. Beaches are composed of whatever material is locally available and coarser beaches have steeper slopes.

- *Waves that break at the shore move sand perpendicular to shore* (toward and away from shore). In *light wave activity*, swash dominates the transport system and sand is moved up the beach face toward the berm. In *heavy wave activity*, backwash dominates the transport system and sand is moved down the beach face away from the berm toward longshore bars. In a natural system, there is a *balance between light and heavy wave activity*, alternating between sand piled on the berm (*summertime beach*) and sand stripped from the berm (*wintertime beach*), respectively.

- *Sand is moved parallel to the shore, too.* Waves breaking at an angle to the shore create a *longshore current that results in a zigzag movement of sediment called longshore drift* (longshore transport). Each year, *millions of tons of sediment are moved from upstream to downstream* ends of beaches. Most of the year, *longshore drift moves southward along both the Pacific and Atlantic shores* of the United States.

- *Erosional-type shores are characterized by headlands, wave-cut cliffs, sea caves, sea arches, sea stacks, and marine terraces* (caused by uplift of a wave-cut bench). Wave erosion increases as more of the shore is exposed to the open ocean, tidal range decreases, and bedrock weakens.

- *Depositional-type shores are characterized by beaches, spits, bay barriers, tombolos, barrier islands, deltas, and beach compartments.* Viewed from ocean side to lagoon side, barrier islands commonly have an ocean beach, dunes, barrier flat, and salt marsh. Deltas form at the mouths of rivers that carry more sediment to the ocean than the longshore current can carry away. *Beach*

starvation occurs when the sand supply is interrupted. *Beach replenishment* (beach nourishment) is an expensive and temporary way to reduce beach starvation.

- *Coasts can be classified as either primary or secondary coasts.* Primary coasts include those developed by *nonmarine processes*; they include land erosion coasts, subaerial deposition coasts, glacial deposition coasts, volcanic coasts, coasts shaped by earth movements, and ice coasts. *Secondary coasts include those where the nonmarine character has been destroyed by marine processes*; they include wave erosion coasts, marine deposition coasts, and coasts formed by marine biological activity.

- *Shorelines can also be classified as emerging or submerging based on their position relative to sea level.* Ancient wave-cut cliffs and stranded beaches well above the present shoreline may indicate a *drop in sea level relative to land.* Old drowned beaches, submerged dunes, wave-cut cliffs, or drowned river valleys may indicate a *rise in sea level relative to land.* Changes in sea level may result from *tectonic processes* causing local movement of the landmass or from *eustatic processes* changing the amount of water in the oceans or the capacity of ocean basins. Melting of continental ice caps during the past 18,000 years has caused a eustatic rise in sea level of about 120 meters (400 feet).

- *Sea level is rising along the Atlantic Coast* about 0.3 meter (1 foot) per century, and the average erosion rate is −0.8 meter (−2.6 feet) per year. *Along the Gulf Coast*, sea level is rising 0.3 meter (1 foot) per century, and the average rate of erosion is −1.8 meters (−6 feet) per year. The Mississippi River Delta is eroding at 4.2 meters (13.7 feet) per year, resulting in a large loss of wetlands every year. *Along the Pacific Coast*, the average erosion rate is only −0.005 meter (−0.016 foot) per year. Different shorelines erode at different rates depending on wave exposure, amount of uplift, and type of bedrock.

- *Hard stabilization, such as groins, jetties, breakwaters, and seawalls, is often constructed in an attempt to stabilize a shoreline. Groins* (built to trap sand) and *jetties* (built to protect harbor entrances) widen the beach by trapping sediment on their upstream side, but erosion usually becomes a problem downstream. Similarly, *breakwaters* (built parallel to a shore) trap sand behind the structure, but cause unwanted erosion downstream. *Seawalls* (built to armor a coast) often cause loss of the recreational beach. Eventually, the constant pounding of waves destroys all types of hard stabilization. *Relocation* is a technique that has been successfully used to protect coastal structures.

Key Terms

Backshore (p. 320)
Backwash (p. 321)
Barrier island (p. 327)
Bay barrier (bay-mouth bar) (p. 327)
Beach (p. 321)
Beach compartment (p. 331)
Beach face (p. 321)
Beach replenishment (beach nourishment) (p. 332)
Beach starvation (p. 332)
Berm (p. 321)
Breakwater (p. 341)
Coast (p. 321)
Coastline (p. 320)

Delta (p. 328)
Depositional-type shore (p. 325)
Drowned beach (p. 335)
Drowned river valley (p. 335)
Emerging shoreline (p. 335)
Erosional-type shore (p. 325)
Eustatic sea level change (p. 336)
Foreshore (p. 320)
Groin (p. 340)
Groin field (p. 341)
Hard stabilization (p. 340)
Headland (p. 325)
Jetty (p. 341)
Longshore bar (p. 321)
Longshore current (p. 322)

Longshore drift (longshore transport) (p. 322)
Longshore trough (p. 321)
Marine terrace (p. 327)
Nearshore (p. 320)
Offshore (p. 321)
Primary coast (p. 333)
Relocation (p. 344)
Rip current (p. 323)
Rip-rap (p. 340)
Sea arch (p. 326)
Sea cave (p. 326)
Sea stack (p. 326)
Seawall (p. 343)
Secondary coast (p. 333)

Shepard, Francis (p. 333)
Shore (p. 320)
Shoreline (p. 320)
Spit (p. 327)
Stranded beach deposit (p. 335)
Submerged dune topography (p. 335)
Submerging shoreline (p. 335)
Summertime beach (p. 322)
Swash (p. 321)
Tombolo (p. 327)
Wave-cut bench (p. 321)
Wave-cut cliff (p. 326)
Wintertime beach (p. 322)

Questions and Exercises

1. To help reinforce your knowledge of beach terminology, construct and label your own diagram similar to Figure 11-1 from memory.

2. Describe differences between summertime and wintertime beaches. Explain why these differences occur.

3. What variables affect the speed of longshore currents?

4. What is longshore drift, and how is it related to a longshore current?

5. How is the flow of water in a stream similar to a longshore current? How are the two different?

6. Why does the direction of longshore current sometimes reverse in direction? Along both U.S. coasts, what is the primary direction of annual longshore current?

7. Describe the formation of rip currents. What is the best strategy to ensure that you won't drown if you are caught in a rip current?

8. Discuss the formation of such erosional features as wave-cut cliffs, sea caves, sea arches, sea stacks, and marine terraces.

9. Describe the origin of these depositional features: spit, bay barrier, tombolo, and barrier island.

10. Describe the response of a barrier island to a rise in sea level. Why do some barrier islands develop peat deposits running through them from the ocean beach to the salt marsh?

11. Discuss why some rivers have deltas and others do not. What are the factors that determine whether a "bird's-foot" delta (like the Mississippi Delta) or a smoothly curved delta (like the Nile Delta) will form?

12. Describe all parts of a beach compartment. What will happen when dams are built across all of the rivers that supply sand to the beach?

13. Define the characteristics of the two major categories of Shepard's classification of coasts, and list the subcategories of each.

14. Compare the causes and effects of tectonic versus eustatic changes in sea level.

15. List the two basic processes by which coasts advance seaward, and list their counterparts that lead to coastal retreat.

16. List and discuss four factors that influence the classification of a coast as either erosional or depositional.

17. Describe the tectonic and depositional processes causing subsidence along the Atlantic Coast.

18. Compare the Atlantic Coast, Gulf Coast, and Pacific Coast by describing the conditions and features of emergence-submergence and erosion-deposition that are characteristic of each.

19. List the types of hard stabilization and describe what each is intended to do.

20. Draw an aerial view of a shoreline to show the effect on erosion and deposition caused by constructing a groin, a jetty, a breakwater, and a seawall within the coastal environment.

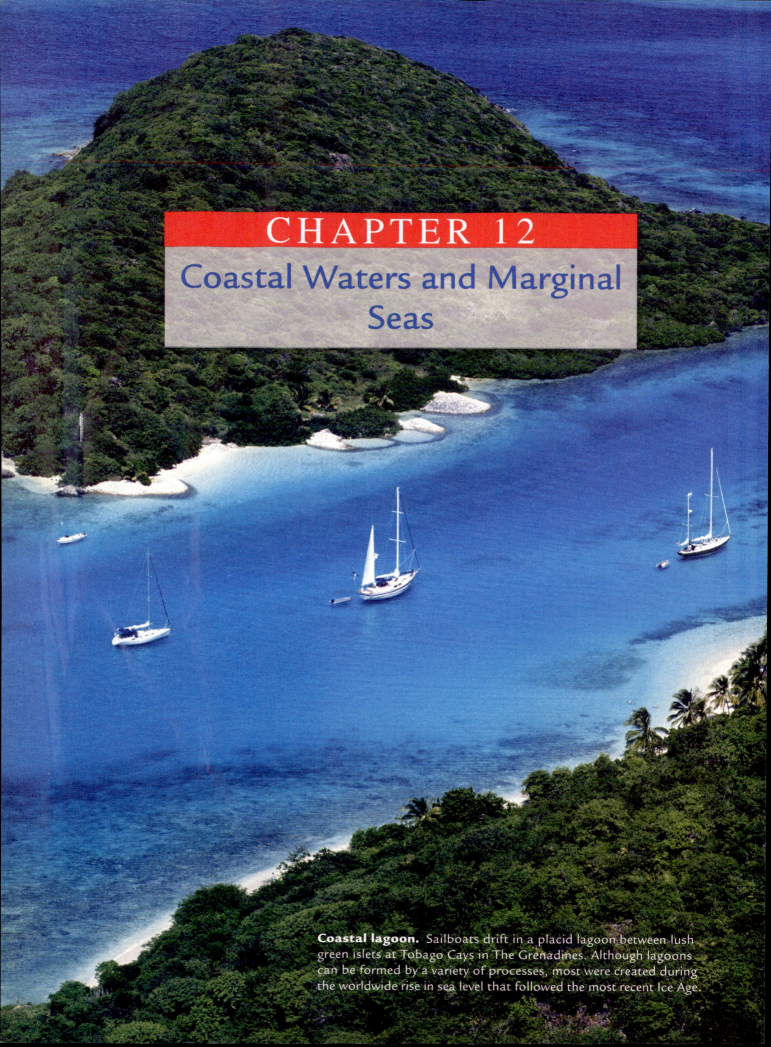

CHAPTER 12
Coastal Waters and Marginal Seas

Coastal lagoon. Sailboats drift in a placid lagoon between lush green islets at Tobago Cays in The Grenadines. Although lagoons can be formed by a variety of processes, most were created during the worldwide rise in sea level that followed the most recent Ice Age.

Key Questions

- How does the coastal ocean vary in terms of salinity, temperature, and currents?
- How are estuaries created and what kinds of estuaries exist?
- Why are coastal wetlands important?
- Why is the circulation pattern in the Mediterranean Sea so unusual?
- What characteristics do marginal seas exhibit?

The answers to these questions (and much more) can be found in the highlighted concept statements within this chapter.

"All the rivers run into the sea, yet the sea is not full."

—From *Ecclesiastes*, 1.7

Coastal waters and their adjoining marginal seas are filled with life, commerce, recreation, and fisheries. Of the world fishery,[1] about 95% is obtained within 320 kilometers (200 miles) of shore. Coastal waters also support about 95% of the total mass of life in the oceans. Further, coastal estuary and wetland environments are among the most biologically productive ecosystems on Earth and serve as nursery grounds for many species of marine organisms that inhabit the open ocean. In addition, these waters are the focal point of most shipping routes, oil and gas production, and recreational activities.

Coastal waters are also the conduits through which land-derived compounds must pass to reach the open ocean. Numerous chemical, physical, and biological processes occur in these environments that tend to protect the quality of the water in the open ocean. Human activities, however, are increasingly altering coastal environments. As we'll see in Chapter 18, "Marine Environmental Concerns," coastal waters are the final destination of much of the waste products of those living on the adjacent land.

Coastal Waters

Coastal waters are those relatively shallow-water areas that adjoin continents or islands. If the continental shelf is broad and shallow, coastal waters can extend several hundred kilometers from land. If it has significant relief or drops rapidly onto the deep-ocean basin, on the other hand, coastal waters will occupy a relatively thin band near the margin of the land. Beyond coastal waters lies the open ocean.

Because of their proximity to land, coastal waters are directly influenced by processes that occur on or near land. River runoff and tidal currents, for example, have a far more significant effect on coastal waters than on the open ocean.

Salinity

Fresh water is less dense than seawater, so river runoff does not mix well with seawater along the coast. Instead, the fresh water forms a wedge at the surface, which creates a well-developed **halocline**[2] (Figure 12-1a). When water is shallow enough, however, tidal mixing causes fresh water to mix with seawater, thus reducing the salinity of the water column (Figure 12-1c). There is no halocline here; instead, the water column is **isohaline** (*iso* = same, *halo* = salt).

Freshwater runoff from the continents generally lowers the salinity of coastal regions compared to the open ocean. Where precipitation on land is mostly rain, river runoff peaks in the rainy season. Where runoff is due mainly to melting snow and ice, on the other hand, runoff always peaks in summer.

Prevailing offshore winds can increase the salinity in some coastal regions. As winds travel over a continent, they usually lose most of their moisture. When these dry winds reach the ocean, they typically evaporate considerable amounts of water as they move across the surface of the coastal waters. The increased evaporation rate increases surface salinity, creating a halocline (Figure 12-1b). The gradient of the halocline, however, is reversed compared to the one developed from the input of fresh water (Figure 12-1a).

Temperature

Sea ice forms in many high-latitude coastal areas where water temperatures are uniformly cold—generally greater than $-2°C$ (28.4°F) (Figure 12-1d). In low-latitude coastal regions, where circulation with the open ocean is restricted, surface waters are prevented from mixing thoroughly, so maximum surface temperature may approach 45°C (113°F) (Figure 12-1e). In both high- and low-latitude coastal waters, **isothermal** (*iso* = same, *thermo* = heat) conditions prevail.

Surface temperatures in mid-latitude coastal regions are coolest in winter and warmest in late summer. A strong **thermocline**[3] may develop from surface water being warmed during the summer (Figure 12-1f) and cooled during the winter (Figure 12-1g). In summer, very-high-temperature surface water may form a relatively thin layer. Vertical mixing reduces the surface temperature by distributing the heat through a greater volume of water, thus pushing the thermocline deeper and making it less pronounced. In winter, cooling increases the density of

---

[1]The term *fishery* refers to fish caught from the ocean by commercial fishers.

[2]Recall that a *halocline* (*halo* = salt, *cline* = slope) is a layer of rapidly changing salinity, as discussed in Chapter 6.

[3]Recall that a *thermocline* (*themo* = heat, *cline* = slope) is a layer of rapidly changing temperature, as discussed in Chapter 6.

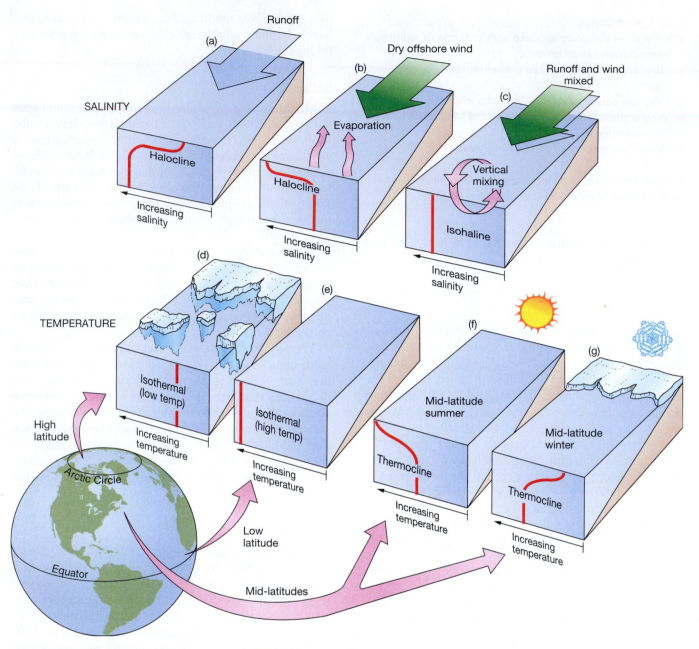

Figure 12–1 Salinity and temperature variation in the coastal ocean.
Changes in coastal salinity (*top row*) can be caused by the input of freshwater runoff (**a**), by dry offshore winds causing a high rate of evaporation (**b**), or by both (**c**). Changes in coastal temperature (*bottom row*) depend on latitude. In high latitudes (**d**), the temperature of coastal water remains uniformly near freezing. In low latitudes (**e**), coastal water may become uniformly warm. In the mid-latitudes, coastal surface water is significantly warmed during summer (**f**) and cooled during the winter (**g**).

surface water, which causes it to sink, thus creating an isothermal water column.

Prevailing offshore winds can significantly affect surface water temperatures. These winds are relatively warm during the summer, so they increase the ocean surface temperature and seawater evaporation. During winter, they are much cooler than the ocean surface, so they absorb heat and cool surface water near shore. Mixing from strong winds may drive the thermoclines in Figures 12-1f and 12-1g deeper and even mix the entire water column, producing isother-

mal conditions. Tidal currents can also cause considerable vertical mixing in shallow coastal waters.

Coastal Geostrophic Currents

Recall from Chapter 8 that *geostrophic* (*geo* = earth, *strophio* = turn) *currents* move in a circular path around the middle of a current gyre. Wind and runoff create geostrophic currents in coastal waters, too, where they are called **coastal geostrophic currents**.

Wind blowing parallel to the coast piles up water along the shore. Gravity eventually pulls this water back toward the open ocean. As it runs downslope away from the shore, the Coriolis effect causes it to curve to the right in the Northern Hemisphere and to the left in the Southern Hemisphere. Thus, in the Northern Hemisphere, the coastal geostrophic current curves *northward* on the western coast and *southward* on the eastern coast of continents. These currents are reversed in the Southern Hemisphere.

A high-volume runoff of fresh water produces a surface wedge of fresh water that slopes away from the shore (Figure 12-2). This causes a surface flow of low-salinity water toward the open ocean, which the Coriolis effect curves to the right in the Northern Hemisphere and to the left in the Southern Hemisphere.

Coastal geostrophic currents are variable because they depend on the wind and the amount of runoff for their strength. If the wind is strong and the volume of runoff is high, then the currents are relatively strong. They are bounded on the ocean side by the steadier boundary currents comprising the open-ocean gyres.

An example of a coastal geostrophic current is the **Davidson Current** that develops along the coast of Washington and Oregon during the winter (Figure 12-2). Heavy precipitation (which produces high volumes of runoff) combines with strong southwesterly winds to produce a relatively strong northward-flowing current. It flows between the shore and the southward-flowing California Current.

The shallow coastal ocean adjoins land and experiences changes in salinity and temperature that are more dramatic than the open ocean. Coastal geostrophic currents can also develop.

Estuaries

An **estuary** (*aestus* = tide) is a partially enclosed coastal body of water in which freshwater runoff dilutes salty ocean water. The most common estuary is a river mouth, where the river empties into the sea. Many bays, inlets, gulfs, and sounds may be considered estuaries, too. All estuaries exhibit large variations in temperature and/or salinity.

The mouths of large rivers form the most economically significant estuaries, because many are seaports, centers of ocean commerce, and important commercial fisheries. Examples include Baltimore, New York, San Francisco, Buenos Aires, London, Cairo, Tokyo, and many others.

Origin of Estuaries

The estuaries of today exist because sea level has risen approximately 120 meters (400 feet) since major continental glaciers began melting 18,000 years ago. As described in Chapter 11, these glaciers covered portions of North America, Europe, and Asia during the Pleistocene Epoch, which is more commonly referred to as the *Ice Age*. Four major classes of estuaries can be identified based on their origin (Figure 12-3):

1. A **coastal plain estuary** forms as sea level rises and floods existing river valleys. These estuaries, such as Chesapeake Bay in Maryland and Virginia, are called *drowned river valleys* (see Figure 11-18).

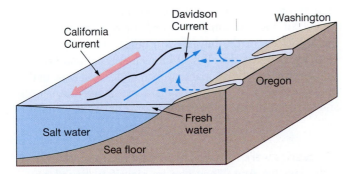

Figure 12-2 Davidson coastal geostrophic current.

The Davidson Current is a coastal geostrophic current that flows north along the coast of Washington and Oregon. During the winter, runoff produces a freshwater wedge (*light blue*) that thins away from shore. This causes a surface flow of low-salinity water toward the open ocean, which is acted upon by the Coriolis effect, curving to the right.

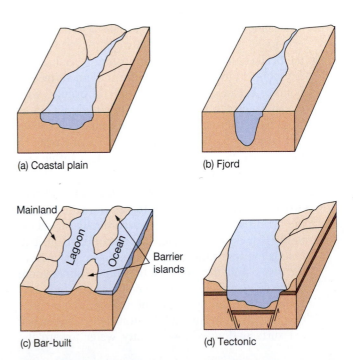

Figure 12-3 Classifying estuaries by origin.

Diagrammatic views of the four types of estuaries based on origin. **(a)** Coastal plain estuary. **(b)** Glacially carved fjord. **(c)** Bar-built estuary. **(d)** Tectonic estuary.

2. A **fjord**[4] forms as sea level rises and floods a glaciated valley. Water-carved valleys have V-shaped profiles, but fjords are U-shaped valleys with steep walls. Commonly, a shallowly submerged glacial deposit of debris (called a *moraine*) is located near the ocean entrance, marking the farthest extent of the glacier. Fjords are common along the coasts of Alaska, Canada, New Zealand, Chile, and Norway (Figure 12-4a).

3. A **bar-built estuary** is shallow and is separated from the open ocean by sand bars that are deposited parallel to the coast by wave action. Lagoons that separate *barrier islands* from the mainland are bar-built estuaries. They are common along the U.S. Gulf and East Coasts, including Laguna Madre in Texas and Pamlico Sound in North Carolina (see Figure 11-9).

4. A **tectonic estuary** forms when faulting or folding of rocks creates a restricted downdropped area into which rivers flow. San Francisco Bay is in part a tectonic estuary (Figure 12-4b), formed by movement along faults including the San Andreas Fault.

Water Mixing in Estuaries

Generally, freshwater runoff moves across the upper layer of the estuary toward the open ocean, whereas denser seawater moves in a layer just below toward the head of the estuary. Mixing takes place at the contact between these water masses.

Estuaries can be classified based on the way freshwater and seawater mix, as shown in Figure 12-5:

1. **Vertically mixed estuary**—A shallow, low-volume estuary where the net flow always proceeds from the head of the estuary toward its mouth. Salinity at any point in the estuary is uniform from surface to bottom because river water mixes evenly with ocean water at all depths. Salinity simply increases from the head to the mouth of the estuary, as shown in Figure 12-5a. Salinity lines curve at the edge of the estuary because the Coriolis effect influences the inflow of seawater.

2. **Slightly stratified estuary**—A somewhat deeper estuary in which salinity increases from the head to the mouth at any depth, as in a vertically mixed estuary. However, two water layers can be identified. One is the less saline, less dense upper water from the river, and the other is the more saline, more dense deeper water from the ocean. These two layers are separated by a zone of mixing. The circulation that develops in slightly stratified estuaries is a net surface flow of low-salinity water toward the

ocean and a net subsurface flow of seawater toward the head of the estuary (Figure 12-5b), which is called an **estuarine circulation pattern**.

3. **Highly stratified estuary**—A deep estuary in which upper-layer salinity increases from the head to the mouth, reaching a value close to that of open-ocean water. The deep-water layer has a rather uniform open-ocean salinity at any depth throughout the length of the estuary. An estuarine circulation pattern is well developed in this type of estuary (Figure 12-5c). Mixing at the interface of the upper water and the lower water creates a net movement from the deep-water mass into the upper water. Less-saline surface water simply moves from the head toward the mouth of the estuary, growing more saline as water from the deep mass mixes with it. Relatively strong haloclines develop at the contact between the upper and lower water masses.

4. **Salt wedge estuary**—An estuary in which a wedge of salty water intrudes from the ocean beneath the river water. This kind of estuary is typical of the mouths of deep, high-volume rivers. No horizontal salinity gradient exists at the surface because surface water is essentially fresh throughout the length of—and even beyond—the estuary (Figure 12-5d). There is, however, a *horizontal* salinity gradient at depth and a very pronounced vertical salinity gradient—a halocline—at any location throughout the length of the estuary. This halocline is shallower and more highly developed near the mouth of the estuary.

The mixing pattern within an estuary may vary with location, season, or tidal conditions. In addition, mixing patterns in real estuaries are rarely as simple as the models presented here. For example, Chesapeake Bay, which is one of the most intensely studied coastal bodies of water on the planet, often exhibits complex and poorly understood mixing patterns. In some cases, Chesapeake Bay exhibits upstream surface flow accompanied by downstream deep flow, which is exactly the opposite flow pattern than would normally be expected. In other cases, periods of downstream flow have been observed throughout all depths of the bay. In still other cases, even more complex patterns develop, such as a surface and bottom flow in one direction separated by a mid-depth flow in the opposite direction, or landward flows along the shores and seaward flows in the central portions of the estuary.

> Estuaries were formed by the rise in sea level after the last Ice Age. They can be classified based on origin as coastal plain, fjord, bar built, or tectonic. Estuaries can also be classified based on mixing as vertically mixed, slightly stratified, highly stratified, or salt wedge.

---

[4]The Norwegian word *fjord* is pronounced "FEE-yord" and means a long, narrow sea inlet bordered by steep cliffs.

Figure 12–4 Estuaries.
(a) A Norwegian fjord, which is a deep glacially formed estuary that has been flooded by the sea.
(b) Aerial view of San Francisco Bay in California, which is a tectonic estuary that was created by faulting.

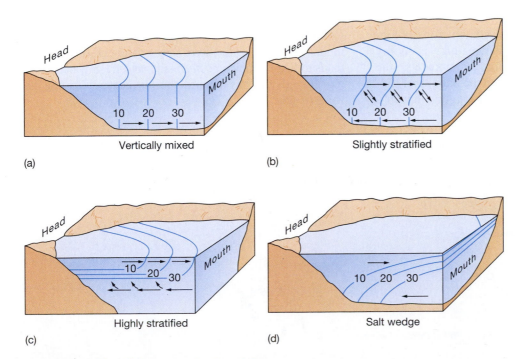

Figure 12–5 Classifying estuaries by mixing.
The basic flow pattern in an estuary is a surface flow of less dense fresh water toward the ocean and an opposite flow in the subsurface of salty seawater into the estuary. Numbers represent salinity in ‰; arrows indicate flow directions. **(a)** Vertically mixed estuary. **(b)** Slightly stratified estuary. **(c)** Highly stratified estuary. **(d)** Salt wedge estuary.

Estuaries and Human Activities

Estuaries are important breeding grounds and protective nurseries for many marine animals, so the ecological well-being of estuaries is vital to fisheries and coastal environments worldwide. Nevertheless, estuaries support shipping, logging, manufacturing, waste disposal, and other activities that can potentially damage the environment.

Estuaries are most threatened where human population is large and expanding, but they can be severely damaged where populations are still modest, too. Development in the Columbia River estuary, for example, demonstrates how a relatively small population can damage an estuary.

Columbia River Estuary The Columbia River, which forms most of the border between Washington and Oregon, has a long salt-wedge estuary at its entrance to the Pacific Ocean (Figure 12-6). The strong flow of the river and tides drive a salt wedge as far as 42 kilometers (26 miles) upstream and raise the river's water level over 3.5 meters (12 feet). When the tide falls, the huge flow of freshwater [up to 28,000 cubic meters (1,000,000 cubic feet) per second] creates a freshwater wedge that can extend hundreds of kilometers into the Pacific Ocean.

Most rivers create floodplains along their lower courses, which have rich soil that can be used for growing crops. In the late 19th century, farmers and dairymen moved onto the floodplains along the Columbia River. Eventually, protective dikes were built to prevent the annual flooding. Flooding brings new nutrients, however, so the dikes deprived the floodplain of the nutrients necessary to sustain agriculture.

The river has been the principal conduit for the logging industry, which dominated the region's economy through most of its modern history. Fortunately, the river's ecosystem has largely survived the additional sediment caused by clear cutting by the logging industry.

The construction of over 250 dams along the river and its tributaries, on the other hand, has permanently altered the river's ecosystem. Many of these dams, for example, do not have salmon ladders, which help fish "climb" in short vertical steps around the dams to reach their spawning grounds at the headwaters of their home streams.

Even though the dams have caused a multitude of problems, they do provide flood control, electrical power, and a dependable source of water, all of which have become necessary to the region's economy. To aid shipping operations, the river receives periodic dredging of sediment, which brings an increased risk for pollution. If these kinds of problems have developed in such sparsely populated areas as the Columbia River estuary, then larger environmental effects must exist in more highly populated estuaries, such as Chesapeake Bay.

Chesapeake Bay Estuary Chesapeake Bay, which formed by the drowning of the Susquehanna River (Figure 12-7), is a large coastal plain estuary that is about 320 kilometers (200 miles) long and 50 kilometers (30 miles) wide at its widest point. Most of the fresh water entering the bay comes from its western margin via rivers that drain the slopes of the Appalachian Mountains. About 15 million people live near this estuary.

Chesapeake Bay is a slightly stratified estuary that experiences large seasonal changes in salinity, temperature, and dissolved oxygen. Figure 12-7a shows the estuary's average surface salinity, which increases oceanward. The salinity lines are oriented virtually north–south in the middle of the bay because of the Coriolis effect. The Coriolis effect causes flowing water to curve to the right in the Northern Hemisphere, so seawater entering the bay tends to hug the bay's *eastern* side, and fresh water flowing through the bay toward the ocean tends to hug its *western* side.

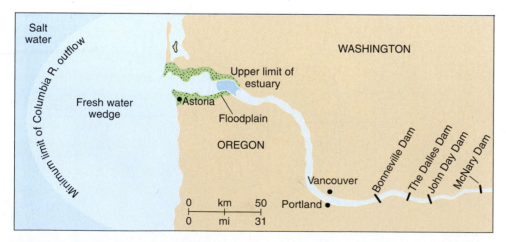

Figure 12–6 Columbia River estuary.

The long estuary at the mouth of the Columbia River has been severely affected from interference by floodplains that have been diked, by logging activities, and—most severely—by hydroelectric dams. The tremendous outflow of the Columbia River creates a large wedge of low-density fresh water that remains traceable far out at sea.

Figure 12–7 Chesapeake Bay.

(a) Map of Chesapeake Bay, showing average surface salinity (*blue lines*) in ‰. The purple area in the middle of the bay represents anoxic (oxygen-poor) waters. (b) Profile along length of Chesapeake Bay showing dissolved oxygen concentration (in ppm) during July–August 1980, indicating deep anoxic waters (*purple*). (c) Comparison profile showing normal dissolved oxygen concentrations (in ppm) during July 1950.

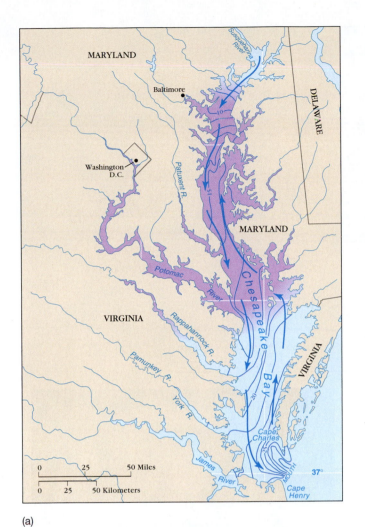

(a)

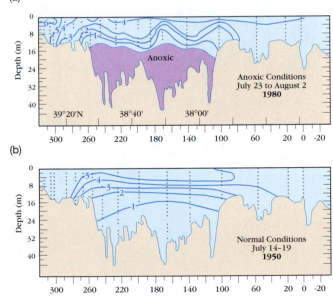

(b)

(c)

With maximum river flow in the spring, a strong halocline (and *pycnocline*[5]) develops, preventing the fresh surface water and saltier deep water from mixing. Beneath the pycnocline, which can be as shallow as 5 meters (16 feet), waters may become **anoxic** (*a* = without, *oxic* = oxygen) from May through August, as dead organic matter decays in the deep water (Figure 12-7b). Major kills of commercially important blue crab, oysters, and other bottom-dwelling organisms occur during this time.

The degree of stratification and extent of mortality of bottom-dwelling animals have increased since the early 1950s. Increased nutrients from sewage and agricultural fertilizers have been added to the bay during this time, too, which has increased the productivity of microscopic algae (algal blooms). When these organisms die, their remains accumulate as organic matter at the bottom of the bay and promote the development of anoxic conditions. In drier years with less river runoff, however, anoxic conditions aren't as widespread or severe in bottom waters because fewer nutrients are supplied.

Coastal Wetlands

Wetlands are ecosystems in which the water table is close to the surface, so they are typically saturated most of the time. Wetlands can border either fresh water or coastal environments. Coastal wetlands occur along the margins of estuaries and other shore areas that are protected from the open ocean and include swamps, tidal flats, coastal marshes, and bayous.

The two most important types of coastal wetlands are **salt marshes** and **mangrove swamps**. Both are intermittently submerged by ocean water and both have oxygen-poor mud and accumulations of organic matter called *peat deposits*. Marshes support a variety of grasses and are known to occur from the Equator to latitudes as high as 65 degrees (Figure 12-8a and b). Mangroves are restricted to latitudes below 30 degrees (Figure 12-8a and c).

Wetlands are some of the most highly productive ecosystems on Earth and provide enormous economic benefits when left alone. Salt marshes, for example, serve as nurseries for over half the species of commercially important fishes in the southeastern United States. Other

[5]Recall that a *pycnocline* (*pycno* = density, *cline* = slope) is a layer of rapidly changing density, as discussed in Chapter 6. A pycnocline is caused by a change in temperature and/or salinity with depth.

(a)

(b)

(c)

Figure 12–8 Salt marshes and mangrove swamps.

(a) Map showing the distribution of salt marshes (higher latitudes) and mangrove swamps (lower latitudes). (b) Salt marsh along San Francisco Bay at Shoreline Park, California. (c) Mangrove trees on Lizard Island, Great Barrier Reef, Australia.

fishes, such as flounder and bluefish, use marshes for feeding and protection during the winter. Fisheries of oysters, scallops, clams, eels, and smelt are located directly in marshes, too. Mangrove ecosystems are important nursery areas and habitats for commercially valuable shrimp, prawn, shellfish, and fish species. Both marshes and mangroves also serve as important stopover points for many species of waterfowl and migrating birds.

Wetlands are amazingly efficient at cleansing polluted water. Just 0.4 hectare (1 acre) of wetlands, for example, can filter up to 2,760,000 liters (730,000 gallons) of water each year, cleaning agricultural runoff, toxins, and other pollutants long before they reach the ocean. Wetlands remove inorganic nitrogen compounds (from sewage and fertilizers) and metals (from groundwater polluted by land sources), which become attached to clay-sized particles in the wetland mud. Some nitrogen compounds trapped in sediment are decomposed by bacteria that release the nitrogen to the atmosphere as gas and many of the remaining nitrogen compounds fertilize plants, further increasing the productivity of wetlands. As marsh plants die, their remains either accumulate as peat deposits or are broken up to become food for bacteria, fungi, and fish.

Serious Loss of Valuable Wetlands

Despite all the benefits they provide, over half of the nation's wetlands have vanished. Of the original 87 million hectares (215 million acres) of wetlands that once existed in the conterminous United States (excluding Alaska and Hawaii), only about 43 million hectares (106 million acres) remain. Wetlands have been filled in and developed for housing, industry, and agriculture, because people want to live near the oceans and because they often view wetlands as unproductive, useless land that harbors diseases.

Other countries have experienced similar losses of wetlands, too. In fact, scientists estimate that 50% of wetlands worldwide have been destroyed in the past century. The Philippines, for example, has lost 70% of its original mangrove cover.

To help prevent the loss of remaining wetlands, the U.S. Environmental Protection Agency established an Office of Wetlands Protection (OWP) in 1986. At that time, wetlands were being lost to development at a rate of 121,000 hectares (300,000 acres) per year! Currently, the rate of wetland loss has slowed to 8100 hectares (20,000 acres) per year and the agency's goal is to minimize the loss of wetlands to the point that there is no net loss of wetlands in the U.S. The OWP actively enforces regulations against wetlands pollution and identifies the most valuable wetlands so that they may be protected or restored.

A rise in sea level is predicted to exacerbate the loss of wetlands. Even using a conservative estimate of sea level rise over the next 100 years of 50 centimeters (20 inches), it is estimated that 38% to 61% of existing U.S. coastal wetlands would be lost. Some of this wetland loss, however, would be partially offset by new wetland formation on former upland areas, although even under ideal circumstances not all lost wetlands would be replaced.

> Coastal wetlands such as salt marshes and mangrove swamps are highly productive areas that serve as important nurseries for many marine organisms and act as a filter for polluted runoff.

Lagoons

Landward of barrier islands lie protected, shallow bodies of water called **lagoons** (see Figure 12-3c). Lagoons form in a bar-built type of estuary. Because of restricted circulation between lagoons and the ocean, three distinct zones can usually be identified within a lagoon. A *freshwater*

zone lies near the mouths of rivers that flow into the lagoon. A *transitional zone* of brackish[6] water occurs near the middle of the lagoon. A *saltwater zone* lies close to the entrance (Figure 12-9a).

Salinity within a lagoon is highest near the entrance and lowest near the head (Figure 12-9b). In latitudes that have seasonal variations in temperature and precipitation, ocean water flows through the entrance during a warm, dry summer to compensate for the volume of water lost through evaporation, thus increasing the salinity in the lagoon. Lagoons actually may become hypersaline[7] in arid regions, where the flow of seawater cannot keep pace with the lagoon's surface evaporation. During the rainy season, the lagoon becomes much less saline as freshwater runoff increases.

Tidal effects are greatest near the entrance to the lagoon (Figure 12-9c) and diminish inland from the saltwater zone until they are nearly undetectable in the freshwater zone.

Laguna Madre

Laguna Madre is located along the Texas coast between Corpus Christi and the mouth of the Rio Grande

---

[6]Brackish water is water with salinity between that of fresh water and seawater.

[7]Hypersaline conditions are created when water becomes excessively salty.

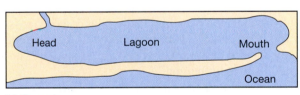

(a) Geometry

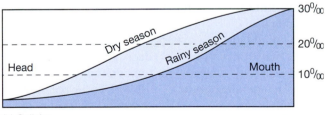

(b) Salinity

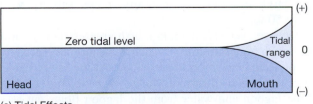

(c) Tidal Effects

Figure 12–9 Lagoons.

Typical geometry **(a)** salinity **(b)** and tidal effects **(c)** of a lagoon.

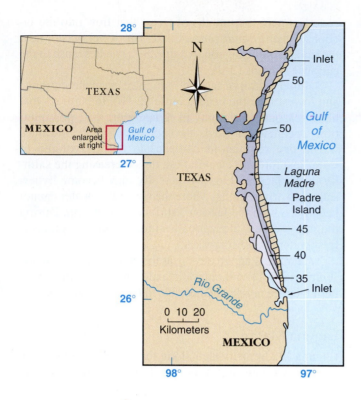

Figure 12–10 Laguna Madre summer surface salinity.

Map showing geometry of Laguna Madre, Texas, and typical summer surface salinity (in ‰).

(Figure 12-10). This long, narrow body of water is protected from the open ocean by Padre Island, a barrier island 160 kilometers (100 miles) long. The lagoon probably formed about 6000 years ago as sea level approached its present height.

The tidal range of the Gulf of Mexico in this area is about 0.5 meter (1.6 feet). The inlets at each end of Padre Island are quite narrow (Figure 12-10), so there is very little tidal interchange between the lagoon and the open sea.

Laguna Madre is a hypersaline lagoon and much of it is less than 1 meter (3.3 feet) deep. As a result, there are large seasonal changes in temperature and salinity. Water temperatures reach 32°C (90°F) in the summer and fall below 5°C (41°F) in winter. Salinities range from 2‰ when infrequent local storms provide large volumes of fresh water to over 100‰ during dry periods. High evaporation generally keeps salinity well above 50‰.[8]

Because even salt-tolerant marsh grasses cannot withstand such high salinities, the marsh has been replaced by an open sand beach on Padre Island. At the inlets, ocean water flows in as a surface wedge *over* the denser water of the lagoon and water from the lagoon flows out as a *subsurface* flow, which is exactly the opposite of the classic estuarine circulation.

[8]Recall that normal salinity in the open ocean averages 35‰.

Marginal Seas

At the margins of the ocean are relatively large semi-isolated bodies of water called **marginal seas**. Most of these seas result from tectonic events that have isolated low-lying pieces of ocean crust between continents, such as the Mediterranean Sea, or are created behind volcanic island arcs, such as the Caribbean Sea. These waters are shallower than and have varying degrees of exchange with the open ocean, depending on climate and geography; as a result, salinities and temperatures are substantially different from those of typical open ocean seawater. Let's examine some of the more important (and unusual) marginal seas that border the Atlantic, Pacific, and Indian Oceans.

Marginal Seas of the Atlantic Ocean

The Atlantic Ocean has several large marginal seas, including the Mediterranean Sea, the Caribbean Sea, and the Gulf of Mexico.

The Mediterranean Sea The Mediterranean (*medi* = middle, *terra* = land) **Sea** is actually a number of small seas connected by narrow necks of water into one larger sea. It is the remnant of the ancient Tethys Sea that existed when all the continents were combined about 200 million years ago (see Figure 3-39). It is over 4300 meters (14,100 feet) deep, and is one of the few inland seas in the world underlain by oceanic crust. Thick salt deposits and other evidence on the floor of the Mediterranean suggest that it nearly dried up about 6 million years ago, only to refill with a large salt water waterfall (Box 12-1).

The Mediterranean is bounded by Europe and Asia Minor on the north and east and Africa to the south (Figure 12-11a). It is surrounded by land except for very shallow and narrow connections to the Atlantic Ocean through the Strait of Gibraltar, and to the Black Sea through the Bosporus, which is roughly 1.6 kilometers (1 mile) wide. In addition, the Mediterranean Sea has a human-made passage to the Red Sea via the Suez Canal, a waterway 160 kilometers (100 miles) long that was completed in 1869. The Mediterranean Sea has a very irregular coastline, which divides it into subseas such as the Aegean Sea and Adriatic Sea, each of which has a separate circulation pattern.

An underwater ridge called a **sill**, which extends from Sicily to the coast of Tunisia at a depth of 400 meters (1300 feet), separates the Mediterranean into two major basins. This sill restricts the flow between the two basins, resulting in strong currents that run between Sicily and the Italian mainland through the Strait of Messina (Figure 12-11a).

Mediterranean Circulation Atlantic Ocean water enters the Mediterranean as a surface flow through the Strait of Gibraltar to replace water that rapidly evaporates in the very arid eastern end of the sea. The water level in

BOX 12–1   Research Methods in Oceanography

WHEN A SEA WAS DRY: CLUES FROM THE MEDITERRANEAN

The Mediterranean Sea is surrounded by land (Figure 12A) except for its shallow connection to the Atlantic Ocean through the Strait of Gibraltar, which is only about 14 kilometers (9 miles) wide. Analysis of sea floor sediments from the Mediterranean Sea suggest that it must have nearly dried up at least once (and perhaps several times) in its history.

About 6 million years ago, a drop in sea level or tectonic activity at the Strait of Gibraltar cut off circulation from the Atlantic Ocean—in effect, creating a dam. With the inflow of the Atlantic Ocean eliminated, the arid climate and resulting high evaporation rates caused the Mediterranean Sea to nearly evaporate in just a few thousand years. As the seawater evaporated, the dissolved substances began to precipitate out of the water, leaving thick layers of evaporite minerals on the sea floor. Up to 4000 meters (13,100 feet) of salt can be found in parts of the Mediterranean, suggesting that the basin may have partially filled and dried up several times during this period. At the same time, unusual gravel deposits washed in from the continents and shallow-water carbonate algal mats called *stromatolites*

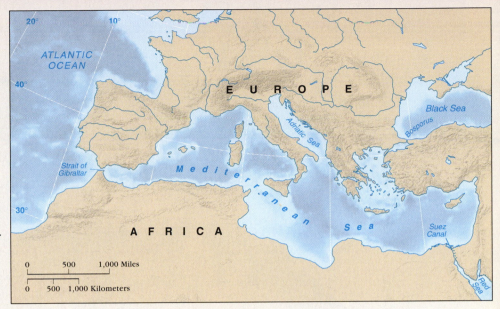

Figure 12A The Mediterranean Sea.

(*stromat* = covering, *lithos* = stone) also formed. Other supporting evidence for the drying of the Mediterranean Sea includes changes in climate, fossil evidence, and even deep notches cut into the surrounding river valleys. Eventually, most of the water evaporated, leaving a hot, salty, desiccated basin floor far below sea level.

About a half million years later, erosion, further tectonic activity, or a rise in sea level caused the dam at Gibraltar to breach and the Mediterranean started to refill with seawater. The waterfall that spilled into the Mediterranean was probably the largest ever to exist and is estimated to have been 1000 times larger than the flow of all rivers in the world. At that rate, the Mediterranean would have again been full of seawater in only 100 years. Nonetheless, the clues to its history of drying out are preserved in its sea floor sediments.

---

the eastern Mediterranean is generally 15 centimeters (6 inches) lower than at the Strait of Gibraltar. The surface flow follows the northern coast of Africa throughout the length of the Mediterranean and spreads northward across the sea (Figure 12-11a).

The remaining Atlantic Ocean water continues eastward to Cyprus. During winter, it sinks to form what is called *Mediterranean Intermediate Water*,[9] which has a temperature of 15°C (59°F) and a salinity of 39.1‰. This water flows westward at a depth of 200 to 600 meters (660 to 2000 feet) and returns to the North Atlantic as a *subsurface* flow through the Strait of Gibraltar (Figure 12-11b).

By the time Mediterranean Intermediate Water passes through Gibraltar, its temperature has dropped to 13°C (55°F) and its salinity to 37.3‰. It is still denser than even Antarctic Bottom Water and much denser than water at this depth in the Atlantic Ocean, so it moves down the continental slope. While descending, it mixes with Atlantic Ocean water and becomes less dense. At a depth of about 1000 meters (3300 feet) its density equals that of the surrounding Atlantic Ocean, so it spreads in all directions (Figure 12-11b). It has been detected in deep waters as far north as Iceland.

Circulation between the Mediterranean Sea and the Atlantic Ocean is typical of closed, restricted basins where evaporation exceeds precipitation. Low-latitude restricted basins such as this always lose water rapidly to evaporation, so surface flow from the open ocean must replace it. Evaporation of inflowing water from

---

[9]For more on subsurface and surface current circulation patterns, see Chapter 8.

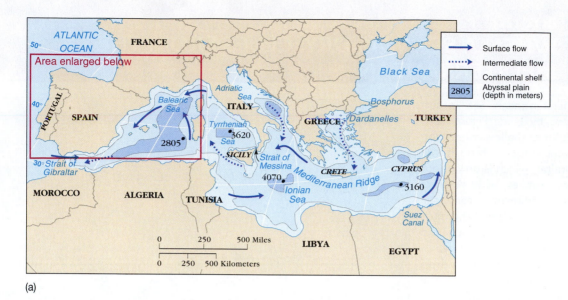

(a)

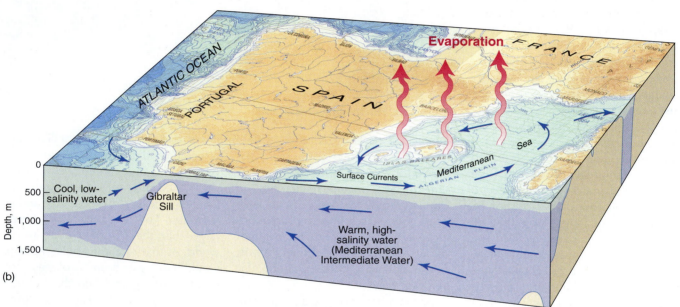

(b)

Figure 12–11 Mediterranean Sea bathymetry and circulation.

(a) Map of the Mediterranean Sea region showing its subseas, depths, sills (underwater ridges), surface flow, and intermediate flow. **(b)** Diagrammatic view of Mediterranean circulation in the Gibraltar Sill area.

the open ocean increases the sea's salinity to very high values. This denser water eventually sinks and returns to the open ocean as a subsurface flow.

This circulation pattern, which is called **Mediterranean circulation**, is opposite that of most estuaries, which experience estuarine circulation where fresh water flows at the surface into the open ocean and salty water flows below the surface into the estuary. In estuaries, however, fresh water input exceeds water loss to evaporation, whereas evaporation exceeds input in the Mediterranean.

STUDENTS SOMETIMES ASK ...
How can Mediterranean Intermediate Water sink if it's so warm?

While it is true that warm water has low density, remember that *both* salinity and temperature affect seawater density. In the case of the Mediterranean Intermediate Water, it has high enough salinity to increase its density despite being warm. Once its density increases enough, it sinks beneath the surface and retains its temperature and salinity characteristics as it flows out through the Strait of Gibraltar into the North Atlantic.

High evaporation rates in the Mediterranean Sea cause it to have a shallow inflow of surface seawater and a subsurface high-salinity outflow—a circulation pattern opposite that of most estuaries.

The Caribbean Sea The **Caribbean Sea** is separated from the Atlantic Ocean by an island arc called the Antillean Chain. Composed of the islands of Cuba, Hispaniola, Puerto Rico, and Jamaica, the *Greater Antilles* form the northern boundary of the Caribbean Sea. The *Lesser Antilles* extend in an arc from the Virgin Islands to the continental shelf of South America. The deepest connection between the Caribbean and the Atlantic Ocean is the Anegada Passage east of the Virgin Islands, with a maximum depth near 2300 meters (7550 feet); many other channels approach that depth. The Caribbean Sea is divided into four major basins from east to west: the Venezuela, Colombia, Cayman, and Yucatán basins, all of which reach depths in excess of 4000 meters (13,100 feet) (Figure 12-12a).

Circulation Patterns The *Guiana Current*, entering the Caribbean through channels between the Lesser Antilles islands, represents a portion of the *South Equatorial Current* that moves northwest along the Guiana Coast of South America. It has a temperature between 26°C and 28°C (78.8°F and 82.4°F) and a salinity between 35.0‰ and 36.5‰. This relatively thin mass of water passes into the Caribbean Sea through the shallow channels north and south of St. Lucia Island and mixes in a 1:3 ratio with *North Atlantic Water*. This current becomes the *Caribbean Current*, which travels about 250 kilometers (155 miles) north of the Venezuelan coast, continues generally west over the deepest portion of the Caribbean Sea, and finally turns north and passes through the Yucatán Strait into the Gulf of Mexico. Surface velocities as high as 4.5 kilometers (2.8 miles) per hour have been measured in the main axis of the Caribbean Current, but typical surface velocities usually average less than half that speed.

The easterly component of the trade winds blowing along the coast of Venezuela and Colombia sets up a surface flow away from the coast that produces shallow *upwelling* (movement of deep water to the surface). Most of the water rising to the surface comes from depths of less than 250 meters (820 feet). The upwelling water is cold and contains high concentrations of nutrients, which, in the presence of sunlight, result in relatively high biological productivity at the surface.

Water Masses There are four readily identifiable water masses in the Caribbean Sea. Two are relatively warm surface masses found above 200 meters (660 feet) depth, and two are deeper masses characterized by lower temperature. The salinity of *Caribbean Surface Water* is determined by the rate of evaporation versus the amount of precipitation and runoff; it is generally above 36‰ during the winter, partially because of the upwelling of high-salinity water. Moving northward, the surface salinity decreases to values generally less than 35.5‰.

Extending from the southeast to the northwest of the Caribbean Sea near the Yucatán Strait is a thin, sheetlike, high-salinity layer called the *Subtropical Underwater*. Located at depths as shallow as 50 meters (164 feet) in the southeast, it dips to a depth of 200 meters (660 feet) near the Yucatán Strait. The maximum salinity within this sheet follows the axis of flow for the Caribbean Current and exceeds 37‰ in the Yucatán Strait; salinity decreases away from the flow axis. Directly beneath the Subtropical Underwater, following the main flow axis, is a low-salinity water mass called the *Subtropical Intermediate Water*. Identification of this water mass is based on a salinity minimum that falls below 34.7‰ in the southeast and becomes less detectable at the Yucatán Strait (Figure 12-12b).

Deep below the surface, *North Atlantic Deep Water* enters the Caribbean primarily through the Anegada Passage between the Virgin Islands and the Leeward Islands of the Lesser Antilles, and through the Windward Passage between Cuba and Hispaniola. Characterized by a salinity slightly less than 35‰ and a temperature just above 2°C (35.6°F), this water spreads out as *Caribbean Bottom Water* and can be identified by an oxygen maximum layer that reaches values in excess of 5 parts per million.

The Gulf of Mexico The **Gulf of Mexico** is tectonically much less complex than the adjoining Caribbean Sea. Surrounded by a wide continental shelf, this relatively broad basin reaches a maximum depth in excess of 3600 meters (11,800 feet). The Gulf of Mexico is connected to the Caribbean Sea by the Yucatán Strait, which reaches a maximum depth of 1900 meters (6200 feet); its only connection with the Atlantic Ocean is through the Straits of Florida, which reach depths approaching 1000 meters (3300 feet).

Taken together, the Caribbean Sea and the Gulf of Mexico have an unusual geometry of a series of deep basins that are connected by passageways similar to the Mediterranean Sea. This relationship, plus the fact that they exist in an area of high evaporation, is why these two bodies of water are often collectively referred to as the "American Mediterranean."

Circulation Patterns The Caribbean Current passing through the Yucatán Strait loops clockwise, producing a dome of water in the Gulf of Mexico that stands 10 centimeters (4 inches) higher than the Atlantic water southeast of Florida. This elevated sea level causes an intense flow known as the *Florida Current* to pass through the Straits of Florida. It joins with water carried north by the *Antilles Current* and flows north along the coast of Florida (see Figure 12-12a).

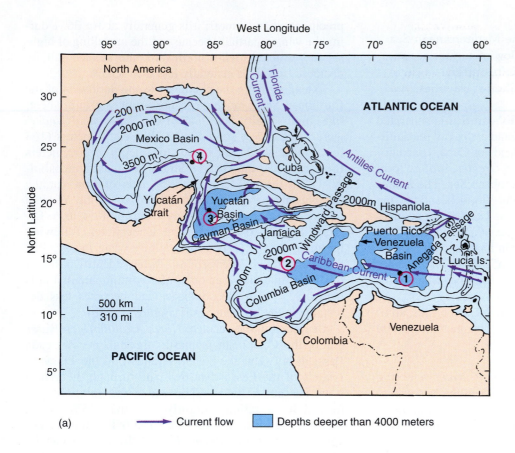

(a)

⟶ Current flow   ▢ Depths deeper than 4000 meters

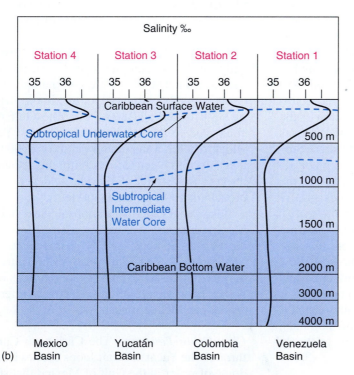

(b)

Figure 12–12 Bathymetry, surface circulation, and salinity profiles in the Caribbean Sea and the Gulf of Mexico.

(a) Location map showing major surface currents and bathymetry. Circled numbers represent stations shown in part (b). **(b)** Vertical salinity profiles and water masses at four stations marked on the location map in part (a). The central cores of the Subtropical Underwater and the Subtropical Intermediate Water can be identified on the profiles as maxima and minima, respectively.

The *Loop Current* is a significant feature of the Gulf of Mexico's surface circulation. Figure 12-13a shows the relationship of the current flow direction to the topography of the 20°C (68°F) isotherm surface in the southeastern Gulf of Mexico. Much of the surface water entering the Gulf of Mexico through the Yucatán Strait loops around the temperature contour in a clockwise flow and heads toward the Straits of Florida. After passing through the Yucatán Strait, surface water characteristics can be identified during the winter to a depth of 90 meters (295 feet) and, during the summer, to 125 meters (410 feet), which marks the depth at which seasonal temperature changes extend. Generally, the surface temperature just off the Yucatán coast ranges from 24° to 27°C (75° to 81°F), whereas temperatures along the northern Gulf Coast range between 18° and 21°C (64° and 70°F) (Figure 12-13b).

Water Masses   The Subtropical Underwater is still identifiable as a salinity maximum north of the Yucatán Strait at a depth of 100 to 200 meters (330 to 660 feet). The boundary between the upper water (containing the surface water and Subtropical Underwater) and the deep water is marked by the 16°C (61°F) isotherm at a depth of about 200 meters (660 feet). Intermediate Water entering through the Yucatán Channel below 850 meters (2800 feet) can be identified by a salinity minimum throughout the Gulf of Mexico. The core of the Intermediate Water can be identified throughout the Gulf at depths as shallow as 550 meters (1800 feet), but its identity is lost passing through the Straits of Florida. Salinity and temperature increase very slightly toward the bottom of the basin.

Figure 12–13 Gulf of Mexico Loop Current.

(a) Location map showing depth of 20°C (68°F) temperature surface and clockwise flow of Gulf of Mexico Loop Current. Block diagram of the Loop Current (*inset*) shows Loop Current flow around the "dome" of water that directly overlies the "bowl" representing the depth of the 20°C temperature surface; the vertical scale is greatly exaggerated. **(b)** February 26, 1988 NOAA 9 satellite infrared image of sea-surface temperature in the Gulf of Mexico. The 20°C temperature surface that forms the "bowl" in part (a) intersects the ocean surface where the two lightest shades of blue meet. The Loop Current flows clockwise around the warm water that "fills" the "bowl," then moves off through the Straits of Florida. Note that the warm water that forms the "dome" reaches temperatures in excess of 26°C (79°F) (*red color*).

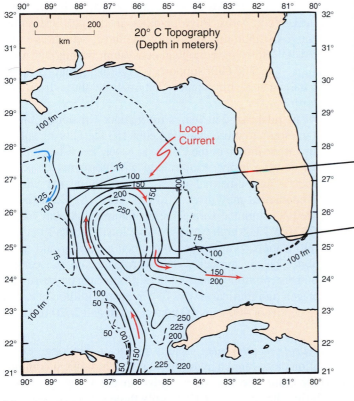

(a)

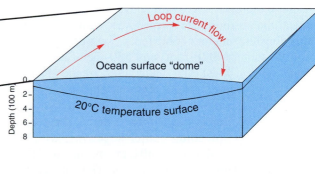

(b)

Marginal Seas of the Pacific Ocean

Two of the most important marginal seas in the Pacific Ocean are the Gulf of California (which lies between Baja California and mainland Mexico) and the Bering Sea (which lies between Alaska and Russian Siberia).

The Gulf of California   The narrow, northwest–southeast trending **Gulf of California**[10] extends from near the Tropic of Cancer at its open southern end to the north at the mouth of the Colorado River, which is the main river system draining into it (Figure 12-14a). About a half-dozen smaller rivers also empty into the Gulf from the east, carrying water from the Sierra Madre Occidental across a broad coastal plain that forms its western margin. From this coastal plain, which is on the North American Plate, a continental shelf extends as far as 50 kilometers (30 miles) from shore. The shelf terminates at an average depth of about 100 meters (330 feet). The western side of the Gulf, which is on the Pacific Plate, does not have a wide shelf; instead, it is characterized by steep rocky slopes.

The Gulf of California is one of the most recently created seas on Earth. Tectonically, it was formed as the East Pacific Rise spreading center migrated northward, rifting Baja California away from mainland Mexico about 6 million years ago. Today, active sea floor spreading still occurs along segments of the mid-ocean ridge that underlie the central portion of the Gulf and continue to enlarge the sea.

The Colorado River has developed a significant delta at the northern end of the Gulf, with depths rarely exceeding 200 meters (660 feet). Two exceptions are the 1500-meter (4920-foot) and 550-meter (1800-foot) basins on the west and east sides, respectively, of Angel de la Guarda Island. Depths gradually increase to the south through a series of basins, the deepest of which is 3700 meters (12,140 feet) deep. These basins are created by spreading center segments, which are separated by sills that generally extend up to 400 meters (1300 feet) above the basin floor. The sills represent east–west trending offsets (transform faults/fracture zones) of spreading centers.

The character of the northern Gulf of California has changed dramatically since the completion of Hoover Dam in 1935. Before the dam was built, the Colorado River provided an annual average flow of almost 18 billion cubic meters (23 billion cubic yards) of water, which carried 147 million metric tons (161 million short tons) of suspended sediment per year to the Gulf. Since 1935, however, the average annual flow has been reduced to less than 8 billion cubic meters (10 billion cubic yards) of water and about 14 million metric tons (15 million short tons) of sediment. In recent years, no Colorado River water (or sediment) makes it to the Gulf; instead, the water is diverted to supply the water needs of many southwest municipalities, depriving the Gulf of an important source of land-based nutrients. The drainage systems on the eastern side of the basin have relatively small flow volumes, and many are intermittent owing to the arid nature of their drainage basins.

Circulation Patterns   Seasonal winds control surface circulation in the Gulf of California. A low-pressure atmospheric system in summer located over the northern end of the peninsula develops winds that drive the surface water from the Pacific into the Gulf. This flow produces upwelling along the steep rocky coast of the Baja peninsula. During the winter months, the low-pressure system is located on the mainland east of the Gulf. These winter winds produce upwelling along the mainland side. Upwelling produces a bloom of rich **plankton** (*planktos* = wandering), which are microscopic, free-floating organisms that abound in nutrient-rich waters. **Phytoplankton** (*phyto* = plant, *planktos* = wandering) are microscopic algae that photosynthesize and are the base of most marine food webs. Phyroplankton in the gulf are represented by diatoms and dinoflagellates, which are the basis of a thriving biological ecosystem throughout most of the year. The Gulf supports a large fish population and is the nursery for whales that migrate there from the North Pacific.

The tidal range increases from about 1 meter (3.3 feet) in the south to more than 10 meters (33 feet) during spring tides at the mouth of the Colorado River. The tidal currents that develop in the north, along with convective mixing, produce an isothermal water column during the winter. Temperatures may drop as low as 16°C (61°F), as compared to summer surface temperatures that may reach 30°C (86°F). A high rate of evaporation produces a marked stratification, with surface water being warmer and more saline. The water below the thermocline in the central southern portion of the Gulf of California possesses an oxygen minimum as low as 0.01 parts per million between the depths of 400 and 800 meters (1300 and 2620 feet).

Hydrothermal Vents   A joint Mexican–American expedition discovered a hydrothermal vent biological community along the mid-ocean ridge in the Guaymas Basin during the summer of 1980, only three years after the first hydrothermal vent biocommunity was discovered in the Galápagos Rift in the eastern Pacific. In 1982, the vent biocommunity was observed and sampled during a dive of the submersible *Alvin* (Figure 12-14b). Since that time, many other hydrothermal vent biocommunities in the Gulf have been identified and observed along segments of active spreading centers. Because the newly forming mid-ocean ridge is so close to areas of high sediment input, many hydrothermal vents are covered by sediment.

The Bering Sea   The **Bering Sea**, on the northern margin of the Pacific Ocean, extends to 66 degrees north latitude

---

[10]The Gulf of California is also known as the Sea of Cortez or the Vermillion Sea.

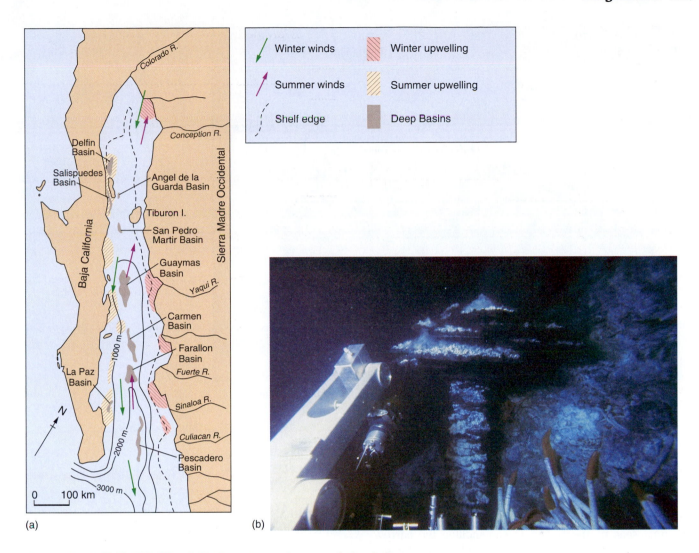

(a)   (b)

Figure 12–14 Gulf of California bathymetry and seasonal circulation.

(a) Map showing location and major rivers that drain into the Gulf of California. Basins within the Gulf increase in depth from north to south. Winds that reverse on a seasonal basis produce winter upwelling on the east side of the Gulf and summer upwelling on the west side. **(b)** Pagodalike structure of a hydrothermal vent atop a sediment mound that covers a spreading center in the Guaymas Basin. Tubeworms and *Alvin's* mechanical "arm" can be seen in the foreground.

and has the shape of a triangle with a curved base formed by the volcanic island arc of the Aleutian Islands (Figure 12-15). A broad continental shelf with depths less than 200 meters (660 feet) extends off the Siberian and Alaskan coasts, but most of the rest of the basin is deeper than 1000 meters (3300 feet). The deepest part of the basin is located in the western half of the sea, and fully 90% of the sea is either less than 200 meters (660 feet) in depth or more than 1000 meters (3300 feet) in depth. Except where it is cut by the Bering Canyon at the end of the Alaska Peninsula, the continental shelf drops off very abruptly into the deep basin at slopes of up to 5 degrees.

Circulation Patterns The major surface flow of water into the Bering Sea occurs between the Komandorski Islands of Russia and Attu Island, which is the most wester-

ly of the Aleutian Islands of Alaska. Here the *Alaskan Current*, flowing in a westerly direction south of the Aleutian chain, converges with northward-moving water of the western Pacific and flows through the passages into the Bering Sea. A small counterclockwise gyre is set up north of the Komandorski Islands, while a clockwise rotation develops to the north of the eastern end of the Aleutians. The main flow continues to the east until it reaches the broad Alaskan shelf and then follows this shelf edge to the north. A portion of this northward flow passes between St. Lawrence Island and the east Siberian coast before crossing the Bering Strait into the Arctic Ocean. Tidal currents dominate the shallow Alaskan shelf region, but a persistent northward-flowing current resulting from the runoff of fresh water from the Alaskan coast flows at speeds up to 10 kilometers (6 miles) per hour along the coast and through the Bering Strait (Figure 12-15).

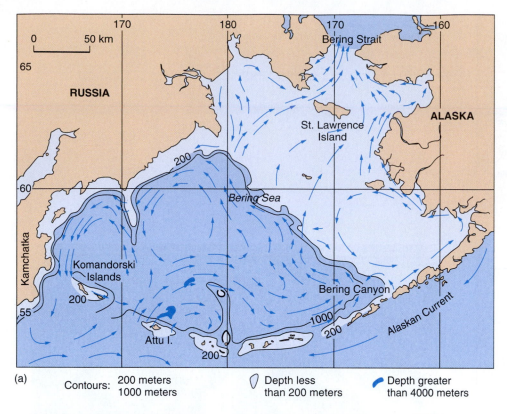

Figure 12–15 Bering Sea bathymetry and surface circulation.

Location map showing bathymetry and surface circulation in the Bering Sea. Note that much of the Bering Sea is either shallower than 200 meters (660 feet) or deeper than 1000 meters (3300 meters).

Marginal Seas of the Indian Ocean

Marginal seas of the Indian Ocean include the narrow Red Sea and the broader Arabian Sea and Bay of Bengal.

The Red Sea   The **Red Sea** extends more than 1900 kilometers (1200 miles) north of the narrow Strait of Bab-el-Mandeb (Gate of Tears) to the northern tip of the Gulf of Suez, the western branch of the Red Sea that separates the Sinai Peninsula from the African mainland. Forming the eastern boundary of the Sinai Peninsula is the Gulf of Aqaba, which is an eastern branch at the northern end of the Red Sea (Figure 12-16).

Extending from 12 degrees to 30 degrees north latitude, the Red Sea lies in a highly arid region and is characterized by surface waters of unusually high temperature and salinity. Broad reef-covered shelves no more than 50 meters (164 feet) deep drop off sharply to a gently sloping surface at about 500 meters (1640 feet), which eventually leads into a central trough with depths from 1500 meters (4920 feet) to more than 2300 meters (7500 feet).

Geologic evidence indicates the Red Sea formed primarily during the past few million years by intracontinental rifting. Similar to the origin of the Gulf of California, the Red Sea is in the initial stages of ocean basin formation, which will continue as the plates containing the Arabian Peninsula and the African mainland rift apart. New oceanic crust and an active spreading center underlie the Red Sea.

Circulation Patterns   Sill depth at the Strait of Bab-el-Mandeb is only 125 meters (410 feet) compared to a maximum depth of more than 1000 meters (3300 feet) over parts of the Red Sea. Across this shallow sill, the basic circulation is dominated by a high rate of evaporation, which exceeds 200 centimeters (79 inches) per year. Because of this water loss, surface water from the Indian Ocean flows through the Gulf of Aden into the Red Sea. As this surface flow moves north, its density increases as evaporation increases its salinity. The dense water sinks and returns as a subsurface flow to the sill and out into the Gulf of Aden. This outflowing warm, saline water sinks rapidly until it finds its equilibrium depth (based on density) and then spreads out into the Indian Ocean. This pattern of circulation (Figure 12-16b) is similar to that of the Mediterranean Sea (see Figure 12-11b).

Because of the arid conditions in the region, the surface water in the Red Sea reaches a salinity of 42.5‰ and a temperature of 30°C (86°F) during the summer months. Below a depth of 200 meters (660 feet), a uniform mass of water extends to the bottom throughout most of the Red Sea. This deep-water mass has a temperature of 21.7°C (71°F) and a salinity of 40.6‰.

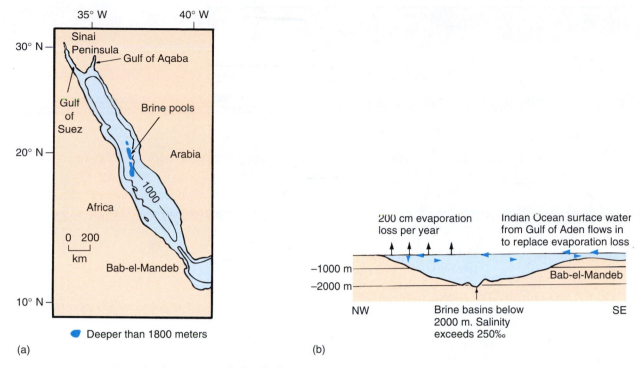

Figure 12–16 Red Sea bathymetry and circulation.
(a) Map showing location and bathymetry of the Red Sea, including location of brine pools. (b) Cross-section view of the mouth of the Red Sea, showing circulation between the Red Sea and the Indian Ocean.

Brine Pools In 1966, investigators aboard the Woods Hole Oceanographic Institution research vessel *Chain* studied a series of deep basins in the central Red Sea that had been previously noted to contain extremely high-salinity and high-temperature water masses. At about 21 degrees north latitude, two major basins were found—the Discovery Deep to the south and the Atlantis II Deep to the north. These and other similar basins contain *brine pools* with temperatures in excess of 36°C (96.8°F) and salinities as high as 257‰! These brine pools, because of their high salinity, have densities great enough to keep them in their basins and prevent them from mixing with overlying surface waters.

The brine pools are formed from hydrothermal circulation of seawater in the hot, porous oceanic crust. The sediments associated with the brine pools have concentrations of salts and metals that give them a great potential for future mining. In addition, there is also enrichment in the crust beneath the sediment, thereby enhancing the economic potential of this area. The Discovery Deep contains the first hydrothermal springs to be discovered. This discovery provided the impetus to initiate the search that led to the discovery of the hydrothermal vents along the Galápagos Rift in 1977.

The Arabian Sea The **Arabian Sea** is the northward extension of the Indian Ocean between Africa and India (Figure 12-17).

Circulation Patterns Surface currents in the Arabian Sea are dominated by *monsoon winds*[11] that blow from the northeast from November until March, when the *southwest monsoon* begins to develop. The air carried onto the continent during this southwest monsoon contains large quantities of water and produces heavy precipitation in the coastal regions.

During the *northeast monsoon* (winter), the surface current moves south along the west coast of India and turns west at about 10 degrees north latitude. Here, some of the surface water flows into the Gulf of Aden and the rest turns south along the Somali coast to converge with the *North Equatorial Current*. When the southwest monsoon begins in the summer, the North Equatorial Current disappears, and a portion of the *South Equatorial Current* flows north along the Somali coast as the *Somali Current*. This strong seasonal current flows with velocities in excess of 11 kilometers (7 miles) per hour and continues along the coast of Arabia and India in a clockwise pattern until it reaches 10 degrees north latitude. Here, it becomes the *Southwest Monsoon Current* and replaces the North Equatorial Current. Because of the alignment of winds relative to the African and Arabian coasts, significant upwelling occurs during the southwest monsoon.

[11]For more about Indian Ocean monsoon circulation, see Chapter 8.

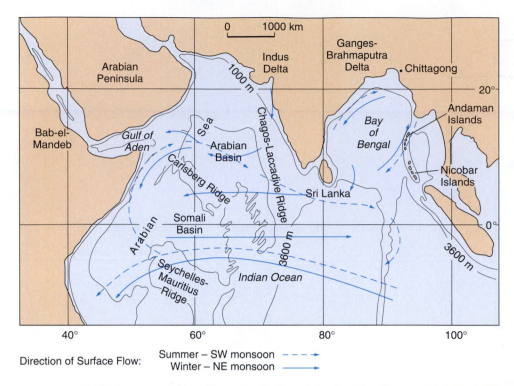

Figure 12–17 Bathymetry and surface circulation of the Arabian Sea and the Bay of Bengal.

Map showing location, bathymetry, and surface circulation of the Arabian Sea and the Bay of Bengal, both of which are affected by seasonal monsoon winds.

Surface salinities north of 5 degrees north latitude in the Arabian Sea are generally above 36‰ during the northeast monsoon. The Somali coastal region surface salinity may fall below 35.5‰ because of dilution by the South Equatorial Current and upwelling during the southwest monsoon. During the rainy season, surface salinities of less than 35‰ can be found as a result of dilution due to precipitation and runoff. Surface temperatures in the central region reach a maximum of 28°C (82°F) in June and a minimum temperature of 24°C (75°F) in February.

Below 200 meters (660 feet), the salinity decreases to values near 35‰ until an abrupt salinity maximum of 35.4‰ to 35.5‰ develops at depths above 1000 meters (3300 feet). This salinity maximum represents the flow of water into the Arabian Sea from the Persian Gulf and the Red Sea. The temperature of the Red Sea water ranges from 9°C to 10°C (48°F to 50°F). Red Sea water can also be identified by the low concentration of dissolved oxygen (0.45 parts per million) at about 790 meters (2600 feet) off the Somali coast. Interestingly, hydrogen sulfide has been observed on the continental slope at depths where this oxygen minimum occurs in the northern Arabian Sea.

Bay of Bengal The **Bay of Bengal** is a large body of water bounded by India and the island of Sri Lanka on the

west, and by the Malay Peninsula and the Andaman-Nicobar Island Ridge to the east (Figure 12-17). A significant continental influence is exerted on this body of water by the runoff from the Ganges and Brahmaputra rivers at the extreme north end of the bay.

Circulation Patterns The currents in the Bay of Bengal are also dominated by the monsoon wind system. During the southwest monsoon, a clockwise rotation is established within the Bay; this circulation is accompanied by upwelling along the east Indian coast. With the development of the northeast monsoon in November, the circulation reverses and forms a counterclockwise gyre. At Chittagong in southeast Bangladesh, a seasonal change in the sea level of 1.2 meters (4 feet) occurs as a result of monsoon wind changes.

The surface salinity in the Bay of Bengal seldom exceeds 34‰. During the southwest monsoon, particularly during late summer when the rainfall is the greatest, the runoff from the Ganges, Brahmaputra, and other rivers along the coast of Burma and India dilute the water and reduce surface salinity. As a result, salinity values as low as 18‰ can be observed at the extreme north end of the Bay. The major influence of this dilution is observed along the Indian coast; it is less significant away from shore.

? **STUDENTS SOMETIMES ASK** ...

Which nation on Earth would experience the greatest impact of a rise in sea level?

Of all the nations on Earth, Bangladesh, with a population of over 127 million, is under the greatest threat if a significant sea level rise occurs. This is because more than 80% of Bangladesh is built on a low-lying delta within a few meters of sea level, and it is regularly devastated by storm surges associated with tropical cyclones, particularly during the southwest monsoon. Even a small rise in sea level (like that predicted due to global warming) will severely impact the country. Other low-lying small islands will certainly be impacted, but they don't have the population size that Bangladesh does.

> The characteristics of marginal seas are influenced by their tectonic origin, location, circulation, and physical processes, all of which affect water temperature and salinity.

Chapter in Review

- *Coastal waters support about 95% of the total mass of life in the oceans*, and they are important areas for commerce, recreation, fisheries, and the disposal of waste. *The temperature and salinity of the coastal ocean vary over a greater range than the open ocean* because the coastal ocean is shallow and experiences river runoff, tidal currents, and seasonal changes in solar radiation. *Coastal geostrophic currents are produced from freshwater runoff and coastal winds.*

- *Estuaries are semienclosed bodies of water where freshwater runoff from the land mixes with ocean water.* Estuaries are *classified by their origin* as coastal plain, fjord, bar built, or tectonic. Estuaries are also *classified by their mixing patterns* of fresh and salt water as vertically mixed, slightly stratified, highly stratified, and salt wedge. *Typical circulation in an estuary consists of a surface flow of low-salinity water toward its mouth and a subsurface flow of marine water toward its head.*

- *Estuaries provide important breeding and nursery areas for many marine organisms* but often suffer from human population pressures. The *Columbia River Estuary*, for example, has degraded from agriculture, logging, and the construction of dams upstream. In *Chesapeake Bay*, an anoxic zone occurs during the summer that kills many commercially important species.

- *Wetlands are some of the most biologically productive regions on Earth. Salt marshes and mangrove swamps are important examples of coastal wetlands.* Wetlands are ecologically important because they remove land-derived pollutants from water before it reaches the ocean. Nevertheless, human activities continue to destroy wetlands.

- *Long offshore deposits called barrier islands protect marshes and lagoons.* Some lagoons have restricted circulation with the ocean, so water temperatures and salinity may vary widely with the seasons.

- *Marginal seas are relatively large semi-isolated bodies of marine water often formed by tectonic processes that isolate patches of ocean crust or that are forming new ocean basins.* For example, the Mediterranean is a remnant of the ancient Tethys Sea, which has been closing up as the plates containing Africa and Europe move closer together. Both the Gulf of California and the Red Sea are geologically young seas produced by sea floor spreading rifting apart adjoining landmasses.

- *Marginal seas typically are shallower than the open ocean and have varying degrees of restricted circulation.* Circulation in the Mediterranean Sea is characteristic of restricted bodies of water in areas where evaporation greatly exceeds precipitation. Called *Mediterranean circulation, it is the reverse of estuarine circulation.* In other marginal seas, salinities may fall below those of normal seawater because of large influxes of river runoff.

- *Seasonal wind patterns like those found in the Gulf of California cause coastal upwelling and high biological productivity.* An extensive shelf on the northeast side and a deep basin to the southwest are characteristic of the *Bering Sea* north of the Aleutian Islands. Many important commercial fisheries exist in this sea. *The Arabian Sea* to the west of India and the *Bay of Bengal* to the east have circulation patterns controlled by monsoon winds.

Key Terms

Anoxic (p. 355)
Arabian Sea (p. 367)
Bar-built estuary (p. 352)
Bay of Bengal (p. 368)
Bering Sea (p. 364)
Caribbean Sea (p. 361)
Coastal geostrophic current
(p. 350)
Coastal plain estuary (p. 351)
Coastal water (p. 349)
Davidson Current (p. 351)

Estuarine circulation pattern
(p. 352)
Estuary (p. 351)
Fjord (p. 352)
Gulf of California (p. 364)
Gulf of Mexico (p. 361)
Halocline (p. 349)
Highly stratified estuary
(p. 352)
Isohaline (p. 349)
Isothermal (p. 349)

Lagoon (p. 357)
Mangrove swamp (p. 355)
Marginal sea (p. 358)
Mediterranean circulation
(p. 360)
Mediterranean Sea (p. 358)
Phytoplankton (p. 364)
Plankton (p. 364)
Red Sea (p. 366)
Salt marsh (p. 355)
Salt wedge estuary (p. 352)

Sill (p. 358)
Slightly stratified estuary
(p. 352)
Tectonic estuary (p. 352)
Thermocline (p. 349)
Vertically mixed estuary
(p. 352)
Wetland (p. 355)

Questions and Exercises

1. For coastal oceans where deep mixing does not occur, discuss the effect that offshore winds and freshwater runoff will have on salinity distribution. How will the winter and summer seasons affect the temperature distribution in the water column?

2. How does coastal runoff of low-salinity water produce a coastal geostrophic current?

3. Based on their origin, draw and describe the four major classes of estuaries.

4. Describe the difference between vertically mixed and salt wedge estuaries in terms of salinity distribution, depth, and volume of river flow. Which displays the more classical estuarine circulation pattern?

5. Discuss factors that cause the surface salinity of Chesapeake Bay to be greater along its east side, and why periods of summer anoxia in deep water are becoming increasingly severe with time.

6. Name the two types of coastal wetland environments and the latitude ranges where each will likely develop. How do wetlands contribute to the biology of the oceans and the cleansing of polluted river water?

7. What factors lead to a wide seasonal range of salinity in Laguna Madre?

8. List the evidence that exists to support the idea that the Mediterranean Sea completely dried up in the geologic past.

9. Describe the circulation between the Atlantic Ocean and the Mediterranean Sea, and explain how and why it differs from estuarine circulation.

10. Describe how coastal upwelling in the Gulf of California is related to seasonal winds.

11. Compare and contrast the circulation between the Red Sea and the Indian Ocean to that between the Mediterranean Sea and the Atlantic Ocean.

12. Explain why Red Sea water that flows into the Indian Ocean at a depth of 125 meters (410 feet) sinks to 1000 meters (3300 feet) before spreading throughout the Arabian Sea.

13. Describe the relationship between surface circulation and monsoon winds in the Bay of Bengal.

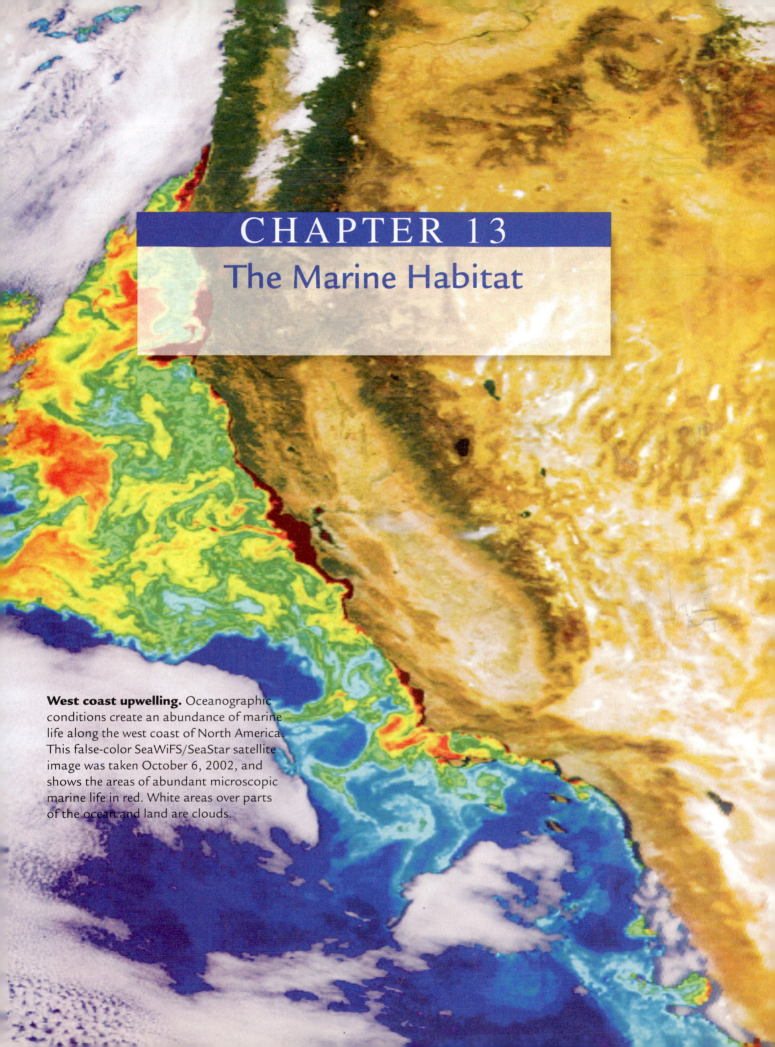

CHAPTER 13
The Marine Habitat

West coast upwelling. Oceanographic conditions create an abundance of marine life along the west coast of North America. This false-color SeaWiFS/SeaStar satellite image was taken October 6, 2002, and shows the areas of abundant microscopic marine life in red. White areas over parts of the ocean and land are clouds.

Key Questions

- How are living things classified?
- How are marine organisms classified?
- How many marine species exist?
- What is osmosis, and how does it affect marine organisms?
- What adaptations do marine organisms have that allow them to live in the ocean?
- What are the main divisions of the marine environment?

The answers to these questions (and much more) can be found in the highlighted concept statements within this chapter.

"I never dreamed that islands, about fifty or sixty miles apart, and most of them in sight of each other, formed of precisely the same rocks, placed under a quite similar climate, rising to a nearly equal height, would have been differently tenanted."

—Charles Darwin, commenting about species diversity in the Galápagos Islands (1837)

A wide variety of organisms inhabits the marine environment. These organisms range in size from microscopic bacteria and algae to the blue whale, which is as long as three buses lined up end to end. Marine biologists have identified over 250,000 marine species, and this number is constantly increasing as new organisms are discovered.

Most marine organisms live within the sunlit surface waters of the ocean. Strong sunlight supports photosynthesis by marine algae, which either directly or indirectly provides food for the vast majority of marine organisms. All marine algae live near the surface because they need the sunlight; most marine animals live near the surface because this is where food can be obtained. In shallow-water areas close to land, sunlight reaches all the way to the ocean floor, resulting in an abundance of marine life.

There are advantages and disadvantages to living in the marine environment. One advantage is that there is an abundance of water available, which is necessary for supporting all types of life. One disadvantage is that maneuvering in water, which has high density and impedes movement, can be difficult. The individual success of species depends on their ability to find food, avoid predators, and cope with the many physical barriers to their movement.

Classification of Living Things

All living things[1] belong to one of three domains (branches) of life: Archaea, Bacteria, and Eukarya (Figure 13-1). The domain **Archaea** (*archaeo* = ancient) is a group of sim-

ple microscopic bacteria-like creatures that includes methane producers and sulfur oxidizers that inhabit deep-sea vents and seeps,[2] as well as other forms—many of which prefer environments with extreme temperatures and/or pressures. The domain **Bacteria** (*bakterion* = rod) includes simple life forms with cells that usually lack a nucleus, including purple bacteria, green nonsulfur bacteria, and cyanobacteria (blue-green algae). The domain **Eukarya** (*eu* = good, *karuon* = nut) includes complex organisms—protoctists, fungi, and multicellular plants and animals—with cells that usually contain a nucleus.

Within the three domains of life, a system of five kingdoms was first proposed by ecologist and biologist Robert H. Whittaker in 1959. These five kingdoms of life are Monera, Protoctista, Fungi, Plantae, and Animalia.

Kingdom **Monera** (*monos* = single) includes some of the simplest organisms. These organisms are single celled but lack a discrete nucleus, so their nuclear material is spread throughout the cell. Included in this kingdom are the cyanobacteria (blue-green algae), heterotrophic bacteria, and archaea. Recent discoveries have shown bacteria to be a much more important part of marine ecology than previously believed. These organisms are found throughout the breadth and depth of the oceans.

Kingdom **Protoctista** (*proto* = first, *ktistos* = to establish) includes single- and multicelled organisms that have a nucleus, so they represent a higher stage of evolutionary development than Kingdom Monera. Examples of organisms in Kingdom Protoctista include various types of marine *algae* (aquatic photosynthetic organisms that can be either single-celled microscopic or multi-celled macroscopic) and single-celled animals called **protozoa** (*proto* = first, *zoa* = animal).

Kingdom **Fungi** (*fungus* = [probably from Greek *sp(h)ongos* = sponge]) includes 100,000 species of mold and lichen, though less than one-half of 1% of them are sea dwellers. Fungi exist throughout the marine environment, but they are much more common in the intertidal zone, where they live with cyanobacteria or green algae to form lichen. Other fungi remineralize organic matter and function primarily as decomposers in the marine ecosystem.

Kingdom **Plantae** (*planta* = plant) comprises the multicelled plants, all of which photosynthesize. Only a few species of true plants—such as surf grass (*Phyllospadix*) and eelgrass (*Zostera*)—inhabit shallow coastal environments. In the ocean, photosynthetic marine algae occupy the ecological niche of land plants. However, certain plants are vital parts of coastal ecosystems, including mangrove swamps and salt marshes.

Kingdom **Animalia** (*anima* = breath) comprises the multicelled animals. Organisms from kingdom Animalia range in complexity from the simple sponges to complex vertebrates (animals with backbones), which also include humans.

---

[1]See the discussion about living things (including a working definition of life) in Chapter 2.

[2]A *seep* is an area where water trickles out of the sea floor.

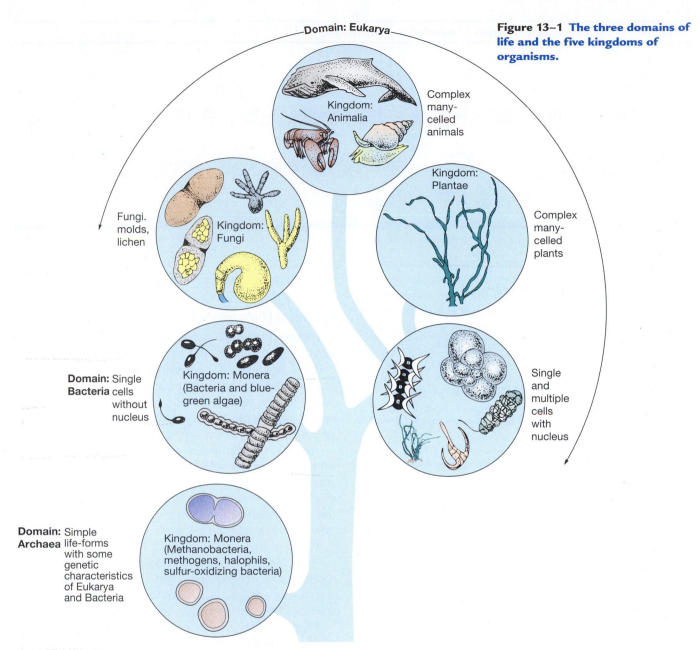

Domain: Eukarya

Kingdom: Animalia

Complex many-celled animals

Kingdom: Plantae

Complex many-celled plants

Fungi, molds, lichen

Kingdom: Fungi

Domain: Single **Bacteria** cells without nucleus

Kingdom: Monera (Bacteria and blue-green algae)

Single and multiple cells with nucleus

Domain: Simple **Archaea** life-forms with some genetic characteristics of Eukarya and Bacteria

Kingdom: Monera (Methanobacteria, methogens, halophils, sulfur-oxidizing bacteria)

Common ancestral community of primitive cells

In an effort to determine the relationships of all living things on Earth, Swedish botanist Carl von Linné, who Latinized his name to Carolus Linnaeus, developed a system in 1758 that is the basis of the modern scientific system of classification used today. The systematic classification of organisms—called **taxonomy** (*taxis* = arrangement, *nomia* = a law)—involves placing organisms into the following increasingly specific groupings:

- Kingdom
- Phylum (Division for plants)
- Class
- Order
- Family   *ex: cat or Dolphin*
- Genus
- Species

All organisms that share a common category (for instance, a family—such as the cat or dolphin family) have certain characteristics and evolutionary similarities. In some cases, subdivisions of these categories are also used, such as subphylum (Table 13-1). The categories assigned to an individual species must be agreed upon by an international panel of experts.

A species is the fundamental unit of classification. **Species** (*species* = a kind) consist of populations of genetically similar, interbreeding (or potentially interbreeding) individuals that share a collection of inherited characteristics whose combination is unique.

Every type of organism has a unique scientific name that includes its genus and species, which is italicized with the first letter of the generic (genus) name capitalized—for example, *Delphinus delphis*. Most organisms

TABLE 13–1 **Taxonomic classification of selected organisms.**

Category	Human	Common dolphin	Killer whale	Bat star	Giant kelp
Kingdom	Animalia	Animalia	Animalia	Animalia	Protoctista
Phylum	Chordata	Chordata	Chordata	Echinodermata	Phaeophyta
Subphylum	Vertebrata	Vertebrata	Vertebrata		
Class	Mammalia	Mammalia	Mammalia	Asteroidea	Phaeophycae
Order	Primates	Cetacea	Cetacea	Valvatida	Laminariales
Family	Hominidae	Delphinidae	Delphinidae	Oreasteridae	Lessoniaceae
Genus	Homo	Delphinus	Orcinus	Asterina	Macrocystis
Species	sapiens	delphis	orca	miniata	pyrifera

also have one or more common names. *Delphinus delphis*, for instance, is the common dolphin, and *Orcinus orca* is the killer whale or orca. If the scientific name is referred to repeatedly within a document, it is often shortened by abbreviating the genus name to its first letter. Thus, *Delphinus delphis* becomes *D. delphis*.

? STUDENTS SOMETIMES ASK. ...

Why are there scientific names for all organisms? Wouldn't it be easier just to know the common name of an organism?

Each individual species has a unique scientific name, so the scientific name identifies one particular species more clearly than the common name. Common names are often used for more than one species of organisms, which can be confusing. "Dolphin," for example, is used to describe dolphins, porpoises, and even a type of fish! Most people would shudder at being served dolphin in a restaurant, but dolphin *fish* (also called mahi-mahi) is often a featured menu item.

Common names can also be confusing because there can be more than one for the same species, and because they vary from language to language. Scientific names, on the other hand, are Latin based, so they are the same in all languages. This allows a Chinese scientist, for example, to communicate effectively with a Greek scientist about a particular organism. Thus, scientific names are useful, descriptive (if you know a bit about Latin terms and word roots), and unambiguous.

> Living things can be classified into one of three domains and five kingdoms, each of which is split into increasingly specific groupings of phylum, class, order, family, genus, and species.

Classification of Marine Organisms

Marine organisms can be classified according to where they live (their habitat) and how they move (their mobility). Organisms that inhabit the water column can be classified as either *plankton* (floaters) or *nekton* (swimmers). All other organisms are *benthos* (bottom dwellers).

Plankton (Floaters)

Plankton (*planktos* = wandering) include all organisms—algae, animals, and bacteria—that drift with ocean currents. An individual organism is called a **plankter**. Just because plankters drift does not mean they are unable to swim. In fact, many plankters can swim but either move only weakly or move only vertically. As such, they cannot determine their horizontal position within the ocean.

Among plankton, the microscopic single-celled photosynthetic algae are called **phytoplankton** (*phyto* = plant, *planktos* = wandering), and those organisms not capable of photosynthesis (mostly animals) are called **zooplankton** (*zoo* = animal, *planktos* = wandering). Representative members of each group are shown in Figure 13-2.

Plankton also include bacteria. It has recently been discovered that free-living **bacterioplankton** are much more abundant than previously thought. Having an average diameter of only one-half of a micrometer[3] (0.00002 inch), they were missed in earlier studies because they are so small.

Plankton are unbelievably abundant and important within the marine environment. In fact, *most of Earth's biomass—the mass of living organisms—consists of plankton adrift in the oceans.* Even though 98% of marine *species* are bottom dwelling, the vast majority of the ocean's *biomass* is planktonic.

Plankton range greatly in size. They include large floating animals and algae, such as jellyfish and *Sargassum*,[4] which are called **macroplankton** (*macro* = large, *planktos* = wandering) and measure 2 to 20 centimeters (0.8 to 8 inches). Plankton also include bacterioplankton, which are so small that they can be removed from the water only with special microfilters. These very tiny floaters are called **picoplankton** (*pico* = small, *planktos* = wandering) and measure 0.2 to 2 micrometers (0.000008 to 0.00008 inch).

---

[3]One micrometer (also known as one *micron*) is one-millionth of a meter and is designated by the symbol μm.

[4]*Sargassum* is a floating type of brown macro marine algae commonly referred to as a "seaweed" that is particularly abundant in the Sargasso Sea.

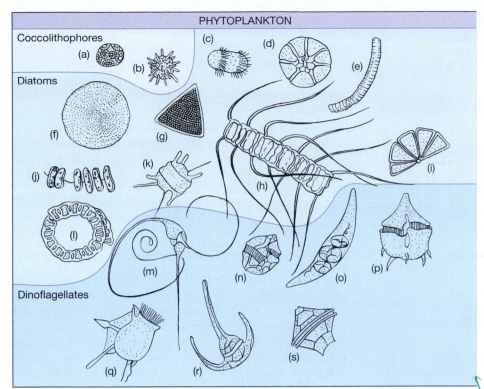

PHYTOPLANKTON

Coccolithophores (a) (b) (c) (d) (e)

Diatoms (f) (g) (h) (i) (j) (k) (l)

Dinoflagellates (m) (n) (o) (p) (q) (r) (s)

Figure 13–2 Phytoplankton and zooplankton (floaters).

Not drawn to scale; typical maximum dimension is in parentheses.
Phytoplankton: (a) and **(b)** Coccolithophoridae (15 μm, or 0.0006 in.). **(c)–(l)** Diatoms (80 μm, or 0.0032 in.): **(c)** *Corethron;* **(d)** *Asteromphalus;* **(e)** *Rhizosolenia;* **(f)** *Coscinodiscus;* **(g)** *Biddulphia favus;* **(h)** *Chaetoceras;* **(i)** *Licmophora;* **(j)** *Thalassiorsira;* **(k)** *Biddulphia mobiliensis;* **(l)** *Eucampia.* **(m)–(s)** Dinoflagellates (100 μm, or 0.004 in.): **(m)** *Ceratium recticulatum;* **(n)** *Goniaulax scrippsae;* **(o)** *Gymnodinium;* **(p)** *Goniaulax triacantha;* **(q)** *Dynophysis;* **(r)** *Ceratium bucephalum;* **(s)** *Peridinium.*
Zooplankton: (a) Fish egg (1 mm, or 0.04 in.). **(b)** Fish larva (5 cm, or 2 in.). **(c)** Radiolaria (0.5 mm, or 0.02 in.). **(d)** Foraminifer (1 mm, or 0.04 in.). **(e)** Jellyfish (30 cm, or 12 in.). **(f)** Arrowworms (3 cm, or 1.2 in.). **(g)** and **(h)** Copepods (5 mm, or 0.2 in.). **(i)** Salp (10 cm, or 4 in.). **(j)** Doliolum (10 cm, or 4 in.). **(k)** Siphonophore (30 cm, or 12 in.). **(l)** Worm larva (1 mm, or 0.04 in.). **(m)** Fish larva (5 cm, or 2 in.). **(n)** Tintinnid (1 mm, or 0.04 in.). **(o)** Foraminifer (1 mm, or 0.04 in.). **(p)** Dinoflatellate (*Noctiluca*) (1 mm, or 0.04 in.).

how are these "single celled"?

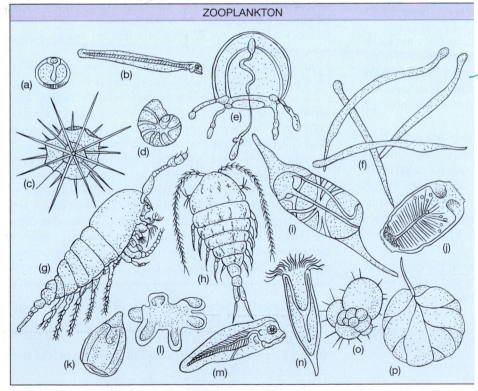

ZOOPLANKTON

(a) (b) (c) (d) (e) (f) (g) (h) (i) (j) (k) (l) (m) (n) (o) (p)

Plankton can be classified as either phytoplankton, zooplankton, or bacterioplankton. They can also be classified according to the portion of their life cycle spent as plankton. Organisms that spend their entire lives as plankton are **holoplankton** (*holo* = whole, *planktos* = wandering). Many organisms that spend their adult lives as nekton or benthos spend their juvenile and/or larval stages as plankton (Figure 13-3). These organisms are called **meroplankton** (*mero* = a part, *planktos* = wandering).

Nekton (Swimmers)

Nekton (*nektos* = swimming) include all animals capable of moving independently of the ocean currents, by

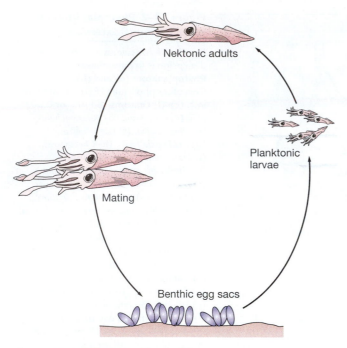

Figure 13–3 Life cycle of a squid.

Squid are meroplankton because they are planktonic only during their larval stage. Adult squid are nekton and their egg sacks are benthos.

swimming or other means of propulsion. They are capable not only of determining their own positions within the ocean but also, in many cases, of long migrations. Nekton include most adult fish and squid, marine mammals, and marine reptiles (Figure 13-4). When you go ocean swimming, you become nekton, too.

Although nekton move freely, they are unable to move throughout the breadth of the ocean. Gradual changes in temperature, salinity, viscosity, and availability of nutrients effectively limit their lateral range. The deaths of large numbers of fish, for example, can be caused by temporary horizontal shifts of water masses in the ocean. Water pressure normally limits the vertical range of nekton.

Fish may appear to exist everywhere in the oceans, but they are more abundant near continents and islands and in colder waters. Some fish, such as salmon, ascend freshwater rivers to spawn. Many eels do just the reverse, growing to maturity in fresh water and then descending the streams to breed in the great depths of the ocean.

Benthos (Bottom Dwellers)

The term **benthos** (*benthos* = bottom) describes organisms living on or in the ocean bottom. **Epifauna** (*epi* = upon, *fauna* = animal) live on the surface of the sea floor, either attached to rocks or moving along the bottom. **Infauna** (*in* = inside, *fauna* = animal) live buried in the sand, shells, or mud. Some benthos, called **nektobenthos**, live on the bottom but also swim or crawl through the water above the ocean floor. Examples of benthos are shown in Figure 13-5.

The shallow coastal ocean floor contains a wide variety of physical and nutritive conditions, which has allowed a great number of animal species to develop. Moving across the bottom from the shore into deeper water, the *number* of benthos species per square meter may remain relatively constant, but the *biomass* of benthos organisms decreases. In addition, the shallow coastal areas are the only locations where large marine algae (often called "seaweeds") attached to the bottom are found, because these are the only areas of the sea floor that receive sufficient sunlight.

STUDENTS SOMETIMES ASK...

What is the difference between kelp, seaweed, and marine algae?

In common usage, these terms all refer to large, branching, photosynthetic marine organisms that contain various pigments, which give these organisms their color. However, differences between the terms do exist.

Early mariners probably coined the term *seaweed* because they thought these organisms were a nuisance. They clogged harbors, entangled vessels, and washed up on the beach in useless bunches after storms. Although they could be eaten, they could not be consumed in large quantities. Historically, all marine algae (except microscopic species) had similarities to weeds on land, so they have become known collectively as "seaweeds."

It turns out, however, that these organisms are vital to coastal ecosystems, so marine biologists prefer to call them "marine algae." To differentiate them from the microscopic planktonic species of algae, large marine algae are often called "marine macro algae." The branching types of brown algae (phylum Phaeophyta) are called "kelp."

Today marine macro algae have many uses. They are used as a thickener and emulsifier in many products (such as toothpaste) and foods (such as ice cream). Cooking recipes sometimes call for small quantities of "sea vegetable," and nori, a type of red alga, is used to wrap sushi. Marine algae are used as fertilizer, and some species have recently been touted as health foods. Marine algae are discussed in Chapter 14, "Biological Productivity and Energy Transfer."

Throughout most of deeper parts of the sea floor, animals live in perpetual darkness, where photosynthetic production cannot occur. They must feed on each other, or on whatever outside nutrients fall from the productive zone near the surface.

The deep-sea bottom is an environment of coldness, stillness, and darkness. Under these conditions, life progresses slowly, and organisms that live in the deep sea usually are widely distributed because physical conditions vary little on the deep-ocean floor, even over great distances.

Hydrothermal Vent Biocommunities In 1977, the first biocommunity at a hydrothermal vent was discovered in the Galápagos Rift off South America,

Figure 13–4 Nekton (swimmers).

Not drawn to scale; typical maximum dimension is in parentheses. **(a)** Bluefin tuna (2 m, or 6 ft.).
(b) Bottlenose dolphin (4 m, or 13 ft.). **(c)** Nurse shark (3 m, or 10 ft.). **(d)** Barracuda (1 m, or 3.3
ft.). **(e)** Stripped bass (0.5 m, or 1.6 ft.). **(f)** Sardine (15 cm, or 6 in.). **(g)** Deep-ocean fish (8 cm, or
3 in.). **(h)** Squid (1 m, or 3.3 ft.). **(i)** Anglerfish (5 cm, or 2 in.). **(j)** Lantern fish (8 cm, or 3 in.).
(k) Gulper (15 cm, or 6 in.).

demonstrating that high concentrations of deep-ocean
benthos are possible. The primary limiting factor for life
on the deep-ocean floor is probably the availability of
food because food is abundant at these hydrothermal
vents. Archaeon (bacteria-like organisms) produce food
not by photosynthesis, for no sunlight is available, but by
chemosynthesis. The size of individuals and the total biomass in the hydrothermal communities far exceed that

previously known for the deep-ocean benthos. These biocommunities are discussed in Chapter 16, "Animals of
the Benthic Environment."

Marine organisms can be classified according to
their habitat and mobility as plankton (floaters),
nekton (swimmers), or benthos (bottom dwellers).

Figure 13–5 Benthos (bottom dwellers): Representative intertidal and shallow subtidal forms.

Not drawn to scale; typical maximum dimension is in parentheses. **(a)** Sand dollar (8 cm, or 3 in.). **(b)** Clam (30 cm, or 12 in.). **(c)** Crab (30 cm, or 12 in.). **(d)** Abalone (30 cm, or 12 in.). **(e)** Sea urchins (15 cm, or 6 in.). **(f)** Sea anemone (30 cm, or 12 in.). **(g)** Brittle star (20 cm, or 8 in.). **(h)** Sponge (30 cm, or 12 in.). **(i)** Acorn barnacles (2.5 cm, or 1 in.). **(j)** Snail (2 cm, or 0.8 in.). **(k)** Mussels (25 cm, or 10 in.). **(l)** Gooseneck barnacles (8 cm, or 3 in.). **(m)** Sea star (30 cm, or 12 in.). **(n)** Brain coral (50 cm, or 20 in.). **(o)** Sea cucumber (30 cm, or 12 in.). **(p)** Lamp shell (10 cm, or 4 in.). **(q)** Sea lily (10 cm, or 4 in.). **(r)** Sea squirt (10 cm, or 4 in.).

Distribution of Life in the Oceans

It is difficult to describe the extent to which the marine environment is inhabited because the ocean is immense, so little is known about marine organisms, and some populations fluctuate greatly each season. What is known, however, is that there are over 1,750,000 species[5] in marine and terrestrial environments and only about 14% of all known species live in the ocean (Figure 13-6).

Why Are There So Few Marine Species?

If the ocean is such a prime habitat for life and if life originated there, then why do so few of the world's known species live in the ocean? The disparity may arise because

---

[5]Various methods of estimation place the number of all species on Earth, known and still unknown, between 3 million and 100 million. Of the unknown species on Earth, a large number probably inhabit the oceans.

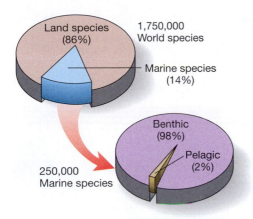

Figure 13–6 Distribution of species on Earth.

Of the 1,750,000 known species on Earth, 86% inhabit land environments and 14% inhabit the ocean. Of the 250,000 known marine species, 98% inhabit the benthic environment and live in or on the ocean floor, while only 2% inhabit the pelagic environment and live within the water column as either plankton or nekton.

the marine environment is more stable than the terrestrial environment. The relatively uniform conditions of the open ocean do not pressure organisms to adapt. The need for adaptation is thought to be responsible for the creation of new species. In addition, temperatures are not only stable but also relatively low below the sunlit surface waters. The rate of chemical reaction is slowed, which may further reduce the tendency for variation to occur.

The less stable environment on land, on the other hand, presents many opportunities for natural selection to produce new species to inhabit varied new niches. At least 75% of all land species, for example, are insects that have evolved the capability of inhabiting very restricted environmental niches. If insects are ignored, the sea possesses over 45% of the remaining species living in marine and terrestrial environments.

Figure 13-6 also shows that only 2% or about 5000 of the 250,000 known marine species inhabit the **pelagic** (*pelagios* = of the sea) **environment** and live within the water column. The other 98% inhabit the **benthic** (*benthos* = bottom) **environment** and live either in or on the sea floor. These numbers are minimums, however, because recent discoveries indicate that many more species may inhabit the benthic environment than previously thought.

Why do most marine species inhabit the benthic environment? The ocean floor contains numerous benthic environments (such as rocky, sandy, muddy, flat, sloped, irregular, and mixed bottoms) that create different habitats to which organisms have adapted. On the other hand, most of the pelagic environment—especially that below the sunlit surface waters—is a watery world that is quite uniform from one region to the next and does not experience extreme environmental variability to which organisms need to be adapted in order to survive.

Marine species represent 14% of the total number of known species on Earth. The benthic environment, which has large environmental variability, is home to 98% of the 250,000 known marine species.

Adaptations of Organisms to the Marine Environment

The ocean environment—particularly its temperature—is far more stable than the terrestrial environment. As a result, ocean-dwelling organisms have not developed highly specialized regulatory systems to adjust to sudden changes that might occur within their environment. Marine organisms can be adversely affected by quite small changes in temperature, salinity, turbidity, pressure, or other environmental variables.

Water constitutes over 80% of the mass of **protoplasm** (*proto* = first, *plasm* = something molded) the substance of living matter. In fact, over 65% of your body's weight—and 95% of a jellyfish's weight—is water (Table 13-2). Water carries dissolved within it the gases and minerals organisms need to survive. Water is also a raw material in the photosynthesis of food by marine phytoplankton.

TABLE 13-2 Percent water content of selected organisms.

Organism	Percent water content
Jellyfish	95
Lobster	79
Scallop	78
Herring	67
Human	65

Land plants and animals have developed complex "plumbing systems" to retain water and to distribute it throughout their bodies. The inhabitants of the open ocean do not risk atmospheric *desiccation* (drying out), however, because they live in an environment of abundant water.

Need for Physical Support

One basic need of all plants and animals is for simple physical support. Land plants, for example, have vast root systems that anchor the plant securely to the ground. Land animals have skeletons and combinations of appendages—legs, arms, fingers, and toes—to support their entire weight.

In the ocean, water physically supports marine plants and animals. Organisms such as photosynthetic phytoplankton, which must live in the upper surface waters of the ocean, depend primarily upon buoyancy and frictional resistance to sinking to maintain their desired position.

Still, maintaining position can be difficult, so some organisms have developed special adaptations to increase their efficiency. These adaptations are discussed in this and succeeding chapters.

Water's Viscosity

Viscosity (*viscos* = sticky) is a substance's *internal resistance to flow*. Recall from Chapter 3 that a substance with high resistance to flow (high viscosity)—such as toothpaste—does not flow easily. Conversely, a substance that has low viscosity—such as water—flows more readily. Viscosity is strongly affected by temperature. Tar, for example, must be heated to decrease its viscosity before it can be spread onto roofs or roads.

The viscosity of ocean water increases as salinity increases and temperature decreases. Thus, single-celled organisms that float in colder, higher-viscosity waters have less need for extensions to help them maintain their positions near the surface. Figure 13-7 shows, for example, that a warm-water species of copepod (small crustacean) has ornate, featherlike appendages, whereas a cold-water variety has no such appendages.

The Importance of Organism Size   The major requirements of phytoplankton are that they stay in the upper portion of ocean water where solar radiation is available, that they have available necessary nutrients, that they efficiently take in these nutrients from surrounding waters, and that they expel waste materials. Their ingenious size and shape help single-celled phytoplankton satisfy these requirements without needing specialized multiple cells.

Phytoplankton cannot propel themselves, so they use frictional resistance to maintain their general position near the surface of the water. Frictional resistance to sinking increases as an organism's surface area in contact with surrounding water increases. Figure 13-8 shows the surface area to volume ratio of three different cubes, illustrating that the ratio increases as an organism's size decreases. For instance, cube (a) has twice the surface area per unit of volume of cube (b) and four times the surface area per unit of volume as cube (c). If the cubes were plankton, cube (a)

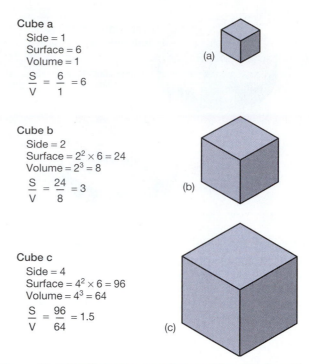

Cube a
Side = 1
Surface = 6
Volume = 1
$$\frac{S}{V} = \frac{6}{1} = 6$$

(a)

Cube b
Side = 2
Surface = $2^2 \times 6 = 24$
Volume = $2^3 = 8$
$$\frac{S}{V} = \frac{24}{8} = 3$$

(b)

Cube c
Side = 4
Surface = $4^2 \times 6 = 96$
Volume = $4^3 = 64$
$$\frac{S}{V} = \frac{96}{64} = 1.5$$

(c)

Figure 13–8 Surface area to volume ratios of cubes of different sizes.

As the linear dimension of a cube increases, the ratio of surface area to volume decreases. Thus, smaller bodies have a higher surface area to volume ratio, which allows them to more easily stay afloat and efficiently exchange nutrients and wastes.

would have four times the resistance to sinking per unit of mass as cube (c), so cube (a) would need to exert far less energy to stay afloat. Single-celled organisms, which make up the bulk of photosynthetic marine life, clearly benefit from being as small as possible. They are so small, in fact, that one needs a microscope to see them!

Photosynthetic cells take in nutrients from surrounding water and expel waste through their cell membranes. The efficiency of both functions increases with a higher surface area to volume ratio. Thus, if cubes (a) and (c) in Figure 13-8 were planktonic algae, cube (a) could take in nutrients and dispose of waste four times more efficiently than cube (c). This is why cells in all plants and animals are microscopic, regardless of the overall size of the organism.

Figure 13–7 Water temperature and appendages.

(a) Copepod (*Oithona*) displays the ornate plumage characteristic of warm-water varieties. **(b)** Copepod (*Calanus*) displays the less ornate appendages found on temperate and cold-water forms.

(a)   (b)

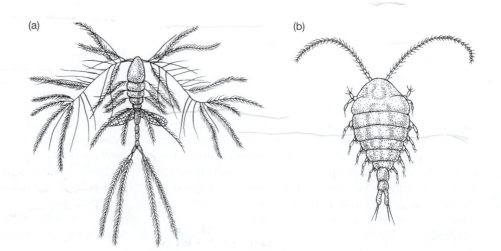

Diatoms—one of the most important groups of phytoplankton—often have unusual appendages, needlelike extensions, or even rings (Figure 13-9) to increase their surface area, thus preventing them from sinking below the sunlight surface waters. Other planktonic marine organisms—particularly warm-water species—use similar strategies to stay afloat.

Some small organisms produce a tiny droplet of oil, which lowers their overall density and increases buoyancy. Accumulations of vast amounts of these organisms in sediment can produce offshore oil deposits if their droplets of oil are combined, sufficiently matured, and trapped in a reservoir.

Despite adaptations to remain in the upper layers of the ocean, organisms still have a higher density than seawater, so they tend to sink, if ever so slowly. This is not a serious handicap, however, because wind causes considerable mixing and turbulence near the surface. Turbulence, in turn, keeps these organisms positioned to bask in the solar radiation needed to photosynthesize, producing the energy used by essentially all other members of the marine community.

Viscosity and Streamlining As organisms increase in size, viscosity ceases to enhance survival and instead becomes an obstacle. This is particularly true of large organisms that swim freely in the open ocean. They must pursue prey or flee predators, yet the faster they swim, the more the viscosity of water impedes their progress. Not only must water be displaced ahead of the swimmer, but water also must move in behind it to occupy the space that the animal has vacated.

Figure 13-10 shows the advantage of **streamlining**, which is having a shape that offers the least resistance to fluid flow. Streamlining allows marine organisms to overcome water's viscosity and move more easily through water. A streamlined shape usually consists of a flattened

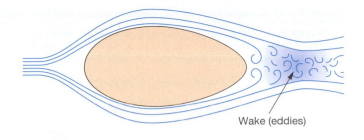

Wake (eddies)

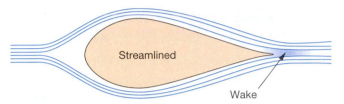

Streamlined

Wake

Figure 13–10 Streamlining.

To move through water efficiently, a body must produce as little stress as possible as it moves through and displaces the water. After the water has moved past the body, it must flow in behind the body with as little eddy action as possible.

body, which presents a small cross section at the front end and a gradually tapering back end to reduce the wake created by eddies. It is exemplified in the shape of free-swimming fish (and in marine mammals such as whales and dolphins).

Temperature

Figure 13-11, which compares extremes in land and ocean surface temperatures, shows that ocean temperatures have a far narrower range than temperatures on land. For example, the minimum surface temperature of the open ocean is seldom much below −2°C (28.4°F) and the maximum surface temperature seldom exceeds 32°C (89.6°F), except in some shallow-water coastal regions, where the temperature may reach 40°C (104°F). On land, however, extremes in temperatures have ranged from −88°C (−127°F) to 58°C (136°F), which represents a temperature range more than four times greater than that experienced by the ocean.

Further, the ocean has a smaller daily, seasonal, and annual temperature range than that experienced on land, which provides a stable environment for marine organisms. The reasons for this are fourfold:

1. Recall from Chapter 6 that the heat capacity of water is much higher than that of land, which causes land to heat up by a greater amount and much more rapidly than the ocean.

2. The warming of the ocean is reduced substantially because of evaporation, a cooling process that stores excess heat as latent heat.

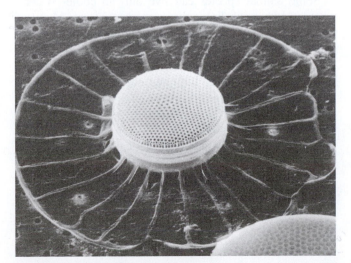

Figure 13–9 Warm-water diatom.

Scanning electron micrograph of the warm-water diatom *Planktoniella sol*, which has a prominent marginal ring to increase its surface area and prevent it from sinking (diameter = 60 microns).

Figure 13–11 Comparison of extremes in ocean and land surface temperatures.

In the open ocean, the maximum temperature range is limited to only 34°C (61°F) and increases to 42°C (76°F) in the coastal ocean. On land, the maximum temperature range is 146°C (263°F), over four times the temperature range in the open ocean.

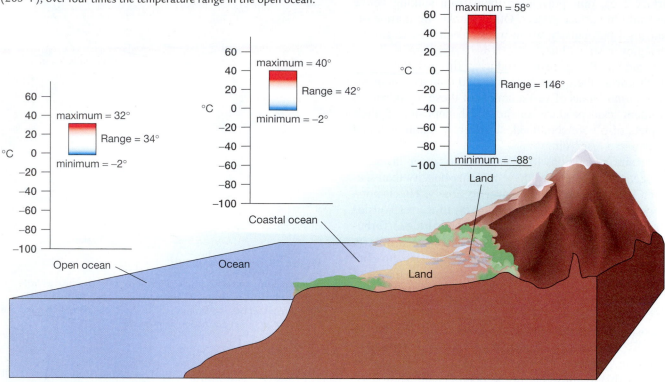

3. Radiation received at the surface of the ocean can penetrate several tens of meters deep and distribute its energy throughout a very large mass. In contrast, solar radiation absorbed by land heats only a very thin surface layer.

4. Unlike solid land surfaces, water has good mixing mechanisms such as currents, waves, and tides that allow heat from one area to be transported to other areas.

In addition, the small daily and seasonal temperature variations are confined to ocean surface waters and decrease with depth, becoming insignificant throughout the deeper parts of the ocean. At ocean depths that exceed 1.5 kilometers (0.9 mile), for example, temperatures hover around 3°C (37.4°F) year-round, regardless of latitude.

Comparing Cold- and Warm-Water Species Cold water is denser and has a higher viscosity than warm water. These factors, among others, profoundly influence marine life, resulting in the following differences between warm-water and cold-water species in the marine environment:

- Floating organisms are physically smaller in warm waters than in colder waters. Small organisms expose more surface area per unit of body mass, which helps them more easily maintain their position in the lower viscosity and density of warm seawater.

- Warm-water species often have ornate plumage to increase surface area, which is strikingly absent in the larger cold-water species (see Figures 13-7 and 13-9).

- Warmer temperatures increase the rate of biological activity, which more than doubles with an increase of 10°C (18°F). Tropical organisms apparently grow faster, have a shorter life expectancy, and reproduce earlier and more frequently than those in colder water.

- There are more *species* in warm waters, but the total *biomass* of plankton in colder, high-latitude waters greatly exceeds that of the warmer tropics.

Some animal species can live only in cooler waters, whereas others can live only in warmer waters. Many of these organisms can withstand only very small temperature changes and are called **stenothermal** (*steno* = narrow, *thermo* = temperature). Stenothermal organisms are found predominantly in the open ocean at depths where large temperature ranges do not occur.

Other species are little affected by different temperatures and can withstand large and even rapid changes in temperature. These organisms are called **eurythermal** (*eury* = wide, *thermo* = temperature) and are found predominantly in shallow coastal waters—where the largest temperature ranges are found—and in surface waters of the open ocean.

Salinity

The sensitivity of marine animals to changes in their environment varies from organism to organism. Those that inhabit estuaries, for example, such as oysters, must be able to withstand considerable fluctuations in salinity.

The daily rise and fall of the tides forces salty ocean water into river mouths and draws it out again, changing the salinity considerably. During floods, the salinity in estuaries can reach extremely low levels. The coastal organisms that can tolerate large changes in salinity are known as **euryhaline** (*eury* = broad, *halo* = salt).

Marine organisms that inhabit the open ocean, on the other hand, are seldom exposed to a large variation in salinity. They have adapted to a constant salinity and can tolerate only very small changes. These organisms are called **stenohaline** (*steno* = narrow, *halo* = salt).

Extraction of Salinity Components   Some organisms extract minerals from ocean water—particularly *silica* (SiO_2) and *calcium carbonate* ($CaCO_3$)—to construct the hard parts of their bodies, which serve as protective coverings. In doing so, they reduce the amount of dissolved material in ocean water. For example, phytoplankton (including diatoms) and microscopic protozoans such as radiolarians and silicoflagellates extract silica from seawater. Coccolithophores, foraminifers, most mollusks, corals, and some algae that secrete a calcium carbonate skeletal structure extract calcium carbonate. *to make their skeleton*

Diffusion   Molecules of soluble substances, such as nutrients, move through water from areas of *high* concentration to areas of *low* concentration until the distribution of the substance is uniform (Figure 13-12a). This process is called **diffusion** (*diffuse* = dispersed) and is caused by random motion of molecules.

The outer membrane of a living cell is permeable to many molecules. Organisms may take in nutrients they need from the surrounding water by diffusion of the nutrients through their cell walls. Because nutrients are usually plentiful in seawater, they pass through the cell wall into the interior, where nutrients are less concentrated (Figure 13-12b).

After a cell uses the energy stored in nutrients, it must dispose of waste. Waste passes out of a cell by diffusion, too. As the concentration of waste materials becomes greater within the cell than in the water surrounding it, these materials pass from within the cell into the surrounding fluid. The waste products are then carried away by circulating fluid that services cells in higher animals, or by the water that surrounds simple one-celled organisms.

Osmosis   When water solutions of unequal salinity are separated by a semipermeable membrane (such as skin or the membrane around a living cell), water molecules (but not dissolved ions) diffuse through the membrane. Water molecules always move from the *less* concentrated solution into the *more* concentrated solution in a process called **osmosis** (*osmos* = to push) (Figure 13-13a).

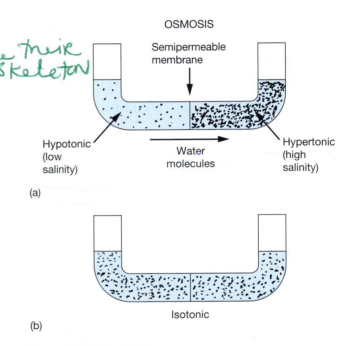

OSMOSIS

Semipermeable membrane

Hypotonic (low salinity)

Water molecules

Hypertonic (high salinity)

(a)

Isotonic

(b)

Figure 13–13 Osmosis.

(a) The semipermeable membrane that separates two water solutions of different salinities allows water molecules (but not dissolved substances) to pass through it by the process of osmosis. Water molecules will diffuse through the membrane from the less concentrated (hypotonic) solution (*left*) into the more concentrated (hypertonic) solution (*right*). **(b)** If the salinity of the two solutions is the same (isotonic), there is no net movement of water molecules.

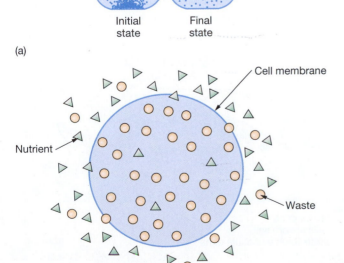

DIFFUSION

Initial state

Final state

(a)

Cell membrane

Nutrient

Waste

(b)

Figure 13–12 Diffusion.

(a) A water-soluble substance that is added to the bottom of a container of water (*initial state*) will eventually become evenly distributed throughout the water by diffusion (*final state*). **(b)** Nutrients (*triangles*) are in high concentration outside the cell and diffuse into the cell through the cell membrane. Waste particles (*circles*) are in high concentration inside the cell and diffuse out of the cell through the membrane.

Osmotic pressure is the pressure that must be applied to the more concentrated solution to prevent water molecules from passing into it.

Osmosis causes water to move through an organism's skin (its semipermeable membrane) and affects both marine and freshwater organisms. If the salinity of an organism's body fluid equals that of the ocean, it is **isotonic** (*iso* = same, *tonos* = tension), has equal osmotic pressure, and no net transfer of water will occur through the membrane in either direction (Figure 13-13b).

If seawater has a lower salinity than the fluid within an organism's cells, seawater will pass through the cell walls into the cells (always toward the more concentrated solution). This organism is **hypertonic** (*hyper* = over, *tonos* = tension), which means it has body fluids that are saltier than the surrounding seawater.

If the salinity within an organism's cells is less than that of the surrounding seawater, water from the cells will pass through the cell membranes out into the seawater (toward the more concentrated solution). This organism is **hypotonic** (*hypo* = under, *tonos* = tension) relative to the water outside its body.

In essence, osmosis is diffusion that produces a net transfer of water molecules through a semipermeable membrane from the side with the *greatest concentration* of water molecules to the side with the *lesser concentration* of water molecules.

During osmosis, three things can occur simultaneously across the cell membrane:

1. There is a net transfer of water molecules moving through the semipermeable membrane toward the side with the lower concentration of water.
2. Nutrient molecules or ions move from where they are more concentrated into the cell, where they are used to maintain the cell.
3. Waste molecules move from within the cell to the surrounding seawater.

Molecules or ions of all the substances in the system are passing through the membrane in both directions. A net transport of molecules of a given substance always occurs from the side on which they are most highly concentrated to the side where the concentration is less, until equilibrium is attained.

The body fluids of marine invertebrates (those without backbones) such as worms, mussels, and octopi, and the seawater in which they live, are nearly isotonic. As a result, these organisms have not had to evolve special mechanisms to maintain their body fluids at a proper concentration. This gives them an advantage over their fresh water relatives, whose body fluids are hypertonic.

? STUDENTS SOMETIMES ASK. …

Why do my fingers get all wrinkly when I stay in the water for a long time?

It's caused by water molecules passing through your skin (a semipermeable membrane) due to osmosis. The water molecules outside your body—contained in either pure water or seawater—flow into your skin cells in an attempt to dilute the dissolved particles inside those cells. After a short time in water, your skin cells contain many more water molecules than before, which hydrates your skin and causes it to look wrinkly. The effect is most obvious on appendages such as fingers and toes because they have a high surface area per unit of volume. It's only a temporary condition, though, because your body absorbs and excretes the excess water molecules after you leave the water.

An Example of Osmosis: Marine Fish versus Freshwater Fish Marine fish have body fluids that are only slightly more than one-third as saline as ocean water, possibly because they evolved in low-salinity coastal waters. They are, therefore, hypotonic (less salty) compared to the surrounding seawater.

This salinity difference means that saltwater fish, without some means of regulation, would lose water from their body fluids into the surrounding ocean and eventually dehydrate. This loss is counteracted, however, because marine fish drink ocean water and excrete the salts through special chloride-releasing cells located in their gills. These fish also help maintain their body water by discharging a very small amount of very highly concentrated urine (Figure 13-14a).

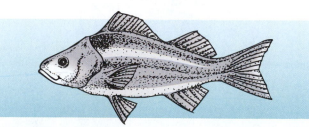

(a) MARINE FISH
(Hypotonic)

Drink large quantities of water
Secrete salt through special cells
Small volume of highly concentrated urine

(b) FRESHWATER FISH
(Hypertonic)

Do not drink
Cells absorb salt
Large volume of dilute urine

Figure 13–14 Salinity adaptations of marine and fresh water fish.

Osmotic processes cause these two types of fish to have different adaptations to the environment.

Freshwater fish are hypertonic (internally more saline) compared to the very dilute water in which they live. The osmotic pressure of the body fluids of such fish may be 20 to 30 times greater than that of the fresh water that surrounds them, so freshwater fish risk rupturing cell walls from excessive quantities of water taken on through osmosis.

To prevent this, freshwater fish do not drink water and their cells have the capacity to absorb salt. They also excrete large volumes of very dilute urine to reduce the amount of water in their cells (Figure 13-14b).

> Osmosis produces a net transfer of water molecules through a semipermeable membrane from the side with the greatest concentration of water molecules to the side with the lesser concentration.

Dissolved Gases

The amount of gases that can dissolve in seawater increases as the temperature of seawater decreases, so cold water dissolves more gas than warm water. As a result, vast phytoplankton communities develop in high latitudes during summer, when solar energy becomes available for photosynthesis. These cold waters contain an abundance of dissolved gases, specifically carbon dioxide (which phytoplankton need for photosynthesis) and oxygen (which all organisms need to metabolize their food).

Most animals that live in the ocean—except air-breathing marine mammals and certain fishes—must extract dissolved oxygen from seawater. How do they do it? Most marine animals have specially designed fibrous respiratory organs called **gills** that exchange oxygen and carbon dioxide directly with seawater. Most fish, for instance, take water in through their mouths (which gives them the appearance of "breathing" underwater), pass it through their gills to extract oxygen, and then expel it through the gill slits on the sides of their bodies (Figure 13-15). Most fish need at least 4.0 parts per million (ppm) of dissolved oxygen in seawater to survive for long periods, and even more for activity and rapid growth. That's why aquariums need a pump to continually resupply oxygen to the water in the tank.

During low-oxygen conditions, most marine animals with gills cannot simply breathe air at the surface. Their adaptations allow them to use only the oxygen dissolved in water. If dissolved oxygen levels become low enough (such as after an algae bloom when decomposition consumes dissolved oxygen), many marine organisms will suffocate and die.

Gill structure and location vary among animals of different groups. In fishes, gills are located at the rear of the mouth and contain capillaries. In higher aquatic invertebrates, they protrude from the body surface and contain extensions of the vascular system. In mollusks, they are inside the mantle cavity. In aquatic insects, they occur as projections from the walls of the air tubes. In amphibians, gills are usually present only in the larval stage. In higher vertebrates (including humans), they occur as rudimentary, nonfunctional gill slits, which disappear during embryonic development.

Water's High Transparency

Water—including seawater—has relatively high transparency compared to many other substances, allowing sunlight to penetrate to a depth of about 1000 meters (3300 feet) in the open ocean. The actual depth depends on the amount of turbidity (suspended sediment) in the water, the amount of plankton in the water, latitude, time of day, and the season.

Because of water's high transparency, many marine organisms have developed keen eyesight, which helps them locate and capture prey. To combat keen-eyed predators, many marine organisms such as jellyfish are themselves nearly transparent, which helps them blend into

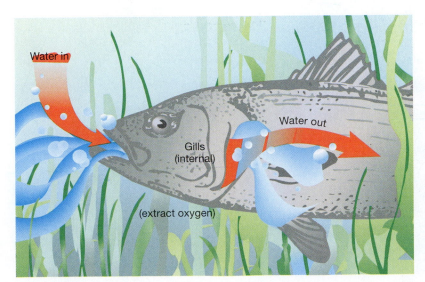

Figure 13–15 **Gills on fish.**
Water is taken in through the mouth, passed across the gills to extract dissolved oxygen, and exits through the gill slits.

Water in

Water out

Gills
(internal)

(extract oxygen)

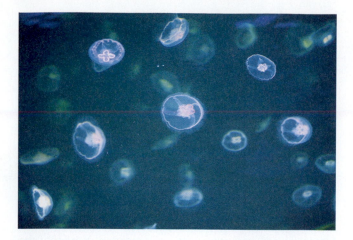

Figure 13–16 Jellyfish.

Jellyfish are nearly transparent, which makes them difficult for predators to see.

their environment (Figure 13-16). In fact, almost all open-ocean animals not otherwise protected by teeth, toxins, speed, or small size have some degree of invisibility. Only at depths where sunlight never penetrates is transparency uncommon. Not only can transparency help organisms elude their predators, it can also help organisms stalk their prey.

Other organisms hide by using their coloration pattern as camouflage (Figure 13-17a). Still others use **countershading** to blend into their environment, which means they are dark colored on top and light colored on bottom (Figure 13-17b). Many fish—especially flat fish—have countershading so they cannot be easily seen against the dark background of deep water or the ocean floor and they blend into the sunlight when viewed from below. Other organisms undertake a daily vertical migration to deeper, darker parts of the ocean to avoid becoming prey (Box 13-1).

(a)

Figure 13–17 Camouflage and countershading.

(a) The head and eye of a well-camouflaged rock fish.
(b) Halibut on a dock in Alaska show countershading.

(b)

Figure 13–18 Use of color by tropical fish.

Many tropical fish such as this spotfin butterfly fish have bright colors and bold patterns, which can allow them to blend into the environment by using disruptive coloration. Alternatively, they may also stand out so that they can advertise their identity, sex, or weaponry.

In contrast to species trying to blend into their environment, many species of tropical fish display bright colors (Figure 13-18). Why would they be brightly colored if it makes them easily seen by predators? The bright markings of tropical fish may be an example of **disruptive coloration**, where large, bold patterns of contrasting colors tend to make an object blend in when viewed against an equally variable, contrasting background. Zebras use this principle to evade predators, tigers use it to conceal themselves from their prey, and military uniforms are camouflaged in a similar way.

Even considering disruptive coloration, many tropical fish still don't easily blend into their environment. Perhaps the bright colors and distinctive markings that make tropical fish more apparent make it easier to advertise their identity, to attract mates, or to display weaponry such as spines or poison. Scientists have yet to agree on why tropical fish have such vivid colors, but there must be some biological advantage or the fish wouldn't have them.

Pressure

Water pressure increases about 1 kilogram per square centimeter (1 atmosphere or 14.7 pounds per square inch) with every 10 meters (33 feet) of water depth. Humans are not well adapted to the high pressures that exist below the surface.[6] Even when diving to the bottom of the deep end of a swimming pool, one can feel the dramatic increase in pressure in one's ears.

In the deep ocean, water pressure is on the order of several hundred kilograms per square centimeter (several hundred atmospheres, or several tons per square inch). How do deep-water marine organisms withstand pressures that can easily kill humans? Most marine organisms lack large compressible air pockets inside their bodies. They do not have lungs, ear canals, or other passageways as we do, so these organisms don't feel the high pressure pushing in on their bodies. Water is nearly incompressible, moreover, so their water-filled bodies have the same amount of pressure pushing outward and they are unaffected by the high pressures found in deep-ocean environments.

A few species appear to be extremely tolerant of pressure changes. In fact, some marine species that are found in nearshore areas can also be found at depths of several kilometers.

[6]For a discussion of the physiologic problems associated with human diving, see Box 1–3 in Chapter 1.

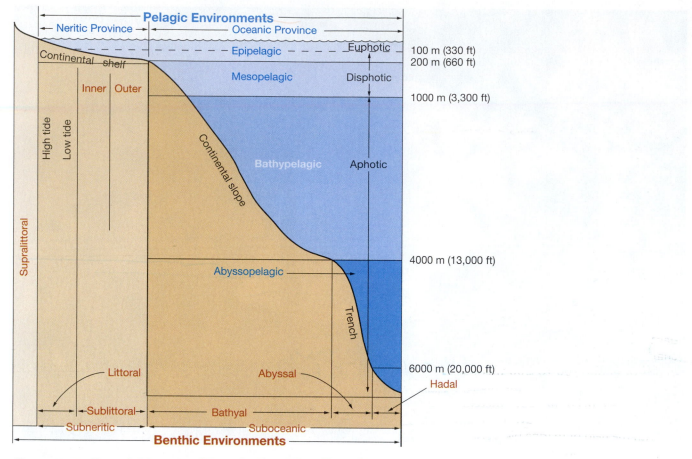

Figure 13–19 Oceanic biozones of the pelagic and benthic environments.
Pelagic environments are shown in blue and benthic environments are shown in brown. Subdivisions of pelagic and benthic environments are both based on depth, not necessarily on distance from shore. Sea floor features and sunlight zones are shown in black lettering.

STUDENTS SOMETIMES ASK...

If animals that have large air pockets inside are affected by the extreme pressures at depth, how are sperm whales able to dive so deeply?

All marine mammals have lungs and breathe air and certain ones have special adaptations that allow them to make extremely deep dives. Sperm whales, for example, can dive to more than 2800 meters (9200 feet) and stay submerged for more than one hour during their search for food! They are able to use small amounts of oxygen very efficiently and have a collapsible rib cage, which forces air out and collapses the lungs, thereby closing the air cavities inside their bodies. Marine mammals and their adaptations are discussed in Chapter 15, "Animals of the Pelagic Environment."

The ocean's physical support, viscosity, temperature, salinity, sunlit surface waters, dissolved gases, high transparency, and pressure create conditions to which marine organisms are superbly adapted.

Divisions of the Marine Environment

The oceans can be divided into two main environments. The *ocean water itself is the pelagic environment*, where floaters and swimmers play out their lives in a complex food web. The *ocean bottom is the benthic environment*, where marine algae and animals that do not float or swim (or at least not very well) spend their lives.

Pelagic (Open Sea) Environment

The pelagic environment can be divided into **biozones** that have distinctive biological characteristics, as shown in Figure 13–19.

The pelagic environment is divided into neritic and oceanic provinces (Figure 13–19). The **neritic** (*neritos* = of the coast) **province** extends from the shore seaward and includes all water less than 200 meters (660 feet) deep. Seaward of the neritic province is the **oceanic province**, where depth increases beyond 200 meters (660 feet). The oceanic province is further subdivided into four biozones:

BOX 13–1   Research Methods in Oceanography

A FALSE BOTTOM: THE DEEP SCATTERING LAYER (DSL)

Within the mesopelagic layer in most parts of the ocean, there exists a curious feature called the **deep scattering layer (DSL).** It was discovered when the U.S. Navy was testing sonar equipment to detect enemy submarines early in World War II. On many of the sonar recordings, a mysterious sound-reflecting surface appeared that was much too shallow to be the ocean floor. It was often referred to as a "false bottom" (see Figure 4–1). What was even more surprising was that the depth of the deep scattering layer changed with time. It was at a depth of about 100 to 200 meters (330 to 660 feet) during the night, but was as deep at 900 meters (3000 feet) during the day.

With the help of marine biologists, sonar specialists were able to determine that sonar signals were reflecting off densely packed concentrations of marine organisms (Figure 13A). Investigation with plankton nets and submersibles revealed that the DSL contained many different organisms, including copepods (which constitute

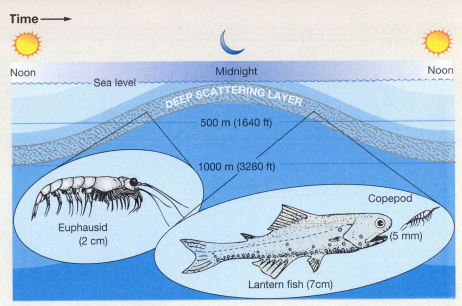

Figure 13A   Daily movement and organisms of the deep scattering layer.

a large proportion of planktonic animals), euphausids (krill), and lantern fish (family Myctophidae)

The daily movement of the deep scattering layer is caused by the vertical migration of marine organisms that feed in the highly productive surface waters but must protect themselves from being seen by predators.

Organisms within the DSL ascend only at night to feed under cover of darkness and then migrate to deeper (darker) water to hide during the day.

Today, fish-finding sonar is still used on commercial and sport fishing trips to track the movement of fish below the surface, often quite successfully.

---

1. The **epipelagic** (*epi* = top, *pelagic* = of the sea) **zone** from the surface to a depth of 200 meters (660 feet)

2. The **mesopelagic** (*meso* = middle, *pelagios* = of the sea) **zone** from 200 to 1000 meters (660 to 3280 feet)

3. The **bathypelagic** (*bathos* = depth, *pelagios* = of the sea) **zone** from 1000 to 4000 meters (3300 to 13,000 feet)

4. The **abyssopelagic** (*a* = without, *byssus* = bottom, *pelagios* = of the sea) **zone**, which includes all the deepest parts of the ocean below 4000 meters (13,000 feet)

- The **euphotic** (*eu* = good, *photos* = light) **zone** extends from the surface to a depth where enough light still exists to support photosynthesis. This rarely is deeper than 100 meters (330 feet).

- The **disphotic** (*dis* = apart from, *photos* = light) **zone** has small but measurable quantities of light. It extends from the euphotic zone to a depth where light no longer exists—usually about 1000 meters (3300 feet).

- The **aphotic** (*a* = without, *photos* = light) **zone** has no light, so it exists below about 1000 meters (3300 feet).

The single most important factor that determines the distribution of life in the oceanic province is the availability of sunlight. Thus, in addition to the four biozones, the distribution of life in the ocean is also divided into zones based on the availability of sunlight as follows:

Epipelagic Zone   The upper half of the epipelagic zone is the only place in the ocean where there is sufficient light to support photosynthesis. The boundary between the epipelagic and mesopelagic zones, at 200 meters (660

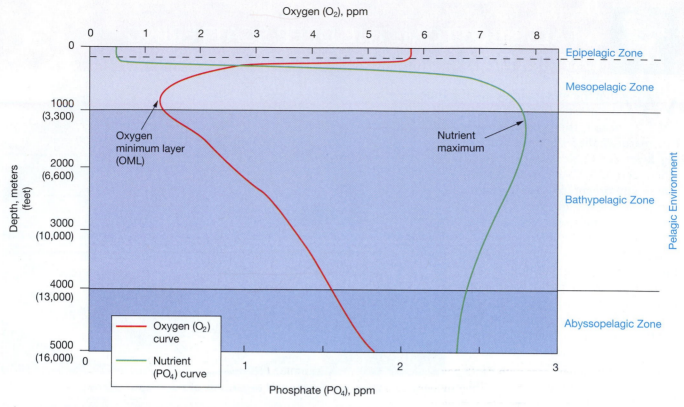

Figure 13–20 **Abundance of dissolved oxygen and nutrients with depth.**

In surface water, oxygen is abundant due to mixing with the atmosphere and plant photosynthesis, and nutrient content (phosphate) is low due to uptake by algae. At deeper depths, oxygen decreases and produces an oxygen minimum layer (OML), which coincides with a nutrient maximum. Below that, nutrients remain high and oxygen increases as it is replenished with high-oxygen cold water from polar regions (ppm = parts per million).

feet), is also where the level of dissolved oxygen begins to decrease significantly (Figure 13–20, *red curve*).

Oxygen decreases at this depth because no photosynthetic algae live below about 150 meters (500 feet), and dead organic tissue descending from the biologically productive upper waters is decomposing by bacterial oxidation. Nutrient content also increases abruptly below 200 meters (600 feet) (Figure 13–20, *green curve*) and it is the approximate bottom of the mixed layer, seasonal thermocline, and surface water mass.

Mesopelagic Zone A dissolved **oxygen minimum layer (OML)** occurs at a depth of about 700 to 1000 meters (2300 to 3280 feet) (Figure 13–20). The intermediate-water masses that move horizontally in this depth range often possess the highest levels of nutrients in the ocean.

Sunlight from the surface is very, very dim, so fish in the mesopelagic zone have unusually large eyes that are capable of detecting light levels 100 times lower than humans can sense.

Bioluminescent (*bio* = life, *lumen* = light, *esc* = becoming) organisms, which can produce light biologically and "glow in the dark," inhabit the mesopelagic zone, too. Examples of bioluminescent organisms include certain species of shrimp, squid, and fish. Approximately 80% of bioluminescent organisms have light-producing cells

called **photophores** (*photo* = light, *phoros* = bearing), which are glandular cells containing luminous bacteria surrounded by dark pigments. Some photophores contain lenses to amplify the light radiation.

Bioluminescence is produced when molecules of *luciferin* (*lucifer* = light-bringing) are excited and emit photons of light in the presence of oxygen, similar to how fireflies and glowworms produce light. Only a 1% loss of energy is required to produce this illumination.

Bathypelagic Zone and Abyssopelagic Zone The aphotic (lightless) bathypelagic and abyssopelagic zones represent over 75% of the living space in the oceanic province. Many completely blind fish exist in this region of total darkness and all are small, bizarre-looking, and predaceous.

Many species of shrimp that normally feed on **detritus**[7] become predators at these depths, where the food supply is greatly reduced compared to surface waters. Animals that live in these deep zones feed mostly upon one another. They have evolved impressive warning devices and unusual apparatuses to make them extremely efficient predators (Figure 13–21). Many also have sharp teeth and extremely large mouths relative to their body size.

---

[7]*Detritus* is a catchall term for dead and decaying organic matter, including waste products.

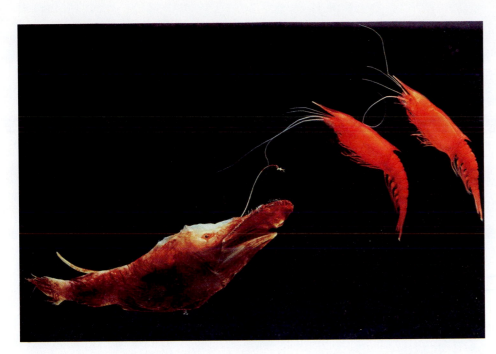

Figure 13–21 Deep-sea anglerfish and shrimp.

An anglerfish (*Lasiognathus saccostoma*) attracts two deep-water shrimp with its bioluminescent "lure." The fish is about 10 centimeters (4 inches) long.

Oxygen content increases with depth below the oxygen minimum layer because it is replenished by deep currents originating in polar regions as cold surface water high in oxygen. The abyssopelagic zone is the realm of the bottom-water masses, which commonly move in the direction opposite the deep-water masses in the bathypelagic zone.

Benthic (Sea Bottom) Environment

The benthic, or sea floor, environment is divided into two main units that correspond to the neritic and oceanic provinces of the pelagic environment (see Figure 13–19):

- The **subneritic province** extends from the spring high tide shoreline to a depth of 200 meters (660 feet), approximately encompassing the continental shelf.
- The **suboceanic province** includes the benthic environment below 200 meters (660 feet).

The transitional region from land to sea floor above the spring high-tide line is called the **supralittoral** (*supra* = above, *littoralis* = the shore) **zone**. Commonly called the *spray zone*, it is covered with water only during periods of extremely high tides and when tsunami or large storm waves break on the shore.

The subneritic zone is subdivided into the littoral and sublittoral zones. The *intertidal zone* (the zone between high and low tides) coincides with the **littoral** (*littoralis* = the shore) **zone**. The **sublittoral** (*sub* = below, *littoralis* = the shore) **zone**, or shallow subtidal zone, extends from low-tide shore line out to a depth of 200 meters (660 feet).

The sublittoral zone consists of inner and outer regions. The **inner sublittoral zone** extends to the depth at which marine algae no longer grows attached to the ocean bottom [approximately 50 meters (160 feet)], so it varies depending on the amount of solar radiation that penetrates the surface water. All photosynthesis seaward of the inner sublittoral zone is carried out by floating microscopic algae.

The **outer sublittoral zone** extends from the inner sublittoral zone out to a depth of 200 meters (660 feet) or the shelf break, which is the seaward edge of the continental shelf.

The suboceanic province is subdivided into bathyal, abyssal, and hadal zones. The **bathyal** (*bathus* = deep) **zone** extends from a depth of 200 to 4000 meters (660 to 13,000 feet) and corresponds generally to the continental slope.

The **abyssal** (*a* = without, *byssus* = bottom) **zone** extends from a depth of 4000 to 6000 meters (13,000 to 20,000 feet) and includes over 80% of the benthic environment. The ocean floor of the abyssal zone is covered by soft oceanic sediment, primarily abyssal clay. Tracks and burrows of animals that live in this sediment can be seen in Figure 13–22.

The **hadal** (*hades* = hell)[8] **zone** extends below 6000 meters (20,000 feet), so it consists only of deep trenches along the margins of continents. Animal communities that are found in these deep environments have been isolated from each other, often resulting in unique adaptations.

> The pelagic environment includes the water column and the benthic environment includes the sea bottom. Subdivisions of pelagic and benthic environments are based on depth, which influences the amount of sunlight.

---

[8]The inhospitable high-pressure environment of the hadal zone is aptly named.

Figure 13–22 Benthic organisms produce tracks on the ocean floor.

As benthic organisms—such as this sea urchin (*below*) and brittle star (*above*)—move across or burrow through the ocean bottom, they often leave tracks in the sediment on the ocean floor.

Chapter in Review

- *A wide variety of organisms lives in the ocean*, ranging in size from microscopic bacteria and algae to blue whales. *All living things belong to one of three major domains (branches) of life: Archaea*, simple microscopic bacteria-like creatures; *Bacteria*, simple life forms consisting of cells that usually lack a nucleus; and *Eukarya*, complex organisms (including plants and animals) consisting of cells that have a nucleus.

- *Organisms are further divided into five kingdoms: Monera*, single-celled organisms without a nucleus; *Protoctista*, single- and multi-celled organisms with a nucleus; *Fungi*, mold and lichen; *Plantae*, many-celled plants; and *Animalia*, many-celled animals. *Classification of organisms involves placing individuals within the kingdoms into increasingly specific groupings of phylum, class, order, family, genus, and species*, the last two of which denote an organism's scientific name. Many organisms also have one or more common names.

- *Marine organisms can be classified into one of three groups based on habitat and mobility. Plankton are free-floating forms with little power of locomotion; nekton are swimmers; and benthos are bottom dwellers. Most of the ocean's biomass is planktonic.*

- *Only about 14% of all known species inhabit the ocean, and over 98% of marine organisms are benthic.* The *marine environment—especially the pelagic environment—is much more stable than the terrestrial environment*, so there is less pressure on marine organisms to diversify.

- *Marine organisms are well adapted to life in the ocean.* Those organisms that have established themselves on land have had to develop complex systems for support and for acquiring and retaining water.

- The *algae* that must stay in surface water to receive sunlight and the *small animals* that feed on them *lack an effective means of locomotion.* To keep from sinking below sunlit surface waters, they depend on their *small size* and other adaptations to *increase their ratio of surface area to body mass*, which gives them high frictional resistance to sinking. Their small size also allows them to efficiently absorb nutrients and dispose of wastes. Many nektonic organisms have developed *streamlined bodies so that they can overcome the viscosity of seawater* and more easily move through it.

- *Surface temperature of the world ocean does not vary on a daily, seasonally, or yearly basis as much as on land. Organisms living in warm water tend to be individually smaller, have ornate plumage, comprise a greater number of species, and constitute a much smaller total biomass than organisms living in cold water.* Warm-water organisms also tend to live shorter lives and reproduce earlier and more frequently than cold-water organisms.

- *Osmosis is the passing of water molecules through a semipermeable membrane from a region of higher concentration to a region of lower concentration.* If the body fluids of an organism and ocean water are separated by a membrane that allows water molecules to pass through, the organism may become severely dehydrated from osmosis. Many marine invertebrates are essentially *isotonic*—the salinity of their body fluids is similar to that of ocean water. Most marine vertebrates are *hypotonic*—the salinity of their body fluids is lower than that of ocean water, so they tend to lose water through osmosis. Freshwater organisms are essentially all *hypertonic*—the salinity of their body fluids is greater than the water in which they live, so they tend to gain water through osmosis.

- *Most marine animals extract oxygen through their gills.* Many marine organisms have well-developed eyesight because water is so transparent. *To avoid being seen and consumed by predators*, many marine organisms are transparent, camouflaged, countershaded, or disruptively colored. Unlike humans, most *marine organisms are unaffected by the high pressure* at depth because they do not have large internal air pockets that can be compressed.

- *The marine environment is divided into pelagic (open sea) and benthic (sea bottom) environments.* These regions are further divided based on depth and have varying physical conditions to which marine life is adapted.

Key Terms

Abyssal zone (p. 391)
Abyssopelagic zone (p. 389)
Animalia (p. 372)
Aphotic zone (p. 389)
Archaea (p. 372)
Bacteria (p. 372)
Bacterioplankton (p. 374)
Bathyal zone (p. 391)
Bathypelagic zone (p. 389)
Benthic environment (p. 379)
Benthos (p. 376)
Bioluminescent (p. 390)
Biomass (p. 374)
Biozone (p. 388)
Countershading (p. 385)
Deep scattering layer (DSL) (p. 389)
Detritus (p. 390)

Diffusion (p. 383)
Disphotic zone (p. 389)
Disruptive coloration (p. 387)
Epifauna (p. 376)
Epipelagic zone (p. 389)
Eukarya (p. 372)
Euphotic zone (p. 389)
Euryhaline (p. 383)
Eurythermal (p. 382)
Fungi (p. 372)
Gill (p. 385)
Hadal zone (p. 391)
Holoplankton (p. 375)
Hypertonic (p. 384)
Hypotonic (p. 384)
Infauna (p. 376)
Inner sublittoral zone (p. 391)
Isotonic (p. 384)

Littoral zone (p. 391)
Macroplankton (p. 374)
Meroplankton (p. 375)
Mesopelagic zone (p. 389)
Monera (p. 372)
Nektobenthos (p. 376)
Nekton (p. 375)
Neritic province (p. 388)
Oceanic province (p. 388)
Osmosis (p. 383)
Osmotic pressure (p. 384)
Outer sublittoral zone (p. 391)
Oxygen minimum layer (OML) (p. 390)
Pelagic environment (p. 379)
Photophore (p. 390)
Phytoplankton (p. 374)
Picoplankton (p. 374)

Plankter (p. 374)
Plankton (p. 374)
Plantae (p. 372)
Protoctista (p. 372)
Protoplasm (p. 379)
Protozoa (p. 372)
Species (p. 373)
Stenohaline (p. 383)
Stenothermal (p. 382)
Streamlining (p. 381)
Sublittoral zone (p. 391)
Subneritic province (p. 391)
Suboceanic province (p. 391)
Supralittoral zone (p. 391)
Taxonomy (p. 373)
Viscosity (p. 380)
Zooplankton (p. 374)

Questions and Exercises

1. List the three major domains of life and the five kingdoms of organisms. Describe the fundamental criteria used in assigning organisms to these divisions.

2. Describe the lifestyles of plankton, nekton, and benthos. Why is it true that plankton account for a much larger percentage of the ocean's biomass than the benthos and nekton?

3. List the subdivisions of plankton and benthos and the criteria used for assigning individual species to each.

4. List the relative number of species of animals found in the terrestrial, pelagic, and benthic environments, and discuss the factors that may account for this distribution.

5. Discuss the major differences between marine algae and land plants, and explain the reasons why there is greater complexity in land plants.

6. Determine the surface to volume ratio of an organism whose average linear dimension is (a) 1 centimeter (0.4 inch); (b) 3 centimeters (1.1 inches); and (c) 5 centimeters (2 inches). Which one is better able to resist sinking, and why?

7. Discuss some adaptations other than size that are used by organisms to increase their resistance to sinking.

8. Changes in water temperature significantly affect the density, viscosity of water, and ability of water to hold gases in solution. Discuss how decreased water temperature changes these variables and how these changes affect marine life.

9. List differences between cold- and warm-water species in the marine environment.

10. What do the prefixes *eury-* and *steno-* mean? Define the terms *eurythermal/stenothermal* and *euryhaline/stenohaline*. Where in the marine environment will organisms displaying a well-developed degree of each characteristic be found?

11. Describe the process of osmosis. How is it different from diffusion? What three things can occur simultaneously across the cell membrane during osmosis?

12. What is the problem requiring osmotic regulation faced by hypotonic fish in the ocean? How have these animals adapted to meet this problem?

13. How does water temperature affect the water's ability to hold gases? How do marine organisms extract the dissolved oxygen from seawater?

14. How does the depth of the deep scattering layer vary over the course of a day? Why does it do this? Which organisms comprise the DSL?

15. Construct a table listing the subdivisions of the pelagic and benthic environments and the physical factors used in assigning their boundaries.

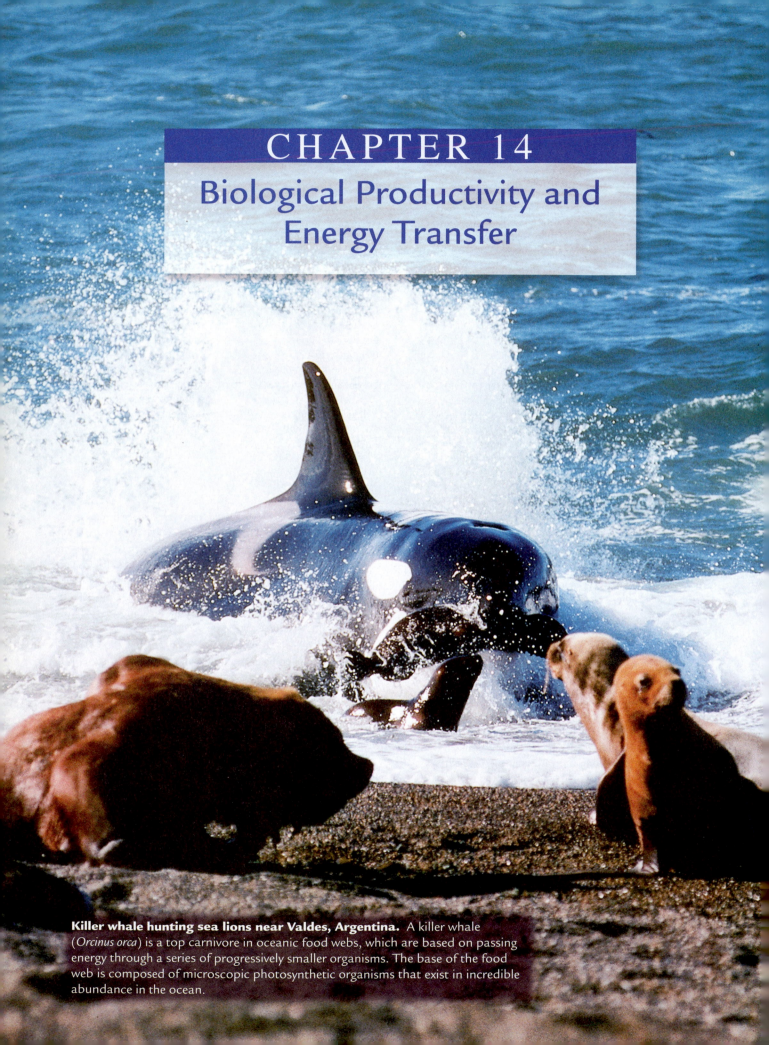

CHAPTER 14
Biological Productivity and Energy Transfer

Killer whale hunting sea lions near Valdes, Argentina. A killer whale (*Orcinus orca*) is a top carnivore in oceanic food webs, which are based on passing energy through a series of progressively smaller organisms. The base of the food web is composed of microscopic photosynthetic organisms that exist in incredible abundance in the ocean.

Key Questions

- What is primary productivity?
- What are the key limiting factors in the marine environment that control photosynthesis?
- What types of marine organisms photosynthesize?
- How does productivity differ in polar, tropical, and temperate regions?
- What is the efficiency of energy transfer between various organisms?
- How does a food chain differ from a food web?

The answers to these questions (and much more) can be found in the highlighted concept statements within this chapter.

"We cannot cheat on DNA. We cannot get round photosynthesis. We cannot say I am not going to give a damn about phytoplankton. All these tiny mechanisms provide the preconditions of our planetary life."

—Barbara Ward, *Who Speaks for Earth?* (1973)

Producers are plants and algae that photosynthesize their own food from carbon dioxide, water, and sunlight. Their ability to capture solar energy and bind it into their food sugars is the basis for all nutrition in the ocean (except near hydrothermal vents, where chemosynthesis[1] is the major source of "food" energy). The ocean's producers are the foundation of the ocean food web. The major primary producers of the oceans are marine algae. They capture nearly all of the solar energy used to support the entire marine biological community.

Large species of marine algae play only a minor role in the production of energy for the ocean population as a whole. Instead, marine organisms depend primarily on microscopic marine algae that are scattered throughout the near-surface sunlit waters of the world's oceans. They represent the largest community of **biomass**[2] in the marine environment—the *phytoplankton*.

In the total darkness of the deep sea, where no measurable sunlight penetrates, certain archaeon (bacteria-like organisms) oxidize hydrogen sulfide or methane to synthesize food. This method of food production supports a vast array of unusually large deep-sea benthos.

The chemical energy stored by surface phytoplankton and deep-sea archaeon is passed to the various populations of animals that inhabit the oceans through a series of feeding relationships called food chains and food webs.

Primary Productivity

Primary productivity is the amount of carbon fixed by organisms through the synthesis of organic matter using energy derived from solar radiation (**photosynthesis**) or chemical reactions (**chemosynthesis**). Although chemosynthesis supports deep-sea vent and seep biocommunities, it is much less significant than photosynthesis in worldwide marine primary production.

Photosynthetic Productivity

Photosynthesis is a chemical reaction in which energy from the Sun is stored in organic molecules (Figure 14-1). The

---

[1]Hydrothermal vent biocommunities are discussed in more detail in Chapter 16, "Animals of the Benthic Environment."

[2]Biomass is the mass of living organisms.

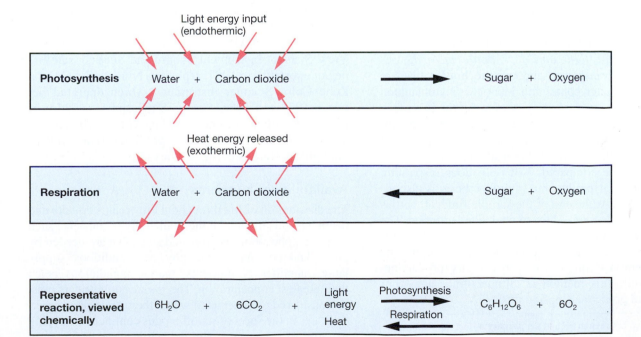

Figure 14–1 Photosynthesis (*top*), respiration (*middle*), and representative reactions viewed chemically (*bottom*).

total amount of organic matter produced by photosynthesis per unit of time is the **gross primary production** of the oceans. Algae use some of this organic matter for their own maintenance, through cellular respiration. What remains is **net primary production**, which is defined as gross primary production minus cellular respiration and is manifested as growth and reproduction products.[3] Net primary production supports the rest of the marine population.

Gross primary production has two components: new production and regenerated production. **New production** results from nutrients brought in from outside the local ecosystem by processes such as upwelling. **Regenerated production** results from nutrients that are recycled within the ecosystem. As the ratio of new production to gross primary production increases, so does the ecosystem's ability to support animal populations that are depended on for fisheries, such as pelagic fishes and benthic scallops.

> Primary productivity is the amount of carbon (organic matter) produced by microbes, algae, and plants mostly through photosynthesis but also includes chemosynthesis.

Measurement of Primary Productivity

Various properties of the ocean can be measured to give an approximation of the amount of primary productivity. One of the most direct at-sea methods is to capture plankton in cone-shaped nylon **plankton nets** (Figure 14-2). These fine mesh nets—which resemble windsocks at airports—filter plankton from the ocean as they are towed at a specific depth by research vessels. Analysis of the amounts and types of organisms captured reveals much about the productivity of the area.

In the 1920s, the *Gran method* was developed, which uses bottles to measure the amount of oxygen produced by phytoplankton and thus indicate the amount of organic carbon synthesized (Figure 14-3). This method, which is also called the *light and dark bottle method*, uses paired sets of transparent and opaque bottles that are filled with seawater containing identical phytoplankton samples and are suspended at various depths in surface waters for at least 24 hours. In the opaque bottles, phytoplankton consumes equal amounts of oxygen through respiration, so these bottles act as a standard to which the other bottles are compared. Not only does respiration occur in the clear bottles, but photosynthesis is also allowed to occur. At the end of the sample time, all bottles are retrieved and analyzed for oxygen concentration. The net increase in oxygen observed in the shallow clear bottles (less the amount of respiration denoted by the opaque bottles) is proportional to primary production. Unfortunately, this method is cumbersome, time consuming, and prone to sampling errors.

Figure 14–2 Plankton nets.

These large, cone-shaped, fine mesh plankton nets being washed are lowered into the water and towed behind a research vessel to collect plankton.

Other methods include laboratory analysis using radioactive carbon and using orbiting satellites to determine chlorophyll levels based on seawater surface color. The **SeaWiFS** (Sea-viewing Wide Field-of-View Sensor) instrument aboard the SeaStar satellite began operating in 1997, replacing Nimbus-7's Coastal Zone Color Scanner instrument, which operated between 1978 and 1986. SeaWiFS measures the color of the ocean with a radiometer and provides global coverage of estimated ocean chlorophyll levels as well as land productivities every two days.

Availability of Nutrients

The distribution of life throughout the ocean's breadth and depth depends mainly on the availability of nutrients such as nitrate, phosphorous, iron, and silica that are needed by phytoplankton. Where the physical conditions supply large quantities of nutrients, marine populations reach their greatest concentration. The *sources* of nutrients must be considered to understand where these areas are found.

Water in the form of runoff erodes continents, carrying material to the oceans, and depositing it as sediment at the margins of the continents. Runoff also dissolves and

---

[3]The difference between gross and net primary production is similar to the difference between your gross pay (your earnings before taxes) and your net pay (the amount you take home after taxes).

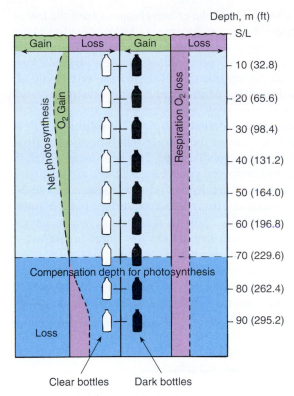

Depth, m (ft)

Figure 14–3 Gran method for determining the compensation depth for photosynthesis.

Two sets of bottles are lowered into the ocean to determine phytoplankton production and consumption of oxygen (O_2). The clear bottles (*left*) let in light needed for photosynthesis; the dark bottles (*right*) do not and provide a control group. Comparison of the two sets of bottles reveals how deeply light can penetrate to support photosynthesis (the compensation depth for photosynthesis) as well as an estimate of the amount of productivity.

transports substances such as nitrates and phosphates,[4] which are the basic nutrients for phytoplankton. The major nutrients used by organisms in metabolic processes are listed in Table 14-1.

Through photosynthesis, phytoplankton combine these nutrients with carbon dioxide and water to produce the carbohydrates, proteins, and fats that the rest of the ocean's biological community depends upon for food.

The continents are the major sources of these nutrients, so the greatest concentrations of marine life are found along the continental margins. The concentration of marine life decreases, however, as the distance from the continental margins into the open sea increases. The vast depth of the world's oceans and the great distance between the open ocean and the coastal regions where nutrients are concentrated account for these differences.

Recent studies in the waters near Antarctica and the Galápagos Islands reveal that photosynthetic production is low even though the concentration of all nutrients—except iron—is high.[5] Production is high only in regions of shallow water downcurrent from islands or landmasses where iron from rocks and sediments was dissolved into the water to significant levels.

Availability of Solar Radiation

Photosynthesis cannot proceed unless light energy (solar radiation) is available. Despite the atmosphere's thickness

---

[4]Nitrates and phosphates are the basic ingredients in garden and farm fertilizers.

[5]The idea of fertilizing the ocean with iron to stimulate productivity and increase the amount of CO_2 gas absorbed by the ocean is discussed in Chapter 7.

TABLE 14–1 **Major nutrients used by organisms in metabolic processes.**

Nutrient (symbol)	Metabolic use(s)
Nitrogen[a] (N)	Production of proteins and nucleic acids
Phosphorous (P)	Production of nucleic acids and teeth/bones/shells
Sodium (Na)	Body fluids; osmotic regulation
Magnesium (Mg)	Osmotic balance; chlorophyll production
Sulfur (S)	Production of proteins; cell division
Chlorine (Cl)	Nerve discharge; osmotic regulation; ATP formation
Potassium (K)	Nerve discharge; osmotic regulation; enzyme activation
Calcium (Ca)	Production of shells/bones/coral/teeth
Iodine (I)	Production of thyroid hormone
Silicon (Si)	Construction of hard supporting structures such as tests
Iron (Fe)	Electron transport; nitrogen assimilation
Copper (Cu)	Electron transport
Zinc (Zn)	Nucleic acid replication and transcription

[a] Before nitrogen can be assimilated, it must be "fixed" by being converted into nitrate compounds such as ammonia (NH_4OH).

of more than 80 kilometers (50 miles), its high transparency allows sunlight to penetrate it quite readily, so land-based plants almost always have an abundance of solar radiation to conduct photosynthesis.

In the clearest ocean water, however, solar energy may be detected to depths of only about 1 kilometer (0.6 mile) and, even then, the amount reaching these depths is inadequate for photosynthesis. Photosynthesis in the ocean, therefore, is restricted to the uppermost surface waters and those areas of the sea floor where the water is shallow enough to allow light to penetrate.

The depth at which net photosynthesis becomes zero is called the **compensation depth for photosynthesis**. The **euphotic** (*eu* = good, *photos* = light) **zone** extends from the surface down to the compensation depth for photosynthesis, which is approximately 100 meters (330 feet) in the open ocean. Near the coast, the euphotic zone may extend to less than 20 meters (66 feet) because the water contains more suspended inorganic material (turbidity) or microscopic organisms that limit light penetration.

How do the two factors necessary for photosynthesis—the supply of nutrients and the presence of solar radiation—differ between coastal areas and the open ocean? In the open ocean (far from continental margins), solar energy extends deeper into the water column, but concentration of nutrients is low. In coastal regions, on the other hand, light penetration is much less, but the concentration of nutrients is much higher. Because the coastal zone is much more productive, nutrient availability must be the most important factor affecting the distribution of life in the oceans.

Margins of the Oceans

If the stability of the ocean environment is ideal for sustaining life, why are the richest concentrations of marine organisms in the very margins of the oceans, where conditions are the most *un*stable? In the coastal ocean, for instance,

- Water depths are shallow, allowing much greater seasonal variations in temperature and salinity than in the open ocean.
- The thickness of the water column varies in the nearshore region in response to tides that periodically cover and uncover a thin strip of land along the margins of the continents.
- Waves breaking in the surf zone release large amounts of energy that has been carried for great distances across the open ocean.

Each of these conditions stresses organisms. Over the billions of years of geologic time, however, new species have evolved by natural selection to fit every imaginable biological niche—even in environments that pose difficulties for organisms. In fact, many organisms have adapted to live under adverse conditions—such as coastal environments—as long as nutrients are available.

Along continental margins, some areas have more abundant life than others. What characteristics create such an uneven distribution of life? Again, only those basic requirements for the production of food need be considered. For ex-

ample, areas that have the greatest biomass have the lowest water temperatures, too, because cold water dissolves larger amounts of oxygen and carbon dioxide than warm water. Carbon dioxide, in particular, stimulates phytoplankton growth and phytoplankton growth, in turn, profoundly affects the distribution of all other life in the oceans.

Upwelling and Nutrient Supply **Upwelling** is a flow of deep water toward the surface that brings water from depths below the euphotic zone. This deep water is rich in nutrients and dissolved gases because there are no phytoplankton at these depths to consume these compounds. When chilled water from below the surface rises, it hoists nutrients from the depths to the surface, where phytoplankton thrive and make food for larger organisms—copepods, fish, and on up to whales.

Highly productive areas of *coastal upwelling* are found along the western margins of continents, where surface currents are moving toward the Equator (Figure 14-4). Ekman transport (see Chapter 8, "Ocean Circulation") moves surface water away from these coasts, so nutrient-rich water from depths of 200 to 1000 meters (660 to 3300 feet) constantly rises to replace it.

> Photosynthetic productivity is limited in the marine environment by the amount of sunlight and the supply of nutrients. Upwelling greatly enhances the conditions for life by lifting cold, nutrient-rich water to the sunlit surface.

Light Transmission in Ocean Water

The graph in Figure 14-5 shows that most solar energy falls in the range of wavelengths called **visible light**. This radiant energy from the Sun powerfully affects three major components of the oceans:

1. The major wind belts of the world, which produce ocean currents and wind-driven ocean waves, ultimately derive their energy from solar radiation. Wind belts and ocean currents strongly influence world climates.

2. A thin layer of warm water at the ocean surface, created by solar heating, overlies the great mass of cold water that fills most of the ocean basins. This is the "life layer" where most sea life exists.

3. Photosynthesis can occur only where sunlight penetrates the ocean water, so phytoplankton and most animals that eat them must live where the light is, in the relatively thin layer of sunlit surface water.

The Electromagnetic Spectrum The Sun radiates a wide range of wavelengths of electromagnetic radiation. Together they comprise the **electromagnetic spectrum**, which is shown in the upper part of Figure 14-5. Only a very narrow portion of the electromagnetic spectrum is visible to humans as visible light. We call it "visible" light because our electromagnetic sensors—our eyes—are adapted

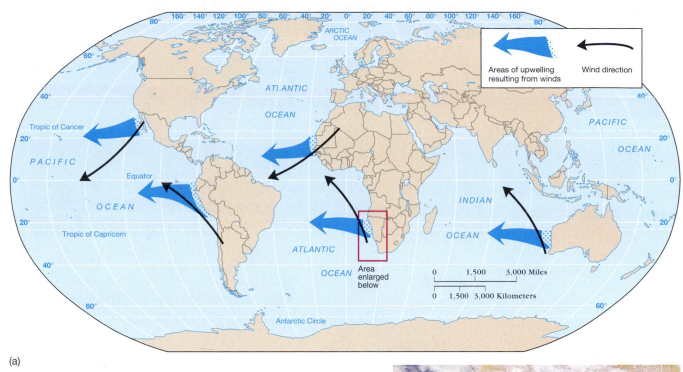

(a)

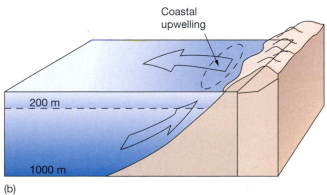

(b)

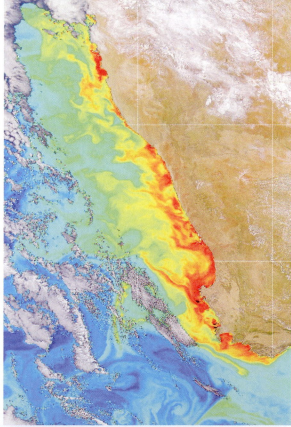

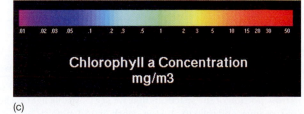

Chlorophyll a Concentration mg/m3

(c)

Figure 14–4 Coastal upwelling.

(a) Coastal winds (*black arrows*) cause Ekman transport, which drives surface water away from the west coasts of continents (*blue arrows*).
(b) Block diagram showing how coastal upwelling is created by surface water that moves away from shore, bringing cold, nutrient-rich water to the surface. (c) SeaWiFS image of chlorophyll concentration along the southwest coast of Africa (February 21, 2000). High chlorophyll concentrations indicate high phytoplankton biomass, which is caused by coastal upwelling. Concentration is reported in milligrams of chlorophyll per cubic meter (mg/m^3).

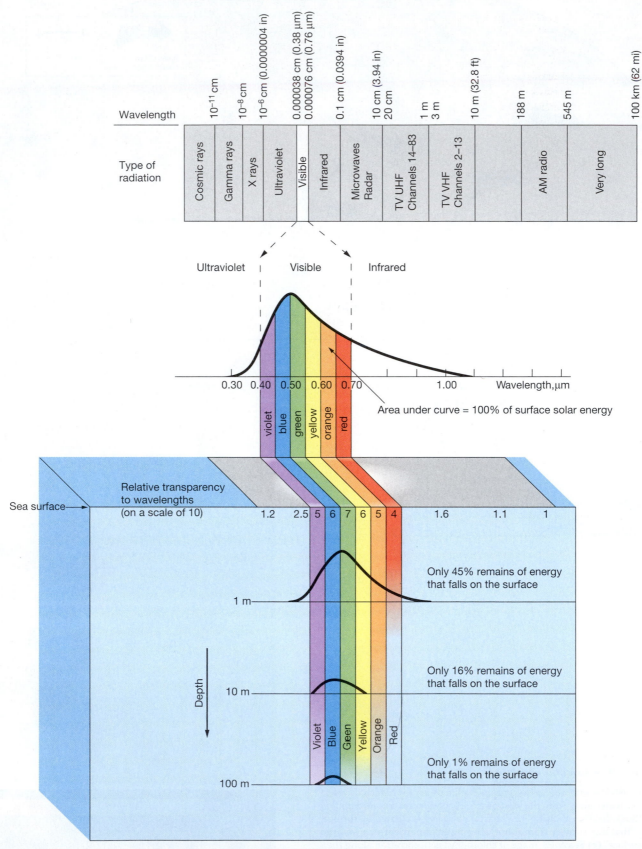

Figure 14–5 **The electromagnetic spectrum and transmission of visible light in seawater.**

The spectrum (*top*) runs from extremely short cosmic rays (*left side*) with progressively increasing wavelength shown toward the right. The narrow portion of the spectrum that we see as visible light is shown passing through seawater (*bottom*), which absorbs the longer wavelengths (red, orange, and yellow) above a depth of 100 meters.

to detect only the wavelengths in the visible region. In essence, our eyes "tune into" the visible light wavelengths, just as a radio "tunes into" specific radio waves.

Visible light can be further divided by wavelength into violet, blue, green, yellow, orange, and red energy levels. Together, these different wavelengths produce white light. The shorter wavelengths of energy to the left of visible light (for example, X rays and gamma rays) damage tissue in high enough doses. The longer wavelengths of energy to the right of visible light (for example, infrared, microwaves, and radio waves) are used for heat transfer and communication.

The Color of Objects   Light from the Sun includes all the visible colors. Most of the light we see is reflected from objects. All objects absorb and reflect different wavelengths of light, and each wavelength represents a color in the visible spectrum. Vegetation, for example, absorbs most wavelengths except green and yellow, which they reflect, so most plants look green. Similarly, a red jacket absorbs all wavelengths of color except red, which is reflected.

The lower part of Figure 14-5 shows how the ocean selectively absorbs the longer-wavelength colors (red, orange, and yellow) of visible light. The true colors of objects can be observed in natural light only in the surface waters, because only there can all wavelengths of the visible spectrum be found. Red light is absorbed within the upper 10 meters (33 feet) of the ocean, and yellow is completely absorbed before a depth of 100 meters (330 feet). Thus, the shorter-wavelength portion of the visible spectrum (violet, blue, and some green light) is all that can be transmitted to greater depths, and even then, their intensity is low. In the open ocean, sunlight strong enough to support photosynthesis occurs only with the euphotic zone to a depth of 100 meters (330 feet) and no sunlight penetrates below a depth of about 1000 meters (3300 feet).

A Secchi disk, like the one shown in Figure 14-6, is used to measure the depth to which visible light penetrates the ocean. The **Secchi** (pronounced "SECK-ee") **disk** is named after its inventor, Angelo Secchi, an Italian astronomer who first used the device to measure water clarity of lakes in 1865. It consists of a round flat disk about 30 centimeters (12 inches) in diameter attached to a line that is marked off at regular intervals. After the disk is lowered into the ocean, the depth at which it can last be seen indicates the water's clarity. Both the amount of microorganisms in the water and the water's turbidity—the amount of suspended material—increase the degree of light absorption, thus decreasing the depth to which visible light can penetrate the ocean.

Water Color and Life in the Oceans   The color of the ocean ranges from deep indigo (blue) to yellow-green. Why are some areas of the ocean blue, whereas others appear green? Ocean color is influenced by (1) the amount of turbidity from runoff and (2) the amount of photosynthetic pigment, which increases with increasing biological production.

Figure 14–6 Secchi disk.

A Secchi disk is used to measure the depth of penetration of sunlight, and thus indicate the clarity of the water.

Coastal waters and upwelling areas are biologically very productive and almost always yellow-green in color because they contain large amounts of yellow-green microscopic marine algae and suspended particles. These materials disperse solar radiation so that the wavelengths for greenish or yellowish light are scattered most.

Water in the open ocean—particularly in the tropics—lacks productivity (and turbidity), so it is usually a clear, indigo blue color. Water molecules disperse solar radiation so that the wavelength for blue light is scattered most. The atmosphere scatters blue light, too, which is why clear skies are blue.

Satellites can easily detect differences in ocean color, which is related to the water's productivity. Figure 14-7 is a SeaStar satellite/SeaWiFS instrument view of worldwide productivity, showing that highly productive areas, which are called **eutrophic** (*eu* = good, *tropho* = nourishment), are found in shallow-water coastal regions, areas of upwelling, and high-latitude regions. Areas of low productivity, on the other hand, which are called **oligotrophic** (*oligo* = few, *tropho* = nourishment), are found in the open oceans of the tropics.

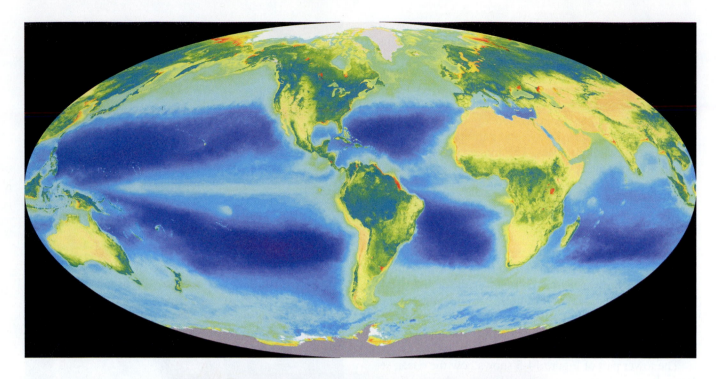

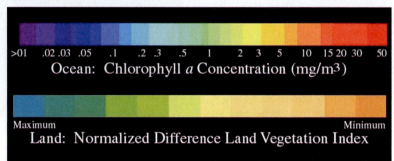

Figure 14–7 World productivity.

SeaWiFS image of world productivity (September 1997–August 2000). The SeaWiFS instrument aboard the SeaStar satellite detects changes in seawater color caused by changing concentrations of chlorophyll, which varies with the photosynthetic productivity. Ocean concentrations are reported in milligrams of chlorophyll per cubic meter (mg/m^3).

STUDENTS SOMETIMES ASK...

My friends are planning an ocean-fishing trip. To get to good fishing grounds that haven't been fished out, one of my friends has suggested that we go far offshore. Should we expect to find more fish there?

Generally, no. Most people fish coastal waters because they have high nutrient levels and are highly productive, whereas few people fish the open ocean because it has a low nutrient level and correspondingly low productivity. Far offshore, there just isn't enough food for many fish to live (see, for example, the distribution of phytoplankton shown in Figure 14-7). Thus, the farther you go from land, the less likely there would be good fishing, especially past the continental shelf. However, there are some exceptions to this, such as shallowly submerged banks and areas of upwelling far from shore.

Photosynthetic Marine Organisms

Many types of marine organisms photosynthesize. Mostly, they are represented by microscopic marine algae, but also include larger forms of algae and seed-bearing plants.

Seed-Bearing Plants (Anthophyta)

The only members of kingdom Plantae that exist in the marine environment belong to the highest group of plants, the seed-bearing Anthophyta, and they occur exclusively in shallow coastal areas. Eelgrass (*Zostera*), for example, is a grasslike plant with true roots that exists primarily in the quiet waters of bays and estuaries from the low-tide

Figure 14–8 *Surf grass (Phyllospadix).*

zone to a depth of 6 meters (20 feet). Surf grass (*Phyllospadix*) (Figure 14-8), which is also a seed-bearing plant with true roots, is typically found in the high-energy environment of exposed rocky coasts from the intertidal zones down to a depth of 15 meters (50 feet).

Other seed-bearing plants are found in salt marshes and include grasses (mostly of the genus *Spartina*), whereas mangrove swamps contain primarily mangroves (genera *Rhizophora* and *Avicennia*). All of these plants are important sources of food and protection for the marine animals that inhabit nearshore environments.

Macroscopic (Large) Algae

Various types of marine macro algae (the "seaweeds") are typically found in shallow waters along the ocean margins. These algae are usually attached to the bottom, but a few species float. Algae are classified in part on the color of the pigment they contain (Figure 14-9).

Brown Algae   The brown algae of the phylum Phaeophyta (*phaeo* = dusky, *phytum* = a plant) include the largest members of the attached (not free-floating) species of marine algae. Their color ranges from very light brown to black. Brown algae occur primarily in temperate and cold-water areas. Their sizes range widely.

Figure 14–9 **Macroscopic algae.**

(a) Brown algae, *Sargassum*. This attached form is similar to the floating form that is the namesake of the Sargasso Sea. **(b)** Brown algae, *Macrocystis*, a major component of kelp beds. **(c)** Green algae, *Codium fragile*, also known as sponge weed or dead man's fingers.
(d) Red algae, *Lithothamnion*, an encrusting form that produces a pink coating on these rounded cobbles.

The smallest is a black encrusting patch of *Ralfsia* of the upper and middle intertidal zones. The largest is bull kelp (*Pelagophycus*), which may grow in water deeper than 30 meters (100 feet) and extend to the surface. Other types of brown algae include *Sargassum* (Figure 14-9a) and *Macrocystis* (Figure 14-9b).

Green Algae Although green algae of phylum Chlorophyta (*chloro* = green, *phytum* = a plant) are common in freshwater environments, they are not well represented in the ocean. Most marine species are intertidal or grow in shallow bay waters. They contain the pigment chlorophyll, which gives them their green color. They grow only to moderate size, seldom exceeding 30 centimeters (12 inches) in the largest dimension. Forms range from finely branched filaments to thin sheets.

Various species of sea lettuce (*Ulva*), a thin membranous sheet only two cell layers thick, are widely scattered throughout colder-water areas. Sponge weed (*Codium*), a two-branched form more common in warm waters, can exceed 6 meters (20 feet) in length (Figure 14-9c).

Red Algae Red algae of phylum Rhodophyta (*rhodo* = red, *phytum* = a plant) are the most abundant and widespread of marine macroscopic algae. Over 4000 species occur from the very highest intertidal levels to the outer edge of the inner sublittoral zone. Many are attached to the bottom, either as branching forms or as forms that encrust surfaces (Figure 14-9d). They are very rare in fresh water. Red algae range from just barely visible to the unaided eye to 3 meters (10 feet) long. While found in both warm and cold waters, the warm-water varieties are relatively small.

The color of red algae varies considerably depending on its depth in the intertidal or inner sublittoral zones. In upper, well-lighted areas, it may be green to black or purplish. In deeper-water zones, where less light is available, it may be brown to pinkish red.

The bulk of marine photosynthetic productivity occurs within the surface layer of the ocean to a depth of 100 meters (330 feet), which corresponds to the depth of the euphotic zone. At this depth, the amount of light is reduced to 1% of that available at the surface. A red alga, however, has been documented growing at a depth of 268 meters (880 feet) on a seamount near San Salvadore, Bahamas. Available light at this depth was only 0.05% of that available at the surface.

Microscopic (Small) Algae

Microscopic algae produce food either directly or indirectly for over 99% of marine animals. Most microscopic algae are phytoplankton—photosynthetic organisms that live in the upper surface waters and drift with currents—although some live on the bottom in the nearshore environment, where sunlight reaches the shallow ocean floor.

Golden Algae The golden algae of phylum Chrysophyta (*chrysus* = golden, *phytum* = a plant) contain the orange-yellow pigment *carotin*. They consist of diatoms and coccolithophores, both of which are described in Chapter 5, "Marine Sediments," and they store food as carbohydrates and oils.

Diatoms The **diatoms** (*diatoma* = cut in half) are a class of algae that are contained in a microscopic shell called a **test** (*testa* = shell). The tests are composed of opaline silica ($SiO_2 \cdot nH_2O$) and are important geologically because they accumulate on the ocean bottom, producing **diatomaceous earth**. Some deposits of diatomaceous earth that have been elevated above the water surface by tectonic forces are mined and used in filtering devices and numerous other applications (see Box 5–2). Diatoms are a very productive group of marine algae in coastal waters.

The tests of diatoms have a variety of shapes, but all have a top and bottom half that fit together (Figure 14-10a). The single cell is contained within this test, and it exchanges nutrients and waste with the surrounding water through holes in its test.

Coccolithophores **Coccolithophores** (*coccus* = berry, *lithos* = stone, *phorid* = carrying) are covered with small calcareous plates called **coccoliths**, made of calcium carbonate ($CaCO_3$) (Figure 14-10b). The individual plates are about the size of a bacterium, and the entire organism is too small to be captured in plankton nets. Coccolithophores contribute significantly to calcareous deposits in temperate and warmer oceans.

? STUDENTS SOMETIMES ASK...
Why does a red tide glow blue-green at night?

Many of the species of dinoflagellates that produce red tides (most notably those of genus *Gonyaulax*) also have bioluminescent capabilities—that is, they can produce light organically. When the organisms are disturbed, they emit a faint blue-green glow. When waves break during a red tide at night, the waves are often spectacularly illuminated by millions of bioluminescent dinoflagellates. During these times, one can easily observe marine animals moving through the water because their bodies are silhouetted by bioluminescent dinoflagellates that light up as they pass over the animal's body.

Dinoflagellates The **dinoflagellates** (*dino* = whirling, *flagellum* = a whip) belong to the phylum Pyrrophyta (*pyrros* = fire, *phytum* = a plant) (Figures 14-10c and d). They possess **flagella** (small, whiplike structures) for locomotion, giving them a slight capacity to move into areas that are more favorable for photosynthetic productivity. Dinoflagellates are rarely important geologically because

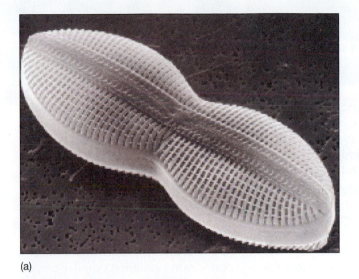

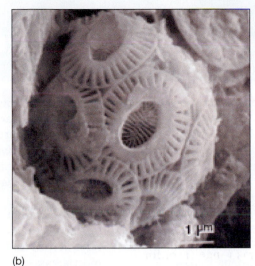

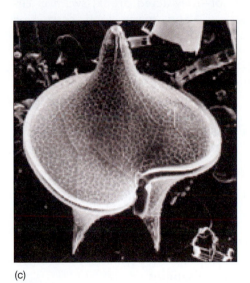

(a)

(b)

(c)

(d)

Figure 14–10 Microscopic algae.

(a) Peanut-shaped diatom *Diploneis* (length = 50 microns, or 0.002 inch). (b) Coccolithophore *Emiliania huxleyi*, showing disk-shaped calcium carbonate ($CaCO_3$) plates—called coccoliths—that cover the organism (bar scale = 1 micron, or 0.00004 inch). (c) Dinoflagellate *Protoperidinium divergnes* (length = 70 microns, or 0.003 inch). (d) Leaflike tropical dinoflagellate *Heterodinium whittingae* (length = 100 microns, or 0.004 inch).

their tests are made of cellulose, which is biodegradable and not preserved as deposits on the sea floor.

Dinoflagellates sometimes exist in such great abundance that they color surface waters red, producing the phenomenon known as a **red tide** (Box 14-1)—which has nothing to do with tidal phenomena and is more accurately called a **harmful algal bloom (HAB)**. In addition, many of the 1100 species undergo structural changes in response to changes in their environment (Box 14-2).

Marine photosynthetic organisms include seed-bearing plants (such as surf grass), macroscopic algae (seaweeds), and microscopic algae (diatoms, coccolithophores, and dinoflagellates).

Regional Productivity

Primary photosynthetic production in the oceans varies dramatically from place to place (see Figure 14-7). Typical units of photosynthetic production are in weight of carbon (*grams of carbon*) per unit of area (*square meter*) per unit of time (*year*), which is abbreviated as $gC/m^2/yr$. Values range from as low as 1 $gC/m^2/yr$ in some areas of the open ocean to as much as 4000 $gC/m^2/yr$ in some highly productive coastal estuaries (Table 14-2). This variability is the result of the uneven distribution of nutrients throughout the photosynthetic zone and seasonal changes in the availability of solar energy.[6]

[6]For a review of Earth's seasons, see Chapter 7.

BOX 14–1   People and Ocean Environment

RED TIDES: WAS ALFRED HITCHCOCK'S *THE BIRDS* BASED ON FACT?

Conditions in the oceans sometimes stimulate the productivity of certain dinoflagellates. During these times, up to 2 million dinoflagellates may be found in 1 liter (about 1 quart) of water, giving the water a reddish color and causing what is known as a red tide (Figure 14A). Red tides are by no means a new phenomenon. In fact, the Old Testament makes reference to waters turning blood red, which is most likely how the Red Sea got its name.

Although many red tides are harmless to marine animals and humans, they can still be responsible for mass die-offs of marine organisms. When huge numbers of dinoflagellates die, oxygen is removed from seawater during decomposition and many types of marine life literally suffocate to death. In other cases, dinoflagellates that are responsible for many red tides produce toxins that can spread to many different types of organisms—including humans (Figure 14B). *Ptychodiscus* and *Gonyaulax*, for example, are two common genera of dinoflagellates in red tides that produce water-soluble toxins. Certain filter-feeding shellfish called bivalves—various clams, mussels, and oysters—then strain the dinoflagellates from the water for food. *Ptychodiscus* toxin kills fish and shellfish. *Gonyaulax* toxin is not poisonous to shellfish, but it concentrates in their tissues and is poisonous to humans who eat the shellfish, even after the shellfish are cooked. This malady is called *paralytic shellfish poisoning (PSP)*.

The symptoms of PSP in humans are similar to those of drunkenness—incoherent speech, uncoordinated movement, dizziness, and nausea—and can occur only 30 minutes after ingesting contaminated shellfish.

There is no known antidote for the toxin, which attacks the human nervous system, but the critical period usually passes within 24 hours. At least 300 fatal and 1750 nonfatal cases of PSP have been documented worldwide.

April through September are particularly dangerous months for red tides in the Northern Hemisphere. In most areas, quarantines exist to prohibit harvesting those shellfish that feed on toxic microscopic organisms.

Worldwide, increasing numbers of mysterious poisonings are implicating species of toxic dinoflagellates that can spread throughout the marine food web and to humans. For instance, domoic acid—a toxin produced by a diatom (*Pseudonitzschia*)—was first recognized in 1987 as the poison that infected over 100 people who ate contaminated mussels from Prince Edward Island, Canada. Four of the victims died and 10 suffered permanent memory loss, which led researchers to call poisoning by domoic acid *amnesic shellfish poisoning.*

Domoic acid poisoning may also be responsible for birds attacking humans in California's Monterey Bay in 1961. The incident—reported to have inspired Alfred Hitchcock's 1963 movie, *The Birds* (Figure 14C)—apparently happened after seagulls consumed fish loaded with domoic acid produced by a bloom of diatoms. Infected by the poison, the birds smashed into structures during flight and pecked eight people. In the same area in September 1991, brown pelicans and Brandt's cormorants exhibited similar behavior—they acted drunk, swam in circles, and made loud squawking sounds. Research revealed that toxic diatoms (*Pseudonitzschia australis*) were ingested by unaffected anchovies that were subsequently eaten by birds, causing more than 100 birds to wash up dead at the shore. In 1998, another toxic diatom bloom in Monterey Bay caused the death of 400 California sea lions that ate sardines and anchovies contaminated with domoic acid. The brains of dead sea lions were found to have lesions and sea lions that did not die showed symptoms of permanent neurological disability.

Red tides are affecting larger areas and more marine species (including fish, dolphins, and humpback whales) are succumbing to

Figure 14A Red tide.

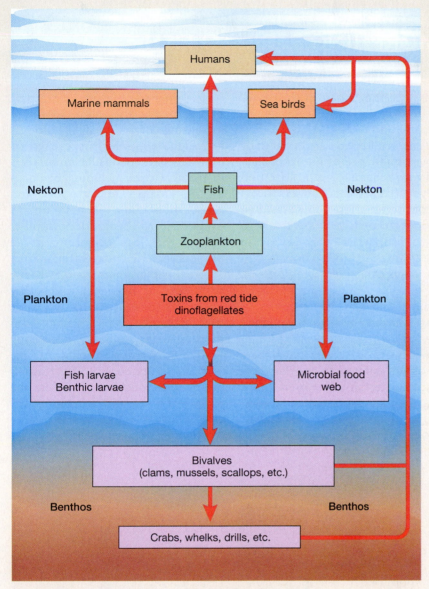

Figure 14B **Routes through which dinoflagellate toxins spread to marine organisms and humans.**

the toxins. In fact, toxic species of dinoflagellates have been implicated in several group strandings of marine mammals. Human activities have been shown to contribute to red tides when excess nutrients in the form of fertilizer, sewage, and animal waste make their way into coastal waters, producing ocean *eutrophication* (*eu* = good, *tropho* = nourishment, *ation* = action). Eutrophication is the enrichment of waters by a previously scarce nutrient and can result in dangerous phytoplankton blooms. In addition, red tides appear to be occurring more frequently around the globe. It is also possible, however, that the system for reporting them has simply improved.

Figure 14C **Actress Tippi Hedren fights off a seagull in Hitchcock's 1963 classic, *The Birds*.**

BOX 14–2   People and Ocean Environment

PFIESTERIA: A MORPHING PERIL TO FISH AND HUMANS

In 1995, the death of 15 million fish—mostly Atlantic menhaden (*Brevoortia tyrannus*)—in North Carolina's Neuse Estuary (Figure 14D) was traced to an outbreak of *Pfiesteria piscicida*, a toxic dinoflagellate that caused a similar fish kill in nearby Pamlico Sound in 1991. This and related dinoflagellate species kill millions of fish yearly in coastal waters off North Carolina and in Chesapeake Bay by undermining the ability of fish to reproduce and resist disease.

P. piscicida is known to change into as many as 24 distinct forms (Figure 14E) within a life cycle that can be initiated by the presence of fish. In water that is at least 26°C (79°F), cysts lie dormant in sediments and produce nontoxic *zoospores* (*zoo* = animal, *spora* = a seed) that feed on other microorganisms present in the water. When fish are present, however, the nontoxic zoospores become toxic and attach to fish, where they drug the fish, destroy their skin, and weaken their resistance to disease-causing bacteria and fungi. The spores then feed on the material oozing from sores that develop on the skin of infected fish. Toxic zoospores reproduce asexually and also produce gametes that merge to produce planozygotes that resemble the zoospores. When the fish die, the zoospores and planozygotes change into amoebae that gorge on the dead fish. In colder waters, amoebae rise from the sediment when fish are sensed, attack the fish with toxins, and return to the bottom to feed on the fish when it dies. Unable to perform photosynthesis on their own, zoospores can save chloroplasts from algae they ingest in their non-toxic form and use them for weeks to generate food.

Pfiesteria outbreaks are initiated by an overabundance of nutrients such as nitrogen that allow a proliferation of algae that serve as a food sources for zoospores. These zoospores then reproduce rapidly and are ready to attack fish when they arrive in coastal waters.

Humans can suffer from *Pfiesteria*, too, but not from eating *Pfiesteria*-contaminated fish. Harm comes to humans from getting toxin-laden water on their skin or even breathing air over the toxic waters. Researchers and others exposed to the *Pfiesteria*-infested water often reported symptoms such as shortness of breath, itchy or burning eyes, headaches, and forgetfulness.

The *Pfiesteria* outbreaks are part of a generally worsening pattern of increasing harmful algae blooms and ciguatera, which is caused by dinoflagellate toxins accumulating in tropical fish, including barracuda, red snapper, and grouper, and causes more human illness than any other form of seafood poisoning. Factors that lead to harmful algae blooms include poor water quality in streams that enter the ocean due to runoff of nutrients and other chemicals, destruction of wetlands that naturally filter water from these streams, and the increasing human population in coastal areas. Sickness in our coastal waters must be taken seriously, as it affects the quality of life for fish in coastal waters, which is linked to the quality of life for humans.

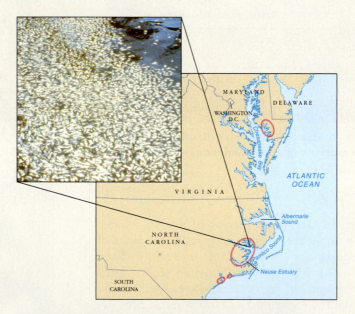

Figure 14D Fish kills related to *Pfiesteria* outbreaks.

Up to 100,000 fish died in Pamlico Sound (*inset*) during an outbreak of the dinoflagellate *Pfiesteria* in 1991. Since then, smaller fish kills related to *Pfiesteria* have occurred in Chesapeake Bay and along the North Carolina coast (*red circled areas*).

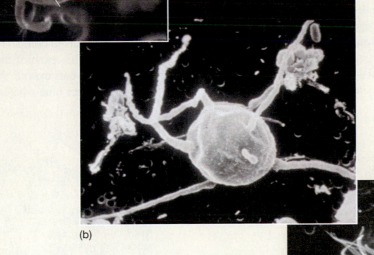

Figure 14E **Scanning electron micrographs of stages of** *Pfiesteria piscicida.*

Pfiesteria undergoes numerous transformations during the stages of its life cycle, some of which are shown here. **(a)** The toxic zoospore stage, when *Pfiesteria* releases toxins that kill fish. **(b)** The amoeba stage, when *Pfiesteria* feed on dead fish. **(c)** The cyst stage, when *Pfiesteria* is covered with bristles and settles to the bottom after a fish kill.

(a)

(b)

(c)

About 90% of the biomass generated in the euphotic (sunlit) zone of the open ocean is decomposed into inorganic nutrients before descending below this zone. The remaining 10% of this organic matter sinks into deeper water, where all but about 1% of it is decomposed. The 1% that reaches the deep-ocean floor accumulates there. The process of removing material from the euphotic zone to the sea floor is called a **biological pump**, because it "pumps" carbon dioxide and nutrients from the upper ocean and concentrates them in deep-sea waters and sea floor sediments.

Throughout much of the subtropical gyres, a permanent **thermocline** (and resulting **pycnocline**[7]) develops. It forms a barrier to vertical mixing, so it prevents the resupply of nutrients to the sunlit surface layer. In the mid-latitudes, a

[7]Recall that a *thermocline* is a layer of rapidly changing temperature and a *pycnocline* is a layer of rapidly changing density. Development of ocean thermoclines and pycnoclines is discussed in Chapter 6.

TABLE 14–2 **Values of net primary productivity for various ecosystems.**

	Ecosystem	Primary range (gC/m²/yr)	Productivity average (gC/m²/yr)
Oceanic	Algae beds and coral reefs	1000–3000	2000
	Estuaries	500–4000	1800
	Upwelling zone	400–1000	500
	Continental shelf	300–600	360
	Open ocean	1–400	125
Land	Freshwater swamp and marsh	800–4000	2500
	Tropical rainforest	1000–5000	2000
	Mid-latitude forest	600–2500	1300
	Cultivated land	100–4000	650

thermocline develops only during the summer season and, in polar regions, a thermocline does not usually develop. The degree to which waters develop a thermocline profoundly affects the patterns of biological production observed at different latitudes.

Productivity in Polar Oceans

Polar regions such as the Arctic Ocean's Barents Sea, which is off the northern coast of Europe, experiences continuous darkness for about three months of winter and continuous illumination for about three months during summer. Diatom productivity peaks there during May (Figure 14-11a), when the Sun rises high enough in the sky so that there is deep penetration of sunlight into the water. As soon as diatoms develop, zooplankton—mostly small crustaceans (Figure 14-11d)—begins feeding on them. The zooplankton biomass peaks in June and con-

tinues at a relatively high level until winter darkness begins in October.

In the Antarctic region—particularly at the southern end of the Atlantic Ocean—productivity is somewhat greater. This is caused by the upwelling of North Atlantic Deep Water, which forms on the opposite side of the ocean basin where it sinks and moves southward below the surface. Hundreds of years later, it rises to the surface near Antarctica, carrying with it high concentrations of nutrients (Figure 14-11b). When the summer Sun provides sufficient radiation, there is an explosion of biological productivity.

Blue whales, the largest of all whales (see Figure 15–23), eat mostly zooplankton and time their migration through temperate and polar oceans to coincide with maximum zooplankton productivity. This enables the whales to develop and support calves that can exceed 7 meters (23 feet) in length at birth. The mother blue whale suckles the calf with

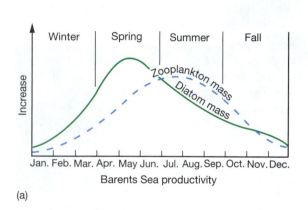

(a)

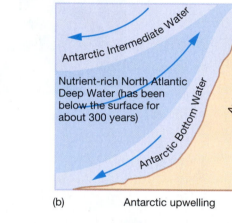

(b)   Antarctic upwelling

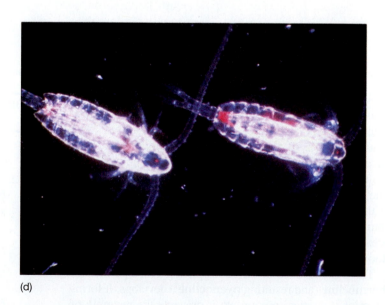

(c)   (d)

Figure 14–11 Productivity in polar oceans.

(a) A springtime increase of diatom mass is followed closely by an increase in zooplankton abundance. (b) The continuous upwelling of North Atlantic Deep Water keeps Antarctic waters rich in nutrients (c) Polar water shows nearly uniform temperature with depth (an isothermal water column). (d) Copepods of the genus *Calanus*, each about 8 millimeters (0.3 inch) in length.

rich, high-fat milk for six months. By the time the calf is weaned, it is over 16 meters (50 feet) long. In two years, it will be 23 meters (75 feet) long, and after about three years, it will weigh 55 metric tons (60 short tons)! This phenomenal growth rate gives some indication of the enormous biomass of small copepods and krill upon which these large mammals feed.[8]

Density and temperature change very little with depth in polar regions (Figure 14-11c), so these waters are **isothermal** (*iso* = same, *thermo* = temperature) and there is no barrier to mixing between surface waters and deeper, nutrient-rich waters. In the summer, however, melting ice creates a thin, low-salinity layer that does not readily mix with the deeper waters. This stratification is crucial to summer production, because it helps prevent phytoplankton from being carried into deeper, darker waters. Instead, they are concentrated in the sunlit surface waters where they reproduce continuously.

Nutrient concentrations (phosphates and nitrates) are usually adequate in high-latitude surface waters, so the availability of solar energy limits photosynthetic productivity in polar areas more than the availability of nutrients.

Productivity in Tropical Oceans

Perhaps surprisingly, productivity is low in tropical regions of the open ocean. Because the Sun is more directly overhead, light penetrates much deeper into tropical oceans than temperate and polar waters, and solar energy is available year round, but productivity is low in tropical regions of the open ocean because a permanent thermocline produces a stratification (layering) of water masses. This prevents mixing between surface waters and nutrient-rich deeper waters, effectively eliminating any supply of nutrients from deeper waters below (Figure 14-12).

At about 20 degrees north and south latitude, phosphate and nitrate concentrations are commonly less than $\frac{1}{100}$ of their concentrations in temperate oceans during winter. In fact, nutrient-rich waters in the tropics lie below 150 meters (500 feet), with the highest concentrations between 500 and 1000 meters (1640 and 3300 feet). So, productivity in tropical regions is limited by the lack of nutrients (unlike polar regions, where productivity is limited by the lack of sunlight).

Generally, primary production in tropical oceans occurs at a steady but rather low rate. The total annual production of tropical oceans is only about half of that found in temperate oceans.

Exceptions to the general pattern of low productivity in topical oceans include the following:

- *Equatorial upwelling.* Where trade winds drive westerly equatorial currents on either side of the Equator, Ekman transport causes surface water to diverge toward higher latitudes (see Figure 8–9). This surface water is replaced by nutrient-rich water from depths

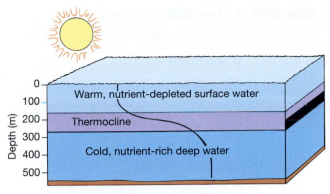

Figure 14–12 Productivity in tropical oceans.
Although tropical regions receive adequate sunlight year-round, a permanent thermocline prevents the mixing of surface and deep water. As phytoplankton consume nutrients in the surface layer, productivity is limited because the thermocline prevents replenishment of nutrients from deeper water. Thus, productivity remains at a steady, low level.

of up to 200 meters (660 feet). Equatorial upwelling is best developed in the eastern Pacific Ocean.

- *Coastal upwelling.* Where the prevailing winds blow toward the Equator and along western continental margins, surface waters are driven away from the coast. They are replaced by nutrient-rich waters from depths of 200 to 900 meters (660 to 2950 feet). This upwelling promotes high primary production along the west coasts of continents (see Figure 14-4), which can support large fisheries.

- *Coral reefs.* Organisms that comprise and live among coral reefs are superbly adapted to low nutrient conditions, similar to the way certain organisms are adapted to desert life on land. Symbiotic algae living within the tissues of coral and other species allow coral reefs to be highly productive ecosystems. Coral reefs also tend to retain and concentrate what little nutrients exist. Coral reef ecosystems are discussed further in Chapter 16, "Animals of the Benthic Environment."

STUDENTS SOMETIMES ASK...
The number and variety of tropical species on land is astounding. I don't understand how the tropical oceans can have such low productivity.

Life on land does not necessarily correspond to life in the ocean! Tropical rain forests support an amazing diversity of species and an enormous biomass. In the tropical ocean, however, a strong, permanent thermocline limits the availability of nutrients that are necessary for the growth of phytoplankton. Without abundant phytoplankton, not much else can live in the ocean. In fact, these areas are often considered biological deserts. It is ironic that the clear blue water of the tropics so prominently displayed in tourist brochures indicates seawater that is biologically quite sterile!

---

[8]As an analogy, consider how many ants you would have to eat as a child to grow to adult size!

Productivity in Temperate Oceans

Productivity is limited by available sunlight in polar regions and by nutrient supply in the low-latitude tropics. In temperate (mid-latitude) regions, a combination of these two limiting factors controls productivity as shown in Figure 14-13a (which shows the pattern for the Northern Hemisphere; in the Southern Hemisphere, the seasons are reversed).

Winter Productivity in temperate oceans is very low during winter, even though nutrient concentration is *highest* at this time (Figure 14-13a). The water column is isothermal, too, similar to polar regions, so nutrients are well distributed throughout the water column. Figure 14-11b (*winter*) shows, however, that the Sun is at its lowest position above the horizon during winter, so a high percentage of the available solar energy is reflected, leaving

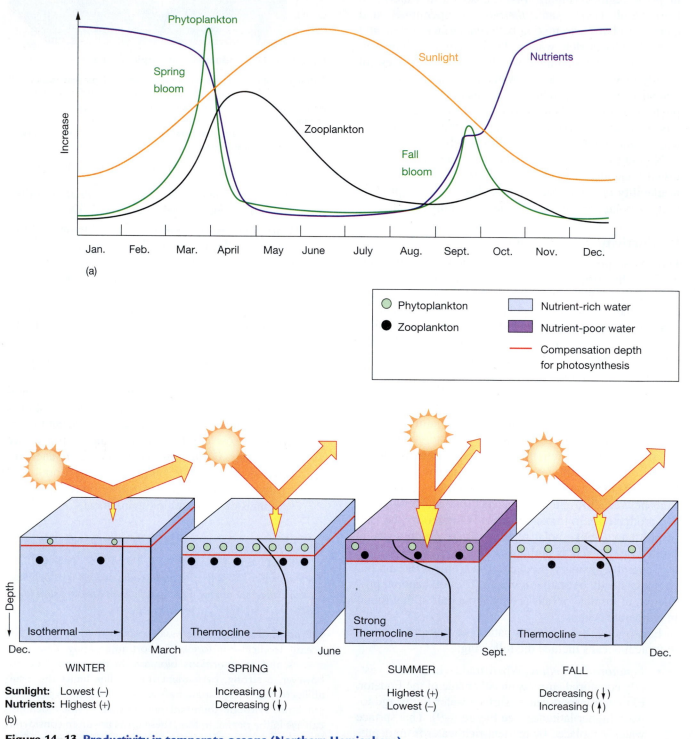

Figure 14–13 Productivity in temperate oceans (Northern Hemisphere).

(a) Relationship among phytoplankton, zooplankton, amount of sunlight, and nutrient levels for surface waters in northern temperate latitudes. **(b)** The seasonal cycle of sunlight affects the presence and depth of the thermocline, which affects the availability of nutrients. This, in turn, affects the abundance of phytoplankton and other organisms such as zooplankton that rely on phytoplankton for food.

only a small percentage to be absorbed into surface waters. As a result, the compensation depth for photosynthesis is so shallow that phytoplankton do not grow much. The absence of a thermocline, moreover, allows algal cells to be carried down beneath the euphotic zone for extended periods by turbulence associated with winter waves.

Spring The Sun rises higher in the sky during spring [Figure 14-13b (*spring*)], so the compensation depth for photosynthesis deepens. A **spring bloom** of phytoplankton occurs (Figure 14-13a) because solar energy and nutrients are available, and a seasonal thermocline develops (due to increased solar heating) that traps algae in the euphotic zone (Figure 14-13b). This creates a tremendous demand for nutrients in the euphotic zone, so the supply becomes limited, causing productivity to decrease sharply. Even though the days are lengthening and sunlight is increasing, productivity during the spring bloom is limited by the lack of nutrients. In most areas of the Northern Hemisphere, therefore, phytoplankton populations decrease in April due to insufficient nutrients and because their population is being consumed by zooplankton (grazers).

Summer The Sun rises even higher in the summer [Figure 14-13b (*summer*)], so surface waters in temperate parts of the ocean continue to warm. A strong seasonal thermocline is created at a depth of about 15 meters (50 feet). The thermocline, in turn, prevents vertical mixing, so nutrients depleted from the surface waters cannot be replaced by those from deeper waters. Throughout summer, the phytoplankton population remains relatively low (Figure 14-13a). Even though the compensation depth for photosynthesis is at its maximum, phytoplankton can actually become scarce in late summer.

Fall Solar radiation diminishes in the fall as the Sun moves lower in the sky [Figure 14-13b (*fall*)], so surface temperatures drop and the summer thermocline breaks down. Nutrients return to the surface layer as increased wind strength mixes surface waters with deeper waters. These conditions create a **fall bloom** of phytoplankton, which is much less dramatic than the spring bloom (Figure 14-13a). The fall bloom is very short-lived because sunlight (not nutrient supply, as in the spring bloom) becomes the limiting factor as winter approaches to repeat the seasonal cycle.

Figure 14-14 compares the seasonal variation in phytoplankton biomass of tropical, north polar, and north temperate regions, where the total area under each curve represents photosynthetic productivity. The figure shows the dramatic peak in productivity in polar oceans during the summer; the steady, low rate of productivity year-round in the tropical ocean; and the seasonal pattern of productivity that occurs in temperate oceans. It also shows that the highest overall productivity occurs in temperate regions.

> In polar regions, productivity peaks during the summer and is limited by sunlight. In tropical regions, productivity is low year-round and is limited by nutrients. In temperate regions, productivity peaks in the spring and fall and is limited by a lack of solar radiation in the winter and a lack of nutrients in the summer.

Energy Flow

Energy flow is not a *cycle* but a *unidirectional flow* that begins with a constant input of solar energy and ends with a high level of **entropy** (*en* = in, *trope* = transformation), a state when energy is so randomly distributed it can no longer do work.

Energy Flow in Marine Ecosystems

A **biotic community** is the assemblage of organisms that live together within some definable area. An **ecosystem**

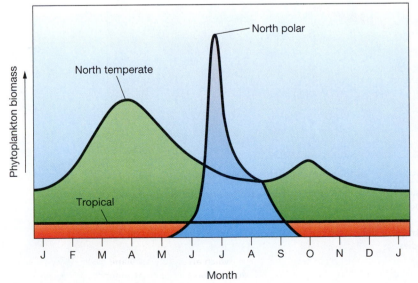

Figure 14–14 Comparison of productivity in tropical, temperate, and polar oceans (Northern Hemisphere).

Seasonal variation in phytoplankton biomass, where the total area under each curve represents annual photosynthetic productivity.

includes the biotic community plus the environment with which it exchanges energy and chemical substances. A kelp forest biotic community, for instance, includes all organisms living within or near the kelp and receiving some benefit from it. A kelp forest ecosystem, on the other hand, includes all those organisms plus the surrounding seawater, the hard substrate onto which the kelp is attached, and the atmosphere where gases are exchanged.

In an algae-supported biotic community (Figure 14-15), energy enters the system when algae absorb solar radiation. Photosynthesis converts this solar energy into chemical energy, which is used for the algae's respiration. This chemical energy is also passed on to the animals that consume the algae for their growth and other life functions. The animals expend mechanical and heat energy, which are progressively less recoverable forms of energy. Finally, the residual energy becomes biologically useless as entropy increases.

Generally, three basic categories of organisms exist within an ecosystem: **producers**, **consumers**, and **decomposers**. Algae, plants, archaeons, and certain bacteria are called **autotrophic** (*auto* = self, *tropho* = nourishment) producers because they can nourish themselves through chemosynthesis or photosynthesis. Consumers and decomposers are called **heterotrophic** (*hetero* = different, *tropho* = nourishment) organisms because they depend on the organic compounds produced by the autotrophs for their food supply.

Consumers may be categorized as **herbivores** (*herba* = grass, *vora* = eat), which feed directly on plants or algae; **carnivores** (*carni* = meat, *vora* = eat), which feed only on other animals; **omnivores** (*omni* = all, *vora* = eat), which feed on both; and **bacteriovores** (*bacterio* = bacteria, *vora* = eat), which feed only on bacteria.

Decomposers such as bacteria break down organic compounds that comprise **detritus** (*detritus* = to lessen) —dead and decaying remains and waste products of organisms—for their own energy requirements. In the decomposition process, compounds are released that are again available for use by algae and plants as nutrients.

Symbiosis

Symbiosis (*sym* = together, *bios* = life) occurs when two or more organisms associate in a way that benefits at least one of them. Symbiotic relationships are classified as commensalism, mutualism, or parasitism.

In **commensalism** (*commensal* = sharing a meal, *ism* = process), a smaller or less dominant participant benefits without harming its host, which affords subsistence or protection to the other. Remoras, for example, attach them-

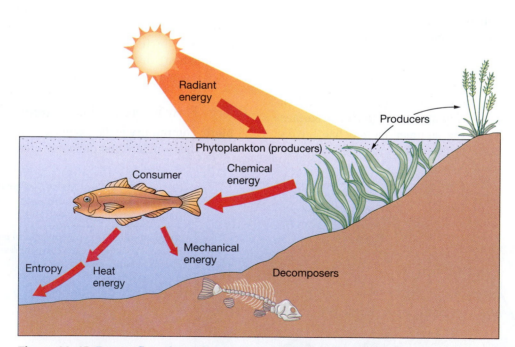

Figure 14–15 Energy flow through a photosynthetic marine ecosystem.

Energy enters a marine ecosystem as radiant solar energy and is converted to chemical energy through photosynthesis by producers. Metabolism in the fish (a consumer) then releases the chemical energy for conversion to mechanical energy. Energy also is lost from the biotic community (the algae and fish) as heat, which increases the entropy of the ecosystem. Decomposers work to break down the remaining energy after an organism dies.

(a)

(b)

(c)

Figure 14–16 **Symbiosis.**

(a) Commensalism occurs when an organism benefits without harming its host, such as this remora attached to a Caribbean grouper. **(b)** Mutualism occurs when both participants benefit, such as this clown fish and sea anemone. **(c)** Parasitism occurs when one participant benefits at the expense of the other, such as this isopod that has attached itself to the head of a blackbar soldierfish.

fish are hosts to isopods, which attach to the fish and derive their nutrition from the body fluids of the fish, thereby robbing the host of some of its energy supply (Figure 14-16c). Usually, the parasite does not rob enough energy to kill the host because if the host dies, so does the parasite.

Biogeochemical Cycling

Unlike the noncyclic, unidirectional flow of energy through a biotic community, the flow of nutrients depends on **biogeochemical cycles**.[9] That is, matter does not dissipate (as energy does) but is *cycled* from one chemical form to another by the various members of the community.

Figure 14-17 shows the biogeochemical cycling of matter within the marine environment. The chemical components of organic matter enter the biological system through photosynthesis (or, less commonly, through chemosynthesis at hydrothermal vents). These chemical components are passed on to animal populations (consumers) through feeding. When organisms die, some of the material is used and reused within the euphotic zone, while some sinks as detritus. Some of this detritus feeds organisms living in deep water or on the sea floor, while some undergoes bacterial or other decomposition processes that convert organic remains into useable nutrients (nitrates and phosphates). When upwelling hoists these nutrients to the surface again, they can be used by algae and plants to begin the cycle anew.

Of the many biogeochemical cycles, three cycles are especially important: carbon, nitrogen, and phosphorous. A fourth cycle, the silicon cycle, also has important implications in the marine environment.

Carbon, Nitrogen, and Phosphorus Cycles

The element carbon is the basic component of all organic compounds (including carbohydrates, proteins, and fats). There is no scarcity of carbon for photosynthetic production: only about 1% of the total carbon in the sea is involved in photosynthetic productivity.[10] Thus, carbon does not limit productivity.

selves to a shark or other fish to obtain food and transportation, generally without harming its host (Figure 14-16a).

In **mutualism** (*mutuus* = borrowed, *ism* = process), both participants benefit. For example, the stinging tentacles of the sea anemone protect the clown fish and the clown fish serves as bait to draw other fish within reach of the anemone's tentacles (Figure 14-16b).

In **parasitism** (*parasitos* = person who eats at someone else's table, *ism* = process), one participant (the parasite) benefits at the expense of the other (the host). Many

---

[9]Biogeochemical cycling is so named because it involves *biological* and *geological* (Earth processes), and *chemical* components.

[10]The remainder of carbon atoms are in seawater as bicarbonate ions or are bound into calcium carbonate shells.

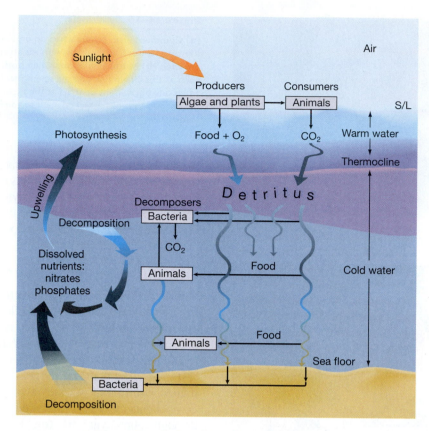

Figure 14–17 **Biogeochemical cycling of matter.**

The chemical components of organic matter enter the biological system through photosynthesis and are passed on to consumers through feeding. Detritus sinks and feeds organisms living below the surface or undergoes decomposition, which returns nutrients to the water that can be hoisted to the surface by upwelling.

Nitrogen and phosphorous, however, often limit marine productivity and, as a result, are among the most studied elements in chemical oceanography. Comparatively, nitrogen compounds involved in photosynthesis may be 10 times the total nitrogen compound concentration that can be measured as a yearly average. This level implies that the soluble nitrogen compounds must be completely recycled up to 10 times per year. Available phosphates may be turned over up to four times per year.

The ratio of carbon to nitrogen to phosphorus in the tissues of algae is in the proportion of 105 : 15 : 1 (C : N : P), which is called the *Redfield ratio* after marine chemist Alfred Redfield who first discovered it. This ratio is also observed in the zooplankton that feed on diatoms and in most ocean water samples taken worldwide. Moreover, phytoplankton take up nutrients in the ratio in which they are available in the ocean water and pass them on to zooplankton in the same ratio. When these plankton and animals die, carbon, nitrogen, and phosphorus are restored to the water in this same ratio.

The Carbon Cycle The carbon cycle (Figure 14-18) involves the uptake of carbon (as carbon dioxide) by algae and plants for use in photosynthesis. Carbon is returned to the ocean through respiration of algae, animals, and microbes as well as by the breakdown of dissolved organic materials.

The Nitrogen Cycle The nitrogen cycle (Figure 14-19) shows how nitrogen is cycled in the marine environment. Nitrogen is essential in producing amino acids, which are the building blocks of proteins. These photosynthetic products are consumed by animals and free-living microbes. They are then passed on to decomposing bacteria in the form of dead algae tissue, dead animal tissue, and excrement. The decomposing bacteria gain energy from breaking down these compounds. This breakdown liberates inorganic compounds, such as nitrates, that are basic nutrients.

What makes the nitrogen cycle complex is that many types of different bacteria break down nitrogen compounds. Although some bacteria consume dissolved organic matter or convert organic compounds into inorganic substances, other **nitrogen-fixing bacteria** can bind molecular nitrogen (N_2) into the useful nutrient nitrate (NO_3). Conversely, **denitrifying bacteria** convert nitrates into molecular nitrogen.

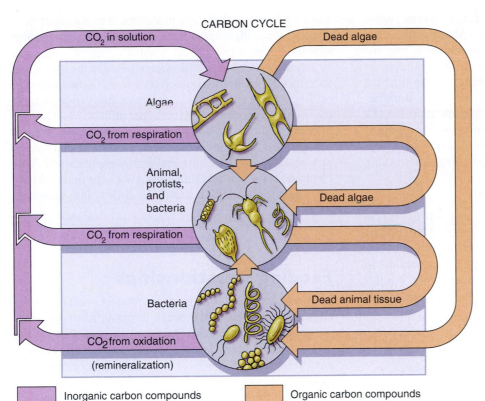

CARBON CYCLE

CO$_2$ in solution

Dead algae

Algae

CO$_2$ from respiration

Animal, protists, and bacteria

Dead algae

CO$_2$ from respiration

Bacteria

Dead animal tissue

CO$_2$ from oxidation

(remineralization)

Inorganic carbon compounds

Organic carbon compounds

Figure 14–18 The carbon cycle.

A large supply of inorganic carbon (*purple*) exists in the oceans, but only about 1% of it is involved in photosynthetic productivity (*orange*). Through photosynthesis, algae store carbon, which is passed on to animals that eat them. Bacteria decompose dead algae and animals to release carbon. All release carbon in the form of carbon dioxide through respiration.

Figure 14–19 The nitrogen cycle.

The total nitrogen fixed into organic molecules (*orange*) at any given time may be 10 times as great as the yearly average of soluble nitrogen compounds dissolved in seawater. Therefore, each nitrogen atom must be recycled biogeochemically about 10 times per year. Also, the decomposition of organic nitrogen compounds back into useable forms of nitrate (NO$_3$) requires three steps of bacterial decomposition. As a result, nitrogen can often limit biological productivity.

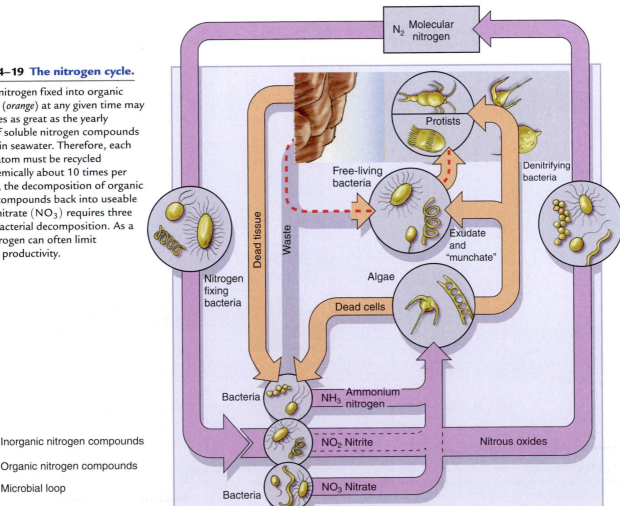

NITROGEN CYCLE

N$_2$ Molecular nitrogen

Protists

Denitrifying bacteria

Free-living bacteria

Dead tissue

Waste

Exudate and "munchate"

Algae

Nitrogen fixing bacteria

Dead cells

Bacteria

NH$_3$ Ammonium nitrogen

NO$_2$ Nitrite

Nitrous oxides

Bacteria

NO$_3$ Nitrate

Inorganic nitrogen compounds

Organic nitrogen compounds

- - - - Microbial loop

Studies suggest that nitrogen availability limits photosynthetic activity in temperate waters. Mostly, this is because the conversion of organic nitrogen back into the nutrient nitrate requires three stages of bacterial conversion, which can take up to three months. By the time the conversion is completed, the nitrate will have sunk beneath the euphotic zone. Because of the development of a strong thermocline during summer months, the nitrate is unlikely to be returned to the euphotic zone until the thermocline weakens, which allows upwelling and mixing during fall and winter months.

The Phosphorus Cycle The phosphorous cycle (Figure 14-20) is simpler than the nitrogen cycle, primarily because there is a single step of bacterial action required to convert organic phosphorus compounds to nutrient form. Therefore, phosphorus is more readily available and is less likely to limit photosynthesis except in some tropical ecosystems.

The Silicon Cycle

A fourth cycle, the silicon cycle (not shown), is also important because diatoms and other groups of silica-secreting marine microorganisms use this element to construct their tests. As a result, when silicon is in low concentration in seawater, diatom production is usually limited. One such region that experiences this phenomenon is the upwelling zone that occurs in the equatorial eastern Pacific Ocean.

Silicon concentrations in seawater range from barely measurable up to about 400 parts per million. Fluctuations in concentration roughly coincide with those observed in nitrogen and phosphorus but the amount of silicon fluctuates much more than nitrogen and phosphorus do. This condition is probably because silica does not undergo bacterial decay; instead, it is dissolved directly into seawater.

Feeding Relationships

As producers make food (organic matter) available to the consuming animals of the ocean, it passes from one feeding population to the next. Only a small percentage of the energy taken in at any level is passed on to the next because energy is consumed and lost at each level. As a result, the producers' biomass in the ocean is many times greater than the mass of the top consumers, such as sharks or whales.

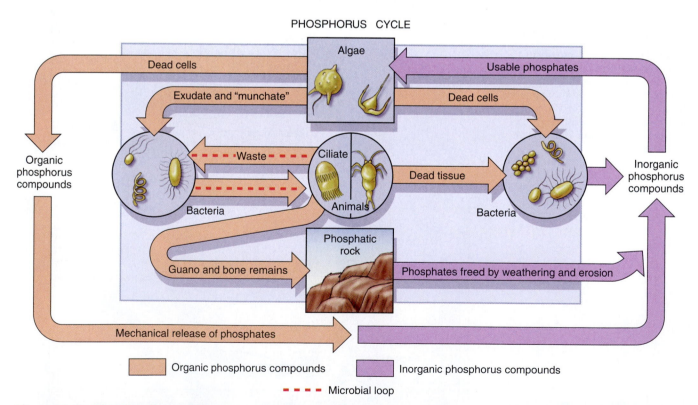

PHOSPHORUS CYCLE

Figure 14–20 The phosphorus cycle.

Each phosphorus atom may need to be recycled up to four times per year to maintain biological productivity in the oceans. Yet, phosphorus is seldom depleted to where it limits biological productivity. Organic phosphorus is quickly returned as a usable nutrient to algae through breakdown and the single step of bacterial decomposition.

Trophic Levels

Chemical energy stored in the mass of the ocean's algae (the "grass of the sea") is transferred to the animal community mostly through feeding. Zooplankton are *herbivores*, like cows,[11] so they eat diatoms and other microscopic marine algae. Larger herbivores feed on the larger algae and marine plants that grow attached to the ocean bottom near shore.

? STUDENTS SOMETIMES ASK ...
How likely am I to see a whale during a whale-watching trip?

It depends on the area and the time of the year, but generally your chances are quite low because large marine organisms comprise such a small proportion of marine life. In fact, it has been estimated that large nektonic organisms like whales comprise only *one-tenth of 1%* of all biomass in the sea! With your knowledge of food pyramids, it should be no surprise that the majority of the ocean's biomass is comprised of phytoplankton. Perhaps commercial boat operators should conduct *plankton*-watching trips! Everyone would be guaranteed to see dozens of different species—all that would be required is a plankton net and a microscope.

The herbivores are then eaten by larger animals, the *carnivores*. They, in turn, are eaten by another population of larger carnivores, and so on. Each of these feeding stages is called a **trophic** (*tropho = nourishment*) **level**.

Generally, individual members of a feeding population are larger—but not too much larger—than the organisms they eat. There are conspicuous exceptions, however, such as the blue whale. At 30 meters (100 feet) long, it is possibly the largest animal that has ever existed on Earth, yet it feeds on krill, which have a maximum length of only 6 centimeters (2.4 inches).

---

[11]In fact, zooplankton might be considered the miniature drifting "cows of the sea."

The transfer of energy from one population to another is a continuous *flow* of energy. Small-scale recycling and storage interrupts the flow, which slows the conversion of potential (chemical) energy to kinetic energy, then to heat energy, and finally to entropy.

Transfer Efficiency

The transfer of energy between trophic levels is very inefficient. The efficiencies of different algal species vary, but the average is only about *2%*, which means that 2% of the light energy absorbed by algae is ultimately synthesized into food and made available to herbivores.

The **gross ecological efficiency** at any trophic level is the ratio of energy passed on to the next higher trophic level divided by the energy received from the trophic level below. The ecological efficiency of herbivorous anchovies, for example, would be the energy consumed by carnivorous tuna that feed on the anchovies divided by the energy contained in the phytoplankton that the anchovies consumed.

Figure 14–21 shows that some of the chemical energy taken in as food by herbivores is excreted as feces and the rest is assimilated. Of the assimilated chemical energy, much is converted through respiration to kinetic energy for maintaining life, and what remains is available for growth and reproduction. Thus, only about 10% of the food mass consumed by herbivores is available to the next trophic level.

Figure 14–22 shows the passage of energy between trophic levels through an entire ecosystem, from the solar energy assimilated by phytoplankton through all trophic levels to the ultimate carnivore—humans. Because energy is lost at each trophic level, it takes thousands of smaller marine organisms to produce a *single* fish that is so easily consumed during a meal!

The efficiency of energy transfer between trophic levels depends on many variables. Young animals, for example, have a higher growth efficiency than older animals. In

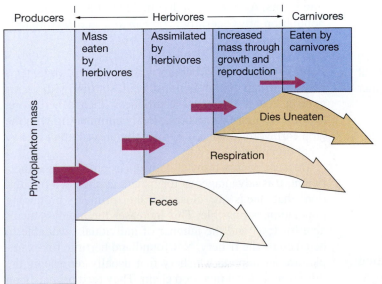

Figure 14–21 Passage of energy through a trophic level.

As food mass initially produced by phytoplankton passes from herbivores to carnivores, a large percentage is excreted by feces, used during respiration, or dies uneaten. Thus, only about 10% of the food mass consumed by herbivores is available for consumption by carnivores.

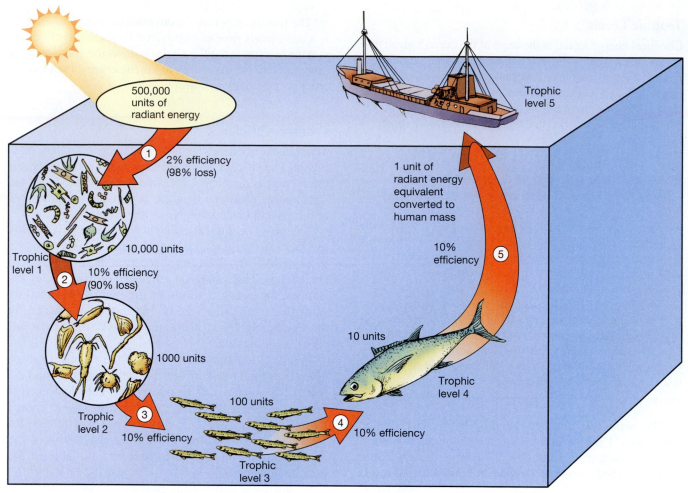

Figure 14–22 Ecosystem energy flow and efficiency.

For every 500,000 units of radiant energy input available to the producers (phytoplankton), only one unit of mass is added to the fifth trophic level (humans). Average phytoplankton transfer efficiency is 2% (98% loss), and all other trophic levels average 10% efficiency (90% loss).

addition, when food is plentiful, animals expend more energy in digestion and assimilation than when food is scarce.

Most ecological efficiencies in natural ecosystems average about 10% but range between 6% and 15%. There is some evidence, however, that ecological efficiencies in populations important to present fisheries may run as high as 20%. The true value of this efficiency is of practical importance because it determines the size of the fish harvest that can be safely taken from the oceans without damaging the ecosystem. Fisheries are discussed further in Chapter 17, "Marine Resources."

> The transfer of energy between various trophic levels operates at low efficiencies, averaging only 2% for marine algae and 10% for most consumer levels.

Food Chains, Food Webs, and the Biomass Pyramid

The loss of energy between each feeding population limits the number of feeding populations in an ecosys-

tem. If there were too many levels, there would not be enough energy to support the organisms in higher and higher trophic levels. In addition, each feeding population necessarily must have less mass than the population it eats. As a result, individual members of a feeding population are generally *larger in size* and *less numerous* than their prey.

Food Chains A **food chain** is a sequence of organisms through which energy is transferred, starting with an organism that is the primary producer, then a herbivore, then one or more carnivores, finally culminating with the "top carnivore," which is not usually preyed upon by any other organism.

Because energy transfer between trophic levels is inefficient, it is advantageous for fishers to choose a population that feeds as close to the primary producing population as possible. This increases the biomass available for food and the number of individuals available to be taken by the fishery. Newfoundland herring, for example, are an important fishery that usually represents the third trophic level in a food chain. They feed primarily on small crustaceans (copepods) that feed, in turn, upon diatoms (Figure 14–23a).

Food Webs Feeding relationships are rarely as simple as that of the Newfoundland herring. More often, top carnivores in a food chain feed on a number of different animals, each of which has its own simple or complex feeding relationships. This constitutes a **food web**, as shown in Figure 14–23b for North Sea herring.

Animals that feed through a food web rather than a food chain are more likely to survive because they have alternative foods to eat should one of their food sources diminish in quantity or even disappear. Newfoundland herring, on the other hand, eat only copepods, so the disappearance of copepods would catastrophically affect their population.

Conversely, Newfoundland herring are more likely to have a larger biomass to eat, because they are only two steps removed from the producers, whereas North Sea herring are three steps removed in some of the food chains within their web.

> A food chain is a linear feeding relationship among producers and one or more consumers. A food web is a branching network of feeding relationships among many different organisms.

Biomass Pyramid The ultimate effect of energy transfer between trophic levels can be seen in the oceanic **biomass pyramid** in Figure 14–24. *The number of individuals* and *total biomass* decrease at successive trophic levels because the amount of available energy decreases. The figure also shows that organisms *increase in size* at successive tropic levels up the food web.

Microbes in the Marine Environment

Although it has long been widely accepted that zooplankton are the primary grazers of phytoplankton, it is becoming clear that free-living bacteria and archaea may consume up to 50% of the production of phytoplankton. This consumption by microbes, however, relies on unusual sources of organic matter:

1. *Phytoplankton exudates*. As phytoplankton age, they leak some of their cytoplasm directly into the ocean.
2. *Phytoplankton munchate*. As phytoplankton are unmeticulously eaten by zooplankton, cytoplasm is spilled into the ocean.
3. *Zooplankton excretions*. The liquid excretions of zooplankton are dissolved into ocean water.

After consuming this dissolved organic matter, the microbes may reenter the conventional food web by being consumed by microscopic flagellates. However, some investigators believe that much of this consumption does not reenter the conventional food web, but is trapped in the **microbial loop** because there are no zooplankton that are efficient at grazing on flagellates. Microbial loops are shown on the nitrogen and phosphorous cycles shown in Figures 14–19 and 14–20 (*red dashed lines*).

Cyanobacteria Nitrogen-fixing cyanobacteria convert molecular nitrogen (N_2) to the nutrient nitrate (NO_3) and may be important producers in oligotrophic open ocean waters.

Viruses Viruses (Figure 14–25) are extremely small particles about 0.2 to 0.02 microns (0.000008 to 0.0000008 inch) long that have been recently discovered to play an

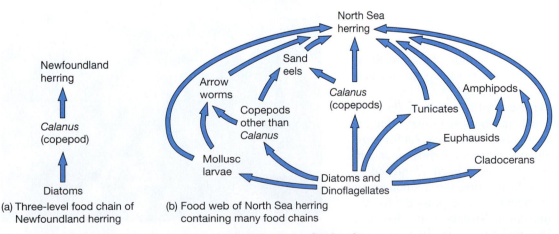

(a) Three-level food chain of Newfoundland herring

(b) Food web of North Sea herring containing many food chains

Figure 14–23 Comparison between a food chain and a food web.

(a) A food chain, showing the passage of energy along a single path, such as from diatoms to copepods to Newfoundland herring in three trophic levels. **(b)** A food web, showing multiple paths for food sources of the North Sea herring, which may be at the third or fourth trophic level.

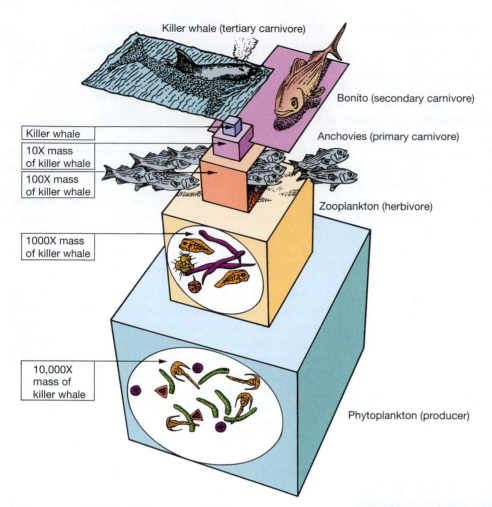

Killer whale (tertiary carnivore)

Bonito (secondary carnivore)

Anchovies (primary carnivore)

Killer whale

10X mass
of killer whale

100X mass
of killer whale

Zooplankton (herbivore)

1000X mass
of killer whale

10,000X
mass of
killer whale

Phytoplankton (producer)

Figure 14–24 Oceanic biomass pyramid.

A huge mass of phytoplankton constitutes the base of the oceanic biomass pyramid. At each step up the pyramid, there are larger organisms but fewer individuals and a smaller total biomass because transfer efficiency between steps averages only 10%.

important role in many marine biological processes. Viruses consist of genetic material surrounded by a protein or lipid coat and are parasitic, so they must depend on a host cell for their metabolism.

The abundance of viruses in the ocean is about 10 billion per 1 liter (about 1 quart) of seawater, which is from 5 to 25 times that of larger microbes. As for most groups of marine organisms, viruses are most abundant in coastal waters. They have also been identified as having an important role in limiting the abundance of microbes because they appear to be responsible for 8% to 34% of microbe mortality. Viruses may also play important roles in biogeochemical cycling and the overall biodiversity and species distribution of many larger organisms (by infection of microbes, phytoplankton, and zooplankton).

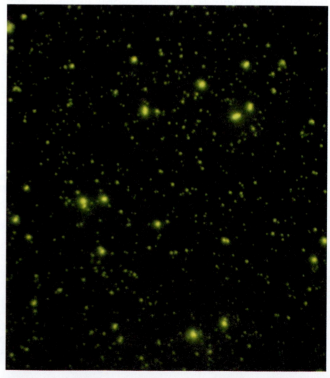

Figure 14–25 Marine viruses.

Fluorescent image showing marine viruses (*small dots*) and either archaea or bacteria (*large dots*). Most viruses shown are less than 0.2 micron (0.000008 inch) in diameter.

Chapter in Review

- *Microscopic planktonic algae that photosynthesize represent the largest biomass in the ocean.* They are the ocean's *primary producers*—the foundation of the ocean's food web. Organic biomass is also produced near deep-sea hydrothermal springs through *chemosynthesis*, in which bacteria-like organisms trap chemical energy by the oxidation of hydrogen sulfide.

- *The availability of nutrients and the amount of solar radiation limits photosynthetic productivity* in the oceans. Nutrients—such as *nitrate, phosphorous, iron, and silica*—are most abundant in coastal areas, due to runoff and upwelling. The depth at which net photosynthesis is zero is the *compensation depth for photosynthesis*. Algae cannot live successfully below this depth, which may be less than 20 meters (65 feet) in turbid coastal waters or as much as 100 meters (330 feet) in the open ocean.

- *Marine life is most abundant along continental margins,* where nutrients and sunlight are optimal. It decreases with distance from the continents and with increased depth. In addition, *cool water typically supports more abundant life than warm water* because cool water can dissolve more of the gases necessary for life (oxygen and carbon dioxide). *Areas of upwelling bring cold, nutrient-rich water to the surface* and have some of the highest productivities.

- *Ocean water selectively absorbs the colors of the visible spectrum.* Red and yellow light are absorbed at relatively shallow depths, whereas *blue and green light are the last to be removed.* Ocean water of *low biologic productivity* scatters the short wavelengths of visible light, producing a *blue color.* Turbidity and photosynthetic algae in *more productive ocean water* scatter more green light wavelengths, which produces a *green color.*

- *There are many different types of photosynthetic marine organisms.* The seed-bearing Anthophyta are represented by a few genera of *nearshore plants* such as eelgrass (*Zostera*), surf grass (*Phyllospadix*), marsh grass (*Spartina*), and mangrove trees (genera *Rhizophora* and

Avicennia). *Macroscopic algae* include *brown algae* (Phaeophyta), *green algae* (Chlorophyta), and *red algae* (Rhodophyta). *Microscopic algae* include *diatoms, coccolithophores* (Chrysophyta), and *dinoflagellates* (Pyrrophyta, which are responsible for *red tides*).

- *A thermocline is generally absent* in high-latitude (*polar*) *areas,* so *upwelling* can readily occur. The *availability of solar radiation limits productivity* more than the availability of nutrients. In low-latitude (*tropical*) regions, however, a *strong thermocline* may exist year-round, so the *lack of nutrients limits productivity,* except in areas of upwelling or near coral reefs. In *temperate regions, productivity peaks in the spring and fall* and is limited by lack of solar radiation in the winter and lack of nutrients in the summer.

- *Radiant energy captured by algae is converted to chemical energy* and passed through the different trophic levels of the *biotic community.* It is expended as *mechanical and heat energy* and ultimately reaches a high state of *entropy,* where it is biologically useless. Upon death, organisms are *decomposed* to an inorganic form that algae can use again for nutrients.

- *Marine ecosystems are composed of populations of organisms called producers* (which photosynthesize or chemosynthesize), *consumers* (which eat producers), and *decomposers* (which break down detritus). Animals can be categorized as *herbivores* (eat plants), *carnivores* (eat animals), *omnivores* (eat both), or *bacteriovores* (eat bacteria). Some organisms live closely together in *symbiotic relationships.* The organisms of a biotic community *cycle nutrients* and other chemicals from one form to another.

- On average, *only about 10% of the mass taken in at one feeding level is passed on to the next.* As a result, the *size of individuals increases* but the *number of individuals decreases* with each *trophic level* of a *food chain* or *food web.* Overall, *the total biomass of populations decreases the higher they are in the biomass pyramid.*

Key Terms

Autotrophic (p. 414)
Bacteriovore (p. 414)
Biogeochemical cycle (p. 415)
Biological pump (p. 409)
Biomass (p. 395)
Biomass pyramid (p. 421)
Biotic community (p. 413)
Carnivore (p. 414)
Chemosynthesis (p. 395)
Coccolith (p. 404)
Coccolithophore (p. 404)
Commensalism (p. 414)

Compensation depth for photosynthesis (p. 398)
Consumer (p. 414)
Decomposer (p. 414)
Denitrifying bacteria (p. 418)
Detritus (p. 404)
Diatom (p. 404)
Diatomaceous earth (p. 404)
Dinoflagellate (p. 404)
Ecosystem (p. 413)
Electromagnetic spectrum (p. 398)

Entropy (p. 413)
Euphotic zone (p. 398)
Eutrophic (p. 401)
Fall bloom (p. 413)
Flagella (p. 404)
Food chain (p. 420)
Food web (p. 420)
Gross ecological efficiency (p. 419)
Gross primary production (p. 396)
Harmful algal bloom (HAB) (p. 405)

Herbivore (p. 414)
Heterotrophic (p. 414)
Isothermal (p. 411)
Microbial loop (p. 421)
Mutualism (p. 414)
Net primary production (p. 396)
New production (p. 396)
Nitrogen-fixing bacteria (p. 418)
Oligotrophic (p. 401)
Omnivore (p. 414)
Parasitism (p. 414)

Photosynthesis (p. 395)
Plankton net (p. 396)
Primary productivity (p. 395)
Producer (p. 414)
Pycnocline (p. 409)

Red tide (p. 405)
Regenerated production
   (p. 396)
SeaWiFS (p. 396)
Secchi disk (p. 401)

Spring bloom (p. 413)
Symbiosis (p. 414)
Test (p. 404)
Thermocline (p. 409)
Trophic level (p. 419)

Upwelling (p. 398)
Virus (p. 421)
Visible light (p. 398)

Questions and Exercises

1. How does gross primary production differ from net primary production? What are the two components of gross primary production, and how do they differ?

2. An important variable in determining the distribution of life in the oceans is the availability of nutrients. How are the following variables related: proximity to the continents, availability of nutrients, and the concentration of life in the oceans?

3. Another important determinant of productivity is the availability of solar radiation. Why is biological productivity relatively low in the tropical open ocean, where the penetration of sunlight is greatest?

4. Discuss the characteristics of the coastal ocean where unusually high concentrations of marine life are found.

5. Why does everything in the ocean at depths below the shallowest surface water take on a blue-green appearance?

6. Compare the macroscopic algae in terms of color, maximum depth in which they grow, common species, and size.

7. The golden algae include two classes of important phytoplankton. Compare their composition and the structure of their tests and explain their importance in the geologic fossil record.

8. Discuss and compare the contributions of the Pyrrophyta genera *Ptychodiscus* and *Gonyaulax* to red tide development.

9. Describe how a biological pump works. What percentage of organic material from the euphotic zone accumulates on the sea floor?

10. Compare the biological productivity of polar, temperate, and tropical regions of the oceans. Consider seasonal changes, the development of a thermocline, the availability of nutrients, and solar radiation.

11. Generally, the productivity in tropical oceans is rather low. What are three environments that are exceptions to this, and what factors contribute to their higher productivity?

12. Describe the flow of energy through the biotic community and include the forms into which solar radiation is converted. How does this flow differ from the manner in which matter is moved through the ecosystem?

13. What are the three types of symbiosis, and how do they differ?

14. What is the average efficiency of energy transfer between trophic levels? Use this efficiency to determine how much phytoplankton mass is required to add *1 gram* of new mass to a killer whale, which is a third-level carnivore. Include a diagram that shows the different trophic levels and the relative size and abundance of organisms at different levels. How would your answer change if the efficiency were half the average rate, or twice the average rate?

15. Describe the advantage that a top carnivore gains by eating from a food web as compared to a single food chain.

CHAPTER 15
Animals of the Pelagic Environment

Reef sharks. Sharks like these Caribbean reef sharks (*Carcharhinus perezi*) are efficient predators that are uniquely adapted to the marine environment.

Key Questions

- How are marine organisms able to stay above the ocean floor?
- What types of fins do fish have, and how are they used?
- What kinds of adaptations do fish have that live in the deep ocean?
- How are pelagic organisms adapted for seeking prey?
- What are the advantages of schooling?
- Which marine mammals live in the ocean?
- Why do gray whales make the longest yearly migration of any mammal?

The answers to these questions (and much more) can be found in the highlighted concept statements within this chapter.

"If you were to make little fishes talk, they would talk like whales."

—Oliver Goldsmith (1773)

Pelagic organisms live suspended in seawater (not on the ocean floor) and comprise the vast majority of the ocean's **biomass**.[1] Phytoplankton live within the sunlit surface waters of the ocean and are the food source for nearly all other marine life, so most marine animals live in surface waters too so they can be close to their food supply. The immense depth of the oceans makes it challenging for these organisms to remain afloat.

Phytoplankton depend primarily on their small size to provide a high degree of frictional resistance to sinking. Most animals, however, are more dense than ocean water, and have less surface area per unit of body mass. Therefore, they tend to sink more rapidly than phytoplankton.

To remain in surface waters where the food supply is greatest, pelagic marine animals must increase their buoyancy or continually swim. Animals apply one or both of these strategies in a wonderful variety of adaptations and lifestyles.

Staying Above the Ocean Floor

Some animals increase their buoyancy to remain in near-surface waters. They may have containers of gas, which significantly reduce their average density, or they may have soft bodies void of hard, high-density parts. Larger animals with bodies denser than seawater must exert more energy to propel themselves through the water.

Gas Containers

Air is approximately 1000 times less dense than water at sea level, so even a small amount of air inside an organism can dramatically increase its buoyancy. Some ani-

mals, such as the cephalopods (*cephalo* = the head, *podium* = a foot), have rigid gas containers in their bodies. For instance, the genus *Nautilus* have an external shell, whereas the cuttlefish *Sepia*[2] and deep-water squid *Spirula* have an internal chambered structure (Figure 15-1). These animals are neutrally buoyant, which means the amount of air in their bodies regulates their density, so they can remain at a particular depth.

Because the pressure in their air chambers is always 1 kilogram per square centimeter (1 atmosphere, or 14.7 pounds per square inch), the *Nautilus* must stay above a depth of approximately 500 meters (1640 feet) to prevent collapse of its chambered shell. The *Nautilus*, therefore, rarely ventures below about 250 meters (800 feet).

Some slow-moving fish use an internal organ called a **swim bladder** (Figure 15-2) to achieve neutral buoyancy. Very active swimmers (such as tuna) or fish that live on the bottom do not usually have a swim bladder because they don't have a problem maintaining their positions in the water column.

A change in depth either expands or contracts the swim bladder, so fish must remove or add gas to maintain a constant volume. In some fish, a pneumatic duct connects the swim bladder to the esophagus, so these fish can quickly add or remove gases through the duct. In other fish without the pneumatic duct, the gases of the swim bladder must be added or removed more slowly by an in-

[2]Many species of cephalopod have an inking response. The ink of the cuttlefish *Sepia* was used as writing ink (with the brand name Sepia) before alternatives were developed.

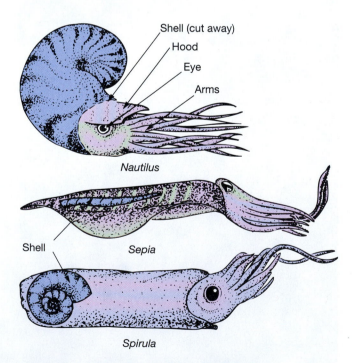

Nautilus

Shell (cut away)
Hood
Eye
Arms

Shell
Sepia

Spirula

Figure 15–1 Gas containers in cephalopods.

The *Nautilus* has an external chambered shell, while *Sepia* and *Spirula* have rigid internal chambered structures that can be filled with gas to provide buoyancy.

[1]Remember that *biomass* is the mass of living organisms.

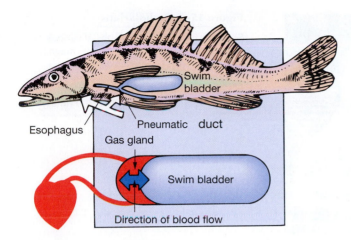

Figure 15–2 Swim bladder.

Some bony fishes have a swim bladder, which is connected to the esophagus by the pneumatic duct, allowing air to be added or removed rapidly. In fish with no pneumatic duct, all gas must be added or removed through the blood, which requires more time.

terchange with the blood, so they cannot withstand rapid changes in depth.

The composition of gases in the swim bladders of shallow-water fishes is similar to that of the atmosphere. At the surface, the concentration of oxygen in the swim bladder is about 20% and as depth increases, the oxygen concentration increases to more than 90%. Fish with swim bladders have been captured from as deep as 7000 meters (23,000 feet), where the pressure is 700 kilograms per square centimeter (700 atmospheres, or 10,300 pounds per square inch). Pressure this high compresses the gas to a density of 0.7 gram per cubic centimeter.[3] This is approximately the same density as fat, so many deep-water fish have special organs for buoyancy that are filled with fat instead of compressed gas.

Floating Heterotrophs (Zooplankton)

Organisms floating at the surface range in size from microscopic to relatively large, such as the familiar jellyfish. These floating organisms—collectively called zooplankton—comprise the second largest biomass in the ocean after the phytoplankton. Microscopic forms have a hard shell or test. Many larger forms have soft, gelatinous bodies with little if any hard tissue, which reduces their density and allows them to stay afloat.

Microscopic Zooplankton Microscopic zooplankton are incredibly abundant in the ocean. They are *primary consumers* because they eat microscopic phytoplankton, the primary producers. Thus, many zooplankton are herbivores. Others are omnivores because they eat other zooplankton in addition to phytoplankton. Most types of

microscopic zooplankton have adaptations to increase the surface area of their bodies (or shells) so they can remain in the sunlit surface waters near their food source.

Three of the most important groups of zooplankton are the radiolarians, foraminifers (both of which are discussed in more detail in Chapter 5, "Marine Sediments"), and copepods.

Radiolarians (*radio* = a spoke or ray) are single-celled, microscopic organisms that build their hard shells (*tests*) out of silica (Figure 15-3). Their tests have intricate ornamentation including long projections. Although the spikes and spines appear to be a defense mechanism against predators, they increase the test's surface area so the organism won't sink through the water column.

Foraminifers (*foramen* = an opening) are microscopic to (barely) macroscopic single-celled animals. While the most abundant types of foraminifers are planktonic, the most diverse (in terms of number of species) are benthic. Foraminifers produce a hard test made of calcium carbonate (Figure 15-4) that is segmented or chambered with a prominent opening in one end. The tests of both radiolarians and foraminifers are common components of deep-sea sediment.

Copepods (*kope* = oar, *pod* = a foot) are microscopic shrimplike animals of the subphylum Crustacea, which also includes shrimps, crabs, and lobsters. Like other crustaceans, copepods have a hard exoskeleton (*exo* = outside) and a segmented body with jointed legs (Figure 15-5). Copepods probably represent the majority of the ocean's zooplankton biomass and are an important link in many marine food webs. They have special adaptations for filtering their tiny floating food from seawater.

Macroscopic Zooplankton Many types of zooplankton are large enough to be seen without the aid of a microscope. One important group is **krill**, which means "young fry of fish" in Norwegian. Krill, however, are in the subphylum Crustacea (genus *Euphausia*) and resemble mini-shrimp or large copepods (Figure 15-6). There are over 1500 species of krill, most of which achieve a length no longer than 5 centimeters (2 inches). They are abundant near Antarctica and form a critical link in the food web there, supplying food for many organisms from sea birds to the largest whales in the world.

Another important group of macroscopic zooplankton is the **cnidarians** [*cnid* = nematocyst (*nemato* = thread, *cystis* = bladder)], which were formerly known as *coelenterates* (*coel* = hollow, *entreon* = the intestines). All members in this group have soft bodies that are more than 95% water and tentacles armed with stinging cells called *nematocysts*. Macroscopic zooplankton that are cnidarians fall into one of two basic groups: the *hydrozoans* and the *scyphozoans* (jellyfish).

Hydrozoan (*hydro* = water, *zoa* = animal) *cnidarians* are represented in all oceans by the Portuguese man-of-war (genus *Physalia*) and the "by-the-wind sailor" (genus

---

[3]For comparison, note that the density of water is 1.0 gram per cubic centimeter.

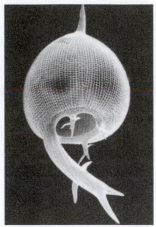

(a) *Euphysetta elegans, x280*

(b) *Anthocyrtidium ophirense, x230*

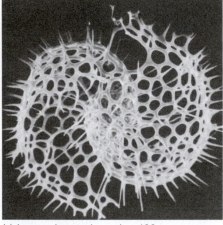

(c) *Larcospira quadrangula, x190*

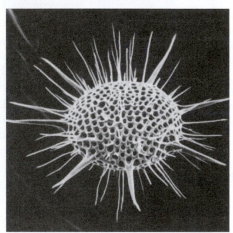

(d) *Heliodiscus asteriscus, x200*

Figure 15–3 Radiolarians.

Scanning electron micrographs of various radiolarians. **(a)** *Euphysetta elegans* (magnified 280 times). **(b)** *Anthocyrtidium ophirense* (magnified 230 times). **(c)** *Larcospira quadrangula* (magnified 190 times). **(d)** *Heliodiscus asteriscus* (magnified 200 times).

Figure 15–4 Foraminifers.

Photomicrograph of various species of foraminifers, the largest of which is 1 millimeter (0.04 inch) long. These pelagic foraminifers were collected from the Ontong Java Plateau in the western Pacific Ocean.

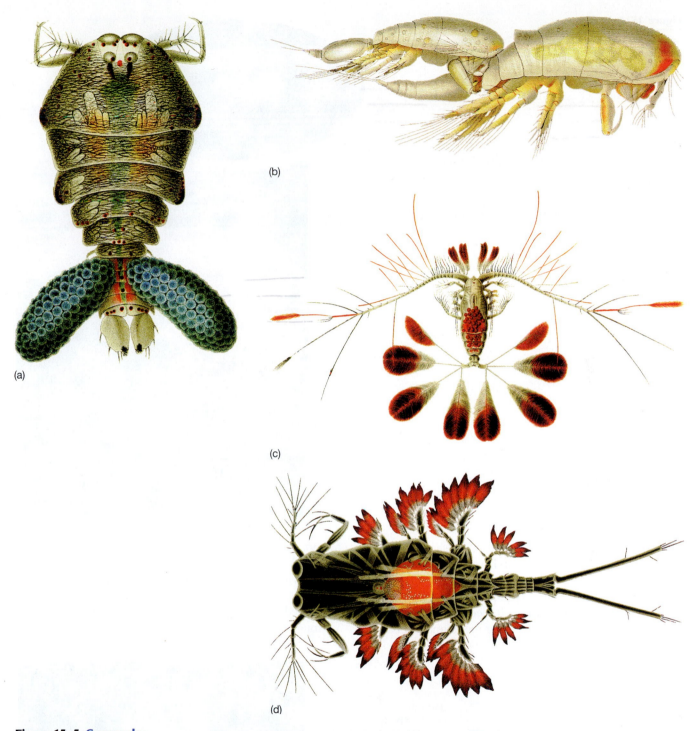

(b)

(a)

(c)

(d)

Figure 15–5 Copepods.

Line drawings of various copepods (shown many times their actual size) from Wilhelm Giesbrecht's 1892 book on the flora and fauna of the Gulf of Naples. **(a)** The adult female *Sapphirina auronitens* carries a pair of lobelike egg sacks. **(b)** Copulating pair of *Oncaea conifera*. **(c)** *Calocalanus pavo* showing elaborate feathery appendages that are characteristic of warm-water species. **(d)** *Copilia vitrea* uses its appendages to cling to large particles in the water column or to larger zooplankton.

Velella). Their gas chambers, called *pneumatophores* (*pneumato* = breath, *phoros* = bearing), serve as floats and sails that allow the wind to push them across the ocean surface (Figure 15-7a,). Sometimes the wind pushes large numbers of these organisms toward a beach, where they can wash ashore and die. In the living organism, a colony of tiny individuals is suspended beneath the float. Portuguese man-of-war tentacles may be many meters long and, because they

are armed with nematocysts, they have been known to inflict a painful and occasionally dangerous neurotoxin poisoning in humans.

Jellyfish, or **scyphozoan** (*skuphos* = cup, *zoa* = animal) cnidarians, have a bell-shaped body with a fringe of tentacles and a mouth at the end of a clapperlike extension hanging beneath the bell-shaped float (Figure 15-7). Ranging in size from nearly microscopic to 2 meters (6.6 feet) in

Figure 15–6 Krill.

Krill that has washed up on a beach in Antarctica and close-up view of *Meganyctiphanes norvegica*, which is about 3.8 centimeters (1.5 inches) long (*inset*).

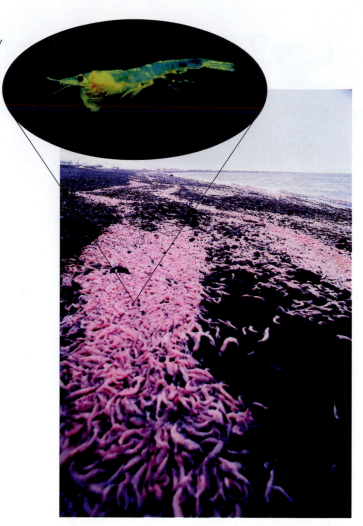

diameter, most jellyfish are less than 0.5 meter (1.6 feet) in diameter. The largest jellyfish can have tentacles as long as 60 meters (200 feet).

Jellyfish move by muscular contraction. Water enters the cavity under the bell and is forced out when muscles that circle the bell contract, jetting the animals ahead in short spurts. To allow the animal to swim generally in an upward direction, sensory organs that are light sensitive or gravity sensitive are spaced around the outer edge of the bell. The ability to orient is important because jellyfish feed by swimming to the surface and sinking slowly through the rich surface waters.

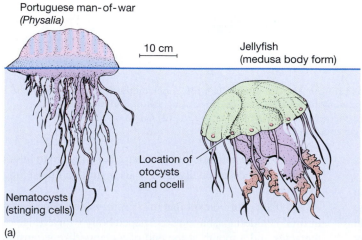

(a)

(b)

Figure 15–7 Planktonic cnidarians.

(a) Portuguese man-of-war (*Physalia*) and jellyfish. **(b)** A medusa jellyfish.

Tunicates (*tunic* = a cape) are barrel-shaped animals with openings at each end: an *incurrent* opening for inflow and an *excurrent* opening for outflow (Figure 15-8a). Although some tunicates are benthic and have a tough covering that may be brightly colored, pelagic forms are usually transparent. They move by a feeble means of jet propulsion, forcing water out the excurrent opening. Tunicates include *salps* (genus *Salpa*), which include solitary forms reaching lengths of 20 centimeters (8 inches) and smaller aggregate forms that produce new individuals by asexual budding. Individual members of an aggregate chain (Figure 15-8b) may be 7 centimeters (2.8 inches) long, while chains may be more than 1 meter (3.3 feet) in length.

The tunicate genus *Pyrosoma* is a luminescent tube-shaped colony of thousands of individuals that have their excurrent openings aligned and facing the inside of a tube, which is closed at one end (Figure 15-8a). As the colony forces water out of their excurrent ends, it provides a slow means of propulsion for the entire colony.

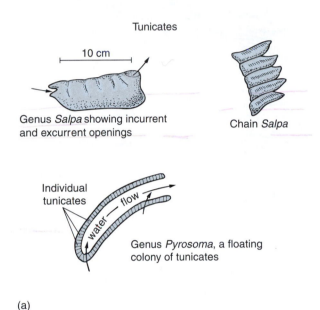

Tunicates

10 cm

Genus *Salpa* showing incurrent and excurrent openings

Chain *Salpa*

Individual tunicates

water — flow

Genus *Pyrosoma*, a floating colony of tunicates

(a)

(b)

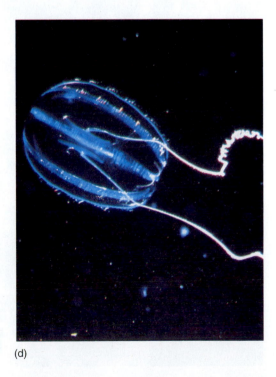

1 cm

Comb rows of cilia

Mouth

Tentacle

Ctenophore *Pleurobranchia*

(c)

(d)

Figure 15–8 Pelagic tunicates and ctenophores.

(a) Body structure and features of tunicates, including chain of tunicates such as *Salpa* and colonial tunicates. (b) Photograph of a chain of salps (*Salpa*). Individual salps are about 7 centimeters (2.8 inches) long. (c) Body structure and features of the ctenophore *Pleurobranchia*. (d) Photograph of the ctenophore *Pleurobranchia*, which is 2 centimeters (0.8 inch) long.

Ctenophores (*cteno* = comb, *phoros* = bearing) (Figure 15-8c) are another type of gelatinous, transparent zooplankton. Their basic body form is spherical with eight rows of hairlike iridescent *cilia* that are spaced evenly around the sphere and resemble miniature combs. Each ctenophore has a pair of tentacles with adhesive pads for catching prey. A common variety of ctenophore is the "sea gooseberry" (genera *Pleurobranchia*) (Figure 15-8d).

Chaetognaths (*chaet* = bristle, *gnath* = jaw) or "*arrowworms*" (Figure 15-9) are an important group of zooplankton that dart through the water like slim arrows to catch prey with their impressive mouthparts. Some grow to lengths in excess of 2.5 centimeters (1 inch). Chaetognaths are voracious feeders that eat primarily small zooplankton; in turn, chaetognaths are fed on by small fish and jellyfish.

> Marine organisms use a variety of adaptations to stay within the sunlit surface waters, such as rigid gas containers, swim bladders, spines to increase their surface area, or soft bodies.

Swimming Organisms (Nekton)

Many larger pelagic animals can swim easily against currents. These organisms include invertebrate squids, fish,

and marine mammals. Because they can swim, some of these organisms undertake long migrations.

Swimming squid include the common squid (genus *Loligo*), flying squid (*Ommastrephes*), and giant squid (*Architeuthis*). Active predators of small fish, the smaller squid varieties have long, slender bodies with paired fins (Figure 15-10). Unlike the less active *Sepia* and *Spirula*

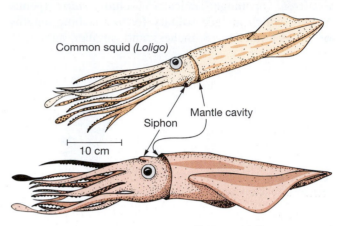

Common squid *(Loligo)*

Siphon — Mantle cavity

10 cm

Flying squid *(Ommastrephes)*

Figure 15–10 Squid.

Squid move by trapping water in their mantle cavity between their soft body and penlike shell. They then jettison the water through their siphon for rapid propulsion.

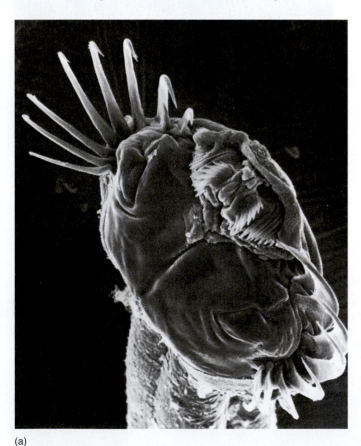

(a)

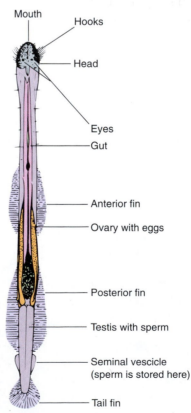

Mouth — Hooks
Head
Eyes
Gut
Anterior fin
Ovary with eggs
Posterior fin
Testis with sperm
Seminal vescicle (sperm is stored here)
Tail fin

(b) General Anatomy of an Arrowworm

Figure 15–9 Chaetognath (arrowworm).

(a) Photomicrograph of the head of a chaetognath (*Sagitta tenuis*) from the Gulf of Mexico, magnified 161 times. **(b)** General anatomy of a chaetognath (arrowworm). Adult chaetognaths reach lengths ranging from 1 to 5 centimeters (0.4 to 2 inches).

shown in Figure 15-1, they have no hollow chambers in their bodies and therefore require more energy to remain in the upper water of the oceans without sinking.

Squid can swim about as fast as any fish their size, and do so by trapping water in a cavity between their soft body and pen-like shell and forcing it out through a siphon. To capture prey, they use two long arms with pads containing suction cups at the ends (Figure 15-10). Eight shorter arms with suckers convey the prey to the mouth, where it is crushed by a mouthpiece that resembles a parrot's beak.

Locomotion in fish occurs when a wave of lateral body curvature passes from the front of the fish to the back. This is achieved by the alternate contraction and relaxation of muscle segments, called *myomeres* (*myo* = muscle, *merous* = parted), along the sides of the body. The backward pressure of the fish's body and fins produced by the movement of this wave provides the forward thrust (Figure 15-11).

Fin Designs in Fish Most active swimming fish use two sets of paired fins—*pelvic fins* and *pectoral* (*pectoralis* = breast) *fins*—to turn, brake, and balance (Figure 15-11). When not in use, these fins can be folded against the body. Vertical fins, both *dorsal* (*dorsum* = back) and *anal*, serve primarily as stabilizers.

The fin that is most important in propelling the high-speed fish is the tail fin, or *caudal* (*cauda* = tail) *fin*. Caudal fins flare vertically to increase the surface area available to develop thrust against the water.[4] The increased surface area also increases frictional drag. The efficiency of the design of a caudal fin depends on its shape and can be expressed mathematically as the **aspect ratio**:

$$\text{Aspect ratio} = \frac{\text{fin height}^2}{\text{fin area}} \qquad (15\text{-}1)$$

---

[4]This is equivalent to humans donning swimming fins on their feet to enable them to swim more efficiently.

In essence, the larger the aspect ratio of a fin, the more efficient it is in providing propulsion. The caudal fin types shown in Figure 15-12 are keyed by letter to the following descriptions, which include aspect ratios:

a. *Rounded fins* (aspect ratio = 1) are flexible and useful in accelerating and maneuvering at slow speeds.

b. & c. *Truncate fins* (b) (aspect ratio = 3) and *forked fins* (c) (aspect ratio = 5) are found on faster fish; the fins are somewhat flexible for better propulsion but are also used for maneuvering.

d. *Lunate fins* (aspect ratio = 7 to 10) are found on fast-cruising fishes such as tuna, marlin, and swordfish; the fins are very rigid and useless for maneuverability but very efficient for propulsion.

e. *Heterocercal* (*hetero* = uneven, *cercal* = tail) *fins* (aspect ratio = 7 to 10) are asymmetrical, with most of their mass and surface area in the upper lobe. The heterocercal fin produces a significant lift to sharks as it is moved from side to side. Sharks need this lift because they have no swim bladder and tend to sink when they stop moving. The pectoral (chest) fins, moreover, are large and flat and positioned so they function like airplane wings to lift the front of the shark's body, balancing the rear lift supplied by the caudal fin. Although the shark gains tremendous lift from this adaptation of its pectoral fins, it sacrifices maneuverability. This is why sharks tend to swim in broad circles—like a circling airplane—and do not make sharp turns while swimming.

Instead of being used simply for maneuvering, the pectoral fins in some fish are modified for highly specialized uses. For example, the flying fish *Exocoetus* (Figure 15-13a) uses greatly enlarged pectoral fins to

Figure 15–11 Swimming motion and general features of fish.

Fish send a wave of body curvature along their bodies to produce a forward thrust (*above*). General features of fish (*below*), with the major fins labeled.

Alternate contraction and relaxation of the myomeres sends a wave of body curvature back along the body to produce a forward thrust.

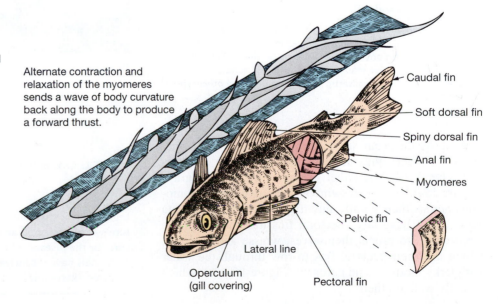

Caudal fin

Soft dorsal fin

Spiny dorsal fin

Anal fin

Myomeres

Pelvic fin

Lateral line

Operculum (gill covering)

Pectoral fin

Rounded

(a)

Truncate

(b)

(d)

(c)

Forked

Lunate
Heteroceral

(e)

Figure 15–12 Caudal fin shapes.

(a) Rounded fin on a queen angel (other examples: sculpin, flounder). **(b)** Truncate fin on a gray angelfish (also on salmon, bass). **(c)** Forked fin on a goatfish (also on herring, yellowtail). **(d)** Lunate fin on a blue marlin (also on bluefish, tuna). **(e)** Heterocercal fin on a silvertip shark (also on many other types of sharks).

glide above the ocean surface for distances up to 400 meters (1300 feet), thus allowing them to evade dolphins and other predators. Other examples include stingrays (Figure 15-13b), which swim by sending undulating motions across their greatly modified pectoral fins; manta rays, which have large pectoral fins that they flap like large wings to propel themselves; wrasse and sculpins, which use their pectoral fins to row through the water with jerky motions; and gurnards (Figure 15-13c), which actually walk on theirs.

? STUDENTS SOMETIMES ASK ...
Do sharks ever get cancer?

Legend has it that sharks don't get cancer, making the creature's cartilage popular in the alternative health market as a cure for the disease. However, researchers who study tumors in animals have recently reported that sharks (and their close relatives, skates and rays) can and do get cancer.

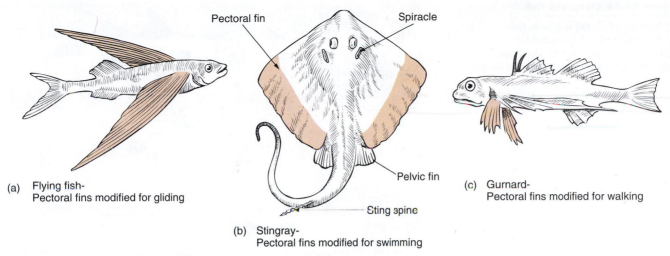

Figure 15–13 Pectoral fin modifications.

(a) Flying fish, with fins modified for gliding. (b) Stingray, with pectoral fins modified for swimming.
(c) Gurnard, with pectoral fins modified for walking.

Fish use their fins for swimming and staying afloat within the water column. The fin that provides the most thrust is the caudal (tail) fin, which can have a variety of shapes depending on the lifestyle of the fish.

Deep-Water Nekton Living below the surface water but still above the ocean floor are deep-water nektonic species—mostly various species of fish—that are specially adapted to the deep-water environment, where it is very still and completely dark. Their food source is either **detritus**—dead and decaying organic matter and waste products that slowly settle through the water column from the surface waters above—or each other. The lack of abundant food limits the *number* of organisms (total biomass) and the *size* of these organisms. As a result, small populations of these organisms exist and most individuals are less than 30 centimeters (1 foot) long. Many have low metabolic rates to conserve energy, too.

These **deep-sea fish** (Figure 15-14) have special adaptations to efficiently find and collect food. They have good sensory devices, for example, such as long antennae or sensitive lateral lines that can detect movement of other organisms within the water column.

Many species have large and sensitive eyes—perhaps 100 times more sensitive to light than our own—that enable them to see potential prey. To avoid being prey, most species are dark so they blend into the environment. Other species are blind and rely on senses such as smell to track down prey.

Other adaptations to the deep-sea environment include large sharp teeth, expandable bodies to accommodate large food items, hinged jaws that can unlock to open widely, and mouths that are huge in proportion to their bodies (Figure 15-15). These adaptations allow deep-sea

fish to ingest species that are larger than they are and to process food efficiently whenever it is captured.

STUDENTS SOMETIMES ASK...

Those deep-sea fish look frightening. Do they ever come to the surface? Are they related to piranhas?

They never come to the surface, and that is fortunate for us because they are vicious predators. They are only distantly related to piranhas (they are within the same group of bony fishes), but their similarly adapted large, sharp teeth may be a good example of *convergent evolution*: the evolution of similar characteristics on different organisms independent of one another yet adapted to the same problem (in this case, a low food supply).

Well over half of deep-sea fish can **bioluminesce**, which means they can produce light organically. Bioluminescent fish produce light by using specially designed structures or cells called *photophores*, some of which contain symbiotic luminescent bacteria. In a world of darkness, the ability to produce light has the following uses:

- Attracting prey (the deep-sea anglerfish in Figure 15-14f uses its specially modified dorsal fin as a bioluminescent lure).
- Staking out territory by constantly patrolling an area.
- Communicating or seeking a mate by sending signals.
- Escaping from predators by using a flash of light to temporarily blind them.

Deep-sea fish have special adaptations such as good sensory devices and bioluminescence that allow them to survive in this still and completely dark environment.

Figure 15–14 **Deep-sea fish.**

(a) A hatchet fish. (b) A lantern fish.
(c) A stomiatoid. (d) A hatchet fish.
(e) A gulper eel. (f) A female deep-sea
anglerfish, with attached parasitic male (g).

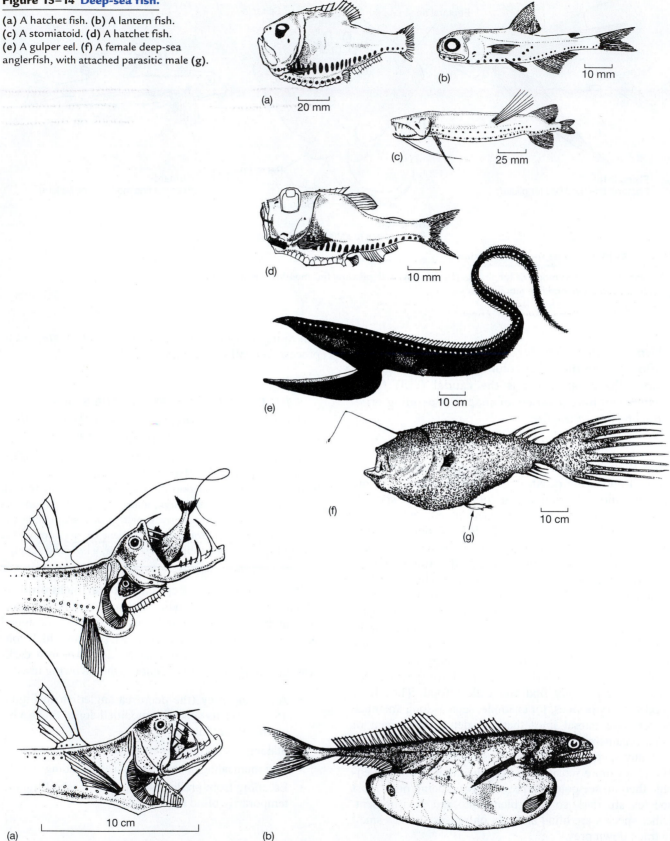

Figure 15–15 **Adaptations of deep-sea fish.**

(a) Large teeth, hinged jaw, and swallowing mechanism of the deep-sea viper fish *Chauliodus sloani*.
(b) Ingestion capability of *Chiasmodon niger*, with a curled-up fish in its stomach that is longer than it is.

BOX 15–1   People and Ocean Environment

SOME MYTHS (AND FACTS) ABOUT SHARKS

Sharks (Figure 15A) are the fish humans most fear. The strength, large size, sharp teeth, and unpredictable nature of sharks are enough to keep some people from *ever* entering the ocean. The publicity generated by the occasional shark attack on humans has led to many myths about sharks.

• *Myth 1: All sharks are dangerous.* Of the more than 350 shark species, about 80% are unable to hurt people or rarely encounter people. The largest shark—also the largest fish in the world—is the whale shark (*Rhincodon typus*), which reaches lengths of up to 15 meters (50 feet) but eats only plankton and so is not considered dangerous.

• *Myth 2: Sharks are voracious eaters that must eat continuously.* Like other large animals, sharks eat periodically depending upon their metabolism and the availability of food. Humans are not a primary food source of any shark, and many large sharks prefer the higher fat content of seals and sea lions.

• *Myth 3: Most people attacked by a shark are killed.* Of every 100 people attacked by sharks, 85 survive. Many large sharks commonly attack by biting their prey to immobilize it before trying to eat it. Consequently, many potential prey escape and survive.

• *Myth 4: Many people are killed by sharks each year.* The chances of a person being killed by a shark are quite low (Table 15A). Sharks kill an average of only 5 to 15 people worldwide each year. Humans, on the other hand, kill as many as 100 million sharks a year (mostly as bycatch from fishing activities). Compared to many other fishes, sharks have low reproduction rates and grow slowly, which may result in many sharks being designated as endangered species in the future.

• *Myth 5: The great white shark is a common, abundant species found off most beaches.* Great white sharks are relatively uncommon predators that prefer cooler waters. At most beaches, great whites are rarely encountered.

• *Myth 6: Sharks are not found in fresh water.* A specialized osmoregulatory system enables some species (such as the bull shark) to cope with dramatic changes in salinity, from the high salinity of seawater to the low salinity of freshwater rivers and lakes.

• *Myth 7: All sharks need to swim constantly.* Some sharks can remain at rest for long periods on the bottom and obtain enough oxygen by opening and closing their mouths to pump oxygen through their gills. Typically, sharks swim very slowly—cruising speeds are less than 9 kilometers (6 miles) per hour—but they can swim at bursts of over 37 kilometers (23 miles) per hour.

• *Myth 8: Sharks have poor vision.* The lens of a shark's eye is up to seven times more powerful than that of a human's. Sharks can even distinguish color.

• *Myth 9: Eating shark meat makes one aggressive.* There is no indication that eating shark meat will alter a person's temperament. The firm texture, white flesh, low fat content, and mild taste of shark meat have made it a favorite seafood in many countries.

• *Myth 10: No one would ever want to enter water filled with sharks.* Long regarded with fear and suspicion, sharks are more recently being viewed with fascination and wonder. Diving tours specializing in close encounters with sharks are becoming increasingly popular.

Figure 15A Great white shark (*Carcharodon carcharias*).

TABLE 15A **Types and number of occurrences in the United States.**

Occurrence to people in the U.S.	Average number per year
Highway fatalities	2611
Struck by lightning	352
Killed by lightning	94
Bit by a squirrel in New York City	88
Killed by a snake bite	15
Bit by a shark	10
Killed by a shark	0.4

Adaptations for Seeking Prey

Several factors affect each species' adaptation for capturing food, including mobility (lunging versus cruising), speed, body length, body temperature, and circulatory system.

Lungers versus Cruisers

Some fish wait patiently for prey and exert themselves only in short bursts as they lunge at the prey. Others cruise relentlessly through the water, seeking prey. A marked difference occurs in the musculature of fishes that use these different styles of obtaining food.

Lungers, such as the grouper in Figure 15-16a, sit and wait for prey to come close by. Lungers have truncate caudal fins for speed and maneuverability, and almost all their muscle tissue is white.

Cruisers, such as the tuna in Figure 15-16b, actively seek prey. Less than half of a cruiser's muscle tissue is white; most is red.

What is the significance of red versus white muscle tissue? Red muscle fibers are 25 to 50 microns in diameter (0.01 to 0.02 inch), whereas white muscle fibers are 135 microns (0.05 inch) and contain lower concentrations of **myoglobin** (*myo* = muscle, *globus* = sphere), a red pigment with an affinity for oxygen. Red fibers, therefore, get a much greater oxygen supply than white fibers. They support a metabolic rate six times that of white fibers, which cruisers need for endurance.

Lungers need very little red tissue because they do not move continually. Instead, they need white tissue, which fatigues much more rapidly than red tissue, for quick bursts of speed to capture prey. Cruisers use white tissue, too, for short periods of acceleration while on the attack.

Speed and Body Size

Fish swim slowly when cruising, fast when hunting for prey, and fastest of all when trying to escape from pred-ators. Generally, the larger the fish, the faster it can swim. For tuna, which are well adapted for sustained cruising and short, high-speed bursts, cruising speed averages about three body lengths per second. They can, however, maintain a maximum speed of about 10 body lengths per second, but only for about one second. Remarkably, yellowfin tuna (*Thunnus albacares*) have been clocked at 74.6 kilometers (46 miles) per hour! Even though this speed is more than 20 body lengths per second, the tuna could maintain it for only a *fraction* of a second.

Theoretically, a 4-meter (13-foot) bluefin tuna (*Thunnus thynnus*) can reach speeds up to about 144 kilometers (90 miles) per hour.[5] Like fish, many of the toothed whales can swim fast, too. For instance, spotted dolphins of the genus *Stenella* have been clocked at 40 kilometers (25 miles) per hour, and killer whales may exceed 55 kilometers (34 miles) per hour during short bursts.

Cold-Blooded versus Warm-Blooded

Fish are mostly **cold-blooded** or **poikilothermic** (*poikilos* = spotted, *thermos* = heat), so their body temperatures are nearly the same as their environment. Usually, these fish are not fast swimmers. The mackerel (*Scomber*), yellowtail (*Seriola*), and bonito (*Sarda*), on the other hand, are indeed fast swimmers. They also have body temperatures that are 1.3, 1.4, and 1.8°C (2.3, 2.5, and 3.2°F), respectively, higher than the surrounding seawater.

The mackerel shark (*Lamna* and *Isurus*) and tuna (*Thunnus*) genera have body temperatures much higher than their environment. Bluefin tuna can maintain a body temperature of 30 to 32°C (86 to 90°F) regardless of the

---

[5]Imagine how difficult it would be to clock a bluefin tuna in the ocean at this speed!

(a)

(b)

Figure 15–16 Feeding styles—lungers and cruisers.

(a) Lungers, such as this tiger grouper, sit patiently on the bottom and capture prey with quick, short lunges. **(b)** Cruisers, such as these yellowfin tuna, swim constantly in search of prey and capture it with short periods of high-speed swimming.

water temperature, which is characteristic of **warm-blooded** or **homeothermic** (*homeo* = alike, *thermos* = heat) organisms. Although these tuna are more commonly found in warmer water, where the temperature difference between fish and water is no more than 5°C (9°F), body temperatures of 30°C (86°F) have been measured in bluefin tuna swimming in 7°C (45°F) water.

Why do these fish exert so much energy to maintain their body temperatures at high levels, when other fish do quite well with ambient body temperatures? It may be that for cruisers, any adaptation (high temperature and high metabolic rate) that increases the power output of their muscle tissue helps them seek and capture prey.

Circulatory System Modifications

A modified circulatory system (Figure 15-17) helps mackerel sharks and tuna maintain their high body temperatures. Most fish have a *dorsal aorta* located just beneath the vertebral column that provides blood to the swimming muscles. Some fish, such as mackerel sharks and tuna, have additional *cutaneous* (*cutane* = skin) *arteries* just beneath the skin on either side of the body. As cool blood flows into red muscle tissue, muscle contractions generate heat that increases the temperature of the blood. A fine network of tiny blood vessels within the muscle tissue is designed to minimize heat loss.

The vessels that return the blood to the *cutaneous vein*, parallel to the cutaneous artery along the side of the fish, are all paired with small vessels carrying blood into the muscle tissue. In this way, the warm blood leaving the tissue helps to heat the cooler blood entering from the cutaneous artery.

> Adaptations of pelagic organisms for seeking prey include mobility (lunging versus cruising), high swimming speed, body length, high body temperature, and a modified circulatory system.

Adaptations to Avoid Being Prey

Obtaining food occupies most of the time of many inhabitants of the open ocean. Some animals are fast and agile, and obtain food through active predation. Other animals, called **filter feeders**, move more leisurely as they filter their small prey from seawater.

Marine animals exhibit a variety of behaviors that serve as defensive mechanisms to help them ward off predators—or to be more successful predators themselves. These include using speed, secreting poisons, and mimicry of other poisonous or distasteful species. Others use transparency, camouflage, or countershading, as discussed in Chapter 13. Still others use symbiotic relationships (see Chapter 14). Let's examine schooling.

Schooling

Although vast populations of phytoplankton and zooplankton may be highly concentrated in certain areas of the ocean, they are not usually referred to as schools. The term **school** is usually reserved for large numbers of fish, squid, or shrimp that form well-defined social groupings.

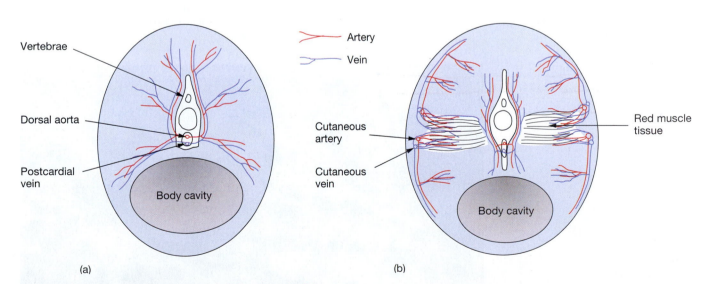

Figure 15–17 Circulatory system modifications in fish.

(a) Most cold-blooded (poikilothermic) fish have major blood vessels arranged so that blood flows to muscle tissue from the dorsal aorta and returns to the postcardial vein beneath the vertebral column. **(b)** Warm-blooded (homeothermic) fishes such as the bluefin tuna have cutaneous arteries and veins that help maintain high blood temperature by using heat energy generated by contracting muscle tissue.

The number of individuals in a school can vary from a few larger predaceous fish (such as bluefin tuna) to hundreds of thousands of small filter feeders (such as anchovies). Within the school, individuals move in the same direction and are evenly spaced. Spacing is probably maintained through visual contact and, in the case of fish, by use of the lateral line system (see Figure 15-11) that detects vibrations of swimming neighbors. The school can turn abruptly or reverse direction as individuals at the head or rear of the school assume leadership positions (Figure 15-18).

What is the advantage of schooling? During spawning, schooling ensures that there will be males to release sperm to fertilize the eggs released into the water or deposited on the bottom by females. The most important function of schooling in small fish, however, is protection from predators.

It may seem illogical that schooling would be protective. For instance, schooling creates tighter groupings of organisms so that any predator lunging into a school would surely catch something, just as land predators run a herd of grazing animals until one weakens and becomes a meal. So, aren't the smaller fish making it easier for the predators by forming a large target?

Scientists who study fish behavior suggest that schooling does indeed serve to protect a group of organisms based on the following strategies that give them "safety in numbers":

1. If members of a species form schools, they reduce the percentage of ocean volume in which a cruising predator might find one of their kind.

2. Should a predator encounter a large school, it is less likely to consume the entire unit than if it encounters a small school or an individual.

3. The school may appear as a single large and dangerous opponent to the potential predator and prevent some attacks.

4. Predators may find the continually changing position and direction of movement of fish within the school confusing, making attack particularly difficult for predators, who can attack only one fish at a time.

In addition, the fact that over 2000 fish species are known to form schools and half of all fishes join schools during a portion of their lives suggests that schooling enhances survival of species, especially for those with no other means of defense. Schooling may also help fish swim greater distances than individuals because each schooling fish gets a boost from the vortex created by the fish swimming in front of it.

> Many pelagic species (especially fish) school to increase their chances of avoiding predators.

Marine Mammals

Marine mammals include some of the largest, best-known, and most charismatic animals in the sea. All organisms in class Mammalia (including marine mammals) share the following characteristics:

- They are warm-blooded.
- They breathe air.
- They have hair (or fur) in at least some stage of their development.
- They bear live young.[6]
- The females of each species have mammary glands that produce milk for their young.

---

[6]This is true except for a few egg-laying mammals of Australia from the subclass Prototheria, which include the duck-billed platypus and the spiny anteater (echidna).

Figure 15–18 Schooling.

A school of soldier fish near a reef in the Maldives, Indian Ocean.

Recent discoveries of ancient whale fossils in Egypt, Pakistan, and China provide strong evidence that marine mammals evolved from mammals on land about 50 million years ago. Some have small, unusable hind legs, suggesting that the land mammal predecessor had no need for its hind legs when it developed a large paddle-shaped structure for a tail used to swim through water. Many anatomical similarities between land mammals and marine mammals are observed, too. Mammals, which originally evolved from organisms that inhabited the sea millions of years ago, may have returned to the sea because of more abundant food sources.

Marine mammals include at least 116 species within the orders Carnivora, Sirenia, and Cetacea.

Order Carnivora

Animals within order **Carnivora** (*carni* = meat, *vora* = eat) have prominent canine teeth, including the cat and dog families on land. Marine representatives or order Carnivora include sea otters, polar bears, and the **pinnipeds** (*pinni* = feather, *ped* = a foot). The name pinniped describes these organisms' prominent skin-covered flippers—which are well adapted for propelling them through water—and includes walruses, seals, sea lions, and fur seals.

Sea otters (Figure 15-19a) lack an insulating layer of blubber but have extremely dense fur. This fur was highly sought for pelts, causing sea otters to be hunted to the brink of extinction in the late 1800s. Fortunately, they have made a remarkable comeback and now inhabit most areas where they were formerly hunted. Sea otters are some of the smallest marine mammals, and seem particularly playful because they continually scratch themselves, which serves to clean their fur and adds an insulating layer of air. They eat various shellfish and crustaceans,

and often use a rock to break open the shells of their food while floating at the surface on their backs. Sea otters commonly inhabit kelp beds.

Polar bears (Figure 15-19b) have massive webbed paws that make them excellent swimmers. The polar bear's fur is thick and each hair is hollow for better insulation. Polar bears also have large teeth and sharp claws, which they use for prying and killing. Their diet consists mainly of seals, which they often capture at holes in the Arctic ice when the seals come up for a breath of air.

Walruses have large bodies and adults (both male and female) have ivory tusks up to 1 meter (3 feet) long (Figure 15-20a), which are used for territorial fighting, for hauling themselves onto icebergs, and sometimes for stabbing their prey.

Seals—also called the *earless seals* or *true seals*—differ from the **sea lions** and **fur seals**—also called the *eared seals*—in the following ways:

- Seals lack prominent ear flaps that are specific to sea lions and fur seals (compare Figures 15-20b and 15-20c).
- Seals have smaller and less-prominent front flippers (called *fore flippers*) than sea lions and fur seals.
- Seals have prominent claws that extend from their fore flippers that sea lions and fur seals lack (Figure 15-21).
- Seals have a different hip structure than sea lions and fur seals. Thus, seals cannot move their rear flippers underneath their bodies as sea lions and fur seals can (Figure 15-21).
- Seals with their smaller front flippers and different hip structure do not move around on land very well, and can only slither along like a caterpillar. Sea lions and fur seals, on the other hand, use their large

(a)

(b)

Figure 15–19 Marine mammals of order Carnivora.

(a) Sea otter. (b) Polar bear.

(a)

(b)

(c)

Figure 15–20 Marine mammals of order Carnivora (pinnipeds).

(a) Walruses. **(b)** Harbor seals. **(c)** California sea lions.

front flippers and their rear flippers, which they can turn under their bodies, to walk easily on land, and can even ascend steep slopes, climb stairs, and do other acrobatic tricks.

- Seals propel themselves through the water using a back-and-forth motion of their rear flippers (similar to a wagging tail), whereas sea lions and fur seals flap their large front flippers.

Order Sirenia

Animals of order **Sirenia** (*siren* = a mythical mermaid-like creature with an enticing voice) include the manatees and dugongs, collectively known as "sea cows." Manatees are concentrated in coastal areas of the tropical Atlantic Ocean, while the dugongs populate the tropical regions of the Indian and Western Pacific Oceans. Both of these animals have a paddlelike tail and rounded front flippers (Figure 15-22). The land-dwelling ancestors of sirenians were elephant-like, and sirenians today have large bodies and manatees have nails on their front flippers. The sirenians have sparse hairs covering their bodies, concentrated around the mouth.

Sirenians eat only shallow-water grasses and are thus the only vegetarian marine mammals. They spend most of their lives in heavily traveled shallow coastal waters, resulting in much competition with humans for space. In addition, there have been many accidents, with boats running over these large animals that move slowly and cannot be easily seen. The populations of manatees and dugongs have been decreasing, and both are considered endangered.

Order Cetacea

The order **Cetacea** (*cetus* = a whale) includes the whales, dolphins, and porpoises (Figure 15-23). The cetacean body is more or less cigar shaped and insulated with a thick layer of blubber. Cetacean forelimbs are modified into flippers that move only at the "shoulder" joint. The hind limbs are vestigial (rudimentary), not attached to the rest of the skeleton, and are usually not visible externally. All cetaceans share the following characteristics:

- An elongated (telescoped) skull
- Blowholes on top of the skull
- Very few hairs
- A horizontal tail fin called a *fluke* that is used for propulsion by vertical movements

These characteristics make cetacean's bodies very streamlined, allowing them to be excellent swimmers.

Modifications to Increase Swimming Speed Cetaceans' muscles are not a great deal more powerful than those of other mammals, so their ability to swim fast must result from modifications that reduce frictional drag. The muscles of a small dolphin, for example, would need to be five times stronger than they are to swim at 40 kilometers (25 miles) per hour in turbulent flow.

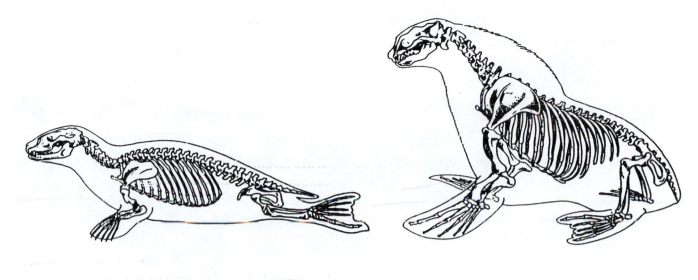

Skeleton of a typical seal, genus *Phoca*

Skeleton of the Steller sea lion

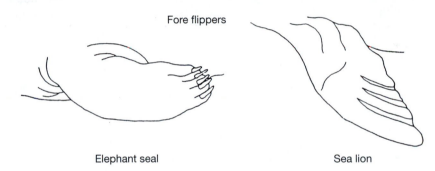

Fore flippers

Elephant seal

Sea lion

Figure 15–21 Skeletal and morphological differences between seals and sea lions.

(a)

(b)

Figure 15–22 Marine mammals of order Sirenia.

(a) Manatees. (b) Dugong.

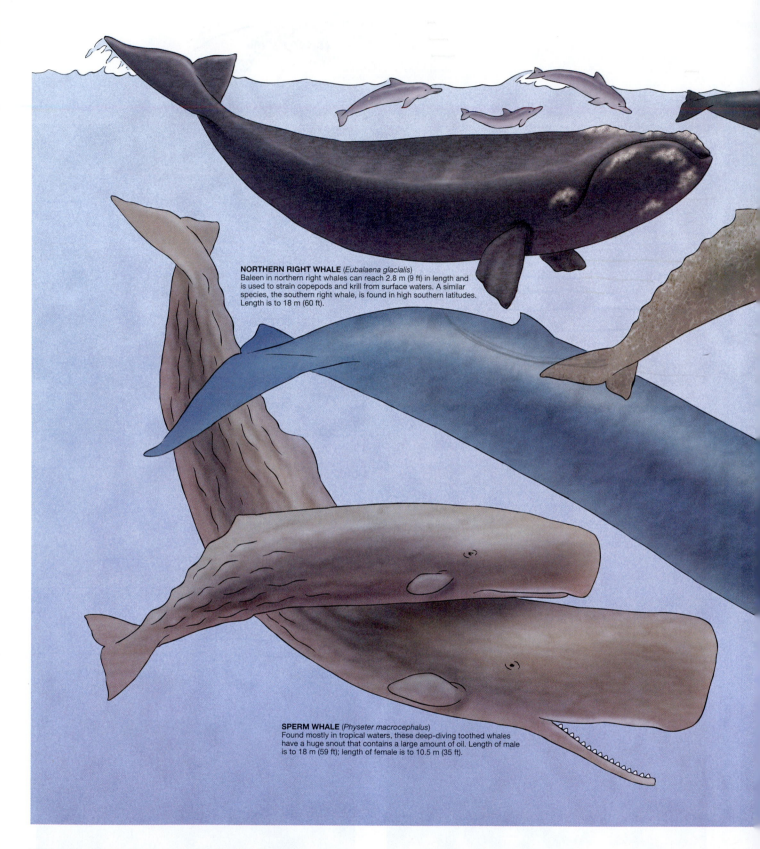

NORTHERN RIGHT WHALE (*Eubalaena glacialis*)
Baleen in northern right whales can reach 2.8 m (9 ft) in length and is used to strain copepods and krill from surface waters. A similar species, the southern right whale, is found in high southern latitudes. Length is to 18 m (60 ft).

SPERM WHALE (*Physeter macrocephalus*)
Found mostly in tropical waters, these deep-diving toothed whales have a huge snout that contains a large amount of oil. Length of male is to 18 m (59 ft); length of female is to 10.5 m (35 ft).

Figure 15–23 Marine mammals of order Cetacea.

A composite drawing of representatives of the two whale suborders, drawn to relative scale. The toothed whales (Odontoceti) form complex social communities and include the bottlenose dolphin, killer whale, narwhal, and sperm whale. The baleen whales (Mysticeti) are the largest of all whales and include the right whale, gray whale, humpback whale, and blue whale.

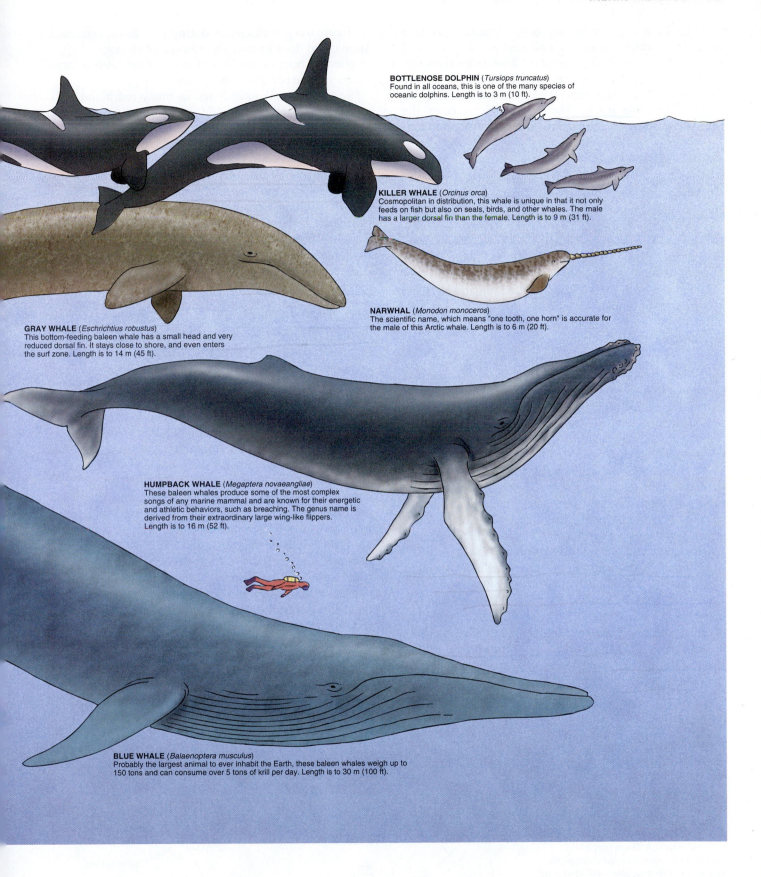

BOTTLENOSE DOLPHIN (*Tursiops truncatus*)
Found in all oceans, this is one of the many species of oceanic dolphins. Length is to 3 m (10 ft).

KILLER WHALE (*Orcinus orca*)
Cosmopolitan in distribution, this whale is unique in that it not only feeds on fish but also on seals, birds, and other whales. The male has a larger dorsal fin than the female. Length is to 9 m (31 ft).

NARWHAL (*Monodon monoceros*)
The scientific name, which means "one tooth, one horn" is accurate for the male of this Arctic whale. Length is to 6 m (20 ft).

GRAY WHALE (*Eschrichtius robustus*)
This bottom-feeding baleen whale has a small head and very reduced dorsal fin. It stays close to shore, and even enters the surf zone. Length is to 14 m (45 ft).

HUMPBACK WHALE (*Megaptera novaeangliae*)
These baleen whales produce some of the most complex songs of any marine mammal and are known for their energetic and athletic behaviors, such as breaching. The genus name is derived from their extraordinary large wing-like flippers. Length is to 16 m (52 ft).

BLUE WHALE (*Balaenoptera musculus*)
Probably the largest animal to ever inhabit the Earth, these baleen whales weigh up to 150 tons and can consume over 5 tons of krill per day. Length is to 30 m (100 ft).

In addition to a streamlined body, cetaceans improve the flow of water around their bodies with a specialized skin structure. Their skin consists of a soft outer layer that is 80% water and has narrow canals filled with spongy material, and a stiffer inner layer composed mostly of tough connective tissue. The soft layer decreases the pressure differences at the skin–water interface by compressing when pressure is high and expanding when pressure is low, reducing turbulence and drag.

Modifications to Allow Deep Diving

Humans can free-dive to a maximum depth of 130 meters (428 feet) and hold their breath in rare instances for up to six minutes. In contrast, sperm whales (*Physeter macrocephalus*) dive deeper than 2800 meters (9200 feet), and northern bottlenose whales (*Hyperoodon ampullatus*) can stay submerged for up to two hours. These remarkable feats require special adaptations, such as special structures that allow them to use oxygen efficiently and an ability to resist nitrogen narcosis.

Cetacean Breathing

Figure 15-24 shows the internal structures that allow cetaceans to remain submerged for extended periods. Inhaled air finds its way to tiny terminal chambers, the *alveoli* (*alveus* = a small hollow). Alveoli are lined with a thin alveolar membrane that is in contact with a dense bed of capillaries. The exchange of gases between the inhaled air and the blood (oxygen in, carbon dioxide out) occurs across the alveolar membrane. Some cetaceans have an exceptionally large concentration of capillaries surrounding the alveoli (Figure 15-24b), which have muscles that move air against the membrane by repeatedly contracting and expanding.

Cetaceans take from one to three breaths per minute while resting, compared with about 15 in humans. Because they hold the inhaled breath much longer, and because of the large capillary mass in contact with the alveolar membrane and the circulation of the air by muscular action, cetaceans can extract almost 90% of the oxygen in each breath, whereas terrestrial mammals extract only 4 to 20%.

To use oxygen efficiently during long dives, cetaceans store it and limit its use. The storage of so much oxygen is possible because prolonged divers have such a large blood volume per unit of body mass.

Some cetaceans have twice as many red blood cells per unit of blood volume and up to nine times as much myoglobin in their muscle tissue as terrestrial animals. As a result, large supplies of oxygen can be chemically stored in **hemoglobin** (*hemo* = blood, *globus* = sphere) within red blood cells and myoglobin within muscles.

Additionally, muscles at the start of a dive that have a significant oxygen supply can continue to function through anaerobic respiration when oxygen is used up. The muscle tissue is relatively insensitive to high levels of carbon dioxide.

Research has shown that cetaceans' swimming muscles can function without oxygen during a dive. This suggests that these muscles and other organs, such as the digestive tract and kidneys, may be sealed off from the circulatory system by constriction of key arteries. The circulatory system would then service only essential components, such as the heart and brain. Because of the decreased circulatory requirements, the heart rate can be reduced by 20 to 50% of normal. Other research has shown, however, that no such reduction in heart rate occurs during dives by the common dolphin (*Delphinus delphis*), the whale (*Delphinapterus leucas*), or the bottlenose dolphin (*Tursiops truncatus*).

Nitrogen Narcosis

Another difficulty with deep and prolonged dives is the absorption of compressed gases into the blood. When humans dive using compressed air—which includes nitrogen and oxygen—they can experience nitrogen narcosis or decompression sickness (the "bends").[7] The effect of **nitrogen narcosis** is similar to drunkenness, and it can occur when a diver either goes too deep or stays too long at depths greater than 30 meters (100 feet).

---

[7]For a discussion of the physiologic problems associated with human diving, see Box 1-3 in Chapter 1.

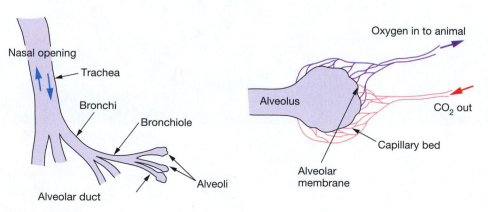

Figure 15–24 Cetacean modifications to allow prolonged submergence.

(a) Basic lung design. Air enters the lung through the trachea, and oxygen is absorbed into the blood through the walls of the alveoli. (b) Oxygen exchange in the alveolus. A dense mat of capillaries receives oxygen through the alveolar membrane, allowing whales to extract as much as 90% of oxygen from each breath.

Nasal opening

Trachea

Bronchi

Bronchiole

Alveolar duct

Alveoli

(a)

Oxygen in to animal

Alveolus

CO₂ out

Capillary bed

Alveolar membrane

(b)

If a diver surfaces too rapidly, the lungs cannot remove the excess gases fast enough, and the reduced pressure may cause small bubbles to form in the blood and tissue. The bubbles interfere with blood circulation, and the resulting *decompression sickness* can cause excruciating pain, severe physical debilitation, or even death.

Cetaceans and other marine mammals do not suffer from these difficulties. By the time a cetacean has reached a depth of 70 meters (230 feet), its rib cage has collapsed under the 8 kilograms per square centimeter (8 atmospheres, or 118 pounds per square inch) of pressure. The lungs within the rib cage also collapse, removing all air from the alveoli. This, in turn, prevents the blood from absorbing additional gases across the alveolar membrane, so nitrogen narcosis cannot occur.

It is possible, however, that a collapsible rib cage is not the main defense against nitrogen narcosis. When enough nitrogen was put into the tissue of a dolphin to give a human a severe case of the "bends," for instance, the dolphin suffered no ill effects. Dolphins, as well as other marine mammals, may have simply evolved an insensitivity to nitrogen gas.

Suborder Odontoceti Members of order Cetacea can be divided into two suborders: Odontoceti (the *toothed whales*) and Mysticeti (the *baleen whales*). Suborder **Odontoceti** (*odonto* = a tooth, *cetus* = a whale) includes the killer whale, sperm whale,[8] porpoises, and dolphins.

Differences between Dolphins and Porpoises Dolphins and porpoises are small toothed whales of suborder Odontoceti. They have similarities in appearance, behavior, and range, and so are easily confused. For instance, both dolphins and porpoises (as well as seals, sea lions, and fur seals) can exhibit a behavior known as "porpoising," which is leaping out of the water while swimming. However, there are several morphological differences between dolphins and porpoises.

Porpoises are somewhat smaller and have a more "stout" (bulky and robust) body shape compared to the more elongated and streamlined dolphins. Generally, porpoises have a blunt snout (rostrum) while dolphins have a longer rostrum. Porpoises have a smaller and more triangular (or, on one species, no) dorsal fin, whereas a dolphin's dorsal fin is sickle shaped or **falcate** (*falcatus* = sickle) and appears hooked and curved backward in profile view.

They also have differences in the shape of their teeth, although it is often difficult to get close enough to see them. The teeth of dolphins end in points, while the teeth of porpoises are blunt or flat (shovel shaped) and resemble our incisors (front teeth). Killer whales have teeth that end in points (Figure 15-25) and thus are members of the dolphin family.

Characteristics of Suborder Odontoceti All toothed whales have prominent teeth, which are used to hold and orient fish and squid, although the killer whale is known to feed on a variety of larger animals, including other whales. The toothed whales form complex and long-lived social groups. Toothed whales have one external nasal opening (blowhole), while baleen whales have two. Although both toothed and baleen whales can emit and receive sounds, the ability to use sound is best developed in toothed whales (particularly sperm whales, which are the most vocal cetaceans).

Despite their lack of vocal cords, toothed whales can produce a variety of sounds—some of which are within the range of human hearing. Sounds are emitted from the blowhole or, in sperm whales, near a special structure called the *museau du singe* ("monkey's muzzle") (Figure 15-26). Contractions of muscles in these structures produce sound that reflects off the front of the skull, which is bowl shaped and resembles a radar dish. The sound is concentrated as it passes through an organ called the **melon** (or the **spermaceti organ** in sperm whales). This organ can form various shapes and sizes of lenses (Figure 15-26a) and acts as an acoustical lens, focusing the sound. Speculations about the sounds' purpose range from **echolocation**—using sound to determine the direction and distance of objects—(clearly true) to a highly developed language (doubtful). In fact, what marine biologists know about cetaceans' use of sound is limited.

All marine mammals have good vision, but conditions often limit its effectiveness. In coastal waters (where suspended sediment and dense plankton blooms make the water turbid) and in deeper waters (where light is limited or absent), echolocation surely assists the pursuit of prey or location of objects.

Using lower-frequency clicks at great distance and higher frequency at closer range, the bottlenose dolphin (*Tursiops truncatus*) can detect a school of fish at distances exceeding 100 meters (330 feet). It can pick out an individual fish 13.5 centimeters (5.3 inches) long at a

Figure 15–25 Jawbone of a killer whale.

The lower jawbone of a killer whale (*Orcinus orca*), showing large teeth that end in points. Thus, killer whales are in the dolphin family.

---

[8]Recent DNA studies suggest that the sperm whale is more closely related to baleen whales than to other toothed whales.

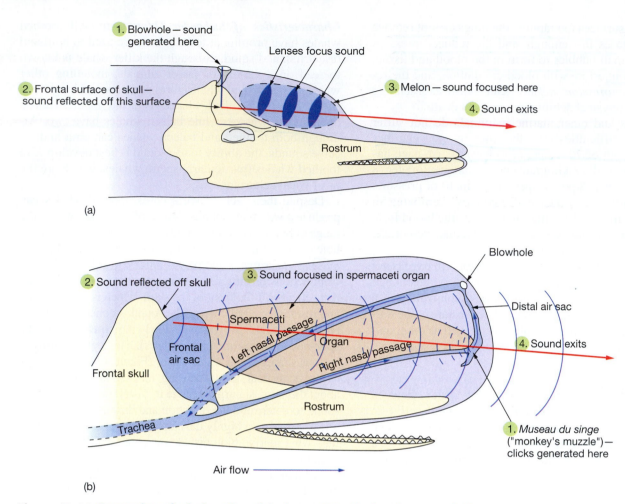

(a)

(b)

Figure 15–26 Generation of echolocation clicks in small toothed and sperm whales.

(a) In small toothed whales, clicks are generated within the blowhole, reflected off the frontal surface of the skull, and then focused by the melon, which can form lenses to focus the sound. **(b)** In the sperm whale, air passes from the trachea through the right nasal passage across the *museau du singe* ("monkey's muzzle"), where clicks are generated. The air passes along the distal air sac, past the closed blowhole, and returns to the lungs along the left nasal passage. The sounds are reflected off the frontal surface of the skull and are focused by the spermaceti organ before leaving the whale.

distance of 9 meters (30 feet). Sperm whales, moreover, can detect their main prey—squid—from distances of up to 400 meters (1300 feet).

To locate an object and determine its distance, toothed whales send sound signals through the water, some of which are reflected from various objects and are returned to the animal and interpreted (Figure 15-27). Because sound penetrates objects, echolocation can produce a three-dimensional image of the object's internal structure and density (which is more than eyesight alone can do). Recent research indicates that the cetaceans may also use a sharp burst of sound to stun their prey before they close in for the kill.

How the reflected sound is received back by toothed whales is not yet fully understood. In most cetaceans, the bony housing of the inner ear is fused to the skull. When submerged, sounds transmitted through the water are picked up by the skull and travel to the hearing structure from many directions, which makes it impossible to accurately locate the source of the sounds. This kind of

hearing structure would not work for an animal that depends on echolocation to find objects in water.

All cetaceans have evolved structures that insulate the inner ear housing from the rest of the skull. In toothed whales, the inner ear is separated from the rest of the skull and surrounded by an extensive system of air sinuses (cavities). The sinuses are filled with an insulating emulsion of oil, mucus, and air and are surrounded by fibrous connective tissue and venous networks. In many toothed whales, it is believed that sound is picked up by the thin, flaring jawbone and passed to the inner ear via the connecting oil-filled body.[9]

How Intelligent Are Toothed Whales? The question of cetacean intelligence is a topic of much debate. Although there may not be a definitive answer, the following facts

---

[9]To simulate this, try pushing the end of a vibrating tuning fork into our chin. The sound is transmitted through your jaw directly to your ear.

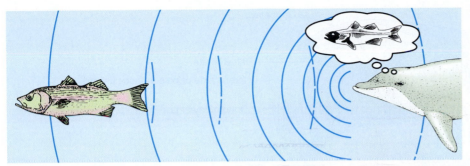

Figure 15–27 Echolocation.

Sounds are generated by toothed whales and bounced off objects in the ocean to determine their size, shape, distance, movement, density—and even internal structure.

about toothed whales (suborder Odontoceti) imply a certain level of intelligence:

- They communicate with each other by using sound.
- They have large brains relative to their body size.
- Their brains are highly convoluted—a characteristic shared by many organisms that are considered to have highly developed intelligence (such as humans and other primates).
- Some dolphins have been reported to assist drowning humans in the wild.
- Some dolphins have been trained to respond to hand signals and do tricks on command (such as retrieve objects).

? STUDENTS SOMETIMES ASK...

In a battle between a killer whale and a great white shark, which one would win?

Although many people who are fascinated with large and powerful wild animals have often wondered which of the two would win such a fight, there was little evidence to settle the dispute until recently. A remarkable video was taken in waters off northern California in 1997, documenting a battle between a 6-meter (20-foot) juvenile killer whale and a 3.6-meter (12-foot) adult great white shark. The video clearly shows the killer whale biting and completely severing off the shark's head! If this is representative of the way these two animals interact in the wild, then the killer whale is the top carnivore in the ocean. It is believed that the killer whale's superior maneuverability and use of echolocation helped it conquer the shark.

Although Odontocetes have remarkable abilities, this does not necessarily imply intelligence. Pigeons, for instance, which are not known for being highly intelligent, have also been trained to retrieve objects by using hand signals. Perhaps many of us would like to think that whales and dolphins are more intelligent than they really are because humans feel an attachment to these playful, seemingly ever-smiling, air-breathing creatures. It is interesting to note that even experts in the field of animal intelligence disagree on how to assess *human* intelligence accurately, let alone that of a marine mammal.

If the large brain that Odontocetes have is not an indication of intelligence, then why is their brain so large?

Leading whale researches don't exactly know, but it might be because Odontocetes need a large brain to process the wealth of information they receive from the sound echoes they transmit. Because intelligence is difficult to measure, perhaps it is best to say that animals of suborder Odontoceti are tremendously well adapted to the marine environment.

Suborder Mysticeti Suborder **Mysticeti** (*mystic* = a moustache, *cetus* = a whale)—also known as the baleen whales—includes the world's largest whales (the blue whale, finback whale, and humpback whale) and the gray whale (a bottom feeder).

Baleen whales are generally much larger than toothed whales because of differences in food sources. Baleen whales eat lower on the food web (including zooplankton such as krill and small nektonic organisms), which are relatively abundant in the marine environment. How are the largest whales in the world able to survive on eating such small prey, especially when these smaller organisms are widely dispersed in the marine environment?

To concentrate small prey and separate them from seawater, baleen whales have parallel rows of **baleen** plates (Figure 15-28a) in their mouths instead of teeth. These baleen plates hang from the whale's upper jaw and, when the whale opens its mouth, the baleen resembles a moustache (except that it is on the *inside* of their mouths), which is why these whales are sometimes called the moustached whales (Figure 15-28b). Baleen is made of flexible keratin—the same as human nails and hair—and can be up to 4.3 meters (14 feet) long[10] (Figure 15-28c). To feed, baleen whales fill their mouths with water that contains their prey items, allowing their pleated lower jaw to balloon in size. The whales force the water out between the fibrous plates of baleen, trapping small fish, krill, and other plankton inside their mouths. Mostly, baleen whales feed at or near the surface, sometimes working together in large groups and surfacing in vertical lunges (Figure 15-29). The gray whale, however, has short baleen slats and feeds by filtering sediment from the shallow bottom of its North Pacific and Arctic feeding grounds, straining benthic organisms such as amphipods and shellfish.

---

[10]Baleen (also called whalebone) was used for such items as buggy whips and corset stays before synthetic materials were substituted.

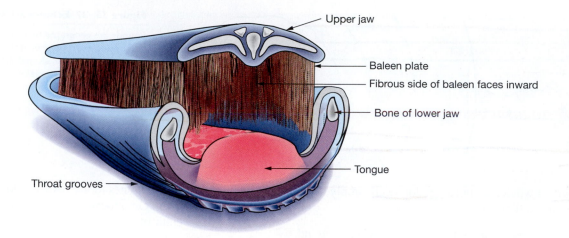

(a)

(b)

Figure 15–28 Baleen.

(a) A diagrammatic cross section through the head of a typical baleen whale. The baleen plates, hanging from the upper jaw, form a sieve that allows these whales to concentrate and eat large quantities of smaller organisms. (b) A rack of baleen from a bottom-feeding gray whale. (c) Individual slat of baleen from a surface-skimming northern right whale.

(c)

Baleen whales include three families:

1. The **gray whale**, which has short, coarse baleen, no dorsal fin, and only two to five ventral grooves on its lower jaw.

2. The **rorqual**[11] **whales**, which have short baleen, many ventral grooves, and are divided into two subfamilies:

   a) The *balaenopterids*, which have long, slender bodies, small sickle-shaped dorsal fins, and flukes with smooth edges (minke, Bryde's, sei, fin, and blue whales)

   b) The *megapterids*, or humpback whales, which have a more robust body, long flippers, flukes with uneven trailing edges, tiny dorsal fins, and tubercles (a row of large bumps) on the head

3. The **right whales**,[12] which have long, fine baleen, broad triangular flukes, no dorsal fin, and no ventral

---

[11]The term *rorqual* refers to the longitudinal grooves on the lower jaw called rorqual folds. The term *rorqual* is from the Old Norse term *raudhr*, meaning "red."

[12]The right whales are so named because they didn't tend to sink when killed, so, from a whaler's point of view, they were the "right" whale to take.

Figure 15–29 Cooperative feeding by humpback whales.

A humpback whale (*Megaptera novaeangliae*) surfaces in a vertical lunge with mouth open during cooperative feeding, where groups of humpback whales work together as a team to capture prey. Baleen plates occur as parallel rows on the upper jaw that are separated by the roof of the mouth (*pink*). The expanded throat grooves can also be seen near the water line.

grooves. The northern right whale is most threatened with extinction. The other members of this family are the Southern Hemisphere's southern right whale and the bowhead whale, which remains near the edge of the Arctic pack ice.

Baleen whales also produce sound, but at much lower frequencies than toothed whales. Gray whales produce pulses (possibly for echolocation) and moans that may maintain contact with other gray whales. Rorqual whales produce moans that last from one to many seconds. These sounds are extremely low in frequency and are probably used to communicate over distances of up to 50 kilometers (31 miles). Blue whales produce sounds that may travel along the SOFAR channel across entire ocean basins.[13] Songs of humpback whales are thought to be a form of sexual display, but it is unclear whether their main purpose is to repel other males or to attract females.

> Marine mammals include orders Carnivora (sea otters, polar bears, and pinnipeds—walrus, seals, sea lions, and fur seals); Sirenia (manatees and dugongs); and Cetacea (whales, dolphins, and porpoises).

An Example of Migration: Gray Whales

Fish, sea turtles, and marine mammals migrate seasonally. Some of the longest migrations known in the open ocean are those of baleen whales, and one of the best-studied is that of the gray whale (*Eschrichtius robustus*, often called the "California" gray whale).

The migratory routes of commercially important baleen whales (including gray whales) have been well known since the mid-1800s. Marine mammals are relatively easy to track because they must surface periodically for air. Gray whales are particularly easy to track because they spend their entire lives in nearshore areas. More recently, radio tracking of individual gray whales has delineated the route and timing of the migration.

Why Gray Whales Migrate   Gray whales undertake a 22,000-kilometer (13,700-mile) round-trip journey every year, the longest known migration of any mammal. Gray whales feed in cold high-latitude waters in the coastal Arctic Ocean and the far northern Pacific Ocean near Alaska; they breed and give birth in warm tropical lagoons along the west coast of Baja California and mainland Mexico (Figure 15-30). Feeding occurs during summer when the long hours of sunlight produce a vast feast of crustaceans and other bottom-dwelling organisms in the shallow Bering, Chukchi, and Beaufort Seas. Without this bountiful food, the whales could not sustain themselves during the long migrations and the mating and calving season, when feeding is minimal.

Initially, gray whales were thought to migrate so far because the physical environment of their cold-water feeding grounds does not meet the needs of young gray whales. Recent research on the physiology of newborn gray whales indicates, however, that gray whale calves can survive in much colder water. So why do they migrate? Perhaps the

---

[13]For more details about the SOFAR channel, see Box 7-2.

Figure 15–30 Gray whale migration route.

Gray whales (*Eschrichtius robustus*) undertake as much as a 22,000 kilometers (13,700 miles) annual migration, the longest of any mammal. They migrate from Arctic summer feeding areas in the Bering and Chukchi Seas to warmer winter breeding and calving lagoons offshore Mexico.

migration is a relic from the Ice Age, when sea level was lower. During that time, the feeding grounds that are so productive today were above sea level. Hence, gray whales could not feast on the abundant food and probably gave birth to smaller calves that could not survive in the cold water. This necessitated the migration to warmer-water regions, which continues to this day despite the abundant food supply.

Alternatively, gray whales may have left the colder waters to avoid killer whales, which are more numerous there and are a major threat to young whales. This may also explain why only those lagoons in Mexico with shallow entrances are used for calving. Killer whales have been seen near the lagoons, however, and have also been observed to feed in extremely shallow water (see Figure 15B).

Timing of Migration The timing of the gray whale migration is closely linked to physical oceanographic conditions. The migration usually begins in September, after the high-latitude summer bloom in productivity has peaked. By this time, the whales have stored enough fat to last them until they return to high-latitude waters. Gray whales have been observed to feed during their migration, however, when the opportunity presents itself. When pack ice begins to form over their feeding grounds along the continental shelf, the whales move south.

First to leave are the pregnant females. They are followed by the nonpregnant mature females, immature fe-

males, mature males, and then immature males. After navigating through passages between the Aleutian Islands, they follow the coast throughout their southern journey. Traveling about 200 kilometers (125 miles) per day, most reach the lagoons of Baja California by the end of January.

In these warm-water lagoons, the pregnant females give birth to 1.8-metric ton (2-short ton) calves. The calves nurse on milk that is almost half butterfat and has the consistency of cheese, allowing the calves to put on weight quickly during the next two months. While the calves are nursing, the mature males breed with the mature females that did not bear calves. Producing large offspring (the gestation period is up to one year) and providing them with fat-rich milk for several months requires enormous amounts of energy, so it is not uncommon for females to mate only once every two or three years.

Late in March, they return north in reverse order (beginning, that is, with the immature males). Most of the whales are back in their high-latitude feeding grounds by the end of June, which coincides with the beginning of the summer bloom in productivity. The whales feed on prodigious quantities of bottom-dwelling organisms to replenish their depleted stores of fat and blubber before their next trip south.

Gray Whales as Endangered Species Although many other whales exist in smaller populations than before whaling (Figure 15-31) and are endangered species, the North

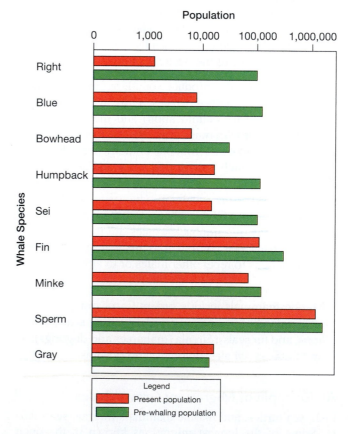

Figure 15–31 World whale populations: present and pre-whaling.

Graph showing present world whale populations (*red*) compared to pre-whaling populations (*green*).

Pacific gray whale was removed from the endangered species list in 1993 when their numbers exceeded 20,000, which surpassed the estimated size of their population prior to whaling. Other populations of gray whales were not so fortunate. For instance, the gray whales that used to inhabit the North Atlantic Ocean were hunted to extinction several centuries ago, and the gray whales that live in the waters near Japan may also have recently gone extinct.

These slow-moving whales of suborder Mysticeti spend most of their lives in coastal waters, which made them easy targets for whalers. In the mid-1800s, they were traced to their birthing and breeding lagoons and were hunted to the brink of extinction. A common strategy was to harpoon a calf and then its mother when she came to its rescue. At this time, gray whales were known as "devilfish" because they would often capsize small whaling

BOX 15–2   People and Ocean Environment
KILLER WHALES: A REPUTATION DESERVED?

Killer whales (*Orcinus orca*) are sleek and powerful predatory animals that inhabit all oceans of the world. Once considered ferocious man-eaters, it turns out they are social animals that don't harm people. Why, then, are they called *killer* whales?

Killer whales prey mainly on fish. When fish are not plentiful, some feed on other dolphins, large whales, penguins, birds, squid, turtles, seals, or sea lions (Figure 15B). Killer whales often hunt in groups, employing various strategies, including deception, to trap their prey. Their distinctive black-and-white coloration—a type of *disruptive coloration*—and white eye patch often confuse their prey as they close in for

the kill. Killer whales have often been called "wolves of the sea" because of their group hunting tactics.

Killer whales often play with their food before killing it. Similar to the way a cat plays with a mouse, killer whales have been known to hone their hunting skills by releasing their prey only to catch it over and over again. Sometimes, they throw their prey completely out of the water. These behaviors appear cruel, so they may have contributed to the whale's common name.

Killer whales have rammed and sunk boats many times and have even been known to tug on divers' fins, but there is only one documented case of a killer whale ever killing a

human. This occurred in 1991 when trainer Keltie Byrne of the marine park Sealand of the Pacific in Victoria, Canada, accidentally slipped into the killer whale pool. As she attempted to climb out, a killer whale pulled her back into the water, where she was tossed about by killer whales in the pool for 10 minutes. Finally, one whale carried her underwater in its mouth for longer than she could hold her breath, causing her to drown. Evidently, the killer whales saw her as a new play object and inadvertently killed her. In the wild, however, there has never been a documented case where killer whales have maliciously attacked and killed a human being.

Figure 15B **Killer whale (*Orcinus orca*) hunting sea lions near Valdes, Argentina.**

boats when the adults came to the aid of their young. By the late 1800s, the number of gray whales had diminished to the point that they were difficult to find during their annual migration. Fortunately, these low numbers also made it difficult for whalers to hunt them successfully.

In 1938, the International Whaling Treaty banned the taking of gray whales, which were thought to be nearly extinct. This protection has allowed them to steadily increase in number to this day and to become the first marine creature to be removed from endangered status. Their repopulation is truly one of the most impressive success stories of how protecting animals can ensure their continued survival. What is perhaps most surprising is how friendly they are now toward people in boats (Figure 15-32) in the same lagoons where they were hunted to near extinction over 150 years ago. Devilfish, indeed!

> Gray whales undertake the longest migration of any mammal, traveling from high latitude Arctic summer feeding areas to low latitude winter birthing and breeding lagoons in Mexico.

Figure 15–32 Gray whale friendly behavior.

A gray whale (*Eschrichtius robustus*) exhibits friendly behavior as it approaches a boat in Scammon's Lagoon, Baja California, Mexico.

Chapter in Review

- *Pelagic animals that comprise the majority of the ocean's biomass remain mostly within the upper surface waters of the ocean*, where their primary food source exists. Those animals that are not planktonic (floating forms such as microscopic zooplankton) depend on *buoyancy* or their *ability to swim* to help them remain in food-rich surface waters.

- *The rigid gas containers in some cephalopods and the expandable swim bladders in some fishes help increase buoyancy.* Other organisms maintain their positions near the surface with *gas-filled floats* (such as those of the Portuguese man-of-war) and *soft bodies that lack high-density hard parts* (such as the jellyfish).

- *Nekton—squid, fish, and marine mammals—are strong swimmers* that depend on their swimming ability to avoid predators and obtain food. Squid swim by trapping water in their body cavities and forcing it out through a siphon. Most fish swim by creating a wave of body curvature that passes from the front of the fish to the back and provides a forward thrust.

- *The caudal (rear) fin provides the most thrust, while the paired pelvic and pectoral (chest) fins are used for maneuvering. The dorsal (back) and anal fins serve primarily as stabilizers.* A rounded caudal fin is flexible and can be used for maneuvering at slow speeds. The lunate fin is rigid and is of little use in maneuvering but produces thrust efficiently for fast swimmers such as tuna.

- *Deep-water nekton feed on detritus and each other.* They have *special adaptations—such as good sensory devices and bioluminescence—that allow them to survive* in this still and completely dark environment.

- *Fish can be categorized a lungers (such as groupers) or cruisers (such as tuna).* Lungers sit motionless and lunge at passing prey. They have mostly white muscle tissue, which fatigues more quickly than red muscle tissue. Cruisers swim constantly in search of prey and possess mostly red, myoglobin-rich muscle tissue.

- *Fish swim slowly when cruising, fast when hunting for prey, and fastest when trying to escape from predators.* Although *most fish are cold-blooded*, the fast-swimming tuna, Thunnus, is homeothermic, meaning it maintains its body temperature well above water temperature.

- *Many marine organisms such as fish, squid, and crustaceans exhibit schooling*, probably because it increases their chances of avoiding predation than swimming alone and serves to preserve the species.

- *Marine mammals are warm-blooded; breathe air; have hair or fur; bear live young; and the females have mammary glands.* Good fossil evidence exists that *marine mammals evolved from land-dwelling animals* about 50 million years ago. Marine mammals belong to orders *Carnivora, Sirenia,* and *Cetacea.*

- *Marine mammals within order Carnivora have prominent canine teeth and include *sea otters, polar bears,* and the *pinnipeds (walruses, seals, sea lions, and fur seals).* Marine mammals of order Sirenia, which include *manatees* and *dugongs* ("sea cows"), have toenails (manatees only), sparse hairs covering their bodies, and are vegetarians.

- The mammals best-adapted to life in the open ocean are those of the *order Cetacea*, which includes *whales,* *dolphins, and porpoises.* Cetaceans have *highly streamlined bodies* so that they are fast swimmers. Other adaptations—such as being able to absorb 90% of the oxygen they inhale, store large quantities of oxygen, reduce the use of oxygen by noncritical organs, and collapse their lungs below depths of 100 meters (330 feet)—*allow them to dive deeply* without suffering the effects of nitrogen narcosis.

- *Cetaceans are divided into suborder Odontoceti (the toothed whales) and suborder Mysticeti (the baleen whales). Odontocetes use echolocation* to find their way through the ocean and locate prey. They emit clicking sounds, and can determine the size, shape, internal structure, and distance of the objects from the nature of the returning signals and the time elapsed.

- *Mysticetes*, which include the largest whales in the world, *separate their small prey from seawater using their baleen plates as a strainer.* Baleen whales include the *gray whale,* the *rorqual whales,* and the *right whales.*

- *Gray whales migrate from their cold-water summer feeding grounds in the Arctic to warm, low-latitude lagoons in Mexico during winter for breeding and birthing purposes.* This behavior may have evolved to allow their young to be born into warm water during the Ice Age, when lower sea level eliminated today's highly productive Arctic feeding areas.

Key Terms

Aspect ratio (p. 433)	Deep-sea fish (p. 435)	Lunger (p. 438)	School (p. 439)
Baleen (p. 449)	Detritus (p. 435)	Melon (p. 447)	Scyphozoan (p. 429)
Bioluminesce (p. 435)	Echolocation (p. 447)	Myoglobin (p. 438)	Sea lion (p. 441)
Biomass (p. 426)	Falcate (p. 447)	Mysticeti (p. 449)	Sea otter (p. 441)
Carnivora (p. 441)	Filter feeder (p. 439)	Nitrogen narcosis (p. 446)	Seal (p. 441)
Cetacea (p. 442)	Foraminifer (p. 427)	Odontoceti (p. 447)	Sirenia (p. 442)
Chaetognath (p. 432)	Fur seal (p. 441)	Pinniped (p. 441)	Spermaceti organ (p. 447)
Cnidarian (p. 427)	Gray whale (p. 450)	Poikilothermic (p. 438)	Swim bladder (p. 426)
Cold-blooded (p. 438)	Hemoglobin (p. 446)	Polar bear (p. 441)	Tunicate (p. 431)
Copepod (p. 427)	Homeothermic (p. 439)	Radiolarian (p. 427)	Walrus (p. 441)
Cruiser (p. 438)	Hydrozoan (p. 427)	Right whale (p. 450)	Warm-blooded (p. 439)
Ctenophore (p. 432)	Krill (p. 427)	Rorqual whale (p. 450)	

Questions and Exercises

1. Discuss why the rigid gas chamber in cephalopods limits the depth to which they can descend. Why do fish with a swim bladder not have this limitation?

2. Draw and describe several different types of microscopic zooplankton and macroscopic zooplankton.

3. Name and describe the different types of fins that fish exhibit. What are the five basic shapes of caudal fins, and what are their uses?

4. What are the two food sources of deep-water nekton? List several adaptations of deep-water nekton that allow them to survive in their environment.

5. What are the major structural and physiological differences between the fast-swimming cruisers and lungers that patiently lie in wait for their prey?

6. Are most fast swimming fish cold-blooded or warm-blooded? What circulatory system modifications do these fish have to minimize heat loss?

7. What are several benefits of schooling?

8. What common characteristics do all organisms in class Mammalia share?

9. Describe marine mammals within the order Carnivora, including their adaptations for living in the marine environment.

10. How can true seals be differentiated from the eared seals (sea lions and fur seals)?

11. Describe the marine mammals within the order Sirenia, including their distinguishing characteristics.

12. List the modifications that are thought to give some cetaceans the ability to (a) increase their swimming speed, (b) dive to great depths without suffering the "bends," and (c) stay submerged for long periods.

13. Describe differences between cetaceans of the suborder Odontoceti (toothed whales) with those of the suborder Mysticeti (baleen whales). Be sure to include examples from each suborder.

14. Describe the process by which the sperm whale produces echolocation clicks.

15. Discuss how sound reaches the inner ear of toothed whales.

16. Describe the mechanism by which baleen whales feed.

17. Discuss reasons why gray whales leave their cold-water feeding grounds during the winter season.

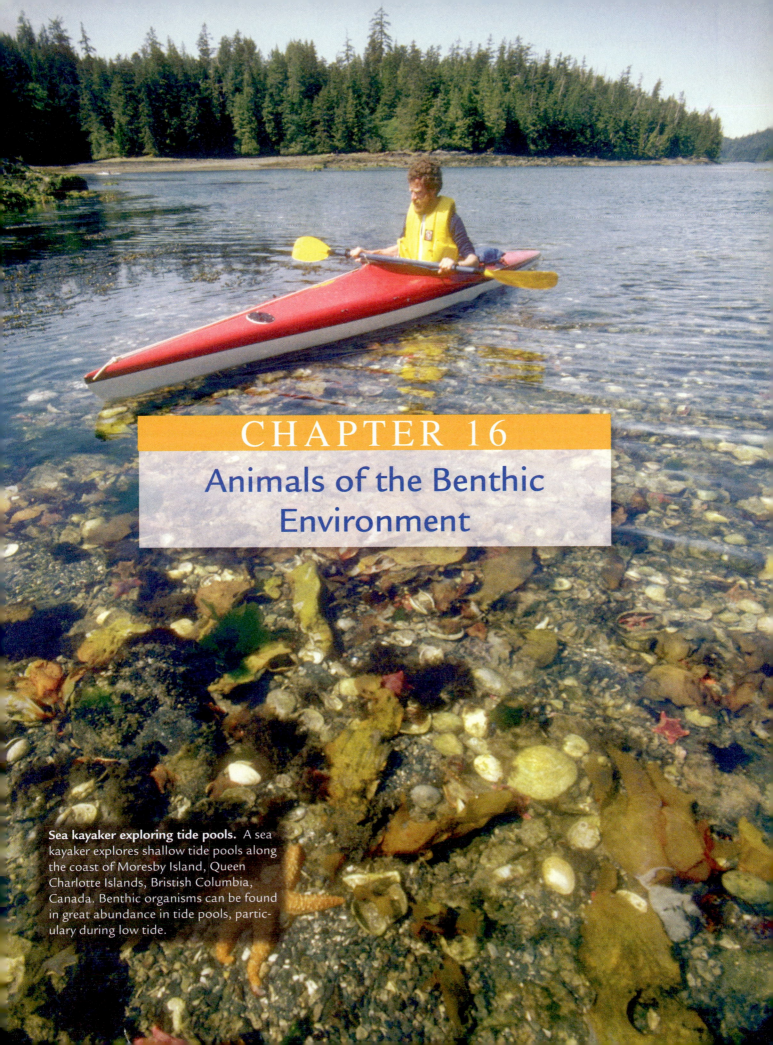

CHAPTER 16

Animals of the Benthic Environment

Sea kayaker exploring tide pools. A sea kayaker explores shallow tide pools along the coast of Moresby Island, Queen Charlotte Islands, Bristish Columbia, Canada. Benthic organisms can be found in great abundance in tide pools, particulary during low tide.

Key Questions

■ What types of organisms occur along rocky shores?
■ What types of organisms occur within sediment covered shores?
■ What are the conditions necessary for coral growth?
■ How are corals able to survive in such nutrient-depleted warm water?
■ What are the environmental conditions of the deep-ocean floor?
■ What types of biocommunities exist on the deep-ocean floor?

The answers to these questions (and much more) can be found in the highlighted concept statements within this chapter.

"From however great a depth we may be able to bring the mud and stones of the bed of the ocean, we shall find them teeming with animal life."

—Sir James Clark Ross (circa 1840)

Of the 250,000 known species that inhabit the marine environment, more than 98% (about 245,000) live in or on the ocean floor. Ranging from the rocky, sandy, and muddy intertidal zone to the muddy deposits of the deepest ocean trenches, the ocean floor provides a tremendously varied environment that is home to a diverse group of specially adapted organisms.

Living at or near the interface of the ocean floor and seawater, an organism's success is closely related to its ability to cope with the physical conditions of the water, the ocean floor, and other members of the biological community. The vast majority of known benthic species live on the continental shelf, where the water is often shallow enough to allow sunlight to penetrate to the ocean bottom.

The number of benthic species found at similar latitudes on opposite sides of an ocean basin depends on how ocean surface currents affect coastal water temperature—one of the most important variables affecting species diversity. The Gulf Stream, for example, warms the European coast from Spain to the northern tip of Norway, giving rise to more than three times the number of benthic species than are found in similar latitudes along the Atlantic coast of North America, where the Labrador Current cools the water as far south as Cape Cod, Massachusetts.

The distribution of benthic **biomass**[1] (Figure 16–1) closely matches the distribution of photosynthetic productivity in surface waters (compare with Figure 14–7). Mostly, life on the ocean floor depends on the productivity of the ocean's surface waters, and great abundances of benthic life are found beneath areas of high productivity and in shallow water.

---

[1]Remember that *biomass* is the mass of living organisms.

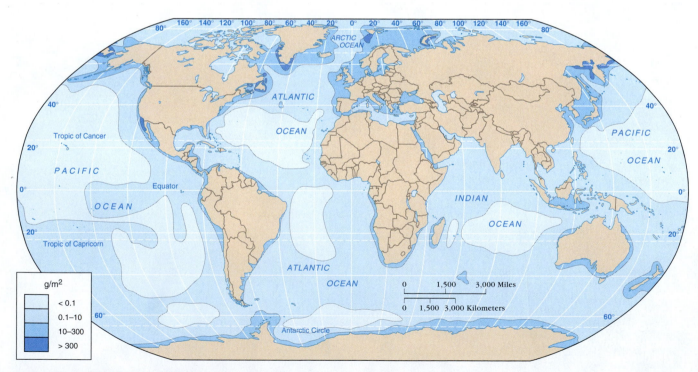

Figure 16–1 Distribution of benthic biomass.

The distribution of benthic biomass (in grams of biomass per square meter) shows that the lowest benthic biomass is beneath the centers of subtropical gyres, and the highest values are in high-latitude continental shelf areas. The pattern is similar to that of surface productivity (see Figure 14–7), suggesting that most of the benthic community receives their food from surface waters.

Rocky Shores

Rocky shorelines teem with organisms that live on the surface of the ocean floor. These organisms are called **epifauna** (*epi* = upon, *fauna* = animal) and are either permanently attached to the bottom (such as marine algae) or they move over it (such as crabs). Table 16–1 lists some of the special adaptations these organisms have to withstand the rigors of life on rocky shores.

A typical rocky shore (Figure 16–2a) can be divided into a **spray zone**, which is above the spring high tide line and is covered by water only during storms, and an **intertidal zone**, which lies between the high and low tidal extremes. Along most shores, the intertidal zone can be clearly separated into the following subzones (Figure 16–2):

- The **high tide zone**, which is relatively dry and is covered only by the highest high tides
- The **middle tide zone**, which is alternately covered by all high tides and exposed during all low tides
- The **low tide zone**, which is usually wet but is exposed during the lowest low tides

The subzones of the intertidal zone can also be delineated based on the populations of organisms that attach themselves to the bottom. Each centimeter of the rocky shore has a significantly different character than the centimeter above and below it, so organisms have evolved to withstand very specific degrees of exposure to the atmosphere. Consequently, the most finely delineated biozones in the marine environment can be found along rocky shores.

Diversity of species that inhabit rocky shores varies widely. Overall, rocky intertidal ecosystems have a moderate diversity of species compared to other benthic environments. The greatest animal diversity is at lower (tropical) latitudes, while the diversity of algae is greater in the mid-latitudes, probably because of better availability of nutrients.[2]

Spray (Supralittoral) Zone

Organisms that live within the spray zone [also known as the **supralittoral** (*supra* = above, *littora* = the seashore) **zone**] must avoid drying out during the long periods when they are above water level. Consequently, many animals, such as the periwinkle snail (genus *Littorina*; Figure 16–2b), have shells, and few species of marine algae are found.

Found among the cobbles and boulders that typically cover the floors of sea caves well above the high tide line are rock lice or sea roaches (isopods of the genus *Ligia*).

[2]As discussed in Chapter 14, "Biological Productivity and Energy Transfer," this increased nutrient supply is the result of the lack of a permanent thermocline in mid-latitude regions.

TABLE 16–1 **Adverse conditions of rocky intertidal zones and organism adaptations.**

Adverse conditions of rocky intertidal zones	Adaptations for adverse conditions	Examples
Drying out during low tide	Ability to seek shelter or withdraw into shells	Sea slugs, snails, crabs
	Thick exterior or exoskeleton to prevent water loss	
Strong wave activity	Strong holdfasts (in algae) to prevent being washed away	Kelp, snails, sea stars, sea urchins
	Strong attachment threads, a muscular foot, multiple legs, or hundreds of tube feet (in animals) to allow them to attach firmly to the bottom	
	Hard structures adapted to withstand wave energy	
Predators occupy area during low tide	Firm attachment	Mussels, sea anemones, sea slugs, octopi, sea stars
	Stinging cells	
	Camouflage	
	Inking response	
	Ability to break off body parts and regrow them later (regenerative capability)	
Difficulty finding mates for attached species	Release of large numbers of egg/sperm into the water column during reproduction	Abalones, sea urchins
Rapid changes in temperature, salinity, pH, and oxygen content	Ability to withdraw into shells to minimize exposure to rapid changes in environment	Snails, barnacles
	Ability to exist in varied temperature, salinity, pH, and low-oxygen environments for extended periods	
Lack of abundant attachment sites	Organisms attach to others	Bryozoans, coral

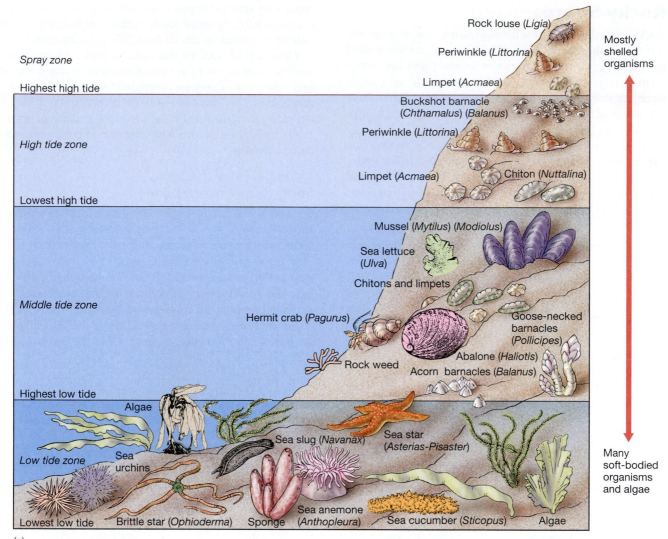

Spray zone

Highest high tide

High tide zone

Lowest high tide

Middle tide zone

Highest low tide

Low tide zone

Lowest low tide

Rock louse (*Ligia*)

Periwinkle (*Littorina*)

Limpet (*Acmaea*)

Buckshot barnacle (*Chthamalus*) (*Balanus*)

Periwinkle (*Littorina*)

Limpet (*Acmaea*)

Chiton (*Nuttalina*)

Mussel (*Mytilus*) (*Modiolus*)

Sea lettuce (*Ulva*)

Chitons and limpets

Hermit crab (*Pagurus*)

Rock weed

Goose-necked barnacles (*Pollicipes*)

Abalone (*Haliotis*)

Acorn barnacles (*Balanus*)

Algae

Sea urchins

Sea slug (*Navanax*)

Sea star (*Asterias-Pisaster*)

Brittle star (*Ophioderma*)

Sponge

Sea anemone (*Anthopleura*)

Sea cucumber (*Sticopus*)

Algae

Mostly shelled organisms

Many soft-bodied organisms and algae

(a)

(b)

(c)

Figure 16–2 The rocky shore.

(a) Zonations and typical organisms of a rocky shore (not to scale). **(b)** Periwinkles (*Littorina*), spray zone to upper high tide zone. **(c)** Rock louse (*Ligia*), spray zone. **(d)** Rough keyhole limpet (*Diodora aspera*) with encrusting red algae (*Lithothamnion*), high tide zone. **(e)** Buckshot barnacles (*Chthamalus*), high tide zone. **(f)** Rock weed (*Fucus filiformes*), middle tide zone. **(g)** Acorn barnacles (*Balanus*), middle tide zone. **(h)** Goose-necked barnacles (*Pollicipes*) and blue mussels (*Mytilus*), middle tide zone. **(i)** Sea anemone (*Anthopleura*), low tide zone.

These scavengers reach lengths of 3 centimeters (1.2 inches) and scurry about at night feeding on organic debris (Figure 16–2c). During the day, they hide in crevices.

A distant relative of the periwinkle snail, the limpet is also found in the spray zone (genus *Acmaea*; Figure 16–2d). Both limpets and periwinkle snails feed on marine algae. The limpet has a flattened conical shell and a muscular foot with which it clings tightly to rocks.

High Tide Zone

Like animals within the spray zone, most animals that inhabit the high tide zone have a protective covering to prevent them from drying out. Periwinkles, for example, have a protective shell and can move between the spray zone and the high tide zone. Buckshot barnacles (Figure 16–2e) have a protective shell, too, but they cannot live above the high tide shoreline because they filter-feed from seawater, and their larval form is planktonic.

The most conspicuous algae in the high tide zone are rock weeds, members of the genus *Fucus* that live in colder latitudes (Figure 16–2f) and *Pelvetia* that live in warmer latitudes. Both have thick cell walls to reduce water loss during periods of low tide.

Rock weeds are among the first organisms to colonize a rocky shore. Later, **sessile** (*sessilis* = sitting on) animal forms—those that are attached to the bottom, such as barnacles and mussels—begin to establish themselves, competing for attachment sites with the rock weeds.

Middle Tide Zone

Seawater constantly bathes the middle tide zone, so more types of marine algae and soft-bodied animals can live there. The total biomass is much greater than in the high tide zone so there is much greater competition for rock space among sessile forms.

Shelled organisms inhabiting the middle tide zone include acorn barnacles (Figure 16–2g); goose-necked barnacles (*Pollicipes*) (Figure 16–2h), which attach themselves to rocks with a long, muscular neck; and various mussels (genera *Mytilus* and *Modiolus*). Mussels attach to bare rock, algae, or barnacles during their planktonic stage and remain in place by means of strong *byssus* threads.

Carnivorous snails and sea stars (such as genera *Pisaster* and *Asterias*) feed upon the mussels. To pry open the mussel shell, sea stars pull on either side with hundreds of tube-like feet. The mussel eventually becomes fatigued and can no longer hold its shell halves closed. When the shell opens ever so slightly, the sea star turns its stomach inside out, slips it through the crack in the mussel shell, and digests the edible tissue inside (Figure 16–3).

The most striking feature of the middle tidal zone along most rocky coasts is a mussel bed that thickens toward the bottom until it reaches an abrupt bottom limit. This is where the physical conditions restrict barnacle growth and in most areas, the boundary is quite pronounced. Protruding from the mussel bed will be numerous goose-necked barnacles, and sea stars browsing on the mussels are concentrated in the lower levels of the bed. Less-conspicuous forms common to mussel beds are varieties of algae, worms, clams, and crustaceans.

Where the rock surface flattens out within the middle tidal zone, tide pools trap water as the tide ebbs. These pools support microecosystems containing a wide variety of organisms. The most conspicuous member of this community is often the sea anemone (see Figure 16–2i), which is a relative of the jellyfish.

Shaped like a sack, anemones have a flat foot disk that provides a suction attachment to the rock surface. Directed upward, the open end of the sack is the mouth, which

Figure 16–3 Sea star feeding on a mussel.

An ochre sea star (*Pisaster*) pulls apart the two halves of a mussel's (*Mytilus*) shell with its tube feet. The star then turns its own stomach inside out and forces it through the opening between the mussel's shells, where it digests the mussel's soft tissue within the mussel's own shell.

leads directly to the gut cavity and is surrounded by rows of tentacles (Figure 16–4). The tentacles are covered with stinging needlelike cells called **nematocysts** (*nemato* = thread, *cystis* = bladder) (Figure 16–4, *inset*), which inject the victim with a potent neurotoxin. Nematocysts are automatically released when any organism brushes against a sea anemone's tentacles.

Hermit crabs (*Pagurus*) inhabit tide pools, too. They have a well-armored pair of claws and upper body but a soft, unprotected abdomen, which they protect by inhabiting an abandoned snail shell (Figure 16–5a). They can often be seen scurrying around the tide pool area or fighting with other hermit crabs for new shells. Their abdomen has even evolved a curl to the right to make it fit properly into snail shells. Once in the snail shell, the crab can protect itself by closing off the shell's opening with its large claws.

In tide pools near the lower limit of the middle tide zone, sea urchins may be found feeding on algae (Figure 16–5b). Sea urchins have a five-toothed mouth centered on the bottom side of their hard spherical shell, consisting of fused calcium carbonate plates perforated to allow tube feet and water to pass through. Resembling a pincushion, the shell of a sea urchin has numerous spines for protection and to scrape out protective holes in rocks.

? STUDENTS SOMETIMES ASK...

I've been at a tide pool and seen sea anemones. When I put my finger on one, it tends to gently grab my finger. Why does it do that?

The sea anemone is trying to kill you and wants to eat you (seriously!) Disguised as a harmless flower, the sea anemone is actually a vicious predator that will attack any unsuspecting animal (even a human) that its stinging tentacles entrap. Fortunately, the skin on your hands is thick enough to resist the stinging nematocyst and its neurotoxin. A couple of people, however, were interested in finding out if the sea anemone grabbed other things with its tentacles, so they put their *tongues* into a sea anemone. After a short time, their throats swelled almost completely closed, and they had to be rushed to a hospital. They lived, but the moral of this story is: *NEVER* put your tongue into a sea anemone!

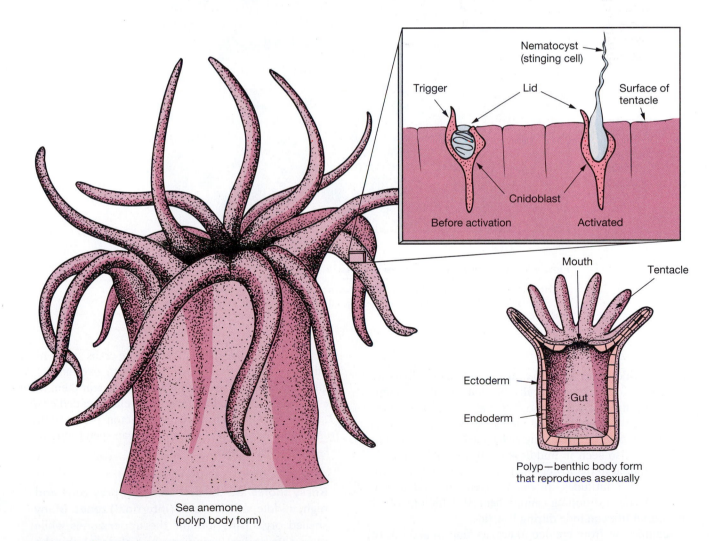

Figure 16–4 Sea anemone.

Sea anemone morphology and detail of its stinging nematocysts (*inset*).

(a)

(b)

Figure 16–5 Hermit crab and sea urchins.

(a) Hermit crab (*Pagurus*) that has taken up residence in a *Maxwellia gemma* shell.
(b) Sea urchins (*Echinus*) that have burrowed into the bottom of a tide pool within the middle tide zone.

Low Tide Zone

The low tide zone is almost always submerged, so an abundance of algae is typically present. A diverse community of animals exists, too, but they are hidden by the great variety of marine algae and surf grass (*Phylospadix*) (Figure 16–6). The encrusting red algae (*Lithothamnion*), which are also seen in middle-zone tide pools, becomes very abundant in the lower tide pools (see Figure 16–2d). In temperate latitudes, moderate-sized red and brown algae provide a drooping canopy beneath which much of the animal life can hide during low tide.

Scampering from crevice to crevice and in and out of tide pools across the full range of the intertidal zone are various species of shore crabs (Figure 16–7). These scav-

engers help keep the shore clean. Shore crabs spend most of the day hiding in cracks or beneath overhangs. At night, they eat algae as rapidly as they can tear them from the rock surface with their large front claws, called *chelae* (*khele* = claw). Their hard exoskeleton prevents them from drying out too quickly so they can spend long periods of time out of water.

Rocky shores are divided into the spray zone and high, middle, and low tide (intertidal) zones. Many shelled organisms inhabit the upper zones while more soft-bodied organisms and algae inhabit the lower zones.

Figure 16–6 Marine algae and surf grass.

Dark-colored sea palms (a brown alga) and green surf grass (*Phylospadix*) are exposed during an extremely low tide in a California low tide zone but provide protection for many organisms.

(a)

(b)

Figure 16–7 Shore crabs.

(a) Coral crab, or queen crab (Bonaire Island, Netherlands Antilles). **(b)** Shore crab, *Pachygrapsis crassipes*. This female is carrying eggs (dark oval structure curled under her abdomen).

Sediment-Covered Shores

The sediment-covered shore ranges from steep boulder beaches, where wave energy is high, to the mud flats of quiet, protected embayments, and includes sandy beaches where wave energy is usually moderate.

Nearly all organisms that inhabit sediment-covered shores are called **infauna** (*in* = inside, *fauna* = animal) because they can burrow into the sediment. Most sediment-covered shores have intertidal zones similar to rocky shores. Also, there is much less species diversity in sediment-covered shores, but the organisms are usually found in great numbers.

The Sediment

Sediment-covered shores include *beaches, salt marshes*, and *mud flats*, which represent progressively lower-energy environments and are consequently composed of progressively finer sediment. The energy level that a shore experiences is related to the strength of waves and longshore currents. Along shores that experience low energy levels, particle size becomes smaller, the sediment slope decreases, and overall sediment stability increases. Thus, the sediment in a fine-grained mud flat is more stable than that of a high-energy sandy beach.

A large quantity of water from breaking waves rapidly sinks into the sand and brings a continual supply of nutrients and oxygen-rich water for the animals that live there. This supply of oxygen also enhances bacterial decomposition of dead tissue. The sediment in salt marshes and mud flats is not nearly so rich in oxygen, however, so decomposition occurs more slowly.

Intertidal Zonation

The intertidal zone of the sediment-covered shore consists of supralittoral, high tide, middle tide, and low tide zones, as shown in Figure 16–8. These zones are best developed on steeply sloping, coarse-sand beaches and are less distinct on the more gentle sloping, fine-sand beaches. On mud flats, the tiny clay-sized particles form a deposit with essentially no slope, so zonation is not possible in this protected, low-energy environment.

The species of animals differ from zone to zone. As in intertidal rocky shores, however, the maximum *number* of species and the greatest *biomass* in intertidal sediment-covered shores are found near the low tide shoreline, and both diversity and biomass decrease toward the high tide shoreline.

Life in the Sediment

Life on and in the sediment requires very different adaptations than on rocky shores. Sandy beaches support fewer species than rocky shores—and mud flats fewer still—but the total *number of individuals* may be as high. In the low tide zone of some beaches and on mud flats, for example, as many as 5000 to 8000 burrowing clams have been counted in only 1 square meter (10.8 square feet).

Burrowing is the most successful adaptation for life in sediment-covered shores, so organisms are much less obvious than in other environments. By burrowing only a few centimeters beneath the surface, they encounter a much more stable environment where they are not bothered by fluctuations of temperature and salinity or the threat of drying out.

Suspension feeding—also called filter feeding—and deposit feeding are two techniques animals of sediment-covered shores commonly use to obtain food from the clear water above or from the sediment itself. In **suspension feeding**, organisms that are buried in sediment use specially designed structures to filter plankton from seawater (Figure 16–9a). Clams, for example, bury themselves in sediment and extend siphons through the surface. They pump in overlying water and filter suspended plankton and other organic matter from it.

In **deposit feeding**, organisms feed on food items that occur as deposits. These deposits include detritus—dead and decaying organic matter and waste products—and the sediment itself, which is coated with organic matter. Some deposit feeders, such as the segmented worm *Arenicola* (Figure 16–9b), feed by ingesting sediment and extracting organic matter from it. Others, such as the amphipod *Orchestoidea* (Figure 16–9c), feed on more concentrated deposits of organic matter (detritus) on the sediment surface.

Another less common method is **carnivorous feeding**. The sand star *Astropecten* (Figure 16–9d), for example, cannot climb rocks the way its sea star relatives can, but it can burrow rapidly into the sand, where it feeds voraciously on crustaceans, mollusks, worms, and other echinoderms.

Sandy Beaches

Most animals at the beach burrow into the sand and are safely hidden from view because there is no stable, fixed surface (as on rocky shores) to which they can attach.

Bivalve Mollusks A **bivalve** (*bi* = two, *valva* = a value) value) is an animal having two hinged shells, such as a clam or a mussel. A **mollusk** is a member of the phylum Mollusca (*molluscus* = soft), characterized by a soft

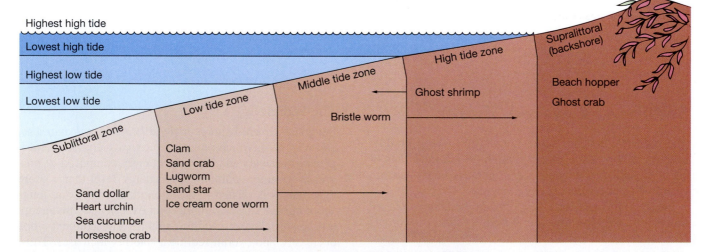

Figure 16–8 Intertidal zonation and typical organisms on a sediment-covered shore.

Intertidal zonation is related to the amount of exposure during low tide. The zonation is best displayed on coarse sand beaches with steep slopes; as the sediment becomes finer and the beach slope decreases, zonation becomes less distinct and disappears entirely on mud flats.

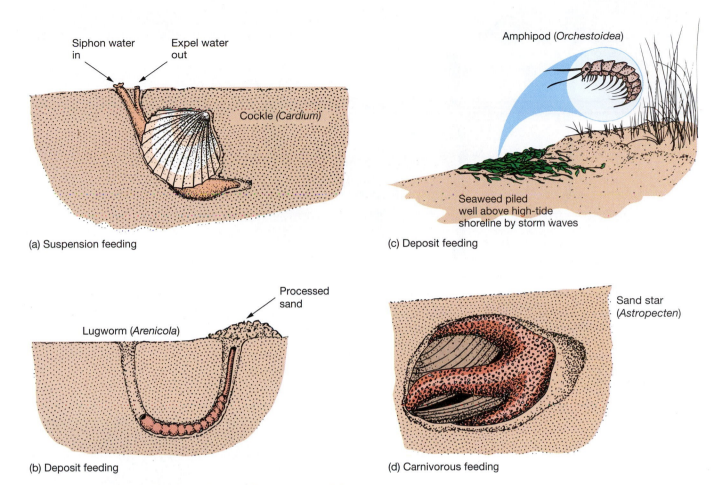

Figure 16–9 Modes of feeding along sediment-covered shores.

(a) Suspension feeding by the clam (cockle) *Cardium*, which uses its siphon to filter plankton and other organic matter that is suspended in the water. **(b)** Deposit feeding by the segmented worm *Arenicola*, which feeds by ingesting sediment and extracting organic matter from it. **(c)** Deposit feeding by the amphipod *Orchestoidea*, which feeds on more concentrated deposits of organic matter (detritus) on the sediment surface. **(d)** Carnivorous feeding of a clam by the sand star *Astropecten*.

body and either an internal or external hard calcium carbonate shell.

Bivalve mollusks are well adapted to life in the sediment. A single foot digs into the sediment to pull the creature down into the sand. Their siphon extends vertically through the sediment for feeding (Figure 16–9a). The method by which clams bury themselves is shown in Figure 16–10.

How deeply a bivalve can bury itself depends on the length of its siphons, which must reach above the sediment surface to pull in water for food (plankton) and oxygen. Indigestible matter is forced back out the siphon periodically by quick muscular contractions. The greatest biomass of clams is burrowed into the low tide region of sandy beaches and it decreases where the sediment becomes muddier.

Annelid Worms   A variety of **annelids** (*annelus* = a ring)—segmented worms—are also well adapted to life in the sediment. The lugworm (*Arenicola* species), for example, lives in a U-shaped burrow (see Figure 16–9b),

the walls of which are strengthened with mucus. The worm moves forward to feed and extends its proboscis (snout) up into the head shaft of the burrow to loosen sand with quick pulsing movements. A cone-shaped depression forms at the surface over the head end of the burrow as sand continually slides into the burrow and is ingested by the worm. As the sand passes through its digestive tract, the organic content is digested, and the processed sand is deposited at the surface.

Crustaceans   Crustaceans (*crusta* = shell)—such as crabs, lobsters, shrimps, and barnacles—include predominately aquatic animals that are characterized by a segmented body, a hard exoskeleton, and paired, jointed limbs. On most sandy beaches, numerous crustaceans called *beach hoppers* feed on kelp cast up by storm waves or high tides. A common genus is *Orchestoidea*, which is only 2 to 3 centimeters (0.8 to 1.2 inches) long (see Figure 16–9c) but can jump more than 2 meters (6.6 feet) high. Laterally flattened, beach hoppers usually spend the day buried in the sand or hidden in kelp. They become

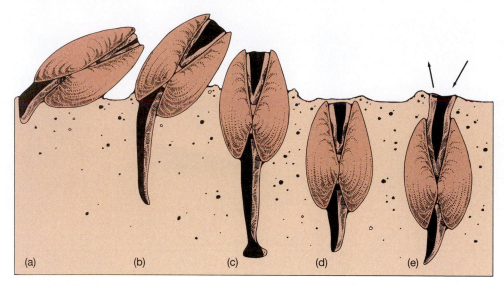

Figure 16–10 How a clam burrows.

A clam burrows into sediment by **(a)** extending its pointed foot into the sediment and **(b)** forcing its foot deeper into the sediment and using this increasing leverage to bring the exposed, shell-clad body toward vertical. When the foot has penetrated deeply enough, a bulbous anchor forms at the bottom **(c)**, and a quick muscular contraction pulls the entire animal into the sediment **(d)**. The siphons are then pushed up above the sediment to pump in water **(e)**, from which the clam extracts food and oxygen.

particularly active at night, when large groups many hop at the same time and form clouds above the piles of kelp on which they feed.

Sand crabs (*Emerita*) (Figure 16–11) are a type of crustacean common to many sandy beaches. Ranging in length from 2.5 to 8 centimeters (1 to 3 inches), they move up and down the beach near the shoreline. They bury their bodies in the sand and leave their long, curved, V-shaped antennae pointing up the beach slope. These little crabs filter food particles from the water, and can be located by looking in the lower intertidal zone for a V-shaped pattern in the swash as it runs down the beach face.

Echinoderms Echinoderms (*echino* = spiny, *derma* = skin) found in beach deposits include the sand star (*Astropecten*) and heart urchins (*Echinocardium*). Sand stars prey on invertebrates that burrow into the low tide region of sandy beaches. The sand star (see Figure 16–9d) is well designed for moving through sediment, with five tapered legs with spines and a smooth back.

Figure 16–11 Sand crab.

A sand crab (*Emerita*) emerges from the sand. They can often be found just beneath the surface of sandy beaches within the lower intertidal zone.

More flattened and elongated than the sea urchins of the rocky shore, heart urchins live buried in the sand near the low tide line. They gather sand grains into their mouths, where the coating of organic matter is scraped off and ingested (Figure 16–12).

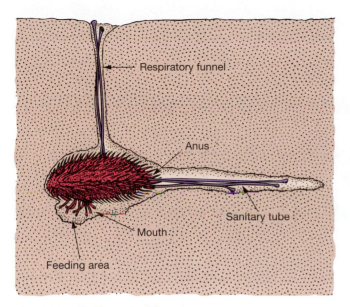

Figure 16–12 Heart urchin.

Feeding and respiratory structures of a heart urchin (*Echinocardium*), which feeds on the film of organic matter that covers sand grains.

Meiofauna Meiofauna (*meio* = lesser, *fauna* = animal) are small organisms that live in the spaces between sediment particles. These organisms, only 0.1 to 2 millimeters (0.004 to 0.08 inch) long, feed primarily on bacteria removed from the surface of sediment particles. Meiofauna include polychaetes, mollusks, arthropods, and nematodes (Figure 16–13) and are found in sediment from the intertidal zone to deep-ocean trenches.

Mud Flats

Eelgrass (*Zostera*) and turtle grass (*Thalassia*) are widely distributed in the low tide zone of mud flats and the adjacent shallow coastal regions. Numerous openings at the surface of mud flats attest to a large population of bivalve mollusks and other invertebrates.

Fiddler crabs (*Uca*) live in burrows that may exceed 1 meter (3 feet) deep in the mud flats. Relatives of the shore crabs, they usually measure no more than 2 centimeters (0.8 inch) across the body. Male fiddler crabs have one small claw and one oversized claw, which is up to 4 centimeters (1.6 inches) long (Figure 16–14). Fiddler crabs get their name because this large claw is waved around as if they were playing an imaginary fiddle. The females have two normal-sized claws. The large claw of the male is used to court females and to fight competing males.

> Sediment-covered shores—including sandy beaches and mud flats—have similar intertidal zonations as rocky shores but contain many organisms that live within the sediment (infauna).

Shallow Offshore Ocean Floor

The shallow offshore ocean floor extends from the spring low tide shoreline to the seaward edge of the continental shelf. It is mainly sediment covered but rocky exposures may occur locally near shore. On rocky exposures, many types of marine algae exist, which have adaptations (such as gas-filled floats) for reaching from the shallow sea floor to near the sunlit surface waters.

The sediment-covered shelf has moderate to low species diversity. Surprisingly, the diversity of benthic organisms is *lowest* beneath upwelling regions. This is because upwelling waters that are rich in nutrients produce high pelagic production, so large amounts of dead organic matter are produced. When this matter rains down on the bottom and decomposes, it consumes oxygen, so the oxygen supply can be locally depleted, thereby limiting benthic populations. However, kelp beds associated with rocky bottoms are a specialized shallow-water community with higher diversity.

(a)

(b)

(c)

Figure 16–13 Scanning electron micrographs of meiofauna.

(a) Nematode head, magnified 804 times. The projections and pit on the right side are sensory structures. **(b)** Amphipod, magnified 20 times. This 3-millimeter-long organism builds a burrow of cemented sand grains. **(c)** A 1-millimeter long polychaete worm (magnified 55 times) with its proboscis (mouth) extended.

Figure 16–14 Fiddler crab.

A male fiddler crab (*Uca*) among eelgrass (*Zostera*) at Cape Hatteras, North Carolina. The fiddler crab uses its large claw for protection and for attracting mates.

In tropical latitudes where conditions are favorable, the great beauty and biological diversity of the coral reef environment is found. Although this ecosystem exists in waters with low nutrient levels, organisms that flourish here have developed adaptations that make the most of available nutrients through a unique system of *symbiosis*.

Rocky Bottoms (Sublittoral)

A rocky bottom within the shallow inner **sublittoral** (*sub* = under, *littora* = the seashore) **zone** is usually covered with various types of marine macro algae. Along the North American Pacific coast, the giant brown bladder kelp (*Macrocystis*) attaches to rocks as deep as 30 meters (100 feet) and is attached to rocky bottoms with a rootlike anchor called a *holdfast* (Figure 16–15a) so strong only large storm waves can break the algae free. The *stipes* and *blades* of the algae are supported by gas-filled floats called *pneumatocysts* (*pneumato* = breath, *cystis* = bladder), which allow the algae to grow upward and extend for another 30 meters (100 feet) along the surface to allow for good exposure to sunlight. Under ideal conditions, *Macrocystis* can grow up to 0.6 meter (2 feet) per day.

The giant brown bladder kelp and bull kelp (*Nereocystis*), another fast-growing kelp, often form beds called **kelp forests** along the Pacific coast (Figure 16–15b). Smaller tufts of red and brown algae are found on the bottom and also live on the kelp blades.

Kelp forests are highly productive ecosystems that provide shelter for a wide variety of organisms living within or directly upon the kelp as epifauna. These organisms are an important food source for many of the animals living in and near the kelp forest, including mollusks, sea stars, fishes, octopus, lobsters, and marine mammals. Surprisingly, very few animals feed directly on the living kelp plant. Among those that do are the large sea hare (*Aplysia*) and sea urchins (Figure 16–15c). The distribution of kelp forests is shown in Figure 16–16.

STUDENTS SOMETIMES ASK...
What is an urchin barren?

An urchin barren is created when the population of sea urchins goes unchecked and they devour entire areas of giant brown bladder kelp (*Macrocystis*), one of the main types of algae in kelp forests. The urchins chew through the holdfast structure that holds the kelp in place and set it adrift. In California, the severe reduction of animals that prey on urchins (such as the wolf eel and sea otter) has upset the natural balance of ocean food webs. Consequently, sea urchins have proliferated and urchin barrens now exist where there were once lush kelp beds.

Lobsters Large crustaceans—including lobsters and crabs—are common along rocky bottoms. The spiny lobsters are named for their spiny covering and have two very large, spiny antennae (Figure 16–17a). These antennae serve as feelers and are equipped with noise making devices near their base that are used in protection. The genus *Panulirus*, which reaches lengths to 50 centimeters (20 inches), is considered a delicacy and lives in water deeper than 20 meters (65 feet) along the European coast. The Caribbean lobster (*Panulirus argus*) sometimes exhibits a remarkable behavior when it migrates single-file across the sea floor in lines that are several kilometers long.

Panulirus interruptus is the spiny lobster of the American West Coast. All spiny lobsters are taken for food, but none are as highly regarded as the so-called true lobsters (genus *Homarus*), which include the American lobster, *Homarus americanus* (Figure 16–17b). Although they are scavengers like their spiny relatives, the true lobsters also feed on live animals, including mollusks, crustaceans, and other lobsters.

Oysters Oysters are thick-shelled sessile (anchored) bivalve mollusks found in estuaries. They grow best where

(a)

(b)

(c)

Figure 16–15 Macrocystis and other kelp forest inhabitants.

(a) Structure of the giant brown bladder kelp (*Macrocystis*). (b) A kelp forest. (c) Sea hares
(*Aplysia californica*) and sea urchins (*Echinus*) in a kelp forest.

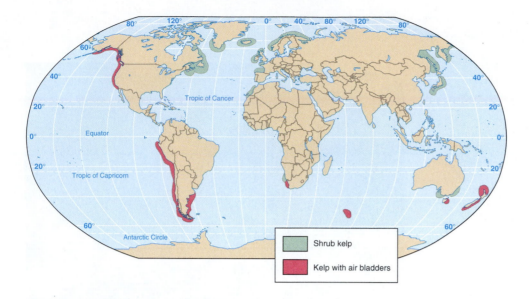

Figure 16–16 Worldwide distribution of kelp forests.

Map showing distribution of kelp forests. Shrub kelp (*green*) includes small smaller species of kelp such as *Sargassum* and rock weed (*Fucus, Pelvetia*). Kelp with air bladders (*red*) includes larger species such as giant brown bladder kelp (*Macrocystis*) and bull kelp (*Nereocystis*).

(a)

(b)

Figure 16–17 Spiny and American lobsters.

(a) Spiny lobster (*Panulirus interruptus*). **(b)** American lobster (*Homarus americanus*).

there is a steady flow of clean water to provide plankton and oxygen.

Oysters are food for sea stars, fishes, crabs, and snails that bore through the shell and rasp away the soft tissue inside (Figure 16–18). In fact, this may be one of the main reasons that oysters have such a thick shell.[3] Oys-

ters also have great commercial importance to humans throughout the world as a food source.

Oyster beds are composed of empty shells of many previous generations that are cemented to a hard substrate or to one another, with the living generation on top. Each female produces many millions of eggs each year, which become planktonic larvae when fertilized. After a few weeks as plankton, the larvae attach themselves to the bottom. As a material upon which to anchor, the oyster larvae prefer (in order): live oyster shells, dead oyster shells, and rock.

---

[3]This is an example of a pattern in nature where a type of armor or defense possessed by one species creates evolutionary pressure in another species for a weapon to defeat it.

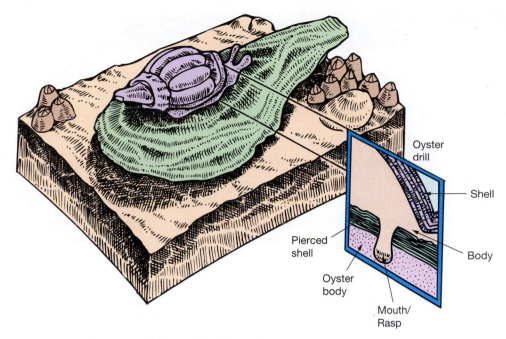

Figure 16–18 An oyster-drill snail feeding on an oyster.

An oyster-drill snail is able to drill through an oyster's shell with a rasp-like mouthpiece (*inset*) to feed on the soft tissue beneath the shell.

Coral Reefs

Individual corals—called **polyps** (*poly* = many, *pous* = foot)—are small benthic marine animals that feed with stinging tentacles and are related to jellyfish. Most species of corals are about the size of an ant, live in large colonies, and construct hard calcium carbonate structures for protection. Coral species are found throughout the ocean, but coral accumulations that are classified as **coral reefs** are restricted to shallow warmer-water regions.

Corals are very temperature sensitive. To survive, they need water where the average monthly temperature exceeds 18°C (64°F) throughout the year (Figure 16–19). If the water exceeds 30°C (86°F), however, many corals die, too. Warmer-than-normal sea-surface temperatures occur during El Niño events, which appear to be related to outbreaks of **coral bleaching** (Box 16–1).

Water warm enough to support coral growth is found primarily within the tropics. Reefs also grow as far north and south as 35 degrees latitude on the western margins of ocean basins, however, where warm-water currents raise average sea surface temperatures (Figure 16–19).

The map in Figure 16–19 also shows the greater diversity of reef-building corals on the western side of ocean basins. More than 50 genera of corals thrive in a broad area of the western Pacific Ocean and a narrow belt of the western Indian Ocean. Fewer than 30 genera, however, occur in the Atlantic Ocean, with the greatest diversity occurring in the Caribbean Sea. This pattern is related to the positions of the continents prior to about 30 million years ago, when the warm equatorial Tethys Sea connected the world's tropical oceans and provided a highway for the worldwide distribution of coral species and reef-associated organisms. With time, tectonic changes in landmass position closed the Tethys Sea and were accompanied by changes in ocean currents and climate, which reduced coral reef biodiversity in areas such as the Atlantic. Further, the presence of numerous tropical islands in the western Pacific favors speciation.

Besides warm water, other environmental conditions that allow for coral growth include the following:

- Strong sunlight (not for the corals themselves, which are animals and can exist in deeper water, but for a symbiotic photosynthetic microscopic algae called **zooxanthellae** that lives within the coral's tissues[4]).
- Strong wave or current action (to bring nutrients and oxygen).
- Lack of turbidity (suspended particles in the water tend to interfere with the coral's filter-feeding capability and absorb radiant energy, so corals are not usually found close to areas where major rivers drain into the sea).
- Salt water (corals die if the water is too fresh, which is another reason coral reefs do not form near the mouths of freshwater rivers).
- A hard substrate for attachment (corals cannot attach to a muddy bottom, so they often build upon the hard skeletons of their ancestors, creating coral reefs that are several kilometers thick).

---

[4]It is zooxanthella (*zoo* = animal, *xanthos* = yellow, *ella* = small) algae that give corals their distinctive bright coloration (which can be many colors besides yellow).

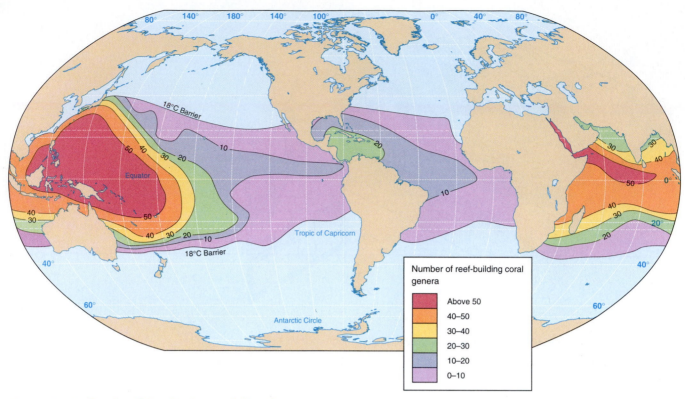

Figure 16–19 Coral reef distribution and diversity.

Coral reef development is restricted to warm tropical waters between the two 18°C (64°F) temperature lines. On the western side of each ocean basin, the coral reef belt is wider and the diversity of coral genera is greater, which is most likely related to surface circulation patterns and the presence of numerous tropical islands that favor speciation.

Coral are small colonial animals with stinging cells that are found primarily in shallow tropical waters and need strong sunlight, wave or current action, lack of turbidity, normal salinity seawater, and a hard substrate for attachment.

Because of changes in wave energy, salinity, water depth, temperature, and other less obvious factors, there is a well-developed vertical and horizontal zonation of the reef slope (Figure 16–20). These zones can be readily identified by the types of coral present and the assemblages of other organisms found in and near the reef.

Symbiosis of Coral and Algae "Coral reefs" are more than just coral. Algae, mollusks, and foraminifers contribute to the reef structure, too. Individual reef-building corals are **hermatypic** (*herma* = secret, *typi* = type) because they have a *mutualistic relationship*[5] with microscopic algae (zooxanthellae) that live within the tissue of the coral polyp. The algae provide their coral host with a

continual supply of food, and the corals provide the zooxanthellae with nutrients. Although coral polyps capture tiny planktonic food with their stinging tentacles, most reef-building corals receive up to 90% of their nutrition from symbiotic zooxanthellae algae. In this way, corals are able to survive in the nutrient-poor waters characteristic of the tropical oceans.

Other reef animals also have a symbiotic relationship with various types of marine algae. Those that derive part of their nutrition from their algae partners are called **mixotrophs** (*mixo* = mix, *tropho* = nourishment) and include coral, foraminifers, sponges, and mollusks (Figure 16–21). The algae not only nourish the coral but also may contribute to their calcification by extracting carbon dioxide from the coral's body fluids.

Coral reefs actually contain up to three times as much algal biomass as animal biomass. Zooxanthellae, for example, account for up to 75% of the biomass of reef-building coral. Nevertheless, zooxanthellae account for less than 5% of the reef's overall algal mass (most of the rest is filamentous green algae).

Because the algae need sunlight for photosynthesis, the greatest depth to which active coral growth extends is 150 meters (500 feet). Water motion is less at these depths, so relatively delicate plate corals can live on the outer slope of the reef from 150 meters (500 feet) up to

---

[5]See Chapter 14, "Biological Productivity and Energy Transfer," for a discussion of the types of symbiosis, including mutualism.

BOX 16–1   People and Ocean Environment

HOW WHITE I AM: CORAL BLEACHING AND OTHER DISEASES

Coral bleaching is the loss of color in coral reef organisms—often in response to an increase in water temperature—that causes them to turn white (Figure 16A). Bleaching occurs when the coral's symbiotic partner, the zooxanthellae algae, is removed or expelled. Once bleached, the coral no longer received nourishment from the algae and if the coral does not regain its symbiotic algae, it will eventually die. Bleaching often occurs in surface waters—the top two or three meters (7 to 10 feet)—but has recently been observed at depths of 30 meters (100 feet) and can occur as quickly as overnight. The coral does not die immediately, but is weakened and does not grow. Recovery from a minor bout of bleaching can take as little as four weeks, but severe bouts can take as long as four years.

Florida's coral reefs have experienced at least eight widespread bleachings since the early 1900s, and at least 70% of the corals along the Pacific Central American coast died due to bleaching associated with the severe El Niño event of 1982–1983. Coral reefs around the Galápagos Islands thrive in ocean water at or below 27°C (81°F). If the water is even 1 or 2°C (2 or 4°F) warmer for an extended period, however, the coral may expel the algae, in effect "bleaching" itself. The warming during 1982–1983 was so severe and long-lived that two species of Panamanian coral became extinct during this El Niño. The bleaching episode of 1987 affected coral reefs worldwide, especially those in Florida and throughout the Caribbean. Since then, widespread bleachings have occurred with increasing frequency and intensity. For instance, the El Niño event of 1997–1998 raised water temperatures several degrees higher than normal and has been blamed for the most geographically widespread bleaching ever recorded, including the equatorial eastern Pacific Ocean, the Yucatán coast, the Florida Keys, and the Netherlands Antilles.

Besides abnormally high surface water temperatures (such as El Niño events), the algae may leave their host due to pollution, elevated ultraviolet radiation levels, changes in salinity, invasion of disease, or a combination of factors. Some researchers believe that excess oxygen builds up in the coral's tissues and becomes toxic when temperatures are excessively warm. Then the algae are expelled or, perhaps, the algae leave with dead tissue. The strong correlation between coral bleaching and elevated water temperature concerns scientists, some of whom believe that coral bleaching may be one of the first oceanic indications of global warming. Whatever the reason, experts agree that bleaching indicates that the coral is experiencing severe environmental stress.

John Porter—a coral reef ecologist at the University of Georgia—and his colleagues study diseases that affect corals. They have been monitoring the health of corals in the Florida Keys since 1995 and have discovered the reappearance of *white plague disease* as well as a dozen new diseases, such as *white band disease, white pox, black band disease, yellow band (yellow blotch) disease, patchy necrosis*, and *rapid wasting disease*.

The cause of most of these diseases is still being investigated, and it is not known if the new diseases are from the invasion of microorganisms—bacteria, viruses, or fungi—or related to environmental stress as coral bleaching is. As human population has increased along the Florida Keys, the coral reefs of the Keys have begun to show signs of stress, thus making them more susceptible to a host of diseases. The increased nutrient levels and water turbidity resulting from soil runoff and improper sewage disposal in the Keys may contribute to the problem, too.

Easter Island

March 1999   **March 2000**

Figure 16A Normal coral (*Pocillopora verrucosa*) near Easter Island (*left*) and bleached coral one year later (*right*) after exceptionally high sea surface temperatures.

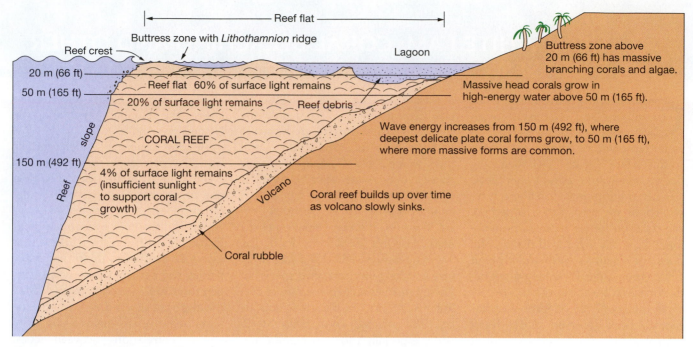

Figure 16–20 **Coral reef zonation.**
Coral reefs exhibit zonation because of the decrease in both wave energy and sunlight intensity with increasing ocean depth. As a result, massive branching corals occur above 20 meters (66 feet), where wave energy is great; corals become more delicate with increasing depth, until a depth of around 150 meters (500 feet), where too little solar radiation is available to allow the survival of their symbiotic zooxanthellae algae.

about 50 meters (165 feet), where light intensity is as low as 4% of the surface intensity (see Figure 16–20).

From 50 meters (164 feet) to about 20 meters (66 feet), water motion from breaking waves increases on the side of the reef facing into the prevailing current flow. Correspondingly, the mass of coral growth and the strength of the coral structure supporting it increase toward the top of this zone, where light intensity is as low as 20% of the surface value.

The reef flat may have a water depth of a few centimeters to a few meters at low tide, so it has at least 60% of the surface light intensity. Many species of colorful reef fish inhabit this shallow water, as well as sea cucumbers, worms, and mollusks. In the protected water of the reef lagoon live gorgonian coral, anemones, crustaceans, mollusks, and echinoderms (Figure 16–22).

> Corals are able to survive in nutrient-depleted warm water by living symbiotically with zooxanthellae algae, which live within the coral's tissues, provide it with food, and give the coral its color.

The Importance of Coral Reefs Coral reefs are some of the largest structures created by living creatures on Earth [the Great Barrier Reef, for example, is over 2000 kilometers (1250 miles) long]. Coral reefs foster a diversity of species that surpasses even that of tropical rain forests. Reefs provide shelter, food, and breeding grounds

for an estimated 35,000 to 60,000 species worldwide, including almost a third of the world's estimated 20,000 species of marine fishes. Other species that inhabit reefs include creatures as diverse as anemones, sea stars, crabs, eels, sea slugs, clams, sharks, and sponges. This is why marine biologists consider coral reefs the most diverse communities in the marine environment.

Coral reefs are also important to the economies of countries that have them. Many of these tropical countries receive over 50% of their gross national product as tourism related to reefs, which provides a much-needed incentive for these countries to protect them. Fisheries associated with reefs supply more than one-sixth of all fish from the sea. Recently, pharmacologists and marine chemists have discovered a storehouse of new medical compounds that fight maladies such as cancer and infections. In addition, reefs help to prevent shoreline erosion and protect coastal communities from waves and storms. The hard calcium carbonate skeletons of coral have even been used in some human bone grafts.

Coral Reefs and Nutrient Levels When human populations increase on land adjacent to coral reefs, the reefs deteriorate. Fishing, trampling, boat collisions with the reef, sediment increase due to development, and removal of reef inhabitants by visitors all damage the reef. One of the more subtle effects is the inevitable increase in the nutrient levels of the reef waters from sewage discharge and farm fertilizers.

(a)

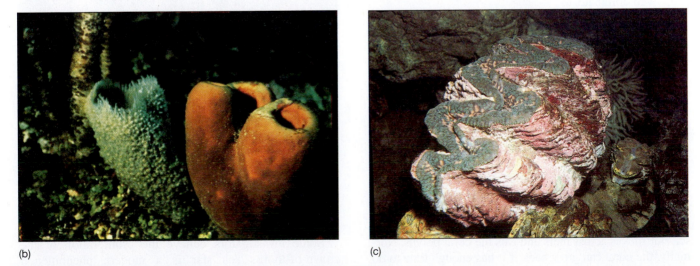

(b)   (c)

Figure 16–21 Coral reef inhabitants that rely on symbiotic algae.

(a) Coral polyps, which are nourished by internal zooxanthellae algae and also by extending their tentacles to capture tiny planktonic organisms from the surrounding water. (b) The blue-gray sponge *Niphates digitalis* (*left*), and the brown sponge *Angelas* (*right*), which contain symbiotic algae or bacteria. (c) A giant clam (*Tridacna gigas*), which depends on symbiotic algae living within its mantle tissue.

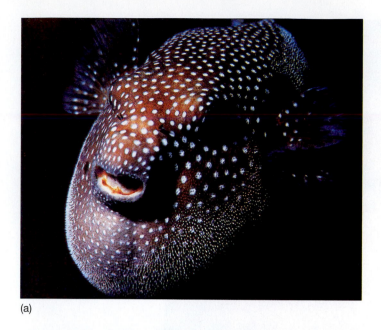

(a)

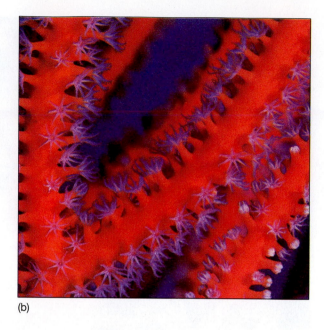

(b)

Figure 16–22 **Non-reef-building inhabitants.**

(a) Coral reefs supply habitat and protection for many fishes, including this puffer (*Arothron*). Puffers usually aren't quick enough to escape a predator, but they can expand their bodies to produce a large, spherical shape that cannot be easily eaten. **(b)** Unlike reef-building corals, some corals do not secrete a hard calcium carbonate structure, such as this soft gorgonian coral, which has feeding polyps (*purple*) extending from its branches.

As nutrient levels increase in reef waters, the dominant benthic community changes:

- At low nutrient levels, hermatypic corals and other reef animals that contain algal symbiotic partners thrive.
- Moderate nutrient levels favor fleshy benthic algae.
- At high nutrient levels, the phytoplankton mass exceeds the benthic algal mass, so benthic filter-feeders such as clams tied to the phytoplankton food web dominate.

Increased phytoplankton biomass reduces the clarity of the water, too, which interferes with the coral's filter-feeding capability. The fast-growing members of the phytoplankton-based ecosystem destroy the reef structure by overgrowing the slow-growing coral and through *bioerosion*, which is erosion of the reef by organisms. Bioerosion by sea urchins and sponges is particularly damaging to many coral reefs.

The Crown-of-Thorns Phenomenon The crown-of-thorns (*Acanthaster planci*) is a sea star (Figure 16–23) that has greatly proliferated and destroyed living coral on many reefs throughout the western Pacific Ocean since 1962. The sea star moves across reefs and eats the coral polyps. Normally, the coral can grow back if it has enough time to do so. Vast numbers of the crown-of-thorns sea stars upset the natural balance, however, decimating coral communities. Initially, divers were employed to smash the crown-of-

Figure 16–23 **Crown-of-thorns sea star.**

Crown-of-thorns sea stars (*Acanthaster planci*) have plagued Australia's Great Barrier Reef.

thorns, but sea stars (which have tremendous regenerating capabilities) can easily produce new individuals from various body parts, so it only made the problem worse.

Some investigators believe the proliferation of the crown-of-thorns sea star is a modern phenomenon brought about by the activities of humans, although there is little supporting evidence. Studies suggest that during the last 80,000 years, the crown-of-thorns sea star has

Figure 16–24 Spawning staghorn corals.

These staghorn corals (*Acropora*) on Australia's Great Barrier Reef spawn by releasing millions of egg and sperm directly into seawater.

been even more abundant on the reefs studied than it is today. Thus, the sea star may be an integral part of the reef ecology in this region, and their increase may be part of a long-term natural cycle rather than a destructive event triggered by human actions.

All reef corals can reproduce asexually by budding off new individuals. In addition, corals can also reproduce sexually either by *spawning* (releasing egg and sperm cells into water) (Figure 16–24) or by *brooding* (releasing larvae into water); most reproduce sexually by spawning. Remarkably, corals on Australia's Great Barrier Reef reproduce through spawning at a rate about 100 times greater than corals in the Caribbean Sea. Thus, Caribbean Sea reefs can be expected to recover from human-induced or naturally occurring disturbances much more slowly than those of the Great Barrier Reef.

The Deep-Ocean Floor

The vast majority of the ocean floor lies submerged below several kilometers of water. Less is known about life in the deep ocean than about life in any of the shallower nearshore environments because it is difficult and expensive to investigate the deep sea. Just to obtain samples from the deep-ocean floor requires a specially designed submersible or a properly equipped research vessel that has a spool of high-strength cable at least 12 kilometers (7.5 miles) long.

Collecting samples by submersible or with a biological dredge is a time-consuming process. Because the supply of oxygen is limited, manned submersibles can stay down for only 12 hours, and it may take eight of those hours to descend and ascend. To send a dredge to the deep-ocean floor and retrieve it from the depths takes about 24 hours.

Robotics and remotely operated vehicles (ROVs) are making it possible to observe and sample even the deepest reaches of the ocean more easily. Because they are unmanned, ROVs are cheaper to operate and can stay beneath the surface for months if necessary. These developments should lead to further discoveries in one of Earth's least-known habitats.

The Physical Environment

The deep-ocean floor includes bathyal, abyssal, and hadal zones.[6] Here, the physical environment is much different than at the surface, but it is quite stable and homogeneous. Light is present in only the lowest concentrations down to a maximum of 1000 meters (3300 feet) and absent below this depth. Everywhere the temperature rarely exceeding 3°C (37°F) and falls as low as −1.8°C (28.8°F) in the high latitudes. Salinity remains at slightly less than 35‰.[7] Oxygen content is constant and relatively high. Pressure exceeds 200 kilograms per square centimeter (200 atmospheres or 2940 pounds per square inch) on the oceanic ridges, exceeds 300 to 500 kilograms per square centimeter (300 to 500 atmospheres or 4410 to 7350 pounds per square inch) on the deep-ocean abyssal plains, and exceeds 1000 kilograms per square centimeter (1000 atmospheres or 14,700 pounds per square inch) in the deepest trenches.[8] Bottom currents are generally slow but more variable than once believed. **Abyssal storms** created by

[6]The bathyal, abyssal, and hadal zones are described in Chapter 13, "The Marine Habitat."

[7]Remember that average surface seawater salinity is 35 parts per thousand (‰).

[8]The pressure is 1 atmosphere (1 kilogram per square centimeter) at the ocean surface and increases by 1 atmosphere for each 10 meters (33 feet) of depth. Thus, a pressure of 1000 atmospheres is 1000 times that at the ocean's surface.

warm- and cold-core eddies of surface currents affect certain areas, lasting several weeks and causing bottom currents to reverse and/or increase in speed.

A thin layer of sediment covers much of the deep-ocean floor. On abyssal plains and in deep trenches, sediment is composed of mudlike abyssal clay deposits. The accumulation of oozes—composed of dead planktonic organisms that have sunk through the water column—occurs on the flanks of oceanic ridges and rises. On the continental rise, there may be some coarse sediment from nearby land sources. Sediment may be absent on steep areas of the continental slope. It may also be absent near the crest of the mid-ocean ridge and along the slopes of seamounts and oceanic islands, where it had not had enough time to accumulate on newly formed ocean floor.

Food Sources and Species Diversity

Because of the lack of light, photosynthetic primary production cannot occur. Except for the chemosynthetic productivity that occurs around hydrothermal vents, all benthic organisms receive their food from the surface waters above. Only about 1% to 3% of the food produced in the euphotic (sunlit) zone reaches the deep-ocean floor, so the scarcity of food that drifts down from the sunlit surface waters—not low temperature or high pressure—limits deep-sea benthic biomass. However, some variability in the supply of food is caused by seasonal phytoplankton blooms at the surface. Figure 16–25 shows the food sources for deep-sea organisms.

Many of the organisms that inhabit the deep sea have special adaptations to help them detect food using chemical clues. Once food is found, these organisms are efficient at consuming it (Box 16–2).

For many years, it was believed that the species diversity of the deep-ocean floor was quite low compared with shallow-water communities. Researchers studying sediment-dwelling animals in the North Atlantic, however, discovered an unexpectedly large diversity of species. An area of 21 square meters (225 square feet) contained 898 species, of which 460 were new to science. After analyzing 200 samples, new species were being discovered at a rate that suggested millions of deep-sea species!

It turns out that deep-sea species diversity—especially for small infaunal deposit feeders—rivals tropical rain forests. It also appears, however, that the distribution of deep-sea life is patchy and depends to a large degree on the presence of certain microenvironments.

> The deep-ocean floor is a stable environment of darkness, cold water, and high pressure but still supports life. The food source for most deep-sea organisms is from the sunlit surface waters.

Deep-Sea Hydrothermal Vent Biocommunities

An active hydrothermal (*hydro* = water, *thermo* = heat) vent field on the ocean floor was visited for the first time in 1977 during a dive of the submersible *Alvin*. The field exists in complete darkness in water below 2500 meters (8200 feet) in the Galápagos Rift, near the Equator in the eastern Pacific

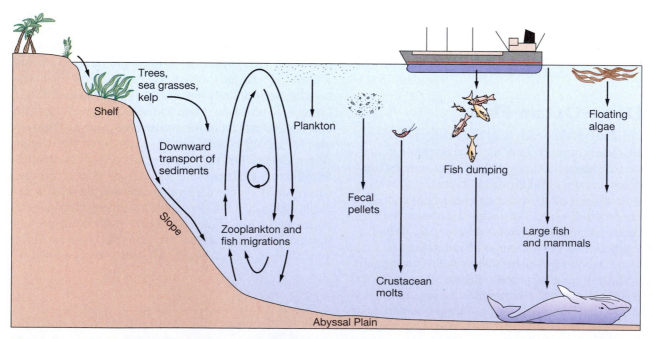

Figure 16–25 Food sources for deep-sea organisms.

Most deep-sea organisms obtain their food from surface water after it slowly settles through the water column to the sea floor. The food supply is usually limited, except when large fish or mammals (such as whales) sink to the bottom.

BOX 16–2   Research Methods in Oceanography

HOW LONG WOULD YOUR REMAINS REMAIN ON THE SEA FLOOR?

What happens to people who are buried at sea? How long do their remains remain on the sea floor? How long do the remains of a large organism such as a whale remain on the sea bottom? Oceanographers who study deep-sea biocommunities have conducted experiments in the deep sea to help answer these questions.

One experiment was conducted along the sea floor in the Philippine Trench at a depth of 9600 meters (31,500 feet) in 1975. Several whole fish were placed on the sea floor and an underwater camera positioned above them took a picture every few minutes to observe how long they remained there (Figure 16B). *Hirondellea gigas*, a scavenging benthic amphipod (a shrimplike animal), discovered the bait after only a few hours. The bait was swarming with amphipods after nine hours and it was stripped of flesh in as little as 16 hours! Other studies obtained similar results, suggesting that organisms the size of humans would have their soft tissue devoured within a day on the deep-ocean floor.

For deep-sea organisms, large food falls represent an unpredictable but intense nutrient supply. Deep-sea scavengers such as amphipods, hagfish, and sleeper sharks use special chemoreceptive sensory devices to identify and quickly locate food on the sea floor. Whale carcasses, for example, can support a thriving ecosystem of benthic organisms, including some species that inhabit hydrothermal vents.

To test how long a whale's remains lasted on the sea floor, researchers used two dead juvenile gray whales that washed up on the beach in southern California in 1996 and 1997. With permission from the National Marine Fisheries Service, the 5000-kilogram (5.5-ton) whales were intentionally weighted and sunk in the San Diego Trough. Researchers in a deep-diving submersible visited the carcasses at regular intervals and found that the gray whales were completely stripped of flesh in four months! Other studies indicate that even blue whales, which can weigh 25 times more than a gray whale, are stripped of their flesh in as little as six months.

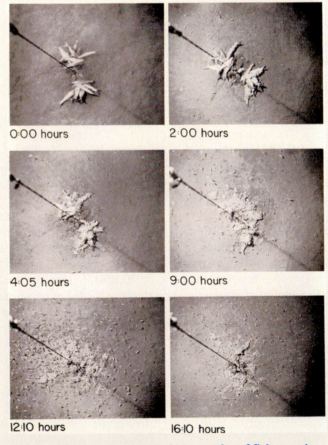

0:00 hours

2:00 hours

4:05 hours

9:00 hours

12:10 hours

16:10 hours

Figure 16B Time-sequence photography of fish remains on the deep-ocean floor.

Figure 16–26 *Alvin* **approaches a hydrothermal vent community.**
Schematic view of a hydrothermal vent area, showing lava pillows and a black smoker that spews hot
(350°C, or 662°F), sulfide-rich water from a chimney. Organisms (counterclockwise from *Alvin*)
include the grenadier fish (or rattail fish), octocoras, a sea anemone, white brachyuran crabs, large
clams (*Calypotogena*), and tube worms (*Riftia*).

Ocean (Figures 16–26 and 16–27). Water temperature near
the vents was 8 to 12° C (46 to 54° F), whereas normal water
temperature at these depths is about 2° C (36°F).

These vents supported the first known **hydrothermal
vent biocommunities**, consisting of organisms that were
unknown to science and unusually large for those depths.
The most prominent were tube worms over 1 meter (3.3
feet) long (Figure 16–29a), giant clams up to 25 centimeters (10 inches) long, large mussels, two varieties of
white crabs, and extensive microbial mats. These biocommunities have up to 1000 times more biomass than
the rest of the deep-ocean floor. In a region of scarce nutrients and small populations of organisms, these hydrothermal vents are truly the oases of the deep ocean.

In 1979, south of the tip of Baja California at 21 degrees north latitude on the East Pacific Rise, tall underwater chimneys were discovered belching hot vent water
(350° C or 662° F) so rich in metal sulfides that it colored
the water black. These chimney vents, which are composed primarily of sulfides of copper, zinc, and silver,
came to be called **black smokers**.

The most important members of these hydrothermal
vent biocommunities are microscopic **archaea** (*archaeo* =
ancient), which are simple bacteria-like organisms that

thrive on sea floor chemicals. Through **chemosynthesis**
(*chemo* = chemistry, *syn* = with, *thesis* = an arranging) of hydrogen sulfide, archaea manufacture carbohydrates from carbon dioxide and water (Figure 16–28) and
form the base of the food web for vent organisms. Although
some animals feed directly on archaea and larger prey,
many of them depend primarily on a symbiotic relationship
with archaea. The tube worms and giant clams, for instance,
depend entirely on sulfur-oxidizing archaea that live symbiotically within their tissues (Figure 16–29b).

In 1981, humans in a submersible first visited the Juan
de Fuca Ridge biocommunity offshore of Oregon. Although vent fauna at this site are less abundant than at the
Galápagos Rift and on the East Pacific Rise, the metallic
sulfide deposits from the vents aroused much interest because they are the only active hydrothermal vent deposits
in U.S. waters.

In 1982, the first hydrothermal vents beneath a thick
layer of sediment were discovered during a submersible
dive in the Guaymas Basin of the Gulf of California. In
this region, a spreading center is actively working to rift
apart the sea floor as it is being covered with sediment.
Sediment samples recovered in this region were high in
sulfide and saturated with hydrocarbons, which may have

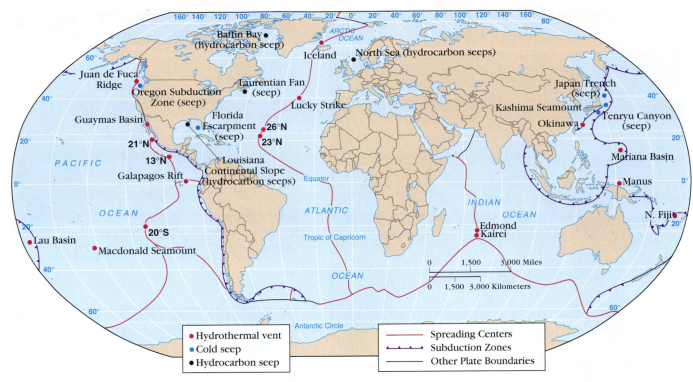

Figure 16–27 Vents and seeps known to support deep-sea biocommunities.

Location map of selected hydrothermal vents (*red*), cold seeps (*blue*), and hydrocarbon seeps (*black*). Major plate boundaries are also shown.

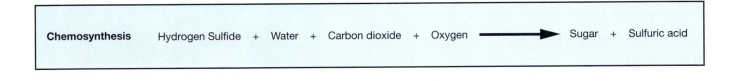

| Chemosynthesis | Hydrogen Sulfide | + | Water | + | Carbon dioxide | + | Oxygen | ⟶ | Sugar | + | Sulfuric acid |

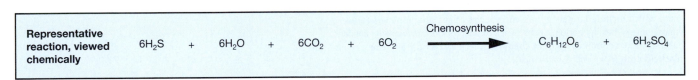

| Representative reaction, viewed chemically | $6H_2S$ | + | $6H_2O$ | + | $6CO_2$ | + | $6O_2$ | $\xrightarrow{\text{Chemosynthesis}}$ | $C_6H_{12}O_6$ | + | $6H_2SO_4$ |

Figure 16–28 Chemosynthesis (*top*) and representative reaction viewed chemically (*bottom*).

entered the food chain through bacteria. The abundance and diversity of life discovered here may exceed that of the Galápagos Rift and along the East Pacific Rise.

Like the Guaymas Basin, the Mariana Basin of the western Pacific has a small spreading center beneath a sediment-filled basin. A research dive in a submersible in 1987 revealed many new species of hydrothermal vent organisms (Figure 16–29c). Subsequent exploration has revealed numerous hydrothermal vent biocommunities in other parts of the Pacific Ocean (see Figure 16–27) and additional new species.

In 1985, the first active hydrothermal vents with associated biocommunities in the Atlantic Ocean were dis-

covered at depths below 3600 meters (11,800 feet) near the axis of the Mid-Atlantic Ridge between 23 and 26 degrees north latitude. The predominant fauna of these vents consists of shrimp that have no eye lens but can detect levels of light emitted by the black smoker chimneys that are invisible to the human eye (Figure 16–30).

In 1993, a hydrothermal vent community was discovered on a flat-topped volcano rising to 1525 meters (5000 feet)—well above the walls of the Mid-Atlantic Ridge rift valley. Called the "Lucky Strike" vent field, it is about 1000 meters (3300 feet) shallower than most other sites. It is the only Mid-Atlantic Ridge site to possess the mussels common

(a)

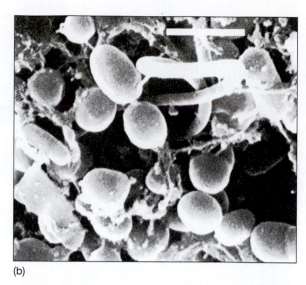

(b)

Figure 16–29 Chemosynthetic life.

(a) Tube worms up to 1 meter (3.3 feet) long are found at the Galápagos Rift and other deep-sea hydrothermal vents. **(b)** Sulfur-oxidizing archaea (enlarged 20,000 times; white bar at top = 1 micron) that live symbiotically within the tissue of tube worms, clams, and mussels found at hydrothermal vents. **(c)** A hydrothermal vent biocommunity from the Mariana Back-Arc Basin, which included a new genus and species of sea anemone (*Marianactis bythios*); a new family, genus, and species of gastropod (*Alviniconcha hessleri*; the first known conch to contain chemosynthetic bacteria); and the galatheid crab (*Munidopsis marianica*).

(c)

at many other vent sites and is the only location where a new species of pink sea urchin has been found.

In August 2000, Japanese investigators discovered the Indian Ocean's first hydrothermal vent biocommunity, which is associated with black smokers spewing water up to 365°C (689°F). The vents are covered by shrimp similar to those found in Atlantic fields while sea anemones mark the ambient temperature boundaries beyond. In between are clusters of animals similar to those found at other hydrothermal vents.

?STUDENTS SOMETIMES ASK...
Are the mussels found at hydrothermal vents edible?

Not for humans. The microbes that form the base of the food web use hydrogen sulfide gas (which has a characteristic "rotten egg" odor) as a source of energy. Sulfide, which is poisonous at low levels to most organisms, tends to concentrate in the tissues of these organisms. Although organisms within the hydrothermal biocommunities can ingest sulfide and have mechanisms of getting rid of it, hydrothermal vent organisms are toxic to humans. Even if they were edible, they would be expensive to harvest because they live at such great depths.

Figure 16–30 Atlantic Ocean hydrothermal vent organisms.

Swarm of particulate-feeding shrimp, the predominant animals observed at hydrothermal vents near 26 degrees north latitude on the Mid-Atlantic Ridge.

Life Span of Hydrothermal Vents   Because the hot-water plumbing of the sea floor is controlled by unpredictable volcanic activity associated with mid-ocean ridge spreading centers, a vent may remain active for only limited periods—years or sometimes decades. For instance, a hydrothermal vent field called the Coaxial Site along the Juan de Fuca Ridge offshore of Washington that had been active was revisited a few years later and found to be inactive. Inactive sites such as this one are identified by an accumulation of large numbers of dead hydrothermal vent organisms. When the vent becomes inactive and the hydrogen sulfide that serves as the source of energy for the community is no longer available, organisms of the community die if they cannot move elsewhere.

Other sites indicate an increase in volcanic activity. For example, at a site along the East Pacific Rise known as Nine-Degrees North, a large number of tube worms were cooked by lava flowing into their midst in what has been described as a "tube-worm barbecue." The discovery of newly formed and ancient vent areas along spreading centers indicates that hydrothermal vents can suddenly appear or cease to operate. Moreover, areas of active venting may lie hundreds of kilometers apart.

Hydrothermal vent organisms are well adapted to the temporary nature of hydrothermal vents. Most have high metabolic rates, for example, which cause them to mature rapidly so they can reproduce while the vent is still active.

Studies of several hydrothermal vent sites suggest that species diversity is low. In fact, only a little more than 300 animal species have been identified to date. Many species, however, are common to widely separated hydrothermal vent fields. Although hydrothermal vent animals typically release drifting larvae into the water, it is not clear how the larvae are able to survive the journey to hydrothermal vents that lie at such great distances from one another.

One idea, called the "*dead whale hypothesis*," suggests that when large animals die they may sink to the deep-ocean floor, decompose, and provide an energy source in stepping stone fashion for the larvae of hydrothermal vent organisms. The organisms settle and grow here, then breed and release their own larvae, some of which make it to the next hydrothermal vent field. Other researchers believe that deep-ocean currents are strong enough to transport drifting larvae to new sites. Still others have suggested that the rift valleys of mid-ocean ridges act as passageways along which drifting larvae traverse to inhabit new vent fields. By whatever means they travel, they colonize new hydrothermal vents soon after the vents are created. In 1989, for example, a newly formed hydrothermal vent along the Juan de Fuca Ridge had no life forms. By 1993, however, tube worms and other life forms had already established themselves.

Hydrothermal Vents and the Origin of Life   Life is thought to have begun in the oceans, and environments similar to those of the hydrothermal vents must have been present in the early history of the planet. The uniformity of conditions and abundant energy of the vents, therefore, has led some scientists to propose that hydrothermal vents would have provided an ideal habitat for the origin of life. In fact, hydrothermal vents may represent one of the oldest life-sustaining environments, because hydrothermal activity occurs wherever there are both volcanoes and water. The presence of archaea, which have ancient genetic makeup, helps support this idea.

Low-Temperature Seep Biocommunities

Three additional submarine seep environments—locations where water trickles out of the sea floor—have been found that chemosynthetically support biocommunities similar to hydrothermal vent communities.

Hypersaline Seeps   In 1984, a hypersaline seep was studied in water depths below 3000 meters (9800 feet) at the base of the Florida Escarpment in the Gulf of Mexico (Figure 16–31a). The water from this seep had a salinity of 46.2% but its temperature was not warmer than normal. Researchers discovered a **hypersaline seep biocommunity** similar in many respects to hydrothermal vent communities. The seeping water appears to flow from fractures at the base of a limestone escarpment (Figure 16–31b) and move out across the clay deposits of the abyssal plain at a depth of about 3200 meters (10,500 feet).

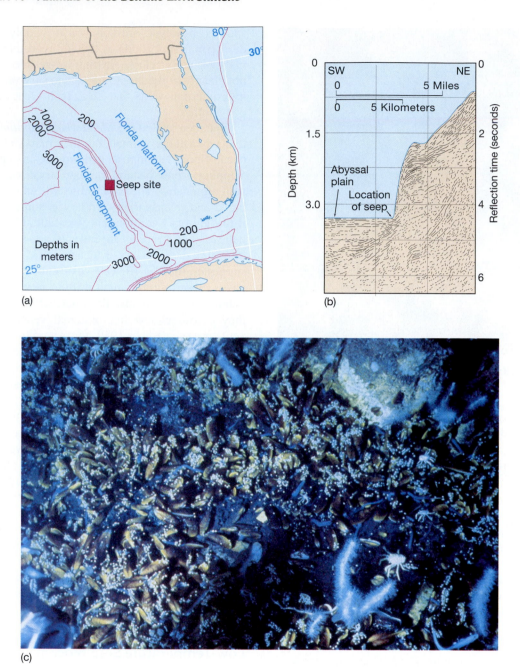

(a)

(b)

(c)

Figure 16–31 Hypersaline seep biocommunity at the base of the Florida Escarpment.

(a) Map showing location of hypersaline seep. **(b)** Seismic reflection profile of Florida Escarpment and abyssal sediments at its base. Arrow marks location of seep. **(c)** Florida Escarpment seep biocommunity of dense mussel beds. White dots are small gastropods on mussel shells. Tube worms (*lower right*) are covered with hydrozoans and galatheid crabs.

The hydrogen sulfide–rich waters support a number of white microbial growths called mats, which conduct chemosynthesis in a fashion similar to archaea at hydrothermal vents. These and other chemosynthetic archaea may provide most of the sustenance for a diverse community of animals that includes sea stars, shrimp, snails, limpets, brittle stars, anemones, tube worms, crabs, clams, mussels, and a few species of fish (Figure 16–31c).

Hydrocarbon Seeps   Also observed in 1984 were dense biological communities associated with oil and gas seeps on the Gulf of Mexico continental slope (Figure 16–32). Trawls at depths of between 600 and 700 meters (2000 and 2300 feet) recovered fauna similar to those observed at hydrothermal vents and at the hypersaline seep in the Gulf of Mexico. Subsequent investigations identified seeps with associated communities down to depths of 2200 meters (7300 feet) on the continental slope.

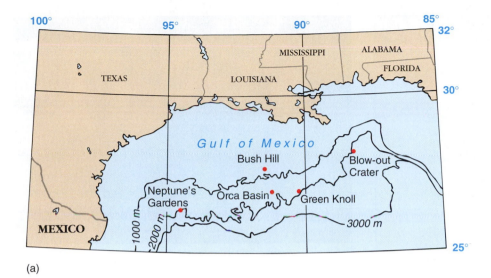

(a)

(b)

(c)

Figure 16–32 Hydrocarbon seeps on the continental slope in the Gulf of Mexico.

(a) Map showing locations of known hydrocarbon seeps with associated biocommunities.
(b) Chemosynthetic mussels and tube worms from the Bush Hill seep. **(c)** Alaminos Canyon site
(Neptune's Gardens) was discovered in 1990 and also contains mussels and tube worms.

Carbon-isotope analysis indicates that these **hydro-carbon seep biocommunities** are based on chemosynthesis that derives its energy from hydrogen sulfide and/or methane. Microbial oxidation of methane produces calcium carbonate slabs found here and at other hydrocarbon seeps (see Figure 16–27).

Subduction Zone Seeps In 1984, a **subduction zone seep biocommunity** was discovered during one of *Alvin's* dives to study folding of the sea floor in a subduction zone. The seep is located near the Cascadia subduction zone of the Juan de Fuca Plate at the base of the continental slope off the coast of Oregon (Figure 16–33a). The trench is filled with sediments, which are folded into a ridge at the seaward edge of the slope. At the crest of this ridge, water slowly flows from the 2-million-year-old folded sedimentary rocks into a thin overlying layer of soft sediment on the sea floor. Eventually, the water is released from the sediment through seeps on the ocean bottom.

At a depth of 2036 meters (6678 feet), the seeps produce water that is only about 0.3°C (0.5°F) warmer than seawater at that depth. The vent water contains methane that is probably produced by decomposition of organic material in the sedimentary rocks. Microbes oxidize the methane, chemosynthetically producing food for them-

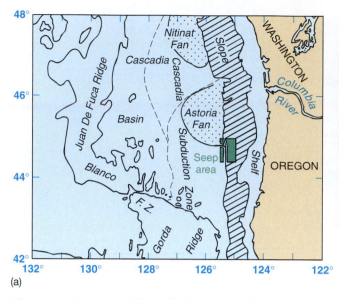

(b)

Figure 16–33 Subduction zone seep biocommunities.

(a) Map showing sea floor features and the location of vent communities off the coast of Oregon. These communities are associated with the Cascadia subduction zone, where sediment filling the trench is folded into a ridge with vents at its crest. **(b)** Giant white clams (*Calyptogena soyoae*) half buried in methane-rich mud at 1100 meters (3600 feet) near the Japan Trench. The clams host sulfide-oxidizing microorganisms, which oxidize methane and supply the clams with food.

selves and the rest of the community, which contains many of the same genera found at other vent and seep sites (Figure 16–33b).

Since the detection of subduction zone seeps, similar communities have been discovered in other subduction zones, including the Japan Trench and the Peru-Chile Trench. All these subduction zone seeps are located on the landward side of the trenches at depths from 1300 to 5640 meters (4265 to 18,500 feet).

> Hydrothermal vent biocommunities occur near black smokers and rely on chemosynthetic archaea for food. Other deep-sea biocommunities that depend on chemosynthesis exist around hypersaline, hydrocarbon, and subduction zone seeps.

The Deep Biosphere

Beneath the sea floor lies yet another frontier: a watery realm known as the **deep biosphere**. Only recently have scientists even considered that life might exist deep within Earth and in 2002, researchers made the first expedition to study life in this environment. Cores were drilled up to 420 meters (1380 feet) deep into the sea floor off Peru in water depths between 150 and 5300 meters (490 and 17,400 feet), which allowed the researchers to discover a host of microbes. These microbes live within circulating fluids that pass through the porous sea floor.

The abundance of microbes in the cores suggests that as much as two-thirds of Earth's entire bacterial biomass might exist in the deep biosphere. Further, if other bodies in the solar system have similar subsurface conditions, it is likely they might harbor microbes, too. As such, Earth's deep biosphere will continue to be an area of active research.

Chapter in Review

- *Over 98% of the 250,000 known marine species live in diverse environments within or on the ocean floor. Species diversity of these benthic organisms depends on their ability to adapt to the conditions of their environment*, particularly temperature. With few exceptions, the *biomass of benthic organisms closely matches that of photosynthetic productivity in surface waters above*.

- *Many adverse conditions exist in the intertidal zone of rocky shores, but organisms have adapted so they can densely populate these environments. Influenced by the tides, rocky shores can be divided into a high tide zone (mostly dry), a middle tide zone (equally wet and dry), and a low tide zone (mostly wet). The intertidal zone is bounded by the supralittoral (covered only by storm waves) and the sublittoral, which extends below the low tide shoreline.

- *Each of these zones contains characteristic types of life.* Periwinkle snails, rock louses, and limpets can be found in the supralittoral zone. Sessile organisms can be found in the high tide zone, especially buck-shot barnacles. *Algae become more abundant in the middle tide zone, and the diversity and abundance of the flora and fauna increase toward the lower inter-tidal zone.* Acorn barnacles, goose-necked barna-cles, mussels, and sea stars are commonly found in the middle tide zone, as well as sea anemones, fish-es, hermit crabs, and sea urchins. The low tide zone in temperate latitudes has a variety of moderately sized red and brown algae that provide a drooping canopy for animal life.

- *Many varieties of burrowing infauna are common along sediment-covered shores* (that is, beaches, salt marshes, and mud flats). Compared with rocky shores, however, *the diversity of species in sediment-covered shores is less.* As with the rocky shore, *the diversity of species and abundance of life on the sediment-covered shore increases toward the low tide shoreline.*

- In more protected segments of the shore, wave energy is lower, so sand and mud are deposited. Sand deposits are usually well oxygenated compared to mud de-posits. *The intertidal region of sediment-covered shores has high, middle, and low tide zones, similar to rocky shores.*

- Common *methods for feeding* on sediment-covered shores include *suspension feeding* (filtering planktonic organisms from the water), *deposit feeding* (ingesting sediment and detritus), and *carnivorous feeding* (prey-ing directly upon other organisms). Organisms charac-teristic of *sandy beaches* include bivalve (two-shelled) mollusks, lugworms, beach hoppers, sand crabs, sand stars, and heart urchins. Organisms characteristic of *mud flats* include eelgrass, turtle grass, bivalve mol-lusks, and fiddler crabs.

- *Attached to the rocky sublittoral bottom just beyond the shoreline is a band of algae that often creates kelp forests.* Kelp forests are the home of many organisms, including other varieties of algae, mollusks, sea stars, fishes, octopus, lobsters, marine mammals, sea hares, and sea urchins.

- *Spiny lobsters* are common to rocky bottoms in the Caribbean and along the West Coast, and the *American lobster* is found from Labrador to Cape Hat-teras. *Oyster beds* found in estuarine environments consist of individuals that attach themselves to the bot-tom or to the empty shells of previous generations.

- *Coral reefs consist of large colonies of coral polyps and many other species that need warm water and strong sunlight to live.* Coral reefs are usually *found in nutrient-poor tropical waters.* Reef-building corals and other mixotrophs are *hermatypic*, containing symbiotic algae (zooxanthellae) in their tissues. Delicate varieties are found at 150 meters, and they become more mas-sive near the surface, where wave energy is higher. *The potentially lethal "bleaching" of coral reefs is caused by the removal or expulsion of symbiotic algae*, proba-bly under stress of elevated temperature.

- *The physical conditions of the deep-ocean floor* are much different from those of shallow water. There is *no light* and the water is *uniformly cold. The primary food source is from the surface waters above, which limits biomass.* Species diversity in the deep ocean, however, is much higher than was previously thought.

- *Primary production in hydrothermal vent communities near black smokers is due to chemosynthesis.* Some ev-idence suggests that *hydrothermal vents may have been some of the first regions where life became established on Earth*, despite the short life span of individual vents. Chemosynthesis has also been identified in low-tem-perature seep biocommunities near *hypersaline, hydro-carbon*, and *subduction zone seeps.*

Key Terms

Abyssal storm (p. 479)	Crustacean (p. 467)	Hypersaline seep biocommunity (p. 485)	Polyp (p. 473)
Annelid (p. 467)	Deep biosphere (p. 488)	Infauna (p. 465)	Sessile (p. 462)
Archaea (p. 482)	Deposit feeding (p. 466)	Intertidal zone (p. 459)	Spray zone (p. 459)
Biomass (p. 458)	Echinoderm (p. 468)	Kelp forest (p. 470)	Subduction zone seep biocommunity (p. 487)
Bivalve (p. 466)	Epifauna (p. 459)	Low tide zone (459)	Sublittoral zone (p. 470)
Black smoker (p. 482)	Hermatypic (p. 474)	Meiofauna (p. 469)	Supralittoral zone (p. 459)
Carnivorous feeding (p. 466)	High tide zone (p. 459)	Middle tide zone (p. 459)	Suspension feeding (p. 466)
Chemosynthesis (p. 482)	Hydrocarbon seep biocommunity (p. 487)	Mixotroph (p. 474)	Zooxanthellae (p. 473)
Coral bleaching (p. 473)	Hydrothermal vent biocommunity (p. 482)	Mollusk (p. 466)	
Coral reef (p. 473)		Nematocyst (p. 463)	

Questions and Exercises

1. What are some adverse conditions of rocky intertidal zones? What are some organisms' adaptations to some of those adverse conditions? Which conditions seem to be the most important in controlling the distribution of life?

2. Draw a diagram of the zones within the rocky-shore intertidal region and list characteristic organisms of each zone.

3. Describe how sandy and muddy shores differ in terms of energy level, particle size, sediment stability, and oxygen content.

4. Describe the two types of feeding styles (other than predation) that are characteristic of rocky, sandy, and muddy shores. One of these feeding styles is rather well represented in all three types of shores; name it, and for each type of shore, give an example of an organism that uses it.

5. How does the diversity of species on sediment-covered shores compare with that of the rocky shore? Suggest at least one reason why this occurs.

6. In which intertidal zone of a steeply sloping, coarse sand beach would you find each of the following organisms: clams; beach hoppers; ghost shrimp; sand crabs; and heart urchins?

7. Discuss the dominant species of kelp, their epifauna, and animals that feed on kelp in Pacific coast kelp forests.

8. Describe the environmental conditions required for development of coral reefs.

9. Describe the zones of the reef slope, the characteristic coral types, and the physical factors related to zonation.

10. What is coral bleaching? How does it occur? What other diseases affect corals?

11. As one moves from the shoreline to the deep-ocean floor, what changes in the physical environment are experienced?

12. Where does the food come from to supply organisms living on the deep-ocean floor? How does this affect benthic biomass?

13. Describe the characteristics of hydrothermal vents. What evidence suggests that hydrothermal vents have short life spans?

14. What is the "dead whale hypothesis"? What other ideas have been suggested to help explain how organisms from hydrothermal vent biocommunities populate new vent sites?

15. What are the major differences between the conditions and biocommunities of the hydrothermal vents and the cold seeps? How are they similar?

CHAPTER 17
Marine Resources

Fishing vessel hauling in a catch. Fishing vessels are using increasing sophisticated techniques to catch fish and, as a result, many fish stocks are declining.

Key Questions

■ Who owns the ocean and the sea floor?
■ What problems do fisheries have?
■ What types of organisms are being farmed in mariculture operations?
■ What energy resources exist in the ocean?
■ What geologic resources exist in the ocean?
■ What chemical resources exist in the ocean?

The answers to these questions (and much more) can be found in the highlighted concept statements within this chapter.

"In the ocean depths, there exists mines of zinc, iron, silver, and gold which would be quite easy to exploit."

—Captain Nemo in Jules Verne's *Twenty Thousand Leagues Under the Sea* (1870)

Since ancient times, it was believed that the marine environment could provide society with an endless supply of food, minerals, and other resources. While this has been generally true in the past, rapid population growth and recent advances in technology have resulted in increased human exploitation of marine resources to the point that the oceans are in danger of being altered forever. Examples of marine resources that are being depleted include marine fisheries and offshore petroleum deposits.

In this chapter, we will discuss the legal framework of who owns the oceans before examining some of the more important marine resources. The pollution that has resulted from the removal of resources and the disposal of wastes into the oceans is addressed in Chapter 18, "Marine Environmental Concerns."

Laws and Regulations

Who owns the ocean? Who owns the sea floor? If a company wanted to drill for oil offshore between two different countries, would it have to obtain permission from either country? Extensive exploitation of the ocean floor for minerals and petroleum is well under way, necessitating laws that unambiguously answer these questions. In the future, exploration will occur beyond the jurisdiction of the country with the nearest coastline. Furthermore, overfishing and pollution are worsening. Are these kinds of problems covered by long-established laws? The answer is yes ... and no.

Mare liberum and the Territorial Sea

In 1609 Hugo Grotius, a Dutch jurist and scholar whose writings eventually helped formulate international law, urged freedom of the seas to all nations in his treatise *Mare liberum* (*mare* = sea, *liberum* = free), which was premised on the assumption that the sea's major known

resource—fish—exists in inexhaustible supply. Nevertheless, controversy continued over whether nations could control a *portion* of an ocean, such as the ocean adjacent to a nation's coastline.

Dutch jurist Cornelius van Bynkershoek attempted to solve this problem in *De dominio maris* (*dominio* = domain, *maris* = sea), published in 1702. It provided for national domain over the sea out to the distance that could be protected by cannons from the shore, an area called the **territorial sea**. Just how far from shore did the territorial sea extend? The British had determined in 1672 that cannon range extended 1 league (3 nautical miles) from shore. Thus, every country with a coastline maintained ownership over a *three-mile territorial limit* from shore.

Law of the Sea

In response to new technology that facilitated mining the ocean floor, the first **United Nations Conference on the Law of the Sea**, held in 1958 in Geneva, Switzerland, established that prospecting and mining of minerals on the continental shelf was under the control of the country that owned the nearest land. Because the continental shelf is that portion of the sea floor extending from the coastline to where the slope markedly increases, the seaward limit of the shelf is subject to interpretation. Unfortunately, the continental shelf was not well defined in the treaty, which led to disputes. In 1960, the second United Nations Conference on the Law of the Sea was also held in Geneva, but it made little progress toward an unambiguous and fair treaty concerning ownership of the coastal ocean.

Meetings of the third Law of the Sea Conference were held during 1973–1982. A new Law of the Sea treaty was adopted by a vote of 130 to 4, with 17 abstentions. Most developing nations that could benefit significantly from the treaty voted to adopt it. The United States, Turkey, Israel, and Venezuela opposed the new treaty because it made sea floor mining unprofitable. The abstaining countries included the Soviet Union, Great Britain, Belgium, the Netherlands, Italy, and West Germany, all of which were interested in sea floor mining, too. The treaty was ratified by the required sixtieth nation in 1993, establishing it as international law. Negotiations removed the objections of nations interested in sea floor mining, and the United States signed the revised treaty in 1994. However, the United States has not ratified the treaty although it has publicly indicated its intention to do so.

The primary components of the treaty are as follows:

1. **Coastal nations jurisdiction.** The treaty established a uniform 12-mile (19-kilometer) territorial sea and a 200-nautical-mile (370-kilometer) **exclusive economic zone (EEZ)** from all land (including islands) within a nation. Each of the 151 coastal nations has jurisdiction over mineral resources, fishing, and pollution regulation within its EEZ. If the continental shelf (defined geologically) exceeds the 200-mile EEZ, the EEZ is extended to 350 nautical miles (648 kilometers) from shore.

2. **Ship passage.** The right of free passage for all vessels on the high seas is preserved. The right of free passage is also provided within territorial seas and through straits used for international navigation.

3. **Deep-ocean mineral resources.** Private exploitation of sea floor resources may proceed under the regulation of the International Seabed Authority (ISA), within which a mining company will be strictly controlled by the United Nations. This provision, which caused some industrialized nations to oppose ratification, required mining companies to fund two mining operations—their own and one operated by the regulatory United Nations. Recently, this portion of the law was modified to eliminate some of the regulatory components, thus favoring free market principles and development by private companies. Still, this portion of the Law of the Sea has been one of the most contentious issues in international law.

4. **Arbitration of disputes.** A United Nations Law of the Sea tribunal will arbitrate any disputes in the treaty or disputes concerning ownership rights.

The Law of the Sea puts 42% of the world's oceans under the control of coastal nations. The EEZ of the United States consists of about 11.5 million square kilometers (4.2 million square miles) (Figure 17–1), which is about 30% more than the entire land area of the U.S. and its territories. This huge offshore area is widely believed to have tremendous economic potential.

> Ownership of the ocean and sea floor is regulated by the internationally ratified Law of the Sea, which gives nations control of waters immediately adjacent to their coasts.

Ecosystems and Fisheries

Since well before the beginning of recorded history, humans have used the sea as a source of food. Over the last several decades, **fisheries** (fish caught from the ocean by commercial fishers) have provided about 16% of the protein consumed by humans.

Fisheries harvest from the **standing stock** of a population, which is the mass present in an ecosystem at a given time. Successful fisheries leave enough individuals from the standing stock to repopulate the ecosystem after fisheries have made their harvest. **Overfishing** occurs when harvesting of fish stocks takes place so rapidly that the majority of the population is sexually immature and is therefore unable to reproduce. Predictably, overfishing results in the decline of marine fish populations and the reduction of a fishery's **maximum sustainable yield (MSY)** (the maximum fishery *biomass*[1] that can be removed yearly and still be sustained by the fishery ecosystem). Worldwide, about 30% of fish stocks are now depleted or overfished, while another 47% of fish stocks

---

[1] Remember that *biomass* is the mass of living organisms.

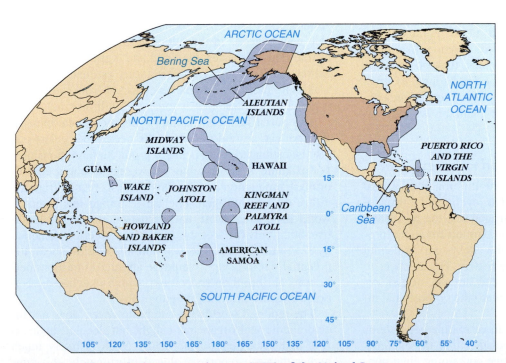

Figure 17–1 The exclusive economic zone (EEZ) of the United States.

The exclusive economic zone extends from shore to a distance of 200 nautical miles (370 kilometers) from the continent or islands. If the continental shelf (defined geologically) exceeds the 200-mile EEZ, the EEZ is extended to 350 nautical miles (648 kilometers) from shore.

are being fished at their biological limit. In U.S. waters, 80% of 191 commercial stocks are fully exploited or overfished. Overfishing can only be reversed by curtailing fish harvests.

Fish Recruitment and Survival

Recent large fluctuations in fish stocks that were not clearly related to fishing pressures such as overfishing convinced many fisheries biologists of the need for detailed study of fisheries ecology. One object of the studies was to produce a reliable measure of the survival rates of *larval* and *juvenile* fishes to determine how many young adult fish were added to a fishery each year. Other findings of these studies include the following:

- *Fish recruitment* (addition of young adult fish to the fishery) was found to depend on survival at critical stages in the life cycle of a fish. The mortality rate is very high for eggs and larvae. Mortality decreases in juvenile fish and reaches its lowest level in adults. Much less than 1% of larvae survive to become breeding adults.

- *Survival of fish larvae* depends on the availability of appropriate food, particularly after it uses up its yolk reserves. For instance, anchovy larvae require at least 20 phytoplankton cells per milliliter (0.03 ounce) of seawater within two and a half days, and the minimum cell size must be at least 40 microns (0.002 inch) in diameter.

- Once a fish survives the larval stage, *it still must survive a juvenile existence* before it can be recruited into the fishery. Juvenile fish may die from predation, disease, parasitism, genetic defects, and the effects of pollution. Availability of adequate food is still important because rapid growth is the best insurance against predation; if juvenile fish grow faster than their adult predators, they become too large to be suitable prey. The food source of juvenile fish may change as they grow, so that they are typically feeding on larger plankton as they increase in size.

Primary Productivity Effect on Fisheries

In most ecosystems, nitrogen is the limiting nutrient that curtails primary productivity. The cycling of nitrogen[2] within and the influx of new nitrogen into the ecosystem thus can be used to estimate the maximum amount of protein that can be removed from the ecosystem by fishing while still maintaining a healthy fishery. For a fishery to maintain (or increase) its size, the amount of fish removed must be equal to (or less than) the proportional influx of new nitrogen.

Recall from Chapter 14 that most marine ecosystems are supported by nutrients recycled within the *euphotic* (*eu* = good, *photos* = light) zone containing algal populations and by nutrients added from surrounding and deeper waters. In effect, the potential amount of fish that can be harvested without stressing the ecosystem can be determined by the amount of nitrogen input. Studies suggest that influx of new nitrogen accounts for less than 10% of the primary productivity in open-ocean *oligotrophic* (*oligo* = few, *tropo* = nourishment) waters, where-as it accounts for about 50% in areas of **upwelling**. Figure 17–2 shows the potential fishery that can result due to the presence of a large nutrient input from upwelling.

World Fishery

Figure 17–3 shows that the world marine fishery is drawn from five ecosystems. In descending order, they are (1) nontropical shelves, (2) tropical shelves, (3) upwellings, (4) coastal and coral systems, and (5) open ocean. The largest proportion of the marine fishery is found in highly productive shallow shelf and coastal waters, whereas low productivity open ocean areas comprise only 3.8% of the total. Nearly 21% of the world's total catch is from very highly productive upwelling areas, which represent only about 0.1% of the ocean surface area.

Figure 17–4 shows the world total fish production in marine waters since 1950. After increasing steadily for 35 years, the world catch of ocean fish finally leveled off in the 1990s. In 1998, in fact, it dipped to 79.2 million metric tons (87.1 million short tons).

How much fish can the oceans produce? The National Research Council estimated in 1999 that the maximum catch that might be expected from the oceans is about 100 million metric tons (110 million short tons). This amount might be exceeded, however, if new species—such as Antarctic krill, squid, and pelagic crabs—become significant new fisheries in the future. Based on current yields of traditional fisheries, the Food and Agriculture Organization (FAO) of the United Nations estimates that the maximum catch that might be expected from the oceans is about 120 million metric tons (132 million tons).

The **potential world fishery**—the theoretical mass of fish that can be removed from the oceans per year—can also be estimated using primary production relationships. For example, if one assumes an average of about 20% of primary production on a worldwide basis results from the influx of new nitrogen, one can estimate the potential world fishery. Such estimates must be considered tentative, but if only 0.1% of this primary production sustained by new nitrogen found its way into the world fishery, the potential world fishery would be about 100 million metric tons (110 million short tons), which matches the National Research Council's figure.

The reported worldwide catch, however, does not include additional organisms caught but deemed undesirable and discarded, so the actual catch is likely much higher than the reported value. Unfortunately, this suggests that the amount of biomass removed by fishing exceeds that produced by the influx of new nutrients, which means that the oceans are being overfished.

---

[2]For more details on the nitrogen cycle, see Chapter 14.

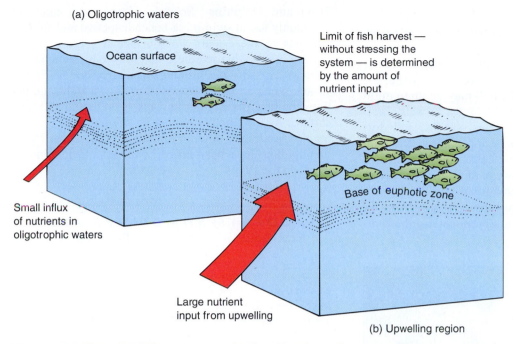

(a) Oligotrophic waters

Ocean surface

Limit of fish harvest —
without stressing the
system — is determined
by the amount of
nutrient input

Small influx
of nutrients in
oligotrophic waters

Base of euphotic zone

Large nutrient
input from upwelling

(b) Upwelling region

Figure 17–2 Potential fishery as a result of nutrient input from upwelling.

(a) A small amount of nutrient (nitrogen) input (*small red arrow*) into an oligotrophic open-ocean
ecosystem produces few fish. **(b)** A large input of nutrients (*large red arrow*) into the euphotic zone of an
upwelling region produces high productivity and a much greater fishery.

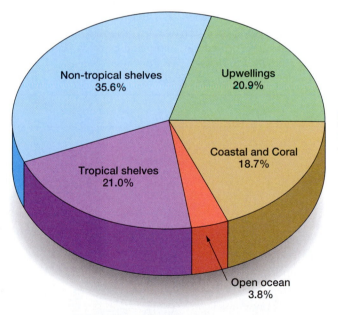

Non-tropical shelves
35.6%

Upwellings
20.9%

Coastal and Coral
18.7%

Tropical shelves
21.0%

Open ocean
3.8%

Figure 17–3 Marine fishery ecosystems.

Pie diagram showing the relative contribution of various ecosystems to
the total world marine fishery.

**Figure 17–4 World total fish production in
marine waters since 1950.**

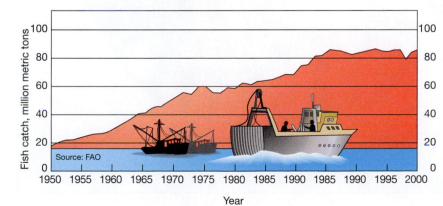

Source: FAO

Fish catch, million metric tons

Year

Incidental Catch

Incidental catch or **bycatch** includes any marine organisms that are caught incidentally by fishers seeking commercial species. On average, close to one-fourth of the catch is discarded, although for some fisheries, such as shrimp, the incidental catch may be up to eight times larger than the catch of the target species. Incidental catch includes birds, turtles, sharks, and dolphins, as well as many species of non-commercial fish. In most cases, these animals die before they are thrown back overboard, even though some of them are protected by United States and international law. Each year, an estimated 18 to 36 million metric tons (20 to 40 million short tons) of bycatch is produced by the fishing industry.

Figure 17–5 Spotted dolphin (*Stenella attnuata*), which are commonly associated with yellowfin tuna.

Tuna and Dolphins Schools of yellowfin tuna are commonly found swimming beneath spotted and spinner dolphins in the eastern Pacific Ocean (Figure 17–5). Fishers commonly used these dolphins to locate tuna and set a *purse seine net* around the entire school (Figure 17–6). When an underwater line is drawn tight, the net traps the tuna underwater as well as the dolphins at the surface. As the catch is hauled in, the dolphins become trapped, leading to dolphin mortality. In fact, over 7 million dolphins have been killed by the tuna fishing industry since the 1950s.

? STUDENTS SOMETIMES ASK ...
Is tuna labeled as "dolphin-safe" really safe to dolphins?

For the time being, yes. Tuna and tuna products harvested in the eastern tropical Pacific Ocean can only be labeled as "dolphin-safe" in the U.S. if no nets were intentionally set on dolphins during the fishing trip and no dolphins were killed or seriously injured during the set in which the tuna were caught. The U.S. National Marine Fisheries Service has developed an extensive monitoring, tracking, and verification program to ensure that only tuna and tuna products that meet the definition of dolphin-safe are indeed labeled as "dolphin-safe." This program along with consumer awareness has helped reduce yearly dolphin mortality to less than 2000 since 1998, down from 133,000 in 1986. Recently, however, changes have been proposed to the definition of "dolphin-safe" to address international trade concerns and allow more tuna to be imported from international sources that harm dolphins during netting of tuna.

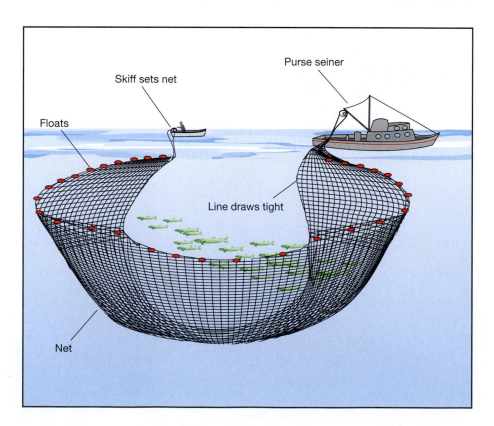

Figure 17–6 Purse seiner.

Purse seine nets are set around schools of yellowfin tuna, often trapping dolphins at the surface as bycatch.

The problem of dolphin deaths caused by tuna fishing was graphically presented to the world in March 1988 through video footage of dolphins struggling in tuna fishing nets taken by biologist Samuel F. La Budde. In 1990, under intense public outcry and the largest consumer boycott in U.S. history, the United States tuna canning industry declared that it would not buy or sell tuna caught by methods that kill or injure dolphins. In 1992, a special addendum was added to the **Marine Mammals Protection Act**, further protecting dolphins. The policy of the U.S. tuna canning industry has greatly reduced the practice of using dolphins to locate tuna. In addition, purse seine nets have been modified so that dolphins can be released safely.

Driftnets Another means of netting tuna and other species is by use of **driftnets** or **gill nets**, which are made of monofilament fishing line that is virtually invisible and cannot be detected by most marine organisms. Depending on the size of the holes in the net, it is highly effective at catching anything large enough to become entangled in it. As a result, driftnets often have high amounts of bycatch.

Up until 1993, Japan, Korea, and Taiwan had the largest driftnet fleets, deploying as many as 1500 fishing vessels into the North Pacific and setting over 48,000 kilometers (30,000 miles) of driftnets in one day. Although driftnetting was supposed to be restricted to specific fisheries, some fishers who claimed to be fishing for squid were involved in illegally taking large quantities of salmon and steelhead trout. Driftnetters were also targeting immature tuna in the South Pacific, which could result in the reduced abundance of South Pacific tuna. In addition, tens of thousands of birds, turtles, dolphins, and other species were killed annually in these nets.

A prohibition against the importation of fish caught in driftnets was included in the International Convention of Pacific Long Driftnet Fishing, signed in 1989. Although driftnets are now banned by international law, some fishers continue to use them illegally.

Fisheries Management

Fisheries management is the organized effort directed at regulating fishing activity with the goal of maintaining a long-term fishery. It includes assessing ecosystem health, determining fish stocks, analyzing fishing practices, and setting and enforcing catch limits. Unfortunately, however, fisheries management has historically been more concerned with maintaining human employment than preserving a self-sustaining marine ecosystem. For example, fisheries such as the anchovy, cod, flounder, haddock, herring, and sardine are suffering from overfishing in spite of being managed (Box 17–1).

One of the problems with fisheries management results from the fact that some fisheries encompass the waters of many different countries, including a variety of ecosystems. Some species of commercial fish, for instance, are raised in coastal estuaries and migrate long distances across international waters to their preferred environment. Fishing limits are difficult to enforce internationally, and if human interference occurs at any location where the fish exist, these species can be severely reduced in number.

Regulation of Fishing Vessels A major regulatory failure has been the absence of restrictions on the number of fishing vessels. Figure 17–7 shows that the number of decked (large) fishing vessels in the world more than doubled between 1970 and 1995. The slight decrease in 2000 is more a reflection of decreasing fish stocks than any limitation on fishing vessels. Many of these larger vessels use nets that can hold up to 27,000 kilograms (60,000 pounds) of fish in one haul. In addition, there are more than 1.6 million smaller nondecked fishing vessels, mostly in Asia and Africa.

The increase in fishing vessels has resulted in an increased fishing effort, which often leads to overfishing. In some locations, the fish are becoming so scarce that the fishing effort costs more than what the catch is worth! In 1995, for instance, the world fishing fleet spent $124 billion to catch $70 billion worth of fish. To make up for the shortfall, many governments have given subsidies to fishers, which compounds the problem by maintaining (or even *increasing*) the number of fishing vessels.

Fisheries in the Northwest Atlantic The effects of inadequate fisheries management are illustrated by the history of fisheries in the northwest Atlantic. Under management by the International Commission for the Northwest Atlantic Fisheries, the fishing capacity of the international fleet increased 500% from 1966 to 1976; the total catch, however, rose by only 15%. This was a significant decrease in the catch per unit of effort, which is a good indication that the fishing stocks of the Newfoundland–Grand Banks area were being overexploited. The biologists who recommended the total allowable fishing quotas for the major species within this region complained that enforcement was lacking and that countries' quotas were being bartered in a game of international politics such that they exceeded the total allowable catch set by the commission.

The difficulty of enforcing regulations by the international commission was largely responsible for Canada's unilateral decision to extend its right to control fish stocks for a distance of 200 nautical miles (370 kilometers) from its shores beginning January 1, 1977. The United States followed with a similar action only two months later.

Claiming coastal waters, however, proved to have limited effectiveness. After essentially all coastal nations assumed regulatory control over their coastal waters, things continued to deteriorate and overfishing became even more of a problem. The Canadian government, in fact, had to shut down the Grand Banks fishery

BOX 17–1 People and Ocean Environment

A CASE STUDY IN FISHERIES MISMANAGEMENT: THE PERUVIAN ANCHOVETA FISHERY

The waters off the coast of Peru have historically been filled with life, particularly a small silvery fish called the *anchoveta* or **anchovy** (*Engraulis ringens*), which swims through the water with its mouth open to catch its food (Figure 17A, *inset*). Human disruption of the anchovy ecosystem combined with periodic oceanographic changes resulted in depletion of anchovy stocks in 1972.

Off the coast of Peru, winds parallel to the coast from the southeast produce Ekman transport, which moves surface water away from shore and causes the upwelling of cold, nutrient-rich water from below (see Figure 8–11) that creates high productivity and an abundance of phytoplankton. Phytoplankton support a large population of zooplankton, which in turn are the food source for the anchovies. Anchovies are a vital link in the food web for larger marine animals such as birds, large fishes, marine mammals, and squid.

Recall from Chapter 8 that the waters offshore Peru periodically experience warm surface water, a deeper thermocline, and a reduction of upwelling during El Niño years. This reduction of upwelling severely affects productivity, limiting the number of anchovies and the abundance of many other marine organisms that feed on them. In 1972, a severe El Niño affected the area, decreasing the total biomass of anchovies to about 10% of its former level. Normally, anchovy populations had rebounded in the years following El Niño events, but failed to do so after the 1972 El Niño. What had prevented the repopulation of anchovies?

The sharp reduction of anchovies after 1972 may have resulted from a combination of the effects of El Niño and the establishment of the anchoveta fishing industry, which began in 1957. In 1960, the Intituto del Mar del Peru was created with the aid of the United Nations to study the anchoveta fishery and initiate a management program to avoid problems like those experienced by the Japan herring and California sardine fisheries, which had collapsed as a result of overexploitation. Still, the production of anchovies increased through the 1960s (Figure 17A).

Biologists had estimated that 10 million metric tons (11 million short tons) was the maximum sustainable yield of the anchoveta (Figure 17A, *green line*). However, this amount was exceeded in 1968, 1970, and 1971. Anchovy production peaked in 1970 at 12.3 million metric tons (13.5 million short tons), representing nearly one-quarter of the *entire* catch of fish from the sea worldwide! The reported amounts of anchovy catch may in fact be higher if processing losses, spoilage, and underreporting are considered.

In addition, there was a lack of understanding of the anchoveta fishery with respect to natural cycles. In January 1972, for example, the anchovy catch was set at 1.2 million metric tons (1.3 million short tons), which

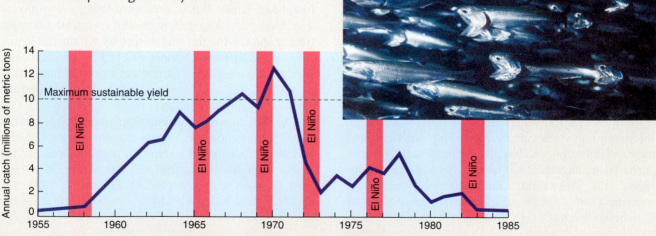

Figure 17A Peruvian anchovies (*inset*) and annual catch, 1955–1985.

Graph showing the annual catch of Peruvian anchovies, which severely declined after a peak production in 1970. Also shown are El Niño years (*red bars*) and estimated maximum sustainable yield (*green line*).

was reached easily in waters close to shore. The abundance of anchovies in nearshore waters raised expectations of a record harvest—in spite of the strengthening El Niño. As a result, the allotment for February 1972 was raised to 1.8 million metric tons (2.0 million short tons). The warm water of the 1972 El Niño, however, began to influence surface water temperatures. The combination of warm water plus an increased fishing effort severely reduced the abundance of anchovies, and in March, the fishery collapsed, leading to the loss of many fishers' livelihood.

Further, it wasn't realized that anchovies play a vital role in the health of the entire ecosystem. Studies have revealed that even with upwelling, ammonia (NH_3) from decomposition of large quantities of anchovy excrement is required to support enough phytoplankton to allow the anchovy population to expand. When anchovy populations are small and scattered, they do not provide much nutrient enrichment. When the population is larger, the whole region receives additional enrichment as part of a natural cycle. It will require a gradual increase in the anchovy population over many years to supply enough nutrients throughout the entire region for the ecosystem's health to be restored.

In the meantime, anchovies are gradually being replaced by sardines, which are a more commercially valuable variety of fish because sardines are used for direct human consumption (unlike anchovies, most of which are ground up and converted to fish meal that is used to feed poultry and hogs). Currently, Peru's fishing activity emphasizes species conversion, so the anchovy population may recover and be viable in the future if it is managed properly. Because the area was severely overfished immediately prior to an El Niño, however, it may never reach previous levels of anchovy abundance.

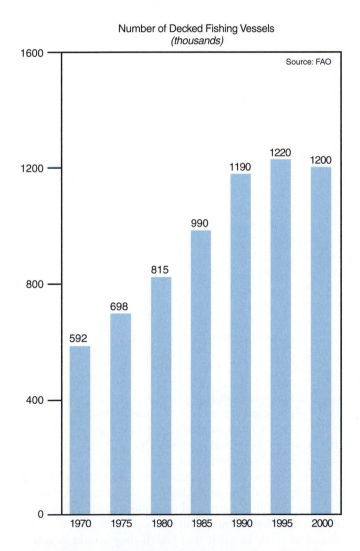

Number of Decked Fishing Vessels
(thousands)

Source: FAO

Figure 17–7 Number of decked (large) fishing vessels in the world (in thousands).

off Newfoundland, which resulted in the loss of about 40,000 jobs and subsequent government outlays of more than $3 billion in welfare. This cost far exceeds the value of the fishery, which generated no more than $125 million during one of its best years. A similar situation has developed on Georges Bank, which lies in Canadian and U.S. waters.

If fisheries are to remain viable in the future, marine ecosystems must be better studied to document the natural relationships among organisms in marine food webs, the critical environmental factors for the health of the fishery, and the effects of removing so many organisms as fishery catch. In addition, critical fish habitats must be protected and fishing limits must be upheld despite political factors. Only after such obstacles are overcome can a fishery be successfully managed.

To ensure sustainable fisheries and a healthy marine environment in U.S. waters, there has been a recent push to institute *ecosystem-based fishery management*, which employs a more comprehensive approach to understanding fish stocks including analysis of such variables as fish habitat, migration routes, and predator-prey interactions.

Seafood Choices

Consumer demand has driven some fish populations to the brink of disappearing. However, consumers can help by making wise choices in the fish they consume by purchasing only those fish that are from healthy, thriving fisheries. Certainly, some types of seafood carry less environmental impact than others because of differences in abundance, how they're caught, and how well fishing is managed. Table 17–1 lists some recommendations for seafood (both fish and shellfish) in three categories: safe to eat (*green*), use caution (*yellow*), and seafood to avoid (*red*).

TABLE 17–1 **Recommended seafood choices.**

Green: Abundant, relatively well-managed species	Yellow: Some concerns about a species' status, fishing methods, and/or management	Red: A species with problems such as severe depletion, overfishing, or poor management
Arctic char	Mahimahi, *longline-caught*	Yellowfin and bigeye tuna steak (Ahi), *longline-caught*
Spanish mackerel, *U.S. Atlantic and Gulf of Mexico*	Pacific cod	Atlantic flounders and soles (including grey sole)
Farmed clams, mussels, and scallops (including littleneck clams and bay scallops)	Snow, blue, and king crabs	Atlantic cod
White sturgeon caviar, *farmed*	Pacific flounder and soles	Groupers
Mahimahi, *troll-caught*	Black sea bass	Shrimp
Farmed oysters	Sea scallops	Skate
Alaska salmon	American ("Maine") lobster	Monkfish
Alaska halibut	Squid (calamari)	Sharks
Tilapia, *U.S.-farmed*	Swordfish	Farmed salmon (including Atlantic)
Stone crab	Canned tuna	Chilean seabass (toothfish)
Yellowfin, bigeye, albacore tuna, *pole/troll-caught*		Snappers
Striped bass		Caspian Sea caviar (beluga, osetra, sevruga)
Catfish		

The fishing industry suffers from overfishing, practices that produce a large amount of unwanted bycatch, and a lack of adequate fisheries management.

Mariculture

Marine aquaculture, or **mariculture** (*mari* = sea, *cultus* = to till), has been conducted for years throughout the Far East, making major contributions to the available food supply in that part of the world. Only recently has it expanded to other areas, mostly because of declining harvests of natural fisheries (despite increases in technology that give fishers the ability to catch wild fish more easily). Although mariculture is also fraught with various problems, it may offer a better way to serve consumer demand for marine products.

Figure 17–8 shows how aquaculture has increased dramatically in recent years. In 2000, worldwide combined marine and freshwater aquaculture accounted for about 35.6 million metric tons (38.5 short tons), up from 13.1 million metric tons (14.4 million short tons) in 1990. In 1990, all types of aquaculture represented only 16% of the total world fishery, while in 2000, it represented 37%

of the total. The figure also shows that mariculture typically represents about 40% of total aquaculture worldwide while inland fresh waters constitute the other 60%.

Organisms chosen for mariculture should be popular marine products that command a high price, be easy and inexpensive to grow, and be able to reach marketable size within a year or less. Candidates also should be hardy and resistant to disease and parasites. Severe economic problems can result if the chosen organism is not able to reproduce or cannot be brought to sexual maturity in captivity.

Fish

Mariculture with fish is difficult, so only a few countries have significant programs. Salmon, yellowtail, tuna, puffers, and a few other fish species are being raised in Japan on an experimental basis. Norway has pioneered the successful mariculture of Atlantic salmon in suspended pens (Figure 17–9), which has been emulated in other countries. In Washington, Oregon, and British Columbia, for example, it has replaced the failed attempt at salmon ranching, where young salmon were released into the ocean to grow large. It was hoped they would return to where they entered the ocean as wild salmon do; however, few did. This new mariculture activity is controversial in the Pacific Northwest because of the introduction of

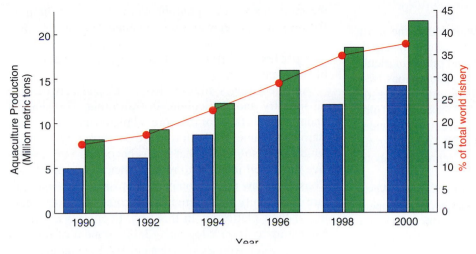

Figure 17–8 World marine and inland aquaculture production.

Bar graph showing the amount of marine (*blue*) and inland (*green*) aquaculture production. Note that mariculture comprises about 40% of total aquaculture while inland aquaculture comprises the other 60%. *Red line* shows how aquaculture is responsible for a steadily increasing amount of the total world fishery.

Figure 17–9 Salmon pen mariculture.

Mariculture salmon are raised in large floating pens like these in coastal British Columbia, Canada.

Atlantic salmon into Pacific waters. Although the technology does not involve releasing the salmon from their pens, some have managed to escape, fueling concern about the effect the escaped Atlantic salmon might have on indigenous populations.

In Alaska, good returns of salmon released by Alaska Fish and Game hatcheries have recently been reported. In some Canadian lakes, researchers have added algal nutrients (nitrates and phosphates) to develop a larger biomass of plankton, which in turn gives young salmon an abundance of food before they go to sea. This rather simple modification—which had no apparent ill effects on the ecology of the lakes—increased salmon production in the lakes and salmon returns from the open ocean.

Some estuary fish farming involving mullet and milkfish has been practiced successfully in the Far East for centuries. These fishes are hardy and tolerate salinity ranging from that of fresh water to full seawater. Unfortunately, they cannot be spawned and raised to sexual maturity in an artificial environment, so the fry must be collected from a natural nursery environment. Because the fish do not have to be artificially fed, however, the costs of operating the farms are relatively low. Typically, these mariculture operations can be quite profitable because they can produce yields of about 2.5 metric tons per hectare (1.1 short tons per acre).

STUDENTS SOMETIMES ASK...
Does aquaculture relieve some of the demand for wild fish?

Even though global production of farmed fish and shellfish has nearly tripled in the past 10 years and now provides over one-third of all fish directly consumed by humans, some types of aquaculture have actually *increased* the demand for wild fish. This is because the farming of carnivorous species such as salmon, tuna, cod, and seabass require large inputs of wild fish for feed. Some aquaculture systems further reduce wild fish supplies through habitat modification, the collection of wild fish for initially stocking aquaculture operations, and other negative ecological impacts including waste disposal, exotic species introduction, and pathogen invasions, all of which probably have contributed to the collapse of fishery stocks worldwide. If the aquaculture industry is to sustain its contribution to world fish supplies, it must reduce wild fish inputs in feed and adopt more ecologically sound management practices.

Crustaceans

Shrimp and prawn are, on a worldwide basis, the most valuable commodity produced by mariculture. In 2000, for example, the total value of shrimp and prawn mariculture production was almost $7 billion. Asian nations lead the way, with China producing about one-quarter of world production; Ecuador, Taiwan, Indonesia, Thailand, and the Philippines also are major producers. Worldwide, shrimp mariculture exceeds 450,000 metric tons (495,000 short tons) per year, which represents about 22% of the world supply.

The brightest spot in shrimp mariculture in the Western Hemisphere is Ecuador, where a very profitable operation is producing 45,000 metric tons (49,500 short tons) per year. The success in Ecuador—where the cost of labor and dealing with government regulation is low—contrasts with high costs in the United States, where a similar operation might have to cope with more than 35 regulating agencies and high labor costs. Nonetheless, it is clear that shrimp can be produced in mariculture operations at much lower cost than they can be provided by fishing. As a result, the world shrimp fishing fleet has been negatively impacted by the shrimp mariculture industry. The largest problem shrimp mariculture faces, however, is the production of larvae, which are susceptible to bacterial and viral infections.

Although lobsters are an attractive species for mariculture operations, raising them presents several problems. For instance, American lobsters (*Homarus americanus*) (see Figure 16–17b) can easily live through their complete life cycle in captivity, but they have a cannibalistic nature. To be successfully grown, these lobsters would have to be compartmentalized and maintained at an optimum temperature of about 20°C (68°F) for at least two years. In addition, artificial feeding would also be necessary. Encouraging results, however, have been achieved in rearing American lobster in the warm effluent from an electrical generating plant in Bodega Bay, California and similar projects are being tried in the coastal waters of Maine. The spiny lobster (*Panulirus interruptus*) (see Figure 16–17a) is not so easily reared through its long, complicated larval development, so as a result, less effort has been expended in developing it for commercial mariculture.

Bivalves

Cultivation of *bivalves* (*bi* = two, *valva* = a valve) such as oysters and mussels is one of the most successful forms of mariculture. Commercially, no bivalves have yet been reared from the larval stage to market size in captivity because sufficient phytoplankton to feed them cannot be economically supplied. Instead, hatchery-produced juveniles are allowed to mature in natural environments such as bays, estuaries, and other protected coastal waters where the mollusks feed on phytoplankton carried to them by the tides and other water movements. This simple procedure results in yields ranging from 25 to 2500 metric tons per hectare (10 to 1000 short tons per acre) of edible meat per year.

An interesting symbiosis has developed between the oil industry and mariculture in California. The shallow underwater surfaces of offshore oil platforms in the Santa Barbara Channel in California are colonized by mussels, which grow so rapidly that it was becoming very expensive for the oil industry to remove them. Ecomar Marine Consulting saw the oil companies' nuisance as a mariculture opportunity and offered the oil companies free cleaning of their platforms. In this way, Ecomar was able to harvest up to 1600 kilograms (3500 pounds) of mussels per day, which were sold to restaurants. Ecomar also cultures up to 50,000 scallops and oysters annually on trays suspended from oil platforms.

Algae

About 17% of mariculture production involves marine macro algae (seaweed), which has many uses. The most important commercial product is **algin**, which is made from the mucus that makes marine algae slick. The algin from seaweeds is collected and purified, making it available for many diverse applications such as an emulsifier and thickener in paint, printer's ink, and various food items including salad dressings. It is also in ice cream and many other frozen foods to prevent the formation of ice crystals. It is even used in the production of beer and wine. Because seaweeds have an abundance of trace minerals and high fiber, they are also eaten directly. For example, dried *nori* (a red algae) is used primarily as a wrap for sushi; the Japanese alone consume 150,000 metric tons (165,000 short tons) of it a year.

The life cycle of most of these algae is very complicated, and most of the successful operations are conducted in Japan, where spore-producing algae are cultivated in laboratories. When the algae release their spores into the water, local growers come to the laboratories to dip devices (typically poles or nets) into the water, giving the algae spores an attachment surface. Then, the attachment surfaces are submerged in estuarine waters, where the spores grow into adults using naturally available nutrients. In addition, marine algae can be harvested directly from the sea and dried on beaches (Figure 17–10).

Mariculture provides 37% of the total world fishery and includes such species as fish, crustaceans (shrimp and lobster), bivalves (oysters and mussels), and algae.

Figure 17–10 Algae mariculture.

Algae harvested from the ocean is dried on this beach in Lombok, Indonesia.

Energy Resources

The oceans contain a variety of renewable energy resources that, if properly harnessed, could provide society with vast amounts of power. The Sun, which imparts large amounts of radiant energy to Earth, is the ultimate source of all of these resources (except for tidal power, which relies on the pull of the Moon and the Sun to lift water). This energy imparted by the Sun is stored in the ocean and the atmosphere.

Can this type of solar energy be harnessed for use by humans? The potential for extracting energy from the movement and heat distribution patterns of the ocean and the atmosphere is attractive because

- It can be achieved without significant air or water pollution.
- The amount of energy available at any time is far greater than that in fossil fuels (coal, oil, natural gas) and nuclear fuel (uranium).
- The energy is renewable as long as the Sun continues to radiate energy.
- It can be collected any time because it depends on solar heat stored in the oceans and atmosphere over time, not on energy directly from the Sun.

Let's examine some of the main sources of renewable energy in the oceans.

Power from Offshore Winds

As discussed in Chapter 7, the uneven heating of Earth by the Sun drives various small- and large-scale winds. These winds, in turn, can be harnessed to turn windmills or turbines and generate electricity. Wind farms consisting of hundreds of large turbines mounted on tall towers can be found on land where the wind blows constantly, thereby taking advantage of this renewable, clean energy source. Similar facilities could be built offshore, where the wind blows harder and more steadily than on land.

In the U.S., persistent westerly winds off the coast of New England are an attractive target for a large *offshore wind power system (OWPS)* (Figure 17–11a). This floating system consists of dozens of individual turbines that could generate enough electricity to meet the needs of large areas of the U.S. North Atlantic coast. Across wind-swept northern Europe, hundreds of high-powered turbines are being planned in offshore wind farms and about 100 sea-based turbines are already operating (Figure 17-11b).

Power from Currents

Winds also drive ocean surface currents (see Chapter 8), which carry much more energy than winds because of the higher density of water as compared to air. As such, ocean currents can be used to drive turbans and generate electricity, too.

One location that has received much consideration as a site for harnessing power from ocean currents is the Florida–Gulf Stream Current System, which is a fast, western intensified surface current that runs along the east coast of the U.S. In fact, researchers have determined that at least 2000 million watts (2000 megawatts[3]) of electricity could be recovered from this ocean current system along the southeast coast of Florida.

Various devices have been proposed to extract the energy in ocean currents. For example, a series of large hydroturbines could be placed within a current and anchored to the ocean floor (Figure 17–12). Since each hydroturbine generates up to 43 megawatts of electricity, a grid of 1800 square kilometers (700 square miles) with 242 units could generate as much as 10,000 megawatts annually—the equivalent to the energy contained in about one-third of the present domestic inventory of crude oil in the United States. Other devices proposed for extracting energy from ocean currents include underwater windmills and a series of underwater parachutes attached to a conveyer belt. However, placing any moving machinery into the marine environment is problematic because seawater tends to corrode most materials.

Power from Waves

Moving water has a huge amount of energy, which is why there are so many hydroelectric power plants on rivers. Even greater energy exists in ocean waves (see Chapter 9),

---

[3]Each megawatt of electricity is enough to serve the energy needs of about 800 average U.S. homes.

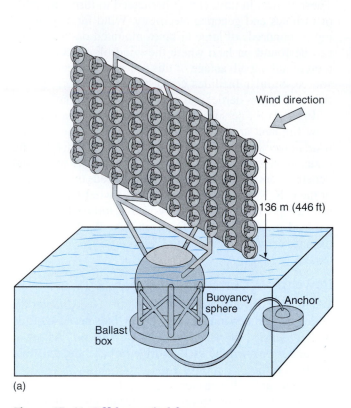

(a)

(b)

Figure 17–11 Offshore wind farms.

(a) An offshore wind power system (OWPS), which contains a large array of individual turbines that are driven by the wind and generate electricity. (b) Offshore wind turbines form a part of a wind farm in the North Sea off the coast of Blyth in the U.K.

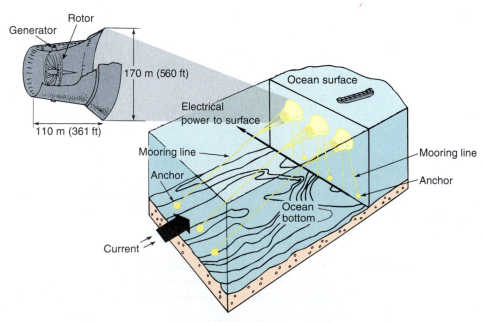

Figure 17–12 Power from ocean currents.

A prototype of an ocean current power system, which uses a series of large hydroturbines anchored to the ocean bottom. As a surface current flows past the turbines, it turns the rotors and generates electricity.

but significant problems must be overcome for the power to be harnessed efficiently. For example, a serious obstacle to the use of any device to harness wave energy is the monumental engineering problem of preventing the devices from being destroyed by the wave force they are built to harness.

Another key disadvantage of wave energy is that the system produces significant power only when large storm waves break against it, so the system could serve only as a power supplement. In addition, a series of one hundred or more of these structures along the shore would be required. Structures of this type could have a significant impact on the environment, with negative effects on marine organisms that rely on wave energy for dispersal, transporting food supplies, or removing wastes. Also, harnessing wave energy might alter the transport of sand along the coast, causing erosion in areas deprived of sediment.

Still, the immense power contained in waves could be used for generating electricity. Offshore wave generating plants would be able to tap into the higher wave energy found offshore, but they are more likely to be damaged in large waves and more difficult to maintain. The most promising locations for coastal power generation from waves are where waves refract (bend) and converge, such as at headlands, which tend to focus wave energy (see Figure 9–19b). Using this advantage, an array of wave power plants might extract up to 10 megawatts of power per kilometer (0.6 mile) of shoreline.

Internal waves are a potential source of energy, too. Along shores that have favorable ocean floor topography for focusing wave energy, internal waves may be effectively concentrated by refraction. This energy could power an energy-conversion device that would produce electricity.

In November 2000, the world's first commercial wave power plant began generating electricity. The small plant, called *LIMPET 500* (Land Installed Marine Powered Energy Transformer), is located on Islay, a small island off the west coast of Scotland. The plant was constructed at a cost of about $1.6 million and allows waves to compress air in a partially submerged chamber that, in turn, rotates a turbine for the generation of power (Figure 17–13). As waves recede, air is sucked back into the chamber and rotates the turbine in the other direction, which also generates power. Under peak operating capacity, the facility is capable of producing 500 kilowatts of power, which is capable of lighting about 400 homes. Economic conditions in the future may lead to the construction of larger wave plants that are capable of using this renewable source of energy.

Figure 17–14 shows the average wave height experienced along coastal regions and indicates the sites most favorable for wave energy generation (*red areas*). The map shows that west-to-east movement of storm systems in the mid-latitudes between 30 and 60 degrees north or south latitude causes the western coasts of continents to be struck by larger waves than eastern coasts. Thus, more wave energy is generally available along western than eastern shores. Furthermore, some of the largest waves (and greatest potential for wave power) are associated with the prevailing westerly wind belt in the mid-latitude Southern Hemisphere.

Power from Tides

Throughout history, ocean tides (see Chapter 10) have been used as a source of power. During high tide, water can be trapped in a basin and then harnessed to do work

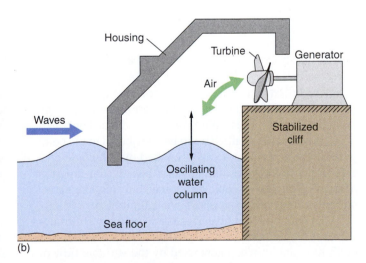

(a)   (b)

Figure 17–13 How a wave power plant works.

(a) LIMPET 500, the world's first commercial wave power plant. **(b)** Cutaway view of LIMPET 500, showing how the power plant generates electricity. As a wave advances into the power plant, it causes water to surge into the structure, forcing air out through a turbine. As the wave recedes, air is sucked back through the turbine. Electricity is generated as air moves both ways past the turbine as waves advance and recede.

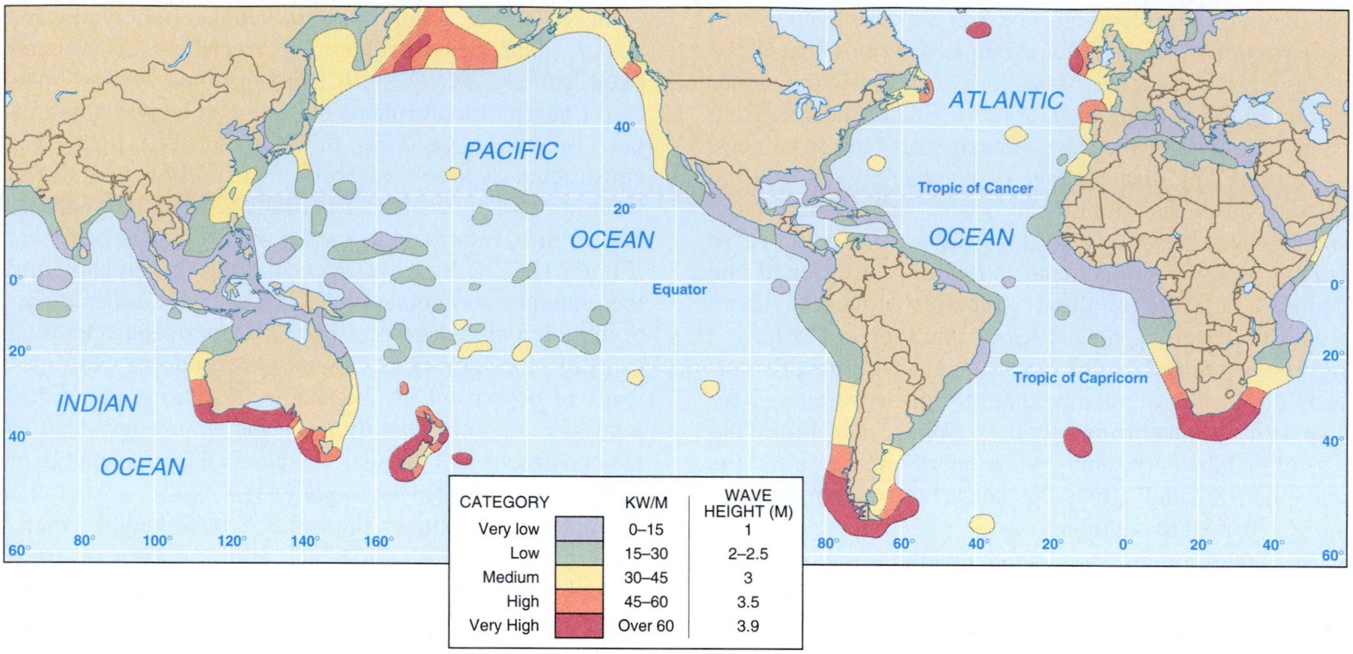

Figure 17–14 Global coastal wave energy resources.

Distribution of coastal wave energy shows that more wave energy is available along western shores of continents, especially in the Southern Hemisphere. KW/M is kilowatts per meter (for example, every meter of "red" shoreline is a potential site for generating over 60 kilowatts of electricity); average wave height is in meters.

as it flows back to the sea. In the 12th century, for example, water wheels driven by the tides were used to power gristmills and sawmills. During the 17th and 18th centuries, much of Boston's flour was produced at a tidal mill. Today, tidal water trapped in bays and estuaries can be used to turn turbines and generate electrical energy.

Tidal power has long been considered a renewable resource with vast potential. The initial cost of building a tidal power-generating plant may be higher than a conventional thermal power plant, but the operating costs would be less because it does not use fossil fuels or radioactive isotopes to generate electricity.

One disadvantage of tidal power, however, is the periodicity of the tides, allowing power to be generated only during a portion of a 24-hour day. People operate on a solar period, but tides operate on a lunar period, so the energy available from the tides would coincide with need only part of the time. Power would have to be distributed to the point of need at the moment it was generated, which could be a great distance away, resulting in an expensive transmission problem. The power could be stored, but this alternative presents a large and expensive technical problem.

To generate electricity effectively, electrical turbines (generators) need to run at a constant speed, which is difficult to maintain when generated by the variable flow of tidal currents in two directions (flood tide and ebb tide). Specially designed turbines that allow both advancing and receding water to spin their blades are necessary to solve the problem of generating electricity from the tides.

Another disadvantage of tidal power is unwanted environmental effects resulting from the modification of tidal current flow. In addition, a tidal power plant would likely interfere with many traditional uses of coastal waters, such as transportation and fishing.

Tidal Power Plants Currently, there are only a few small tidal power plants in the world. One successful tidal power plant is operating in the estuary of La Rance River near the English Channel in northern France (Figure 17–15). The estuary has a surface area of approximately 23 square kilometers (9 square miles), and the tidal range[4] is 13.4 meters (44 feet). Usable tidal energy increases as the area of the basin increases and as the tidal range increases.

The power-generating barrier was built across the estuary a little over 3 kilometers (2 miles) upstream to protect it from storm waves. The barrier is 760 meters (2500 feet) wide and supports a two-lane road (Figure 17–15). Water passing through the barrier powers 24 electricity-generating units that operate beneath the power plant. At peak operating capacity, each unit can generate 10 megawatts of electricity.

To generate electricity, the plant needs a sufficient water height between the estuary and the ocean—which only occurs about half of the time. Annual power production of about 540 million kilowatt-hours can be increased to 670 million kilowatt-hours by using the turbine generators as pumps to move water into the estuary at proper times.

---

[4]Recall from Chapter 10 that the *tidal range* is the vertical difference between high and low tides.

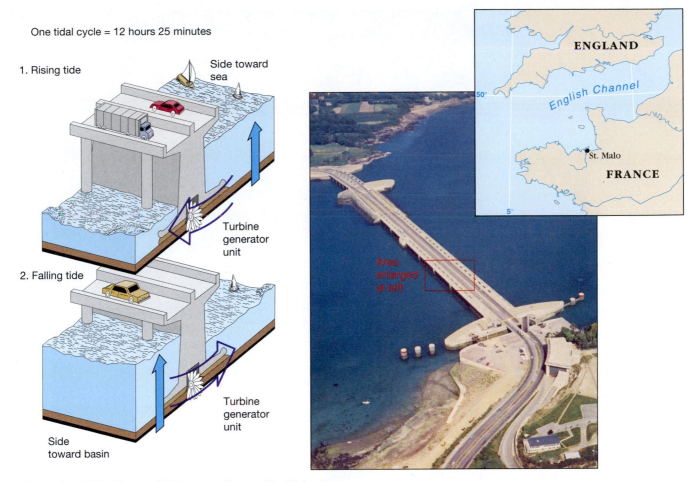

One tidal cycle = 12 hours 25 minutes

1. Rising tide

Side toward sea

Turbine generator unit

2. Falling tide

Turbine generator unit

Side toward basin

ENGLAND

English Channel

50°

St. Malo

FRANCE

5°

Area enlarged at left

Figure 17–15 La Rance tidal power plant at St. Malo, France.

Electricity is generated at the La Rance tidal power plant at St. Malo, France when water from a rising tide (1) flows into the estuary and turns turbines; electricity is also generated when water from a falling tide (2) exits the estuary and turns turbines in the other direction.

Within the Bay of Fundy, which has the largest tidal range in the world, the Canadian province of Nova Scotia has constructed a small tidal power plant that has generated up to 40 million kilowatt-hours per year since its completion in 1984. It is the only North American tidal power plant in operation today, and is built on the Annapolis River estuary, an arm of the Bay of Fundy (see Figure 10–24), where maximum tidal range is 8.7 meters (26 feet).

Some engineers think a tidal power plant could be made to generate electricity continually if it were located on the Passamaquoddy Bay near the U.S.–Canadian border at the south end of the Bay of Fundy. Although a tidal power plant across the Bay of Fundy has often been proposed, it has never been built. Potentially, the usable tidal energy seems great compared to the La Rance plant, because the flow volume is over 100 times greater.

China and Russia operate tidal power plants, too, and a few other countries are experimenting with various tidal power machines. For example, Norway and the U.K. are currently testing devices that harness swift coastal currents driven by the tides.

Whether or not tidal power stations are ever constructed on a large scale, this potential source of energy will likely receive increased attention as fossil fuels are depleted and the cost of generating electricity by conventional means increases. Worldwide, many sites have the potential for tidal power generation (Figure 17–16).

Power from Thermal Energy

Remarkably, the energy source with the greatest amount of energy potential is the heat stored in the warm surface layer of the oceans. It exists because of water's unique thermal properties (see Chapter 6).

The ocean occupies 90% of Earth's surface in the *tropics* (the region between the Tropics of Cancer and Capricorn). What makes this warm tropical surface water such an abundant source of energy is the presence of much colder water beneath the *thermocline*. For instance, the temperature difference between tropical surface water and water below the thermocline can be 20°C (36°F) or more. A temperature difference of at least 17°C (31°F) is enough to provide a source of energy that can be converted to electricity using an **ocean thermal energy conversion (OTEC)** system (Figure 17–17).

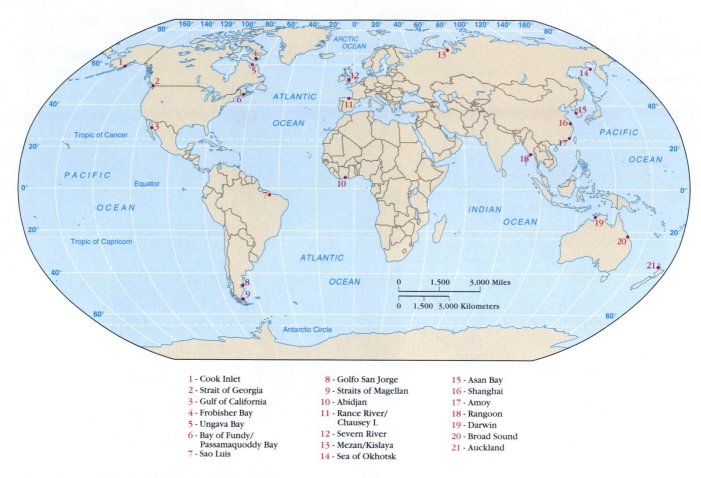

1 - Cook Inlet	8 - Golfo San Jorge	15 - Asan Bay
2 - Strait of Georgia	9 - Straits of Magellan	16 - Shanghai
3 - Gulf of California	10 - Abidjan	17 - Amoy
4 - Frobisher Bay	11 - Rance River/	18 - Rangoon
5 - Ungava Bay	Chausey I.	19 - Darwin
6 - Bay of Fundy/	12 - Severn River	20 - Broad Sound
Passamaquoddy Bay	13 - Mezan/Kislaya	21 - Auckland
7 - Sao Luis	14 - Sea of Okhotsk	

Figure 17–16 Sites with high potential for tidal power generation.

Twenty-one locations occur worldwide where tidal ranges are great enough to create potential for generating electricity. Where a large area exists for storing water behind a dam, a tidal range of 3 meters (10 feet) suffices; where smaller storage areas exist, greater tidal ranges are required.

An OTEC system works like a typical home refrigeration system but in the opposite direction and on a much larger scale. It converts heat energy stored in the ocean by using warm surface water to heat a fluid (such as liquefied propane gas or ammonia), which evaporates (Figure 17–17, *inset*). The vapor is directed through tubes and its increased volume allows it to turn a turbine, which powers an electrical generator that produces electricity. After passing through the turbine, cold water pumped up from the deep ocean condenses the fluid again, making it ready for heating by warm surface water to repeat the cycle.

By the late 1980s, a few small OTEC plants were tested using technology developed by the Japanese. In the U.S., only southern Florida and Hawaii have the necessary temperature difference to allow OTEC systems to successfully operate. In Florida, for example, an offshore warm surface current called the Gulf Stream provides an adequate temperature difference with colder deep water below.

An experimental OTEC plant on the island of Hawaii's Kona Coast began operating in the early 1980s, using cold seawater brought up from below the thermocline at a depth of 610 meters (2000 feet). Although the plant pro-

duced up to 210 kilowatts of electricity, it eventually shut down due to high operating expenses and problems associated with biofouling (corrosion due to marine organisms). Today, the *Natural Energy Laboratory of Hawaii Authority (NELHA)* still pumps cold seawater to the surface, where it is used for conducting biotechnology research and producing various commercial mariculture products such as algae, pearl oysters, American (Maine) lobster, and a variety of cold-water fish including halibut, black cod, and even seahorses. NELHA is also experimenting with using the cold seawater to chill roots and stimulate growth in crops by channeling the cold seawater through pipelines below the soil. Another benefit is the enhancement of fishing grounds due to artificial upwelling, which results from cold, deep, nutrient-rich water being brought to the surface.

Ocean energy resources include power from offshore winds, surface currents, waves, tides, and thermal energy.

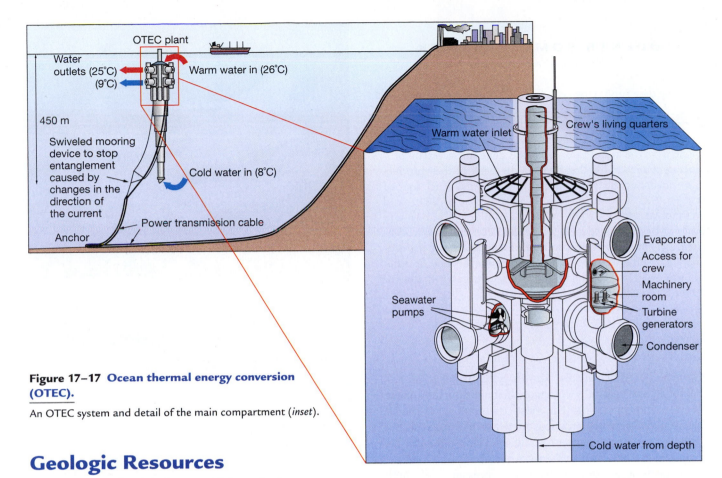

Figure 17–17 Ocean thermal energy conversion (OTEC).

An OTEC system and detail of the main compartment (*inset*).

Geologic Resources

The sea floor is rich in potential geologic resources. Much of it, however, is not easily accessible so the recovery of these resources involves technological challenges and high cost. Nevertheless, let's examine some of the most appealing exploration targets.

Petroleum

The ancient remains of microscopic organisms, buried within marine sediments before they could decompose, are the source of today's **petroleum** (oil and natural gas) deposits. Of the nonliving resources extracted from the oceans, more than 95% of the economic value is in petroleum products.

The percentage of world oil produced from offshore regions has increased from trace amounts in the 1930s to over 30% in the 2000s. Most of this increase results from continuing technological advancements employed by offshore drilling platforms (Figure 17–18). Major offshore reserves exist in the Persian Gulf, in the Gulf of Mexico, off southern California, in the North Sea, and in the East Indies. Additional reserves are probably located off the north coast of Alaska and in the Canadian Arctic, Asian seas, Africa, and Brazil. With almost no likelihood of finding major new reserves on land, future offshore petroleum exploration will continue to be intense, especially in deeper waters of the continental margins. However, a major drawback to offshore petroleum exploration is the inevitable oil spills caused by inadvertent leaks or blowouts during the drilling process.

Figure 17–18 Offshore drilling rig.

Constructed on tall stilts, rigs like this one in the Gulf of Mexico off Texas are important for extracting petroleum reserves from beneath the continental shelves.

Not any time soon. However, from an economic perspective, when the world runs completely out of oil—a finite resource—is not as relevant as when production begins to taper off. When this happens, we will run out of the *abundant* and *cheap* oil on which all industrialized nations depend. Several oil-producing countries are already past the peak of their production—including the United States and Canada, which topped out in 1972. Current estimates indicate that by 2010—and perhaps as soon as 2005—more than half of all known and likely to-be-discovered oil will be gone. After that, it will be increasingly more costly to produce oil and prices will rise dramatically—unless demand declines proportionately or other sources such as extra-heavy oil, tar sands, gas hydrates, or other sources become readily available.

(a)

Gas Hydrates

Gas hydrates, which are also known as **clathrates** (*clathr* = a lattice), are unusually compact chemical structures made of water and natural gas. The most common type of natural gas is methane, which produces **methane hydrate**. Gas hydrates occur beneath permafrost areas on land and under the ocean floor, where they were discovered in 1976.

In deep-ocean sediments, where pressures are high and temperatures are low, water and natural gas combine in such a way that the gas is trapped inside a latticelike cage of water molecules. Vessels that have drilled into gas hydrates have retrieved cores of mud mixed with chunks or layers of gas hydrates that fizzle and evaporate quickly when they are exposed to the relatively warm, low-pressure conditions at the ocean surface. Gas hydrates resemble chunks of ice, but ignite when lit by a flame because methane and other flammable gases are released as gas hydrates vaporize (Figure 17–19).

Most oceanic gas hydrates are created when bacteria break down organic matter trapped in sea floor sediments, producing methane gas with minor amounts of ethane and propane. These gases can be incorporated into gas hydrates under high pressure and low temperature conditions. Most ocean floor areas below 525 meters (1720 feet) provide these conditions, but gas hydrates seem to be confined to continental margin areas, where high productivity surface waters enrich ocean floor sediments below with organic matter.

Studies of the deep-ocean floor reveal that at least 50 sites worldwide may contain extensive gas hydrate deposits. Research suggests that at various times in the geologic past, changes in sea level or sea floor instability have allowed large quantities of methane to be released. The release of methane from the sea floor can affect global climate as methane—a potent greenhouse gas—increases in the atmosphere. Methane seeps also support a rich community of organisms, many of which are species new to science.

(b)

Figure 17–19 Gas hydrates.

(a) A sample retrieved from the ocean floor shows layers of white icelike gas hydrate mixed with mud. **(b)** Gas hydrates evaporate when exposed to surface conditions and release natural gas, which can be ignited.

Some estimates indicate that as much as 20 quadrillion cubic meters (700 quadrillion cubic feet) of methane are locked up in sediments containing gas hydrates. This is equivalent to about twice as much carbon as Earth's coal, oil, and conventional gas reserves combined (Figure 17–20), so gas hydrates may potentially be the world's largest source of usable energy. One of the major drawbacks in exploiting reserves of gas hydrate is that they rapidly decompose at surface temperatures and pressures. In the future, however, these vast sea floor reserves may be used to power modern society.

Sand and Gravel

The offshore sand and gravel industry is second in economic value only to the petroleum industry. Sand and gravel, which includes rock fragments that are washed

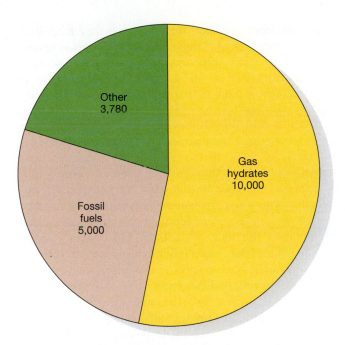

Figure 17–20 Organic carbon in Earth reservoirs.
Pie diagram showing that gas hydrates contain twice as much organic carbon as all fossil fuels combined. Values in billions of tons of carbon; "other" includes sources such as soil, peat, and living organisms.

out to sea and shells of marine organisms, are mined by offshore barges using a suction dredge. This material is primarily used as an aggregate in concrete, as a fill material in grading projects, and for beach replenishment.

Offshore deposits are a major source of sand and gravel in New England, New York, and throughout the Gulf Coast. Many European countries, Iceland, Israel, and Lebanon also depend heavily on such deposits.

Some offshore sand and gravel deposits are rich in valuable minerals. Gem quality diamonds, for example, are recovered from gravel deposits on the continental shelf offshore South Africa and Australia, where waves reworked them during low stands of the sea. Sediments rich in tin have been mined offshore southeast Asia from Thailand to Indonesia. Platinum and gold have been found in deposits offshore of gold mining areas throughout the world, and some Florida beach sands are rich in titanium. The largest unexplored potential for metallic minerals in offshore sand deposits may exist along the west coast of South America, where rivers have transported Andean metallic minerals.

Phosphorite (Phosphate Minerals)

Phosphorite is a sedimentary rock consisting of various phosphate minerals containing the element phosphorus, an important plant nutrient. Consequently, phosphate deposits can be used to produce phosphate fertilizer. Although there is currently no commercial phosphorite mining occurring in the oceans, the marine reserve is estimated to exceed 45 billion metric tons (50 billion short tons). Phosphorite occurs in the ocean at depths of less

than 300 meters (1000 feet) on the continental shelf and slope and is typically associated with upwelling, which hoists deep, phosphorus-enriched water to the surface.

Some shallow sand and mud deposits contain up to 18% phosphate. Many phosphorite deposits occur as nodules, with a hard crust formed around a nucleus. The nodules may be as small as a sand grain or as large as 1 meter (3.3 feet) in diameter and may contain over 25% phosphate. For comparison, most land sources of phosphate have been enriched to more than 31% by groundwater leaching.

Metal Sulfides

Metal sulfides are rich deposits of copper, lead, zinc, and silver that are sometimes found along the margins of tectonic plates and originate during plate tectonics processes (see Chapter 3).

One of the most common environments where metal sulfides form is at *divergent plate boundaries*, where ocean water enters fractures in oceanic crust and leaches metals from the rock. This metal-enriched hot water rises to the sea floor near the axis of the mid-ocean ridge and comes in contact with cold seawater, which causes the metal sulfides to precipitate. Metal sulfide ores have been discovered in spreading centers such as those in the Red Sea and along the Mid-Atlantic Ridge, the East Pacific Rise, and the Juan de Fuca Ridge. The Troodos Massif on the island of Cyprus in the Mediterranean Sea is an ancient example of a rich copper sulfide ore that formed along an ancient mid-ocean ridge and was later thrust above sea level (see Box 3–2).

Metal sulfide deposits are also found along *convergent plate boundaries*, where descending metal-rich ocean crust melts and fluids rich in metals rise. These metals precipitate into overlying volcanic rock and create ore bodies such as those found in the Andes Mountains. In addition, similar hydrothermal mineralization processes occur at submerged *fore-arc volcanoes* and associated *back-arc spreading centers*.[5] Many of these metal sulfide deposits are economically more promising than those at mid-ocean ridges because of higher precious metal content (because of larger magmatic input) and shallower water depths.

Manganese Nodules and Crusts

Manganese nodules are rounded, hard, golf-to tennis-ball-sized lumps of metals that contain significant concentrations of manganese, iron, and smaller concentrations of copper, nickel, and cobalt, all of which have a variety of economic uses. In the 1960s, mining companies began to assess the feasibility of mining manganese nodules from the deep-ocean floor (Figure 17–21). The map in Figure 17–22 shows that vast areas of the sea floor contain manganese nodules, particularly in the Pacific Ocean.

The formation of manganese nodules has puzzled oceanographers since they were first discovered in 1872 by

---

[5]For more details about fore-arc volcanoes and back-arc spreading centers, see Chapter 4.

Figure 17–21 Mining manganese nodules.

Manganese nodules can be collected by dredging the ocean floor. This metal dredge is shown unloading nodules onto the deck of a ship.

the British research vessel HMS *Challenger*.[6] For example, if manganese nodules are truly a type of *hydrogenous* (*hydro* = water, *generare* = to produce) *sediment* and

precipitate from seawater, then how can they have such high concentrations of manganese (which occurs in seawater at concentrations often too small to measure accurately)? Furthermore, why are the nodules on *top* of ocean floor sediment and not buried by the constant rain of sedimentary particles?

Unfortunately, nobody has definitive answers to these questions. Perhaps the creation of manganese nodules is the result of one of the slowest chemical reactions known—on average, they grow at a rate of about 5 millimeters (0.2 inch) per *million* years. Recent research suggests the formation of manganese nodules may be aided by bacteria and an as-yet-unidentified marine organism that intermittently lifts and rotates them. Other studies reveal that the nodules don't form continuously over time but in spurts that are related to specific conditions such as a low sedimentation rate of lithogenous clay and strong deepwater currents. Interestingly, the larger the nodules are, the faster they grow. The mystery of manganese nodules is awaiting someone to investigate what is considered the most interesting unresolved problem in marine chemistry.

Technologically, mining the deep-ocean floor for manganese nodules is possible. However, the political issue of determining international mining rights at great distances from land has hindered exploitation of this resource. Additionally, environmental concerns about mining the deep-ocean floor and removing this essentially nonrenewable resource have not been fully addressed.

Of the five metals commonly found in manganese nodules, cobalt is the only metal deemed "strategic" (es-

[6]For more information about the many accomplishments made during the HMS *Challenger* Expedition, see Chapter 1.

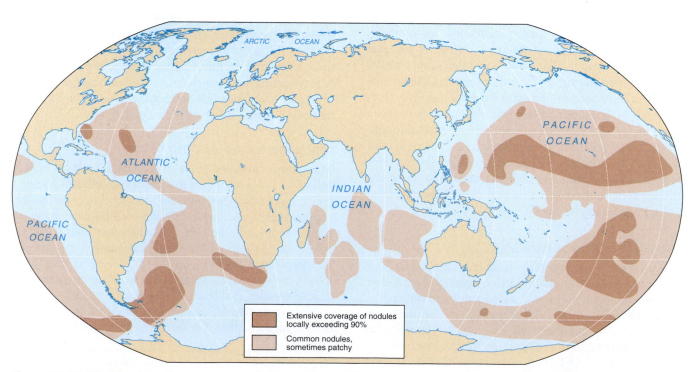

Figure 17–22 Distribution of manganese nodules on the sea floor.

sential to national security) for the United States. It is required to produce dense, strong alloys with other metals for use in high-speed cutting tools, powerful permanent magnets, and jet engine parts. At present, the United States must import all of its cobalt from large deposits in southern Africa. However, the United States has considered deep-ocean nodules and **crusts** (hard coatings on other rocks) as a more reliable source of cobalt.

In the 1980s cobalt-rich manganese crusts were discovered on the upper slopes of islands and seamounts that lie relatively close to shore and within the jurisdiction of the United States territories in the western equatorial Pacific. The cobalt concentrations in these crusts are half again as rich as in the best African ores and at least twice as rich as in the deep-sea manganese nodules. However, interest in mining these deposits has faded because of lower metal prices from land-based sources.

> Ocean geologic resources include petroleum (oil and natural gas), gas hydrates, sand and gravel (including deposits of valuable minerals), phosphorite, metal sulfides, and manganese nodules and crusts.

Chemical Resources

Some of the ocean's most important chemical resources include fresh water, evaporative salts, and drugs from the sea.

Fresh Water

Earth's growing population uses fresh water in greater volumes each year. As fresh water becomes scarcer, several countries have begun to use the ocean as a source of water. **Desalination**, or salt removal from seawater, can provide fresh water for business, home, and agricultural use. Today, there are more than 12,500 desalination plants in existence worldwide that provide about 1% of the world's drinking water. The majority of these plants are located in the Middle East, the Caribbean, and the Mediterranean. The United States produces only about 10% of the world's desalted water, with the majority produced in Florida and California. More than half of the world's desalination plants use distillation to purify water, while the remaining plants use various membrane processes.

Distillation Distillation is shown schematically in Figure 17–23. Salt water is boiled and the resulting water vapor is passed through a cooling condenser, where it condenses and is collected as fresh water. This simple procedure is very efficient at purifying seawater. For instance, distillation of 35‰ seawater produces fresh water with a salinity of only 0.03‰, which is about 10 times fresher than bottled water so it needs to be mixed with less pure water to make it taste better. Distillation is expensive, however, because it requires large amounts of heat energy to boil the salt water. Because of water's high latent heat of vaporization, it takes 540 calories to convert only 1 gram (0.04 ounce) of water at the boiling point to the vapor state.[7] Increased efficiency, such as using the waste heat from a power plant, is required to make distillation practical on a large scale.

Solar humidification or **solar distillation** does not require supplemental heating and has been used successfully in small-scale agricultural experiments in arid regions such as Israel, West Africa, and Peru. Solar

---

[7]Assuming 100% efficiency, it takes a whopping *540,000 calories* of heat energy to make a half-liter (about 1 pint) bottle of distilled water.

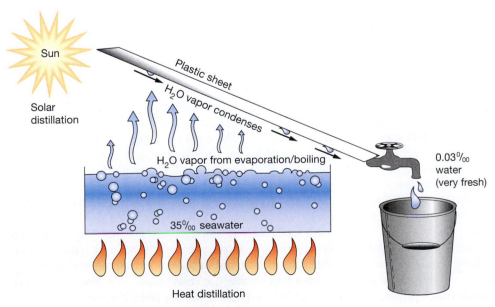

Figure 17–23 Distillation.

The process of distillation requires boiling saltwater (heat distillation) or using the Sun's energy to evaporate seawater (solar distillation). In either case, the water vapor is captured and condensed, which produces very pure water.

Sun

Solar distillation

Plastic sheet

H_2O vapor condenses

H_2O vapor from evaporation/boiling

35‰ seawater

Heat distillation

0.03‰ water (very fresh)

humidification is similar to distillation in that saltwater is evaporated in a covered container, but the water is heated by direct sunlight instead (Figure 17–23). Salt water in the container evaporates, and the water vapor that condenses on the cover runs into collection trays. The major difficulty lies in effectively concentrating the energy of sunlight into a small area to speed evaporation.

Membrane Processes **Electrolysis** can be used to desalinate seawater, too. In this method, two volumes of fresh water—one containing a positive electrode and the other a negative electrode—are placed on either side of a volume of seawater. The seawater is separated from each of the fresh water reservoirs by semipermeable membranes. These membranes are permeable to salt ions but not to water molecules. When an electrical current is applied, positive ions such as sodium ions are attracted to the negative electrode, and negative ions such as chloride ions are attracted to the positive electrode. In time, enough ions are removed through the membranes to convert the seawater to fresh water. The major drawback to electrolysis is that it requires large amounts of energy.

Reverse osmosis (*osmos* = to push) may have potential for large-scale desalination. In osmosis, water molecules naturally pass through a thin, semipermeable membrane from a fresh water solution to a salt water solution. In reverse osmosis, water on the salty side is highly pressurized to drive water molecules—but not salt and other impurities—through the membrane to the fresh water side (Figure 17–24). A significant problem with reverse osmosis is that the membranes are flimsy, become clogged, and must be replaced frequently. Advanced composite materials may help eliminate these problems because they are sturdier, provide better filtration, and last up to 10 years. Worldwide, at least 30 countries located in arid climates are operating reverse osmosis units. For example, Santa Barbara, California, operates a reverse osmosis plant that produces up to 34 million liters (9 million gallons) daily, which supplies up to 60% of its municipal water needs. This method is also used in many household water purification units and aquariums.

Other Methods of Desalination Seawater selectively excludes dissolved substances as it freezes—a process called **freeze separation**. As a result, the salinity of sea ice (once it is melted) is typically 70% lower than seawater. To make this an effective desalination technique, though, the water must be frozen and thawed multiple times, with the salts washed from the ice between each thawing. Like electrolysis, freeze separation requires large amounts of energy, so it may be impractical except on a small scale.

Yet another way to obtain fresh water is to melt naturally formed ice. Imaginative thinkers have proposed towing large icebergs to coastal waters off countries that need fresh water. Once there, the fresh water produced

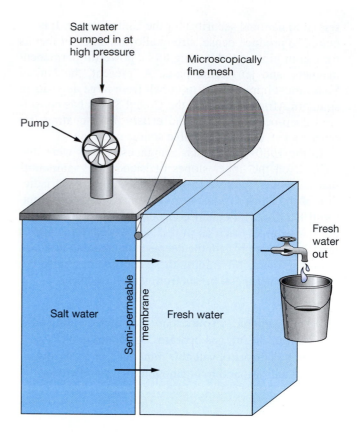

Figure 17–24 Reverse osmosis.

The process of reverse osmosis involves applying pressure to salt water and forcing it through a semipermeable membrane, thus removing the salts and producing fresh water.

as the icebergs melt could be captured and pumped ashore. Studies have shown that towing large Antarctic icebergs to arid regions would be technologically feasible and, for certain Southern Hemisphere locations, economically feasible, too.

Other novel approaches to desalination include crystallization of dissolved components directly from seawater, solvent demineralization using chemical catalysts, and even making use of salt-eating bacteria!

Evaporative Salts

When seawater evaporates, the salts increase in concentration until they can no longer remain dissolved, so they precipitate out of solution and form **salt deposits** (Figure 17–25). The most economically useful salts are *gypsum* and *halite*. Gypsum is used in plaster of Paris to make casts and molds and is the main component in gypsum board (wallboard or sheet rock).

Halite—common table salt—is widely used for seasoning, curing, and preserving foods. It is also used to de-ice roads, in water conditioners, in agriculture, and in the clothing industry for dying fabric. Additionally, halite is used in the production of chemicals such as sodium hydroxide (to make soap products), sodium hypochlorite (for disinfectants, bleaching agents, and PVC piping), sodium chlorate (for herbicides, matches, and fire-

Figure 17–25 Mining sea salt.

A salt mining operation at San Ignacio Lagoon, Baja California, Mexico. Low-lying areas near the lagoon are allowed to flood with seawater, which evaporates in the arid climate and leaves deposits of salt that are then collected.

works), and hydrochloric acid (for use in chemical applications and for cleaning scaled pipes). The manufacture and use of salt is one of the oldest chemical industries.[8]

Drugs from the Sea

Although many drugs are extracted from land plants,[9] scientists have only recently turned to the sea as a source of pharmaceuticals. In particular, tropical coral reefs have high biodiversity and contain a vast array of marine creatures, some of which contain chemical compounds that have unique biochemical properties useful in combating and curing human diseases. On coral reefs, marine chemists seek sponges and other soft-bodied animals that use chemicals rather than shells or spines to protect themselves. Chemicals are then extracted from the sampled organisms and are tested for their effectiveness against bacteria, fungi, and viruses and their potential for inhibiting inflammation, pain, and cancer cell growth.

Examples of recently discovered pharmaceuticals from the sea include the following:

- Acyclovir, which is an antiviral compound derived from a Caribbean sea sponge that can be used in fighting herpes infections of the skin and nervous systems. It is the first compound from a marine organism that was approved for human use.

[8]An interesting historical note about salt is that part of a Roman soldier's pay was in salt. That portion was called the *salarium*, from which the word *salary* is derived. If a soldier did not earn it, he was not worth his salt.

[9]For example, the active chemical compound in aspirin was initially discovered in the bark and leaves of the willow tree.

- Pseudopterosins, which are a group of compounds derived from sea fans (relatives of coral) (Figure 17–26) that are useful as anti-inflammatory drugs. These compounds can be used to fight maladies such as psoriasis, sunburn, and arthritis, all of which involve inflammation. They are also used in an internationally marketed skin-care product that reduces swelling and retards degeneration of the skin.
- Eleutherobin, which is derived from a soft coral and is effective in limiting tumor growth by disrupting cell division.
- Anticancer, antiviral, and antitumor agents derived from bryozoans (small mosslike encrusting invertebrates) and tunicates (small gelatinous chordates). The bryozoan *Bugula neritina*, for example, produces a potent chemical effective against cancers of the blood that is the active ingredient in Bryostatin 1, which is being tested in clinical trials as a treatment for leukemia.
- Antiviral compounds derived from cyanobacteria and sponges that are useful for combating HIV, the causative agent of AIDS.
- Antibiotic and cancer-fighting compounds from bacteria in deep ocean sediments that inhibit the growth of some cancer cells from human colon, lung, and breast tissues.

Because of increasing bacterial resistance to antibiotics, drugs from marine organisms may soon be among society's most valuable marine resources.

Ocean chemical resources include desalination of seawater to produce fresh water, evaporative salts, and a host of newly discovered drugs from the sea.

Figure 17–26 Sea fan (*Pseudopterogorgia elisabethae*).

Chemical compounds called pseudopterosins that have anti-inflammatory properties have been extracted from this species of sea fan. This is one of many examples of how marine organisms are supplying society with drugs from the sea.

Chapter in Review

- *International efforts at creating a law of the sea* date back to a 1609 doctrine that the ocean was free to all nations. In 1702, the *territorial sea* was established so that coastal nations could control the oceans to 3 nautical miles from shore. The present law of the sea is detailed in the United Nations Convention on the *Law of the Sea*, which establishes an *exclusive economic zone (EEZ)* within 200 nautical miles (370 kilometers) from shore. It also describes the *right of free passage*, rules for *mining deep-ocean mineral resources*, and *arbitration of disputes*.

- *Marine fisheries harvest standing stocks of populations from various ecosystems*, particularly shallow shelf and coastal waters and areas of upwelling. *Overfishing occurs when adult fish are harvested faster than they can reproduce* and results in the decline of fish populations as well as a reduction of a fishery's *maximum sustainable yield (MSY)*. Many fishing practices capture unwanted *bycatch*. Despite the *management of fisheries, many fish stocks worldwide are declining*. Wise *seafood choices* can help reverse the decline in fish populations.

- *Maine aquaculture (mariculture) projects have, in some cases, indicated potential for increasing the amount of food taken from the ocean*. Although the aquaculture industry struggles with a variety of problems, it *supplies over one-third of the total world fishery*. Some of the most successful mariculture operations raise various fish, shrimp and prawn, bivalves (oysters and mussels), and marine algae.

- *Renewable, nonpolluting energy can be extracted from the motions and patterns of heat distribution in the atmosphere and the oceans*. The most significant sources of marine energy resources include *power from offshore winds, surface currents, waves, tides, and thermal gradients*. Although these sources can be harnessed to turn a turbine and generate electricity, they all have *significant problems to overcome* before they become practical sources of energy. One of the most promising ways to extract ocean energy is by *ocean thermal energy conversion (OTEC)*, which is a process that uses the temperature difference between warm tropical surface water and the cold water below the thermocline to produce electricity.

- *The most valuable geologic resource from the ocean today is petroleum*, which is recovered from below the continental shelves and used as a source of energy. *Gas hydrates* include vast deposits of icelike material that may some day be used as a source of energy. Other important resources include *sand and gravel* (including deposits of valuable minerals), *phosphorite, metal sulfides*, and *manganese nodules and crusts*.

- *Ocean chemical resources include fresh water, evaporative salts, and drugs from the sea. Desalination* of ocean water to provide fresh water for business, home, and agricultural use is of growing interest. Solar distillation (solar humidification), electrolysis, freeze separation, and reverse osmosis are methods currently used to desalinate seawater. *Evaporative salt deposits are created when seawater evaporates*, producing salts such as gypsum and halite. Researchers have recently identified a host of *chemical compounds produced by marine organisms* that have unique properties useful in combating and curing certain human diseases.

Key Terms

Algin (p. 502)
Anchovy (p. 498)
Bycatch (p. 496)
Clathrate (p. 510)
Crusts (p. 513)
Desalination (p. 513)
Distillation (p. 513)
Driftnet (p. 497)
Electrolysis (p. 514)
Exclusive economic zone (EEZ) (p. 492)

Fisheries management (p. 497)
Fishery (p. 493)
Freeze separation (p. 514)
Gill net (p. 497)
Incidental catch (p. 496)
Manganese nodule (p. 511)
Mariculture (p. 500)
Marine Mammals Protection Act (p. 497)
Maximum sustainable yield (MSY) (p. 493)

Metal sulfide (p. 511)
Methane hydrate (p. 510)
Ocean thermal energy conversion (OTEC) (p. 507)
Overfishing (p. 493)
Petroleum (p. 509)
Phosphorite (p. 511)
Potential world fishery (p. 494)
Reverse osmosis (p. 514)
Salt deposit (p. 514)
Solar distillation (p. 513)

Solar humidification (p. 513)
Standing stock (p. 493)
Territorial sea (p. 492)
United Nations Conference on the Law of the Sea (p. 492)
Upwelling (p. 494)

Questions and Exercises

1. Discuss possible reasons why less-developed nations believe that the open ocean is the common heritage of all, whereas the more developed nations believe that the open ocean's resources belong to those who recover them.

2. What are the two critical survival stages that must occur before young adults can be recruited to a fishery? What factor is most important in increasing the chance of survival at each stage?

3. Explain the relationship between the influx of nutrients (nitrogen) into an ecosystem and the amount of the fishery that can safely be removed each year.

4. Explain why moderate rates of upwelling over long periods of time produce larger fisheries than either high or low rates of upwelling.

5. When a species is overfished, what changes are there in the standing stock and the maximum sustainable yield? What are some problems with fisheries management?

6. What do bivalve and algae cultivation have in common that helps make those organisms good choices for mariculture operations?

7. Discuss some environmental problems that might result from developing facilities for conversion of wave energy to electrical energy.

8. Explain how a tidal power plant works, using an estuary that has two high and two low tides a day as an example. Why does potential for usable tidal energy increase with an increase in the tidal range?

9. Discuss at least two positive and two negative factors related to tidal power generation.

10. Explain how an ocean thermal energy conversion (OTEC) unit generates electricity (a labeled diagram might help).

11. Discuss the present importance and the future prospects for the production of petroleum; sand and gravel; phosphorite; and manganese nodules and crusts.

12. What are gas hydrates, where are they found, and why are they important?

13. Describe how the origin of metal sulfide deposits in the lithosphere is related to plate tectonics processes.

14. Compare and contrast the following seawater desalination methods: distillation, solar humidification, and reverse osmosis.

15. For developing new drugs from the sea, why would researchers want to find coral reef organisms that have no obvious means of protection? List some recently discovered pharmaceuticals from the sea.

CHAPTER 18
Marine Environmental Concerns

Oil tanker *Prestige* broken in two off the Spanish coast. The oil tanker *Prestige* ruptured its hull in heavy seas north of Spain and broke in two on November 19, 2002, leaking an estimated 26.5 million liters (7 million gallons) of fuel oil. The oil spill damaged rich fishing grounds and tainted kilometers of Spanish beaches.

Key Questions
- How is marine pollution defined?
- What are the main types of marine pollution?
- What other marine environmental concerns exist?
- What are the current laws on ocean dumping?
- What can people do to prevent marine pollution?

The answers to these questions (and much more) can be found in the highlighted concept statements within this chapter.

"Most people think of oceans as so immense and bountiful that it's difficult to imagine any significant impact from human activity. Now we've begun to recognize how much of an impact we do have."

—Jane Lubchenco, marine ecologist (2002)

Earth's rapidly expanding human population has put an ever-increasing stress on the marine environment. Although the ocean supplies society with many resources, it is also the final destination of much of the waste products of those living on the adjacent land. The ocean has a tremendous ability to assimilate waste materials, yet negative results are beginning to be felt worldwide. Recently, the effects of cumulative stresses on the oceans have become large enough for humans to finally acknowledge the finiteness and fragility of the world environment.

Marine pollution is just one of many environmental concerns. Other concerns include invasion of harmful non-native species, eutrophication, ozone depletion, and habitat destruction. Let's examine these environmental concerns and what can be done about reducing or eliminating them.

What Is Pollution?

Pollution is *any harmful substance*, but how do scientists determine which substances are harmful? For example, a substance may be esthetically unappealing to people yet is not harmful to the environment. Conversely, certain types of pollution cannot be easily detected by humans, yet they can do harm to the environment. A substance may not be immediately harmful, but it may cause harm years, decades, or even centuries later. Also, to whom must this harm be done? For instance, some marine species thrive when exposed to a particular compound that is quite toxic to other species. Interestingly, natural conditions in coastal waters, such as dead seaweed on the beach, may be considered "pollution" by some people. It should be remembered, however, that although nature may produce conditions that we dislike, it does not pollute. The amount of a pollutant is also important: If a sub-

stance that causes pollution is present in extremely tiny amounts, can it still be characterized as a pollutant? All of these questions are difficult to answer.

The World Health Organization defines pollution of the marine environment as follows:

> The introduction by man, directly or indirectly, of substances or energy into the marine environment, including estuaries, which results or is likely to result in such deleterious effects as harm to living resources and marine life, hazards to human health, hindrance to marine activities, including fishing and other legitimate uses of the sea, impairment of quality for use of sea water and reduction of amenities.

It is often difficult to determine the degree to which pollution affects the marine environment. Most areas were not studied sufficiently before they were polluted, so scientists do not have an adequate baseline from which to determine how pollutants have altered the marine environment. The marine environment is affected by decade- to century-long cycles, too, so it is difficult to determine whether a change is due to a natural biologic cycle or any number of introduced pollutants, many of which combine to produce new compounds.

STUDENTS SOMETIMES ASK...
Is dilution the solution to ocean pollution?

That's certainly a catchy phrase, but the implications are controversial. It suggests that the oceans can be used to dispose society's wastes, as long as the wastes are diluted to the point they no longer threaten marine organisms (which is often difficult to determine). Because the oceans are vast and consist of a good solvent (water), they appear ideally suited to this disposal strategy. In addition, the oceans have good mixing mechanisms (currents, waves, and tides), which dilute many forms of pollution.

Air pollution was once viewed in a similar manner. Disposal of pollutants into the atmosphere was thought to be acceptable as long as they were dispersed widely and high enough—so tall smokestacks were constructed. Over time, however, pollutants such as nitric acid and acid sulfates increased in the atmosphere to the point that acid rain is now a problem. The ocean, like the atmosphere, has a finite *holding capacity* for pollutants and even experts disagree on exactly how much that is.

As disposal sites on land begin to fill, the ocean is increasingly evaluated as an area for disposal of society's wastes. One thing that we can *all* do is to limit the amount of waste we generate, alleviating some of the problem of where to put waste. It is likely, however, that the ocean will continue to be used as a dumping ground in the foreseeable future. Despite many new disposal techniques, a long-term "solution" to ocean pollution hasn't appeared as yet.

Waste disposal facilities on land (such as landfills) have limited capacities that are already being exceeded in many cases. Should additional waste be discarded in the open ocean? Unlike coastal areas, the open ocean has mixing mechanisms (waves, tides, and currents) that distribute pollutants over a wide area—including an entire ocean basin. Diluting pollutants often renders them less harmful. On the other hand, do we really want to distribute a pollutant across an entire ocean without knowing what the long-term effects might be?

Some experts believe that we should not dump *anything* in the ocean, while others believe that the ocean can be a repository for many of society's wastes, as long as proper monitoring is conducted. Unfortunately, there are no easy answers and the issues are complex. What is clear is that more research is needed to assess the impact of pollutants in the ocean.

> Marine pollution is difficult to define but includes any human-induced substance that is harmful to the marine environment.

Predicting the Effects of Pollution on Marine Organisms

To date, the most widely used technique for determining the concentration of pollutants that negatively affect the living resources of the ocean is the **standard laboratory bioassay** (*bio* = biologic, *essaier* = to weigh out). Regulatory agencies such as the Environmental Protection Agency (EPA) use a bioassay that determines the concentration of a pollutant that causes 50% mortality among the test organisms. If a pollutant exceeds a 50% mortality rate, then concentration limits are established for the discharge of the pollutant into coastal waters. One shortcoming of the bioassay is that it does not predict the long-term effect of pollution on marine organisms. Another is that it does not take into account how pollutants may combine with other chemicals, creating new types of pollutants.

Marine ecologists have found it desirable to measure a pollutant's effects at these four **levels of biological response** that require increasingly long periods of observation:

1. **Biochemical–cellular: minutes to hours.** Exposure to organic pollutants such as hydrocarbons and PCBs produce immediate dysfunction in metabolic processes, causing death.
2. **Organism: hours to months.** Pollution stress may create changes in physiological processes such as metabolic rate or digestive efficiency that severely reduce the energy available for growth and reproduction.
3. **Population dynamics: months to decades.** These responses manifest as changes in the abundance and distribution of a species, population structure on an age class basis, growth rates within age classes, productiveness, and incidence of disease.
4. **Community dynamics and structure: years to decades.** Communities, as assemblages of popu-

lations, become less diverse, and more-resistant populations increase in numbers.

Much more research is needed to identify early warning signs of stress at each level of biological response. Because the degree of system complexity and the time needed to measure a response increase exponentially from the biochemical–cellular level to the community level, the predictive difficulties also increase at each level.

Main Types of Marine Pollution

Marine pollution comes from substances such as petroleum, sewage, and various chemical compounds, all of which can have severe deleterious effects on marine organisms, particularly in coastal environments. Coastal waters are more polluted than the open ocean because more pollution is dumped into coastal waters, and coastal waters are not as well circulated as the open ocean.

Petroleum

Major oil (**petroleum**) spills into the ocean are a fact of our modern oil-powered economy. Some oil spills are the result of loading/unloading accidents, collisions, or tankers running aground, such as the 1989 spill from the *Exxon Valdez* in Prince William Sound, Alaska (Box 18-1). Others are intentionally created, such as the oil that was spilled during the Persian Gulf War in 1991. Still others are caused by the blowout of undersea oil wells during drilling or pumping. The largest such spill occurred in June 1979 in the Gulf of Mexico, when the Petroleus Mexicanos (PEMEX) oil-drilling station Ixtoc #1 in the Bay of Campeche off the Yucatán peninsula, Mexico, blew out and caught fire. Before it was capped nearly 10 months later, it spewed 530 million liters (140 million gallons) of oil into the Gulf of Mexico, some of which washed up along the coast of Texas (Figure 18-1).

Oil is a **hydrocarbon**, which means it is composed of the elements *hydrogen* and *carbon*. Hydrocarbons are organic substances, so they can be broken down or *biodegraded* by microorganisms. Because hydrocarbons are biodegradable, many experts consider oil to be among the *least* damaging pollutants introduced into the ocean! In fact, natural undersea oil seeps have occurred for millions of years, and the ocean ecosystem seems unaffected or even enhanced by them (because oil is a source of energy).

Data from the *Exxon Valdez* oil spill is a case in point. The oil spill released almost 44 million liters (11.6 million gallons) of oil into a pristine wilderness area in Alaska. The affected waters were expected to have a long, slow recovery, but the fisheries that closed in 1989 bounced back with record takes in 1990. Ten years after the spill, several key species have rebounded to the point where their numbers are now greater than before the spill (Figure 18-2).

Still, oil is a complex mixture of hydrocarbons and other substances, including the elements oxygen, nitrogen,

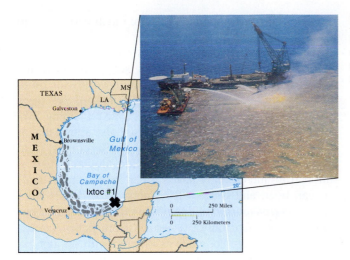

Figure 18–1 Blowout from the Ixtoc #1 oil well, Gulf of Mexico.

Location of the 1979 Bay of Campeche blowout and oil slick that affected the Texas coast. The well blew out, caught fire, and flowed for 10 months, spilling 530 million liters (140 million gallons) of oil into the Gulf of Mexico. The accident produced the world's largest oil spill from an oil well.

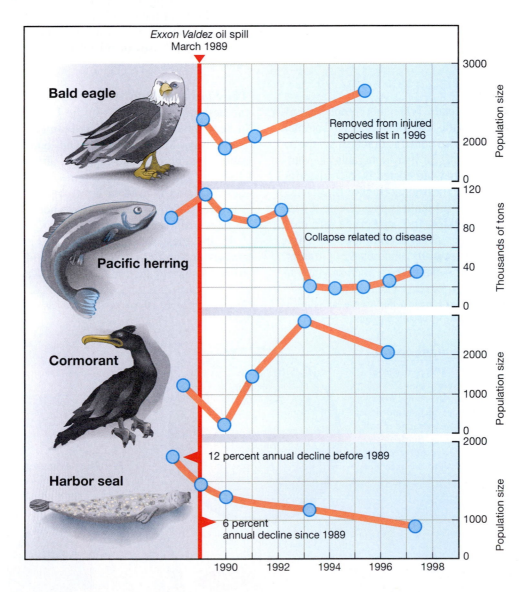

Figure 18–2 Recovery of organisms affected by the *Exxon Valdez* oil spill.

The populations of several key organisms in the Prince William Sound area of Alaska have rebounded after the 1989 *Exxon Valdez* oil spill. The bald eagle is so numerous that it was removed from the endangered species list in 1996. The collapse of the Pacific herring's population is thought to be unrelated to effects of the oil spill.

Figure 18–3 A bird covered by oil from the *Exxon Valdez* **oil spill.**

When marine organisms are covered by oil from an oil spill, their feathers or fur loose their insulation properties, resulting in high fatality rates. Some marine organisms such as this cormorant were rescued and cleaned of oil.

sulfur, and various trace metals. When this complex chemical mixture combines with seawater—another complex chemical mixture, which also contains organisms—the results are usually devastating for marine organisms. Many are killed outright when they are coated by oil, rendering their insulating feathers or fur useless (Figure 18-3).

In addition, spills from oil tankers may release a wide variety of petroleum products (not just crude oil), each of which contains different concentrations of **toxic** (*toxicum* = poison) **compounds**[1] and behaves differently in the environment. For example, refined oil such as

---

[1]A *toxic compound* is a poisonous substance that has the capability of causing injury or death, especially by chemical means.

fuel oil is rich in compounds that are much more toxic to the environment than crude oil.

Although spills from oil tankers receive much media attention, they are not the primary source of oil to the oceans. Figure 18-4 shows that 47% of worldwide oil to the oceans is caused by underwater *natural oil seeps* (many of which occur in U.S. waters); the remaining 53% comes from *human sources*. The figure also shows that of human-caused oil to the oceans, 72% comes from *petroleum consumption*, which includes nontank vessels, runoff from increasingly paved urban areas, and individual car, boat, and watercraft owners; 12% comes from *petroleum transportation*, including refining and distribution activities; and only 3% comes from *petroleum extraction*, which are activities associated with oil and gas exploration or production. Remarkably, the overwhelming majority of the petroleum that enters the oceans due to human activity is a result of small but frequent and widespread releases of oil related to activities that consume petroleum.

The *Florida* Spill in West Falmouth Harbor One of the best-studied oil spills in the United States occurred in September 1969 near West Falmouth Harbor in Buzzards Bay, Massachusetts. The barge *Florida* came ashore, ruptured, and spilled about 680,000 liters (180,000 gallons) of No. 2 fuel oil, which is similar to diesel oil and is used in home heating. The oil spread northward into Wild Harbor, where the most severe damage occurred (Figure 18-5).

In the most severely oiled areas, nearly all marsh grasses and most intertidal and subtidal[2] animals were killed. A sharp reduction in *species diversity* (the number

---

[2]The intertidal zone extends from high to low tide; the subtidal zone is below that.

Figure 18–4 Sources of oil to the oceans.

Of worldwide oil to the oceans, 47% comes from natural seeps while 53% comes from human sources. Of all human sources, 72% comes from petroleum consumption activities such as individual car and boat owners, non-tank vessels, and runoff from increasingly paved urban areas. Surprisingly, petroleum transportation and extraction account for only 28% of human-caused oil to the oceans.

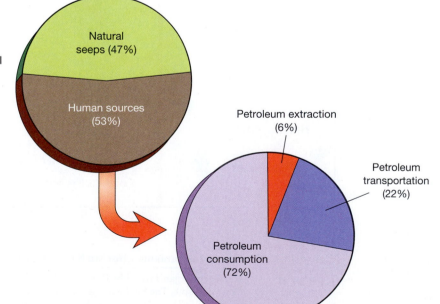

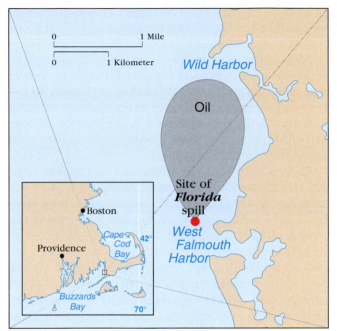

Figure 18–5 *Florida* **oil spill at West Falmouth Harbor, Massachusetts.**

When the barge *Florida* came ashore and ruptured, currents carried its load of No. 2 fuel oil northward into Wild Harbor, where the most severe damage occurred.

visible damage after 10 years. After 20 years, there was virtually no oil in the subtidal sediments, and the intertidal marsh sediments were more than 99% oil-free. At the most heavily oiled site in Wild Harbor, however, enough oil was still present in beaches at a depth of 15 centimeters (6 inches) to kill animals that burrow into the sediments.

Despite the extensive damage caused by the oil spill at Wild Harbor in 1969, recovery can occur much more quickly than some researchers thought possible. Evidently, natural processes effectively biodegrade and remove oil from the marine environment, although it can take at least a couple of decades. Studies of other areas affected by oil spills indicate that most return to normal conditions within a few years and seem to experience no long-term damage.

Unfortunately, another fuel oil spill occurred in Buzzards Bay on April 27, 2003. This time, a barge ruptured and spilled nearly 57,000 liters (15,000 gallons) of oil, which spread to local beaches and tarred hundreds of seabirds.

of different species present) was accompanied by a rapid increase in the population of polychaete worms[3] that have a high tolerance of oil. In fact, a single small red polychaete worm species, *Capitella capiata*, accounted for up to 99.9% of the organisms collected in samples of the most severely oiled locations during the first year. Species diversity did not increase appreciably until well into the third year after the spill.

Marsh grasses and animals reentered the area from three to five years after the spill. Remarkably, there was no

The *Argo Merchant* Spill off Nantucket Island The *Argo Merchant* sank after running aground on Fishing Rip Shoals 40 kilometers (25 miles) southeast of Nantucket Island, Massachusetts, in December 1976, spilling 29 million liters (7.7 million gallons) of No. 6 fuel oil (Figure 18-6). Fortunately, the winds were such that no oil came ashore. The surface slick moved eastward out to sea and was gone within a month of the spill.

[3]Polychaete worms are close relatives of segmented earthworms found on land.

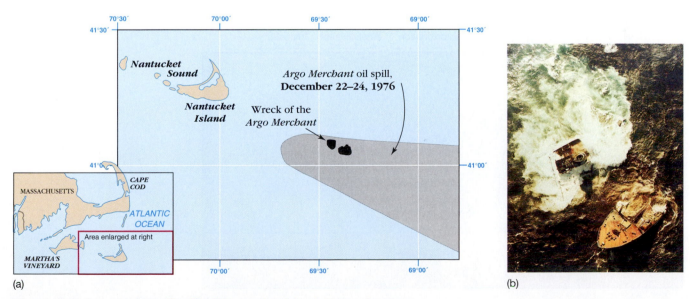

(a) (b)

Figure 18–6 *Argo Merchant* **oil spill off Nantucket Island, Massachusetts.**

(a) Map showing the ship's grounding site (*black*) and oiled area of the ocean (*gray*). **(b)** The *Argo Merchant*, after breaking up and spilling much of its cargo into the Atlantic Ocean southeast of Nantucket Island, Massachusetts.

Although the oil was not visible for long, it significantly damaged many organisms, such as planktonic (drifting) fish eggs of cod and pollock (Figure 18-7). In one study of fish eggs collected shortly after the spill, 60% of cod eggs and 94% of pollock eggs were oil fouled. In addition, 20% of cod eggs and 46% of pollock eggs were dead or dying.

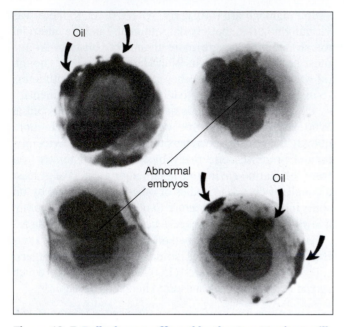

Figure 18–7 Pollock eggs affected by the *Argo Merchant* spill.

The pollock eggs at the upper left and lower right show outer membranes contaminated with oil (*arrows*); the other two eggs have abnormally developed embryos. Each egg is about 1 millimeter (0.04 inch) in diameter.

In comparison, only 4% of cod eggs spawned in a laboratory under natural conditions were dead or dying at a similar stage of development.

Most of the oil floated as a surface slick, so contamination in subsurface water samples did not exceed more than 250 parts per billion. Other than the damage to fish eggs and other plankton, little direct evidence of major biological damage was obtained. A large number of dead and oiled birds, however, did wash ashore at Nantucket Island and Martha's Vineyard.

Each season, female pollock and cod spawn about 225,000 and 1 million eggs each, respectively, in an area from New Jersey to Greenland. Because of the huge number of eggs and the large spawning area, it is unlikely that a single oil spill would significantly affect these fisheries. Still, fishing communities of the northwest Atlantic are pleased when oil exploration on Georges Bank yielded poor results because the pollock and cod fisheries—both overexploited already—would probably not survive the level of pollution brought about by even a small oil spill.

Cleaning Oil Spills   When oil enters the ocean, it initially floats because oil is less dense than water and forms a slick at the surface, where it starts to break down through natural processes (Figure 18-8). The volatile, lighter components of crude oil evaporate over the first few days, leaving behind a more viscous substance that aggregates into tar balls and eventually sinks. The tarry oil also coats suspended particles, which settle to the seafloor, too.

If the floating oil hasn't dispersed, it can be collected with specially designed skimmers or absorbent materials. The collected oil (or oiled materials), however, must still be disposed of elsewhere. Waves, winds, and currents serve to further disperse an oil slick and mix the remaining oil with water to make a frothy emulsion called *mousse*. In ad-

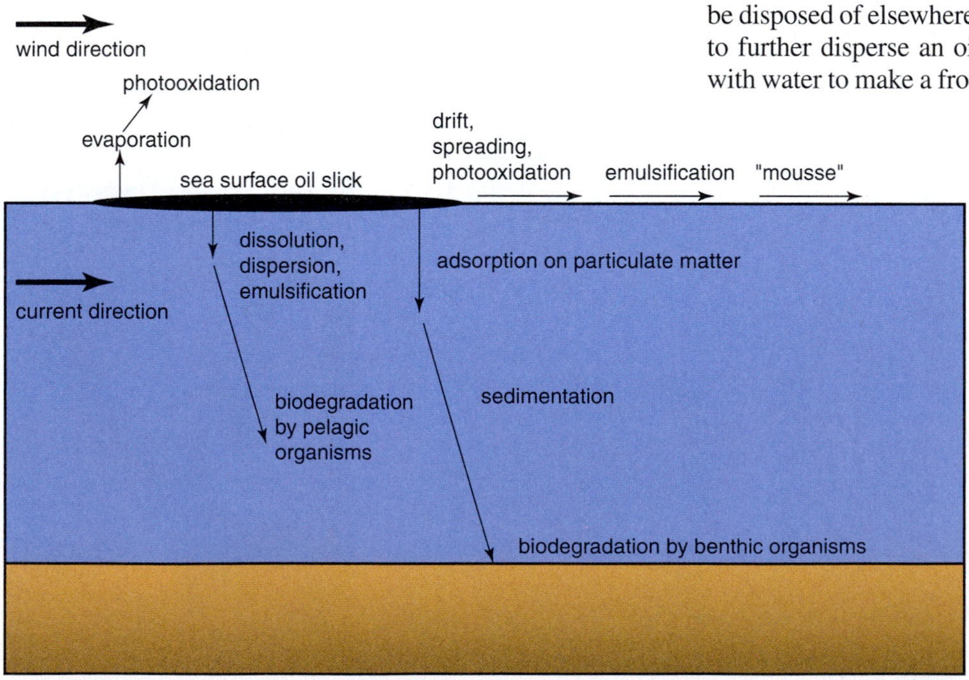

Figure 18–8 Processes acting on oil spills.

After an oil spill enters the ocean, it is acted upon by various natural processes that break up the spill. The lighter components evaporate, while the heavier components form tar balls or coat suspended particles and sink. The remaining dispersed oil photo-oxidizes or can mix with water, creating a frothy substance called "mousse."

dition, bacteria and photo-oxidation act to break down the oil into compounds that dissolve in water.

Microorganisms such as bacteria and fungi naturally biodegrade oil, so they can be used to help clean oil spills—a method called **bioremediation** (*bio* = biologic, *remedium* = to heal again). Virtually all marine ecosystems harbor naturally occurring bacteria that degrade hydrocarbons. Although certain types of bacteria and fungi can break down particular kinds of hydrocarbons, none is effective against all forms. In 1980, however, microbiologist A. M. Chakrabarty isolated a microorganism capable of breaking down nearly two-thirds of the hydrocarbons in most crude oil spills.

Releasing bacteria directly into the marine environment is one form of bioremediation. For example, a strain of oil-degrading bacteria was released into the Gulf of Mexico to test its effectiveness in cleaning up about 15 million liters (4 million gallons) of crude oil spilled after an explosion disabled the tanker *Mega Borg* in 1990. Preliminary results indicate that the bacteria reduced the amount of oil with no negative effects on the ecology due to the bacteria.

Providing conditions that stimulate the growth of naturally occurring oil-degrading bacteria is another form of bioremediation. Exxon, for example, spent $10 million dollars to spread fertilizers rich in phosphorus and nitrogen on Alaskan shorelines to boost the growth of indigenous oil-eating bacteria after the *Exxon Valdez* spill (Box 18-1). The resulting cleanup rate was more than twice that under natural conditions.

Preventing Oil Spills One of the best ways to protect areas from oil spills is to prevent spills from occurring in the first place. Because our society relies on petroleum products, however, oil spills are a likely occurrence in the future (Figure 18-9), especially as petroleum reserves beneath the continental shelves of the world are increasingly exploited.

After the *Exxon Valdez* oil spill in 1989, Congress enacted the 1990 Oil Pollution Act to phase in double hulls by 2015 for all oil tankers traveling in the United States. The double hull houses two layers: an inner hull can prevent oil

spillage if damage should occur to the outer hull. Studies of hull designs during groundings and collisions indicate that double-hull designs are overall more effective at reducing oil spills. However, analysis of the *Exxon Valdez* spill suggests that even a double-hulled tanker would not have prevented the disaster. (Currently, only 10% of the tankers operating in Prince William Sound have double hulls.) Tanker designs are also being modified to limit the amount of oil spilled should there be a hull rupture.

In February 1999, the Japanese-owned freighter M/V *New Carissa* ran aground just offshore of Coos Bay, Oregon, with nearly 1,500,000 liters (400,000 gallons) of tar-like fuel oil aboard and began leaking oil through cracks in its hull. When the ship washed into the surf zone and an approaching storm threatened to tear it apart, federal and state authorities decided to ignite the vessel and its fuel rather than risk a larger oil spill (Figure 18-10). This was the first time that oil on a ship in U.S. waters was intentionally burned to prevent an oil spill. Eventually, the ship split in two and about half of its oil burned, limiting the amount of oil spilled into the ocean. Most of the remaining oil was sunk with the wrecked ship a month later when it was towed offshore and sank in water 3 kilometers (1.9 miles) deep by Naval gunfire and a torpedo.

Sewage Sludge

Sewage treated at a facility typically undergoes **primary treatment**, where solids are allowed to settle and dewater, and **secondary treatment**, where it is exposed to bacteria-

Figure 18–10 The *New Carissa* on fire off the Oregon coast.

When the freighter M/V *New Carissa* ran aground in shallow water offshore Coos Bay, Oregon, in 1999 and began leaking oil, it was intentionally set on fire to prevent further oil from spilling into the ocean.

Figure 18–9 Who is at fault?

BOX 18–1 People and Ocean Environment

THE *EXXON VALDEZ* OIL SPILL: NOT THE WORST SPILL EVER

A large percentage of oil enters the oceans from spills by tankers and transportation operations. One of the most publicized oil spills was from the supertanker *Exxon Valdez*, which occurred in Prince William Sound, Alaska, and caused the largest oil spill in U.S. territorial waters.

Crude oil produced from the North Slope of Alaska is carried by pipeline to the southern port of Valdez, Alaska, where it is loaded onto supertankers like the *Exxon Valdez*, which are capable of holding almost 200 million liters (53 million gallons) when full. On March 29, 1989, the tanker left Valdez with a full load of crude oil and was headed toward refineries in California. She was only 40 kilometers (25 miles) out of Valdez when the ship's officers noted icebergs from nearby Columbia Gla-cier within the shipping channel. While maneuvering around the icebergs, the ship ran aground on a shallowly submerged rocky outcrop known as Bligh Reef (Figure 18A), rupturing

eight of the ship's 11 cargo tanks. About 22% of her cargo—almost 44 million liters (about 11.6 million gallons) of oil—spilled into the pristine waters of Prince William Sound, where it subsequently spread into the Gulf of Alaska and fouled over 1775 kilometers (1100 miles) of shoreline.

The U.S. Fish and Wildlife Service reported that at least 994 sea otters and 34,434 birds were killed outright by the spill. By some official estimates, however, the actual kill could have been 10 times that amount, because not all dead organisms are recovered. Due to the remote location and the size of the affected area, the exact total death toll will never be determined.

Immediately after the spill, Exxon spent over $2.5 billion in cleanup efforts and another $900 million in subsequent years for restoration. Absorbent materials and skimming devices were used to remove oil from the water, whereas super hot water (60°C or 140°F) sprayed through

high-pressure hoses was used to clean oil from the rocky beaches. The hot water removed the oil but also killed most shoreline organisms. Analysis of the cleanup effort, when compared to areas that were left to biodegrade naturally, reveals that the beaches that were left alone recovered more quickly and more completely than the cleaned beaches.

As large and damaging as the *Exxon Valdez* spill was, it ranks as only the *fifty-third* largest oil spill worldwide (Table 18A). The world's largest oil spill occurred because of intentional dumping by the Iraqi army during their invasion of Kuwait during the 1991 Persian Gulf War. By the time the Iraqi were driven out of Kuwait and the leaking oil wells and sabotaged production facilities were brought under control, more than 908 million liters (240 million gallons) of oil had spilled into the Persian Gulf (Figure 18B)—more than *20 times* the amount spilled by the *Exxon Valdez*.

Figure 18A The 1989 *Exxon Valdez* oil spill, Prince William Sound, Alaska.

Map showing location of the oil spill (*left*), the *Exxon Valdez* on Bligh Reef (*upper right*), and spilled oil coating a beach (*lower right*).

TABLE 18A **The world's largest oil spills.**

Rank	Date	Location	Source of spill	Size of spill (million liters)	(million gallons)
1	1/1991	Kuwait, Saudi Arabia	Oil terminals, tankers	908	240
2	6/1979	Gulf of Mexico	Ixtoc #1 oil well	530	140
3	3/1992	Uzbekistan	Oil well	333	88
4	2/1983	Iran	Oil well	303	80
5	8/1983	Near coast of South Africa	*Castillo de Bellver* tanker	299	79
6	3/1978	Near coast of France	*Amoco Cadiz* tanker	261	69
53	3/1989	Prince William Sound, Alaska	*Exxon Valdez* tanker	44.0	11.6

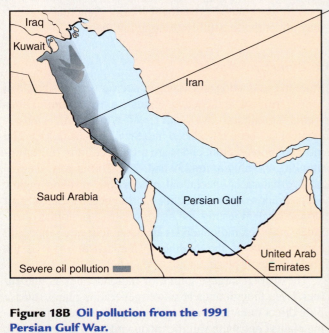

Figure 18B Oil pollution from the 1991 Persian Gulf War.

Map showing location of the spilled oil, which was confined to the northwest coast of the Persian Gulf by currents and southeasterly winds, and a Saudi Arabian government official examining some of the damage (*inset*).

killing chlorine. **Sewage sludge** is the semisolid material that remains after such treatment. It contains a toxic brew of human waste, oil, zinc, copper, lead, silver, mercury, pesticides, and other chemicals. Since the 1960s, at least 500,000 metric tons (550,000 short tons) of sewage sludge has been dumped into the coastal waters of southern California and more than 8 million metric tons has been dumped in the New York Bight between Long Island and the New Jersey shore.

Although the Clean Water Act of 1972 prohibited the dumping of sewage into the ocean after 1981, the high cost of treating and disposing of sewage sludge on land resulted in extension waivers being granted to many municipalities. In the summer of 1988, however, nonbiodegradable debris including medical waste—probably carried by heavy rains into the ocean through storm drains—washed up on Atlantic coast beaches and adversely affected the tourist business. Although this event was completely unrelated to sewage disposal at sea, it focused public awareness on ocean pollution and helped pass new legislation to terminate sewage disposal at sea.

New York's Sewage Sludge Disposal at Sea Sewage sludge from New York and Philadelphia has traditionally been transported offshore by barge and dumped in the ocean at sites totaling 150 square kilometers (58 square miles) within the New York Bight Sludge Site and the Philadelphia Sludge Site (Figure 18-11).

The water depth is about 29 meters (95 feet) at the New York Bight Sludge Site and about 40 meters (130 feet) at the Philadelphia Sludge Site. The water column in such shallow water is relatively uniform, so even the

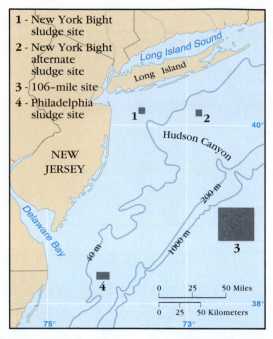

Figure 18–11 Atlantic sewage sludge disposal sites.

More than 8 million metric tons of sewage sludge was dumped by barge annually at the New York Bight Sludge Site (1) and the Philadelphia Sludge Site (4). After 1986, the new dump site is the larger and deeper-water 106-mile site (3).

smallest sludge particles reach the bottom without undergoing much horizontal transport, and the ecology of the dump site can be severely affected. At the very least, such a concentration of organic and inorganic matter seriously disrupts the chemical cycling of nutrients. Greatly reduced species diversity results, and in some locations the environment becomes devoid of oxygen (*anoxic*).

In 1986, the shallow-water sites were abandoned and sewage was subsequently transported to a deep-water site 171 kilometers (106 miles) out to sea (Figure 18-11). The deep-water site is beyond the continental shelf break, so there is usually a well-developed density gradient that separates low-density, warmer surface water from high-density, colder deep water. Internal waves moving along this density gradient can horizontally transport particles at rates 100 times greater than they sink.

Local fishermen reported adverse effects on their fisheries soon after deep-water dumping began. Also, concern was expressed that the sewage could be transported great distances in eddies of the Gulf Stream (see Chapter 8), even as far as the coast of the United Kingdom. This program was terminated in 1993, and municipalities must now dispose of their sewage on land.

Boston Harbor Sewage Project Some 48 different communities that comprise the greater Boston area have, until recently, used an antiquated sewage system to dump sludge and partially treated sewage at the entrance to Boston Harbor. Tidal currents often swept the sewage back into the bay and at other times, the system became overloaded and dumped raw sewage directly into the bay,

making Boston Harbor one of the most polluted bays in the country.

A court-ordered cleanup of Boston Harbor in the 1980s resulted in a new sewage system that came on-line in 1998. It treats all sewage with bacteria-killing chlorine and carries it through a tunnel 15.3 kilometers (9.5 miles) long into deeper waters offshore (Figure 18-12a), which prevents it from returning to the bay. Since the system started, there has been a dramatic improvement in the bay's water quality. To pay for the $4 billion system, however, the average annual sewage bill for a Boston-area household is now about $1200.

Some fear that the project will degrade the environment in Cape Cod Bay and Stellwagen Bank (Figure 18-12b), which is an important whale habitat. The area was recently designated a National Marine Sanctuary, which may affect the feasibility of dumping there.

Radioactive Waste

Since the production and testing of nuclear weapons began in 1944, artificial radioactive atoms called **radionuclides** (*radio* = radioactivity, *nucleos* = a little nut, *idus* = small) have been introduced into the oceans as fallout. The most common radionuclides are tritium, cesium-137, strontium-90, and plutonium-239 and -240. Fortunately, the dispersive nature of the fallout process has resulted in low concentrations of radionuclides in the oceans. Since most atomic weapon testing and usage occurred in northern latitudes, however, radionuclides are about four times higher in concentration in the Northern Hemisphere than in the Southern Hemisphere.

The second largest source of radionuclides is nuclear fuel. Even though nuclear power generating plants release little direct radiation to the oceans, nuclear fuel reprocessing plants contribute significant quantities. Such plants are found at Sellafield and Dounreay in Great Britain and Cape de la Hague, France. The Irish Sea, which has been receiving reprocessing discharges since the early 1950s, is the most radioactive sea in the world.

Radioactive Waste Disposal in Sea Floor Sediments
Nuclear waste that remains from the production of nuclear weapons and from commercial power generation must be safely disposed of until it is no longer dangerous, which typically takes several hundred thousand years. In fact, the U.S. Environmental Protection Agency (EPA) has established strict criteria for maximum levels of radioactivity deemed safe for disposal sites. Several countries—including the United States—have seriously considered high-level nuclear waste disposal in sea floor sediments.

The disposal of nuclear waste has so far been focused on land sites because the materials remain accessible. From 1976 until 1986, however, research was conducted into disposal of high-level nuclear waste within the sea floor of the deep sea. Given the requirements that the waste must be isolated from human activities, exposure by natural erosion, and shaking from earthquakes, the

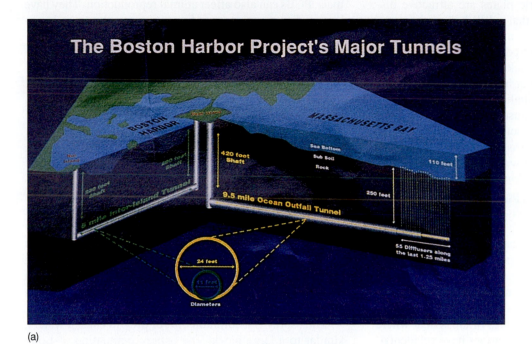

(a)

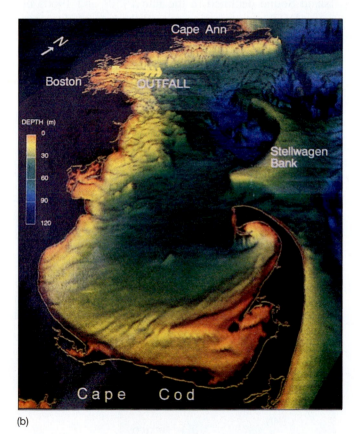

(b)

Figure 18–12 Boston Harbor sewage project.

(a) Diagrammatic view of the Boston Harbor Project tunnels that transport sewage to an outfall 15 kilometers (9.5 miles) offshore at a depth of 76 meters (250 feet) beneath the ocean floor. (b) Bathymetric map of the coastal ocean in the Boston–Cape Cod area, showing the proximity of the outfall to Stellwagen Bank. Depths in meters, vertical exaggeration = 100×.

centers of oceanic lithospheric plates are attractive disposal sites. Proposals specify that a drill-hole in the deep-ocean floor[4] could be used for the disposal of canisters containing nuclear waste hundreds of meters beneath the sea floor. The plan calls for stacking the canisters within the drill hole, which would later be filled with mud.

In 1986, the United States halted research into seabed disposal of nuclear waste because of budget constraints. After a lengthy evaluation of potential sites nationwide, Yucca Mountain in Nevada was selected as the future repository site for U.S. nuclear waste. However, concerns about the safety of the Yucca Mountain site have recently been raised. In addition, it appears that the site is not large enough to hold all the nuclear waste that will be generated by the year 2020.

Given the political realities that may be faced as work proceeds toward land-based disposal, deep-sea disposal will most likely be considered again. The seabed disposal program did produce encouraging results, and, in fact, Japan is currently investigating disposing of its nuclear waste in the Mariana Trench. Studies of areas near where nuclear weapons and nuclear submarines have accidentally been lost at sea indicate that fine-grained abyssal clays on the deep-ocean floor tend to hold fast to certain radioactive elements, effectively isolating them from the environment. It would seem that careful ocean disposal, if conducted safely, might be a disposal strategy with merit. However, environmental consequences must be carefully considered before ocean floor disposal is attempted.

DDT and PCBs

The pesticide **DDT** (dichlorodiphenyltrichloroethane) and the industrial chemicals called **PCBs** (polychlorinated biphenyls, which are also known as *organochlorides*) are now found throughout the marine environment. They are persistent, biologically active chemicals that have been introduced into the oceans entirely as a result of human activities. Because of their toxicity, long life, and propensity for being accumulated in food chains, these and other chemicals have been classified as *persistent organic pollutants (POPs)* capable of causing cancer, birth defects, and other grave harm.

DDT was widely used in agriculture during the 1950s and improved crop production throughout developing countries for several decades. However, its extreme effectiveness as an insecticide and persistence as a toxin in the environment eventually resulted in a host of environmental problems including devastating effects on marine food chains.

PCBs are industrial chemicals that were once widely used as a liquid coolant and insulation in industrial equipment such as power transformers, where they were released into the environment. PCBs have been shown to cause liver cancer and harmful genetic mutations in ani-

mals. PCBs can also affect animal reproduction: They have been indicated as causes of spontaneous abortions in sea lions and the death of shrimp in Escambia Bay, Florida.

DDT and Eggshells U.S. production of DDT was almost completely banned in 1971. By that time, though, 2 billion kilograms (4.4 billion pounds) had already been manufactured, most by the United States. Since 1972, the use of DDT was banned in the United States by the EPA. Worldwide, the pesticide is banned from agricultural use, but it continues to be used in limited quantities for public health purposes.

The danger of excessive use of DDT and similar pesticides first became apparent in the marine environment when it affected marine bird populations. During the 1960s, there was a serious decline in the brown pelican population of Anacapa Island off southern California (Figure 18-13). High concentrations of DDT in the fish eaten by the birds had caused them to produce eggs with excessively thin shells.

The osprey is a common bird of prey in coastal waters, similar to a large hawk. The osprey population of Long Island Sound declined in the late 1950s and 1960s because DDT contamination caused them to produce eggs with thin shells, too. Since the ban on DDT, the osprey, brown pelican, and many other species affected by the chemical are making remarkable comebacks.

Figure 18–13 Survival of brown pelicans threatened by DDT.

Brown pelicans (*Pelecanus occidentalis*) that breed on Anacapa Island offshore southern California were found to have high levels of DDT, which decreased the thickness of their eggshells. Since DDT has been banned, healthy pelicans have returned to these waters, such as this one in breeding plumage (*inset*).

---

[4]These holes are similar to the ones drilled by the Ocean Drilling Program's drill ship *JOIDES Resolution* to recover deep-sea sediment cores; see Box 5-1.

DDT and PCBs Linger in the Environment   DDT and PCBs (which were banned in 1977) generally enter the ocean through the atmosphere and river runoff. They are concentrated initially in the thin slick of organic chemicals at the ocean surface, and then they gradually sink to the bottom, attached to sinking particles. A study off the coast of Scotland indicated that open-ocean concentrations of DDT and PCBs are 10 and 12 times less, respectively, than in coastal waters. Long-term studies have shown that DDT residue in mollusks along the U.S. coasts peaked in 1968.

Although most countries have banned their use, DDT and PCBs are so pervasive in the marine environment that even Antarctic marine organisms contain measurable quantities of them. There has been no agriculture or industry in Antarctica to introduce them directly, so they must have been transported from distant sources by winds and ocean currents.

? STUDENTS SOMETIMES ASK ...

I've heard that some organizations want to lift the ban on DDT. Why would they want to do that?

Since production of DDT was banned in 1971, outbreaks of malaria have dramatically increased because DDT was the most effective and readily available pesticide used to kill mosquitoes that transmit malaria. According to the World Health Organization, malaria infects up to 500 million people a year—mostly in tropical regions—and kills as many as 2.7 million, including at least one child every 30 seconds. In addition, drug-resistant strains of malaria have begun to show up worldwide. This resurgence of malaria has caused many health organizations to call for an exception to the ban on DDT—in spite of its well-documented perseverance and negative effects on the environment—so that it can be used selectively to spray houses in malaria-prone areas like tropical Africa and Indonesia.

Mercury and Minamata Disease

The metal **mercury**, which is a silvery liquid at room temperature, has many industrial uses. When it enters the ecosystem, however, mercury forms an organic compound (*methyl mercury*) that is generally toxic to most living things.

A chemical plant built on Minamata Bay, Japan, in 1938, produced acetaldehyde, which requires mercury in its manufacture. Mercury was discharged into Minamata Bay, where bacteria degraded it into methyl mercury that was later ingested and concentrated in the tissues of larger marine organisms. The first ecological changes in Minamata Bay were reported in 1950 and human effects were noted as early as 1953. The mercury poisoning that is now known as **Minamata disease** became epidemic in 1956, when the plant was only 18 years old. Minamata disease is a degenerative neurologic disorder that affects the human nervous system and causes sensory disturbances including blindness and tremor, brain damage, birth defects, paralysis, and even death. This mercury poisoning was the first major human disaster resulting from ocean pollution. However, the Japanese government did not declare mercury as the cause of the disease until 1968. The plant was immediately shut down, but over 100 people were known to suffer from the disease by 1969 (Figure 18-14), almost half of whom died. A second acetaldehyde plant was closed in 1965 in Niigata, Japan, because it, too, was discharging mercury that poisoned people. Between 1965 and 1970, 47 fishing families contracted Minamata disease. Today the concentration of mercury in Minamata Bay is no longer unusually high, indicating that there has been enough time for the mercury to be widely dispersed within the marine environment.

Bioaccumulation and Biomagnification   During the 1960s and 1970s methyl mercury contamination in seafood received considerable attention. Certain marine organisms

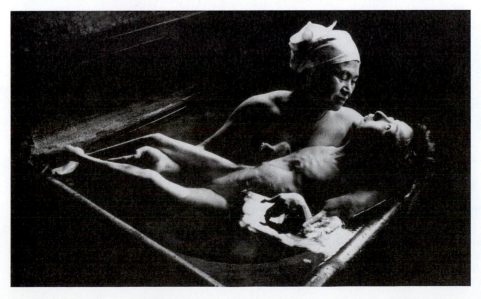

Figure 18–14   A victim of Minamata disease.

concentrate within their tissues many substances found in minute concentrations in seawater in a process called **bioaccumulation**. When animals eat other animals, some of these substances (including toxic chemicals) move up food chains and become concentrated in the tissues of larger animals in a process called **bioamplification**. Because the amount of mercury in the ocean has been increasing (mostly from the mercury in disposable batteries), some seafood such as tuna and swordfish were thought to contain unusually high amounts of mercury.

Studies done on the amount of seafood consumed by various human populations helped establish safe levels of mercury in fish to be marketed. To establish these levels, three variables were considered:

1. The rate at which each group of people consumed fish
2. The mercury concentration in the fish consumed by that population
3. The minimum ingestion rate of mercury that induces disease symptoms

These three variables help establish a maximum allowable mercury concentration that will safeguard people from mercury poisoning, as long as they don't exceed the recommended intake of fish.

Figure 18-15 shows the relative risk of contracting Minamata disease for people in the United States, Sweden, and Japan, including those from the Minamata fishing community. The graph shows that the risk increases with increased consumption of fish and that the higher the mercury concentration of the fish, the greater the risk.

Scientists have determined that the minimum level of mercury consumption that causes poisoning symptoms is 0.3 milligrams per day over a 200-day period. Figure 18-15 shows that for people in the United States, which have an average daily consumption of 17 grams of fish per day, mercury poisoning symptoms occur when mercury concentrations in fish exceed 20 parts per million (ppm). Using a safety factor of 10 times, the maximum concentration of mercury in fish that can be safely consumed by people in the U.S. is 2.0 ppm.

The U.S. Food and Drug Administration (FDA) doubled the safety factor and established a limit of mercury concentration for fish at 1 ppm. Based on consumption rates, this limit has adequately protected the health of U.S. citizens because essentially all tuna and most swordfish fall below this concentration. Still, the FDA issued an advisory in 2001 stating that pregnant women, women of childbearing age, nursing mothers, and young children should avoid eating certain kinds of fish that may contain high levels of methyl mercury, such as swordfish, sharks, king mackerel, and tilefish.

Figure 18-15 shows that for people in Sweden and Japan, the mercury concentration of fish deemed to be at a safe level is lower because these populations eat more

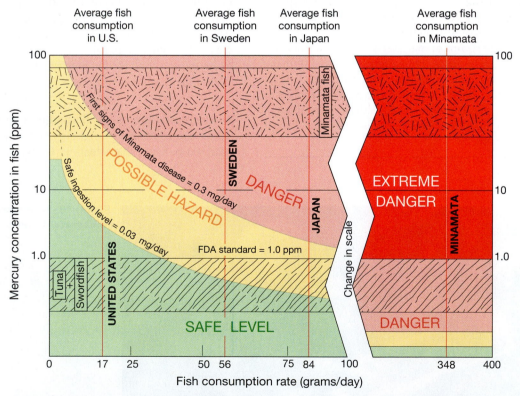

Figure 18–15 Mercury concentrations in fish versus consumption rates for various populations.

Graph showing the relative risk of contracting Minamata disease based on the amount of fish consumed and the concentration of mercury in fish for people in the U.S., Sweden, and Japan—including Minamata. It shows that the safe ingestion level of 0.03 milligrams of mercury per day is within the Food and Drug Administration's standard safety level of 1.0 ppm for fish and that tuna and most swordfish are safe to consume.

fish. The graph also shows the extreme danger to which residents of Minamata were inadvertently subjected when they ate so much of the highly contaminated fish from Minamata Bay.

Non-Point-Source Pollution and Trash

Non-point-source pollution—also called *"poison runoff"*—is any type of pollution entering the surface water system from sources other than underwater pipelines. Mostly, non-point-source pollution arrives at the ocean via runoff from storm drains, many of which have labels indicating they lead directly to the ocean (Figure 18-16).

Because non-point-source pollution comes from many different locations, it is difficult to pinpoint where it originates, although the *cause* of the pollution may be readily apparent. Trash that is washed down a storm drain to the ocean is one such example. Others include pesticides and fertilizers from agriculture and oil from automobiles that is washed to the ocean whenever it rains. In fact, the amount of road oil and improperly disposed oil regularly discharged each year into U.S. waters as non-point-source pollution is as much as 26 times the amount of the *Exxon Valdez* oil spill!

? STUDENTS SOMETIMES ASK ...

Don't storm drains receive treatment before emptying into the ocean?

Contrary to popular belief, water (and any other material) that goes down a storm drain does *not* receive any treatment before being emptied into a river or directly into the ocean. Sewage treatment plants receive enough waste to process without the additional runoff from storms, so it is important to monitor carefully what is disposed into storm drains. For instance, some people discharge used motor oil into storm drains, thinking that it will be processed by a sewage plant. A good rule of thumb is this: *Don't put anything down a storm drain that you wouldn't put directly into the ocean itself.*

Trash enters the ocean as a result of ocean dumping, too. Private, recreational, research, commercial, military, and other vessels all dump various forms of waste into the ocean. Much of this material sinks or biodegrades and does not accumulate at the surface.

Plastic, however, floats and is not biodegradable, so it can remain in the marine environment indefinitely (Box 18-2). Plastic waste has strangled marine organisms and birds that have been caught in plastic netting and packing straps (Figure 18-17). Marine turtles have been killed when they ingested plastic bags, evidently mistaking them for jellyfish or other transparent plankton on which they typically feed. Thus, plastic is one of the few substances that is illegal to dump anywhere in the ocean.

The increasing amount of non-point-source pollution that enters the oceans has resulted in increasing numbers of beach closings. In 2000, for example, there were 11,270 U.S. beach

Figure 18–16 A labeled storm drain that leads to the ocean.

(a)

(b)

Figure 18–17 Floating plastic strangles marine life.

(a) A female northern elephant seal (*Mirounga angustirostris*) with a plastic packing strap around its neck. **(b)** A herring gull (*Larus argentatus*) with a plastic six-pack ring around its neck.

BOX 18–2 People and Ocean Environment

FROM A TO Z IN PLASTICS: THE MIRACLE SUBSTANCE?

Even though plastic products have been used for over a century, their commercial development occurred during World War II when shortages of rubber and other materials created great demand for alternative products. Plastic products are *lightweight, strong, durable,* and *inexpensive,* so they have many advantages over other materials. Today, everything from airplane parts to zippers is made of plastic (Figure 18C). We wear plastics, drive in plastics, cook in plastics, and even carry plastic components inside us as artificial parts. The convenience of plastic items intended for one-time use has also contributed to the popularity of plastics.

However, what was once thought of as a miracle substance has several disadvantages. Disposing of plastics has already strained the capacity of land-based solid-waste disposal systems. Plastic waste is now an increasingly abundant component of oceanic flotsam (floating refuse). In fact, plastics constitute the vast majority of floating trash in all oceans

worldwide. Unfortunately, the very same properties that make plastics so advantageous make them unusually persistent and damaging when released into the marine environment:

- They are *lightweight*, so they float and concentrate at the surface.
- They are *strong*, so they entangle marine organisms.
- They are *durable*, so they don't biodegrade easily, causing them to last almost indefinitely.
- They are *inexpensive*, so they are mass-produced and used in almost everything.

Small pellets called "*nurdles*" ranging in size from a BB to a pea are used to produce plastic products (Figure 18D). They are transported in bulk aboard commercial vessels and are found throughout the oceans, probably due to spillage at loading terminals. In coastal waters, plastic products used in fishing are commonly thrown overboard by recreational and commercial vessels. Plastic trash also finds its way into the open-ocean waters from non-point-source pollution by careless people on land.

Even though plastic material does not sink or biodegrade, ocean currents eventually wash it onto the beaches of islands and continents. Thus, even remote beaches are littered with plastic pellets and plastic trash. Studies conducted between 1984 and 1987 showed that the plastic pellet content of beaches throughout the world is increasing. For instance, some Bermuda beaches have up to 10,000 pellets per square meter (10.7 square feet), and some beaches on Martha's Vineyard in Massachusetts yielded 16,000 plastic spherules per square meter (10.7 square feet).

Plastic trash is not limited to beaches. In the northern Sargasso Sea, researchers have counted more than 10,000 plastic pieces and 1500 pellets per square kilometer (0.4 square mile). Between 1972 and 1987, the concentration of plastic pellets doubled, suggesting that the amount of floating plastic trash in the ocean has doubled, too. Indeed, equipment deployed from research vessels often returns entangled in plastic trash such as 6-pack rings, styrofoam, and fishing lines, nets, and floats.

What can be done to limit the amount of plastic in the marine environment? People can limit their use of disposable plastic, recycle plastic material, and dispose of their plastic trash properly, including not dumping any plastic at sea. If these simple guidelines are followed, it would greatly reduce the amount of plastics in the world's oceans.

Figure 18C Plastic products.

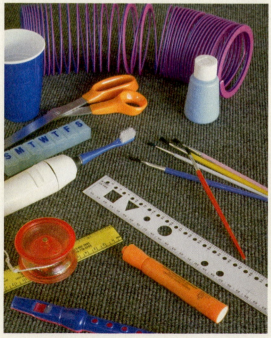

Figure 18D Plastic pellets ("*nurdles*") found at a beach.

closings and advisories issued, which is nearly double that of the previous year. Part of the reason for the unusually high number lies in heavy rains that many states experienced that year. In addition, there has been increased monitoring, better testing standards for bacteria and other pathogens, and more thorough reporting.

Waste Heat

Coastal electrical generating plants use seawater to cool and condense steam. After passing through a heat-exchange system, the seawater is then returned to the ocean up to 11°C (20°F) warmer. This **waste heat** is a form of *thermal pollution* that has the potential to negatively affect coastal environments. Although the heat can create a microenvironment suitable for some warm-water organisms, it can stress other marine organisms, particularly those that are *stenothermal*.[5]

To minimize the coastal environmental impact of waste heat, power plants are dumping their waste heat in deeper waters further offshore. In addition, some power plants are being designed to pump colder water from deeper offshore, so that when the water is heated and released, it more closely matches normal sea surface temperatures.

> Examples of marine pollution include petroleum from oil spills, sewage sludge, radioactive waste, chemicals such as DDT and PCBs, mercury, non-point-source pollution such as road oil and trash, and waste heat.

Other Concerns

Many other environmental concerns affect the marine environment. Among them are the proliferation of non-native species, ocean eutrophication, ozone depletion, and habitat destruction.

Non-Native Species

Non-native species (also called *exotic, alien,* or *invasive species*) are those that originate in a particular area but are introduced into new environments, either by the deliberate or accidental actions of humans. Because they inhabit new areas where they lack predators or other natural controls, non-native species can wreak ecological havoc by outcompeting and dominating native populations. Non-native species can also introduce new parasites and/or diseases. In some cases, non-native species completely transform ecosystems. In the U.S. alone, more than 7000 introduced species (not counting microorganisms) have been documented, of which about 15% cause ecological or economic damage. In fact, invasive species

cause an estimated $137 billion in loss and damages in the U.S. each year.

The seaweed *Caulerpa taxifolia*, which is native in tropical waters in limited patches, is one example of an invasive, non-native marine species. It is ideal as a decorative alga in saltwater aquariums because it is hardy, fast growing, and not edible by most fish. However, when it is introduced into suitable new habitats (probably as a result of dumping of household saltwater aquariums), it becomes a dominant and persistent species that displaces native seaweeds and other marine life. A cold-tolerant clone of *Caulerpa taxifolia* produced for the aquarium industry was first introduced into the Mediterranean Sea in 1984 where it has overwhelmed aquatic ecosystems and continues to spread. This clone has also been found in the lagoons of southern California (Figure 18-18), where it was first reported in 2000, but eradication efforts appear to have been successful. Because *Caulerpa taxifolia* can regenerate from very small fragments, however, repeated surveys are being conducted in these lagoons to eliminate all remaining occurrences of the seaweed. Recently, a new infestation of *Caulerpa taxifolia* has been reported near Sydney, Australia.

Another example of a non-native aquatic species is the **zebra mussel** (*Dreissena polymorpha*), which has invaded the Great Lakes region. After entering North America more than a decade ago (probably in the ballast water[6] of a freighter from Europe), these mussels have proliferated rapidly in waters of eastern Canada and the U.S. In doing so, they've driven out native mussels, altered the ecology of freshwater lakes and streams, and blocked the water-carrying pipes of power plants and many other industrial facilities. Although they are exceedingly hardy organisms, researchers are working to identify predators, parasites, and infectious microbes that can kill zebra mussels but leave native populations unharmed.

Other notable examples of harmful non-native aquatic species include the Atlantic comb jelly *Mnemiopsis leidyi*, which was transported in ballast water to the Black Sea and has done extensive damage; the Atlantic cordgrass *Spartina alterniflora*, which has invaded soft-bottom coasts of California and Washington; the water hyacinth *Eichhornia crassipes*, which infests tropical estuaries and other water bodies; and the European green crab *Carcinus maenas*, which has invaded the Pacific Coast and is altering coastal food webs.

Eutrophication

Ocean **eutrophication** (*eu* = good, *tropho* = nourishment, *ation* = action) is the artificial enrichment of waters by a previously scarce nutrient that can result in dangerous phytoplankton blooms. Large areas of ocean eutrophication are associated with extensive *hypoxic* (*hypo* = under, *oxid* = oxygen) **dead zones** that often occur near the mouths of major rivers after large spring runoffs. Oxygen

---

[5]Recall from Chapter 13 that the term *stenothermal* (*steno* = narrow, *thermo* = temperature) describes organisms that can withstand only very small temperature changes.

---

[6]Ballast water is taken into the hold of a ship to enhance stability and then released in a port when it is no longer needed.

Figure 18–18 Invasion of *Caulerpa taxifolia*.

The seaweed *Caulerpa taxifolia* is thought to have been illegally dumped from an aquarium into southern California coastal lagoons. The alga was eradicated largely because of good public awareness and quick action.

levels within these dead zones drop from above 5.0 parts per million (ppm) to below 2.0 ppm, which is lower than most marine animals can tolerate. Some of the more mobile marine organisms can flee the area, but it kills many bottom-dwelling organisms that cannot swim or crawl.

One of the most prominent dead zones is the one that forms each summer near the mouth of the Mississippi River in the Gulf of Mexico (Figure 18-19). In 1999, in fact, it reached a record size of 20,000 square kilometers (7700 square miles)—about the same size as the state of New Jersey. Smaller dead zones have occurred each summer in the region for decades, but they have increased dramatically in size after the record-breaking Midwest floods in 1993.

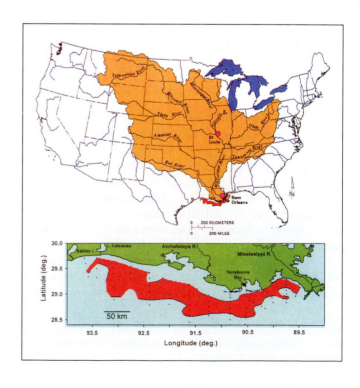

Figure 18–19 The Gulf of Mexico dead zone.

Map (*above*) showing Mississippi River drainage basin (*yellow*) and enlargement (*below*) showing extent of the 1999 Gulf of Mexico dead zone (*red*). The dead zone is a low-oxygen region related to excess nutrients from agricultural runoff from the Mississippi River that stimulates algal blooms. When the algae die, decomposition consumes vast amounts of oxygen, which creates an anoxic dead zone that often affects a multitude of marine life.

The Gulf's dead zone appears to be related to runoff of nutrients from agricultural activities that are washed down the Mississippi River, eventually reaching the Gulf, where they trigger algal blooms. Once these algae die and rain down on the sea floor, bacteria feed on them and on fecal matter, depleting the water of oxygen along the bottom. Similar nutrient pollution has degraded more than half of U.S. estuaries.

Although many other coastal waters experience summer-time hypoxic conditions, the Gulf's dead zone is the largest in the Western Hemisphere. Proposals to combat the spread of the Gulf's dead zone include controlling nutrient runoff from agriculture, preserving and utilizing wetlands that filter runoff before it enters the Gulf, planting buffer strips of trees and grasses between farm fields and streams, altering the times when fertilizers are applied, improving crop rotation, and enforcing existing clean water regulations.

Ozone Depletion and Phytoplankton Production

Atmospheric scientists first established in 1985 that **ozone** (the molecule O_3) in the Antarctic *stratosphere* (part of the upper atmosphere) was being significantly depleted by the actions of human-made chemicals called **chlorofluorocarbons (CFCs)**. Stratospheric ice particles and aerosols catalyze a CFC chemical reaction that produces chlorine monoxide (ClO), which combines with ozone and destroys it. Cold temperatures (as low as $-78°C$ or $-108°F$) also have been shown to accelerate the reaction, which occurs in the presence of sunlight. During each Antarctic spring, the presence of cold temperatures and sunlight can reduce O_3 by as much as 50%, which creates a hole in the South Pole stratospheric ozone layer about the size of Antarctica (Figure 18-20). In the Arctic, low stratospheric temperatures do not last as long, so the Artic ozone hole does not usually develop to the extent that it does in the Southern Hemisphere.

Because ozone protects Earth from ultraviolet (UV) radiation, decreased ozone thickness has resulted in a 10% increase in ultraviolet B (UVB) radiation—the wavelength of radiation that most damages DNA.

In fact, exposure to increased UVB radiation has been linked to the recent increase in human skin cancer.

How will increased UV radiation affect life in the oceans, where it is absorbed in surface waters, thereby providing some degree of protection for marine organisms? A recent study conducted in waters near Antarctica showed that phytoplankton production was reduced by as

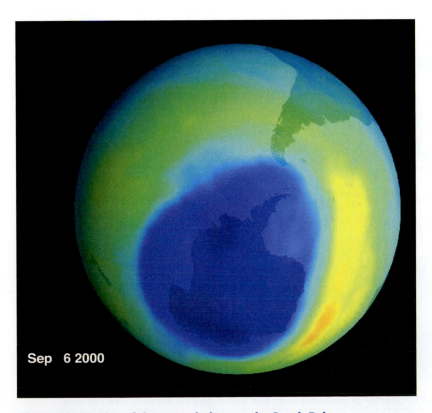

Sep 6 2000

Figure 18–20 Size of the ozone hole over the South Pole.

The Southern Hemisphere ozone hole in September 2000. The area over Antarctica, shown in violet, has the lowest ozone concentration.

much as 12%, which is a very significant cause for concern because Antarctic productivity supports a huge biomass of other organisms.

Concentrations of ozone-depleting compounds in the atmosphere peaked in 1993–1994 and have been decreasing ever since, leading to progressively smaller yearly Antarctic ozone holes. Very short-lived gases such as HCFC-123 that pose only a minor risk to the ozone layer are being phased in for use in refrigeration, so it is possible that ozone depletion can be entirely halted within the next decade.

Coastal Population Increase and Habitat Destruction

The migration of the U.S. population toward coastal towns and villages has occurred in the last half of the 20th century and continues today, with over 80% of the American population now living within 50 miles of the coast or Great Lakes. In fact, 14 of the country's 20 largest urban corridors are along the nation's coast and a major potion of U.S. economic infrastructure is near or on the ocean. Globally, the situation is similar: Over 50%—some 3.2 billion people—live along a coastline today, and this figure is expected to rise to 75% by 2025.

The large-scale movements of populations to coastal areas have been coupled with a significant increase in economic activity and industrialization along the coastline—such as oil and gas exploration, mining, fish farming, tourism, development of ports, marines and coastal defenses—putting enormous pressure on coastal areas. As a result, many coastal habitats have been reduced or completely lost to urbanization (Figure 18-21). Those that remain are plagued by pollution and other destructive human activities. For instance, nearly one-third of U.S. coastal waters used for the harvest of oysters, clams, and mussels are classified as "harvest-limited" due to contamination from pollution.

Highly productive coastal habitats are important for many marine species, which use these habitats for breeding or mating purposes. For example, a recent study of lemon sharks—which have low reproduction rates— found that they always return to the same lagoon to give birth. To conserve this and other threatened species, marine biologists need to identify and help protect traditional nursery grounds as well as the sharks themselves.

Coastal habitats are also important for preserving biodiversity. In fact, a recent study reported that the greatest threat to the 2500 species in the U.S. listed as either imperiled or endangered is habitat destruction, followed by invasion by non-native species, pollution, overexploitation, and disease.

> Other marine environmental concerns include the invasion of non-native species, eutrophication (which produces "dead zones"), ozone depletion by CFCs, and coastal habitat destruction.

Figure 18–21 Coastal population increase and resulting habitat destruction.
Coastal development such as that shown here on North Myrtle Beach, South Carolina, negatively affects coastal habitats such as the tidal marshes and lagoons seen behind the houses.

Laws Governing Marine Waters

The United States has long been concerned with legal protection of the marine environment, so it has been a leader in enacting laws protecting coastal and offshore waters.

Marine Pollution Control in the United States

In the 19th century, increasing amounts of pollution in U.S. waters led to the **Rivers and Harbors Act** of 1899, which provided a straightforward prohibition against dumping refuse into the navigable waters of the United States. Today, most current regulations on waste disposal within the 200-nautical-mile (370-kilometer) *Exclusive Economic Zone* (see Figure 17-1) are contained in two major statutes. The **Federal Water Pollution Control Act** (Clean Water Act) of 1948 and its amendments address point sources of municipal and industrial waste and spills of oil and hazardous materials. The **Marine Protection, Research, and Sanctuaries Act** (Ocean Dumping Act) of 1972 controls dumping of wastes at sea, at-sea research, and the establishment of marine sanctuaries.

Although the regulations are quite complex, these laws: (1) prohibit the dumping of materials known to be harmful, and (2) specify the criteria under which other materials may be dumped. The basic premise of each seems to be that land disposal is preferable to marine disposal. If it cannot be proved to the satisfaction of the regulating agencies that dumping of a material will not adversely affect human health, it will not be allowed to be dumped in the ocean. In fact, the 1991 **Ocean Dumping Ban Act** banned all sewage and industrial-waste dumping in U.S. offshore waters.

In 2000, the U.S. congress approved the **Oceans Act of 2000**, which establishes a 16-member **U.S. Commission on Ocean Policy** to undertake a thorough review of U.S. ocean and coastal programs and activities. Further, the commission will make recommendations for a coordinated and comprehensive national ocean policy, including changes to U.S. law to improve management, conservation, and use of ocean and coastal resources. Four working groups have been established to help the commission address the following issues: governance; research, education, and marine operations; stewardship; and investment and development.

Marine Protected Areas In an effort to protect marine resources, the United States has established **Marine Protected Areas (MPAs)**. MPAs are areas of the ocean in which some or all of the natural and cultural resources are protected. All MPAs focus on conservation, but the level of protection they offer varies widely. Most are open to fishing, although some exclude specific type of fishing gear or protect certain species; others restrict certain activities such as ocean dumping or drilling for oil and gas.

MPAs include 328 sites in the 35 U.S. coastal states, territories, and commonwealths. Of these, 251 are federally managed, including

- National Marine Sanctuaries (14 sites) (Figure 18-22)
- National Wildlife Refuge System (162 sites)
- National Park Service/National Seashore (39 sites)
- National Marine Fisheries Service (36 sites, including 23 Federal Fisheries Management Zone sites and 13 Federal Threatened/Endangered Critical Habitat and Species Protected Area sites)

MPAs provide a more focused, ecosystem-based approach to resource management. As such, they can help protect and restore various components of the nation's marine environment, including natural ecosystems, biodiversity, habitat, and endangered and threatened species. MPAs also preserve and protect important historical and cultural resources. One example is the *U.S.S.* Monitor *National Marine Sanctuary*, which protects a Civil War ship that sank off North Carolina in 1862.

Full protection of all species and habitats occurs only within special types of MPAs known as **marine reserves** (sometimes called *no-take zones, fully protected areas*, or *ecological reserves*). Currently, less than 1% of U.S. waters are fully protected in marine reserves.

International Efforts

Internationally, marine pollution is regulated by the 1973 **MARPOL Convention** (the International Convention for the Prevention of Pollution from Ships), which was modified in 1978. It specifies procedures for preventing operational or accidental causes of pollution of the marine environment by substances such as oil, noxious liquids carried in bulk, harmful substances carried in package form, sewage from ships, garbage from ships, and even air pollution while at sea.

Although consistent worldwide marine protection is difficult, one important program is the United Nation's **Action Plan for the Human Environment**, which was initiated in 1972. It established 12 worldwide environmental units and determined procedures for assessing and monitoring various sources of pollution. The Mediterranean Sea was identified as one of the most critically polluted environmental units, largely because of its geographic setting and large coastal population. In fact, marine pollution in the Mediterranean Sea comes from essentially every conceivable source—domestic sewage, industrial discharge, pesticides, and petroleum. As a result of international cooperation, an action plan that addresses pollution issues was ratified by the majority of nations with Mediterranean coasts.

Success of this program depends greatly on the existence of a regional scientific community interested in environment improvement work and government officials that respect the scientific community and the importance of their work. These conditions are met in only a few environ-

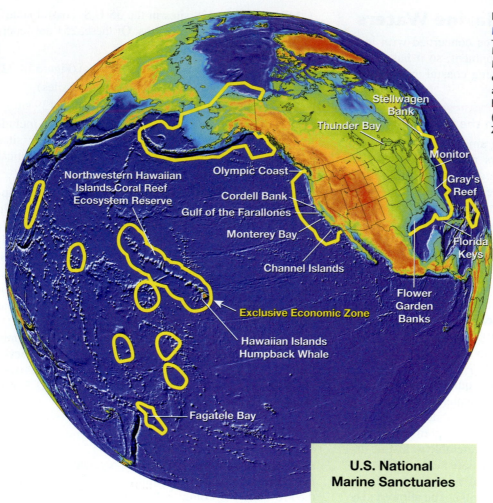

**U.S. National
Marine Sanctuaries**

Figure 18–22 **U.S. National Marine Sanctuaries.**

Map showing the 14 U.S. National Marine Sanctuaries, which were established to protect natural and/or cultural resources, and the boundary of the 200-nautical mile (370-kilometer) Exclusive Economic Zone (*yellow line*).

mental units. In addition, political tensions and frequent wars have had a dramatic negative effect in some regions.

? STUDENTS SOMETIMES ASK ...

All this ocean pollution is distressing. Is there anything I can do about it?

Yes, there are many things you can do to help protect the ocean, some of which are listed in this book's Afterword. They all involve making intelligent choices and minimizing your impact on the environment. Non-point-source pollution, for example, is something that the general public is directly responsible for, so one of the best methods of prevention may be educating people. Once people understand the impact our choices have on the environment, then the solution is up to *all* of us.

Current Laws on Ocean Dumping

What are the laws regulating ocean dumping? Under the MARPOL agreement and U.S. federal law, it is illegal for any vessel to discharge into the ocean plastics or garbage containing plastics. Figure 18-23 shows additional restrictions on dumping nonplastic waste. It shows that certain types of materials such as glass, metal, rags, and food can be *legally* dumped in the ocean—as long as they are dumped far enough away from shore or ground up small enough.

Ocean dumping is regulated by U.S. and international laws that prohibit the dumping of materials known to be harmful and specify the criteria under which other materials may be dumped.

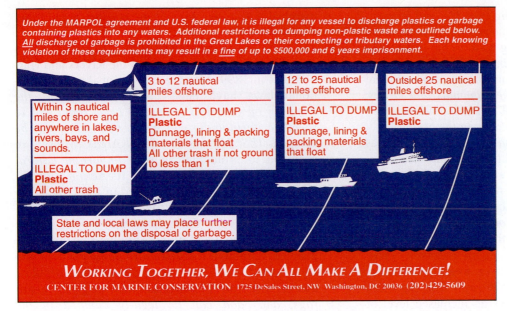

Under the MARPOL agreement and U.S. federal law, it is illegal for any vessel to discharge plastics or garbage containing plastics into any waters. Additional restrictions on dumping non-plastic waste are outlined below. _All_ discharge of garbage is prohibited in the Great Lakes or their connecting or tributary waters. Each knowing violation of these requirements may result in a _fine_ of up to $500,000 and 6 years imprisonment.

Within 3 nautical miles of shore and anywhere in lakes, rivers, bays, and sounds.

ILLEGAL TO DUMP
Plastic
All other trash

3 to 12 nautical miles offshore

ILLEGAL TO DUMP
Plastic
Dunnage, lining & packing materials that float
All other trash if not ground to less than 1"

12 to 25 nautical miles offshore

ILLEGAL TO DUMP
Plastic
Dunnage, lining & packing materials that float

Outside 25 nautical miles offshore

ILLEGAL TO DUMP
Plastic

State and local laws may place further restrictions on the disposal of garbage.

WORKING TOGETHER, WE CAN ALL MAKE A DIFFERENCE!
CENTER FOR MARINE CONSERVATION 1725 DeSales Street, NW Washington, DC 20036 (202)429-5609

Figure 18–23 Current law regulating ocean dumping.

Many types of trash can be legally dumped in the ocean, as long as it is ground fine enough and not composed of plastic. In fact, plastic is the only substance that cannot be dumped anywhere in the ocean.

Chapter in Review

- *Although marine pollution seems easily defined, an all-encompassing definition is quite detailed.* It is often difficult to establish the degree to which pollution affects ocean areas. The most widely used technique for determining the effects of pollution is the *standard laboratory bioassay*, which determines the concentration of pollution that causes a *50% mortality rate* among marine organisms. The debate continues about whether society's wastes should be dumped in the ocean.

- *Oil is a complex mixture of hydrocarbons and other substances*, most of which are naturally *biodegradable*. Thus, *many experts consider oil to be among the least damaging of all substances* introduced into the marine environment. In areas that have experienced oil spills, *recovery can be as rapid as a few years.* Still, *oil spills can cover large areas and kill many animals.*

- *Oil pollution reduces species diversity and persists longer on muddy bottoms than on sandy or rocky bottoms.* Oil spills that do not come ashore, such as that of the *Argo Merchant,* do considerably less environmental damage than those that do. The greatest damage from the *Argo Merchant* spill was probably to plankton, especially pollock and cod eggs. After the much-publicized *Exxon Valdez* oil spill in Alaska, *many novel approaches were used to clean the spilled oil,* including oil-eating bacteria (*bioremediation*).

- *Millions of tons of sewage sludge have been dumped offshore in coastal waters.* Although 1972 legislation required an end to dumping of sewage in the coastal ocean by 1981, exceptions continue to be made. In-creased public concern resulted in *new legislation to prohibit sewage dumping in the ocean.*

- *Most of the radioactive waste in the ocean comes from fallout* resulting from the testing of nuclear explosives since 1944. However, *areas with the greatest concentrations of radionuclides entering the ocean are near fuel-reprocessing plants.* After nearly five decades of production of high-level nuclear waste, disposal of this material still represents an unsolved and complex problem. Although the presently favored option is land disposal, research has been done to identify *deep-ocean sea floor sediments as the most suitable marine disposal sites.*

- *DDT and PCBs are persistent, biologically hazardous chemicals that have been introduced into the ocean by human activities.* DDT pollution produced a decline in the Long Island *osprey population* in the 1950s and the brown *pelican population* of the California coast in the 1960s. *Virtual cessation of DDT use* in the Northern Hemisphere in 1972 allowed the *recovery of both populations.* The DDT thinned the eggshells and reduced the number of successful hatchings. PCBs have been implicated in causing health problems in sea lions and shrimp.

- *Mercury poisoning was the first major human disaster resulting from ocean pollution.* It is now called *Minamata disease* after the bay in Japan where it first occurred in 1953. *Mercury accumulates in the tissues of many large fish,* most notably tuna and swordfish, and works its way up the food web. To prevent mercury poisonings in the United States, *stringent mercury contamination levels in fish have been established by the FDA.*

- *Plastics are lightweight, strong, durable, and inexpensive.* Unfortunately, these same properties make them a relentless source of floating trash in the ocean. *The amount of plastic accumulating in the oceans has increased dramatically.* Certain forms of plastic are known to be *lethal to marine mammals and turtles,* and national and international legislation has been enacted to *ban the disposal of plastic in the oceans.*
- The introduction of *non-native species* (such as *Caulerpa taxifolia* and zebra mussels) into new environments can cause *severe ecological or economic damage. Eutrophication is a result of runoff carrying nutrients to the ocean, which causes algae blooms and resulting low-oxygen conditions* such as the Gulf of Mexico hypoxic dead zone that occurs each spring. *Human-made CFCs are reducing the concentration of Earth's stratospheric ozone layer, which protects organisms from damaging UVB rays.* The Southern Hemisphere *ozone hole* allows an increased level of UVB radiation to reach Earth, where it is having a *negative impact on phytoplankton* in the Antarctic

marginal ice zone. *Increasing coastal populations have led to coastal habitat destruction.*
- *The U.S. government has been concerned about coastal pollution* since it passed the Rivers and Harbors Act in 1899. More recent national and international laws have been passed to *prohibit the dumping of materials known to be harmful* and to *specify criteria under which other materials may be dumped. U.S. Marine Protected Areas (MPAs)* are designed to protect areas with *important natural and/or cultural resources.* With its *Action Plan for the Human Environment,* initiated in 1972, the United Nations Environment Program established *12 environmental units* throughout the world, within which assessment and monitoring have begun or are planned. Although it is *illegal to dump plastics into the ocean,* current laws on ocean dumping stipulate that *certain types of materials such as glass, metal, rags, and food can be legally dumped in the ocean*—as long as they are dumped far enough away from shore or ground up small enough.

Key Terms

Action Plan for the Human
  Environment (p. 539)
Bioaccumulation (p. 532)
Bioamplification (p.532)
Bioremediation (p. 525)
Caulerpa taxifolia (p. 535)
Chlorofluorocarbons (CFCs)
  (p. 537)
DDT (p. 530)
Dead zone (p. 535)
Eutrophication (p. 535)
Federal Water Pollution Control
  Act (p. 539)

Hydrocarbon (p. 520)
Levels of biological response
  (p. 520)
Marine Protected Areas (MPAs)
  (p. 539
Marine Protection, Research,
  and Sanctuaries Act (p. 539)
Marine reserves (p. 539)
MARPOL Convention (p. 539)
Mercury (p. 531)
Minamata disease (p. 531)
Non-native species (p. 535)

Non-point-source pollution
  (p. 533)
Ocean Dumping Ban Act
  (p. 539)
Oceans Act of 2000 (p. 539)
Ozone (p. 537)
PCBs (p. 530)
Petroleum (p. 520)
Pollution (p. 519)
Primary treatment (p. 525)
Radionuclides (p. 528)
Rivers and Harbors Act (p. 539)

Secondary treatment (p. 525)
Sewage sludge (p. 527)
Standard laboratory bioassay
  (p. 520)
Toxic compound (p. 522)
U.S. Commission on Ocean
  Policy (p. 539)
Waste heat (p. 535)
Zebra mussel (p. 535)

Questions and Exercises

1. Without consulting the textbook, define pollution. Then consider these items and determine if each one is a pollutant based on your definition (and refine your definition as necessary):
   a. Dead seaweed on the beach
   b. Natural oil seeps
   c. A small amount of sewage
   d. Warm water dumped into the ocean

2. Discuss the different levels of biological response to the effects of pollution. Also, discuss the symptoms by which these effects can be recognized and the time typically required for them to manifest themselves.

3. Why would many marine pollution experts consider oil among the *least* damaging pollutants in the ocean?

4. Describe the effect of oil spills on species diversity and recovery of bottom-dwelling organisms based on the experience at Wild Harbor.

5. Compare oil spills that wash ashore to those that do not, such as the *Argo Merchant,* in terms of destruction to marine life.

6. Discuss techniques used to clean oil spills. Why is it important to begin the cleanup immediately?

7. When and where was the world's largest oil spill? How many times larger was it than the *Exxon Valdez* oil spill?

8. How would dumping sewage in deeper water off the East Coast help reduce the negative effects to the ocean bottom?

9. What components of the nuclear energy generation industry introduce the most radioactive wastes into the oceans?

10. Discuss the animal populations that clearly suffered from the effects of DDT and the way in which this negative effect was manifested.

11. What causes Minamata disease? What are the symptoms of the disease in humans?

12. What is non-point-source pollution and how does it get to the ocean? What other ways does trash get into the ocean?

13. What properties contributed to plastics being considered a miracle substance? How do those same properties cause them to be unusually persistent and damaging in the marine environment?

14. What are non-native species? Why can they be so damaging to ecosystems?

15. Describe what the Gulf of Mexico dead zone is, how it forms, and why it is so deadly to marine organisms.

16. Describe the effect of CFCs in terms of stratospheric ozone depletion and the resulting increased intensity of UVB radiation.

17. Discuss the laws governing pollution in marine waters.

Afterword

At the end of our journey through this book together, it seems appropriate to examine people's perceptions of the ocean. Many people describe the ocean as "powerful," "awe inspiring," "moving," "serene," "abundant," and "majestic." Others call it "vast," "infinite," or "boundless." These are all appropriate descriptions, because even though humans have been exploring and studying the ocean for centuries, the ocean still holds many secrets. The oceans are constantly surprising researchers with unusual features, animals, and geologic wonders that exist in its watery world.

Despite the ocean's impressive size, it is beginning to feel the effects of human activities. Every ocean contains floating plastic trash, for instance, and even remote beaches are littered with trash. Organisms living in the ocean also feel the effect of humankind's use of the ocean. Whaling in the 19th and 20th centuries, for example, pushed many great whale populations to near extinction. Restrictions on whaling and the development of substitutes for whale products helped whales survive this threat, but now they face destruction of their feeding and breeding grounds. Overfishing, however, more than any other human activity, has altered the marine ecosystem, suggesting that the ocean is not quite as "vast," "infinite," or "boundless" as most people believe.

Whether it is coral bleaching, increased concentrations of greenhouse gases in the lower atmosphere, depletion of upper-atmosphere ozone, the increased frequency of red tides and "dead zones," or polluting the ocean with petroleum, plastics, sewage, chemicals, or toxin-laden sediment, human-induced changes in the environment are broad and far-reaching. All of these problems may be symptoms of a worldwide pathology (sickness) that will require major changes in human behavior before it can be cured.

Marine Sanctuaries and Marine Reserves

In 1972, the U.S. Congress began establishing national **marine sanctuaries** to protect vital pockets of the ocean from further degradation. Today, 14 national marine sanctuaries cover more than 6200 square kilometers (2400 square miles) in areas such as the Florida Keys, the Stellwagen Bank off Massachusetts, the Channel Islands off southern California, the Flower Garden Banks in the Gulf of Mexico, and the Hawaiian Islands. Many activities that degrade the marine environment, however, are still allowed in marine sanctuaries, such as fishing, recreational boating, and even mining of some resources.

Many who recognize the importance of preserving vital marine habitats are calling for governments to set aside large "no-take" refuges called **marine reserves** in which fishing and other activities are prohibited. Such action would allow the recovery of heavily overfished stocks and protect sea floor communities decimated by trawling the sea floor with nets. Currently, however, less than 1% of U.S. waters are fully protected in marine reserves. Although the fishing industry has opposed the creation of marine reserves, research indicates that marine reserves provide benefits to surrounding regions by boosting populations of fish outside their borders. Other countries are also realizing the economic benefit of fully protecting ocean resources. A healthy coral reef, for example, can be worth more as a tourist draw than it might be as a source of seafood.

Although the ocean is feeling the effects of humans, it is very resilient and has a tremendous ability to withstand change. Studies reveal that once human impact is reduced, the ocean often returns to a near pristine state because natural processes in the ocean tend to disperse and eventually remove many types of pollutants. The vast majority of ocean water, moreover, is still relatively unpolluted, except in shallow-water coastal areas near large population centers and near the mouths of major rivers.

What Can I Do?

Because the ocean is vast and capable of absorbing many substances, it has been used as a dumping ground for many of society's wastes. Even today, humans are adding pollutants to the ocean at staggering rates. What can each of us do to help? Some ways to help the environment in general and the ocean in particular include the following:

- Minimize your impact on the environment. Reduce the amount of waste you generate by making wise consumer choices. Avoid products with excessive packaging and support companies that have good environmental records. Reuse and recycle items, then help close the recycling loop by buying goods made of recycled materials. Do simple things that make a positive impact on the environment (Box Aft-1).

- Become politically aware. Many ocean-related issues come before the public and require a majority of voters to approve a proposal before they are enacted into law. This "majority rules" is true for local as well as national and international issues. Within our lifetimes, for instance, we may very well decide whether to spend large amounts of money to add finely ground iron to the ocean to reduce the amount of carbon dioxide in the atmosphere. Many political issues in the future will involve the ocean.

TEN SIMPLE THINGS YOU CAN DO TO PREVENT MARINE POLLUTION[1]

There are many simple things you can do every day to help prevent marine pollution, such as the following:

1. **Snip six-pack rings.** Six-pack rings entangle many marine organisms, so snip each circle with a scissors before you toss them into the garbage. If you find any at a beach, pick them up, snip them, and discard them properly.

2. **Don't overfertilize your lawn and use a clean laundry detergent.** Lawn fertilizers contain nitrates and phosphates, which cause algae blooms when runoff from land enters the ocean. Detergents also contain phosphates, so read the detergent label to find one that is low in phosphate or phosphate free. Using a smaller amount of detergent than recommended also helps.

3. **Clean up after your pet.** Dog and cat feces that wash into a stream and eventually to the ocean also provide nitrates and phosphates, which create algae blooms.

4. **Make sure that your car doesn't leak oil.** Oil that leaks from automobiles is responsible for a large percentage of the oil that gets into the ocean. Annually, the amount of oil that enters the ocean as runoff from road sources is greater than a major oil spill. If you change your car's oil yourself, be sure to recycle the used oil at an appropriate recycling center.

5. **Drive less and carpool more.** Reducing the amount of gasoline you use reduces the amount of oil that must be transported across the ocean, which minimizes the potential for oil spills.

6. **Take your own bags to the grocery store.** Although paper bags are biodegradable, plastic bags are not and are becoming an increasing problem in the ocean, especially for animals like sea turtles that eat plastic bags when they mistake them for jellyfish.

7. **Don't release balloons.** Balloons that are released far from the ocean can still wind up there, where they deflate, quickly loose their color, and resemble drifting jellyfish that marine animals can ingest.

8. **Don't litter.** Any material that is carelessly discarded on land can become non-point-source pollution when it washes down a storm drain, into a stream, and eventually into the ocean.

9. **Pick up trash at the beach or volunteer for an organized beach clean-up.** It is important to remove trash that washes up at the beach so that it can't endanger marine organisms. Trash arrives at the beach from non-point-source pollution, ships, recreational boaters, beachgoers, and other sources. It is truly surprising (and somewhat horrifying) to discover what winds up at the beach.

10. **Inform and educate others.** Many people are unaware that their actions have a negative influence on the environment—especially the marine environment.

Nobody made a greater mistake than he who did nothing because he could only do a little.

—Edmund Burke (circa 1790)

---

[1]With thanks to The Earthworks Group, 1989, *50 Simple Things You Can Do to Save the Earth*, Berkeley, CA: Earthworks Press.

• Educate yourself about how the ocean works. A recent poll indicates that over 90% of the American public consider themselves scientifically illiterate. With our society becoming more scientifically advanced, people need to understand how science operates. Science is not meant to be comprehended by an elite few. Rather, science is for everyone. By studying oceanography, you have begun to understand how the ocean works. It is our hope that you will be a life-long student of the ocean.

In the end, we will conserve only what we love. We love only what we understand. We will understand only what we are taught.
    —Baba Dioum, Senegalese conservationist (1968)

Figure Aft-1 Sunset at the ocean.

Appendix I

Metric and English Units Compared

How many inches are there in a mile? How many cups in a gallon? How many pounds in a ton? In our daily lives, we often need to convert between units. Worldwide, the metric system of measurement is the most widely used system. Besides the United States, only *two other countries in the world*—Liberia and Myanmar (formerly Burma)—still use English units as their primary system of measurement. The metric system has many advantages over the English system. It is simple, logical, and easy to convert between units. For those of us in the United States, it is only a matter of time before the change to the metric system occurs.

On December 23, 1975, U.S. President Gerald R. Ford signed the Metric Conversion Act of 1975. It defined the metric system as the International System of Units (officially called the *Système International d'Unités*, or SI) as interpreted in the United States by the secretary of commerce. The Trade Act of 1988 and other legislation declared the metric system the preferred system of weights and measures for U.S. trade and commerce, called for the federal government to adopt metric specifications, and mandated the Commerce Department to oversee the program. All of this legislation comes 200 years after Benjamin Franklin first proposed that the United States adopt the metric system.

Although the metric system has not become the system of choice for most Americans' daily use, and there is great resistance to using it, the United States must change over in order to remain competitive in world markets. Many of the sciences are leading the way in this changeover. Oceanographers all over the world have been using the metric system for years.

The English System

The English system actually consists of two related systems—the U.S. Customary System[1] (used in the United States and dependencies) and the British Imperial System (used in Great Britain). Great Britain has now largely converted to the metric system. The basic unit of length in the English system is the *yard*; the basic unit of weight (not mass[2]) is the *pound*.

In the English system, the units of length were initially based arbitrarily on dimensions of the body. The yard as a measure of length, for example, can be traced to the early Saxon kings. They wore a sash around their waists that could be used as a convenient measuring device. Thus, the word *yard* comes from the Saxon word *gird* (like a girdle), in reference to the circumference of a person's waist. The circumference of a person's waist varies from person to person, however, and even varies from time to time on the *same* person, so it is of limited use. It had to be standardized to be useful, so King Henry I of England decreed that the yard should be the distance from the tip of his nose to the end of his thumb!

Romans initially defined the mile (*mille passuum* = 1000 paces) as an even 5000 feet. The early Tudor rulers in England, however, defined a *furlong* ("furrow-long") as 220 yards, based on the length of agricultural fields. To facilitate the conversion between miles and furlongs, Queen Elizabeth I declared in the 16th century that the traditional Roman mile of 5000 feet would be forever replaced by one of 5280 feet, making the mile exactly 8 furlongs. Today, a furlong is an archaic unit of measurement (except in horse races), but a mile is still 5280 feet.

The Metric System

The metric system of weights and measures was devised in France and adopted there in 1799. It is based on a unit of length called the *meter* and a unit of mass called the *kilogram*. Originally defined as one ten-millionth the distance from the North Pole to the Equator, the meter is now defined as the distance light travels through a vacuum in $\frac{1}{299,792,458}$ of a second. The kilogram was originally related to the volume of one cubic meter of water. It is now defined as the mass of the International Prototype Kilogram, a platinum-iridium cylinder kept at Sèvres (near Paris), France. Other metric units can be defined in terms of the meter and the kilogram.

Fractions and multiples of the metric units are related to each other by powers of 10, allowing conversion from one unit to a multiple of it simply by shifting the decimal point. The lengthy arithmetical operations required with English units can be avoided. Even the names of the metric units indicate how many are in a larger unit. For instance, how many cents are there in a dollar? It is the same as the number of *centi*grams in a gram. The prefixes listed in the following table, "Common Prefixes for Basic Metric Units," have been accepted for designating multiples and fractions of the meter, the gram (which equals $\frac{1}{1000}$ of a kilogram), and other units.

In spite of many fears that changing to the metric system will be an economic burden on businesses in the

---

[1] The names of the units and the relationships between them are generally the same in the U.S. Customary System and the British Imperial System, but the sizes of the units differ, sometimes considerably.

[2] The basic unit of mass in the English system is the *slug*, although it is rarely used.

Common Prefixes for Basic Metric Units

Factor	Name	Prefix Name
$1 \times 10^{12} = 1,000,000,000,000$	trillion	tera
$1 \times 10^{9} = 1,000,000,000$	billion	giga
$1 \times 10^{6} = 1,000,000$	million	mega
$1 \times 10^{3} = 1,000$	thousand	kilo
$1 \times 10^{2} = 100$	hundred	hecto
$1 \times 10^{1} = 10$	ten	deka
$1 \times 10^{-1} = 0.1$	tenth	deci
$1 \times 10^{-2} = 0.01$	hundredth	centi
$1 \times 10^{-3} = 0.001$	thousandth	milli
$1 \times 10^{-6} = 0.000001$	millionth	micro
$1 \times 10^{-9} = 0.000000001$	billionth	nano
$1 \times 10^{-12} = 0.000000000001$	trillionth	pico

United States, the conversion has already begun. Examples include 35-millimeter film, 2-liter soft drink bottles, 750-milliliter wine bottles, computers with gigabytes of memory, and 10 kilometer runs. Cars are now manufactured metrically in order to compete with other countries. Most people in the United States will experience the increasing use of the metric system in their daily lives. Some ways to ease yourself into this unfamiliar (yet sensible) system of units include noticing distances on road signs in kilometers, using a meterstick instead of a yardstick, and charting your weight in kilograms.

Temperature

The *Celsius (centigrade) temperature scale* was invented in 1742 by a Swedish astronomer named Anders Celsius. The scale was designed so that the freezing point of pure water—the temperature at which it changes state from a liquid to a solid—is set as 0 degrees. Similarly, the boiling point of pure water—the temperature at which it changes state from a liquid to a gas—is set at 100 degrees. Doing so made the liquid range of water—0 degrees to 100 degrees—an even 100 units, giving the scale its name: centigrade (*centi* = 100, *grade* = graduations).

The *Fahrenheit temperature scale*, on the other hand, sets water's freezing and boiling points at 32 degrees and 212 degrees, respectively, for a spread of 180 units. This scale was devised by a German-born physicist named Gabriel Daniel Fahrenheit, who invented the mercury thermometer in 1714. He initially designed the scale so that normal body temperature was set at 100° F. However, because of inaccuracies in early thermometers or how the scale was initially devised, normal human body temperature is now known to be 98.6°F (37°C).

Scientific Notation

Scientists often indicate very large or very small numbers by using *scientific notation*, which also simplifies arithmetic operations on such numbers. In scientific notation, one integer is placed to the left of the decimal, and an exponent (a power of 10) tells which direction and how far the decimal is moved to write a number in its long form. If the exponent is positive, the decimal is moved *to the right* that number of places; if the exponent is negative, the decimal is moved *to the left* that number of places. For example,

Scientific notation	Move the decimal	Long form
2.13×10^{5}	to the right 5 places	213,000
2.13×10^{-5}	to the left 5 places	0.0000213

The "Factor" column of the table "Common Prefixes for Basic Metric Units" shows other examples of numbers in scientific notation.

To add or subtract numbers written in scientific notation, they must first be converted to the same power:

Addition example		Subtraction example	
2.1×10^{3} $=$	0.021×10^{5}	3.4×10^{4} $=$	3.4×10^{4}
$+1.0 \times 10^{5}$	$+1.000 \times 10^{5}$	-2.0×10^{3}	-0.2×10^{4}
	1.021×10^{5}		3.2×10^{4}

To multiply or divide numbers written in scientific notation, the exponents are added or subtracted:

Multiplication example	Division example
6.04×10^{2}	3.0×10^{3}
$\times\ 2.10 \times 10^{4}$	$\div\ 1.5 \times 10^{2}$
12.684×10^{6}	2.0×10^{1}

To raise a number expressed as a power by a power, the powers are multiplied:

Raise a number example
$(10^{5})^{2} = 10^{10}$

The following tables show conversion units between various units.

Conversion Tables

Length

1 micrometer (μm)	0.001 millimeter
	0.0000394 inch
1 millimeter (mm)	1,000 micrometers
	0.1 centimeter
	0.001 meter
	0.0394 inch
1 centimeter (cm)	10 millimeters
	0.01 meter
	0.394 inch
1 meter (m)	100 centimeters
	39.4 inches
	3.28 feet
	1.09 yards
	0.547 fathom
1 kilometer (km)	1,000 meters
	1,093 yards
	3,280 feet
	0.62 statute mile
	0.54 nautical mile
1 inch (in)	25.4 millimeters
	2.54 centimeters
1 foot (ft)	12 inches
	30.5 centimeters
	0.305 meter
1 yard (yd)	3 feet
	0.91 meter
1 fathom (fm)	6 feet
	2 yards
	1.83 meters
1 statute mile (mi)	5,280 feet
	1,760 yards
	1,609 meters
	1.609 kilometers
	0.87 nautical mile
1 nautical mile (nm)	1 minute of latitude
	6,076 feet
	2,025 yards
	1,852 meters
	1.15 statute miles
1 league (lea)	5,280 yards
	15,840 feet
	4,805 meters
	3 statute miles
	2.61 nautical miles

Area

1 square centimeter (cm^2)	0.155 square inch
	100 square millimeters
1 square meter (m^2)	10,000 square centimeters
	10.8 square feet
1 square kilometer (km^2)	100 hectares
	247.1 acres
	0.386 square mile
	0.292 square nautical mile
1 square inch (in^2)	6.45 square centimeters
1 square foot (ft^2)	144 square inches
	929 square centimeters

Volume

1 cubic centimeter (cc; cm^3)	1 milliliter
	0.061 cubic inch
1 liter (l)	1,000 cubic centimeters
	61 cubic inches
	1.06 quarts
	0.264 gallon
1 cubic meter (m^3)	1,000,000 cubic centimeters
	1,000 liters
	264.2 gallons
	35.3 cubic feet
1 cubic kilometer (km^3)	0.24 cubic mile
	0.157 cubic nautical mile
1 cubic inch (in^3)	16.4 cubic centimeters
1 cubic foot (ft^3)	1,728 cubic inches
	28.32 liters
	7.48 gallons

Mass

1 gram (g)	0.035 ounce
1 kilogram (kg)	2.2 pounds
	1,000 grams
1 metric ton (mt)	2,205 pounds
	1,000 kilograms
	1.1 U.S. short tons
1 pound[a] (lb)	16 ounces
	454 grams
	0.454 kilogram
1 U.S. short ton (ton; t)	2,000 pounds
	907.2 kilograms
	0.91 metric ton

[a]The pound is a weight unit, not a mass unit, but is often used as such.

Pressure

1 atmosphere (atm) (at sea level)	760 millimeters of mercury
	14.7 pounds per square inch
	29.9 inches of mercury
	33.9 feet of fresh water
	33 feet of seawater

Speed

1 centimeter per second (cm/s)	0.0328 foot per second
1 meter per second (m/s)	2.24 statute miles per hour
	1.94 knots
	3.28 feet per second
	3.60 kilometers per hour
1 kilometer per hour (kph)	27.8 centimeters per second
	0.62 mile per hour
	0.909 foot per second
	0.55 knot
1 statute mile per hour (mph)	1.61 kilometers per hour
	0.87 knot
1 knot (kt)	1 nautical mile per hour
	51.5 centimeters per second
	1.15 miles per hour
	1.85 kilometers per hour

Oceanographic data

Velocity of sound in 34.85‰ seawater	4,945 feet per second
	1,507 meters per second
	824 fathoms per second
Seawater with 35 grams of dissolved substances per kilogram of seawater	3.5 percent
	35 parts per thousand
	35,000 parts per million
	35,000,000 parts per billion

Temperature

Exact formula	Approximation (easy way)
$°C = \dfrac{(°F - 32)}{1.8}$	$°C = \dfrac{(°F - 30)}{2}$
$°F = (1.8 \times °C) + 32$	$°F = (2 \times °C) + 30$

Some useful equivalent temperatures

100°C = 212°F	(boiling point of pure water)
40°C = 104°F	(heat wave conditions)
37°C = 98.6°F	(normal body temperature)
30°C = 86°F	(very warm—almost hot)
20°C = 68°F	(room temperature)
10°C = 50°F	(a warm winter day)
3°C = 37°F	(average temperature of deep water)
0°C = 32°F	(freezing point of pure water)

In degrees centigrade

Thirty is hot, twenty is pleasing;

Ten is not, and zero is freezing.

Appendix II

Geographic Locations

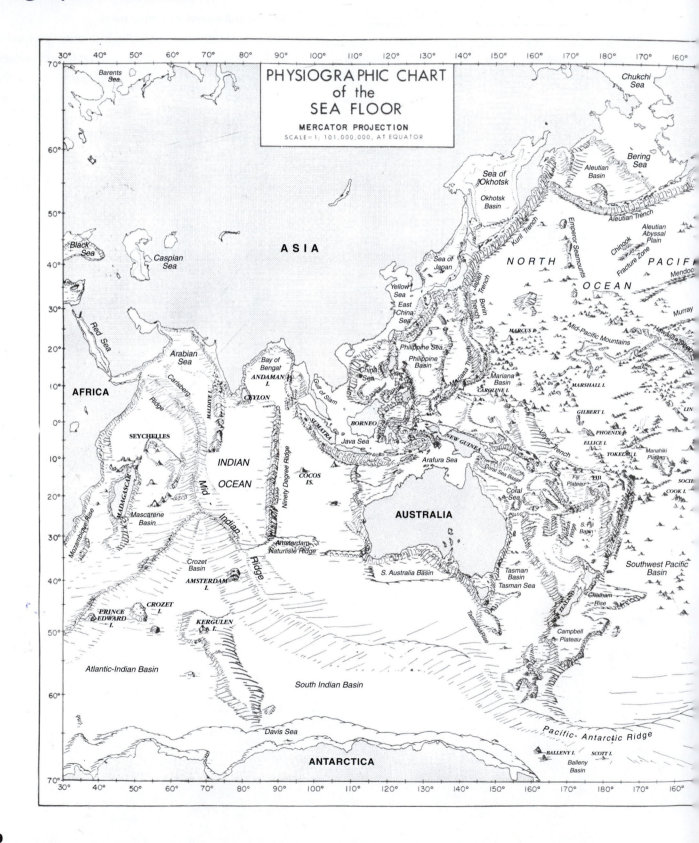

Appendix III

Latitude and Longitude on Earth

Suppose that you find a great fishing spot or a shipwreck on the sea floor. How would you remember where it is—far from any sight of land—so that you might return? How do sailors navigate a ship when they are at sea? Even using sophisticated equipment such as the global positioning system (GPS), which uses satellites to determine location, how is this information reported?

To solve problems like these, a navigational grid—a series of intersecting lines across a globe—is used. Once starting places are fixed for a grid, a location can be identified based on its position within the grid. Some cities, for example, use the regular numbering of streets as a grid. Even if you have never been to the intersection of 5th Avenue and 42nd Street in New York City, you would be able to locate it on a map, or know which way to travel if you were at 10th Avenue and 42nd Street. Although many grid (or coordinate) systems are used today, the one most universally accepted uses latitude and longitude.

A series of north–south and east–west lines that together comprise a grid system can be used to locate points on Earth's surface whether at sea or on land. The north–south lines of the grid are called *meridians* and extend from pole to pole (Figure A3-1). The meridian lines converge at the poles and are spaced farthest apart at the Equator. The east–west lines of the grid are called *parallels* because they are parallel to one another. The longest parallel is the *Equator* (so called because it divides the globe into two equal hemispheres), and the parallels at the poles are a single point (Figure A3-1).

Latitude and Longitude

Latitude is the angular distance (in degrees of arc) measured north or south of the starting line (the Equator) from the center of Earth. All points that lie along the same parallel are an identical distance from the Equator, so they all have the same latitude. The latitude of the Equator is 0 degrees, and the North and South Poles lie at 90 degrees north and 90 degrees south, respectively. Figure A3-2 shows that the latitude of New Orleans is an angular distance of 30 degrees north from the Equator.

Longitude is the angular distance (in degrees of arc) measured east or west of the starting line from the center of Earth. All meridians are identical (none is longer or shorter than the others, and they all pass through both poles), so the starting line of longitude has been chosen arbitrarily. Before it was agreed upon by all nations, different countries used different starting points for 0 degrees longitude. Examples include the Canary Islands, the Azores, Rome, Copenhagen, Jerusalem, St. Petersburg, Pisa, Paris, and Philadelphia. At the 1884 International Meridian Conference, it was agreed that the meridian that passes through the Royal Observatory at Greenwich, England, would be used as the zero or starting point of longitude. This zero degree line of longitude is also called the *prime meridian*, and the longitude for any place on Earth is measured east or west from this line. Longitude can vary from 0 degrees along the prime meridian to 180 degrees (either east or west), which is halfway around the globe and is known as the *International Date Line*. In Figure A3-2, New Orleans is 90 degrees west of the prime meridian.

A particular latitude and longitude defines a unique location on Earth, provided the direction from the starting point is given, too. For instance, 42 degrees north and 120 degrees west defines a unique spot on Earth because it is different from 42 degrees south and 120 degrees west and from 42 degrees north and 120 degrees east.

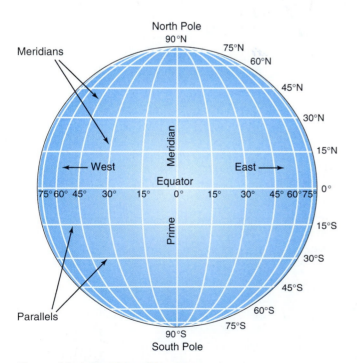

Figure A3–1 Earth's grid system.

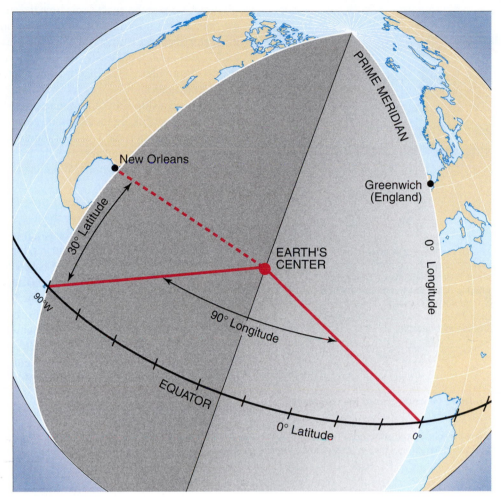

Figure A3–2 **New Orleans is located at 30 degrees north latitude and 90 degrees west longitude.**

A degree of latitude or longitude can be divided into smaller units. One degree (°) of arc (angular distance) is equal to 60 minutes (′) of arc. One minute of arc is equal to 60 seconds (″) of arc. When using a small-scale map or a globe, it may be difficult to estimate latitude and longitude to the nearest degree or two. When using a large-scale map, however, it is often possible to determine the latitude and longitude to the nearest fraction of a minute.

Determination of Latitude and Longitude

Today, latitude and longitude can be determined very precisely by using satellites that remain in orbit around Earth in fixed positions. How did navigators determine their position before this technology was available?

Latitude was determined from the positions of particular stars. Initially, navigators in the Northern Hemisphere measured the angle between the horizon and the North Star (Polaris), which is directly above the North Pole. The latitude north of the Equator is the angle between the two sightings (Figure A3-3). In the Southern Hemisphere, the angle between the horizon and the Southern Cross was used because the Southern Cross is directly overhead at the South Pole. Later, the angle of the sun above the horizon, corrected for the date, was also used.

There was no method of determining longitude until one based on time was developed at the end of the 18th century. As Earth turns on its rotational axis, it moves through 360 degrees of arc every 24 hours (one complete rotation

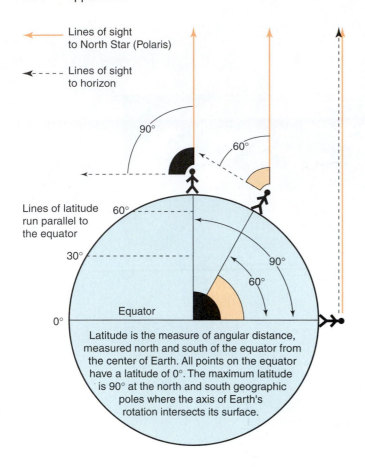

Lines of sight to North Star (Polaris)

Lines of sight to horizon

Lines of latitude run parallel to the equator

Latitude is the measure of angular distance, measured north and south of the equator from the center of Earth. All points on the equator have a latitude of 0°. The maximum latitude is 90° at the north and south geographic poles where the axis of Earth's rotation intersects its surface.

Figure A3–3 Determining latitude based on the North Star (Polaris).

Latitude can be determined by noting the angular difference between the horizon and the North Star, which is directly over the North Pole of Earth.

on its axis) (Figure A3-4a). Earth, therefore, rotates through 15 degrees of longitude per hour (360° ÷ 24 hours = 15 degrees/hour). As a result, a navigator needed only to know the time at the prime or Greenwich meridian (Figure A3-4b) at the exact same time the sun was at its highest point (local noon) at the ship's location. In this way, a navigator aboard a ship could calculate the ship's longitude each day at noon. That's why the development of John Harrison's chronometer (see Box 1-1 in Chapter 1) was so crucial for navigation.

Suppose a ship sets sail west across the Atlantic Ocean from Europe, checking its longitude each day at noon local time. One day when the sun is at the noon position (noon local time), the chronometer reads 16:18 hours (0:00 is midnight and 12:00 is noon). What is the ship's longitude (Figure A3-4c)?

If the clock is keeping good time (which Harrison's chronometer did), then we know that the ship is 4 hours and 18 minutes behind (west of) Greenwich time. To determine the longitude, we must convert time into longitude. Each hour represents 15 degrees of longitude based on the rotation of Earth. Thus, the four hours represents 60 degrees of longitude (4 hours × 15 degrees of longitude = 60 degrees). One degree is divided into 60 minutes of arc, so Earth rotates through $\frac{1}{4}$ degree (15′) of arc per minute of time. Thus, 18 minutes of time multiplied by 15 minutes of arc per minute of time equals 270 minutes of arc. To convert 270 minutes of arc into degrees, it must be divided by 60 minutes per degree, which gives 4.5 degrees of longitude. Therefore, the answer is 60 degrees plus 4.5 degrees, or 64.5 degrees west longitude.

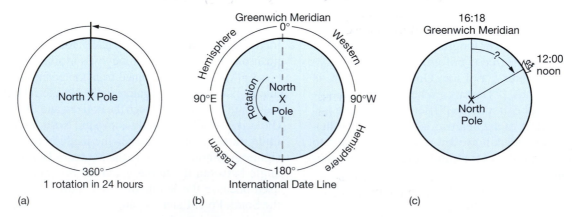

(a)   (b)   (c)

Figure A3–4 Determining longitude based on time.

View of Earth as seen from above the North Pole. (a) Earth rotates 360 degrees of arc every 24 hours. (b) The Greenwich Meridian is set as 0 degree longitude, which divides the globe into Eastern and Western Hemispheres. The International Date Line is 180 degrees from the Greenwich Meridian. (c) An example of how a ship at sea can determine its longitude using time.

Appendix IV

A Chemical Background: Why Water Has 2 H's and 1 O

An **element** (*elementum* = a first principle) is a substance comprised entirely of like particles that cannot be broken into smaller particles by chemical means. The **atom** (*a* = not, *tomos* = cut) is the smallest particle of an element that can combine with similar particles of other elements to produce compounds. The periodic table of elements shown in Figure A4-1 lists the elements and describes their atoms. A **compound** (*compondre* = to put together) is a substance containing two or more elements combined in fixed proportions. A **molecule** (*molecula* = a mass) is the smallest particle of an element or compound that, in the free state, retains the characteristics of the substance.

As an illustration of these terms, consider **Sir Humphrey Davey's** use of electrical dissociation to break the compound water into its component elements, hydrogen and oxygen. Atoms of the elements hydrogen (H) and oxygen (O) combine in the proportion 2 to 1, respectively, to produce molecules of water (H_2O). As an electric current is passed through the water, the molecules dissociate into hydrogen atoms that collect near the cathode (negatively charged electrode) and oxygen atoms that collect near the anode (positively charged electrode). Here they combine to form the diatomic gaseous molecules of the elements hydrogen (H_2) and oxygen (O_2). Because there are twice as many hydrogen atoms as oxygen atoms in a given volume of water, twice as many molecules of hydrogen gas (H_2) as oxygen gas (O_2) are formed. Further, the volume of gas under identical conditions of temperature and pressure is proportional to the number of gas particles (molecules) present, so two volumes of hydrogen gas are produced for each volume of oxygen gas.

A Look at the Atom

Building on earlier discoveries, the Danish physicist **Niels Bohr** (1884–1962) developed his theory of the atom as a small solar system in which a positively charged nucleus takes the place of the sun and the planets that orbit around it are represented by negatively charged electrons. Although this theory has since been diagrammatically altered, it is still commonly used to demonstrate the arrangement of electrons and nuclear particles in the atom.

Bohr's earliest concern was with the atom of hydrogen, which he considered to consist of a single positively charged **proton** (*protos* = first) in its nucleus orbited by a single negatively charged **electron** (*electro* = electricity). Since the mass of the electron is only about $\frac{1}{1840}$ the mass of the proton, it will be considered negligible in our discussion of atomic masses.

According to Bohr, the number of protons—or units of positive charge in the nucleus—coincides with the **atomic number** of the element. Thus hydrogen, with an atomic number of 1, has a **nucleus** (*nucleos* = a little nut) containing a single proton. Helium, the next heavier element, having an atomic number of 2, contains two nuclear protons, and so forth. The atomic number also indicates the number of electrons in a normal atom of any element, because this number is equal to the number of protons in the nucleus (Figure A4-2).

An **isotope** (*isos* = equal, *topos* = place) is an atom of an element that has a different **atomic mass** than other atoms of the same element. As we will see, chemical properties of atoms are determined by the electron arrangement that surrounds the nucleus. This arrangement, in turn, is determined by the number of protons in the nucleus. Because some isotopes have different atomic masses but identical chemical characteristics, there must be a nuclear particle in the nucleus other than the proton that can influence the atomic mass of an atom but does not affect the electron structure surrounding the nucleus (Figure A4-3).

Nuclear research has discovered the additional particle, the **neutron** (*neutr* = neutral), which was postulated by **Ernest Rutherford** in 1920 and was first detected by his associate **James Chadwick** in 1932. It has a mass very similar to that of the proton but no electrical charge. This characteristic of being electrically neutral has made it one of the particles most utilized by nuclear physicists, who continue to explore the atomic nucleus and to make new discoveries of nuclear particles. We will not consider these particles, as knowledge of them is not necessary for our understanding of the chemical nature of atoms.

An atom of a given element can be changed to an ion of that element by adding or taking away one or more electrons, and it is changed to an isotope of the element by adding or taking away one or more neutrons. Adding or taking away one or more protons will change the atom to an atom of a different element.

The atom can be divided into two parts—the nucleus, which contains neutrons and protons, and the **electron cloud** surrounding the nucleus that is involved in chemical reactions. The random motion of electrons makes it impossible to determine the precise location

Periodic Table of the Elements

Atomic number
Symbol of element
Atomic weight
Name of element

| 1 |
| **H** |
| 1.0080 |
| Hydrogen |

Legend:
- Inert Gas (red)
- Gas (blue)
- Liquid (yellow)

Transitional Elements

Heavy Metals

Nonmetals

Light Metals

	I A	II A	III B	IV B	V B	VI B	VII B		VIII B		I B	II B	III A	IV A	V A	VI A	VII A	VIII A
1	1 **H** 1.0080 Hydrogen																	2 **He** 4.003 Helium
2	3 **Li** 6.939 Lithium	4 **Be** 9.012 Beryllium											5 **B** 10.81 Boron	6 **C** 12.011 Carbon	7 **N** 14.007 Nitrogen	8 **O** 15.994 Oxygen	9 **F** 18.998 Fluorine	10 **Ne** 20.1863 Neon
3	11 **Na** 22.990 Sodium	12 **Mg** 24.31 Magnesium											13 **Al** 26.98 Aluminum	14 **Si** 28.09 Silicon	15 **P** 30.974 Phosphorus	16 **S** 32.064 Sulphur	17 **Cl** 35.453 Chlorine	18 **Ar** 39.948 Argon
4	19 **K** 39.102 Potassium	20 **Ca** 40.08 Calcium	21 **Sc** 44.96 Scandium	22 **Ti** 47.90 Titanium	23 **V** 50.94 Vanadium	24 **Cr** 52.00 Chromium	25 **Mn** 54.94 Manganese	26 **Fe** 55.85 Iron	27 **Co** 58.93 Cobalt	28 **Ni** 58.71 Nickel	29 **Cu** 63.54 Copper	30 **Zn** 65.37 Zinc	31 **Ga** 69.72 Gallium	32 **Ge** 72.59 Germanium	33 **As** 74.92 Arsenic	34 **Se** 78.96 Selenium	35 **Br** 79.909 Bromine	36 **Kr** 83.80 Krypton
5	37 **Rb** 85.47 Rubidium	38 **Sr** 87.62 Strontium	39 **Y** 88.91 Yttrium	40 **Zr** 91.22 Zirconium	41 **Nb** 92.91 Niobium	42 **Mo** 95.94 Molybdenum	43 **Tc** (99) Technetium	44 **Ru** 101.1 Ruthenium	45 **Rh** 102.90 Rhodium	46 **Pd** 106.4 Palladium	47 **Ag** 107.870 Silver	48 **Cd** 112.40 Cadmium	49 **In** 114.82 Indium	50 **Sn** 118.69 Tin	51 **Sb** 121.75 Antimony	52 **Te** 127.60 Tellurium	53 **I** 126.90 Iodine	54 **Xe** 131.30 Xenon
6	55 **Cs** 132.91 Cesium	56 **Ba** 137.34 Barium	57 TO 71	72 **Hf** 178.49 Hafnium	73 **Ta** 180.95 Tantalum	74 **W** 183.85 Tungsten	75 **Re** 186.2 Rhenium	76 **Os** 190.2 Osmium	77 **Ir** 192.2 Iridium	78 **Pt** 195.09 Platinum	79 **Au** 197.0 Gold	80 **Hg** 200.59 Mercury	81 **Tl** 204.37 Thallium	82 **Pb** 207.19 Lead	83 **Bi** 208.98 Bismuth	84 **Po** (210) Polonium	85 **At** (210) Astatine	86 **Ra** (222) Radon
7	87 **Fr** (223) Francium	88 **Ra** 226.05 Radium	89 TO 103															

Lanthanide series

57 **LA** 138.91 Lanthanum	58 **Ce** 140.12 Cerium	59 **Pr** 140.91 Praseodymium	60 **Nd** 144.24 Neodymium	61 **Pm** (147) Promethium	62 **Sm** 150.35 Samarium	63 **Eu** 157.25 Europium	64 **Gd** 158.92 Gadolinium	65 **Tb** 158.92 Terbium	66 **Dy** 162.50 Dysprosium	67 **Ho** 164.93 Holmium	68 **Er** 167.26 Erbium	69 **Tm** 168.93 Thulium	70 **Yb** 173.04 Ytterbium	71 **Lu** 174.97 Lutetium

Actinide series

89 **Ac** (227) Actinium	90 **Th** 232.04 Thorium	91 **Pa** (231) Protactinium	92 **U** 238.03 Uranium	93 **Np** (237) Neptunium	94 **Pu** (242) Plutonium	95 **Am** (243) Americium	96 **Cm** (247) Curium	97 **Bk** (249) Berkelium	98 **Cf** (251) Californium	99 **Es** (254) Einsteinium	100 **Fm** (253) Fermium	101 **Md** (256) Mendelevium	102 **No** (256) Nobelium	103 **Lw** (257) Lawrencium

Figure A4–1 The periodic table of the elements.

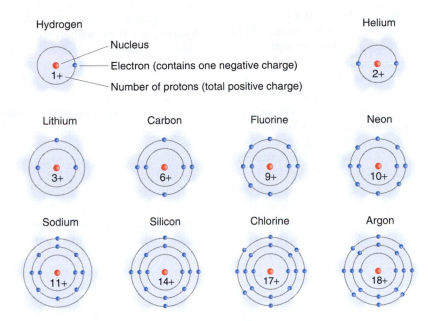

Figure A4–2 Bohr-Stoner orbital models for atoms.

Each atom is composed of a positively charged nucleus with negatively charged electrons around it. The nucleus, which occupies very little space, contains most of the mass of the atom. The atomic number is equal to the number of protons (positively charged nuclear particles) in the nucleus. Note that the first shell holds only two electrons and other shells can hold eight (or more) electrons.

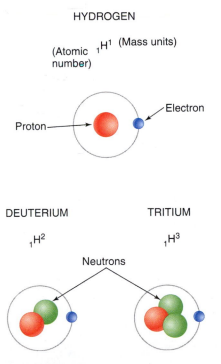

Figure A4–3 Hydrogen isotopes.

Isotopes are atoms of an element that have different atomic masses. The hydrogen atom $_1H^1$ accounts for 99.98% of the hydrogen atoms on Earth and has a nucleus that contains only one proton. Deuterium $_1H^2$ contains one neutron and one proton in its nucleus and combines with oxygen to form "heavy water" with a molecular mass of 20. Tritium $_1H^3$ is a very rare radioactive isotope of hydrogen that has a nucleus containing one proton and two neutrons.

of electrons in this cloud at any instant, but it is possible to estimate the most probable position of an electron in this cloud. The regions in which there would more likely be particular electrons can be viewed as concentric spheres, or shells, that surround the nucleus (see Figure A4-2).

Chemical Bonds

In considering the chemical reactions in which atoms are involved, we will be concerned primarily with the distribution of electrons in the outer shell. When atoms combine to form compounds, they usually do so by forming one of two bonds:

1. **Ionic** (*ienai* = to go) **bonds**, where electrons are either gained or lost
2. **Covalent** (*co* = with, *valere* = to be strong) **bonds**, where electrons are shared between atoms

Ionic bonding produces an ion, an electrically charged atom that no longer has the properties of a neutral atom of the element it represents. A positively charged ion, a **cation** (*kation* = something going down), is produced by the loss of electrons from the outer shell, the positive charge being equal to the number of electrons lost. A negatively charged ion, an **anion** (*anienai* = to go up), is produced by the gain of electrons in the outer shell of an atom, and its charge is equal to the number of electrons gained.

An example of a compound formed by ionic bonding is common table salt, which is sodium chloride (Figure A4-4). In the formation of this compound, the sodium atom, which has one electron in its outer shell, loses this electron and forms a sodium ion with a positive electrical charge of one. The chlorine atom, which contains seven electrons in its outer shell, completes its shell by gaining an electron and becoming a chloride ion with a negative charge of one. These two ions are held in close proximity by an electrostatic attraction between the two ions of equal and opposite charge.

Moreover, it is the tendency of an individual atom to assume the outer-shell electron content of the inert gases such as helium (2), neon (8), and argon (8) that make it chemically reactive (see Figure A4-2). Normally, if an atom can assume this configuration by either sharing one or two electrons with an atom of another element or by losing or gaining one or two electrons, the elements are highly reactive. By contrast, the elements are less reactive if three or four electrons must be shared, gained or lost to achieve the desired configuration.

Valence (*valentia* = capacity) is the number of hydrogen atoms with which an atom of a given element can combine with either ionic or covalent bonds. Elements with lower valences of one or two, for example, combine chemically in a more highly reactive manner than do those with higher valences of three or four or, in certain instances, more than four. Although those elements with higher valences do not react as violently as low-valence elements, they have a greater combining power and can gather about them larger numbers of atoms of other elements than can the atoms with lower valence values.

An example of a covalent bond is the sharing of electrons by hydrogen and oxygen atoms in the water molecule (Figure A4-5). In the formation of this molecule, both hydrogen atoms and the oxygen atom assume the inert gas configuration they seek by sharing electrons.

Figure A4–4 Ionic bonds in sodium chloride (table salt).

Ions of sodim (left) are chlorine (right) combine to form sodim chloride, producing common table salt. Notice the contrast in the outer shells of the ions of sodium (NA^+) and chlorine (Cl^-) with the atoms of these element shown in figure A4-2.

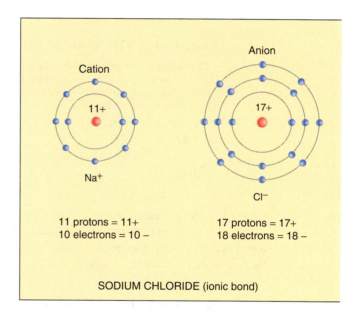

Figure A4–5 Covalent bonds in water.

Water (H_2O) is composed of 2 atoms of hydrogen and 1 atom of oxygen, held together by covalent bonds.

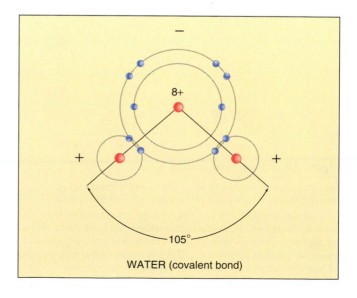

Appendix V

Careers in Oceanography

Many people think a career in oceanography consists of swimming with marine animals at a marine life park or snorkeling in crystal-clear tropical waters studying coral reefs. In reality, these kinds of jobs are extremely rare and there is intense competition for them. Most oceanographers use science to answer questions about the ocean, such as the following:

- What is the role of the ocean in limiting the greenhouse effect?
- What kinds of pharmaceuticals can be found naturally in marine organisms?
- How does sea floor spreading relate to the movement of tectonic plates?
- What economic deposits are there on the sea floor?
- Can rogue waves be predicted?
- What is the role of longshore transport in the distribution of sand on the beach?
- How does a particular pollutant affect organisms in the marine environment?

Preparation for a Career in Oceanography

Preparing yourself for a career in oceanography is probably one of the most interesting and rewarding (yet difficult) paths to travel. The study of oceanography is typically divided into four different academic disciplines (or subfields) of study:

- *Geological oceanography* is the study of the structure of the sea floor and how the sea floor has changed through time; the creation of sea floor features; and the history of sediments deposited on it.
- *Chemical oceanography* is the study of the chemical composition and properties of seawater; how to extract certain chemicals from seawater; and the effects of pollutants.
- *Physical oceanography* is the study of waves, tides, and currents; the ocean–atmosphere relationship that influences weather and climate; and the transmission of light and sound in the oceans.
- *Biological oceanography* is the study of oceanic life forms and their relationships to one another; adaptations to the marine environment; and developing ecologically sound methods of harvesting seafood.

Other disciplines include ocean engineering, marine archaeology, and marine policy. Oceanography is an *interdisciplinary* science because it utilizes all the disciplines of science as they apply to the oceans. Some of the most exciting work and best employment opportunities combine two or more of these disciplines.

Individuals working in oceanography and marine-related fields need a strong background (typically an undergraduate degree) in at least one area of basic science (for example, geology, physics, chemistry, or biology) or engineering. In almost all cases, mathematics is required as well. Marine archaeology requires a background in archaeology or anthropology; marine policy studies require a background in at least one of the social sciences (such as law, economics, or political science).

The ability to speak and write clearly—as well as critical thinking skills—are prerequisites for any career. Fluency in computers—preferably PC systems, not Macintosh—is rapidly becoming a necessity. Because many job opportunities in oceanography require trips on research vessels (Figure A5-1), any shipboard experience

Figure A5–1 Oceanographers at work on a research vessel.

Oceanographers deploy a sonar device on a cable using the A-frame on Scripps Institution of Oceanography's R/V *Melville*.

is also desirable. Mechanical ability (the ability to fix equipment while on board a vessel without having to return to port) is a plus. Depending on the type of work that is required, other traits that may be desirable include the ability to speak one or more foreign languages; certification as a scuba diver; the ability to work for long periods of time in cramped conditions; physical stamina; physical strength; and a high tolerance to motion sickness.

Oceanography is a relatively new science (with much room left for discovery), so most people enter the field with an advanced degree (master's or doctorate). Work as a marine technician, however, usually requires only a bachelor's degree or applicable experience. It does take a large commitment to achieve an advanced degree but, in the end, the journey itself is what makes all the hard work worthwhile.

Job Duties of Oceanographers

There has been and will continue to be enormous expansion in the number of ocean-related jobs. Job opportunities for oceanographers exist with scientific research institutions (universities), various government agencies, and private companies that are engaged in searching for economic sea floor deposits, investigating areas for sea farming, and evaluating natural energy production from waves, currents, and tides. The duties of oceanographers vary from job to job but can be generally described as follows:

- *Geological oceanographers* and *geophysicists* explore the ocean floor and map submarine geological structures. Studies of the physical and chemical properties of rocks and sediments give us valuable information about Earth's history. The results of their work help us understand the processes that created the ocean basins and the interactions between the ocean and the sea floor.

- *Chemical oceanographers* and *marine geochemists* investigate the chemical composition of seawater and its interaction with the atmosphere and the sea floor. Their work may include analysis of seawater components, desalination of seawater, and studying the effects of pollutants. They also examine chemical processes operating within the marine environment and work with biological oceanographers to study living systems. Their study of trace chemicals in seawater helps us understand how ocean currents move seawater around the globe and how the ocean affects climate.

- *Physical oceanographers* investigate ocean properties such as temperature, density, wave motions, tides, and currents. They study ocean–atmosphere relationships that influence weather and climate, the transmission of light and sound through water, and the ocean's interactions with its boundaries at the sea floor and the coast.

- *Biological oceanographers*, *marine biologists*, and *fisheries scientists* study marine plants and animals.

They want to understand how marine organisms develop, relate to one another, adapt to their environment, and interact with their environment. They develop ecologically sound methods of harvesting seafood and study biological responses to pollution. New fields associated with biological oceanography include marine biotechnology (the use of natural marine resources in the development of new industrial and biomedical products) and molecular biology (the study of the structure and function of bioinformational molecules—such as DNA, RNA, and proteins—and the regulation of cellular processes at the molecular level). Because marine biology is the most well known field of oceanography (and because the larger marine animals have such wide appeal), it is currently the most competitive, too.

- *Marine* and *ocean engineers* apply scientific and technical knowledge to practical uses. Their work ranges from designing sensitive instruments for measuring ocean processes to building marine structures that can withstand ocean currents, waves, tides, and severe storms. Subfields include acoustics; robotics; electrical, mechanical, civil, and chemical engineering; as well as naval architecture. They often use highly specialized computer techniques.

- *Marine archaeologists* systematically recover and study material evidence, such as shipwrecks, graves, buildings, tools, and pottery remaining from past human life and culture that is now covered by the sea. Marine archaeologists use state-of-the-art technology to locate underwater archaeological sites.

- *Marine policy experts* combine their knowledge of oceanography and social sciences, law, or business to develop guidelines and policies for the wise use of the ocean and coastal resources.

Other job opportunities for oceanographers include work as science journalists specializing in marine science, teachers at various grade levels, and aquarium and museum curators.

Sources of Information

- Consult the catalog of any college or university that offers a curriculum in oceanography or marine science.

- The Oceanography Society publishes an excellent brochure entitled *Careers in Oceanography and Marine-Related Fields*. The Oceanography Society can be contacted at 4052 Timber Ridge Drive, Virginia Beach, VA 23455.

- The National Sea Grant College Program of NOAA publishes a comprehensive brochure entitled *Marine Science Careers: A Sea Grant Guide to Ocean Opportunities*, which includes interviews with working oceanographers. The Sea Grant College can be reached c/o NOAA, SSMC3 Room 11606, 1315 East-West Highway, Silver Springs, MD 20910.

One of the true pleasures of teaching is that some of your students become so interested in the subject matter that they pursue a career in it. One of Al Trujillo's former students, Joe Cooney (Figure A5-A), works as a facilities manager for the Hubbs–SeaWorld Research Institute's Leon Raymond Hubbard, Jr. Marine Fish Hatchery in Carlsbad, California. His job is to maintain the equipment used to raise sea bass for release to the ocean to restore the former abundance of this sportfish in southern California waters. Joe writes,

After completing coursework in oceanography at a community college, I wanted more information about what types of jobs were available and what degree would be most valuable. I was not certain if I wanted to be involved in work like marine biologists I had seen, or look more towards physical oceanography as a career. Fortunately, I had the opportunity to get a summer job at a marine fish hatchery, which led to an offer of a permanent position. I am now involved daily in a marine sciences working environment where my primary responsibility is to monitor and maintain the seawater systems for the hatchery.

Unlike a traditional fish hatchery that uses water diverted from a river to provide large amounts of clean water at virtually no cost, a saltwater hatchery must pump in its entire water supply from the sea. Also, conditions must be maintained so that open-ocean fish will live and reproduce, and that their larvae will survive and grow.

Water is pumped to recirculating and experimental systems, as well as to traditional flow-through raceways where juvenile fish are briefly held before being released into the ocean. Incoming water is diverted throughout a hatchery building via computer-controlled valves, pumps, filters and other automated systems. Some water is temperature regulated and its gas content manipulated for experimental use. Controlled light and temperature conditions are maintained in larger tanks to achieve regulated spawning for year-round fish production. Much like the automated system that runs a car engine, sensors throughout the hatchery's system allow a computer to maintain desired conditions, and signal an alarm if conditions vary from prescribed limits. The system contains many components, and maintenance of it requires knowledge of everything from computer hardware to large industrial pumps. In addition, the saltwater is very corrosive and constant maintenance is needed to keep every-

thing running smoothly. Much of what is done is experimental, which presents daily challenges to design, build, and run systems that will achieve the desired results. Both the animal caretakers and the machines must work correctly to successfully produce fish.

Lacking an advanced college degree, I was able to draw on past work experience to get the position that I have. In order to be considered for most jobs in marine science, a degree is a virtual necessity. However, experience in any marine-related field is good not only to make you more valuable to an employer, but also to allow you to better choose your goals. My position has exposed me to current developments, which in turn reinforces my desire to continue working in this field and obtain further schooling.

As humans place increasing demands on the ocean, I believe there will be an even larger need for more people to work in many different marine-related areas. Even in such specialized operations such as aquaculture, public aquariums, and research facilities, there are many jobs in animal care, research, lab work, and system operation. The ocean is only recently being studied extensively, and I am looking forward to the investigation into what is found, which should continue for a long time to come.

Figure A5–A Joe Cooney at work in the Hubbs–Sea World Research Institute's Leon Raymond Hubbard, Jr. Marine Fish Hatchery in Carlsbad, California.

- The Scripps Institution of Oceanography at the University of California, San Diego, publishes an informative brochure aimed at perspective students entitled *Preparing for a Career in Oceanography*. General information about Scripps can be obtained by contacting the Scripps Communication Office at the Scripps Institution of Oceanography, University of California San Diego, 9500 Gilman Drive, Department 0233, La Jolla, CA 92093-0233.

- Sea World manages marine and adventure parks and sometimes hires educators and animal trainers—especially those interested in training marine mammals. Their Education Department publishes a booklet on working in a marine life park entitled *The SeaWorld/Bush Gardens Guide to Zoological Park Careers*, available from SeaWorld of California Education Department, 1720 South Shores Road, San Diego, CA. 92109-7995.

- For more information about graduate school, The National Academy of Sciences has published a book entitled *Careers in Science and Engineering: A Student Planning Guide to Graduate School and Beyond* (National Academy Press, 1996). The National Academy of Sciences can be contacted at 2101 Constitution Ave., NW Washington, DC. 20418. Their toll-free telephone number is (800) 624-6242.

Some Web sites that contain oceanography career information include the following:

- The International Oceanographic Foundation at *http://www.rsmas.miami.edu/iof/*

- The U.S. Navy's Web site on careers in oceanography at *http://www.cnmoc.navy.mil/educate/career-o.html* or *http://www.oc.nps.navy.mil/careers.html*

- The Office of Naval Research's Web site, which includes The Oceanography Society's brochure entitled *Careers in Oceanography and Marine-Related Fields* at *http://www.onr.navy.mil/onr/careers/default.htm*

- A comprehensive list of information about careers in oceanography, marine science, and marine biology is available through the Scripps Institution of Oceanography Science Library's Web site at *http://scilib.ucsd.edu/sio/guide/career.html*

- The popular "So You Want to Become a Marine Biologist" Web site at *http://www.siograddept.ucsd.edu/Web/To_Be_A_Marine_Biologist.html*

- A listing of marine laboratories and institutions is available at *http://life.bio.sunysb.edu/marinebio/mblabs.html*

- Woods Hole Oceanographic Institution has developed a Web site devoted to the achievements of women oceanographers, including biographies and unique perspectives of women scientists at *http://www.womenoceanographers.org/*

Appendix VI

Taxonomic Classification of Common Marine Organisms

The following taxonomic classification shows common marine organisms within the five-kingdom system (some authorities place the prokaryotic archaebacteria in their own sixth kingdom). Important marine phyla and classes are also listed; exclusively nonmarine phyla and extinct groups have been omitted. Total number of known species shown in parentheses where accurate counts are available.

Kingdom Monera

Organisms without nuclear membranes; nuclear material is spread throughout cell; predominantly unicellular and colonial—including the true bacteria (eubacteria) and cyanobacteria (blue-green algae).

Phylum Schizophyta   Smallest known cells; true bacteria (1500 species).

Phylum Cyanophyta   Blue-green algae; chlorophyll *a*, carotene and phycobilin pigments (200 species).

Kingdom Protocista (Protista)

Organisms with nuclear material confined to a nucleus by a membrane; includes unicellular protozoans and unicellular and multicellular (macroscopic) algae.

Phylum Chrysophyta   Golden-brown algae; includes diatoms, coccolithophores, and silicoflagellates; chlorophyll *a* and *c*, xanthophyll, and carotene pigments (6000+ species)

Phylum Pyrrophyta   Dinoflagellate algae; chlorophyll *a* and *c*, xanthophyll, and carotene pigments (1100 species).

Phylum Chlorophyta   Green algae; chlorophyll *a* and *b* and carotene pigments (7000 species).

Phylum Phaeophyta   Brown algae; chlorophyll *a* and *c*, xanthophyll, and carotene pigments (1500 species).

Phylum Rhodophyta   Red algae; chlorophyll *a*, carotene, and phycobilin pigments (4000 species).

Phylum Protozoa   Nonphotosynthetic, heterotrophic protists (27,400 species).

Class Mastigophora   Flagellated; dinoflagellates (5200 species).
Class Sarcodina   Ameboid; foraminiferans and radiolarians (11,500 species).
Class Ciliophora   Ciliated (6000 species).

Kingdom Fungi

Simple, multicellular, generally heterotrophic organisms.

Phylum Mycophyta   Fungi, lichens; most fungi are decomposers found on the ocean floor, whereas lichens inhabit the upper intertidal zones (3000 species of fungi; 160 species of lichen).

Kingdom Plantae (Metaphyta)

Multicellular autotrophic (photosynthetic) plants.

Phylum Tracheophyta   Vascular plants with roots, stems, and leaves that are serviced by special cells that carry food and fluids (287,200 species).

Class Angiospermae   Flowering plants with seeds contained in a closed vessel (275,000 species but mostly nonmarine; marine species include eelgrass, surfgrass, manatee grass, turtle grass, salt marsh grasses, mangroves).

Kingdom Animalia (Metazoa)

Multicellular heterotrophic animals.

Phylum Porifera   Sponges; spicules are the only hard parts in these sessile animals that do not possess tissue (10,000 species).

Class Calcarea   Calcium carbonate spicules (50 species).
Class Desmospongiae   Skeleton may be composed of siliceous spicules or spongin fibers or be nonexistent (9500 species).
Class Sclerospongiae   Coralline sponges; massive skeleton composed of calcium carbonate, siliceous spicules, and organic fibers (7 species).
Class Hexactinellida   Glass sponges; six-rayed with siliceous spicules (450 species).

Phylum Cnidaria (Coelenterata)   Radially symmetrical, two-cell, layered body wall with one opening to gut cavity; polyp (asexual, sexual, benthic) and medusa (sexual, pelagic) body forms (10,000 species).

Class Hydrozoa   Polypoid colonies such as pelagic Portuguese man-of-war and benthic Obelia common; medusa present in reproductive cycle but reduced in size (3000 species).
Class Scyphozoa   Jellyfish; medusa up to 1 meter (3 feet) in diameter is dominant form; polyp is small if present (250 species).
Class Anthozoa   Corals and anemones possessing only polypoid body form and reproducing asexually and sexually (6500 species).

Phylum Ctenophora   Predominantly planktonic comb jellies; basic eight-sided radial symmetry modified by secondary bilateral symmetry (80 species).

Phylum Platyhelminthes   Flatworms; bilateral symmetry; hermaphroditic (25,000 species).

Phylum Nemertea   Ribbon worms; as long as 30 meters (100 feet); benthic and pelagic (800 species).

Phylum Nematoda   Roundworms; marine forms are primarily free-living and benthic; most 1 to 3 millimeters (less that 0.1 inch) in length (5000 marine species).

Phylum Rotifera Ciliated, unsegmented forms less than 2 millimeters (less than 0.1 inch) in length (1500 species, only a few marine).

Phylum Bryozoa (Ectoprocta) Moss animals; benthic, branching or encrusting colonies; lophophore feeding structure (4500 species).

Phylum Branchiopoda Lamp shells; lophophorate benthic bivalves (300 species).

Phylum Phoronida Horseshoe worms; lophorate tube worms up to 24 centimeters (9 inches) long that live in sediment of shallow and temperate shallow waters (15 species).

Phylum Sipuncula Peanut worms; benthic (325 species).

Phylum Echiura Spoon worms; sausage shaped with spoon-shaped proboscis; burrow in sediment or live under rocks (130 species).

Phylum Pogonophora Tube-dwelling, gutless worms 5 to 80 centimeters (2 to 31 inches) long; absorb organic matter through body wall (100 species).

Phylum Tardigrada Marine meiofauna that have the ability to survive long periods in a cryptobiotic state (diversity is poorly known).

Phylum Mollusca Soft bodies possessing a muscular foot and mantle that usually secretes calcium carbonate shell (75,000 species).

Class Monoplacophora Rare trench-dwelling forms with segmented bodies and limpetlike shells (10 species).

Class Polyplacophora Chitons; oval, flattened body covered by eight overlapping plates (600 species).

Class Gastropoda Large, diverse group of snails and their relatives; shell spiral if present (64,500 species).

Class Bivalvia Bivalves; includes mostly filter-feeding clams, mussels, oysters, and scallops (7500 species).

Class Aplacophora Tusk shells; sand-burrowing organisms that feed on small animals living in sand deposits (350 species).

Class Cephalopoda Octopus, squid, and cuttlefish that possess no external shell except in the genus *Nautilus* (600 species).

Phylum Annelida Segmented worms in which musculature, circulatory, nervous, excretory, and reproductive systems may be repeated in many segments; mostly benthic (10,000 marine species).

Phylum Arthropoda Jointed-legged animals with segmented body covered by an exoskeleton (30,000 marine species).

Subphylum Crustacea Calcareous exoskeleton, two pairs of antennae; cephlon, thorax, and abdomen body parts; includes copepods, ostracods, barnacles, shrimp, lobsters, and crabs (26,000 species).

Subphylum Chelicerata

Class Merostomata Horseshoe crabs (4 species).

Class Pycnogonide Sea spiders.

Subphylum Uniramia Insects; genus *Halobites* is the only true marine insect.

Phylum Chaetognatha Arrowworms; mostly planktonic, transparent and slender; up to 10 centimeters (4 inches) long (50 species).

Phylum Echinodermata Spiny-skinned animals; benthic animals with secondary radial symmetry and water vascular system (6000 species).

Class Asteroidea Starfishes; free-living, flattened body with five or more rays with tube feet used for locomotion; mouth down (1600 species).

Class Ophiuroidea Brittle stars and basket stars; prominent central disc with slender rays; tube feet used for feeding; mouth down (200 species).

Class Echinoidea Sea urchins, sand dollars, and heart urchins; free-living forms without rays; calcium carbonate test; mouth down or forward (860 species).

Class Holothuroidea Sea cucumbers; soft bodies with radial symmetry obscured; mouth forward (900 species).

Class Crinoidea Sea lilies and feather stars; cup-shaped body attached to bottom by a jointed stalk or appendages; mouth up (630 species).

Phylum Hemichordata Acorn worms and pterobranchs; primitive nerve chord; gill slits; benthic (90 species).

Phylum Chordata Notochord; dorsal nerve chord and gills or gill slits (55,000 species).

Subphylum Urochordata Tunicates; chordate characteristics in larval stage only; benthic sea squirts and planktonic thaliaceans and larvaceans (1375 species).

Subphylum Cephalochordata Amphioxus or lancelets; live in coarse temperate and tropical sediment (25 species).

Subphylum Vertebrata Internal skeleton; spinal column of vertebrae; brain (52,000 species).

Class Agnatha Lampreys and hagfishes; most primitive vertebrates with cartilaginous skeleton, no jaws, and no scales (50 species).

Class Chondrichthyes Sharks, skates, and rays; cartilaginous skeleton; 5 to 7 gill openings; placoid scales (625 species).

Class Osteichthyes Bony fishes; cycloid scales; covered gill opening; swim bladder common (30,000 species).

Class Amphibia Frogs, toads, and salamanders; Asian mud flat frogs are the only amphibians that tolerate marine water (2600 species).

Class Reptilia Snakes, turtles, lizards, and alligators; orders Squamata (snakes) and Chelonia (turtles) are major marine groups (6500 species).

Class Aves Birds; many live on and in the ocean but all must return to land to breed (8600 species).

Class Mammalia Warm-blooded; hair; mammary glands; bear live young; marine representatives found in the orders Carnivora (sea otter, polar bear, pinnipeds), Sirenia (manatee, dugong), and Cetacea (whales, dolphins, and porpoises) (4100 species).

Glossary

Abiogenic Not of biological origin.

Abiotic Without life.

Absolute dating The use of radioisotope half-lives to determine the age of rock units in years. Accurate to within 2 or 3%.

Abyssal clay Deep-ocean (oceanic) deposits containing less than 30% biogenous sediment.

Abyssal hill Volcanic peaks rising less than 1 kilometer (0.6 mile) above the ocean floor.

Abyssal hill province Deep-ocean regions, particularly in the Pacific Ocean, where oceanic sedimentation rates are so low that abyssal plains do not form and the ocean floor is covered with abyssal hills.

Abyssal plain A flat depositional surface extending seaward from the continental rise or oceanic trenches.

Abyssal storm Stormlike occurrances of rapid current movement affecting the deep ocean floor. They are believed to be caused by warm- and cold-core eddies of surface currents.

Abyssal zone The benthic environment between 4000 and 6000 meters (13,100 and 19,700 feet).

Abyssopelagic zone Open-ocean (oceanic) environment below 4000 meters (13,100 feet) depth.

Acid A substance that releases hydrogen ions (H^+) in solution.

Acoelomate Without a secondary body cavity (coelom).

Acoustic Thermometry of Ocean Climate (ATOC) The measurement of ocean-wide changes in water properties such as temperature by transmitting and receiving low-frequency sound signals.

Action Plan for the Human Environment In 1972, this United Nations program established 12 worldwide environmental units and determined procedures for assessing and monitoring varrious sources of pollution.

Active arc An island arc that is volcanically active because of its position above a subducting lithospheric plate.

Active margin A continental margin that is tectonically deformed as it collides with another tectonic plate. It is the leading edge of the continent as it moves away from an oceanic spreading center.

Adhesive force An attractive force that exists between two objects composed of different materials (for example, water and glass).

Adiabatic Pertaining to a change in the temperature of a mass resulting from compression or expansion. It requires no addition of heat to or loss of heat from the substance.

Aerobic Respiration that can take place only in the presence of oxygen. Organic compounds are converted to carbon dioxide and water with the release of energy.

Agulhas Current A warm current that carries Indian Ocean water around the southern tip of Africa and into the Atlantic Ocean.

Air mass A large volume of air that has a definite area of origin and distinctive characteristics.

Alaskan gyre A small Pacific Ocean subpolar surface current gyre that rotates counterclockwise south of Alaska.

Albedo The fraction of incident electromagnetic radiation reflected by a surface.

Algae Primarily aquatic, eukaryotic, photosynthetic organisms that have no root, stem, or leaf systems. Can be microscopic or macroscopic.

Algin A common term used for algenic acid, which is the substance that gives the stalks of kelp strength and elasticity that allows them to survive in the high-energy coastal waters. It is used commercially to smoothe such products as jams and cosmetics and is also used in finishing leather and fireproofing.

Alkaline A substance that releases hydroxide ions (OH^-) in solution. Also called *basic*.

Alveoli A tiny, thin-walled, capillary-rich sac in the lungs where the exchange of oxygen and carbon dioxide takes place.

Amino acid One of more than 20 naturally occurring compounds that contain NH_2 and COOH groups. They combine to form proteins.

Ammonia (NH_3) A colorless, pungent gas composed of nitrogen and hydrogen.

Amnesic shellfish poisoning Poisoning caused by domoic acid secreted by a diatom. It has been known to kill birds and humans and obtains its name from the fact that a pervasive symptom in humans is amnesia.

Amphidromic point A nodal or no-tide point in the ocean or sea around which the crest of the tide wave rotates during one tidal period.

Amphineura Class of mollusks with eight dorsal calcareous plates; includes chitons.

Amphipoda Crustacean order containing laterally compressed members such as "beach hoppers."

Amplitude Height of wave crest above or trough below still water.

Anadromous Pertaining to a species of fish that spawns in fresh water, then migrates into the ocean to grow to maturity.

Anaerobic respiration Respiration carried on in the absence of free oxygen (O_2). Some bacteria and protozoans carry on respiration this way.

Anchovy A small silvery fish (*Engraulis rin-gens*) that swims through the water with its mouth open to catch its food. Also called *anchoveta*.

Andesite A fine-grained igneous rock that is the mineralogic equivalent of diorite. Its name is derived from the Andes Mountains of South America where it is common in association with volcanoes produced by the subduction of the Nazca Plate.

Anhydrite A colorless, white, gray, blue, or lilac evaporite mineral (anhydrous calcium sulfate, $CaSO_4$) that usually occurs as layers associated with gypsum deposits.

Animalia Kingdom of many-celled animals. They are members of the domain eukarya.

Anion A negatively charged ion.

Annelida Phylum of elongated segmented worms.

Anomalistic month The time required for the Moon to go from perigee to perigee, 27.5 days.

Anoxic Without oxygen.

Antarctic Bottom Water A water mass that forms in the Weddell Sea, sinks to the ocean floor, and spreads across the bottom of all oceans.

Antarctic Circle The latitude of 66.5° south.

Antarctic Circumpolar Current The eastward-flowing current that encircles Antarctica and extends from the surface to the deep-ocean floor. The largest volume current in the oceans. Also called the *West Wind Drift*.

Antarctic Convergence The zone of convergence along the northern boundary of the Antarctic Circumpolar Current where the southward-flowing boundary currents of the subtropical gyres converge on the cold Antarctic waters.

Antarctic Divergence The zone of divergence separating the westward-flowing East Wind Drift and the easterly flowing West Wind Drift.

Antarctic krill A species of krill found in Antarctic waters and a major food for baleen whales. It is being considered as a possible fishery.

Antarctic Intermediate Water Antarctic zone surface water that sinks at the Antarctic convergence and flows north at a depth of about 900 meters (2950 feet) beneath the warmer upper-water mass of the South Atlantic subtropical gyre.

Antarctic Ocean This ocean consists of the portions of the Atlantic, Indian, and Pacific Oceans south of the Antarctic Convergence that are recognized by oceanographers as having a distinct character. It is also referred to as the *Southern Ocean*.

Antarctic Polar Front   Name applied to the Antartic Convergence.

Anthophyta   Seed-bearing plants.

Anticyclonic flow   The flow of air around a region of high pressure clockwise in the Northern Hemisphere.

Antilles Current   This warm current flows north seaward of the Lesser Antilles from the North Equatorial Current of the Atlantic Ocean to join the Florida Current.

Antinode   Zone of maximum vertical particle movement in standing waves where crest and trough formation alternate.

Aphelion   The point in the orbit of a planet or comet where it is farthest from the Sun.

Aphotic zone   Without light. The ocean is generally in this state below 1000 meters (3300 feet).

Apogee   The point farthest from Earth in the orbit of the Moon or a human-made satellite.

Apparent polar wandering   Assuming little movement of Earth's axis of rotation relative to Earth's surface, the path the poles appear to have followed as a result of motions of lithospheric plates.

Arabian Sea   A sea at the northern margin of the Indian Ocean that lies between India and the Arabian penninsula.

Arabs   Natives of Arabia or members of the Arabian branch of the Semitic peoples.

Aragonite   A form of $CaCO_3$ that is less common and less stable than calcite. Pteropod shells are usually composed of aragonite.

Archaea   One of the three major domains of life. The domain consists of simple microscopic bacteria-like creatures (including methane producers and sulfur oxidizers that inhabit deep-sea vents and seeps) and other microscopic life forms that prefer environments of extreme conditions of temperature and/or pressure.

Archipelago   A large group of islands.

Arctic Circle   The latitude of 66.5° north latitude.

Arctic Convergence   Regions in the North Atlantic and Pacific Oceans where cold water from the subpolar gyres meets warmer water of the subtropical gyres.

Arctic Ocean   A small and shallow body of Arctic polar water that is centered on the North Pole and surrounded by landmasses.

Arrowworm   A member of the phylum Chaetognatha. It averages about 1 centimeter (0.4 inch) in length and is an important member of the plankton.

Aspect ratio   The index of propulsive efficiency obtained by dividing the square of fin height by the area of the fin.

Asthenosphere   A plastic layer in the upper mantle 80 to 200 kilometers (50 to 124 miles) deep that may allow lateral movement of lithospheric plates and isostatic adjustments.

Atlantic salmon   Any of a variety of fish be-longing to the genus *Salmo* and living in the North Atlantic Ocean. Atlantic salmon have been successfully raised in several mariculture operations worldwide.

Atlantic-type margin   See *Passive margin*.

Atoll   A ring-shaped coral reef growing upward from a submerged volcanic peak. It may have low-lying islands composed of coral debris.

Atom   The smallest particle of an element that can combine with similar particles of other elements to produce compounds.

Atomic mass   A number equal to the sum of protons and neutrons in the nucleus of an atom of an element.

Atomic number   A number representing the relative position of an element in the periodic table of elements. It is equal to the number of positive charges in the atom's nucleus.

Augite   A dark mineral, usually black, rich in iron and magnesium. An important constituent of basalt, the rock that is characteristic of the ocean crust.

Autolytic decomposition   The breakdown of organic matter by enzymes activated when an organism dies.

Autonomous underwater vehicles (AUVs)   Preprogrammed devices that are not tethered to surface veessels and can perform a wide variety of tasks on the ocean floor.

Autotomy   Pertaining to the ability of some organisms to slough off certain body parts as a defensive mechanism.

Autotroph   Plant or bacterium that can synthesize organic compounds from inorganic nutrients.

Autumnal equinox   The passage of the Sun across the Equator as it moves from the Northern Hemisphere into the Southern Hemisphere, approximately September 23. During this time, all places in the world experience equal lengths of night and day. Also called the *fall equinox*.

Auxospore   A diatom cell that has shed its frustule to allow growth.

Back-arc spreading center   Spreading centers behind island arcs that result from the tensional stresses created by the seaward migration of the associated oceanic trench.

Backshore   The inner portion of the shore, lying landward of the mean spring-tide high-water line. Acted upon by the ocean only during exceptionally high tides and storms.

Backwash   The seaward flow of water down the foreshore following the upwash of waves.

Bacteria   A domain of prokaryotic organisms.

Bacterioplankton   Bacteria that live as plankton.

Bacteriovore   An organism that feeds primarily on bacteria.

Bar-built estuary   A shallow estuary (lagoon) separated from the open ocean by a bar deposit such as a barrier island. The water in these estuaries usually exhibits vertical mixing.

Barnacle   See *Cirripedia*.

Barrier flat   Lying between the salt marsh and dunes of a barrier island, it is usually covered with grasses and even forests if protected from overwash for sufficient time.

Barrier island   A long, narrow, wave-built island separated from the mainland by a lagoon.

Barrier reef   A coral reef separated from the nearby landmass by open water.

Barycenter   The center of mass of a system.

Basalt   A dark-colored volcanic rock characteristic of the ocean crust. Contains minerals with high iron and magnesium content.

Base   A compound that releases hydroxide ions in aqueous solution.

Bathyal zone   Benthic environment from 200 to 4000 meters (660 to 13,100 feet).

Bathymetry   The measurement of ocean depth.

Bathypelagic zone   Open-ocean environment of approximately 1000 to 4000 meters (3300 to 13,100 feet).

Bathyscaphe   A specially designed deep-diving submersible

Bathysphere   A spherical chamber with windows that can be lowered into the sea so humans can observe marine life and the ocean floor.

Bay   A coastal body of water enclosed by land on three sides and open to the ocean on one side.

Bay barrier   A marine deposit attached to the mainland at both ends and extending entirely across the mouth of a bay, separating the bay from the open water.

Bay of Bengal   A sea at the northern margin of the Indian Ocean that is located between India and the Malay pennisula.

Bay of Fundy   An elongate, funnel-shaped bay between New Brunswick and Nova Scotia, Canada. The world's largest tidal range occurs in the Minas Basin at its northern end.

Beach   Area seaward of the coastline through the surf zone that contains material in transport along the shore and within the surf zone.

Beach compartment   A segment of beach that is provided sediment by a stream at one end and which looses sediment down a submarine canyon at its other end. Some are well developed along the southern California shore.

Beach face   The sloping surface of a beach extending from the berm to the shoreline.

Beach replenishment   The replenishment of a severely eroded beach by adding sand dredged from offshore or transported in from another location. Also called *beach nourishment*.

Beach starvation   The interruption of sediment supply and resulting narrowing of beaches.

Benguela Current   The cold eastern boundary current of the South Atlantic subtropical gyre.

Benthic Pertaining to the ocean bottom.

Benthos The forms of marine life that live on the ocean bottom.

Bering Sea A sea at the northern margin of the Pacific Ocean that is bounded on the south by the Aleutian Islands.

Berm The dry, gently sloping region on the backshore of a beach at the foot of the coastal cliffs or dunes.

Bicarbonate ion (HCO_3^-) An ion that contains the radical group HCO_3^-.

Big bang theory The theoretical explosion that set the universe into its expansive motion about 13.7 billion years ago.

Bioaccumulation The accumulation of minute dissolved substances in seawater in various tissues of a living organism.

Bioamplification The increased amount of a substance, such as a toxic chemical, in various tissues of larger organisms higher in food chains.

Biocommunity A distinctive community of organisms associated with a specific physical environment such as hydrothermal vents.

Bioerosion Erosion of coral and rock surfaces on the ocean floor by organisms such as sponges and sea urchins.

Biogenous sediment Material produced by plants or animals (e.g., coral reefs, shell fragments, and tests of diatoms, radiolarians, foraminifers, and coccolithophores).

Biogeochemical cycle The natural cycling of compounds among the living and nonliving components of an ecosystem.

Biological pump The movement of CO_2 that enters the ocean from the atmosphere through the water column to the sediment on the ocean floor by biological processes—photosynthesis, secretion of shells, feeding, and dying.

Bioluminescence Light produced in living organisms by chemical reaction. Found in bacteria, phytoplankton, and metazoans.

Biomass Total weight of the organisms in a particular habitat, species, or group of species.

Biomass pyramid A diagram illustrating the increase in the average size of individuals and the decrease in the population mass as one moves through trophic levels above the producing population.

Bioremediation A technique of using microbes to assist in cleaning toxic spills.

Biotic community All the organisms that live within some definable area.

Biozone A region of the environment that has distinctive biological characteristics.

Bivalve An organism that possesses two calcium carbonate shells as an external protective covering. Examples are clams and other bivalved mollusks.

Black smoker Submarine hydrothermal vents that have vent water temperatures around 350°C (660°F). These waters are rich in black particles of metallic sulfides.

Bladder kelp Species of kelp with air bladders that serve as floats to help the organism stand erect within the water column.

Body wave A longitudinal or transverse wave that transmits energy through a body of matter.

Boiling point The temperature at which a substance changes state from a liquid to a gas at a given pressure.

Bore A steep-fronted tide crest that moves up some rivers in association with an incoming high tide.

Bosphorus A narrow strait between the Black Sea and the Sea of Marmara through which Mediterranean and Black Sea waters mix.

Boundary current The northward- or southward-flowing currents that form the western and eastern boundaries, respectively, of subtropical surface current gyres.

Brackish Low salinity water created by mixing of seawater and freshwater.

Brazil Current The warm western boundary current of the South Atlantic subtropical gyre.

Breaker zone Region where waves break at the seaward margin of the surf zone.

Breakwater Any artificial structure constructed to protect a coastal region from the force of ocean waves.

Brittle Descriptive term for a substance that is likely to fracture when force is applied to it.

Brine pool Pools of hypersaline water first observed in the marine environment along the spreading center of the Red Sea.

Bryozoa Phylum of colonial animals that often share one coelomic cavity. Encrusting and branching forms secrete a protective housing (zooecium) of calcium carbonate or chitinous material. Possess lophophore feeding structure.

Buffering Of chemical or other processes, any process that reduces impact.

Buoyancy The ability or tendency to float or rise in a liquid.

Buttress zone The reef slope above 20 meters (66 feet) depth exposed to maximum wave energy.

Bycatch Species caught incidentally that are not the intended target of a fishing effort.

Calcareous Containing calcium carbonate.

Calcareous ooze Deep-sea sediment composed of at least 30% by mass calcareous biogenous particles.

Calcite The most common form of $CaCO_3$.

Calcite composition depth (CCD) The depth at which the amount of calcite ($CaCO_3$) produced by the organisms in the overlying water column is equal to the amount of calcite the water column can dissolve. There will be no calcite deposition below this depth, which, in most parts of the ocean, is at a depth of 4.5 kilometers (2.8 miles).

Calcium carbonate ($CaCO_3$) A chalklike substance secreted by many organisms in the form of coverings or skeletal structures.

California Current The slow moving, cold boundary current that forms the eastern side of the North Pacific subtropical gyre.

Calving The formation of icebergs as large masses of ice break away from glaciers that flow into the ocean.

Calorie Unit of energy defined as the amount of heat required to raise the temperature of 1 gram of water 1°C (1.8°F).

Canary Current The cold eastern boundary current of the North Atlantic subtropical gyre.

Capillarity The process by which a fluid (such as water) is drawn up into small interstices or tubes as a result of surface tension.

Capillary tubes Small diameter tubes of glass or other material into which water is drawn because the adhesive attraction of water molecules for the glass or other tubular material exceeds the cohesive attraction of water molecules.

Capillary wave Ocean wave whose wavelength is less than 1.74 centimeters (0.7 inch). The dominant restoring force for such waves is surface tension.

Carapace Chitinous or calcareous shield that covers the cephalothorax of some crustaceans. Dorsal portion of a turtle shell.

Carbohydrate An organic compound consisting of carbon, hydrogen, and oxygen. Sugars and starches are examples.

Carbonate ion (CO_3^{-2}) An ion that contains the radical group (CO_3^{-2}).

Carbon dioxide (CO_2) A gas containing one atom of carbon and two atoms of oxygen.

Carbonic acid An acid (H_2CO_3) that forms when carbon dioxide is dissolved in water and combines with it.

Caribbean Current The warm current that carries equatorial water across the Caribbean Sea into the Gulf of Mexico.

Caribbean Sea A sea at the western margin of the North Atlantic Ocean. It is bounded on the north and east by the Greater and Lesser Antilles Islands.

Caribbean Surface Water The surface water mass of the Caribbean Sea that generally has salinity in excess of 35.5‰ in open water away from coastal freshwater runoff.

Carnivora A taxonomic order of the class Mammalia.

Carnivore An animal that depends solely or chiefly on other animals for its food supply.

Carnivorous feeding The eating on one animal by another.

Carotin An orange-yellow pigment found in plants.

Catadromous Pertaining to a species of fish that spawns at sea, then migrates into a freshwater stream or lake to grow to maturity.

Cation A positively charged ion.

Caudal fin   Tail fin.

Caulerpa taxifolia   A tropical seaweed. A cold-water clone was introduced into the aquarium industry and found its way into coastal waters of the Mediterranean and southern California. It continues to spread in the Mediterranean, but has been eradicated in southern California.

Center of gravity   The point where the entire mass of a body may be considered to be concentrated.

Centigrade temperature scale (°C)   A temperature scale based on the freezing point (0°C = 32°F) and boiling point (100°C = 212°F) of pure water. Also known as the *Celsius scale* after its founder.

Centrifugal force   An apparent force that appears to make an object move away from the center of a curved path it is following. It results from the application of a centripetal force acting against the inertia of the object.

Centripetal acceleration   The acceleration of an object in motion that results from the application of centripetal force to the object.

Centripetal force   A center-seeking force that tends to make rotating bodies move toward the center of rotation.

Cephalopoda   A class of the phylum Mollusca with a well-developed pair of eyes and a ring of tentacles surrounding the mouth. The shell is absent or internal on most members. The class includes the squid, octopus, and *Nautilus*.

Cetacea   An order of marine mammals that includes the whales.

Chaetognath   A member of a phylum of elongate, transparent, wormlike pelagic animals commonly called *arrowworms*.

Chalk   A lithified sedimentary deposit largely composed of coccoliths.

Characteristic period   The period of a seiche (a single-node free-standing wave) based on a given body of water.

Chela   Arthropod appendage modified to form a pincer.

Chemical energy   A form of potential energy stored in the chemical bonds of compounds.

Chemical weathering   The weathering of rock by chemical reactions. Carbonic acid is an important cause of chemical weathering.

Chemosynthesis   The formation of organic compounds from inorganic substances using energy derived from oxidation.

Chesapeake Bay   A bay formed by the flooding of the mouths of the Susquehanna River and its tributaries. It is bounded by the states of Virginia and Maryland.

Chiton   Common name for any member of the Amphineura class of mollusks with eight dorsal plates.

Chloride ion (Cl⁻)   A chlorine atom that has become negatively charged by gaining one electron.

Chlorinity   The amount of chloride ion and

ions of other halogens in ocean water expressed in parts per thousand (‰) by weight.

Chlorofluorocarbons (CFCs)   A human made chemical that is catalized into chlorine monoxide (ClO), which combines with stratospheric ozone and destroys it.

Chlorophyll   A group of green pigments that make it possible for plants to carry on photosynthesis.

Chlorophyta   Green algae; characterized by the presence of chlorophyll and other pigments.

Chondrite   A stony meteorite composed primarily of silicate rock material and containing chondrules (spheroidal granules). They are the most commonly found meteorites.

Chronometer   The first practical marine timepiece that made possible the accurate determination of longitude. It was created by John Harrison and first put into service by the British admiralty in 1762.

Chrysophyta   An important phylum of planktonic algae, including the diatoms. The presence of chlorophyll is masked by the pigment carotin, which gives the plants a golden color.

Cilia   Short hairlike structures common on lower animals. Beating in unison, they may be used for locomotion or to create water currents that carry food toward the mouth of the organism.

Circadian rhythm   Behavioral and physiological rhythms of organisms related to the 24-hour day. Sleeping and waking patterns are an example.

Circular orbital motion   The motion of water particles caused by a wave as the wave is transmitted through water.

Cirripedia   An order of crustaceans with up to six pairs of thoracic appendages that strain food from the water. They are barnacles that attach themselves to a substrate and secrete an external calcareous housing.

Clathrate   A term applied to a chemical structure in which methane is trapped within a crystalline lattice of ice. This condition is found primarily in continental slope sediments below a depth of 500 meters (1640 feet).

Clay   (1) A particle size between silt and colloid. (2) Any of various hydrous aluminum silicate minerals that are plastic, expansive, and have ion exchange capabilities.

Climate   The prevailing or average weather conditions of a region, as determined by measuring temperature and other meteorological conditions over a period of years.

Cnidaria   Phylum that contains some 10,000 species of predominantly marine animals with a sacklike body and stinging cells on tentacles that surround the single opening to the gut cavity. There are two basic body forms. The medusa is a pelagic form represented by the jellyfish. The polyp is a predominantly benthic form found in sea anemones and corals. Also called *Coelenterata*.

Cnidoblast   Stinging cell of the phylum Cnidaria; contains a stinging mechanism (nematocyst) used in defense and capturing prey.

Coal   A black, hard fuel that results from the burial and partial decomposition of organic matter in the absence of oxygen and under elevated temperature.

Coast   A strip of land that extends inland from the coastline as far as marine influence is evidenced in landforms.

Coastal downwelling   When winds blow water toward shore, the water piles up against the shore and sinks beneath the surface.

Coastal geostrophic current   A current that flows along the coast as a result of relatively fresher water flowing seaward from the shore being turned by the Coriolis effect.

Coastal plain estuary   An estuary formed by rising sea level flooding a coastal river valley.

Coastal upwelling   The movement of deeper nutrient-rich water into the surface water mass as a result of windblown surface water moving offshore.

Coastal waters   Those relatively shallow water areas that adjoin continents or islands.

Coastline   Landward limit of the effect of the highest storm waves on the shore.

Coccolith   Tiny calcareous discs averaging about 3 microns (0.0001 inch) in diameter that form the cell wall of coccolithophores.

Coccolithophore   A microscopic planktonic form of algae, encased by a covering composed of calcareous discs (coccoliths).

Coelenterata   Phylum of radially symmetrical animals that includes two basic body forms, the medusa and the polyp. Includes jellyfish (medusoid) and sea anemones (polypoid). Preferred name is now *Cnidaria*.

Coelom   Secondary body cavity (gut cavity is primary cavity). Forms within the mesoderm in higher animals, is lined with peritoneum, and contains vital organs.

Cohesion   An attractive force between like substances.

Cold-blooded   Organisms whose body temperature is the same as their environment.

Cold boundary current   Slow drifting currents on the eastern side of a subtropical gyre that carry cold water toward the Equator.

Cold core ring   A circular eddy of a surface current that contains cold water in its center and rotates counterclockwise.

Cold front   A weather front in which a cold air mass moves into and under a warm air mass. It creates a narrow band of intense precipitation.

Colloid   Substance having particles of a size smaller than clay.

Colonial animal   An animal that lives in groups of attached or separate individuals. Groups of individuals may serve special functions.

Columbia River estuary   An estuary at the border between the states of Washington and

Oregon that has been most adversely affected by the construction of hydroelectric dams.

Comb jelly Common name for members of the phylum Ctenophora. (See *Ctenophora.*)

Commensalism A symbiotic relationship in which one organism benefits at no expense to its host.

Compensation depth for photosynthesis The depth at which light intensity is insufficient to support photosynthetic cells. It is on average about 100 meters (330 feet), where the light intensity is about 1% that at the ocean surface.

Compound A substance containing two or more elements combined in fixed proportions.

Condensation The conversion of water from the vapor to the liquid state. When it occurs, the energy required to vaporize the water is released into the atmosphere. This is about 585 calories per gram of water at 20°C (68°F).

Condensation point The temperature at which the gaseous state of a substance becomes liquid. For water, it is 100°C (212°F) at one atmosphere pressure.

Conduction The transmission of heat by the passage of energy from particle to particle.

Conductivity In marine science this term usually applies to the ease with which an electrical current flows through a sample of seawater. It is measured with a salinometer calibrated to convert conductivity to salinity.

Conjunction An apparent closeness of two or more heavenly bodies. During the new moon, the Sun and Moon are in conjunction on the same side of Earth.

Conservative constituent A constituent of seawater, the concentration of which is changed only by mixing and diffusion. All of the major constituents of seawater salinity are conservative constituents that have very long residence times.

Conservative property A property of surface ocean water that is changed only by mixing and diffusion after the water sinks below the surface.

Conservative tracer A chemical that has a very low rate of chemical reactivity in ocean water. It can be used to trace subsurface water motion over long periods of time.

Constant proportions, principle of A principle that states that the major constituents of ocean-water salinity are found in the same relative proportions throughout the ocean-water volume, independent of salinity.

Constructive interference A form of wave interference in which two waves come together in phase, e.g., crest to crest, to produce a greater displacement from the still water line than that produced by either of the waves alone.

Consumer A heterotrophic organism that consumes an external supply of organic matter.

Continent About one-third of Earth's surface that rises above the deep-ocean floor to be exposed above sea level. Continents are composed primarily of granite, an igneous rock of lower density than the basaltic oceanic crust.

Continental accretion A process by which continents grow by the addition of island arc volcanoes and other continental crustal material that has been transported from some distant location.

Continental arc A volcanic arc confined to the margin of a continent and associated with an ocean trench. The Andes Mountains and the Peru-Chile Trench constitute such a system.

Continental borderland A highly irregular portion of the continental margin that is submerged beneath the ocean and is characterized by depths greater than those normally associated with continental shelves.

Continental crust Crust composed of granite that makes up the contintnts and may have a thickness of up to 50 kilometers (30 miles).

Continental drift A term applied to early theories supporting the possibility that the continents are in motion over Earth's surface.

Continental flood basalt (CFB) Large volume flows of lava that spread out over continents after a plume of hot mantle material breaks through the lithosphere.

Continental margin Extending from the shoreline to the deep-ocean basin, this feature includes the continental shelf, continental slope, and continental rise.

Continental rise A gently sloping depositional surface at the base of the continental slope.

Continental shelf A gently sloping depositional surface extending from the low-water line to a marked increase in slope around the margin of a continent.

Continental slope A relatively steeply sloping surface lying seaward of the continental shelf.

Continental transform fault A transform fault that cuts through a continent. The San Andreas fault is an example.

Convection Heat transfer in a gas or liquid by the circulation of currents from one region to another.

Convection cell A circular-moving loop of matter involved in convective movement.

Convergence Coming together of water masses in polar, tropical, and subtropical regions of the ocean. Along these lines of convergence, the denser mass will sink beneath the others.

Convergent active margin A continental margin such as the Pacific coast of South America where subduction occurs near the continental margin to produce a coast characterized by volcanic and folded mountain ranges subject to significant earthquake activity.

Convergent plate boundary A lithospheric plate boundary where adjacent plates converge, producing ocean trench island arc systems, ocean trench continental volcanic arcs, or folded mountain ranges. Also called *destructive boundary.*

Conveyor belt circulation The combined circulation pattern for surface and deep currents of the world ocean. It is initiated by sinking of cold, relatively high salinity surface water in the North Atlantic Ocean that flows as a deep current into the South Atlantic, Indian and Pacific Oceans. This deep water gradually surfaces, and the return surface flow is from the low salinity North Pacific Ocean through the Indian and Sourh Atlantic to the North Atlantic Ocean—increasing its salinity along the way—to repeat the pattern.

Copepoda An order of microscopic to nearly microscopic crustaceans that are important members of the zooplankton in temperate and subpolar waters.

Coral A group of benthic cnidarians that exist as individuals or in colonies and may secrete external skeletons of calcium carbonate.

Coral bleaching A condition, caused by stresses such as high temperature, that results in hermatypic corals expelling their algal symbionts for their survival. This causes the coral to appear bleached as the algae provide color to coral tissue.

Coral reef A calcareous organic reef composed significantly of solid coral and coral sand. Algae may be responsible for more than half of the $CaCO_3$ reef material. Found in waters where the minimum average monthly temperature is 18°C (64°F).

Corange lines A circular line, circling an amphidromic point and crossing cotidal lines at right angles, that is composed of all points with a given value of tidal range.

Core (1) The innermost portion of Earth, which is composed primarily of iron and nickel. It has a liquid outer portion 2270 kilometers (1410 miles) thick and a solid inner core with a radius of 1216 kilometers (755 miles). (2) A cylinder of sediment and/or rock material.

Core–mantle boundary (CMB) The boundary between Earth's core and mantle at a depth of about 2885 kilometers (1800 miles).

Coriolis effect An effect resulting from Earth's rotation that causes particles in motion to be deflected to the right in the Northern Hemisphere and to the left in the Southern Hemisphere.

Cosmogenous sediment Sediment derived from outer space.

Cotidal lines Lines connecting points where high tides occur simultaneously.

Countershading Protective coloration in an animal or insect, characterized by darker coloring of areas exposed to light and lighter coloring of areas that are normally shaded.

Covalent bond A chemical bond in which atoms combine to form compounds by sharing electrons. The water molecule is an example.

Crest (wave) The portion of an ocean wave that is displaced above the still water line.

Cromwell Current A ribbonlike eastward-flowing current embedded in the South Equatorial Current that flows from Samoa to the Galápagos Islands.

Cruiser Fish such as the bluefin tuna that constantly cruise the pelagic waters in search of food.

Crust Unit of Earth's structure that is composed of basaltic ocean crust and granitic continental crust. The total thickness of the crustal units may range from 5 kilometers (3 miles) beneath the ocean to 50 kilometers (30 miles) beneath the continents.

Crusts Hydrogenous deposits found on the flanks of volcanic islands and seamounts typically rich in iron, manganese, cobalt, copper, and nickel.

Crustacean An organism belonging to the subphylum Crustacea of the phylum Arthropoda that includes barnacles, copepods, lobsters, crabs, and shrimp.

Crystalline rock Igneous or metamorphic rocks. These rocks are made up of crystalline particles with orderly molecular structures.

Ctenophore A member of the phylum of gelatinous organisms that are more or less spheroidal with biradial symmetry. These exclusively marine animals have eight rows of ciliated combs for locomotion, and most have two tentacles for capturing prey.

Curie point The temperature above which thermal agitation prevents spontaneous alignment of magnetic particles with Earth's magnetic field.

Current A physical movement of water.

Cutaneous artery The artery that runs down both sides of some cruiser-type fish to help maintain a constant elevated temperature in the myomere musculature used for swimming.

Cutaneous vein The vein that runs down both sides of some cruiser-type fish to help maintain a constant elevated temperature in the myomere musculature used for swimming.

Cyanobacteria Photosynthetic bacteria that may have been the first organisms to use aerobic respiration.

Cyclone An atmospheric system characterized by the rapid, inward circulation of air masses about a low-pressure center, usually accompanied by stormy, often destructive, weather. Cyclones circulate counterclockwise in the Northern Hemisphere and clockwise in the Southern Hemisphere.

Cyclonic flow In the Northern Hemisphere, any system that has a counterclockwise pattern of rotation.

Cypris Advanced free-swimming larval stage of barnacles. After attaching to substrate, it metamorphoses into an adult.

Datum A horizontal plane or line from which hights or depths are measured. For example, sea level is the datum for measuring continental el-evations and depth of the oceans. For tides, the datum is either mean low tide for diurnal tides or mean lower low tide where tides are semidiurnal or mixed.

Davidson Current A northward-flowing current along the Washington–Oregon coast that is driven by geostrophic effects on a large freshwater runoff.

DDT An insecticide that caused damage to marine bird populations in the 1950s and 1960s. Its use is now banned throughout most of the world.

Dead zone A region of hypoxic conditions that kills off most marine organisms that cannot escape. It is usually the result of eutrophication cause by runoff from land-based fertilizer applications.

Decapoda (1) An order of crustaceans with five pairs of thoracic "walking legs," including crabs, shrimp, and lobsters. (2) Suborder of cephalopod mollusks with 10 arms that includes squids and cuttlefish.

Decay distance The distance over which waves change from a choppy "sea" to uniform swell.

Declination The angular distance of the Sun or Moon above or below the plane of Earth's Equator.

Decomposers Primarily bacteria that break down nonliving organic material, extract some of the products of decomposition for their own needs, and make available the compounds needed for plant production.

Decompression sickness Serious condition that occurs in divers when they ascend too rapidly, causing nitrogen bubbles to form in the blood and tissue, resulting in great pain and sometimes death. Also known as *the bends*.

Deep boundary current Relatively strong deep current flowing across the continental rise along the western margin of an ocean basin.

Deep current Slow moving subsurface currents that carry oxygen to the deep ocean.

Deep-ocean basin The deep part of the ocean beyond the continental margin.

Deep scattering layer (DSL) A layer of marine organisms in the open ocean that scatter signals from an echo sounder. It migrates daily from depths of slightly over 100 meters (330 feet) at night to more than 800 meters (2600 feet) during the day.

Deep-sea fan A large fan-shaped deposit commonly found on the continental rise at the mouth of submarine canyons. They are particularly well developed seaward of such sediment-laden rivers as the Amazon, Indus, or Ganges-Brahmaputra.

Deep-sea fish Any of a large group of fishes that lives within the aphotic zone and has special adaptations for finding food and avoiding predators in darkness.

Deep-sea system System that includes all benthic environments beneath the littoral (sublittoral, bathyal, abyssal, and hadal).

Deep Sea Drilling Project The first large-scale ocean drilling project that was conducted from 1963 to 1983. The drilling vessel used was the *Glomar Challenger*.

Deep water The water beneath the permanent thermocline (and resulting pycnocline) that has a uniformly low temperature.

Deep-water wave Ocean wave traveling in water that has a depth greater than one-half the average wavelength. Its velocity is independent of water depth.

Delta A low-lying deposit at the mouth of a river.

Denitrifying bacterium Bacterium that reduces oxides of nitrogen to produce free nitrogen (N_2).

Density Mass per unit volume of a substance. Usually expressed as grams per cubic centimeter. For ocean water with a salinity of 35‰ at 0°C (32°F), its density is 1.028 grams/cubic centimeter. (See *Sigma*.)

Density (s, or sigma) Density of ocean water.

Density, in situ (s_T, or sigma T) Density of water in place.

Density, potential (s_U, or sigma theta) Density of ocean water with the adiabatic effect removed. It is always less than *in situ* density except at the surface where the adiabatic effect is zero.

Density stratification Layering based on density, where the highest density material occupies the lowest space.

Deposit feeder An organism that feeds on food items that occur as deposits, including detritus and various detritus-coated sediment.

Depositional-type shores Shorelines dominated by processes that form deposits (such as sand bars and barrier islands) along the shore.

Desalination The removal of dissolved ions from ocean water to produce pure water.

Desiccation Process of drying out.

Destructive interference A form of wave interference in which two waves come together out of phase, e.g., crest to trough, and produce a wave with less displacement than the larger of the two waves would have produced alone.

Detritus (1) Any loose material produced directly from rock disintegration. (2) Material resulting from the disintegration of dead organic remains.

Diamond A precious mineral that is the hardest natural substance on Earth, has great brilliance, and is a pure crystalline form of carbon.

Diatom Member of the class Bacillariophyceae of algae; possesses a wall of overlapping silica valves.

Diatomaceous earth A deposit composed primarily of the tests of diatoms mixed with clay. Also called *diatomite*.

Diatom ooze   An oceanic deposit that is composed of at least 30% biogenous diatom particles.

Diffraction   Any bending of a wave around an obstacle that cannot be interpreted as refraction or reflection.

Diffusion   The transfer of material or a property by random molecular movement. The movement is from a region in which the material or the property is high in concentration to regions of low concentration.

Dinoflagellate   Single-celled microscopic organism that may possess chlorophyll and belong to the phylum Pyrrophyta (autotrophic) or may ingest food and belong to the class Mastigophora of the phylum Protozoa (heterotrophic).

Dipolar   Having two poles. The water molecule possesses a polarity of electrical charge with one pole being more positive and the other more negative in electrical charge.

Discontinuity   An abrupt change in a property, such as temperature or salinity, at a line or surface.

Disphotic zone   The dimly lit zone corresponding approximately with the mesopelagic in which there is not enough light to carry on photosynthesis. Sometimes called the *twilight zone.*

Dissolved organic carbon (DOC)   The carbon contained in dissolved organic compounds.

Dissolved oxygen   Oxygen that is dissolved in seawater.

Dissolved oxygen minimum   The depth at which dissolved oxygen concentration in the water column reaches its lowest value. It is usually between 800 and 1000 meters (2600 and 3300 feet).

Distillation   A method of purifying liquids by heating them to their boiling point and condensing the vapor.

Distributary   A small stream flowing away from a main stream. Such streams are characteristic of deltas.

Distributary channel   One of a system of divergent channels leading away from the mouth of a submarine canyon and across the deep-sea fan deposited by turbidity currents.

Diurnal inequality   The difference in the heights of two successive high or low waters during a lunar (tidal) day.

Diurnal tidal pattern   A tide with one high water and one low water during a tidal day. The tidal period is 24 hour 50 minutes.

Divergence   A horizontal flow of fluid from a central region, as occurs in upwelling.

Divergent plate boundary   A lithospheric plate boundary where adjacent plates diverge, producing an oceanic ridge or rise (spreading center). Also called *constructive boundary.*

Doldrums   A belt of light variable winds within 10° to 15° of the Equator, resulting from the vertical flow of low-density air within this equatorial belt. Doldrums is the common name for the *Intertropical Convergence Zone.*

Dolomite   A common rock-forming mineral; a calcium-magnesium carbonate [$CaMg(CO_3)_2$].

Dolphin   (1) A brilliantly colored fish of the genus *Coryphaena*. (2) The name applied to the small, beaked members of the cetacean family Delphinidae.

Dorsal   Pertaining to the back or upper surface of most animals.

Dorsal aorta   For most fish, this is the only major artery that runs the length of the fish through openings in the vertebrae and supplies blood. Some pelagic cruisers also have a cutaneous artery.

Downwelling   In the open or coastal ocean where Ekman transport causes surface waters to converge or impinge on the coast, surface water will be carried down beneath the surface.

Drift bottle   Any equipment used to study current movement by drifting with currents.

Driftnet   Long net set in the sea to passively catch anything that swims into it. Also called a *gill net.*

Drifts   Thick sediment deposit on the continental rise produced where the western boundary undercurrent slows and loses sediment as it changes direction to follow the base of the continental slope.

Drowned beach   An ancient beach now beneath the coastal ocean because of rising sea level.

Drowned river valley   River valley that is inundated by rising sea level to produce a coastal plain estuary like Chesapeake Bay.

Dune   Coastal deposit of sand lying landward of the beach and deriving its sand from onshore winds that transport beach sand inland.

Dynamical tidal theory   The theory of tidal behavior that takes into account friction between the ocean water and the ocean floor, the effects of changing depth of the ocean floor, and the interference of the continents on the passage of tidal waves.

Dynamic topography   A surface configuration resulting from the geopotential difference between a given surface and a reference surface of no motion. A contour map of this surface is useful in estimating the nature of geostrophic currents.

Earthquake   A sudden ground motion or trembling caused by the sudden release of slowly accumulated strain by faulting (movement along a fracture in Earth's crust) or volcanic activity.

East Australian Current   A warm boundary current flowing south along the east coast of Australia.

East Pacific Rise   A fast-spreading divergent plate boundary extending southward from the Gulf of California through the eastern South Pacific Ocean.

East Wind Drift   The coastal current driven in a westerly direction by the polar easterly winds blowing off of Antarctica.

Eastern boundary current   A slow, drifting cold current that flows toward the Equator on the eastern side of a subtropical gyre.

Ebb current   Seaward-flowing current during a decrease in the height of the tide.

Echinoderms   Members of the phylum Echinodermata which includes animals that have bilateral symmetry in larval forms and usually a five-sided radial symmetry as adults. Benthic and possessing rigid or articulating exoskeletons of calcium carbonate with spines, this phylum includes sea stars, brittle stars, sea urchins, sand dollars, sea cucumbers, and sea lilies.

Echo sounder   A device that transmits sound from a ship's hull to the ocean floor where it is reflected back to receivers. Knowing the speed of sound in the water, the depth can be determined from the travel-time of the sound signal.

Echolocation   A sensory system in odontocete cetaceans in which usually high-pitched sounds are emitted and their echoes interpreted to determine the direction and distance of objects.

Ecliptic   The plane of the center of the Earth–Moon system as it orbits around the Sun.

Ecological efficiency   Efficiency with which energy is transferred from one trophic level to the next, or the ratio of the amount of protoplasm added to a trophic level to the amount of food required to produce it.

Ecosystem   All the organisms in a biotic community and the abiotic environmental factors with which they interact.

Ectoderm   Outermost layer of cells in an animal embryo. In vertebrates it gives rise to the skin, nervous system, sense organs, etc.

Ectothermic   Of or relating to an organism that regulates its body temperature largely by exchanging heat with its surroundings; cold-blooded.

Eddy   A current of any fluid forming on the side of or within a main current. It usually moves in a circular path and develops where currents encounter obstacles or flow past one another.

Ekman spiral   A theoretical consideration of the effect of a steady wind blowing over an ocean of unlimited depth and breadth and of uniform viscosity. The result is a surface flow at 45° to the right of the wind in the Northern Hemisphere. Water at increasing depth will drift in directions increasingly to the right until at about 100 meters (330 feet) depth it is moving in a direction opposite to that of the wind. The net water transport is 90° to the wind, and velocity decreases with depth.

Ekman transport   The net transport of surface water set in motion by wind. Because of the Ekman spiral phenomenon, it is theoretically in a direction 90° to the right and 90° to the left of the wind direction in the Northern Hemisphere and Southern Hemisphere, respectively.

El Niño   A southerly flowing warm current that generally develops off the coast of Ecuador around Christmastime. Occasionally it will move farther south into Peruvian coastal waters and cause the widespread death of plankton, fish, and other organisms dependent upon fish for food.

El Niño–Southern Oscillation (ENSO)   The correlation of El Niño events with an oscillatory pattern of pressure change in a persistent high-pressure cell in the southeastern Pacific Ocean and a persistent low-pressure cell over the East Indies.

Electrical conductivity   The ability or power to conduct or transmit electricity.

Electrolysis   A separation process by which salt ions are removed from saltwater through water-impermeable membranes toward oppositely charged electrodes.

Electromagnetic spectrum   The spectrum of radiant energy emitted from stars and ranging between cosmic rays with wavelengths less than 0.000001 micron and very long waves with wavelengths in excess of 100 kilometers (62 miles).

Electron   A negatively charged particle in orbit around the nucleus of an atom.

Electron cloud   The organized assemblage of electrons surrounding the nucleus of an atom.

Electrostatic force   A force caused by electric charges at rest.

Element   One of a number of substances, each of which is composed entirely of like particles—atoms—that cannot be broken into smaller particles by chemical means.

Emerging shoreline   A shoreline resulting from the emergence of the ocean floor relative to the ocean surface. It is usually rather straight and characterized by marine features usually found at a greater depth.

Endoderm   Innermost cell layer of an embryo. Develops into the digestive and excretory systems and forms the lining for the respiratory system, etc., in vertebrates.

Endothermic reaction   A chemical reaction that absorbs energy. For example, energy is stored in the organic products of the chemical reaction photosynthesis.

ENSO   See *El Niño–Southern Oscillation.*

ENSO index   An index showing the relative strength of El Niño and La Niña conditions.

Entropy   A quantity reflecting the degree of uniform distribution of heat energy in a system. It increases with time and represents a state in which energy is unrecoverable for work.

Environment   The sum of all physical, chemical, and biological factors to which an organism or community is subjected.

Epicenter   The point on Earth's surface that is directly above the focus of an earthquake.

Epifauna   Animals that live on the ocean bottom, either attached or moving freely over it.

Epipelagic zone   The upper region of the oceanic province, extending to a depth of 200 meters (660 feet).

Epitheca   The top valve of a diatom test.

Equator   The imaginary great circle around Earth's surface, equidistant from the poles and perpendicular to Earth's axis of rotation. It divides Earth into the Northern Hemisphere and the Southern Hemisphere.

Equatorial   Pertaining to the equatorial region.

Equatorial Countercurrent   Eastward-flowing currents found between the North and South Equatorial Currents in all oceans, but particularly well developed in the Pacific Ocean.

Equatorial current   A warm current in the equatorial region driven in a westerly direction by the trade winds.

Equatorial low   A band of low atmospheric pressure that encircles the globe along the Equator.

Equatorial tide   A semimonthly tide occurring when the Moon is over the Equator. It displays a minimal diurnal inequality.

Equatorial Undercurrent   A thin, ribbonlike east-flowing current embedded in the Pacific South Equatorial Current that flows over 6000 kilometers (3700 miles) from the western Pacific to the Galápagos Islands.

Equatorial upwelling   Water that rises from beneath the surface to replace surface water that moves away from the Equator at the surface due to Ekman transport.

Equilibrium tide theory   A tidal model that considers the ocean to be of uniform nature and depth throughout Earth's surface. It is further assumed that this ocean will respond instantly to the gravitational forces of the Sun and Moon.

Equinox   The times when the Sun is over the Equator, making day and night of equal length throughout Earth. *Vernal equinox* occurs about March 21 as the Sun is moving into the Northern Hemisphere. *Autumnal equinox* occurs about September 21 as the Sun is moving into the Southern Hemisphere.

Erosion   The group of natural processes, including weathering, dissolution, abrasion, corrosion, and transportation, by which material is worn away from Earth's surface.

Erosional-type shores   Shorelines dominated by processes that form erosional features (such as cliffs and sea stacks) along the shore.

Estuarine circulation pattern   A flow pattern in an estuary characterized by a net surface flow of low-salinity water toward the ocean and an opposite net subsurface flow of seawater toward the head of the estuary.

Estuary   A partially enclosed coastal body of water in which salty ocean water is significantly diluted by fresh water from land runoff. Examples of estuaries include river mouths, bays, inlets, gulfs, and sounds.

Eukarya   A domain of organisms that possess cells containing a nucleus, other intracellular bodies such as mitochondria and plantids, and a cell wall more complex than that of prokaryotic cells.

Eukaryotic cells   Cells that possess a nucleus, other intracellular bodies such as mitochondria and plastids, and a cell wall more complex than that of archaea and bacteria.

Euphausiacea   An order of planktonic crustaceans ranging in length from 5 to 30 centimeters (2 to 12 inches). Most possess luminous organs, and some are the principal food for baleen whales.

Euphotic zone   A layer that extends from the surface of the ocean to a depth where enough light exists to support photosynthesis, rarely deeper than 100 meters (330 feet).

Euryhaline   Pertaining to the ability of a marine organism to tolerate a wide range of salinity.

Eurythermal   Pertaining to the ability of a marine organism to tolerate a wide range of temperature.

Eustatic sea level change   A worldwide raising or lowering of sea level.

Eutrophic   Characterized by an abundance of nutrients.

Eutrophication   The enrichment of waters by a previously scarce nutrient.

Evaporation   The physical process of converting a liquid to a gas. Commonly considered to occur at a temperature below the boiling point of the liquid.

Evaporite   A sedimentary deposit that is left behind when water evaporates. Evaporite minerals include gypsum, calcite, and halite.

Evolution   The theory that groups of organisms change with passage of time, mainly as a result of natural selection, so that descendants differ morphologically and physiologically from their ancestors.

Excess volatile   Any of the volatile compounds found in the oceans, sediments, and atmosphere in quantities greater than the chemical weathering of crystalline rock could produce. They are considered to have been produced by volcanic action.

Exclusive Economic Zone (EEZ)   A coastal zone 200 nautical miles wide over which the coastal nation has jurisdiction over mineral resources, fishing, and pollution. If the continental shelf extends beyond 200 miles, the EEZ may be up to 350 nautical miles in width.

Exothermic reaction   A chemical reaction that liberates energy. For example, the energy stored in the products of photosynthesis is released by the chemical reaction respiration.

Extrusive rock   Igneous rock formed from lava that flows out onto Earth's surface and cools rapidly.

Eye of a hurricane   The low-pressure center of a hurricane that is calm while winds in excess of 120 kilometers (74 miles) per hour rotate around it.

Fact Something having real, demonstrable existence. A scientific fact is an occurrence that has been repeatedly confirmed.

Fahrenheit temperature scale (°F) Scale in which the freezing point of water is 32°, boiling point of water is 212°.

Falcate Curved and tapering to a point; sickle-shaped.

Falkland Current A northward-flowing cold current found off the southeastern coast of South America.

Fall bloom A mid-latitude bloom of phytoplankton that occurs during the fall and is limited by the availability of sunlight.

Fall equinox See *autumnal equinox*.

Fallout The radioactive atomic nuclei that were released into the atmosphere by nuclear bomb testing and fell out onto the continents and into the oceans.

Fan A gently sloping, fan-shaped feature normally located near the lower end of a canyon. Also known as a *submarine fan*.

Fast ice Sea ice that is attached to the shore and therefore remains stationary.

Fat A colorless, odorless organic compound consisting of carbon, hydrogen, and oxygen; insoluble in water.

Fathom (fm) A unit of ocean depth commonly used in countries using the English system of units. It is equal to 1.83 meters, or 6 feet.

Fault A fracture or fracture zone in Earth's crust along which displacement has occurred.

Fault block A crustal block bounded on at least two sides by faults. Usually elongate; if it is down-dropped, it produces a graben; if uplifted, it is a horst.

Fauna The animal life of any particular area or of any particular time.

Fecal pellet Excrement of planktonic crustaceans that assist in speeding up the descent rate of sedimentary particles by combining them into larger packages.

Federal Water Pollution Control Act Commonly called the *Clean Water Act*, it was passed in 1948 to control point-source pollution from sources such as municipal and industrial waste disposal and spills of oil and hazardous materials.

Ferrell cell The atmospheric circulation pattern within the lower atmosphere between the subtropical (30° latitude) high-pressure belt and the relatively low-pressure belt at the polar front (60° latitude).

Ferromagnesian mineral Mineral rich in iron and magnesium.

Fetch (1) Area of the open ocean over which the wind blows with constant speed and direction, thereby creating a wave system. (2) The distance across the fetch (wave-generating area) measured in a direction parallel to the direction of the wind.

Filter feeder Animals that possess devices that filter appropriately sized food particles (organisms or organic matter) from ocean water.

Fishery The process by which humans catch fish or other marine organisms.

Fishery assessment The conduct of research on the ecological and economic factors related to a fishery and the application of the knowledge gained to its regulation.

Fishery management The organized effort directed at regulating fishing activity with the goal of maintaining a long-term fishery.

Fissure A long, narrow opening; a crack or cleft.

Fjord A long, narrow, deep, U-shaped inlet that usually represents the seaward end of a glacial valley that has become partially submerged after the melting of the glacier.

Flagellum A whiplike living process used by some cells for locomotion.

Flocculent A loose, open-structured deposit of organic debris composed of tiny particles.

Floe A piece of floating ice other than fast ice or icebergs. May range in dimension from about 20 centimeters (8 inches) across to more than a kilometer.

Flood basalt A basalt plateau composed of from 1 to 2 million cubic kilometers of basalt deposited in a period of from 1 to 2 million years.

Flood current A tidal current associated with increasing height of the tide, generally moving toward the shore.

Flood tide A rising tide.

Flora The plant life of any particular area or of any particular time.

Florida Current A warm current flowing north along the coast of Florida that flows into the Gulf Stream.

Flux The amount of flow per unit of cross-sectional area per unit of time.

Fog The condensation of water in relatively warmer air of high humidity when it moves over water that is much cooler.

Food chain The passage of energy materials from producers through a sequence of herbivores and carnivores.

Food web A group of interrelated food chains.

Foraminifer A member of an order of planktonic and benthic protozoans that possess protective coverings usually composed of calcium carbonate.

Foraminifer ooze An oceanic deposit composed of more than 30% by weight biogenous foraminifer particles.

Forced wave A wave that is generated and maintained by a continuous force such as the gravitational attraction of the Moon.

Fore-arc The ocean floor that lies between an island arc and its associated trench.

Foreshore The portion of the shore lying between the normal high- and low-water marks—the intertidal zone.

Fossil Any remains, trace, or imprint of an organism that has been preserved in rocks.

Fracture zone An extensive linear zone of unusually irregular ocean floor topography, characterized by large seamounts, steep-sided or asymmetrical ridges, troughs, or long, steep slopes. Usually represents an ancient, inactive transform fault zone.

Free wave A wave created by a sudden rather than a continuous impulse that continues to exist after the generating force is gone.

Freeze separation The process by which seawater is converted to fresh water by freezing to exclude the salt ions.

Freezing Cooling a substance below its freezing point. This is one of the methods used in the desalination of seawater.

Freezing point The temperature at which a liquid becomes a solid under any given set of conditions. The freezing point of water is 0°C (32°F) at one atmospheric pressure.

Frequency See *Wave frequency (f)*.

Fringing reef A reef that is directly attached to the shore of an island or continent. It may extend more than 1 kilometer (0.62 mile) from shore. The outer margin is submerged and often consists of algal limestone, coral rock, and living coral.

Frustule The siliceous covering or test of a diatom, consisting of two halves (epitheca and hypotheca).

Fucoxanthin The reddish-brown pigment that gives brown algae its characteristic color.

Full moon The phase of the Moon that occurs when the Sun and Moon are in opposition, on opposite sides of Earth.

Fully developed sea The maximum average size of waves that can be developed for a given wind speed when it has blown in the same direction for a minimum duration over a minimum fetch.

Fungi Any of numerous eukaryotic organisms of the kingdom Fungi, which lack chlorophyll and vascular tissue. They range in size from a single cell to a body mass of branched filamentous structures that often produce specialized fruiting bodies. The kingdom includes yeasts, molds, lichens, and mushrooms.

Fur seal Any of several eared seals of the genera *Callorhinus* or *Arctocephalus*, having thick, soft underfur and closely related to sea lions.

Furrow Parallel, troughlike structure cut into mud waves by bottom currents. Furrows are aligned in the direction of current flow.

Fusion reaction A reaction that takes place within the Sun and other stars by which hydrogen atoms are fused to produce helium atoms with a huge release of energy. It occurs only at temperatures of tens of millions of degrees. As stars age, helium atoms are fused to produce carbon, and successively oxygen, silicon, and iron atoms are produced.

Gabbro Dark-colored intrusive rock that is the mineralogic equivalent of basalt.

Galápagos Rift A divergent plate boundary extending eastward from the Galápagos Islands toward South America. The first deep-sea hydrothermal vent biocommunity was discovered here in 1977.

Galaxy One of the billions of large systems of stars that make up the universe.

Gas hydrate A latticelike compound composed of water and natural gas (usually methane) formed in high pressure and low temperature environments such as those found in deep ocean sediments. Also known as *clathrates* because of their cagelike chemical structure.

Gaseous state A state of matter in which molecules move by translation and only interact through chance collisions.

Gastropoda A class of mollusks, most of which possess an asymmetrical spiral one-piece shell and a well-developed flattened foot. A well-developed head will usually have two eyes and one or two pairs of tentacles. Includes snails, limpets, abalone, cowries, sea hares, and sea slugs.

Geologic history The history of Earth recorded in the rocks of Earth's crust.

Geologic time scale Scale that shows the 4.6-billion-year history of Earth. About 87% of this history prior to 570 million years ago is composed of three eons that are collectively called the "Precambrian." The Phanerozoic eon covers the last 570 million years, and because it is the most recent, more of the details of this eon are known. It is divided into three eras, Paleozoic, Mesozoic, and Cenozoic, which are further subdivided into periods and epochs. Fossils first become abundant in the Cambrian Period of the Paleozoic Era.

Geostrophic current A current that develops out of Earth's rotation and is the result of a near balance between gravitational force and the Coriolis effect.

Gill A thin-walled projection from some part of the external body or the digestive tract; used for respiration in a water environment.

Gill net A net set in the water to passively catch fish by having them swim into the mesh. The mesh is sized to slide over the gills and trap the fish. Also called a *driftnet*.

Glacial deposit Deposit of rock fragments carried to the ocean by glaciers. The deposits form as the glacier ice melts and releases the rock fragments to fall to the ocean floor.

Glacial Epoch The Pleistocene Epoch, the earlier of two divisions of the Quaternary Period of geologic time. During this time, high-latitude continental areas now free of ice were covered by continental glaciers.

Glacier A large mass of ice formed on land by the recrystallization of old compacted snow.

It flows from an area of accumulation to an area of wasting, where ice is removed from the glacier by melting or calving.

Glauconite A group of green hydrogenous minerals consisting of hydrous silicates of potassium and iron.

Global Positioning System (GPS) A system of satellites that transmit microwave signals to Earth, allowing the accurate determination of locations on Earth.

Globigerina ooze An oceanic deposit composed of at least 30% by weight biogenous particles, most of which are the shells of foraminifers belonging to the genus *Globigerina*.

GLORIA Geological Long-Range Inclined ASDIC (Acoustical Side Derection). GLORIA is used for deep-sea side directed sonar surveys.

Gold A soft, heavy, yellow mineral, the native metallic element, Au. It has been mined from placer deposits in the coastal ocean.

Gondwanaland A hypothetical protocontinent of the Southern Hemisphere named for the Gondwana region of India. It included the present continental masses of Africa, Antarctica, Australia, India, and South America.

Graben An elongate, downdropped crustal block that is bounded by faults. The rift valleys along the axes of spreading centers are grabens.

Graded bedding Stratification in which each layer displays a decrease in grain size from bottom to top.

Gradient The rate of increase or decrease of one quantity or characteristic relative to a unit change in another. For example, the slope of the ocean floor is a change in elevation (a vertical linear measurement) per unit of horizontal distance covered. Commonly measured in meters per kilometer.

Grain size The average size of grains in a sample of material. Also known as *fragment* or *particle size*.

Granite A light-colored igneous rock characteristic of the continental crust. Rich in nonferromagnesian minerals such as feldspar and quartz.

Gran Method A technique involving the observation of the change in dissolved oxygen within paired transparent and opaque bottles containing phytoplankton and suspended in the ocean surface water to determine the base of the euphotic zone.

Gravitational force The force of attraction that exists between any two bodies in the universe that is proportional to the product of their masses and inversely proportional to the distance between the centers of their masses.

Gravitational stability The degree to which segments of the water column tend to remain stationary. The portion of the water column with the greatest degree of gravitational stability is called the pycnocline, where water at the

top is much less dense than water at the bottom. This zone of density stratification normally coincides with the thermocline.

Gravity wave A wave for which the dominant restoring force is gravity. Such waves have a wavelength of more than 1.74 centimeters (0.7 inch), and their velocity of propagation is controlled mainly by gravity.

Gray whale Pacific baleen whales of the species *Eschrichtius robustus* that feed in the Chukchi Sea and Bering Sea and breed and calve in the warm waters of lagoons in Baja California.

Greenhouse effect The heating of Earth's atmosphere that results from the absorption by components of the atmosphere such as water vapor and carbon dioxide of infrared radiation from Earth's surface.

Grenadier A rattailed demersal fish that is usually the first to arrive at bait deposited on the ocean floor.

Groin A low, artificial structure projecting into the ocean from the shore to interfere with longshore transportation of sediment. It usually has the purpose of trapping sand to cause the buildup of a beach.

Groin field A series of closely spaced groins.

Gross ecological efficiency The amount of energy passed on from a trophic level to the one above it divided by the amount it received from the one below it.

Gross primary production The total amount of organic material produced by autotrophs.

Grunion A small fish (*Leursthes tenuis*) of coastal waters of California and Mexico that spawns at night along beaches during the high tides of spring and summer. Grunion time their reproductive acitvities to coincide with tidal phenomena and are the only fish that come completely out of water to spawn.

Guano A phosphate deposit formed by the leaching of bird droppings in arid climates.

Guiana Current The current that diverts South Equatorial Current water north along the Guiana coast of South America into the Northern Hemisphere.

Gulf of California A sea on the eastern margin of the Pacific Ocean between the Mexican mainland and the Baja California pennisula.

Gulf of Mexico A sea on the western margin of the North Atlantic Ocean. It is bounded by the United States, the Mexican mainland, and Cuba.

Gulf Stream The high-intensity western boundary current of the North Atlantic Ocean subtropical gyre that flows north off the east coast of the United States.

Guyot A tablemount; a conical volcanic feature on the ocean floor that has had the top truncated to a relatively flat surface.

Gypsum A colorless, white, or yellowish evaporite mineral, $CaSO_4 \cdot 2H_2O$.

Gyre A circular motion. Used mainly in reference to the circular motion of water in each of the major ocean basins centered in subtropical high-pressure regions.

Habitat A place where a particular plant or animal lives. Generally refers to a smaller area than environment.

Hadal zone The deepest ocean environment, specifically ocean trenches deeper than 6 kilometers (3.7 miles).

Hadley Cell The large atmospheric circulation cell that occurs between the Equator and 30° latitude in each hemisphere.

Half-life The time required for half the atoms of a radioactive isotope sample to decay to atoms of another element.

Halite A colorless or white evaporite mineral, NaCl, which occurs as cubic crystals and is used as table salt.

Halocline A layer of water that has a high rate of salinity change in the vertical dimension.

Hard stabilization Any form of artificial structure built to protect a coast or to prevent the movement of sand along a beach. Examples include groins, breakwaters, and seawalls.

Harmful algae bloom (HAB) See *Red tide*.

Harmonic analysis Mathematical analysis of the partial tides to predict the tides in a given region.

Headland A steep-faced irregularity of the coast that extends out into the ocean.

Heat Energy moving from a high-temperature system to a lower-temperature system. The heat gained by the one system may be used to raise its temperature or to do work.

Heat budget (global) The equilibrium that exists on the average between the amounts of heat absorbed by Earth and that returned to space.

Heat capacity The amount of heat required to raise the temperature of 1 gram of a substance 1°C (1.8°F).

Heat energy Energy of molecular motion. The conversion of higher forms of energy such as radiant or mechanical energy to heat energy within a system increases the heat energy within the system and the temperature of the system.

Heat flow (flux) The measurement of heat moving from Earth's interior to the surface.

Hemoglobin A red proteinaceous pigment containing iron and found in red blood cells of vertebrates. It combines with oxygen and carries oxygen from the lungs to tissues.

Herbivore An organism that eats only algae and/or plants.

Hermatypic Pertaining to corals and other reef animals that depend on algal symbionts for their metabolic energy.

Heterocercal fin A tail fin in which the upper lobe is larger than the lower lobe to it, provides an upward thrust. It is found on many species of shark.

Heterotrophs Animals and bacteria that depend on the organic compounds produced by other animals and plants as food. Organisms not capable of producing their own food by photosynthesis.

High slack water The short period of time associated with the peak of high tide when there is no visible flow of water into or out of a bay or river.

High tide zone Shore covered by the highest high tides but not by the lowest high tides.

High water (HW) The highest level reached by the rising tide before it begins to recede.

Higher high water (HHW) The higher of two high waters occurring during a tidal day where tides exhibit a mixed tidal pattern.

Higher low water (HLW) The higher of two low waters occurring during a tidal day where tides exhibit a mixed tidal pattern.

Highly stratified estuary A relatively deep estuary, with a high volume of fresh water flowing seaward at the surface and a large mass of marine water at depth, producing a well-developed halocline.

Holoplankton Organisms that spend their entire life as members of the plankton.

Homeothermic Of or relating to an animal that maintains a precisely controlled body temperature using its own internal heating and cooling mechanisms; warm-blooded.

Horse latitudes The latitude belts between 30° and 35° north and south latitude where winds are light and variable, since the principal movement of air masses at these latitudes is one of vertical descent. The climate is hot and dry, resulting in the creation of the major continental and maritime deserts of the world.

Hotspot The relatively stationary surface expression of a persistent column of molten mantle material rising to the surface.

Hurricane A tropical cyclone in which winds reach velocities above 120 kilometers (74 miles) per hour. Generally applied to such storms in the North Atlantic Ocean, eastern North Pacific Ocean, Caribbean Sea, and Gulf of Mexico. Such storms in the western Pacific Ocean are called *typhoons* and those in the Indian Ocean are known as *cyclones*.

Hydrated To be chemically combined with water or surrounded by water.

Hydration The condition of being surrounded by water molecules such as when sodium chloride dissolves in water.

Hydration sphere A sphere of water molecules that surround ions dissolved in water.

Hydrocarbon An organic compound consisting solely of hydrogen and carbon. Petroleum is a mixture of many hydrocarbon compounds.

Hydrocarbon seep biocommunity Deep bottom-dwelling community associated with a hydrocarbon seep from the ocean floor. The community depends on methane and sulfur-oxidizing bacteria as producers. The bacteria may live free in the water, on the bottom, or symbiotically in the tissues of some of the animals.

Hydrogen bond An intermolecular bond that forms within water because of the dipolar nature of water molecules.

Hydrogenous sediment Sediment that forms from ocean water precipitation or ion exchange between existing sediment and ocean water. Examples are manganese nodules, phosphorite, glauconite, metal sulfides, and various evaporite salts.

Hydrogen sulfide (H_2S) The gas that smells like rotten eggs. Its molecule is composed of two atoms of hydrogen and one atom of sulfur.

Hydrogen sulfide biocommunity A community of organisms that depend on microbes that oxidize hydrogen sulfide for their existence.

Hydrologic cycle The cycle of water exchange among the atmosphere, land, and ocean through the processes of evaporation, precipitation, runoff, and subsurface percolation. Also called the *water cycle*.

Hydrophilic Pertaining to the property of attracting water.

Hydrophobic Pertaining to the property of being impossible or difficult to wet with water.

Hydrothermal spring Vents of hot water found primarily along the spreading axes of oceanic ridges and rises.

Hydrothermal vent Ocean water that percolates down through fractures in recently formed ocean floor is heated by underlying magma and surfaces again through these vents. They are usually located near the axis of spreading along mid-ocean ridges.

Hydrothermal vent biocommunity Deep bottom-dwelling community associated with a hydrothermal vent. The hot water vent is usually associated with the axis of a spreading center, and the community is dependent on sulphur-oxidizing bacteria that may live free in the water, on the bottom, or symbiotically in the tissue of some of the animals of the community.

Hydrozoa A class of cnidarians that characteristically exhibit alternation of generations, with a sessile polypoid colony giving rise to a pelagic medusoid form by asexual budding.

Hypersaline Pertaining to marine water with salinity significantly higher than normal marine salinity.

Hypersaline lagoon Shallow lagoons such as Laguna Madre, which may become hypersaline due to little tidal flushing and seasonal variability in freshwater input. High evaporation rates and low freshwater input can result in very high salinities.

Hypersaline seep Seeps in which the vented water has a salinity much higher than ocean water.

Hypersaline seep biocommunity Bottom-dwelling community of organisms associated with a hypersaline seep.

Hypertonic   Pertaining to the property of an aqueous solution having a higher osmotic pressure (salinity) than another aqueous solution separated by a semipermeable membrane allowing osmosis. The hypertonic fluid will gain water molecules through the membrane from the other fluid.

Hypotheca   The lower valve of a diatom frustule.

Hypothesis   A tentative, testable statement about the general nature of the phenomena observed.

Hypotonic   Pertaining to the property of an aqueous solution having a lower osmotic pressure (salinity) than another aqueous solution separated by a semipermeable membrane allowing osmosis. The hypotonic fluid will lose water molecules through the membrane to the other fluid.

Hypoxic   Pertaining to an environment with low levels of dissolved oxygen.

Hypsographic curve   A cumulative frequency profile representing the statistical distribution of the areas of Earth's solid surfaces at elevations above or below sea level.

Ice Age   The most recent glacial period, which occurred during the Pleistocene Epoch.

Ice floe   See *Floe.*

Ice rafting   The movement of trapped sediment within or on top of ice by flotation.

Ice sheet   An extensive, relatively flat accumulation of ice.

Ice shelf   A thick layer of ice with a relatively flat surface that is attached to and nourished by a continental glacier from one side. The shelf, which is for the most part afloat, may extend above water level by more than 50 meters (164 feet) along its seaward cliff formed by the break-off of large tabular chunks of ice that become icebergs.

Iceberg   A massive piece of glacial ice that has broken from the front of the glacier (calved) into a body of water. It floats with its tip at least 5 meters (16 feet) above the water's surface and at least four-fifths of its mass submerged.

Igneous rock   One of the three main classes into which all rocks are divided (i.e., igneous, metamorphic, and sedimentary). Rock that forms from the solidification of molten or partly molten material (magma).

In situ   In place (i.e., *in situ* density of a sample of water is its density at its original depth).

In situ **density**   See *Density, in situ.*

Incidental catch   Organisms caught during fishing operations that are not the target of the operation. Also called *bycatch.*

Indian Ocean   The portion of the world ocean bounded by Africa, Asia, Australia, and Antarctica. It exists mostly in the Southern Hemisphere and represents 20.5% of the world ocean surface area.

Inertia   Newton's first law of motion. It states that a body at rest will stay at rest and a body in motion will remain in a uniform motion in a straight line unless acted on by some external force.

Infauna   Animals that live buried in the soft substrate (sand or mud).

Infrared radiation   Electromagnetic radiation between the wavelengths of 0.8 micron (0.0003 inch) and about 1000 microns (0.4 inch). It is bounded on the shorter-wavelength side by the visible spectrum and on the long side by microwave radiation.

Inner core   The solid core of Earth with a radius of 1216 kilometers (755 miles) composed mostly of iron and nickel lying beneath the liquid outer core of similar composition.

Inner sublittoral zone   The section of ocean floor from the low-tide shoreline to the point where attached plants stop growing.

Insolation   The rate at which solar radiation is received per unit of surface area at any point at or above Earth's surface.

Integrated Ocean Drilling Program (IODP)   A new drilling program that will replace the Ocean Drilling Program in 2003 with a new drill ship that has riser technology, enabling cores to be collected from deeper within Earth's interior.

Interface   A surface separating two substances of different properties (i.e., density, salinity, or temperature). In oceanography, it usually refers to a separation of two layers of water with different densities caused by significant differences in temperature and/or salinity.

Interface wave   An orbital wave that moves along an interface between fluids of different density. An example is ocean surface waves moving along the interface between the atmosphere and the ocean, which is 1000 times more dense.

Interference pattern   The pattern of wave development that develops from the combined interference effects of more than one wave system as they pass in the open ocean.

Intergovernmental Panel on Climate Change (IPCC)   A group of 200 scientists sponsored by the United Nations Environment Programme and the World Meteorological Organization that published its first report in 1995.

Intermediate water   Water masses that usually form at the Arctic or Antarctic convergences and sink to a depth of 800 to 1000 meters (2600 to 3300 feet) before spreading out beneath the surface-water masses.

Intermolecular bond   A relatively weak bond that forms between molecules of a given substance. The hydrogen bond and the van der Waals bonds are intermolecular bonds.

Internal wave   A wave that develops below the surface of a fluid, the density of which changes with increased depth. This change may be gradual or occur abruptly at an interface.

Intertidal zone   Littoral zone, the foreshore.

The ocean floor covered by the highest normal tides and exposed by the lowest normal tides and the water environment of the tide pools within this region.

Intertropical Convergence Zone (ITCZ)   Zone where northeast trade winds and southeast trade winds converge. Averages about 5° north latitude in the Pacific and Atlantic Oceans and 7° south latitude in the Indian Ocean.

Intraplate feature   Any feature that occurs within a tectonic plate and not along a plate boundary.

Intrusive rock   Igneous rock such as granite that cools slowly beneath Earth's surface.

Invertebrate   Animal without a backbone.

Ion   An atom that becomes electrically charged by gaining or losing one or more electrons. The loss of electrons produces a positively charged cation, and the gain of electrons produces a negatively charged anion.

Ionic bond   A chemical bond resulting from the electrical attraction that exists between cations and anions.

Irminger Current   A warm current that branches off from the Gulf Stream and moves up along the west coast of Iceland.

Iron hypothesis, the   A hypothesis that states that an effective way of increasing productivity in the ocean is to fertilize the ocean by adding the only nutrient that appears to be lacking—iron. Adding iron to the ocean also increases the amount of carbon dioxide removed from the atmosphere.

Iron spherule   Magnetic, iron-rich spherical particle about 30 microns (0.012 inch) in diameter that rains down on Earth as a component of space dust. Iron spherules may be the products of collisions of asteroids.

Irons   A meteorite consisting essentially of iron but may also contain up to 30% nickel.

Island arc   A linear arrangement of islands, many of which are volcanic, usually curved so that the concave side faces a sea separating the islands from a continent. The convex side faces the open ocean and is bounded by a deep-ocean trench.

Isohaline   Of the same salinity.

Isopoda   An order of dorsoventrally flattened crustaceans that are mostly scavengers or parasites on other crustaceans or fish.

Isopycnal   Of the same density.

Isostasy   A condition of equilibrium, comparable to buoyancy, in which the rigid crustal units float on the underlying mantle.

Isostatic adjustment   Adjustment of crustal material due to isostasy.

Isostatic rebound   The upward movement of crustal material due to isostasy.

Isotherm   A line connecting points of equal temperature.

Isothermal   Of the same temperature.

Isotonic   Pertaining to the property of having

equal osmotic pressure. If two such fluids were separated by a semipermeable membrane that allows osmosis to occur, there would be no net transfer of water molecules across the membrane.

Isotope One of several atoms of an element that has a different number of neutrons, and therefore a different atomic mass, than the other atoms, or isotopes, of the element.

Jellyfish (1) A free-swimming, umbrella-shaped medusoid member of the cnidaria class, Schyphozoa. (2) Also frequently applied to the medusoid forms of other cnidarians.

Jet stream An easterly moving air mass at an elevation of about 10 kilometers (6.2 miles). Moving at speeds that can exceed 300 kilometers (186 miles) per hour, the jet stream follows a wavy path in the mid-latitudes and influences how far polar air masses may extend into the lower latitudes.

Jetty A structure built from the shore into a body of water to protect a harbor or a navigable passage from being closed off by the deposition of longshore drift material.

Juan de Fuca Ridge A divergent plate boundary off the Oregon–Washington coast.

Juvenile water Water that is derived directly from magma, being released for the first time at Earth's surface as the magma crystallizes to igneous rock.

K–T event An extinction event marked by the disappearance of the dinosaurs that occurred about 65 million years ago at the boundary between the Cretaceous (K) and Tertiary (T) Periods of geologic time.

Kelp Large varieties of Phaeophyta (brown algae).

Kelp forest community All of the organisms found in the kelp forests of shallow coastal waters where water temperatures are relatively cool.

Kelvin wave Wave that results when a progressive tide wave moves from the open ocean into and out of a relatively narrow body of water during a tidal cycle; the tidal range will be greater on the right side of the narrow body of water during flood tide. This results from the fact that the channel rotates in a counterclockwise direction as Earth rotates, while the wave tends to move in a straight line.

Key A low, flat island composed of sand or coral debris that accumulates on a reef flat.

Killer whale The toothed whale, *Orcinus orca*, which grows to a length of 9 meters (28 feet). It is cosmopolitan and feeds on the widest variety of animals of any whale—from fish to other whales.

Kinetic energy Energy of motion. It increases as the mass or velocity of the object in motion increases.

Knot (kt) Unit of speed equal to 1 nautical mile per hour, approximately 1.15 statute (land) miles per hour.

Krill A common name frequently applied to members of the crustacean order Euphausiacea (euphausids).

Kuroshio Current The fast-flowing western boundary current of the North Pacific Ocean subtropical gyre.

Kuroshio Extension The northeastern extension of the Kuroshio Current that becomes the North Pacific Current.

Kyoto Protocol An agreement by some 60 nations in 1997 developed at a meeting in Kyoto, Japan, by which target reductions were set for carbon dioxide emissions.

La Niña An event where the surface temperature in the waters of the eastern South Pacific falls below normal values. It often occurs immediately after an El Niño event.

Labrador Current A cold current flowing south along the coast of Labrador in the northeastern Atlantic Ocean.

Lagoon A shallow stretch of seawater partly or completely separated from the open ocean by an elongated, narrow strip of land such as a reef or barrier island.

Laguna Madre A hypersaline lagoon behind Padre Island along the south Texas coast.

Lamina A layer.

Laminar flow Flow in which a fluid flows in parallel layers or sheets. The direction of flow at any point does not change with time; nonturbulent flow.

Land breeze The seaward flow of air from the land caused by differential cooling of Earth's surface.

Langmuir circulation A cellular circulation set up by winds that blow consistently in one direction with velocities above 12 kilometers (7.5 miles) per hour. Helical spirals running parallel to the wind direction are alternately clockwise and counterclockwise.

Lantern fish A mesopelagic fish that averages about 13 centimeters (5 inches) in length. The name is derived from the photophores the fish possesses.

Larva An embryo that is on its own before it assumes the characteristics of the adults of the species.

Latent heat The quantity of heat gained or lost per unit of mass as a substance undergoes a change of state (liquid to solid, etc.) at a given temperature and pressure.

Latent heat of condensation The heat energy that must be removed from one gram of a substance to convert it from a vapor at a given temperature below its boiling point. For water, it is 585 calories at 20°C (68°F).

Latent heat of evaporation The heat energy that must be added to 1 gram of a liquid substance to convert it to a vapor at a given temper-

ature below its boiling point. For water, it is 585 calories at 20°C (68°F) and one atmosphere.

Latent heat of freezing The heat energy that must be removed from one gram of a substance at its melting point to convert it to a solid. For water, it is 80 calories.

Latent heat of melting The heat energy that must be added to 1 gram of a substance at its melting point to convert it to a liquid. For water, it is 80 calories.

Latent heat of vaporization The heat energy that must be added to 1 gram of a substance at its boiling point to convert it to a vapor. For water, it is 540 calories.

Lateral line system A sensory system running down both sides of fishes to sense subsonic pressure waves transmitted through ocean water.

Latitude Location on Earth's surface based on angular distance north or south of the Equator. Equator = 0°, North Pole = 90°N; South Pole = 90°S.

Laurasia A protocontinent of the Northern Hemisphere. The name is derived from Laurentia, pertaining to the Canadian Shield of North America, and Eurasia, of which it was composed.

Lava Fluid magma coming from an opening in Earth's surface, or the same material after it solidifies.

Law of gravitation See *Gravitational force*.

Leeuwin Current A warm current that flows south out of the East Indies along the western coast of Australia.

Leeward Direction toward which the wind is blowing or waves are moving.

Lesser Antilles The arc of West Indian Islands from the Virgin Islands south to the islands of coastal Venezuela.

Levee (1) Natural levees are low ridges on either side of river channels that result from deposition during flooding. (2) Artificial levees are built by human beings.

Levels of biological response The effects of pollution show up on these different levels within the biotic community: biochemical–cellular, organism, population dynamics, and community dynamics and structure.

Lichen Organism involving a photosynthetic, mutualistic relationship between an alga and a fungus. The alga is protected by the fungus, which is dependent on the alga for photosynthetically produced food.

Light Electromagnetic radiation that has a wavelength in the range from about 4000 (violet) to about 7700 (red) angstroms and may be perceived by the normal unaided human eye.

Light-year The distance traveled by light during 1 year at a speed of 300,000 kilometers (186,000 miles) per second. It equals 9.8 trillion kilometers or 6.2 trillion miles.

Limestone A class of sedimentary rock composed of at least 50% calcium or magnesium carbonate. Limestone may be either biogenous or hydrogenous.

LIMPET 500 The world's first commercial wave power plant that can generate up to 500 kilowatts of power. It is located on Islay, a small island off the west coast of Scotland, and began generating electricity in November 2000.

Limpet A mollusk of the class Gastropoda that possesses a low conical shell exhibiting no spiraling in the adult form.

Lipid Fats that, along with proteins and carbohydrates, are the principal structural components of living cells.

Liquid state A state of matter in which a substance has a fixed volume but no fixed shape.

Lithify The process by which sediment becomes hardened into sedimentary rock.

Lithogenous sediment Mineral grains derived from the rock of continents and islands and transported to the ocean by wind or running water.

Lithosphere The outer layer of Earth's structure, including the crust and the upper mantle to a depth of about 200 kilometers (124 miles). It is this layer that breaks into the plates that are the major elements of the theory of plate tectonics.

Lithothamnion **ridge** A feature common to the seaward edge of a reef structure, characterized by the presence of the red alga, *Lithothamnion.*

Littoral zone The benthic zone between the highest and lowest normal water marks; the intertidal zone.

Lobster Large marine crustacean used as food. *Homarus americanus* (American lobster) possesses two large chelae (pincers) and is found off the New England coast. *Panulirus* sp. (spiny lobsters or rock lobsters) have no chelae but possess long spiny antennae effective in warding off predators. *Panulirus argus* is found off the coast of Florida and in the West Indies, while *P. interruptus* is common along the coast of southern California.

Longitude Location on Earth's surface based on angular distance east or west of the Greenwich Meridian (0° longitude). 180° longitude is the International Date Line.

Longitudinal wave A wave in which particle vibration is parallel to the direction of energy propagation.

Longshore bar A deposit of sediment that forms parallel to the coast within or just beyond the surf zone.

Longshore current A current located in the surf zone and running parallel to the shore as a result of breaking waves.

Longshore drift The load of sediment transported along the beach from the breaker zone to the top of the swash line in association with the longshore current.

Longshore trough A low area of beach that separated the beach face from the longshore

bar.

Loop Current A clockwise-flowing current followed by water moving north from the Yucatán Strait through the Gulf of Mexico. It becomes the Florida Current when it enters the Florida Straits.

Lophophore Horseshoe-shaped feeding structure bearing ciliated tentacles characteristic of the phyla Bryozoa, Brachiopoda, and Phoronidea.

Low marsh The marsh that is found between mean sea level and neap high-tide level.

Low slack water The zero velocity that occurs when the ebb current reaches low water before reversing as a flood current.

Low tide terrace See *beach face.*

Low tide zone That portion of the intertidal zone that lies between the lowest low-tide shoreline and the highest low-tide shoreline.

Low water (LW) The lowest level reached by the water surface at low tide before the rise toward high tide begins.

Lower high water (LHW) The lower of two high waters occurring during a tidal day where tides exhibit a mixed tidal pattern.

Lower low water (LLW) The lower of two low waters occurring during a tidal day where tides exhibit a mixed tidal pattern.

Lunar bulge A feature of the equilibrium tidal theory in which a bulge of water is directed toward and another bulge is directed away from the tide generating body (the Sun or Moon).

Lunar day The time interval between two successive transits of the Moon over a meridian (approximately 24 hours and 50 minutes of solar time). Also called a *tidal day.*

Lunar month The $29\frac{1}{2}$-*day* period during which the Moon completes a full orbit around Earth.

Lunar tide The part of the tide caused solely by the tide-producing force of the Moon.

Lunger Fish such as groupers that sit motionless on the ocean floor waiting for prey to appear. A quick burst of speed over a short distance suffices to capture the prey.

Lysocline The depth in the ocean at which the carbon dioxide concentration is sufficient to begin dissolving calcium carbonate particles.

Macroplankton Plankton larger than 2 centimeters (0.8 inch) in their smallest dimension.

Macroscopic biogenous sediment Particles large enough to be seen without the aid of a microscope; includes shells, bones, and teeth of large organisms.

Magma Fluid rock material from which igneous rock is derived through solidification.

Magma chamber A reservoir of molten rock that lies underground. Typically found beneath segments of the mid-ocean ridge.

Magnetic anomaly Distortion of the regular pattern of Earth's magnetic field, resulting from

the various magnetic properties of local concentrations of ferromagnetic minerals in Earth's crust.

Magnetic dip The dip of magnetite particles in rock units of Earth's crust relative to sea level. It is approximately equivalent to the latitude at which the rock formed. Also called *magnetic inclination.*

Magnetic field The region surrounding a magnet or celestial object that generates magentism that is influenced by that magnetism.

Magnetic inclination See *Magnetic dip.*

Manganese nodule Concretionary lump containing oxides of iron, manganese, copper, or nickel found scattered in groups over the ocean floor.

Magnetite An iron oxide ((Fe_2O_4)) mineral that is naturally magnetic.

Mangrove swamp A marshlike environment that is dominated by mangrove trees. They are restricted to latitudes below 30°.

Mantle (1) The zone between the core and crust of Earth; rich in ferromagnesian minerals. (2) In pelecypods, the portion of the body that secretes shell material.

Mantle plume Cylindrical columns of molten material that arise from deep within the mantle and erupt onto Earth's surface at relatively fixed locations called *hotspots.*

Marginal ice zone (MIZ) The water marginal to Antarctic pack ice accumulation.

Marginal sea A semienclosed body of water adjacent to a continent.

Mariculture The application of the principles of agriculture to the production of marine organisms.

Marine Mammals Protection Act An act by U.S. Congress in 1972 that specifies rules to protect marine mammals in U.S. waters.

Marine Protected Areas (MPAs) Areas within United States waters where some or all of the marine biologic or cultural resources within them are protected.

Marine Protection, Research and Sanctuaries Also called the *Ocean Dumping Act,* it was passed in the United States in 1972 and controls dumping of waste at sea, marine research, and the establishment of marine sanctuaries.

Marine reserves A type of MPA in which all species and their habitat are fully protected.

Marine terrace A wave-cut bench that has been exposed above sea level.

MARPOL Convention This 1973 Convention (International Convention for the Prevention of Pollution from Ships), which was modified in 1978, specifies procedures for preventing operational or accidental causes of pollution by sewage or garbage from ships and even air pollution.

Marsh An area of soft, wet land. Flat land periodically flooded by salt water, common in portions of lagoons.

Maturity See *Sediment maturity.*

Maximum sustainable yield (MSY) The maximum fishery biomass that can be removed yearly and be sustained by the fishery ecosystem.

Mean high water (MHW) The average height of all the high waters occurring over a 19-year period.

Mean low water (MLW) The average height of the low waters occurring over a 19-year period.

Mean lower low water (MLLW) The average height of the lower of low waters occurring over a 19-year period.

Mean sea level (MSL) The mean surface water level determined by averaging all stages of the tide over a 19-year period, usually determined from hourly height observations along an open coast.

Mean tidal range The difference between mean high water and mean low water.

Mechanical energy Energy manifested as work being done; the movement of a mass some distance.

Mediterranean circulation Circulation characteristic of bodies of water such as the Mediterranean Sea with restricted circulation with the ocean that results from an excess of evaporation as compared to precipitation and runoff. Surface flow is into the restricted body of water with a subsurface counterflow.

Mediterranean Intermediate Water Water that sinks below the surface in the eastern Mediterranean Sea to a depth between 200 and 600 meters (660 and 2000 feet) and flows toward the Strait of Gibraltar. At the time of its formation, it has a temperature of 15°C (59°F) and a salinity of 39.1‰.

Mediterranean Sea A sea on the eastern margin of the North Atlantic Ocean. It lies east of the Strait of Gibraltar between Europe and Africa.

Mediterranean Water The Mediterranean Intermediate Water becomes the Mediterranean Water of the North Atlantic Ocean after it flows across the Gibraltar sill, sinks to a depth of about 900 meters (2900 feet), and spreads out across the North Atlantic Ocean.

Medusa Free-swimming, bell-shaped cnidarian body form with a mouth at the end of a central projection and tentacles around the periphery.

Meiofauna Small species of animals that live in the spaces among particles in a marine sediment.

Melon A fatty organ located forward of the blowhole on certain odontocete cetaceans that is used to focus echolocation sounds.

Melting point The temperature at which a substance is converted from a solid to a liquid. For water at atmospheric pressure, it is 0°C (32°F).

Mercury A silvery-white poisonous metallic element, Hg.

Meridian of longitude Half a great circle terminating at the North and South Poles.

Meroplankton Planktonic larval forms of organisms that are members of the benthos or nekton as adults.

Mesoderm A primitive cell layer of the embryo that develops between the endoderm and ectoderm. In vertebrates, it gives rise to the skeleton, muscles, circulatory and excretory systems, and most of the reproductive system.

Mesopelagic zone That portion of the oceanic province from about 200 to 1000 meters (660 to 3300 feet) depth. Corresponds approximately with the disphotic (twilight) zone.

Mesosaurus An extinct presumably aquatic reptile that lived about 250 million years ago. The distribution of its fossil remains helps support the theory of plate tectonics.

Mesosphere The middle region of Earth between the asthenosphere and the core; the lower mantle.

Metal sulfide A compound containing one or more metals and sulfur.

Metamorphic rock Rock that has undergone recrystallization while in the solid state in response to changes of temperature, pressure, and chemical environment.

Meteor A bright tail or streak that appears in the sky when a meteoroid is heated to incandescence by friction with Earth's atmosphere. It is often called a *shooting star*.

Meteorite The remains of a meteor that enters Earth's atmosphere and falls as a solid to the surface of Earth. They probably originate as fragments of asteroids.

Methane (CH_4) The simplest hydrocarbon.

Methane hydrate A sedimentary deposit of methane trapped in a crystalline latice of ice in marine sediments at depths in excess of 500 meters (1640 feet) on the continental slope.

Microbial loop A cycling of energy and matter among phytoplankton, heterotrophic bacteria, and protozoans in the pelagic ecosystem without this energy and matter being passed on to larger animals.

Microplankton Plankton not easily seen by the unaided eye, but easily recovered from the ocean with the aid of a fine-mesh plankton net.

Microscopic biogenous sediment Particles that are so small they must be viewed through a microscope. They include tests of diatoms, radiolarians, coccolithophores, and foraminifers.

Mid-Atlantic Ridge A slow-spreading divergent plate boundary running north-south and bisecting the Atlantic Ocean.

Mid-ocean ridge A linear volcanic mountain range that extends through all the ocean basins, rising 1 to 3 kilometers (0.6 to 2 miles) above the floor of the deep ocean basins. Averaging 1500 kilometers (930 miles) in width, rift valleys are common along the central axis. They are the source of new ocean floor.

Mid-ocean ridge basalt (MORB) Basalt typical of oceanic ridges and rises. It is depleted in potassium, rubidium, cesium, uranium, and thorium compared to basalt produced at hotspots. This may be evidence that hotspot basalts arise from near the core–mantle boundary while typical spreading center basalts are formed in the upper mantle.

Middle-tide zone That portion of the intertidal zone that lies between the highest low-tide shoreline and the lowest high-tide shoreline.

Migration Long journeys undertaken by many marine species for the purpose of successful feeding and reproduction.

Milky Way Galaxy The galaxy to which the solar system belongs.

Milkfish A fish capable of withstanding a wide range of salinity that is an important object of mariculture in the Far East.

Minamata Bay, Japan The site of the occurrence of human poisoning in the 1950s by mercury contained in marine organisms that were consumed by victims.

Minamata disease A degenerative neurological disorder caused by poisoning with a mercury compound found in seafood obtained from waters contaminated with mercury-containing industrial waste.

Mineral An inorganic substance occurring naturally in Earth and having distinctive physical properties and a chemical composition that can be expressed by a chemical formula.

Mitochondria Rod-shaped bodies in the cytoplasm of a cell in which cell respiration and energy production take place.

Mixed interference A pattern of wave interference in which there is a combination of constructive and destructive interference.

Mixed surface layer The surface layer of the ocean water mixed by wave and tide motions to produce relatively isothermal and isohaline conditions.

Mixed tidal pattern A tide having two high and two low waters per tidal day with a marked diurnal inequality. Such a tide may also show alternating periods of diurnal and semidiurnal components.

Mixotroph An organism that depends on a combination of autotrophic and heterotrophic behavior to meet its energy requirements. Many coral reef species exhibit such behavior.

Mohorovičič discontinuity (Moho) A sharp compositional discontinuity between the crust and mantle of Earth. It may be as shallow as 5 kilometers (3 miles) below the ocean floor or as deep as 60 kilometers (37 miles) beneath some continental mountain ranges.

Mole The weight of a substance in grams numerically equal to its molecular weight (gram molecule). One mole of water (H_2O) is 18 grams.

Molecule The smallest particle of an element or compound that, in the free state, retains the characteristics of the substance.

Mollusca Phylum of soft unsegmented animals usually protected by a calcareous shell and having a muscular foot for locomotion. Includes snails, clams, chitons, and octopuses.

Molt   Periodic shedding of exoskeleton by arthropods to permit growth.

Monera   Kingdom of organisms that do not have nuclear material confined within a sheath but spread throughout the cell. Bacteria and blue-green algae.

Mononodal   Pertaining to a standing wave with only one nodal point or nodal line.

Monsoon   A name for seasonal winds derived from the Arabic word for season, *mausim*. The term was originally applied to winds over the Arabian Sea that blow from the southwest during summer and the northeast during winter.

Moraine   Unsorted material deposited at the margins of glaciers. Many such deposits have become economically important as fishing banks after being submerged by the rising level of the ocean.

Mud   Sediment consisting primarily of silt and clay-sized particles smaller than 0.06 millimeters (0.002 inch).

Mud waves   Wave feature with lengths of 2 to 3 kilometers (1.2 to 2 miles) that bottom currents produce on the surfaces of drifts or ridges.

Mullet   A group of fish that are able to live in both fresh and salt water.

Mutualism   A symbiotic relationship in which both participants benefit.

Mycota   The kingdom of fungi. In the marine environment they are found living symbiotically with algae as lichen in the intertidal zone and as decomposers of dead organic matter in the open sea.

Myoglobin   A red, oxygen-storing pigment found in the muscle tissue.

Myomere   A muscle fiber.

Mysticeti   The baleen whales.

Nadir   The point on the celestial sphere directly opposite the zenith and directly beneath the observer.

Nannoplankton   Plankton less than 50 microns (0.002 inch) in length that cannot be captured in a plankton net and must be removed from the water by centrifuge or special microfilters.

Nansen bottle   A device used by oceanographers to obtain samples of ocean water from beneath the surface.

National Flood Insurance Program (NFIP)   A program financed by the U.S. government that was intended to prevent costly federal aid after a natural disaster but instead encourages building in risk-prone areas.

Natural selection   The process described by Charles Darwin by which the forces of nature select for survival those most fit to live in a given environment.

Nauplius   A microscopic free-swimming larval stage of crustaceans such as copepods, ostracoids, and decapods. Typically has three pairs of appendages.

Neap tide   Tides of minimal range occurring when the Moon is in quadrature (first and third quarter phases).

Nearshore   That zone from the shoreline seaward to the line of breakers.

Nebula   A diffuse mass of interstellar dust and/or gas.

Nebular hypothesis   A model that describes the formation of the Solar System by contraction of a nebula.

Nektobenthos   Those members of the benthos that can actively swim and spend much time off the bottom.

Nekton   Pelagic animals such as adult squids, fish, and mammals that are active swimmers to the extent they can determine their position in the ocean by swimming.

Nematath   A linear chain of islands and/or seamounts that are progressively older in one direction. It is created by passage of a tectonic plate over a hotspot.

Nematocyst   The stinging mechanism found within the cnidoblast of members of the phylum Cnidaria (Coelenterata).

Nepheloid layer   Well-mixed, turbid layer of water at the base of the oceanic water column. It is particularly well developed in the western boundary undercurrent.

Neritic province   That portion of the pelagic environment from the shoreline to a depth of 200 meters (660 feet).

Neritic sediment   That sediment composed primarily of lithogenous particles and deposited relatively rapidly on the continental shelf, continental slope, and continental rise.

Net primary production   The remaining amount of organic material produced by autotrophs after they have met their respiration needs.

Neutral   A state in which there is no excess of either the hydrogen or hydroxide ion.

Neutron   An electrically neutral particle found in the nucleus of most atoms. It has a mass approximately equal to that of a proton.

New ice   Newly formed sea ice.

New moon   The phase of the Moon that occurs when the Sun and the Moon are in conjunction (on the same side of Earth).

New production   Photosynthetic production supported by nutrients supplied from outside the immediate ecosystem by upwelling or other physical transport.

Niche   The ecological role of an organism and its position in the ecosystem.

Nigata, Japan   In the 1960s, the site of mercury poisoning of humans by mercury-contaminated seafood.

Nitrate   A chemical radical (NO_3) that is an important component of nutrients required for biological production.

Nitrogen fixation   Conversion by bacteria of atmospheric nitrogen (N_2) to oxides of nitrogen (NO_2, NO_3) usable by plants in primary production.

Nitrogen-fixing bacteria   Any of the bacteria that convert atmospheric nitrogen (N_2) to oxides of nitrogen (NO_2, NO_3) usable by algae in primary production.

Nitrogen narcosis   A sickness that effects divers. It results from too much N_2 being dissolved in the blood and reducing the flow of O_2 to tissues. The threat of this problem increases with increasing pressure (depth).

Node   The point on a standing wave where vertical motion is lacking or minimal. If this condition extends across the surface of an oscillating body of water, the line of no vertical motion is a nodal line.

Nonconservative constituent   A constituent of seawater, the concentration of which is is affected by processes other than mixing and diffusion. These additional processes may include biological processes, photosynthesis, chemosynthesis, and respiration that affect such constituents as dissolved oxygen, carbon dioxide, and nutrients.

Nonconservative property   A property of ocean water attained at the surface and changed by processes other than mixing and diffusion after the water sinks below the surface. For example, dissolved oxygen content will be altered by biological activity.

Nonferromagnesian mineral   Any of a group of common igneous rock-forming minerals that do not contain iron and magnesium.

Non-native species   Species that are introduced into waters in which they are alien. They often cause severe problems by displacing native species. Also called *exotic, alien,* or *invasive species.*

Non-point-source pollution   Any type of pollution entering the surface water system from sources other than underwater pipelines; also called "*poison runoff.*"

Nor'easter   Severe North Atlantic storms that produce waves up to 6 meters (20 feet) high that strike the east coast of North America north of Cape Hatteras, North Carolina.

North Atlantic Current   The current flowing west to east across the North Atlantic Ocean. The northern limb of the North Atlantic gyre.

North Atlantic Deep Water   A deep-water mass that forms primarily at the surface of the Norwegian Sea and moves south along the floor of the North Atlantic Ocean.

North Atlantic gyre   The clockwise rotating subtropical gyre of the North Atlantic Ocean.

North Equatorial Current   The westward-flowing equatorial segments of North Pacific and North Atlantic Oceans' subtropical gyres.

North Pacific Current   The eastward-flowing northern segment of the subtropical gyre in the North Pacific Ocean.

North Pacific gyre   The clockwise rotating subtropical gyre of the North Pacific Ocean.

North Pacific Intermediate Water (NPIW) A water mass that sinks beneath the surface at the Arctic Convergence and descends to a depth of about 1 kilometer (0.6 mile) where is spreads south toward the Equator.

Northeast Monsoon A northeast wind that blows off the Asian mainland onto the Indian Ocean during the winter season.

Northeast Trade Winds The prevailing wind system that blows from 30°N latitude toward the Equator.

Norwegian Current A warm current that branches off from the Gulf Stream and flows into the Norwegian Sea between Iceland and the British Isles.

Nuclear fuel cycle The recovery of usable radionuclides from spent fuel rods is achieved at plants that specialize in this process. These plants release radioactive waste into the ocean.

Nucleic base One of the four basic units that combine to form nucleotides that are arranged in a single strand in RNA and two strands coiled as a double helix in DNA.

Nucleus (1) A central, membrane-bound mass in eukaryotic cells; containing chromosomes. (2) The central, positively charged part of an atom; containing protons and neutrons.

Nudibranch A member of the mollusk class Gastropoda that has no protective covering as an adult and is often called a sea slug. Respiration is carried on by gills or other projections on the dorsal surface.

Nutrient Any organic or inorganic compound used by plants in primary production. Nitrogen and phosphorus compounds are important examples.

Obduction The reverse of subduction. In the case of ophiolites, the rock is pushed up on the continent instead of subducting beneath it.

Observation Occurrences that can be measured with one's senses.

Ocean The entire body of salt water that covers more than 71% of Earth's surface.

Ocean beach Beach on the open-ocean side of a barrier island.

Ocean current A physical movement of water.

Ocean Drilling Program (ODP) In 1983, this program replaced the Deep Sea Drilling Project to recover cores from the sea floor. It uses the drilling vessel *JOIDES Resolution*.

Ocean Dumping Ban Act This legislation passed in 1991 banned all sewage and industrial-waste dumping in U.S. offshore waters.

Ocean Thermal Energy Conversion (OTEC) A technology by which the temperature difference between surface waters and deep waters in low latitude regions is used to generate electricity.

Ocean trench A long linear depression in the ocean floor associated with subduction on one oceanic tectonic beneath another oceanic plate or continental plate. It lies seaward of a volcanic island arc or chain of continental mountains at the margin of the continent.

Oceanic Common Water The deep-water mass that enters the Indian and Pacific Oceans from the Antarctic Circumpolar Current. It forms from the mixing of North Atlantic Deep Water and Antarctic Bottom Water in the South Atlantic.

Oceanic crust A mass of rock with basaltic composition that is about 5 kilometers (3 miles) thick and underlies ocean basins.

Oceanic province That division of the pelagic environment where the water depth is greater than 200 meters (660 feet).

Oceanic ridge A portion of the global mid-ocean ridge system that is characterized by slow spreading and steep slopes.

Oceanic rise A portion of the global mid-ocean ridge system that is characterized by fast spreading and gentle slopes.

Oceanic sediment Deep-sea sediment in which at least 50% of the mass is composed of particles less then 5 microns (0.0002 inch) in size and in which less than 25% of the mass of particles larger than 5 microns in size are lithogenous particles. It includes organic oozes and abyssal clay.

Oceanic transform fault Transform faults confined to ocean basins.

Oceans Act of 2000 This act established a 16-member United States Commission on Ocean Policy to study and make recommendations regarding United States governance, research, education, marine operations, stewardship, investment, and development of ocean and coastal programs.

Ocelli Light-sensitive organ around the base of many medusoid bells.

Odontoceti Toothed whales.

Offshore The comparatively flat submerged zone of variable width extending from the breaker line to the edge of the continental shelf.

Oligotrophic Areas such as the middle of subtropical gyres where there are low levels of biological production.

Omnivore An animal that feeds on both plants and animals.

Oolite A deposit formed of small spheres from 0.25 to 2 millimeter (0.002 to 0.08 inch) in diameter. They are usually composed of concentric layers of calcite.

Ooze A pelagic sediment containing at least 30% skeletal remains of pelagic organisms, the balance being clay minerals. Oozes are further defined by the chemical composition of the organic remains (siliceous or calcareous) and by their characteristic organisms (e.g., diatom ooze, foraminifer ooze, radiolarian ooze, pteropod ooze).

Opal An amorphous form of silica ($SiO_2 \cdot nH_2O$) that usually contains from 3 to 9% water. It forms the shells of radiolarians and diatoms.

Opposition The separation of two heavenly bodies by 180° relative to Earth. The Sun and Moon are in opposition during full moon.

Orbital wave A wave phenomenon in which energy is moved along the interface between fluids of different density. The wave form is propagated by the movement of fluid particles in orbital paths.

Organic molecule Molecule of a compound that is naturally produced by organisms. ATP, DNA, carbohydrates, and lipids are examples.

Orthogonal lines Lines drawn perpendicular to wave fronts and spaced uniformly so equal amounts of energy are contained by the segments of the wave front lying between any two orthogonal lines in a series. The areas where energy is concentrated as the waves break on the shore can be identified by the convergence of the orthogonal lines.

Orthophosphate Phosphoric oxide (P_2O_5) can combine with water to produce orthophosphates ($3H_2O \cdot P_2O_5$ or H_3PO_4) that may be used by plants as nutrients.

Osmosis Passage of water molecules through a semipermeable membrane separating two aqueous solutions of different concentrations. The water molecules pass from the solution of lower solute concentration into the other.

Osmotic pressure A measure of the tendency for osmosis to occur. It is the pressure that must be applied to a solution to prevent the passage of water molecules into it from a reservoir of pure water.

Osmotic regulation Physical and biological processes used by organisms to counteract the osmotic effects of differences in osmotic pressures of their body fluids and the water in which they live.

Ostracoda An order of crustaceans that are minute and compressed within a bivalve shell.

Otocyst Gravity sensitive organs around the bell of a medusa.

Outer core Composed mostly of iron and nickel, this 2270-kilometer (1400-mile) thick liquid region surrounds the solid inner core. It is overlain by the mantle at about 2885 kilometers (1800 miles) beneath Earth's surface.

Outer sublittoral zone The section of ocean floor from the seaward edge of the inner sublittoral zone to a depth of 200 meters (660 feet). No attached algae grow here.

Outgassing A process, resulting from heating, by which gases and water vapor are released from molten rocks. It has produced Earth's atmosphere and oceans.

Overfishing Occurs when adult fish in a fishery are harvested faster than the natural rate of reproduction can sustain.

Oviparous Pertaining to an animal that releases eggs, which develop and hatch outside its body.

Ovoviviparous Pertaining to an animal that incubates eggs inside the mother until they hatch.

Oxygen compensation depth The depth at which marine plants photosynthesize at a rate which exactly meets their respiration needs (the base of the euphotic zone).

Oxygen minimum layer (OML) A zone of low dissolved oxygen concentration that occurs at a depth of about 700 to 1000 meters (2300 to 3280 feet).

Oyashio Current A cold current flowing south along the coast of Japan that converges with the warm Kuroshio Current.

Ozone (O_3) A triatomic form of oxygen. It is formed and destroyed by ultraviolet radiation in the stratosphere, thus reducing the level of ultraviolet radiation that reaches Earth's surface.

Pacific Decadal Oscillation (PDO) A natural oscillation in the Pacific Ocean that lasts 20 to 30 years and appears to influence sea surface temperatures.

Pacific Ocean The largest ocean basin, it represents 50.1% of the world ocean area. It is bounded on the east by North America and South America, on the west by Asia and Australia, and on the south by Antarctica.

Pacific Ring of Fire An extensive zone of volcanic and seismic activity that coincides roughly with the borders of the Pacific Ocean.

Pacific-type margin See *Active margin.*

Pacific warm pool The mass of warm water trapped in the western equatorial Pacific Ocean that has sea surface temperatures in excess of 28°C (82.4°F).

Pack ice Any area of sea ice other than fast ice. Less than 3 meters (10 feet) thick, it covers the ocean sufficiently to make navigation possible only by icebreakers.

Paleoceanography Branch of oceanography pertaining to the biological and physical character of ancient oceans.

Paleogeography Branch of geography pertaining to the shapes and positions of ancient continents and oceans.

Paleomagnetism The study of Earth's ancient magnetic field.

Pancake ice Circular pieces of newly formed sea ice from 0.3 to 3 meters (1 to 10 feet) in diameter that form in the early fall in polar regions.

Pangaea An ancient supercontinent of the geologic past that contained all Earth's continents.

Panthalassa A large, ancient ocean that surrounded Pangaea.

Paralytic shellfish poisoning (PSP) A form of poisoning that results from humans ingesting shellfish contaminated with toxins secreted by dinoflagellate algae.

Parapodia Flat protuberances on each side of most segments of polychaete worms. Most possess cirri and setae (bristlelike projections); may be modified for special functions such as feeding, locomotion, and respiration.

Parasite An organism that takes its nutrients from the tissues of another organism and benefits at the host's expense.

Parasitism A symbiotic relationship in which the parasite harms the host from which it takes its nutrition.

Partial tide One of the harmonic components comprising the tide at any location. The periods of the partial tides are derived from the various combinations of the angular velocities of Earth, Sun, and Moon relative to one another.

Parts per thousand (‰) A unit of measurement used in reporting salinity of water equal to the grams of dissolved substances in 1000 grams of seawater. One ‰ is equal to 0.1% or 1000 parts per million (ppm).

Passive margin The margin of a continent that is not significantly deformed by tectonic processes because it is the trailing edge of the continent. It does not directly collide with other lithospheric plates. The Atlantic coast of North America is an example.

PCBs (polychlorinated biphenyls) A group of industrial chemicals used in a variety of products; responsible for several episodes of ecological damage in coastal waters and now spreading to all ocean waters.

Peat deposit Partially carbonized organic matter found in bogs marshes that can be used as fertilizer and fuel.

Pedicellariae Minute stalked or unstalked pincerlike structures around the base of spines and dermal branchiae in certain echinoderms, especially Asteroidea, Echinoidea, and Ophiuroidea. They snap shut on debris and small organisms to keep the surface of the echinoderm clean.

Pelagic deposits Deposits of sediment found on the deep ocean floor beyond the continental rise. They typically include abyssal clay, calcareous ooze, and siliceous ooze.

Pelagic environment The open-ocean environment, divided into the neritic province (water depth 0 to 200 meters, or 660 feet) and the oceanic province (water depth greater than 200 meters, or 660 feet).

Pelecypoda A class of mollusks characterized by two fairly symmetrical lateral valves with a dorsal hinge. These filter feeders pump water through the filter system and over gills through posterior siphons. Many possess a hatchet-shaped foot used for locomotion and burrowing. Includes clams, oysters, mussels, and scallops.

Peridotite An ultramafic mantle rock composed primarily of olivine.

Perigee The point on the orbit of an Earth satellite (the Moon) that is nearest Earth.

Perihelion That point on the orbit of a planet or comet around the Sun that is closest to the Sun.

Period See *Wave period* (*T*).

Periwinkle snail Snails of the genus, *Littorina,* characteristically found in the high-tide zone and spray zone.

Permeability A condition that allows the passage of liquids through a substance.

Peru Current The eastern boundary current of the South Pacific subtropical gyre.

Petroleum A naturally occurring liquid hydrocarbon.

pH The negative of the logarithm of the hydrogen ion concentration in an aqueous solution.

pH scale A measure of the acidity or alkalinity of a solution, numerically equal to 7 for neutral solutions, increasing with increasing alkalinity and decreasing with increasing acidity. The pH scale commonly in use ranges from 0 to 14.

Phaeophyta Brown algae characterized by the carotenoid pigment fucoxanthin. Contains the largest marine algae.

Phase A state of matter; solid, liquid, gas.

Phosphate A compound containing the radical PO_4. It is an important component of nutrients required by algae for primary production.

Phosphorite A sedimentary rock composed primarily of phosphate minerals.

Photic zone The upper ocean in which the presence of solar radiation is detectable. It includes the euphotic and disphotic zones.

Photophore One of several types of light-producing organs found primarily on fishes and squids inhabiting the mesopelagic and upper bathypelagic zones.

Photosynthesis The process by which plants produce carbohydrate from carbon dioxide and water in the presence of chlorophyll, using light energy and releasing oxygen.

Phycoerythrin A red pigment characteristic of the Rhodophyta (red algae).

Phytoplankton Photosynthetic drifters. The most important community of primary producers in the ocean.

Phytoplankton exudate Cytoplasm lost directly into ocean by phytoplankton as they age.

Phytoplankton munchate Cytoplasm released into ocean during ingestion of phytoplankton by zooplankton.

Picoplankton Small plankton within the size range of 0.2 to 2.0 microns (0.000008 to 0.00008 inch) in size. Composed primarily of bacteria.

Pillow basalt A basalt exhibiting pillow structure. See *Pillow lava.*

Pillow lava A general term for those lavas displaying discontinuous pillow-shaped masses (pillow structure) caused by the rapid cooling of lava as a result of underwater eruption of lava or lava flowing into water.

Pinniped A group of marine mammals that have prominent flippers; includes the sea lions/fur seals, seals, and walruses.

Plankter Informal term for plankton.

Plankton Passively drifting or weakly swimming organisms that are dependent on currents. Includes mostly microscopic algae, protozoans, and larval forms of higher animals.

Plankton bloom A very high concentration of phytoplankton, resulting from a rapid rate of reproduction as conditions become optimum.

Plankton net Plankton-extracting device that is cone-shaped and typically of synthetic material. It is towed through the water or lifted vertically to extract plankton down to a size of 50 microns (0.02 inch).

Planktonic The condition of an organism being carried about by ocean currents because of lack of sufficient means of locomotion to determine it position on the water column.

Plantae Kingdom of eukaryotic, many celled plants that can only live in the margins of the ocean where they can put down roots. They are most abundant in mangrove swamps and marshes.

Plastic (1) Capable of being shaped or formed. (2) Composed of plastic or plastics.

Plastid A membrane bounded structure in plant cells that contains DNA, pigments and food reserves.

Plate tectonics Theory of global dynamics having to do with the movement of a small number of semirigid sections of Earth's crust, with seismic activity and volcanism occurring primarily at the margins of these sections. This movement has resulted in continental drift and changes in the shape and size of ocean basins.

Pleistocene Epoch The time in Earth history from 1.6 million years ago to 10,000 years ago during which pronounced glacial advances occurred. Also called the *Ice Age*.

Plume Rising jets of molten mantle material that create hotspots when they penetrate Earth's crust.

Plunging breaker Impressive curling breakers that form on moderately sloping beaches.

Pneumatic duct An opening into the swim bladder of some fishes that allows rapid release of air into the esophagus.

Pneumatocyst Gas-filled floats that help keep large kelp plants to grow upward some 30 meters (100 feet) to take advantage of sunlit surface waters.

Pneumatophore An air-containing float found on siphonophores.

Pogonophora A phylum of marine tube worms that have no gut and are found only in water deeper than 20 meters (66 feet).

Poikilotherm This term is descriptive of an organism whose body temperature varies with and is largely controlled by the temperature of its environment; cold-blooded.

Polar Pertaining to the polar regions.

Polar bear A carnivorous bear of the northern polar region that feeds primarily on seals.

Polar Cell The large atmospheric circulation cell that occurs between 60° and 90° latitude in each hemisphere.

Polar easterly winds Cold air masses that move away from the polar regions toward lower latitudes.

Polar front The boundary between the polar easterlies and the westerly wind belts that is centered at about 60° latitude in both hemispheres.

Polar high The region of high atmospheric pressure that surrounds the poles.

Polar ice The accumulation of sea ice in the Arctic polar region.

Polar wandering curve A curve that shows the apparent change in the location of Earth's poles over time.

Polarity Intrinsic polar separation, alignment, or orientation, especially of a physical property (such as magnetic or electrical polarity).

Pollution (marine) The introduction of substances or energy into the marine environment that results in harm to the living resources of the ocean or humans that use these resources.

Polychaeta Class of annelid worms that includes most of the marine segmented worms.

Polynya A nonlinear opening in sea ice.

Polyp A single individual of a colony or a solitary attached cnidarian.

Population A group of individuals of one species living in an area.

Porifera Phylum of sponges. Supporting structure composed of $CaCO_3$ or SiO_2 spicules or fibrous spongin. Water currents created by flagella-waving choanocytes enter tiny pores, pass through canals, and exit through a larger osculum.

Porosity The ratio of the volume of all the empty spaces in a material to the volume of the whole.

Porpoise Cetaceans that resemble dolphins but are usually smaller, more stout, have no beak, and a more triangular (or no) dorsal fin.

Potential density See *Density, potential.*

Potential world fishery The mass of fish that are supported annually by nutrients that originate outside of an ecosystem (world fishery ecosystems) and flow into it through upwelling or other processes. This amount of fish can be removed on an annual basis without degrading the ecosystem.

Precession Regarding the Moon's orbit around Earth, the axis of this orbit slowly changes its direction and describes a complete cone every 18.6 years. This is accompanied by a clockwise rotation of the plane of the Moon's orbit that is completed in the same time interval.

Precipitate To cause a solid substance to be separated from a solution, usually due to change in physical and/or chemical conditions.

Precipitation In a meteorological sense, the discharge of water in the form of rain, snow, hail, or sleet from the atmosphere onto Earth's surface.

Precision depth recorder A sonar system that emits a high frequency sound signal to measure depths to a resolution of about 1 meter (3.3 feet).

Prevailing westerlies The winds that blow out of the southwest in the Northern Hemisphere and out of the northwest in the Southern Hemisphere between the subtropical highs at 30° latitude and the relatively lower pressure polar fronts at 60° latitude.

Primary treatment The treatment of raw sewage to remove solids from water.

Principle of Constant Proportions See *Constant proportions, principle of.*

Primary coast Coast that has been formed recently by nonmarine processes and little modified by marine processes.

Primary crystalline rock Igneous rock.

Primary productivity The amount of organic matter synthesized by organisms from inorganic substances within a given volume of water or habitat in a unit of time.

Prime meridian The 0° meridian of longitude used as a reference for measuring longitude; the Greenwich Meridian.

Prokaryotic cells Cells with no central nucleus. The first cells and present-day bacteria and archaea are examples of prokaryotic organisms.

Producer The autotrophic component of an ecosystem that produces the food that supports the biocommunity.

Productivity See *Primary productivity.*

Progressive wave A wave in which the waveform progressively moves.

Propagation The transmission of energy through a medium.

Protein A complex organic compound made up of large numbers of amino acids. Proteins make up a large percentage of the dry weight of all living organisms.

Protoctista A kingdom of organisms that includes all one-celled forms with nuclear material confined to a nuclear sheath. Includes the animal phylum Protozoa and the phyla of algal plants.

Protoearth Earth early in its development. It may have had a diameter 1000 times greater and a mass 500 times greater than at present.

Proton A positively charged subatomic particle found in the nucleus of atoms that has a mass approximately equivalent to that of a neutron.

Protoplanet The form taken by any planet early in its development.

Protoplasm The complicated self-perpetuating living material making up all organisms. The elements carbon, hydrogen, and oxygen constitute more than 95%; water and dissolved salts make up from 50 to 97% of most plants and animals, with carbohydrates, lipids (fats), and proteins constituting the remainder.

Protozoa A member of the phylum of one-celled animals with nuclear material confined within a nuclear sheath.

Proxigean   Large tidal ranges that occur when spring tides coincide with perigean conditions.

Pseudopodia   Extensions of protoplasm in broad, flat, or long needlelike projections used for locomotion or feeding. Typical of amoeboid forms such as foraminifers and radiolarians.

Pteropoda   An order of pelagic gastropods in which the foot is modified for swimming and the shell may be present or absent.

Pteropod ooze   An oceanic deposit composed of more than 30% biogenous pteropod particles by weight.

Purse seine net   A curtainlike net that can be used to encircle a school of fish. The bottom is then pulled tight much the way a purse string is used to close a baglike purse.

Pycnocline   A layer of water in which a high rate of change in density in the vertical dimension is present.

Pycnogonid   A spiderlike arthropod found on the ocean bottom at all depths. The more commonly observed nearshore varieties are usually less than 1 centimeter (0.4 inch) across, while deeper water varieties may reach spreads of over 1 meter (3.3 feet).

Pyrrophyta   A phylum of microscopic algae that possesses flagella for locomotion—the dinoflagellates.

Quadrature   The state of the Moon during the first and third quarter moon phases (at right angles to one another relative to Earth).

Quarter moon   First and third quarter moon phases, which occur when the Moon is in quadrature about one week after the new moon and full moon phases, respectively. The third quarter moon phase is also known as the *last quarter moon* phase.

Quartz   A very hard mineral composed of silica, SiO_2.

Radiata   A grouping of phyla with primary radial symmetry—phyla Cnidaria and Ctenophora.

Radioactive waste   Radioactive materials that are no longer useful for the purpose they were designed. They must be stored safely away from contact with the biosphere.

Radioactivity   The spontaneous breakdown of the nucleus of an atom resulting in the emission of radiant energy in the form of particles or waves.

Radiolarian   Any member of an order of planktonic and benthic protozoans that possess protective coverings usually made of silica.

Radiolarian ooze   An oceanic deposit containing more than 30% by weight biogenous radiolarian particles.

Radiometric age dating   See *Absolute dating*.

Radionuclide   Nucleus of a radioactive atom.

Radula   Filelike calcium carbonate rasp used by snails to scrape algae off surfaces, drill through shells, and rasp away the tissue of their prey. Plural *radulae*.

Ray   A cartilaginous fish in which the body is dorsoventrally flattened, eyes and spiracles are on the upper surface, and gill slits are on the bottom. The tail is reduced to a whiplike appendage. Includes electric rays, manta rays, and stingrays.

Recreational beach   The area of a beach above shoreline, including the berm, berm crest, and the exposed part of the beach face.

Recruitment (fishery)   The year-class (number of fish or mass of fish) of young adults added to a fishery following each spawning season.

Red clay   See *Abyssal clay*.

Red muscle fiber   Fine muscle fibers rich in myoglobin that are abundant in cruiser-type fishes.

Red Sea   A narrow, newly formed sea at the northern margin of the Indian Ocean between Africa and the Arabian peninsula.

Red Sea Water   A water mass that spills into the Arabian Sea from the Red Sea. It spreads as a thin layer at a depth of about 900 meters (2950 feet) into the northern Indian Ocean.

Redshift   In light emitted from distant galaxies, the shift of absorption lines toward the red end of the spectrum. Shows the speed with which galaxies are moving away from the Milky Way galaxy.

Red tide   A reddish-brown discoloration of surface water, usually in coastal areas, caused by high concentrations of microscopic organisms, usually dinoflagellates. It probably results from increased availability of certain nutrients. Toxins produced by the dinoflagellates may kill fish directly; decaying plant and animal remains or large populations of animals that migrate to the area of abundant plants may also deplete the surface waters of oxygen and cause asphyxiation of many animals.

Reef   A strip or ridge of rocks, sand, coral, or human-made objects that rises to or near the surface of the ocean and creates a navigational hazard.

Reef flat   A platform of coral fragments and sand that is relatively exposed at low tide.

Reef front   The upper seaward face of a reef from the reef edge (seaward margin of reef flat) to the depth at which living coral and coralline algae become rare, 16 to 30 meters (50 to 100 feet).

Reflection   The process in which a wave has part of its energy returned seaward by a reflecting surface.

Refraction   The process by which the part of a wave in shallow water is slowed down to cause the wave to bend and tend to align itself with the underwater contours.

Regenerated production   The portion of gross primary production that is supported by nutrients recycled within an ecosystem.

Relative age dating   The determination of whether certain rock units are older or younger than others by the use of fossil assemblages. It was not possible to tell the actual age of rocks until radiometric dating was developed.

Relict beach   A beach deposit laid down and submerged by a relative rise in sea level. It is still identifiable on the continental shelf, indicating no deposition is presently taking place at that location on the shelf.

Relict sediment   A sediment deposited under a set of environmental conditions that still remains unchanged although the environment has changed and it remains unburied by later sediment. An example is a beach deposited near the edge of the continental shelf when sea level was lower.

Remnant arc   An inactive volcanic arc that has been split away from an active arc by back-arc spreading.

Relocation   A strategy of moving a structure that is threatened by being claimed by the sea.

Remotely operated vehicle (ROV)   An underwater vehicle that is operated remotely and not manned by humans. It can explore and sample hundreds of square kilometers of ocean floor per month.

Renewable energy   Energy source that cannot be depleted because it is renewed by radiant energy from the Sun.

Reservoir   A container in which a substance is stored (e.g., in the hydrologic cycle, water is contained at least temporarily, in reservoirs such as streams, atmosphere, ground water, glaciers, lakes, and the ocean).

Residence time   The average length of time a particle of any substance spends in the ocean. It is calculated by dividing the total amount of the substance in the ocean by the rate of its introduction into the ocean or the rate at which it leaves the ocean.

Resonance tide   A condition in which the natural free period of oscillation and the forced tidal period of oscillation are identical or multiples of one another and in phase. Thus constructive interference produces an increased tidal range.

Respiration   The process by which organisms use organic materials (food) as a source of energy. As the energy is released, oxygen is used and carbon dioxide and water are produced.

Restoring force   A force such as surface tension or gravity that tends to restore the ocean surface displaced by a wave to that of a still water level.

Resultant force   The small force that results on particles of Earth that are moving in identical circular orbits as a result of the rotation of the Earth–Moon system around its barycenter. The centripetal force to keep the particles in their cricular orbits is provided by the gravitational attraction between each particle and the Moon.

Since this gravitational force is only identical to the required centripetal force at the center of Earth, all other particles experience a resultant force that grows out of the difference in the required centripetal force and the provided gravitational force. The horizontal component of this residual force is the tide-generating force.

Reverse osmosis A method of desalinating seawater that involves forcing water molecules through a water-permeable membrane under pressure.

Reversing current The tide current as it occurs at the margins of landmasses. The water flows in and out for approximately equal periods of time separated by slack water where the water is still at high and low tidal extremes.

Rhodophyta Phylum of algae composed primarily of small encrusting, branching, or filamentous plants that receive their characteristic red color from the presence of the pigment phycoerythrin. With a worldwide distribution, they are found at greater depths than other algae.

Rhyolite A volcanic rock equivalent to granite in its mineral composition.

Ridge See *Oceanic ridge.*

Rift valley A deep fracture or break, about 25 to 50 kilometers (15 to 30 miles) wide, extending along the crest of a mid-ocean ridge.

Rifting The movement of two tectonic plates in opposite directions such as along a divergent boundary.

Right whale A whale of the genus *Balaena* that was the favorite target of early whalers.

Rip current A strong narrow surface or near-surface current of short duration and high speed flowing seaward through the breaker zone at nearly right angles to the shore. It represents the return to the ocean of water that has been piled up on the shore by incoming waves.

Rip-rap Any large blocky material used to armor coastal structures.

Ripple Capillary wave 10–15 centimeters (4–6 inches) long. Ripples are found on the sides of furrows cut into mud waves by current action.

Rise See *Oceanic rise* or *Continental rise.*

Rivers and Harbors Act This 1899 Act provided a prohibition against dumping refuse into the navigable waters of the United States.

Rock cycle Natural cycle that describes how igneous, sedimentary, and metamorphic rocks undergo changes through magmatism, erosion, transportation, deposition, lithification, and metamorphism.

Rock lice Isopods belonging to the genus *Ligia* that are common in the supralittoral zone.

Rogue wave An unusually large wave that usually occurs unexpectedly amid other waves of smaller size. Also known as a *superwave.*

Rorqual whale Any of several baleen whales with many ventral grooves; the minke, Baird's, Bryde's, sei, fin, blue, and humpback whales.

Rotary current Tidal current as observed in the open ocean. The tidal crest makes one complete rotation during a tidal period.

Rotary drilling A method of drilling through rock that involves rotating a bit at the end of a "drill string" of steel pipe.

Saffir–Simpson scale A scale of hurricane intensity that divides tropical cyclones into categories based on wind speed and damage.

Salinity A measure of the quantity of dissolved solids in ocean water. Formally, it is the total amount of dissolved solids in ocean water in parts per thousand by weight after all carbonate has been converted to oxide, the bromide and iodide to chloride, and all the organic matter oxidized. It is normally computed from conductivity, refractive index, or chlorinity.

Salinometer An instrument that is used to determine the salinity of seawater by measuring its electrical conductivity.

Salpa Genus of pelagic tunicates that are cylindrical, transparent, and found in all oceans.

Salt Any substance that yields ions other than hydrogen or hydroxyl. Salts are produced from acids by replacing the hydrogen with a metal.

Salt deposit A deposit of salt that occurs in the marine environment when the concentration of the salt ions reaches the level of supersaturation.

Salt marsh A relatively flat area of the shore where fine sediment is deposited and salt-tolerant grasses grow. One of the most biologically productive regions of Earth.

Salt wedge estuary A very deep river mouth with a very large volume of freshwater flow beneath which a wedge of salt water from the ocean invades. The Mississippi River is an example.

San Andreas Fault A transform fault that cuts across California from the northern end of the Gulf of California to Point Arena north of San Francisco.

Sand Particle size of 0.0625 to 2 millimeters (0.002 to 0.08 inch). It pertains to particles that lie between silt and granules on the Wentworth scale of grain size.

Sardine Any of the small fishes with a maximum length of 40 centimeters (16 inches) that are common in the coastal upwellings of the major subtropical gyres.

Sargasso Sea A region of convergence in the North Atlantic lying south and east of Bermuda where the water is a very clear deep blue color and contains large quantities of floating *Sargassum.*

Sargassum A brown alga characterized by a bushy form, substantial holdfast when attached, and a yellow-brown, green-yellow, or orange color. Two species, *S. fluitans* and *S. natans*, make up most of the macroscopic vegetation in the Sargasso Sea.

Scaphopoda A class of mollusk commonly called *tusk shells.* The shell is an elongate cone open at both ends. The conical foot surrounded by threadlike tentacles extends from the larger end to aid the animal in burrowing.

Scarp A linear, steep slope on the ocean floor separating gently sloping or flat surfaces.

Scavenger An animal that feeds on dead organisms.

Schooling Well-defined large groups of fish, squid, and crustaceans that apparently aid them in survival.

Scientific method The principles and empirical processes of discovery and demonstration considered characteristic of or necessary for scientific investigation, generally involving the observation of phenomena, the formulation of a hypothesis concerning the phenomena, experimentation to demonstrate the truth or falseness of the hypothesis, and a conclusion that validates or modifies the hypothesis.

Scuba Acronym for self-contained underwater breathing apparatus. It is a portable device that uses compressed air for breathing underwater.

Scyphozoa A class of cnidarians that includes the true jellyfish, in which the medusoid body form predominates and the polyp is reduced or absent.

Sea (1) A subdivision of an ocean. Two types of seas are identifiable and defined: the *mediterranean seas,* where a number of seas are grouped together collectively as one sea; and *marginal seas* that are connected individually to an ocean. (2) A portion of the ocean where waves are being generated by wind.

Sea anemone A member of the class Anthozoa whose bright color, tentacles, and general appearance resemble a flower.

Sea arch An opening through a headland caused by wave erosion. Usually develops as sea caves are extended from one or both sides of a headland.

Sea breeze A landward flow of air caused by differential heating of Earth's surface.

Sea cave A cavity at the base of a sea cliff; formed by wave erosion.

Sea cow An aquatic, herbivorous mammal of the order Sirenia that includes the dugong and manatee.

Sea cucumber A common name given to members of the echinoderm class Holotheuroidea.

Sea floor spreading A process producing the lithosphere when convective upwelling of magma along the oceanic ridges moves the ocean floor away from the ridge axes at rates between 2 to 12 centimeters (0.8 to 5 inches) per year.

Sea hare A shelless snail common in the low-tide zones of temperate rocky shores.

Sea ice Any form of ice originating from the freezing of ocean water.

Sea lion Any of several eared seals with relatively long necks and large front flippers, especially the California sea lion, *Zalophus californianus* of the northern Pacific. Along with the fur seals, these animals are known as *eared seals.*

Sea MARC   Sea Mapping and Remote Characterization, a side-scan sonar system, can be towed behind a ship to map a strip of ocean floor bathymetry.

Sea otter   A seagoing otter that has recovered from near extinction along the North Pacific coasts. It feeds primarily on abalone and sea urchins.

Sea roach   See *Rock lice.*

Sea smoke   When very cold air moves over warmer water, the bottom air is heated and rises. As it rises it carries evaporated water into the colder upper air, where it condenses to produce the "smoke."

Sea snake   A reptile belonging to the family Hydrophiidae with venom similar to that of cobras. Found primarily in the coastal waters of the Indian Ocean and the western Pacific Ocean.

Sea stack   An isolated, pillarlike rocky island that is detached from a headland by wave erosion.

Sea state   A description of the ocean surface that includes the average height of the highest one-third of the waves observed in a wave train; referred to a numerical code.

Sea turtle   Any turtle of the reptilian order Testudinata; widely found in warm water.

Sea urchin   An echinoderm belonging to the class Echinoidea; possessing a fused test (external covering) and well-developed spines.

Seabeam   The first multibeam echo sounder; is able to map a strip of sea floor up to 60 kilometers (37 miles) wide.

Seaknoll   See *abyssal hill.*

Seal   (1) Any of the several earless seals with a relatively short neck and small front flippers. Also known as *true seals.* (2) A general term that describes any of the various aquatic, carnivorous marine mammals of the families *Phocidae* and *Otariidae* (true seals and eared seals), found chiefly in the Northern Hemisphere and having a sleek, torpedo-shaped body and limbs that are modified into paddlelike flippers.

Seamount   An individual peak extending more than 1000 meters (3300 feet) above the ocean floor.

Seawall   A wall built parallel to the shore to protect coastal property from waves.

SeaWiFS   An instrument aboard the SeaStar satellite launched in 1997 that measures the color of the ocean with a radiometer and provides global coverage of ocean chlorophyll levels as well as land productivities every two days.

Secchi disk   A light colored disk-shaped device that is lowered into water in order to measure the water's clarity.

Secondary coast   Coast that has aged enough that its nonmarine origin has been largely destroyed or hidden by marine biological or physical processes.

Secondary treatment   The exposure of sewage that has received primary treatment to bacteria-killing chlorine.

Sediment   Particles of organic or inorganic origin that accumulate in loose form.

Sediment maturity   A texture of lithogenous sediment, where increasing maturity (caused by increased time of transport) is indicated by decreased clay content, increased sorting, and increased rounding of the grains within the deposit.

Sedimentary rock   A rock resulting from the consolidation of loose sediment, or a rock resulting from chemical precipitation (i.e., sandstone and limestone).

Seep   An area where water of various temperature trickles out of the sea floor.

Seiche   A standing wave of an enclosed or semienclosed body of water that may have a period ranging from a few minutes to a few hours, depending on the dimensions of the basin. The wave motion continues after the initiating force has ceased.

Seismic   Pertaining to an earthquake or Earth vibration, including those that are artificially induced.

Seismic moment magnitude (M_w)   A scale used for measuring earthquake intensity based on energy released in creating very long-period seismic waves.

Seismic reflection profile   Profiles of sediments and rocks beneath the ocean floor generated by producing strong low-frequency sound that penetrates beneath the ocean floor and reflects off boundaries between different types of sediment and rock.

Seismic sea wave   See *Tsunami.*

Seismic surveying   The use of sound-generating techniques to identify features on or beneath the ocean floor.

Semidiurnal tidal pattern   Tide having two high and two low waters per tidal day with small inequalities between successive highs and successive lows. Tidal period is about 12 hours and 25 minutes solar time. Semidaily tide.

Serpulid   A polychaete worm belonging to the family Serpulidae that builds a calcareous or leathery tube on a submerged surface.

Sessile   Permanently attached to the substrate and not free to move about.

Sewage sludge   Semisolid material precipitated by sewage treatment.

Sextant   A double-reflecting handheld instrument used in navigation to measure apparent altitudes of celestial bodies from a moving ship.

Shallow-water wave   A wave on the surface of the water whose wavelength is at least 20 times water depth. The bottom affects the orbit of water particles, and speed (S) is determined by water depth.

Shelf break   The depth at which the gentle slope of the continental shelf steepens appreciably. It

marks the boundary between the continental shelf and continental rise.

Shelf ice   Thick shelves of glacial ice that push out into Antarctic seas from Antarctica. Large tabular icebergs calve at the edge of these vast shelves.

Shoaling   To become shallow.

Shore   Seaward of the coast, extends from highest level of wave action during storms to the low-water line.

Shoreline   The line marking the intersection of water surface with the shore and migrates up and down as the tide rises and falls.

Shoreline of emergence   Shorelines that indicate a lowering of sea level by the presence of stranded beach deposits and marine terraces above it.

Shoreline of submergence   Shorelines that indicate a rise in sea level by the presence of drowned beaches or submerged dune topography.

Side-scan sonar   A method of mapping the topography of the ocean floor along a strip up to 60 kilometers (37 miles) wide using computers and sonar signals that are directed away from both sides of the survey ship.

Sigma (s)   Symbol for *in situ* density.

Sigma-T(s_T) Symbol for *in situ* density.

Sigma theta(s_U) Symbol for potential density.

Silica   Silicon dioxide (SiO_2).

Silicate   A mineral whose crystal structure contains SiO_4 tetrahedra.

Silicate chrondrule   Cosmic particle that is grouped according to two size ranges, 30 microns or 125 microns (0.012 inch or 0.05 inch). Silicate chondrules are believed to form from the collision of asteroids.

Silicate tetrahedron   The basic pattern of silicate structure—a silicon atom surrounded by four oxygen atoms that form a tetrahedron.

Siliceous   A condition of containing abundant silica (SiO_2).

Siliceous ooze   A pelagic deposit that contains at least 30% by weight biogenous siliceous particles.

Sill   A submarine ridge partially separating bodies of water such as fjords and seas from one another or from the open ocean.

Silt   A particle size of 0.008 to 0.0625 millimeter (0.0003 to 0.002 inch). It is intermediate in size between sand and clay.

Siphonophore   A member of an order of hydrozoan cnidarians that form pelagic colonies containing both polyps and medusae. An examples is *Physalia.*

Sirenia   An order of vegetarian marine mammals that includes the dugong and manatee.

Slack water   Occurs when a reversing tidal current changes direction at high or low water. Current speed is zero.

Slick   A smooth patch on an otherwise rippled surface caused by a monomolecular film of organic material that reduces surface tension.

Slightly stratified estuary An estuary of moderate depth in which marine water invades beneath the freshwater runoff. The two water masses mix so that the bottom water is slightly saltier than the surface water at most places in the estuary.

SOFAR channel Sound fixing and ranging channel. This low velocity sound travel zone coincides with the permanent thermocline in low and mid-latitudes.

Solar day The 24-hour period during which Earth completes one rotation on its axis.

Solar distillation A process by which ocean water can be desalinated by evaporation and the condensation of vapor on the cover of a container. The condensate then runs into a separate container and is collected as fresh water. Also called *solar humidification*.

Solar humidification See *Solar distillation*.

Solar system The Sun and the celestial bodies, asteroids, planets, and comets that orbit around it.

Solar tide The partial tide caused by the tide-producing forces of the Sun.

Solar wind The motion of interplanetary plasma or ionized particles away from the Sun.

Solid state A state of matter in which the substance has a fixed volume and shape. A crystalline state of matter.

Solstice The time during which the Sun is directly over one of the tropics. In the Northern Hemisphere the summer solstice occurs on June 21 or 22 as the Sun is over the Tropic of Cancer, and the winter solstice occurs on December 21 or 22 when the Sun is over the Tropic of Capricorn.

Solute A substance dissolved in a solution. Salts are the solute in salt water.

Solution A state in which a solute is homogeneously mixed with a liquid solvent. Water is the solvent for the solution that is ocean water.

Solvent A liquid that has one or more solutes dissolved in it.

Somali Current This current flows north along the Somali coast of Africa during the southwest monsoon season.

Sonar An acronym for sound navigation and ranging. A method by which objects may be located in the ocean.

Sorting A texture of sediment where well-sorted sediment is characterized by having a narrow range of grain sizes.

Sounding Measuring the depth of water beneath a ship.

South Equatorial Current The equatorial segment of the subtropical gyres in the three major ocean basins.

South Atlantic gyre A counterclockwise rotating subtropical gyre that dominates the South Atlantic Ocean.

Southern Ocean See *Antarctic Ocean*.

Southeast trade winds A belt of prevailing winds found between the subtropics (30°N or S latitude) and the Equator.

Southern Oscillation The periodic change in the pressure differential between the Southeastern Pacific high pressure and the Western Pacific equatorial low pressure that occurs in concert with El Niño–Southern Oscillation events.

Southwest Monsoon A southwest wind that develops during the summer season. It blows off the Indian Ocean onto the Asian mainland.

Southwest Monsoon Current During the southwest monsoon season, this eastward-flowing current replaces the west-flowing North Equatorial Current in the Indian Ocean.

Space dust Micrometeoroid space debris.

Spawning A reproductive process in which male and female members of a species release their sperm and eggs, respectively, to accomplish fertilization.

Species A fundamental category of taxonomic classification, ranking below a genus or subgenus and consisting of related organisms capable of interbreeding.

Species diversity The number or variety of species found in a subdivision of the marine environment.

Specific gravity The ratio of density of a given substance to that of pure water at 4°C (39.2°F) and one atmospheric pressure.

Specific heat The quantity of heat required to raise the temperature of 1 gram of a given substance 1°C (1.8°F). For water it is 1 calorie.

Spermaceti organ A large fatty organ located within the head region of sperm whales (*Physeter macrocephalus*) that is used to focus echolocation sounds.

Spermatophyta See *Anthophyta*.

Sperm whale The largest toothed whale (*Physeter macrocephalus*).

Spherule A cosmogenous microscopic globular mass composed of silicate rock material (tektites) or of iron and nickel.

Spicule A minute needlelike calcareous or siliceous form found in sponges, radiolarians, chitons, and echinoderms that acts to support the tissue or provide a protective covering.

Spilling breaker A type of breaking wave that forms on a gently sloping beach which gradually extracts the energy from the wave to produce a turbulent mass of air and water that runs down the front slope of the wave.

Spit A small point, low tongue, or narrow embankment of land commonly consisting of sand deposited by longshore currents and having one end attached to the mainland and the other terminating in open water.

Splash wave A long-wavelength wave created by a massive object or series of objects falling into water; a type of tsunami.

Sponge See *Porifera*.

Spray zone The shore zone lying between the high-tide shoreline and the coastline. It is covered by water only during storms.

Spreading center A divergent plate boundary.

Spreading rate The rate of divergence of plates at a spreading center.

Spring bloom The rapid increase in biological production that occurs in the spring season in temperate and polar regions.

Spring tide Tide of maximum range occurring about every two weeks when the Moon is new or full.

Stack An isolated mass of rock projecting from the ocean off the end of a headland from which it has been detached by wave erosion.

Standard laboratory bioassay A standard assessment technique that determines the concentration of a pollutant that causes 50% mortality among selected test organisms.

Standard seawater Ampules of ocean water for which the chlorinity has been determined by the Institute of Oceanographic Services in Wormly, England. The ampules are sent to laboratories all over the world so equipment and reagents used to determine the salinity of ocean water samples can be calibrated by adjustment until they give the same chlorinity as is shown on the ampule label.

Standing stock (crop) The mass of fishery organisms present in an ecosystem at a given time.

Standing wave A wave, the form of which oscillates vertically without progressive movement. The region of maximum vertical motion is an *antinode*. Between antinodes are *nodes* where there is no vertical motion but maximum horizontal motion.

Steepness See *Wave steepness*.

Stenohaline Pertaining to organisms that can withstand only a small range of salinity change.

Stenothermal Pertaining to organisms that can withstand only a small range of temperature change.

Stick chart A device made of sticks or pieces of bamboo that was used by early navigators at sea.

Still water level Halfway between crest and trough, it is the level that the water would reside if there were no waves. Also known as *zero energy level*.

Storm An atmospheric disturbance characterized by strong winds accompanied by precipitation and often by thunder and lightning.

Storm surge A rise above normal water level resulting from wind stress and reduced atmospheric pressure during storms. Consequences can be more severe if it occurs in association with high tide.

Strait of Gibraltar The narrow opening between Europe and Africa through which the waters of the Atlantic Ocean and Mediterranean Sea mix.

Stranded beach deposit An ancient beach deposit found above present sea level because of a relative lowering of sea level.

Streamlining The shaping of an object so it

produces the minimum of turbulence while moving through a fluid medium. The teardrop shape displays a high degree of streamlining.

Stromatolite A calcium carbonate sedimentary structure in which algal assemblages trap sediment and bind it into forms that are often dome shaped. They are known to form only in shallow-water environments.

Subaerial Beneath the atmosphere.

Subaqueous Beneath water (typically, the ocean).

Subduction A process by which one lithospheric plate descends beneath another. The surface expression of such a process may be an island arc-trench system or a folded mountain range.

Subduction zone A long narrow region beneath Earth's surface in which subduction takes place.

Subduction zone seep biocommunity Seeps of pore water squeezed out of sediments in subduction zone support a distinctive fauna that depend on sulfur-oxidizing carbon fixing bacteria.

Subduction zone seep Seeps of pore water squeezed out of sediments in subduction zones. A distinctive fauna that depends on sulfur-oxidizing carbon fixing bacteria lives in association with the seeps.

Sublimation The transformation of the solid state of a substance to a vapor without going through the liquid phase, and vice versa.

Sublittoral zone That portion of the benthic environment extending from low tide to a depth of 200 meters (660 feet). Some consider it to be the surface of the continental shelf.

Submarine canyon A steep V-shaped canyon cut into the continental shelf or slope.

Submarine fan See *Deep-sea fan*.

Submerged dune topography Ancient coastal dune deposits found submerged beneath the present shoreline because of a rise in sea level.

Submerging shoreline Shoreline formed by the relative submergence of a landmass in which the shoreline is on landforms developed under subaerial processes. It is characterized by bays and promontories and is more irregular than a shoreline of emergence.

Submersible Either a vessel that can take humans beneath the ocean surface or an unmanned vehicle that can be remotely controlled from the surface or programmed to perform functions beneath the ocean surface.

Subneritic province The benthic environment extending from the shoreline across the continental shelf to the shelf break. It underlies the neritic province of the pelagic environment.

Suboceanic province Benthic environments seaward of the continental shelf.

Subpolar Pertaining to the oceanic region that is covered by sea ice in winter. The ice melts away in summer.

Subpolar gyre Oceanic gyres of ocean water

centered at about 60° latitude in both hemispheres. They rotate counterclockwise in the Northern Hemisphere and clockwise in the Southern Hemisphere.

Subpolar low The atmospheric low pressure belt that represents the polar front at approximately 60° latitude in both hemispheres.

Substrate The base on which an organism lives and grows.

Subsurface current A current usually flowing below the pycnocline, generally at slower speed and in a different direction from the surface current.

Subtropical Pertaining to the oceanic region poleward of the tropics (about 30° latitude).

Subtropical convergence The zone of convergence that occurs within all subtropical gyres as a result of Ekman transport driving water toward the interior of the gyres.

Subtropical gyre The trade winds and westerly winds initiated in the subtropical regions of all ocean basins, with the influence of the Coriolis effect, set large regions of ocean water in motion. They rotate clockwise in the Northern Hemisphere and counterclockwise in the Southern Hemisphere, and they are centered in the subtropics.

Subtropical high The atmospheric high-pressure belt at approximately 30° latitude in both hemispheres at which the easterly and trade wind belts are initiated.

Subtropical Underwater A high-salinity water mass centered at a depth of about 50 meters (164 feet) in the southeastern Caribbean Sea. It slopes to a depth of 200 meters (660 feet) as it flows north through the Yucatán Strait.

Sulfur A yellow mineral composed of the element sulfur. It is commonly found in association with hydrocarbons and salt deposits.

Sulfur-oxidizing bacteria Any of the bacteria that support many deep-sea hydrothermal vent and cold water seep biocommunities by using energy released by oxidation to synthesize organic matter chemosynthetically.

Summer solstice In the Northern Hemisphere, it is the instant when that Sun moves north to the Tropic of Cancer before changing direction and moving southward toward the Equator, approximately June 21.

Summertime beach Gentle summer waves move sand from the offshore bar onto the beach to produce a broad sandy berm with an overall steep beach face.

Superwave See *Rogue wave*.

Supralittoral zone The splash or spray zone above the spring high-tide shoreline.

Surf beat An irregular wave pattern caused by mixed interference that results in a varied sequence of larger and smaller waves.

Surf zone The region between the shoreline and the line of breakers where most wave ener-

gy is released.

Surface tension The tendency for the surface of a liquid to contract owing to intermolecular bond attraction.

Surging breaker A compressed breaking wave that builds up over a short distance and surges forward as it breaks. It is characteristic of abrupt beach slopes.

Suspension feeding See *Filter feeding*.

Suspension settling The process by which fine-grained material that is suspended in the water column is slowly deposited on the ocean floor.

Sverdrup (Sv) A unit of volume transport equal to 1 million cubic meters per second.

Swash A thin layer of water that washes up over exposed beach as waves break at the shore.

Swell A free ocean wave by which energy put into ocean waves by wind in the sea is transported with little energy loss across great stretches of ocean to the margins of continents where the energy is released in the surf zone.

Swim bladder A gas-containing, flexible, cigar-shaped organ that aids many fishes in attaining neutral buoyancy.

Symbiosis A relationship between two species in which one or both benefit or neither or one is harmed. Examples are commensalism, mutualism, and parasitism.

Synodic month The period of time that elapses between conjunction of the Sun and Moon (new moon phase) relative to Earth; $29\frac{1}{2}$ days.

Syzygy Either of two points in the orbit of the Moon when the centers of the Sun, Moon and Earth are perfectly aligned during full or new moon phases.

Tablemount A flat-topped seamount; a guyot.

Taxonomy The classification of organisms in an ordered system that indicates natural relationships.

Tectonic estuary An estuary in which its origin is related to tectonic deformation of the coastal region.

Tectonics Deformation of Earth's surface by forces generated by heat flow from Earth's interior.

Tektite See *Spherule*.

Temperate Pertaining to the oceanic region where pronounced seasonal change occurs (about 40° to 60° latitude); mid-latitude.

Temperature A direct measure of the average kinetic energy of the molecules of a substance.

Temperature of maximum density The temperature at which a substance reaches its highest density. For water, it is 4°C (39.2°F).

Temperature–salinity (T–S) diagram A diagram with axes representing temperature and salinity, whereby the density of the water can be determined.

Terrigenous sediment Sediment produced

from or of land. Also called *lithogenous sediment*.

Territorial sea   A 12-nautical-mile-wide strip of ocean adjacent to land over which the coastal nation has control over the passage of ships.

Test   The supporting skeleton or shell (usually microscopic) of many invertebrates.

Tethys Sea   An ancient body of water that separated Laurasia to the north and Gondwanaland to the south. Its location was approximately that of the present Alpine-Himalayan mountain system.

Texture   The general physical apearance of an object.

Theory   A well-substantiated explanation of some aspect of the natural world that can incorporate facts, laws (descriptive generalizations about the behavior of an aspect of the natural world), logical inferences, and tested hypotheses.

Thermal contraction   The reduction in size as a result of temperature being lowered.

Thermocline   A layer of water beneath the mixed layer in which a rapid change in temperature can be measured in the vertical dimension.

Thermohaline circulation   The vertical movement of ocean water driven by density differences resulting from the combined effects of variations in temperature and salinity.

Tidal bore   A steep-fronted wave that moves up some rivers when the tide rises in the coastal ocean.

Tidal bulges   The mounds of water on both sides of Earth caused by the relative positions of the Moon (lunar tidal bulges) and the Sun (solar tidal bulges).

Tidal datum   The tide level from which the heights of high and low tides are measured. It is usually mean low water (MLW) or mean lower low water (MLLW).

Tidal day   See *Lunar day*.

Tidal period   Elapsed time between successive high or low tides.

Tidal range   The difference in height between consecutive high and low waters. The time frame of comparison may also be a day, month, or year.

Tide   Periodic rise and fall of the ocean surface and connected bodies of water resulting from the unequal gravitational attraction of the Moon and Sun on different parts of Earth.

Tide wave   The long-period gravity wave generated by tide-generating forces and manifested in the rise and fall of the tide.

Tide-generating force   The magnitude of the centripetal force required to keep all particles of Earth with identical mass moving in identical circular paths required by the movements of the Earth–Moon system is identical. This force is provided by the gravitational attraction between the particles and the Moon. This gravitational force is identical to the required centripetal force only at the center of Earth. For ocean tides, the

horizontal component of the small force that results from the difference between the required and provided forces is the tide-generating force on that individual particle. These forces are such that they tend to push the ocean water into bulges toward the tide-generating body on one side of Earth and away from the tide-generating body on the opposite side of Earth.

Tin   The bluish white native metallic element, Sn.

Tintinnid   A ciliate protozoan of the family Tintinnidae with a tubular to vase-shaped outer shell.

Tissue   An aggregate of cells and their products developed by organisms for the performance of a particular function.

Tombolo   A sand or gravel bar that connects an island with another island or the mainland.

Topography   The configuration of a surface. In oceanography, it usually refers to sea floor shape.

Total allowable catch   The permissible annual catch of a given species of fish that will not degrade the fishery given the knowledge of the ecosystem from which the fishery is being removed.

Toxic compound   A poisonous substance capable of causing injury or death, especially by chemical means.

Trade winds   The air masses moving from subtropical high pressure belts toward the Equator. They are northeasterly in the Northern Hemisphere and southeasterly in the Southern Hemisphere.

Transform active margin   A continental margin associated with a transform fault such as the San Andreas Fault that cuts across California and causes large earthquakes.

Transform fault   A fault characteristic of oceanic ridges along which they are offset.

Transform plate boundary   The boundary between two lithospheric plates formed by a transform fault.

Transitional crust   Thinned section of continental crust at the trailing edge of a continent created by the breaking apart of an ancient continent over a newly formed spreading center.

Transitional wave   A wave moving from deep water to shallow water that has a wavelength more than twice the water depth but less than 20 times the water depth. Particle orbits are beginning to be influenced by the bottom.

Transverse ridge   A ridge within a fracture zone that contains large concentrations of mantle rock (periodotite) that have been tectonically squeezed up.

Transverse wave   A wave in which particle motion is at right angles to energy propagation.

Trawl   A sturdy bag or net that can be dragged along the ocean bottom or at various depths above the bottom to catch fish.

Trench   A long, narrow, and deep depression

on the ocean floor with relatively steep sides that is caused by plate convergence.

Troodos Massif   An ophiolite ore deposit located in western Cyprus.

Trophic level   A nourishment level in a food chain. Algal producers constitute the lowest level, followed by herbivores and a series of carnivores at the higher levels.

Tropic of Cancer   The latitude of 23.5° north, which is the furthest location north that receives vertical rays of the Sun.

Tropic of Capricorn   The latitude of 23.5° south, which is the furthest location south that receives vertical rays of the Sun.

Tropical   Pertaining to the regions of the tropics (about 23.5° latitude).

Tropical Atmosphere and Ocean (TOA)   A program sponsored by the United States, Canada, Australia, and Japan that monitors the equatorial Pacific Ocean with 70 moored buoys.

Tropical cyclone   See *Hurricane*.

Tropical Ocean–Global Atmosphere (TOGA) This monitoring program was conducted in the equatorial Pacific Ocean from 1985 to 1995 to learn more about the nature of El Niño events.

Tropical tide   A tide occurring twice monthly when the Moon is at its maximum declination north and south of the Equator. It is in the tropical regions where tides display their greatest diurnal inequalities.

Tropics   The region of Earth's surface that exists between the Tropic of Cancer and the Tropic of Capricorn. It is also known as the *torrid zone*.

Troposphere   The lowermost portion of the atmosphere that extends from Earth's surface to 12 kilometers (7 miles). It is where all weather is produced.

Trough   The part of an ocean wave that is displaced below the still water line.

Tsunami   Seismic sea wave. A long-period gravity wave generated by a submarine earthquake or volcanic event. Not noticeable in the open ocean but builds up to great heights in shallow water.

Tube worm   See *Sabellidae* and *Serpulidae*.

Tunicate   A member of the chordate subphylum Urochordata, which includes sacklike animals. Some are sessile (sea squirts) while others are pelagic (salps).

Turbidite deposit   A sediment or rock formed from sediment deposited by turbidity currents and characterized by graded bedding.

Turbidity   A state of reduced clarity in a fluid caused by the presence of suspended matter.

Turbidity current   A gravity current resulting from a density increase brought about by increased water turbidity. Possibly initiated by some sudden force such as an earthquake, the turbid mass continues under the force of gravity down a submarine slope.

Turbulent flow   Flow in which the flow lines

are confused heterogeneously due to random velocity fluctuations.

Typhoon   A severe tropical storm in the western Pacific.

Ultraplankton   Plankton smaller than 5 microns (0.002 inch). Very difficult to separate from seawater.

Ultrasonic   Pertaining to sound frequencies above human range (above 20,000 cycles per second).

Ultraviolet (UV) radiation   Electromagnetic radiation shorter than visible radiation and longer than X rays.

United Nations Conference on the Law of the Sea   A conference first held in 1958 in response to the expansion of exploration for petroleum onto the continental shelves. The third Conference concluded in 1982 and was adopted by 130 nations. It addresses coastal nations jurisdiction, ship passage through waters where coastal nations have jurisdiction, the exploitation of deep-ocean mineral resources, and arbitration of disputes.

U.S. Commission on Ocean Policy   See *Oceans Act of 2000.*

Universe   All that exists. The cosmos.

Upper water   Includes the mixed layer and the permanent thermocline. It is approximately the top 1000 meters (3300 feet) of the ocean.

Upwelling   The process by which deep, cold, nutrient-rich water is brought to the surface, usually by wind divergence of equatorial currents or coastal winds pushing water away from the coast.

Valence   The combining capacity of an element measured by the number of hydrogen atoms with which it will combine.

van der Waals force   Weak attractive force between molecules; a result of the interaction between the nuclear particles of one molecule and the electrons of another.

Vapor   The gaseous state of a substance that is liquid or solid under ordinary circumstances.

Vapor pressure   A measure of the tendency for molecules of a liquid to escape into the gaseous phase. It increases with increased temperature.

Vent   An opening on the ocean floor that emits hot water and dissolved minerals.

Ventral   Pertaining to the lower or under surface.

Vernal equinox   The passage of the Sun across the Equator as it moves from the Southern Hemisphere into the Northern Hemisphere, approximately March 21. During this time, all places in the world experience equal lengths of night and day. Also known as the *spring equinox.*

Vertebrata   Subphylum of chordates that includes those animals with a well-developed brain and a skeleton of bone or cartilage; includes fish,

amphibians, reptiles, birds, and animals.

Vertically mixed estuary   Very shallow estuaries such as lagoons in which fresh water and marine water are totally mixed from top to bottom so that the salinity at the surface and the bottom is the same at most places within the estuary.

Virus   Less than 200 nannometers in length, a virus consist of genetic material surrounded by a protein or lipic coat and is a parasite depending on other organisms for metabolism.

Viscosity   The property of a substance to offer resistance to flow; internal friction.

Viviparous   Pertaining to an animal that gives birth to living young.

Volcanic arc   An arc-shaped row of active volcanoes directly above a subduction zone. Can occur as a row of islands (island arc) or mountains on land (continental arc).

Volcanogenic particles   Sediment particles produced by volcanic eruptions.

Vortex   A revolving flow of fluid in the atmosphere (cyclone) or ocean (current gyre).

Walker Circulation Cell   The pattern of atmospheric circulation that involves the rising of warm air over the East Indies low-pressure cell and its descent over the high-pressure cell in the southeastern Pacific Ocean off the coast of Chile. It is the weakening of this circulation that accompanies an El Niño event.

Walrus   A large marine mammal (*Odobenus rosmarus*) of Arctic regions belonging to the order Pinnipedia and having two long tusks, tough wrinkled skin, and four flippers.

Waning crescent   When the Moon is between third quarter and new moon phases.

Waning gibbous   When the Moon is between full moon and third quarter phases.

Warm-blooded   See *Homeothermic.*

Warm boundary current   An intense current flowing along the western margin or boundary of a subtropical gyre. It flows away from the Equator. The Gulf Stream is an example.

Warm core ring   As related to the Gulf Stream, they are bowl-shaped masses of warm Sargasso Sea water trapped by pinched off meanders that are about 1 kilometer (0.6 mile) deep and a diameter of about 100 kilometers (62 miles). The warm core sits within a mass of cold slope water on the landward side of the Gulf Stream. Similar features may be found associated with subtropical gyre boundary currents throughout the world ocean.

Warm front   A weather front in which a warm air mass moves into and over a cold air mass producing a broad band of gentle precipitation.

Warm-water vent   An opening in Earth's crust along the axes of spreading centers where water with temperatures ranging between 10 and 20°C (50 and 68°F) is vented.

Waste heat   Warm water that has been heated

by cooling steam lines of coastal power generating plants that is returned to the ocean at temperatures up to 11°C (20°F) above ambient temperatures.

Water (H_2O)   The oxide of hydrogen that makes life possible on Earth.

Water mass   A body of water identifiable by its temperature, salinity, or chemical content.

Water vapor   The gaseous state of H_2O. At sea level it forms at the boiling point of 100°C (212°F).

Wave   A disturbance that moves over or through a medium with a speed determined by the properties of the medium.

Wave base   The depth at which circular orbital motion becomes negligible; exists at a depth of $1/2$ wavelength, measured from still water level.

Wave-cut bench   A gently sloping surface produced by wave erosion and extending from the base of the wave-cut cliff out under the offshore region.

Wave-cut cliff   A cliff produced by wave erosion cutting landward.

Wave dispersion   The separation of waves as they leave the sea area by wave size. Larger waves travel faster than smaller waves and thus leave the sea area first to be followed by progressively smaller waves.

Wave frequency (*f*)   The number of waves that pass a fixed point in a unit of time, usually one second.

Wave height (*H*)   Vertical distance between a crest and an adjoining trough.

Wave period (*T*)   The elapsed time between the passage of two successive wave crests past a fixed point.

Wave speed (*S*)   The rate at which a wave travels. It can be calculated by dividing a wave's wavelength (*L*) by its period (*T*).

Wave steepness   Ratio of wave height (*H*) to wavelength (*L*). If a 1:7 ratio is ever exceeded by the wave, then the wave breaks.

Wave train   A series of waves from the same direction.

Wavelength (*L*)   Horizontal distance between two corresponding points on successive waves, such as from crest to crest.

Waning crescent   When the Moon is between third quarter and new moon phases.

Waning gibbous   When the Moon is between full moon and third quarter phases.

Waxing crescent   When the Moon is between new moon and first quarter phases.

Waxing gibbous   When the Moon is between first quarter and full moon phases.

Weather   The state of the atmosphere at a given time and place, with respect to variables such as temperature, moisture, wind velocity, and barometric pressure. The ocean has "weather" also, but the variables change much more slowly than in the atmosphere due to the greater density of ocean water.

Weathering   A process by which rocks are broken by chemical and mechanical means.

Wentworth scale of grain size   A logarithmic scale for size classification of sediment particles.

West Australian Current   This cold current forms the eastern boundary current of the Indian Ocean subtropical gyre. It is separated from the coast by the warm Leeuwin Current except during El Niño events when the Leeuwin Current weakens.

West Wind Drift   See *Antarctic Circumpolar Current*.

Westerly winds   The air masses moving away from the subtropical high pressure belts toward higher latitudes. They are southwesterly in the Northern Hemisphere and northwesterly in the Southern Hemisphere.

Western boundary current   Poleward-flowing warm currents on the western side of all subtropical gyres.

Western boundary undercurrent (WBUC)   A bottom current that flows along the base of the continental slope eroding sediment from it and redepositing the sediment on the continental rise. It is confined to the western boundary

of deep-ocean basins.

Westward intensification   Pertaining to the intensification of the warm western limb of the subtropical gyre currents that is manifested in higher velocity and deeper flow compared to the cold eastern boundary currents that drift leisurely toward the Equator.

Wetlands   Biologically productive regions bordering estuaries and other protected coastal areas. They are usually salt marshes at latitudes greater than 30° and mangrove swamps at lower latitudes.

Whitecap   A wind-generated wave that breaks in the open ocean.

White muscle fiber   Thick muscle fibers with relatively low concentrations of myoglobin that make up a large percentage of the muscle fiber in lunger-type fish.

White smoker   Similar to a black smoker, but emits water of a lower temperature that is white in color.

Wilson cycle   This cycle illustrates the process by which ocean basins form as a result of continental rifting, grow, and then are destroyed by plate tectonic processes.

Wind-driven circulation   Any movement of ocean water that is driven by winds. This includes most horizontal movements in the sur-

face waters of the world's oceans.

Windward   The direction from which the wind is blowing.

Winter solstice   The instant the southward-moving Sun reaches the Tropic of Cancer before changing direction and moving north back toward the Equator, approximately December 22.

Wintertime beach   Heavy winter wave activity removes sand from the beach face and transports it out to the offshore bar, leaving a narrow, rocky berm and a relatively flat beach face.

Zebra mussel   A nonnative species released into U.S. and Canadian waters of the Great Lakes region.

Zenith   That point on the celestial sphere directly over the observer.

Zooplankton   Animal plankton.

Zooplankton excretions   The undigested organic matter excreted by zooplankton; it is an important source on nourishment for microbes in the ocean.

Zooxanthellae   A form of algae that lives as a symbiont in the tissue of corals and other coral reef animals and provides varying amounts of their required food supply.

Credits and Acknowledgments

Cover

Satellite view of the edge of Bahama Bank courtesy of NASA's Earth Observatory. Sand Pattern Image acquired by the Landsat 7 ETM+ satellite. Processed by Institute for Marine Remote Sensing, University of South Florida courtesy of Serge Andrefouet/Frank Muller-Karger. Starfish photo courtesy of Jack and Sue Drafahl
Illustration of chemistry in ocean courtesy of Quade Paul
Big Sur Coastline photo courtesy of David Muench; CORBIS Bettmann.

Frontspiece

By Bruce C. Heezen and Marie Tharp, 1977. Copyright © 1977 Marie Tharp. Reproduced by permission of Marie Tharp.

Introduction

Figure I–1 APT photo. **Figure I–3** Courtesy of Johnson Space Center, NASA. **Figure I–4** F. Stuart Westmorland/Photo Researchers, Inc.

Chapter 1

Opener Courtesy of National Maritime Museum Picture Library, London, England. **Figure 1A** APT photo; Polynesian stick chart courtesy of Glen Foss. **Figure 1B** APT photo. **Figure 1C** Courtesy of U.S. Department of Defense. **Figure 1D left** Smithsonian Institution Libraries © 2001 Smithsonian Institution. **Figure 1D right** Photo courtesy of Corbis–The Bettmann Archive. **Figure 1E** Courtesy of Scripps Institution of Oceanography Archives, University of California, San Diego. **Figure 1–1** From the *Challenger* Report, Great Britain, 1895. **Figure 1–3** Data from Kennish (1994) and the National Geographic Society. **Figure 1–4** Official photograph U.S. Navy. **Figure 1–11 inset** Courtesy of U.S. Navy. **Figure 1–12** Reprinted by permission of The Bettmann Archive. **Figure 1–13 inset** From C. W. Thompson and Sir John Murray, *Report on the Scientific Results of the Voyage of* H.M.S. Challenger, Vol. 1. Great Britain: Challenger Office, 1895, Plate 1. **Figure 1–14** Photo courtesy of The *Fram* Museum, Oslo, Norway. **Figure 1–15** Map courtesy of U.S. Navy; **photo inset** courtesy of Corbis–The Bettmann Archive. **Figure 1–16** Released Naval Historical Center photograph. **Figure 1–17** Photo courtesy of Corbis–The Bettmann Archive. **Figure 1–18** Courtesy of Patrick Aventurier, Liaison Agency, Inc. **Figure 1–19** From G. Wüst, 1935, *The stratosphere of the Atlantic Ocean. Scientific results of the German American Expedition of the Research Vessel Meteor 1925–1927*, Vol. VI, sec. 1., W. J. Emery (translated and edited), Amerind Pub. Co., New Delhi, 1978. **Figure 1–20a** Courtesy of Scripps Institution of Oceanography, University of California, San Diego. **Figure 1–20b** Courtesy of Clyde H. Smith, Peter Arnold, Inc. **Figure 1–20c** Courtesy of E. Hunt Augustus. **Figure 1–21** Courtesy of Jet Propulsion Laboratory, NASA. **Figure 1–22** Photo courtesy of David Knudsen/National Geographic Society. **Figure 1–23** Courtesy of Woods Hole Oceanographic Institution. **Figure 1–25a and b** Courtesy of Woods Hole Oceanographic Institution. **Figure 1–25c** Courtesy of Mote Marine Laboratory and Rutgers University. **Figure 1–26 left** HuttonGetty/Liaison Agency, Inc. **Figure 1–26 right** Courtesy of Woods Hole Oceanographic Institution, Emory Kristof/National Geographic Society. **Figure 1–27a** Courtesy of Glasheen Collection, Mandeville Special Collections Library, University of California, San Diego. **Figure 1–27b** Photo courtesy of NOAA.

Chapter 2

Opener Courtesy of NASA and the Space Telescope Science Institute. **Figure 2A** Courtesy of NASA's Earth Observatory. **Figure 2–1a** Reprinted by permission from Tarbuck, E. J., and Lutgens, F. K., Earth Science, 6th ed. (Fig. 20.17), Macmillan Publishing Company, 1991. **Figure 2–1b** Courtesy of NASA and the Space Telescope Science Institute. **Figure 2–3a** Reprinted by permission from Tarbuck, E. J., and Lutgens, F. K., Earth Science, 6th ed. (Fig. 19.1), Macmillan Publishing Company, 1991. **Figure 2–3b** Reprinted by permission from Tarbuck, E. J., and Lutgens, F. K., Earth Science, 6th ed. (Fig. 19.2), Macmillan Publishing Company, 1991. **Figure 2–4** Courtesy of NASA and the Space Telescope Science Institute. **Figure 2–5** After Tarbuck, E. J., and Lutgens, F. K., The Earth: An Introduction to Physical Geology, 5th ed. (Fig. 1.10), Prentice Hall, 1996. **Figure 2–6** Reprinted by permission from Tarbuck, E. J., and Lutgens, F. K., The Earth: An Introduction to Physical Geology, 4th ed. (Fig. 1.11), Macmillan Publishing Company, 1993. **Figure 2–9b** Courtesy of Scripps Institution of Oceanography, University of San Diego, California. **Figure 2–11** Reprinted with permission from Schopf, J. W., Microfossils of the Early Archean Apex Chert: New Evidence of the Antiquity of Life, Science 260:5108, 640–646. © 1993 American Association for the Advancement of Science. **Figure 2–15** After Tarbuck, E. J., and Lutgens, F. K., The Earth: An Introduction to Physical Geology, 5th ed. (Fig. 1.9), Prentice Hall, 1997. Data from the Geological Society of America. **Table 2–1** After Rubey, W. W. 1951; courtesy of the Geological Society of America.

Chapter 3

Opener APT photo. **Figure 3A** Courtesy of Deeanne Edwards, Scripps Institution of Oceanography, University of California, San Diego. **Figure 3B** Photo courtesy of R. Koski, U.S. Geologic Survey. **Figure 3C** After Tarbuck, E. J., and Lutgens, F. K., *The Earth: An Introduction to Physical Geology*, 4th ed. (Fig. 19.15), Macmillan Publishing Company, 1993. **Figure 3–1** Courtesy of Bildarchiv Preussischer Kulturbesitz, West Berlin. **Figure 3–2** After Dietz, R. S., and Holden, J. C. 1970. Reconstruction of Pangaea: Breakup and dispersion of continents, Permian to present. *Journal of Geophysical Research* 75:26, 4939–4956. **Figure 3–3** From *Continental Drift* by Don and Maureen Tarling, © 1971 by G. Bell & Sons, Ltd. Reprinted by permission of Doubleday & Co., Inc. **Figure 3–4** After Tarbuck, E. J., and Lutgens, F. K., *Earth Science*, 8th ed. (Fig. 7.6), Prentice Hall, 1997. **Figure 3–5** After Tarbuck, E. J., and Lutgens, F. K., *Earth Science*, 8th ed. (Fig. 7.7), Prentice Hall, 1997. **Figure 3–6** After Tarbuck, E. J., and Lutgens, F. K., *Earth Science*, 8th ed. (Fig. 7.4), Prentice Hall, 1997. **Figure 3–7a** Reprinted by permission from Tarbuck, E. J., and Lutgens, F. K., *The Earth: An Introduction to Physical Geology*, 3rd ed. (Fig. 18.8), Merrill Publishing Company, 1990. **Figure 3–7b** Reprinted by permission from Tarbuck, E. J., and Lutgens, F. K., *The Earth: An Introduction to Physical Geology*, 3rd ed. (Fig. 18.9), Merrill Publishing Company, 1990. **Figure 3–8** After Tarbuck, E. J., and Lutgens, F. K., *Earth Science*, 8th ed. (Fig. 7.17), Prentice Hall, 1997. **Figure 3–9** After Tarbuck, E. J., and Lutgens, F. K., *Earth Science*, 8th ed. (Fig. 7.18), Prentice Hall, 1997. **Figure 3–10** After Tarbuck, E. J., and Lutgens, F. K., *The Earth: An Introduction to Physical Geology*, 3rd ed. (Fig. 18.11), Merrill Publishing Company, 1990. **Figure 3–11** After Tarbuck, E. J., and Lutgens, F. K., *Earth Science*, 8th ed. (Fig. 7.20), Prentice Hall, 1997. **Figure 3–12** Reprinted by permission from Tarbuck, E. J., and

Lutgens, F. K., *The Earth: An Introduction to Physical Geology*, 3rd ed. (Fig. 19.16), Merrill Publishing Company, 1990. After *The Bedrock Geology of the World*, by R. L. Larson et al., Copyright © 1985 by W. H. Freeman. **Figure 3–14a** Reprinted by permission from Tarbuck, E. J., and Lutgens, F. K., *The Earth: An Introduction to Physical Geology*, 3rd ed. (Fig. 5.11), Merrill Publishing Company, 1990. Data from National Geophysical Data Center/NOAA. **Figure 3–14b** After Tarbuck, E. J., and Lutgens, F. K., *The Earth: An Introduction to Physical Geology*, 3rd ed. (Fig. 5.11), Merrill Publishing Company, 1990; modified with data from W. B. Hamilton, U.S. Geological Survey. **Figure 3–17a** Courtesy of Bunge, H. Reprinted with permission from *Nature* 397:571b, 203 © 1999, Macmillan Magazines Ltd. (www.nature.com). **Figure 3–17b** Reprinted with permission from Ritsema, J., et al. Complex shear wave velocity structure imaged beneath Africa and Iceland. *Science* 286:5446, 1925–1928. Supp. Fig. 1. © 1999, American Association for the Advancement of Science. **Figure 3–18** Reprinted by permission from Tarbuck, E. J., and Lutgens, F. K., *The Earth: An Introduction to Physical Geology*, 3rd ed. (Fig. 5.19), Merrill Publishing Company, 1990. **Figure 3–20** Reprinted by permission from Tarbuck, E. J., and Lutgens, F. K., *The Earth: An Introduction to Physical Geology*, 3rd ed. (Fig. 6.7), Merrill Publishing Company, 1990. **Figure 3–21** After Tarbuck, E. J., and Lutgens, F. K., *Earth Science*, 8th ed. (Fig. 7.10), Prentice Hall, 1997. **Figure 3–22** Courtesy of Patricia Deen, Palomar College. **Figure 3–24** After Tarbuck, E. J., and Lutgens, F. K., *Earth Science*, 8th ed. (Fig. 7.12), Prentice Hall, 1997. **Figure 3–25** After Tarbuck, E. J., and Lutgens, F. K., Earth Science, 8th ed. (Fig. 7.11), Prentice Hall, 1997. **Figure 3–26** Courtesy of David Sandwell, Scripps Institution of Oceanography, University of California, San Diego. **Figure 3–27** Reprinted by permission from Tarbuck, E. J., and Lutgens, F. K., *Earth Science*, 5th ed. (Fig. 6.11), Merrill Publishing Company, 1988. **Figure 3–28c** Photo courtesy of U.S. Geologic Survey, Cascades Volcano Observatory. **Figure 3–29** After Tarbuck, E. J., and Lutgens, F. K., *Earth Science*, 8th ed. (Fig. 7.15), Prentice Hall, 1997. **Figure 3–30** After Tarbuck, E. J., and Lutgens, F. K., *The Earth: An Introduction to Physical Geology*, 3rd ed. (Figs. 18.22, 18.23), Merrill Publishing Company, 1990. **Figure 3–31** Data from Robert A. Duncan. **Figure 3–32** After Tarbuck, E. J., and Lutgens, F. K., *The Earth: An Introduction to Physical Geology*, 6th ed. (Fig. 19.30), Prentice Hall, 1999. **Figure 3–36** Photo courtesy of NASA. **Figure 3–37** Courtesy of Paul C. Lowman, Jr., and others, NASA/Goddard Space Flight Center, Greenbelt, Maryland. **Figure 3–38** Plate tectonic reconstructions by Christopher R. Scotese, PALEOMAP Project, University of Texas at Arlington. **Figure 3–40** After Dietz, R. S., and Holden, J. C. 1970. The breakup of Pangaea, *Scientific American* 223:4, 30–41. **Figure 3–41** Adapted from Wilson, J. T., *American Philosophical Society Proceedings* 112, 309–320, 1968: Jacobs, J. A., Russell, R. D., and Wilson, J. T., *Physics and Geology*, McGraw–Hill, New York, 1971.

Chapter 4

Opener Courtesy of Walter H. Fl Smith and David T. Sandwell, Science, 1997/National Geophysical Data Center/NOAA. "Primary Funding from National Science Foundation". **Figure 4A** After Gross, M. G., *Oceanography*, 6th ed. (Fig. 16.10), Prentice Hall, 1993. **Figure 4B** Courtesy of David Sandwell, Scripps Institution of Oceanography, University of California, San Diego. **Figure 4C** Courtesy of David Sandwell, Scripps Institution of Oceanography, University of California, San Diego. **Figure 4D** After Heezen, B. C., and Ewing, M. 1952. Turbidity currents and submarine slumps, and the 1929 Grand Banks earthquake, *American Journal of Science* 250:867. **Figure 4–1** Courtesy of Peter A. Rona, Hudson Laboratories of Columbia University. **Figure 4–2 inset** Courtesy of Daniel J. Fornari, Lamont-Doherty Geological Observatory, Columbia University. Reprinted with permission of the American Geophysical Union. **Figure 4–3** Technical specifications courtesy of Jeffrey Booth, Datasonics, Inc. **Figure 4–4** Seismic profile courtesy of the Deep Sea Drilling Project, Scripps Institution of Oceanography, University of California, San Diego; with thanks to Jerry Bode. **Figure 4–5** After Tarbuck, E. J., and Lutgens, F. K., *Earth Science*, 6th ed. (Fig. 10.2), Macmillan Publishing Company, 1991. **Figure 4–6** After Tarbuck, E. J., and Lutgens, F. K., *The Earth: An Introduction to Physical Geology*, 5th ed. (Fig. 19.3), Prentice Hall, 1996. **Figure 4–7** After Plummer, C. C. and McGeary, D., *Physical Geology*, 5th ed. (Fig. 18.6), Wm. C. Brown, 1991. **Figure 4–8** After Tarbuck, E. J., and Lutgens, F. K., *The Earth: An Introduction to Physical Geology*, 5th ed. (Fig. 19.4), Prentice Hall, 1996. **Figure 4–9a** After Tarbuck, E. J., and Lutgens, F. K., *The Earth: An Introduction to Physical Geology*, 5th ed. (Fig. 19.6), Prentice Hall, 1996. **Figure 4–9b** Courtesy of Francis P. Shepard Photographic Archives/Collections; with thanks to G. G. Kuhn. **Figure 4–10** Courtesy of Charles Hollister, Woods Hole Oceanographic Institution. **Figure 4–13** Courtesy of Peter Lonsdale, Scripps Institution of Oceanography, University of California, San Diego. **Figure 4–14** Courtesy of Aluminum Company of America (Alcoa). **Figure 4–15 upper** Courtesy of Br. Robert McDermott, S. J. **Figure 4–15 lower** Courtesy of Woods Hole Oceanographic Institution. **Figure 4–16a** Map produced by S. P. Miller; provided courtesy of K. C. McDonald, University of California, Santa Barbara. **Figure 4–16b** Courtesy of A. E. J. Engel and Scripps Geological Collections, Scripps Institution of Oceanography, University of California, San Diego. **Figure 4–17a** After Tarbuck, E. J., and Lutgens, F. K., *The Earth: An Introduction to Physical Geology*, 5th ed. (Fig. 21.23), Prentice Hall, 1996. Inset photo by Fred N. Spiess, Scripps Institution of Oceanography, University of California, San Diego. **Figure 4–17b** Photo Ifremer, from Manaute Cruise, courtesy of Jean-Marie Auzende. **Figure 4–19** Courtesy of Enrico Bonatti, Lamont–Doherty Geological Observatory of Columbia University.

Chapter 5

Opener Courtesy of Alfred Wegener Institute. **Figure 5A** Courtesy of the Ocean Drilling Program. **Figure 5B** Courtesy of the Ocean Drilling Program. **Figure 5C** Photo courtesy of World Minerals, Inc., Lompoc, California. **Figure 5C, inset** Reprinted by permission from Hallegraeff, G. M., *Plankton: A Microscopic World*, 1988 (p. 20). Courtesy of E. J. Brill, Inc. **Figure 5E** Courtesy of the Ocean Drilling Program, Texas A&M University. **Figure 5–1** Courtesy of the RISE Project Group, F. N. Spiess et al., Scripps Institution of Oceanography, University of California, San Diego. **Figure 5–2** APT photo; with thanks to Steve Prinz and Warren Smith. **Figure 5–3** After Tarbuck, E. J., and Lutgens, F. K., *The Earth: An Introduction to Physical Geology*, 5th ed. (Fig. 5.11), Prentice Hall, 1996. **Figure 5–4a** APT photo. **Figure 5–4b** Annie Griffiths Belt/CORBIS. **Figure 5–4c** APT photo. **Figure 5–4d** APT photo. **Figure 5–6** Courtesy of Walter N. Mack, Michigan State University. **Figure 5–7** After Leinen, M. et al., 1986, Distribution of biogenic silica and quartz in recent deep-sea sediments. *Geology* 14:3, 199–203. **Figure 5–7 inset** Provided by the SeaWiFS Project, NASA/Goddard Space Flight Center and ORBIMAGE. **Figure 5–8** After Tarbuck, E. J., and Lutgens, F. K., *The Earth: An Introduction to Physical Geology*, 4th ed. (Fig. 10.15), Macmillan Publishing Company, 1993. **Figure 5–10a** Reprinted by permission from Hallegraeff, G. M., *Plankton: A Microscopic World*, 1988 (p. 46). Courtesy of E. J. Brill, Inc. **Figure 5–10b** Courtesy of Warren Smith, Scripps Institution of Oceanography, University of California, San Diego. **Figure 5–10c** Photo courtesy of World Minerals, Inc., Lompoc, California (sample from Celite Corporations Diatomite Mine in Lompoc, California). **Figure 5–11a** Reprinted by permission from Hallegraeff, G. M., *Plankton: A Microscopic World*, 1988 (p. 8). Courtesy of E. J. Brill, Inc. **Figure 5–11b** Reprinted by permission from Hallegraeff, G. M., *Plankton: A Microscopic World*, 1988 (p. 16). Courtesy of E. J. Brill, Inc. **Figure 5–11c** Courtesy of Memorie Yasuda, Scripps Institution of Oceanography, University of California, San Diego. **Figure 5–11d** Courtesy of the Deep Sea Drilling Project, Scripps Institution of Oceanography, University of California, San Diego. **Figure 5–12** APT photo. **Figure 5–13** Maps and photos courtesy of Jeff Dravis; permission for use granted by Exxon Production Research Company and American Association for the Advancement of Science. **Figure 5–15** After Leinen, M. et al., 1986, Distribution of biogenic silica and quartz in recent deep-sea sediments. *Geology* 14:3, 199–203. **Figure 5–17** After Biscaye, P. E. et al., 1976; Berger, W. H. et al., 1976; and Kolla V. and Biscaye, P. E., 1976. **Figure 5–18** Courtesy of Scripps Institution of Oceanography, University of California, San Diego. **Figure 5–19** APT photo. **Figure 5–20** Reprinted by permission of The Open University Course Team, *Ocean Chemistry and Deep-Sea Sediments*, Butterworth-Heinemann, 1989. **Figure 5–21** After image by Joseph Holliday, El Camino College. **Figure 5–22** After M. O. Hayes, *Marine Geology* 5:2, 1967. **Figure 5–24** After Sverdrup, H. U. et al., 1942. **Figure 5–24** Courtesy of Susumu Honjo, Woods Hole Oceanographic Institution. **Table 5–1** After Patricia Deen, Palomar College. **Table 5–2** Source: Wentworth, 1922; After Udden, 1898.

Chapter 6

Opener Image courtesy of Quade Paul. **Figure 6A** Peter Arnold, Inc. **Figure 6–1** After Tarbuck, E. J., and Lutgens, F. K., *The Earth: An Introduction to Physical Geology*, 5th ed. (Fig. 2.4), Prentice Hall, 1996. **Figure 6–3** Used with permission from R. W. Christopherson, *Geosystems*, 2nd ed. (Fig. 7-7, p. 186), Macmillan Publishing Company, 1994. **Figure 6–11** Photo by Electron Microscopy Laboratory, ARS, USDA. **Figure 6–15** Data from GEOSECS. **Figure 6–18** Data from GEOSECS. **Figure 6–19** After Tarbuck, E. J., and Lutgens, F. K., *The Earth: An Introduction to Physical Geology*, 4th ed. (Fig. 10.2), Macmillan Publishing Company, 1993. **Figure 6–20** After Pickard, G. L., *Descriptive Physical Oceanography*, Pergamon Press Ltd., 1963. **Figure 6–21** After Sverdrup, H. U. et al., 1942. **Figure 6–23** After Pickard, G. L., *Descriptive Physical Oceanography*, Pergamon Press Ltd., 1963. **Figure 6–24** After Pickard, G. L., *Descriptive Physical Oceanography*, Pergamon Press Ltd., 1963. **Table 6–4** Source: Data from Broecker, 1974.

Chapter 7

Opener Ralph A. Clevenger/Corbis. **Figure 7A** Courtesy National Oceanic and Atmospheric Administration/Department of Commerce. **Figure 7B inset** Photo courtesy of Walter Munk, Scripps Institution of Oceanography, University of California, San Diego. **Figure 7–4** After Gross, M. G., *Oceanography*, 6th ed. (Fig. 5–2), Prentice Hall, 1993. **Figure 7–13** After Gross, M. G., *Oceanography*, 6th ed. (Fig. 5–17), Prentice Hall, 1993. **Figure 7–14** Reprinted by permission from Lutgens, F. K., and Tarbuck, E. J., *The Atmosphere*, 6th ed. (Fig. 8.2), Prentice Hall, 1995. **Figure 7–18 inset** Photo by Bob Stovall, Bruce Coleman, Inc. **Figure 7–19** Courtesy of National Hurricane Center, NOAA. **Figure 7–21a** Photo courtesy of Stephen J. Krasemann. **Figure 7–21b** Courtesy of NASA. **Figure 7–22a** William W. Bacon, Rapho. **Figure 7–22c** Photo courtesy of Josh Landis/National Science Foundation. **Figure 7–22d** Courtesy of NASA's Earth Observatory. **Figure 7–26** After Tarbuck. E. J., and Lutgens, F. K. *The Earth: An Introduction to Physical Geology*, 6th ed. (Fig. 21.9), Prentice Hall, 1999. **Figure 7–27** Data from 1958 to recent from laboratory of Charles Keeling, Scripps Institution of Oceanography, University of San Diego, California. Carbon dioxide concentrations prior to 1958 estimated from air bubbles in polar ice cores (adapted from *Our Changing Climate*, Reports to the Nation on Our Changing Planet, Fall 1997, No. 4, University Corporation for Atmospheric Research). **Figure 7–28** Adapted from *Our Changing Climate*, Reports to the Nation on Our Changing Planet,

Fall 1997, No. 4, University Corporation for Atmospheric Research. **Table 7–6** Source: After Rodhe, H., 1990.

Chapter 8

Opener Courtesy of NASA's Earth Observatory. **Figure 8A** Map courtesy of Eos Transactions AGU 73:34, 361 (1992); inset courtesy of Eos Transactions AGU 75:37, 425 (1994). **Figure 8B** Photo courtesy of Martin Wikelski, University of Illinois. **Figure 8–1a** Courtesy of Douglas Alden, Scripps Institution of Oceanography, University of California, San Diego. **Figure 8–1b** Courtesy of Aanderaa Instruments. **Figure 8–2** Courtesy of NASA. **Figure 8–8a** Courtesy of NASA and NOAA. **Figure 8–16a** Image courtesy of Charles McLain at the Rosenstiel School of Marine and Atmospheric Science, University of Miami. **Figure 8–19** Maps courtesy of the International Research Institute for Climate Prediction, Lamont-Doherty Earth Observatory, Columbia University. **Figure 8–20** Courtesy of Klaus Wolter and NOAA. **Figure 8–22** Courtesy of Paul C. Fiedler at the National Marine Fisheries Service. **Figure 8–27** Courtesy of Woods Hole Oceanographic Institution.

Chapter 9

Opener Courtesy of Surfer Magazine/Bob Barbour Photography. **Figure 9B** Image courtesy of NASA's Earth Observatory and Jacques Descloitres, MODIS Land Rapid Response Team at NASA GSFC. **Figure 9E** Courtesy of California Geology. **Figure 9–1b** Photo courtesy of NASA. **Figure 9–2** After Kinsman, B., 1965. **Figure 9–3b** From *The Tasa Collection: Shorelines*. Published by Macmillan Publishing Co., New York. Copyright © 1986 by Tasa Graphic Arts, Inc. All rights reserved. **Figure 9–10** Courtesy of Jet Propulsion Laboratory, NASA. **Figure 9–12** Official photograph U.S. Navy. **Figure 9–15** After Gross, M. G., *Oceanography*, 6th ed. (Fig. 8–4), Prentice Hall, 1993. **Figure 9–17** From *The Tasa Collection: Shorelines*. Published by Macmillan Publishing, New York. Copyright © 1986 by Tasa Graphic Arts, Inc. All rights reserved. **Figure 9–18a** © Peter Arnold, Inc. **Figure 9–18b** Vince Cavataio/Allsport/ Agency Vandystadt/Photo Researchers, Inc. **Figure 9–18c** Photo © Woody Woodworth, Creation Captured. All rights reserved. **Figure 9–19b** Adapted from Tarbuck, E. J., and Lutgens, F. K., *Earth Science*, 6th ed. (Fig. 11.14), Macmillan Publishing Company, 1991. **Figure 9–21** Photo by Hal Thurman. **Figure 9–23b** Photos courtesy Kyodo News Agency, Japan. **Figure 9–24** CORBIS. **Figure 9–25** After González, F. I., Tsunami!, *Scientific American* 280:5, 59, 1999. **Figure 9–26** Courtesy of Oregon Department of Geology. **Table 9–1** Source: After Bowditch, N. 1958.

Chapter 10

Opener Courtesy of Tides Photography, photographic artist Dick Killam; website address: *www.tidesinhallsharbour.com*. **Figure 10A** Photo courtesy of New Brunswick Department of Tourism. **Figure 10B** Photo by Eda Rogers. **Figure 10–11** Modified from *The Tasa Collection: Shorelines*. Published by Macmillan Publishing Co., New York. Copyright © 1986 by Tasa Graphic Arts, Inc. All rights reserved. **Figure 10–14** After Hauge, C., Tides, currents, and waves. *California Geology*, July 1972. **Figure 10–18** After von Arx, W. S., 1962; original by H. Poincaré 1910, Leçons de Mécanique Céleste, a Gauther-Crofts, Vol. 3. **Figure 10–19** From Defant, A., *Ebb and flow*. The University of Michigan Press, Ann Arbor, 1958. **Figure 10–20** After Hauge, C., Tides, currents, and waves. *California Geology*, July 1972. **Figure 10–24** Photos courtesy of Nova Scotia Department of Tourism.

Chapter 11

Opener Photo courtesy of Joel Arrington/Visuals Unlimited. **Figure 11A** APT photos. **Figure 11B** Photo courtesy of Drew Wilson, Virginian–Pilot © 1999. **Figure 11–2** APT photos. **Figure 11–3a** Photo by John S. Shelton. **Figure 11–3b** After Tarbuck, E. J., and Lutgens, F. K., *The Earth: An Introduction to Physical Geology*, 4th ed. (Fig. 14.8), Macmillan Publishing Company, 1993. **Figure 11–5** Photo by Bruce F. Molnia, Terra-photographics/BPS. **Figure 11–6** Photo by John S. Shelton. **Figure 11–8a** Photo by USDA-ASCS. **Figure 11–8b** Photo © 2003 Andrew Alden; geology.about.com. **Figure 11–9a and b** After Tarbuck, E. J., and Lutgens, F. K., *The Earth: An Introduction to Physical Geology*, 4th ed. (Fig. 14.12), Macmillan Publishing Company, 1993. **Figure 11–9c** Photo by USDA-ASCS. **Figure 11–11a** Photo by GEOPIC®, Earth Satellite Corporation. **Figure 11–11b** Photo courtesy of NASA. **Figure 11–13 a, b, and d** APT photos. **Figure 11–13c** Courtesy of U.S. Geologic Survey. **Figure 11–13e** Photo courtesy of Mrs. Marge Beaver/Photography Plus/NOAA/Department of Commerce. **Figure 11–13f** Photo by James R. McCullagh. **Figure 11–16** After Neumann, J. E., et al., *Sea-level Rise and Global Climate Change: A Review of Impacts to U.S. Coasts*, PEW Center on Global Climate Change, 2000. **Figure 11–20** Photo by John S. Shelton. **Figure 11–21** Photo by John S. Shelton. **Figure 11–22** After Tarbuck, E. J., and Lutgens, F. K., *Earth Science*, 5th ed., Merrill Publishing Company, 1988. **Figure 11–24a** Courtesy of The Fairchild Aerial Photography Collection at Whittier College, California. Flight C–1670, Frame 4 (9/18/31). **Figure 11–24b** Courtesy of The Fairchild Aerial Photography Collection at Whittier College, California. Flight C–14180, Frame 3:61 (10/21/49). **Figure 11–26** APT photos. **Table 11–1** Source: After Shephard, F. P., *Submarine geology*, 3rd ed. (Table 9, p. 127, "Average beach face slopes compared to sediment diameters"), Harper & Row Publishers, Inc., 1973. **Table 11–3** Source: After Shephard, F. P., Coastal classification and changing coastlines, in *Geoscience and Man*, 14, 53–64, 1976.

Chapter 12

Opener Photo courtesy of Rossi, Guido Alberto/Getty Images Inc. - Image Bank. **Figure 12–4a** Photo courtesy of The Stock Market. **Figure 12–4b** Courtesy of NASA. **Figure 12–7** After Officer et al., 1984. **Figure 12–8b** Photo © Eda Rogers. **Figure 12–8a** Photo © R. N. Mariscal/Bruce Coleman, Inc. **Figure 12–11a** Adapted from *Encyclopedia of Oceanography*, edited by Rhodes Fairbridge, © 1966. Reprinted by permission of Dowden, Hutchinson, & Ross, Inc., Stroudsburg, PA. **Figure 12–11b** After Judson, S. et al., *Physical Geology*, 7th ed. Prentice Hall, 1987. **Figure 12–13** Courtesy of NOAA. **Figure 12–14a** Adapted from *Encyclopedia of Oceanography*, edited by Rhodes Fairbridge, © 1966. Reprinted by permission of Dowden, Hutchinson, & Ross, Inc., Stroudsburg, PA. **Figure 12–14b** Courtesy of Robert Brown, Scripps Institution of Oceanography, University of California, San Diego. **Figure 12–15** Adapted from *Encyclopedia of Oceanography*, edited by Rhodes Fairbridge, © 1966. Reprinted by permission of Dowden, Hutchinson, & Ross, Inc., Stroudsburg, PA.

Chapter 13

Opener Photo courtesy of NASA/Goddard Space Flight Center. **Figure 13–7** After Sverdrup, H. U. et al., 1942. **Figure 13–9** Reprinted by permission from Hallegraeff, G. M., *Plankton: A Microscopic World*, 1988 (p. 21). Courtesy of E. J. Brill, Inc. **Figure 13–16** Photo © Ralph Lee Hopkins/Wilderland Images. **Figure 13–16a** Courtesy of Deeanne Edwards, Scripps Institution of Oceanography, University of California, San Diego. **Figure 13–17b** APT photo. **Figure 13–18** Courtesy National Oceanic and Atmospheric Administration/Department of Commerce. **Figure 13–21** Photo by Peter Arnold, Inc.

Chapter 14

Opener Photo by Francois Gohier, courtesy of Photo Researchers, Inc. **Figure 14A** Carleton Ray/Photo Researchers, Inc. **Figure 14C**

Copyright © 1999 by Universal City Studios, Inc. Courtesy of Universal Studios Publishing Rights, a division of Universal Studios Licensing, Inc. All Rights Reserved. Photo Courtesy of the Academy of Motion Picture Arts and Sciences. **Figure 14D inset** Courtesy of North Carolina Division of Water Quality. **Figure 14E upper** Courtesy of JoAnn Burkholder. **Figure 14E middle** Courtesy of JoAnn Burkholder and Howard Glasgow. **Figure 14E lower** Courtesy of JoAnn Burkholder et al., with permission from Nature. **Figure 14–2** Courtesy of Elizabeth Venrick, Scripps Institution of Oceanography, University of California, San Diego. **Figure 14–4c** Provided by the SeaWiFS Project, NASA/Goddard Space Flight Center and ORBIMAGE. **Figure 14–6** Courtesy of Patricia Deen, Palomar College. **Figure 14–7** Provided by the SeaWiFS Project, NASA/Goddard Space Flight Center and ORBIMAGE. **Figure 14–8** APT photo. **Figure 14–9 a, b, and d** APT photos. **Figure 14–9c** Photo by Hal Thurman. **Figure 14–10a** Reprinted by permission from Hallegraeff, G. M., *Plankton: A Microscopic World*, 1988 (p. 43). Courtesy of E. J. Brill, Inc. **Figure 14–10b** Courtesy of Wuchang Wei, Scripps Institution of Oceanography, University of California, San Diego. **Figure 14–10c** Reprinted by permission from Hallegraeff, G. M., *Plankton: A Microscopic World*, 1988 (p. 91). Courtesy of E. J. Brill, Inc. **Figure 14–10d** Reprinted by permission from Hallegraeff, G. M., *Plankton: A Microscopic World*, 1988 (p. 81). Courtesy of E. J. Brill, Inc. **Figure 14–11d** Photo courtesy of Scripps Institution of Oceanography, University of California, San Diego. **Figure 14–16a** Photo © Marty Snyderman. **Figure 14–16b** Photo by Christopher Newbert. **Figure 14–16c** Photo © Marty Snyderman. **Figure 14–25** Courtesy of Jed Fuhrman. **Table 14–2** Data after Strahler, A. H. and Strahler, A. N., *Modern Physical Geography*, 4th ed. (Table 25.1), Wiley, 1992.

Chapter 15

Opener Photo by Bert Yates. **Figure 15A** Photo by Kelvin Aitken, courtesy Peter Arnold, Inc. **Figure 15B** Photo by Francois Gohier, courtesy of Photo Researchers, Inc. **Figure 15–3** Courtesy of Kozo Takahashi, Kyushu University, Fukuoka, Japan. **Figure 15–4** APT photo. **Figure 15–5** From Wilhelm Giesbrecht's 1892 book *Fauna und Flora des Golfes von Neapel und der Angrenzenden Meeres–Abschnitte*, Berlin: Verlag Von R. Friedländer and Sohn. Courtesy of Scripps Institution of Oceanography Explorations, Scripps Institution of Oceanography, University of California, San Diego. **Figure 15–6** Courtesy of Scripps Institution of Oceanography, University of California, San Diego. Photographer: Paige Jennings. **Figure 15–6 inset** Photo by Pam Blades-Eckelbarger, Harbor Branch Oceanographic Institution, Inc. **Figure 15–7b** Photo by Larry Ford. **Figure 15–8 b and d** Photos by James M. King, Graphic Impressions. **Figure 15–9** SEM photomicrograph courtesy of Howard J. Spero, University of South Carolina. **Figure 15–12a** Photo © Marty Snyderman. **Figure 15–12b** Photo © Wayne and Karen Brown. **Figure 15–12c** Photo © Greg Ochocki/The Stock Market. **Figure 15–12d** Photo © Bob Gomel/The Stock Market. **Figure 15–12e** Photo © Valerie Taylor/Peter Arnold, Inc. **Figure 15–14** After Lalli and Parsons, 1993 (Fig. 6.5). **Figure 15–15** After Lalli and Parsons, 1993 (Figs. 6.6b and 6.7a). **Figure 15–16a** Photo © Fred Bavendam/Peter Arnold, Inc. **Figure 15–16b** Photo courtesy of National Marine Fisheries Service. **Figure 15–18** Photo © Thomas Ives/The Stock Market. **Figure 15–19a** © Jeff Foott/Bruce Coleman, Inc. **Figure 15–19b** Photo © Ralph Lee Hopkins/Wilderland Images. **Figure 15–20a** Photo © Ralph Lee Hopkins/Wilderland Images. **Figure 15–20b** Photo © C. Allan Morgan/Peter Arnold, Inc. **Figure 15–20c** APT photo. **Figure 15–22a** Photo © Fred Bavendam/ Peter Arnold, Inc. **Figure 15–22b** D. Fleetham/Animals Animals/ Earth Scenes. **Figure 15–26** APT photo, courtesy SeaWorld of California.

Figure 15–29b and c APT photos, courtesy SeaWorld of California. **Figure 15–30** Photo by Kennan Ward Photography (© 1990). **Figure 15–32** Data from National Marine Fisheries Service. **Figure 15–33** Photo © Ralph Lee Hopkins/Wilderland Images. **Table 15A** Adapted from the International Shark Attack File at http://www.flmnh.ufl.edu/fish/Sharks/isaf/isaf.htm. **Box 15–1** Adapted from the International Shark Attack File at http://www.flmnh.ufl.edu/fish/Sharks/isaf/isaf.htm.

Chapter 16

Opener Photo courtesy of Joel W. Rogers/CORBIS BETTMANN. **Figure 16A** Courtesy of Jerry Wellington, University of Houston. **Figure 16B** Photos courtesy of Robert Hessler, Scripps Institution of Oceanography, University of California, San Diego. Reprinted with permission from Hessler, R. R., Ingram, C. L., Yayanos, A. A., and Burnett, B. R., Scavenging amphipods from the floor of the Philippine Trench, *Deep-Sea Research*, 25, Fig. 1, © 1978, Elsevier Science. **Figure 16–1** After Zenkevitch, L. A. et al., 1971. **Figure 16–2 b, c, e, g, and h** Photos by Hal Thurman. **Figure 16–2d** Photo © Norbet Wu/Peter Arnold, Inc. **Figure 16–2f** Photo © Breck P. Kent. **Figure 16–2i** APT photo. **Figure 16–3** Photo © Joy Sparr/Bruce Coleman, Inc. **Figure 16–5a** Photo by James McCullagh. **Figure 16–5b** Photo by Hal Thurman. **Figure 16–6** Photo by Hal Thurman. **Figure 16–7a** Photo © Fred Bavendam/ Peter Arnold, Inc. **Figure 16–7b** Photo © Eda Rogers. **Figure 16–11** Photo by Hal Thurman. **Figure 16–13** SEM photomicrographs courtesy of Howard J. Spero, University of South Carolina. **Figure 16–14** Photo by Stephen J. Kraseman © Peter Arnold, Inc. **Figure 16–15b** Photo by Mia Tegner. **Figure 16–15c** Photo by James McCullagh. **Figure 16–17a** Photo by B. Kiwala. **Figure 16–17b** Photo by Harold W. Pratt/Biological Photo Service. **Figure 16–19** After Stehli, F. G. and Wells, J. H., Diversity and Age Patterns in Hermatypic Corals, *Systematic Zoology* 20, 115–126, 1971. **Figure 16–21a** Photo by Christopher Newbert. **Figure 16–21b** Photo by C. R. Wilkerson, Australian Institute of Marine Science. **Figure 16–21c** Photo by Ken Lucas/Biological Photo Service. **Figure 16–22** Photos by Christopher Newbert. **Figure 16–23** Courtesy of Deeanne Edwards, Scripps Institution of Oceanography, University of California, San Diego. **Figure 16–24** Courtesy of Terry P. Hughes, James Cook University. **Figure 16–25** After Lalli and Parsons, 1993 (Fig. 8.16). **Figure 16–28** After Ron Johnson, Old Dominion University. **Figure 16–29a** Photo courtesy of Woods Hole Oceanographic Institution. **Figure 16–29b** SEM photomicrograph courtesy of Woods Hole Oceanographic Institution. **Figure 16–29c** Photo courtesy of Robert Hessler, Scripps Institution of Oceanography, University of California, San Diego. **Figure 16–30** Courtesy of Peter A. Rona, NOAA. **Figure 16–31c** Courtesy of C. K. Paull, Scripps Institution of Oceanography, University of California, San Diego. **Figure 16–32 b and c** Courtesy of Charles R. Fisher, Pennsylvania State University. **Figure 16–33b** Photo courtesy of JAMSTEC.

Chapter 17

Opener Photo © Steven Kazlowski/Seapics.com. **Figure 17A inset** Photo courtesy of National Marine Fisheries Service. **Figure 17–3** After Wyatt, 1980. **Figure 17–4** After Pauley and Christensen, 1995. **Figure 17–5** Data from the Food and Agriculture Organization of the United Nations. **Figure 17–6** Photo by W. High, courtesy of National Marine Fisheries Service. **Figure 17–8** Data from the Food and Agriculture Organization of the United Nations. **Figure 17–9** Data from the Food and Agriculture Organization of the United Na-

tions. **Figure 17–10** Photo by Mark C. Burnett, courtesy Photo Researchers, Inc. **Figure 17–11** Photo by M. Grant Gross. **Figure 17–10** APT photo. **Figure 17–11a** Courtesy of Woods Hole Oceanographic Institution. **Figure 17–11b** Photo courtesy of Michel Brigaud/La Mediatheque EDF, Boulogne, France. **Figure 17–12** Courtesy of U.S. Department of Energy. **Figure 17–13** Courtesy of Wavegen. **Figure 17–14** Map constructed from data in U.S. Navy Summary of Synoptic Meteorological Observations (SSMO). Adapted from Sea Secrets, *Sea Frontiers* 33:4, 260–261, International Oceanographic Foundation, 1987. **Figure 17–15** Photo courtesy of Phototeque/Electricite de France. **Figure 17–18** Photo by Earl Roberge, courtesy Photo Researchers, Inc. **Figure 17–19** Courtesy of GEOMAR Research Center, Kiel, Germany. **Figure 17–21** Courtesy of Deep Sea Ventures, Inc. **Figure 17–22** After Cronan, D. S. 1977. Deep sea nodules: Distribution and geochemistry, *in* Glasby, G. P., ed., *Marine Manganese Deposits*, Elsevier Scientific Publishing Co. **Figure 17–25** Courtesy of Patricia Deen, Palomar College. **Figure 17–27** Photo taken by Dr. Russell G. Kerr, Professor of Chemistry and Director of the Center of Excellence in Biomedical and Marine Biotechnology at Florida Atlantic University. **Table 17–1** Source: After Audubon's Living Oceans/Wildlife Conservation Society Special Edition Seafood Wallet Card.

Chapter 18

Opener Photo courtesy of AP/Wide World Photos. **Figure 18A upper** Natalie Forbes/NGS Image Collection. **Figure 18A lower** Photo courtesy of NOAA. **Figure 18B** Bob Jordan/AP/Wide World Photos. **Figure 18C** © Tony Freeman/ Photo Edit. **Figure 18D** APT photo. **Figure 18–1** Photo courtesy of NOAA. **Figure 18–2** After Mitchell, J. G., In the wake of the spill: Ten years after the *Exxon Valdez*, *National Geographic* 195:3, 96–117, 1999. **Figure 18–3** Gary Braasch/CORBIS. **Figure 18–4** Data after U.S. National Research Council, *Oil in the Sea III: Inputs, Fates, and Effects*, 2003, The National Academy of Sciences. **Figure 18–6b** Photo courtesy of U.S. EPA/EMSL. **Figure 18–7** Courtesy of A. Crosby Longwell. **Figure 18–9** Courtesy of John Trever, Albuquerque Journal. **Figure 18–10** AP/World Wide Photos. **Figure 18–12a** Courtesy of Massachusetts Water Resources Authority. **Figure 18–12b** Courtesy of U.S. Geological Survey, Woods Hole, Massachusetts. **Figure 18–13** Photo by Thomas D. Mangelsen, courtesy Peter Arnold, Inc. **Figure 18–14** Minamata, Tokomo is bathed by her mother, ca 1972. Photograph by W. Eugene Smith. Collection, Center for Creative Photography, The University of Arizona. © Aileen M. Smith, Courtesy Black Star, Inc., New York. **Figure 18–16** APT photo. **Figure 18–17a** Photo by Wayne Perryman, National Marine Fisheries Service. **Figure 18–17b** © Joe McDonald. **Figure 18–18** APT photo. **Figure 18–19** Courtesy of the American Geophysical Union. **Figure 18–20** Image from Reuters NewMedia Inc./Corbis. **Figure 18–21** Courtesy of the American Geophysical Union. **Figure 18–22** After NOAA/National Ocean Service at http://mpa.gov/mpaservices/atlas/fig1_nmsmap.html. **Figure 18–23** Reprinted with permission of the Ocean Conservancy (formerly the Center for Marine Conservation). **Table 18A** Data from The Oil Spill Intelligence Report at http://www.cutter.com/osir/biglist.htm.

Afterword
Figure Aft–1 APT photo.

Appendices
AII Reprinted with permission of Hubbard Scientific, Inc. Physiographic Chart of the Sea Floor © Hubbard Scientific. **Figure A5A** APT photo. **Figure A5–1** APT photo.

Index

60-foot rule, 272–275

A
Abalone, 460
ABE. *See* Autonomous Benthic
    Explorer, 33, 36
Abyss, The (film), 38
Abyssal clay, 144, 159, 160, 480
Abyssal hill provinces, 123
Abyssal hills, 123. *See also* Seamounts
Abyssal plains, 121–122, 123. *See also*
    Ocean floor
    volcanic peaks of, 122–123
Abyssal storms, 479–480
Abyssal zone, 391, 479
Abyssopelagic zone, 389, 390–391. *See
    also* Oceanic environment
Acidity, 181
Acmaea, 460, 462
Acorn barnacles (*Balanus*), 460, 461
Acoustic Thermometry of Ocean
    Climate (ATOC), 225–226
Action Plan for the Human
    Environment (1972), 539
Acyclovir, 515
Adhesion, 166
Adriatic Sea, 358
Aegean Sea, 358
Aerobic respiration, 56
Agassiz, Alexander, 26–27
Age of Discovery, 18, 19
Agulhas Current, 229, 236, 256, 257, 278
Agulhas Retroflection, 229, 256
Air masses, 207, 208
Alaska, oil spill, 520, 525, 526
Alaskan Current, 236, 248, 365
Albedo, 194
Albert, Prince I, 27
Aleutian Islands, 93, 365
Aleutian Trench, 93, 96
Alexander the Great, 16, 31, 35
Algae, 372, 460. *See also* Marine
    organisms; Plankton
    biogenous sediment and, 145
    brown, 403–404, 465
    coral/algal symbiosis, 474–476
    distribution, 21
    golden, 404
    green, 403, 404
    heterotrophs and, 55
    macroscopic, 403–404
    mariculture of, 502, 503
    microscopic, 404–405
    organic matter synthesis, 53, 54
    Pfiesteria outbreak and, 408–409
    primary production and, 396
    red, 403, 404, 461, 464
    vs. seaweed, 376
Algal blooms, 355
Algin, 502
Alkaline, 181
Alluvial plains, 333
Alvin, 31, 33, 34, 35, 36, 127, 364,
    480, 482
Amazon River, 312
Amino acids, 53
Amnesic shellfish poisoning, 406
Amphidromic points, 308–309
Amphipod (*Orchestoidea*), 466,
    467–468
Anaerobic bacteria, 55
Anchovies, 248, 422, 440
    management of, 498–499
Andes Mountains, 85, 91, 125
Andesite, 91, 93
Andrew, Hurricane, 210, 214
Andromeda galaxy, 41–42
Anemones, 476

Angle of incidence, 195
Anglerfish (*Lasiognathus saccostoma*),
    377, 391, 436
Anhydrite, 154
Animals. *See also* Marine organisms
    origins, 57
Animalia, 372, 373
Annelid worms, 467
Anoxic, 355
Antarctic Bottom Water, 258, 259
Antarctic Circle, 197
Antarctic circulation, 242–244
    wind drift, 243–244
Antarctic Circumpolar Current, 234, 236,
    242–243, 244, 247, 248, 256, 257
Antarctic Convergence, 242, 243, 244,
    259
Antarctic Divergence, 243, 244
Antarctic Intermediate Water, 258, 259
Antarctic Ocean, 10
    circulation
        Antarctic Circumpolar Current,
            242–243, 244
        Antarctic Divergence, 243, 244
        East Wind Drift, 243–244
        in general, 242
        West Wind Drift, 242–243, 244
    diatom productivity, 410
    sea ice near, 217
    shelf ice, 219
Antarctic Polar Front, 242
Antarctic Subpolar Current, 243
Antarctica, 20
Anthocyrtidium ophirense, 428
Anticyclonic flow, 206
Antifreeze, 190
Antillean Chain, 361
Antilles Current, 244, 361
Antinodes, 284
Aphelion, 306
Aphotic zone, 389
Apogee, 306
Appalachian Mountains, 67
Appendages, 380, 429
Aqualung, 32, 35
Aquarius, 37, 38
Arabian Sea, 367–368
Aragonite, 154
Archaea, 372, 482, 484, 486
Archaeon, 129, 377
Arctic Circle, 197
Arctic Convergence, 244, 259
Arctic Ocean, 10, 107
    exploration of, 24
    productivity in, 410
    sea ice in, 217
Arenicola spp., 466, 467
Argo Merchant, 523–524
Argo-Jason, 33, 36
Argon, 180, 197
Aristotle, 32, 35
Arrowworm, 432
Aspect ratio, 433
Asthenosphere, 79, 80, 83
Atlantic comb jelly (*Mnemiopsis
    leidyi*), 535
Atlantic cordgrass (*Spartina
    alterniflora*), 535
Atlantic Equatorial Countercurrent, 244
Atlantic menhaden (*Brevoortia
    tyrannus*), 408
Atlantic Ocean, 10, 107
    abyssal clay, 160
    age of floor, 75
    circulation, 244–2473
        climatic effects, 247
        Gulf Stream, 244–247
        North and South Atlantic Gyres, 244

coasts, 338
    Atlantic-type, 338
    erosion, 338
downwelling, 241
exploration of, 17–18
marginal seas, 358–363
    Caribbean Sea, 361
    Gulf of Mexico, 361–363
    Mediterranean Sea, 358–361
    pelagic sediment, 160
Atlantis II Deep, 367
Atmosphere. *See also* Gases; Weather
    atmosphere-ocean interactions, 194
    cycling and mass balance, 50–51
    early, 53
    effect of El Ninos, 254
    gas exchange with ocean, 180–181
    greenhouse effect, 219–226
    insulating effects, 48
    oceans and, 48–49
    origin of, 46–47
    physical properties, 197–199
        composition, 197–198
        density, 198
        movement, 199
        pressure, 198–199
        temperature, 198
        water vapor content, 198
    as reservoir, 50
    tides, 300
Atmospheric circulation, 199–200. *See
    also* Ocean circulation
    atmosphere-ocean interaction, 194
    Coriolis effect and, 203
Atmospheric waves, 265
ATOC. *See* Acoustic Thermometry of
    Ocean Climate
Atoll, 101, 102
Atom, 164. *See also* Chemistry
Atomic fusion, 46
Atomic mass, 164
Autonomous Benthic Explorer (ABE),
    33, 36
Autonomous underwater vehicles
    (AUVs), 33, 34, 36
Autotrophs, 54–57, 414
Autumnal equinox, 196
AUVs. *See* Autonomous underwater
    vehicles
Avicennia, 403

B
Back-arc spreading centers, 124–125,
    511
    Mariana Trench and, 126
Backshore, 320
Backwash, 321
Bacon, Sir Francis, 66
Bacteria
    anaerobic, 55
    bacterioplankton, 374
    coccoidal, 56
    fermenting, 54
    general, 372
    life origins and, 55–57
    nitrogen-fixing, 418
    sulfur-oxidizing, 55–56, 129
Bacterioplankton, 374
Bacteriovores, 414
Baffin Bay, 21
Balaenopterids, 450
Balboa, Vasco Núñez de, 19
Baleen, 449, 450
Bangladesh, 369
Bar-built estuary, 351, 352
Barnacles, 460, 461, 462
Barotrauma, 32
Barracuda, 377, 408

Barrier coasts, 333, 334
Barrier flat, 328
Barrier islands, 327–328, 329, 330, 334,
    338, 352. *See also* Shore
Barrier reef, 100–101, 102
Barycenter, 295, 296, 297
Basalt. *See also* Oceanic crust;
    Volcanism
    composition, 92–93
    flood basalts, 123
    formation, 81
    hotspot, 95
    mid-ocean ridge, 95
    ophiolites, 98
    pillow, 98, 99, 127, 128
    serpentine and, 125
Base, 181
Bathtub toys, drifting, 232
Bathyal zone, 391, 479
Bathymetry, 111–116. *See also* Ocean
    floor
    in general, 111
    hypsographic curve, 116–117
    marine provinces, 117–132
        continental margins, 117–121
        deep-ocean basins, 117, 121–125
        mid-ocean ridge, 117, 125–132
    techniques, 111–116
Bathypelagic zone, 389, 390–391
Bathyscaphe, 12
Bathysphere, 31, 35
Bay barrier, 327, 328, 333. *See also* Shore
Bay-mouth bar, 327
Bay of Bengal, 368
Bay of Fundy, 312, 313, 315–316
    tidal power, 507
    tides, 313, 316
Beach, 328. *See also* Coast; Shore
    compartments, 331–333
    composition and slope, 321
    defined, 321
    deposits, 140
    drowned, 335
    dunes, 328, 333, 334
    marine organisms, 466–469
        annelid worms, 467
        bivalve mollusks, 466–467
        crustaceans, 467–468
        echinoderms, 468–469
        meiofauna, 469
    movement of sand on, 321–325
    nourishment, 332, 345
    pollution, 534
    replenishment, 332–333
    sediment and, 465
    starvation, 332, 338
    summertime, 322, 324
    terminology, 320–321
        backshore, 320
        beach face, 321
        berm, 321
        foreshore, 320
        longshore bars, 321
        longshore trough, 321
        nearshore, 320–321
        offshore, 321
        shoreline, 320
        wave-cut bench, 321
    wintertime, 322, 324
Beach hoppers, 467–468
Beaufort Wind Scale, 273
Beebe, William, 31, 35
Benguela Current, 234, 236, 239, 244
Bennington, 275
Benthic animals. *See* Benthos
Benthic environment, 379, 388, 391–392.
    See also Marine environment
    deep-ocean floor

deep biosphere, 488
deep-sea hydrothermal vent
    biocommunities, 480–485
    food sources, 480
    in general, 479
    low-temperature seep
        biocommunities, 485–488
    physical environment, 479–480
    species diversity, 480
distribution of benthic biomass, 458
in general, 458
rocky shores, 459–465
    in general, 459
    high tide zone, 459, 462
    low tide zone, 459, 463–465
    middle tide zone, 459, 462–463
    spray zone, 459–462
sediment-covered shores
    in general, 465
    intertidal zonation, 466
    life in sediment, 466
    mud flats, 469
    sandy beaches, 466–469
    sediment, 465–466
shallow offshore ocean floor
    coral reefs, 473–479
    in general, 469–470
    rocky bottoms, 470–473
Benthos. *See also* Marine organisms
    epifauna, 376
    in general, 376
    hydrothermal vent biocommunities,
        376–377, 480–485
    infauna, 376
    low-temperature seep communities,
        485–488
    nektobenthos, 376
Bering Sea, 364–366
Berm, 320, 321, 324
Big Bang theory, 43
Bioaccumulation, 532
Bioerosion, 478
Biogenous ooze, 157, 160
Biogenous sediment, 139, 158
    composition, 145
    distribution, 145–153
        in general, 145–146
        neritic deposits, 146–149
        pelagic deposits, 149–153
    origin, 144–145
Biogeochemical cycling. *See also*
    Carbon dioxide; Greenhouse effect;
    Mass balance; Photosynthesis;
    Solar energy
    carbon cycle, 416, 417
    in general, 415–416
    nitrogen cycle, 416–418
    phosphorus cycle, 416, 418
    silicon cycle, 418
Biological oceanography, 3
Biological pump, 409
Biological response, levels of, 520
Bioluminescence, 390, 404, 435
Biomagnification, 532
Biomass, 374, 376, 382, 395, 426,
    458, 466
Biomass pyramid, 421, 422
Bioremediation, 525
Biotic community, 413–414
Biozones, 388
Birds, 521, 530, 533. *See also* Marine
    organisms
Birds, The (film), red tide and, 406–407
Bivalves, 406, 466–467. *See also*
    Marine organisms
    mariculture of, 502
Black smokers, 128–129, 154, 482
Blake Nose, 156
Blue moon, 304
Blue whale *(Balaenoptera musculus)*,
    410–411, 419, 445, 449, 451
Bluefin tuna *(Thunnus thynnus)*, 377,
    440, 438

Blue-gray sponge *(Niphates digitalis)*,
    477
Body size, fish, 438
Body waves, 267–268
Boiling point, 167–168, 190. *See also*
    Heat budget
Bonito *(Scomber)*, 422, 438
Bora-Bora, 102
Boston Harbor, sewage project, 528, 529
Bottlenose dolphin *(Tursiops
    truncatus)*, 377, 445, 446, 447
Boundary currents. *See also* Currents
    discussed, 233–235
    eastern, 234, 239
    subtropical gyre, 234–235
    western, 233–234, 239
Boyle, Robert, 20
Brackish water, 174, 357
Brazil Current, 234, 236, 239, 244
Breakers, 282–283
Breakwaters, 341–343
Brine, 184
Brine pools, 367
Brittle star, 392, 460
Brown algae, 403–404, 465
Brown pelicans *(Pelecanus
    occidentalis)*, 530
Brown sponge *(Angelas)*, 477
Buckshot barnacles *(Chthamalus)*, 460,
    461, 462
Buffering, 182
Bugula neritina, 515
Bull kelp *(Nereocystis)*, 470, 472
Bullard, Sir Edward, 66, 67
Buoyancy, 426–427
Burrowing, 466, 468
Bycatch, 496–497. *See also* Fisheries
Byrne, Keltie, 453

C

Cabot, John, 19
Calanus, 410
Calcareous ooze, 145, 148, 149,
    151–153, 157, 159–160
Calcite, 145, 154
Calcite (calcium carbonate) compensation
    depth (CCD), 151, 183
Calcium, 397
Calcium carbonate, 145, 383
    distribution in sediment, 152
Caledonian Mountains, 67
California
    oil and gas resources, 509
    sewage, 527
California Current, 236, 237, 239, 248
Calocalanus pavo, 429
Calorie, 167
Calving, 215, 217
    waves and, 266
Camille, Hurricane, 210
Camouflage, 386
Canary Current, 234, 236, 239, 244,
    245, 247
Cannon, Berry, 38
Cano, Juan Sebastian del, 19
Cape Cod Bay, 528
Cape Hatteras Lighthouse, relocation
    of, 344–345
Capilia vitrea, 429
Capillarity, 166
Capillary waves, 272
Capitella capiata, 523
Carbon, 51
Carbon cycle, 416, 417
Carbon dioxide. *See also*
    Biogeochemical cycling
    atmospheric, 180
    in dry air, 197
    in global warming, 224
    greenhouse effect and, 221–222
    oceans as buffer for, 183
    plants and, 57–58
    in seawater, 4, 180, 181

Carbonate buffering system, 182–183
Carbonate deposits, 146, 149
Carbonates, 154
Caribbean Bottom Water, 361
Caribbean Current, 244, 361
Caribbean Sea, 361, 362
Caribbean Surface Water, 361
Carlsberg Ridge, 121
Carnivora, 441–442
Carnivores, 414, 419
Carnivorous feeding, 466, 467
Cascade Mountains, 85, 92
Caspian Sea, 11
Caudal fins, 433
Caulerpa taxifolia, 535, 536
CCD. *See* Calcite compensation depth
Census of Marine Life (CoML), 30
Centripetal forces, 295–297, 298, 299
Cetacea, 442–451. *See also* Whales
    modifications to allow deep diving,
        446–447
    modifications to increase swimming
        speed, 442, 446
    suborder Mysticeti, 447, 449–451
    suborder Odontoceti, 447–449
CFCs. *See* Chlorofluorocarbons
Chaetognaths, 432
Chain, 367
Chalk, 145, 149
Challenger Deep region, 12, 124
Chauliodus sloani, 436
Chelonia mydas, 72
Chemical bonds. *See also* Hydrogen
    bonds
    covalent, 165
    ion, 164
    ionic, 166
Chemical oceanography, 3
Chemical resources
    drugs from the sea, 515
    evaporative salts, 514–515
    fresh water, 513–514
Chemical weathering, 40
Chemistry
    atom
        atomic mass, 164
        discussed, 164
        electron, 164
        isotope, 52
        neutron, 164
        nucleus, 164
        proton, 164
    chemical bonds, 165–166
    molecule, 164–167
Chemosynthesis, 55, 395, 482, 483
Chesapeake Bay, 352
Chesapeake Bay Estuary, 354–355
Chiasmodon niger, 436
Chicxulub Crater, 156
Chientang River, 312
Chignecto Bay, 313
China
    compass origins, 20
    ocean exploration, 18
Chiton *(Nuttalina)*, 460
Chloride ion, salinity and, 49
Chlorine, 397
Chlorinity, 177
Chlorofluorocarbons (CFCs), 221, 222,
    233, 537
Chlorophyll, 55
Chondrites, 155
Chronometer, 8, 15
Chromosomes, 57
Circular orbital motion, 268–269
Circulation cells
    boundaries, 203–204
    idealized or real?, 204
    in general, 203
    pressure, 203
    wind belts, 203
Circulatory system modifications, 439
Clams, 129, 406, 466, 476, 482

Clathrates, 510
Clay, 139, 158
    abyssal, 144, 159, 160, 480
Clean Water Act (1972), 527
Cliffs, 326, 344
Climate. *See also* Monsoon;
    Precipitation; Weather
    continental drift and, 67–68
    effects of North Atlantic currents, 247
    in general, 205
    heat budget effects, 220
    latitude and, 68
    ocean currents and, 239
    oceanic patterns, 214–215
    ocean's regions, 215
    solar radiation and, 196–197
    thermostatic effects, 170, 171
Climate Variability and Predictability
    (CLIVAR), 30
Cline, Isaac, 213–214
CLIVAR. *See* Climate Variability and
    Predictability
Closed system, 59
Cnidarians, 427
Coast, 320. *See also* Shore
    characteristics of U.S., 337–340
    classification, 333–335
        primary, 333, 334
        secondary, 333, 334
        tectonic and isostatic movements,
            333, 335
    coastal tide currents, 316–317
        reversing current, 316
Coastal development. *See also* Human
    activity
    artificial barriers, 340–344
        breakwater, 341–343
        groin, 340–341, 345
        jetty, 341
        seawall, 343, 344, 345
    discussed, 340–344
Coastal geostrophic currents, 350–351
Coastal nations jurisdiction, 492
Coastal plain estuary, 351
Coastal region, 320–325
    beach composition and slope, 321
    beach terminology, 320–321
    movement of sand on beach, 321–325
Coastal upwelling, 241–242, 398,
    399, 411
Coastal waters. *See also* Estuaries;
    Marginal seas
    circulation
        geostrophic currents, 350–351
        salinity, 349, 350
        temperature, 349–350
    coastal wetlands, 355–357
    estuaries, 351–355
        origin, 351–352
        water mixing in, 352–353
    in general, 349
    lagoons, 357–358
        Laguna Madre, 357–358
    marginal seas, 358–369
        Atlantic Ocean, 358–363
        Indian Ocean, 366–369
        Pacific Ocean, 364–366
    wetlands
        loss of, 355–357
        mangrove swamp, 355, 356
        salt marsh, 328, 355, 356, 465
Coccoidal bacteria, 56
Coccolith ooze, 149
Coccolithophores, 145, 148, 383, 404
Coccoliths, 145, 404, 405
Cockle *(Cardium)*, 467
Cod, 497, 500
Codium fragile, 403, 404
Coelenterates. *See* Jellyfish; Sea
    anemones
Cohesion, 165
Cold front, 207, 208
Cold-blooded, 438–439

Cold-core rings, 246–247
Colonies, 57
Color
　of objects, 401
　of ocean water, 401
Colorado River, 364
Coloration, disruptive, 453
Columbia River Estuary, 354
Columbus, Christopher, 16, 18–19
Comets, 47
CoML. See Census of Marine Life
Commensalism, 414–415
Compass, 20
Computers, use in oceanography, 29
Condensation, latent heat of, 169–170
Condensation point, 168
Conservative constituents, 180
Conshelf, 34
Constructive interference, 276–277, 278
Consumers, 414
Continental accretion, 102
Continental arc, 91, 124
Continental borderland, 118
Continental crust. See also Earth;
　　Oceanic crust; Plate tectonics
　composition, 79, 82
　density, 46, 84
Continental drift. See Plate tectonics
Continental floor basalts, 123
Continental margins, 117–121, 239. See
　　also Marine provinces
　active, 117–118, 119
　continental rise, 120–121
　continental shelf, 118–119
　continental slope, 117, 119–120, 144
　in general, 117
　　Atlantic-type, 338
　　Pacific-type, 338
　passive, 117–118, 119
　submarine canyons, 120
　turbidity currents, 120
Continental rise, 117, 120–121
Continental separation, 102–103
Continental shelf, 117, 118–119
　benthic species, 458
　Bering Sea, 365
　deposits, 140
　shelf break, 118
Continental Shelf Station, 34
Continental slope, 117, 119–120, 144
Continental transform fault, 94
Continental-continental convergence,
　91, 93
Continental crust, 46, 47, 79, 80
Continents. See also Plate tectonics
　effects on tides, 309
　fit of, 65–66, 67
　vs. oceans, 11–12
　as source of marine nutrients, 397
Convection cell. See also Volcanism
　atmospheric, 198, 200
　sea floor spreading, 74
Convergent active margins, 117
Convergent boundaries, 511. See also
　　Plate boundaries
Conveyer-belt circulation, 259–261
Cook Inlet, 312
Cook, Captain James, 20–21
Copepods, 389, 410, 427, 429
Copernicus, 20
Copper, 397
　ophiolites and, 99–100
Coral crab, 465
Coral reefs, 333. See also Marine
　　organisms
　beach and, 321
　coral/algal symbiosis, 473, 474–476, 477
　coast, 335
　debris, 158
　development of, 100–102
　distribution and diversity, 474
　effect of El Nino, 254, 255
　hermatypic corals, 474

importance of, 476
in general, 473–474
mixotrophs, 474
productivity of, 411
threats to
　bioerosion, 478
　coral bleaching, 473, 475
　crown-of-thorns phenomenon,
　　478–479
　diseases, 475
　nutrient increases, 476, 478
　zonation, 476
　zooxanthellae, 473, 474, 475, 477
Core, earth, 46, 47, 79, 80
Cores, 136, 138
　of sediment, 135
Coriolis effect. See also Currents;
　　Ekman spiral; Upwelling; Wind
　in Chesapeake Bay, 354
　effects
　　atmospheric circulation, 203
　　hurricanes, 209
　　ocean currents, 351
　in general, 200–203
　gravity and, 237
　latitude and, 202
　western intensification and, 237–238
Corona Borealis, 42
Cosmas, 17
Cosmogenous sediment, 139, 155, 157
Costa Rica Rift, 98
Cotidal lines, 308–309
Countershading, 386
Cousteau, Jacques-Yves, 32, 34, 35
Covalent bonds, 165
Crabs, 464, 465, 476, 482
Crests, 268
Crown-of-thorns phenomenon, 478–479
Cruisers, 438
Crust. See Continental crust; Oceanic crust
Crustacea, 427
Crustaceans, 467–468. See also
　　Lobsters; Marine organisms
　mariculture of, 502
Crystalline rocks, 50, 51
Ctenophores, 431, 432
Curie point, 71
Currents. See also Coriolis effect;
　　Gyres; Ocean circulation; Surface
　　currents; Turbidity currents;
　　specific currents
　coastal
　　geostrophic, 350–351
　　tide currents, 316–317
　deep, 257–261
　ebb, 316
　equatorial, 23, 244
　geostrophic, 237, 238
　Gulf Stream, 25–26
　longshore, 322, 325
　in general, 229, 230
　measuring, 230–233
　power generation, 503, 504
　reversing, 316
　rip, 322, 323
　seawater density and, 187–188
　turbidity, 120, 122
Cyanobacteria, 56, 149, 421
Cycling, 49–50
Cyclones, 209–214
Cyclonic flow, 206
Cyprus, 99–100, 129

D
Dams, 331
Dark Ages, 43
Darwin, Charles, 23, 100
Dating, 58–61. See also
　　Paleomagnetism
Davidson Current, 351
DDT
　　(dichlorodiphenyltrichloroethane),
　　530–531

Dead Sea, 175
Dead whale hypothesis, 485
Dead zones, 535–537
Death Valley, 155
Decay distance, 276
Deccan Traps, 156
Declination, 196
　of moon and sun, 305–306
Decomposers, 414
Decomposition, on ocean floor, 481
Decompression illness, 32
Deep biosphere, 488
Deep currents, 230
Deep-ocean basin. See also Ocean
　　basins
　abyssal plains, 121–122
　　volcanic peaks of, 122–123
　back-arc spreading centers, 124–125
　flood basalt, 123
　in general, 117
　oceanic ridges and rises, 88, 90, 129
　ocean trenches, 124
　reefs, 100–102
　subduction zone, 74
　volcanoes, 90–91
Deep-ocean floor. See also Ocean floor
　deep biosphere, 488
　deep-sea hydrothermal vent
　　biocommunities, 480–485
　food sources, 480
　in general, 479
　low-temperature seep biocommunities,
　　485–488
　physical environment, 479–480
　species diversity, 480
Deep-ocean mineral resources, 493
Deep scattering layer (DSL), 389
Deep-sea clamm, 21
Deep Sea Drilling Project (DSDP), 29,
　98, 136
Deep-sea fans, 120, 121
Deep time, 61
Deep water, 189
　dissolved oxygen in, 261
　nekton, 435–437
　sources of, 258–259
　waves, 269–271
　worldwide circulation, 259–261
Deltas, 328, 330–331, 333. See also
　　Shore
Denitrifying bacteria, 418
Density
　atmospheric, 198
　earth layers, 46
　ice, 170–171
　pycnocline, 188–189
　seawater, 187–189, 190
　water, 170–173
　waves and water, 265
Deposit feeding, 466, 467
Deposition, U.S. coast, 337, 340
Depositional-type shores, 325
Depth, of oceans, 11–12
Depth measurements, 21, 22
Depth salinity variation, 185, 187
Desalination, 513–514
Desiccation, 379
Destruction, 146
Destructive interference, 277
Detritus, 390, 414, 415, 435
Detroit Seamount, 96
Diatomaceous earth, 145, 147, 148, 404
Diatomaceous ooze, 149
Diatoms. See also Marine organisms
　in Arctic Ocean, 410
　in general, 136, 146, 147–148, 151,
　　381, 404, 405
　oozes, 145, 148
　silicon and, 418
Diaz, Bartholemeu, 18
Dichlorodiphenyltrichloroethane
　　(DDT), 530–531
Diffusion, 383

Dilution, 146
Dinoflagellates
　in general, 404–405
　Pfiesteria and, 408–409
　red tide and, 406–407
Dinosaurs, extinction of, 156–157
Dipolar, 165
Discovery Deep basin, 367
Disphotic zone, 389
Disruptive coloration, 387, 453
Distillation, 513
Distributaries, 330
Disturbing force, 265
Dittmar, William, 22, 177
Diurnal inequalities, 310
Diurnal tidal pattern, 309, 311
Divergent boundaries. See Plate
　　boundaries
Diving, 32
Diving bell, 32, 35
DNA, 53
Doldrums, 203–204, 206, 214. See also
　　Wind
Dolphins, 72, 442, 445, 446, 447
　as incidental catch, 496–497
Doppler flow meter, 230
Dorsal fins, 433
Dot, Hurricane, 212
Downwelling. See also Currents;
　　Upwelling
　Atlantic Ocean, 241
　coastal, 241–242
　discussed, 239–241
　El Nino and, 250
Drake, Sir Francis, 19, 243
Drake Passage, 243
Dredge/dredging, 27, 136, 343, 479
Drift meters, 230–232
Driftnets, 497. See also Fisheries
Droughts, 222
　El Nino and, 254, 255–256
Drowned beaches, 335. See also Beach;
　　Shore
Drowned glacial-erosion coasts, 333,
　334, 335
Drowned river valleys, 335, 339, 351.
　　See also Estuaries
Drugs, from sea, 515
Drumlins, 333
DSDP. See Deep Sea Drilling Project
DSL. See Deep scattering layer
Dugongs, 442, 443
Dunes, 328, 333, 334
Dynamic tide theory, 308–309. See also
　　Tides

E
Earth. See also Continental crust;
　　Mantle; Oceanic crust; Plate
　　tectonics
　atmosphere and oceans origin, 46–47
　changes through time, 102–103
　changes to environment, 57–58
　circumference of, 16
　composition
　　chemical, 79
　　in general, 46
　　core, 46, 47
　　crust, 46, 47
　　　tectonic and isostatic movement,
　　　335–336
　dating techniques, 61
　internal structure, 46, 47
　magnetic field, 70–73
　nonspinning, 199–200
　orbit
　　aphelion, 306
　　perihelion, 306
　origin, 43–46
　rotation of, 48, 302
　seasons, 195–197
　size, 301
　structure

asthenosphere, 83
inner structure, 80–81
isostatic adjustment, 83–84
lithosphere, 81–82
near the surface, 81–83
physical properties, 79–80
tidal influence, 302–304
universe and solar system, 41, 43–46
Earth tides, 300. *See also* Tides
Earthquakes. *See also* Seismic waves;
Volcanism
convergent boundaries, 93–94
in general, 80–81
Grand Banks, 122
plate boundaries and, 84, 88–89
seismic waves and, 80
transform faults and fracture zones
and, 130–131
turbidity currents, 120, 122
worldwide, 77–78
East African Rift Valleys, 87, 89, 107
East Australian Current, 236, 248
East Greenland Current, 217, 236, 241,
244
East Pacific Rise, 88, 89, 128, 129, 364,
482, 485
East Wind Drift, 243–244
Easter Island, 13
Eastern boundary currents, 234, 239
Ebb current, 316
Ebb tide, 302
Echinoderms, 468–469
Echo sounder, 28, 111–112
Echolocation, 447–448, 449
Ecliptic, 305
Ecosystem, 413
Ecosystem-based fishery management,
499
Eddies, 237, 246
Eelgrass *(Zostera)*, 372, 402, 469, 470
EEZ. *See* Exclusive economic zone
Ekman, V. Walfrid, 24, 235
Ekman spiral, 24, 235–236, 237. *See
also* Coriolis effect
Ekman transport and, 236, 237
Ekman transport, 236–237, 248, 398,
399. *See also* Wind
Ekman spiral and, 236, 237
El Nino-Southern Oscillation (ENSO)
events. *See also* Weather
cool phase (La Nina), 249, 250–251, 253
coral reefs and, 473, 475
effect on anchovies, 248, 498–499
effects of, 194, 253–254
ENSO index, 250, 253
examples, 254–256
frequency of, 253
predicting, 256
warm phase (El Nino), 239, 249, 250,
251, 252
Electrical condition, water, 177
Electrolysis, 514
Electromagnetic spectrum, 398,
400–401
Electron, 164
Electrostatic attraction, 166
Elements. *See also* Chemistry;
individual elements
mass balance, 50–51
Eleutherobin, 515
Elevation, atmospheric pressure and, 199
Elliptical orbits, 306
Emerging shorelines, 335–337. *See also*
Shore
Energy flow. *See also* Heat budget
in marine ecosystems, 413–414
symbiosis, 414–415
Energy resources
from currents, 503, 504
in general, 503
from offshore winds, 503
from thermal energy, 507–509
from tides, 505–507

from waves, 503–505
ENSO. *See* El Nino-Southern
Oscillation events
Entropy, 413
EPA. *See* U.S. Environmental
Protection Agency
Epifauna, 376, 459
Epipelagic zone, 389–390
Equator
Coriolis effect and, 202
diverging surface water, 241
region, 214
tides, 307
upwelling, 241, 411
Equatorial Countercurrent, 236, 239,
248, 256, 257
Equatorial Current, 23, 244
Equatorial latitudes, 197
Equatorial low, 203
Equatorial Undercurrent, 248
Equilibrium tides, 300–309
prediction of, 306–309
Eratosthenes, 16
Erik the Red, 17
Erikson, Leif, 17
Erosion. *See also* Sediment
effect of El Nino, 254, 255
shoreline
conditions in U.S., 337, 338, 340
subaerial, 333, 335
wave erosion, 334, 335
weathering, 136, 140
Erosional-type shores, 325–327
Estuaries. *See also* Coastal waters; Shore
circulation, 352
defined, 351
human activity and, 354–355
origin of, 351–352
water mixing in, 352–353
Eukaryotic cells, 57
Euphotic zone, 389
Euphysetta elegans, 428
Europa, 48–49
European green crab *(Carcinus
maenas),* 535
Euryhaline organism, 383
Eurythermal organism, 382
Eustatic changes, 336
Eutrophic area, 401
Eutrophication, 407, 535–537
Evaporation. *See also* Biogeochemical
cycling; Heat budget
affecting seawater salinity, 184, 185
latent heat of, 169
Evaporation latitudes, 171
Evaporative salts, 514–515
Evaporite minerals, 154, 155. *See also*
Minerals
Evaporites, 179
Evolution, 23, 57
Excess volatiles, 50, 51
Exclusive economic zone (EEZ), 492,
493, 539
Exxon Valdez, 520, 525, 526
Eye of the hurricane, 210, 211

F
Falcate, 447
Falkland Current, 236, 244
Fall bloom, 413
Fast ice, 217. *See also* Ice formation
Fastnet Race disaster, 279–280
Fathom, 111
Fault, 94, 285. *See also* Plate
boundaries; Seismic waves
Fecal pellets, 160–161
Federal Water Pollution Control Act
(1948), 539
Feeding relationships, 418–422
biomass pyramid, 421, 422
food chains, 420–421
food webs, 421
in general, 418

microbes in marine environment,
421–422
transfer efficiency, 419–420
trophic levels, 419
Ferrel cell, 203, 205
Fetch, 272
Fiddler crabs *(Uca),* 469, 470
Fiji, 13
Filter feeding, 439, 466, 467
Fin designs, 433–435
Finches, 23
Fire Island, 328
Fish. *See also* Marine organisms; Nekton
body size, 438
circulatory system modifications, 439
cold-blooded vs. warm-blooded,
438–439
deep-sea, 435–437
distribution, 376, 402
fins, 433–435
mariculture of, 500–501
rocky shore, 478
salinity adaptations, 384–385
schooling, 439–440
speed, 438
Fish recruitment, 494
Fisheries, 349, 493. *See also* Mariculture
fish recruitment and survival, 494
incidental catch, 496–497
management, 497–499
primary productivity effect on, 494
seafood choices, 499–500
world fishery, 494–495
world total fish production, 495
Fishing vessels, regulation of, 497
Fissures, 127
Fitzroy, Robert, 23
Fjord, 334, 333, 351, 352, 353. *See also*
Estuaries
Flagella, 404
Flood basalts, 123. *See also* Basalt
Flood current, 316
Flood tide, 302
Flooding, El Nino and, 256
Florida, 522–523
Florida Current, 236, 244, 245, 361
Florida Escarpment, 486
Florida-Gulf Stream Current System, 503
Flow meters, 230, 231
Fluke, 442
Flying fish *(Exocoetus),* 433, 435
Folger, Timothy, 25–26
Food chains, 420–421
Foods webs, 421
Foraminifer ooze, 149
Foraminifers, 145, 148, 383, 427, 428
Forbes, Edward, 21
Forced standing wave, 313
Forced waves, 277
Fore-arc, 125
Foreshore, 320
Forked fins, 433, 434
Fossils, evidence for continental drift
in, 67–68
Fracture zones, 130–132. *See also* Plate
boundaries
Fram, 24–25
Franklin, Benjamin, 25–26
Free oxygen, 53
Free standing wave, 313
Free waves, 277
Freeze separation, 514
Freezing, latent heat of, 170
Freezing point, 167, 190. *See also*
Ice; Water
Frequency, of wave, 268
Freshwater zone, 357
Fringing reefs, 100, 101
Fucus, 461, 462, 472. *See also* Algae
Full moon, 302

Fully developed sea, 275
Fungi, 372, 373
Fur seals, 441
Fusion reaction, 43

G
Gabbro, 98, 99
Gagan, émile, 32, 35
Galápagos Islands, 23, 95, 252
Galápagos Rift, 364, 376, 480, 482
Galatheid crab *(Munidopsis marianica),*
484
Galaxies, 40, 41–42
movement of, 42–43
Galileo, 20
Galveston (Texas) hurricane, 213–214
Gama, Vasco de, 18
Gas chambers, 429
Gas containers, 426–427
Gas hydrates, 510, 511
Gaseous state, 167, 168, 170, 173
Gases. *See also* Atmosphere; *specific
gases*
in early atmosphere, 46
latent heat of vaporization, 169, 173
in seawater, 175
conservative tracers, 233
dissolved gases, 180–181, 385
volcanic, 48, 49, 50
Gastropod (Alviniconcha hessleri), 484
Geochemical balances, 50
Geochemical Ocean Sections
(GEOSECS), 29
Geographical Equator, 241
Geologic resources
gas hydrates, 510, 511
manganese nodules and crusts, 511–513
metal sulfides, 511
petroleum, 509–510, 511
phosphorite, 511
sand and gravel, 510–511
Geologic time scale, 59–61
Geological oceanography, 3
George Washington DeLong, 24
Georges Bank, 497, 499, 524
Geosat, 30, 115
GEOSECS. *See* Geochemical Ocean
Sections
Geostrophic currents, 237, 238. *See
also* Currents
coastal, 350–351
westward intensification, 237–238
Ghost Head Nebula, 45
Giant brown bladder kelp
(Macrocystis), 470, 471, 472
Gill nets, 497
Gills, 385
Glacial ages, 67–68
Glacial-borne sediment, 136, 137, 141,
412
Glacial ice, 335
Glacier Bay National Park, 64
Glaciers. *See also* Ice
deposits, 142
effects, estuaries, 352, 353
evidence for continental drift, 67–68
isostatic rebound and, 335
sea level changes and, 336
Global Positioning System (GPS), 15
Global warming. *See also* Greenhouse
effect; Temperature
changes as result of, 222–223
El Nino-Southern Oscillation and, 253
in general, 219
international agreements, 223–224
Globigerina ooze, 149
Glomar Challenger, 136
GLORIA (Geological Long-Range
Inclined Acoustical instrument),
112, 113
Goiters, 176
Gold, 511
Golden algae, 404

Gondwanaland, 103–106
Goose-necked barnacles *(Pollicipes)*, 460, 461, 462
Gorky, Hurricane, 212
GPS. *See* Global Positioning System
GRACE, 30
Graded bedding, 120, 142
Grain size, 138
Gran method, 396, 397
Grand Banks, 122
Granite, 82, 90–91
Gravel, 510–511
Gravity
 forces of in sun-moon system, 295–297, 298
 sediment deposits and, 141, 142
 tides and, 295
 variation in ocean surface and, 115
Gravity corer, 136
Gravity waves, 272. *See also* Waves
Gray whale *(Eschrichtius robustus)*, 445, 449, 450
 as endangered species, 452–454
 migration of, 451–452
Great Barrier Reef, 101, 476, 479
Great Salt Lake, 175
Great white sharks *Carcharodon carcharias)*, 437, 449
Greeks, ocean exploration, 13–14, 16
Green algae, 403, 404
Green sea turtles *(Chelonia mydas)*, 72
Greenhouse effect. *See also* Biogeochemical cycling; Human activity
 ATOC experiment, 225–226
 changes as result of, 222–223
 contributing gases, 221–222
 earth temperature, and, 194
 efforts regarding increasing gases, 224
 in general, 48, 219–221
 industrialization and, 58
 Kyoto Protocol, 223–224
 ocean's role in reducing, 224
 sea level and, 336–337
Greenland, 24
Groins, 340–341, 345
Gross ecological efficiency, 419
Gross primary production, 396
Grotius, Hugo, 492
Grouper, 408, 438
Grunions *(Leuresthes tenuis)*, 314–315
Guaymas Basin, 364, 482–483
Guiana Current, 361
Guinea Current, 244
Gulf coast, 338
Gulf of Aden, 366
Gulf of Bothnia, 335
Gulf of California, 87, 364, 365
Gulf of Mexico
 circulation patterns, 361, 363
 dead zone, 536–537
 in general, 361
 hydrocarbon seeps, 486–487
 oil blowout, 521
 water masses, 363
Gulf Stream, 25–26, 234, 236, 239, 241, 244–247, 259, 458
Gulfs, tides in, 313
Gulper, 377
Gulper eel, 436
Gurnards, 434, 435
Guyots, 96–97, 123. *See also* Seamounts
Gypsum, 154, 514
Gyres. *See also* Currents; Ocean circulation
 geostrophic currents and, 237
 North Atlantic, 234–235, 236, 244, 245
 South Atlantic, 234–235, 236, 244
 South Pacific, 234–235, 236, 248
 subpolar gyre, 235
 subtropical gyre, 234–235, 236, 239

H
HAB. *See* Harmful algal bloom
Habitat destruction, coastal population increase and, 538
Hadal zone, 391, 479
Hadley cells, 203
Half-life, 59
Halibut, 386
Halite, 154, 514–515
Halocline, 187, 349, 352, 355. *See also* Salinity
Hansa Carrier, 231
Hard stabilization
 alternatives to, 343–344
 breakwaters, 341–343
 in general, 340
 groins and groin fields, 340–341
 jetties, 341
 seawalls, 343, 344
Harmful algal bloom (HAB), 405
Harmonic analysis, 309
Harrison, John, 8, 15
Hatchet fish, 436
Hawaiian Islands, 13, 21
 effect of El Nino, 255
 hotspots, 94, 95–96, 97
 hurricanes, 212
 thermal generation, 508
Hawaiian Islands-Emperor Seamount Chain, 95–96, 97
Headlands, 325–326
Heart urchins *(Echinocardium)*, 468, 469
Heat
 latent, 169–170
 phase change of water, 167
Heat budget. *See also* Boiling point; Solar energy
 climate patterns in oceans, 214–215
 in general, 220
Heat capacity, 168. *See also* Temperature; Water
Heat flow. *See also* Solar energy
 Coriolis effect and atmospheric circulation, 203
 in general, 77
 oceanic, 197
Heavy wave activity, 322
Heliocentric theory of solar system, 7
*Heliodiscus asteriscus, He*428
Hemoglobin, 446
Henry, Prince, the Navigator, 18
Hensen, Victor, 22, 24
Herbivores, 414, 419
Herjolfsson, Bjarni, 17
Hermatypic coral, 474
Hermit crabs *(Pagurus)*, 460, 463, 464
Herodotus, 9, 13
Herring, 379, 420–421. *See also* Fisheries
Herring gull *(Larus argentatus)*, 533
Hess, Harry, 74
Heterocercal fins, 433, 434
Heterotrophs, 54, 414
High slack water, 316
High tide zone, 459, 462. *See also* Shore
Highly stratified estuary, 352, 353
Hilo (Hawaii), tsunami, 287, 289
Himalaya Mountains, 93, 107, 120
Hirondella gigas, 481
HMS Beagle, 23, 100
HMS Challenger, 21–22, 26–27, 30, 111, 154, 512
Holdfast, 470
Holoplankton, 375. *See also* Plankton
Homarus americanus, 470, 472, 502
Homeothermic, 439
Hoover Dam, 364
Horse latitudes, 204, 206, 215
Hotspots, 94–96, 102. *See also* Volcanism
Hoyle, Fred, 43
Hudson Bay, 335
Human activity. *See also* Coastal development; Greenhouse effect

 estuaries and, 354–355
 stresses on oceans from, 5
Humboldt Current. *See* Peru Current
Humpback whale *(Megaptera novaeangliae)*, 445, 449, 450, 451
Hurricanes. *See also* Solar energy; Weather
 in general, 209
 historic destruction, 210, 212–214
 movement, 210
 origin, 209–210
 types of destruction, 210
Hydra, 42
Hydration, 167
Hydration spheres, 53
Hydrocarbon, 520
Hydrocarbon seep biocommunity, 486–488
Hydrogen. *See also* Gases
 activity, 182
 pH and, 182
 in water molecule, 165
Hydrogen bonds, 165, 170, 172. *See also* Chemical bonds
Hydrogenous sediment, 139, 512
 composition and distribution, 153–154
 origin, 153
Hydrologic cycle, 184–185. *See also* Biogeochemical cycling; Water cycle
Hydrothermal springs, 367
Hydrothermal vents. *See also* Volcanism
 biocommunities, 376–377, 480–485
 life span of, 485
 origin of life and, 51, 485
 black smokers, 128–129, 154, 482
 in Gulf of California, 364
 minerals, 53, 128, 154
 white smokers, 128
Hydrozoan, 427, 429
Hyerdahl, Thor, 13
Hypersaline seep biocommunity, 485–486
Hypersaline water, 174, 175
Hypertonic organisms, 384
Hypothesis, 6
Hypotonic organisms, 384
Hypsographic curve, 116–117

I
Ice. *See also* Glaciers; Water
 density of, 170–171
 extra-terrestrial, 48
 freezing point, 167
Ice age, 67–68, 336
Ice coasts, 335
Ice formation
 fast ice, 217
 in general, 171–172
 ice floes, 216
 pack ice, 217
 pancake ice, 216
 polar ice, 217
 salinity and, 172
 sea ice, 183, 184, 215–217
 shelf ice, 219
Ice rafting, 142
Ice sheets, 67
Icebergs, 183, 184, 193, 217–219
 B-15, 218, 219
 calving, 215, 217
Icebreakers, 217
Iceland, 95. *See also* Mid-Atlantic Ridge
ICESat, 30
IDOE. *See* International Decade of Ocean Exploration
Igneous rocks, 71. *See also* Minerals; Rock
Iguanas, 252
IGY. *See* International Geophysical Year
IIOE. *See* International Indian Ocean Expedition
Incidental catch, 496–497

India, collision with Asia, 93
Indian Ocean
 circulation, 256–257
 in general, 10
 marginal seas, 366–368
  Arabian Sea, 367–368
  Bay of Bengal, 368
  Red Sea, 366–367
 pelagic sediment within, 160
Indian Ocean Gyre, 234–235, 236, 256
Indus Fan, 120–121
Infauna, 376, 465
Iniki, Hurricane, 212
Inner core, 79, 80
Inner sublittoral zone, 391
Institute of Oceanographic Services, 177
Integrated Ocean Drilling Program (IODP), 136
Interdisciplinary science, oceanography as, 3
Interface waves, 267, 268
Interference patterns, 276–277
Intergovernmental Panel on Climate Change (IPCC), 223–224
Internal waves, 265, 266
International Decade of Ocean Exploration (IDOE), 29
International Geophysical Year (IGY), 29
International Ice Patrol, 218, 219
International Indian Ocean Expedition (IIOE), 29
International marine pollution control, 539–540
International Meteorological Conference, 26
International Whaling Treaty, 454
Intertidal zonation, 466
Intertidal zone, 391, 459
Intertropical Convergence Zone (ITCZ), 204
Intraplate features, 94
Iodine, 397
IODP. *See* Integrated Ocean Drilling Program
Ionic bond, 166
Ions, 164, 178
IPCC. *See* Intergovernmental Panel on Climate Change
Iridium, 156
Irminger Current, 247
Iron, 53, 397
Iron hypothesis, 224
Irregular coasts, 333
Island arc, 93, 124
Isohaline, 349
Isopycnal, 189
Isostatic adjustment, 83–84
 of Earth's crust, 335–336
Isostatic rebound, 84
Isothermal, 189, 349, 411
Isotonic, 384
Isotopes, 57
Isthmus of Panama, 103
ITCZ. *See* Intertropical Convergence Zone
Iwa, Hurricane, 212

J
Japan Current. *See* Kuroshio Current
Japan Trench, 488
Jason, 36
Jason, Jr., 317
Jason-1, 30
Jeanette, 24
Jellyfish, 379, 385–386, 429–430
Jet stream, 207, 209
Jetties, 341
JGOFS. *See* Joint Global Ocean Flux Study
JOIDES. *See* Joint Oceanographic Institutions for Deep Earth Sampling
Joint Global Ocean Flux Study

(JGOFS), 29
Joint Oceanographic Institutions for Deep Earth Sampling (JOIDES), 136
JOIDES Resolution, 136, 137
JOIDES Resolution Drill Site 977, 116
Juan de Fuca Plate, 92, 487
Juan de Fuca Ridge, 92, 95, 127, 482, 485
Jules' Undersea Lodge, 38
Jupiter, 48–49

K
Kaiko, 33, 35
Keller, Hannes, 32
Kelp, 376, 470
  beds, 469
  forests, 470, 471, 472
Kilauea, 95–96, 128
Killer whale (*Orcinus orca*), 394, 442, 445, 447, 449, 453
Kinetic energy, 167. *See also* Energy flow; Heat budget
Knot, 233
Komandorski Islands, 365
Kon Tiki, 13
Krakatau, 287
Krill, 389, 419, 430
K-T event, 156–157
Kuroshio Current, 236, 237, 239, 247–248
Kuroshio Extension Current, 248
Kyoto Protocol, 224

L
La Budde, Samuel F., 497
La Nina, 249, 250–251, 253. *See also* El Nino-Southern Oscillation events
La Rance power plant, 506, 507
Labrador Current, 236, 241, 244, 458
Lagoons, 328, 348, 352, 357–358
Laguna Madre, 357–358
Lakes, tides in, 313
Lamont-Doherty Earth Observatory, 28, 136
Land breezes, 207. *See also* Wind
Land erosion coasts, 333, 335
Lantern fish, 377, 389, 436
Lapita people, 13
Larcospira quadrangula, 428
Large igneous provinces, 123
Larsen ice shelf, 216
Latent heats, 169–170
  condensation, 169–170
  evaporation, 169
  freezing, 170
  melting, 169, 173. *See also* Water
  vaporization, 169, 173. *See also* Gases
Lateral gene transfer, 57
Latitude
  Coriolis effect and, 202
  climate and, 68
  coastal temperature and, 350
  determining, 13, 14, 15
  magnetic dip and, 70
  salinity variation and, 185–187
  solar radiation received and, 194–195
Laurasia, 103–106
Lava, 71, 123, 127. *See also* Volcanism
Law of the sea, 492–493
Laws, scientific, 6
Leeuwin Current, 236, 256, 257
Leeward and Windward Islands/Puerto Rico Trench, 93
Library of Alexandria, 13, 16
Life origin. *See also* Marine organisms
  autotrophs, 54–57
  evolution and, 57
  first organisms, 54
  in general, 51
  importance of oxygen to, 51, 53
  life on Mars, 52
  organic substrates, 53, 54
  symbiosis and multicellular life, 57
  working definition of, 51
Light transmission, 398–402. *See also*

Solar energy
  electromagnetic spectrum, 398, 400–401
Light wave activity, 322
Light-year, 42
Limestones, 149, 338
Limpet (*Acmaea*), 460, 462
LIMPET 500, 505
Linear sea, 87, 88
Linnaeus, Carolus, 373
Liquid air, 38
Liquid state, 167, 168, 170, 173
Lithogenous sediment
  composition, 138
  deposition rate, 160
  distribution, 140–144, 158
    neritic deposits, 140–142
    pelagic, 144
  in general, 135
  origin, 136–137
  sources of, 142
  texture, 135, 138–140
Lithosphere, 80
  discussed, 81–83
  earthquakes and, 78
  mid-ocean ridge and, 77
  plate tectonics and, 65
  sea floor spreading and, 74, 79
Lithothamnion, 403
Littoral zone, 391
Lituya Bay (Alaska), 288–289
Lobsters, 379, 470, 502
Log-log scale, 138
Longitude, determining, 15–16. *See also* Latitude
Longitudinal waves, 267. *See also* Waves
Longshore bars, 320, 321, 324
Longshore current, 322, 325
Longshore drift, 322, 325
Longshore transport, 322
Longshore trough, 320, 321
Loop Current, 363
Low slack water, 316
Low tide, 307, 309
Low tide terrace, 321
Low tide zone, 459, 464–465
Low-temperature seep communities, 485–488
Lucky Strike vent field, 483–484
Lugworms, 466, 467
Lunar bulge, 300
Lunar day, 301
Lunate fins, 433, 434
Lungers, 438
Lungs, 387, 388, 446
Lysocline, 151

M
Mackerel (*Sarda*), 438
Mackerel shark, 438
Macrocystis, 403
Macroplankton, 374. *See also* Plankton
Macroscopic biogenous sediment, 144
Magellan, Ferdinand, 10, 19
Magma, 81. *See also* Volcanism
  chambers, 86–87
Magnesium, 397
Magnetic anomalies, 73
Magnetic compass, 20
Magnetic dip, 70, 71
Magnetic field, 70
  navigation and, 72
Magnetic inclination. *See* Magnetic dip
Magnetic polarity reversals, 73
Magnetic sea floor stripes, 72, 74
Magnetite, 71, 72
Magnetodynamics, 70
Magnetometer, 73
Magnetoreception, 72
Malvinas Current, 244
Manatees, 442, 443
Manganese nodules, 153–154, 160, 511–513

Mangrove (*Rhizophora, Avicennia*), 403
Mangrove coasts, 333, 334, 335
Mangrove swamps, 355, 356
Manta, 434
Mantle. *See also* Earth; Plate tectonics
  chemical properties, 79, 80
  in general, 46, 47, 81–82
  isostatic adjustment, 83–84
  as source of ocean water, 49–50
Mantle plumes, 94–96
Manus Basin, 129
Maps
  bathymetric, 112, 115, 116
  cotidal world, 308
  dynamic topography, 230, 233
  early world, 9, 17
Mercator projection, 20
Mare liberum, 492. *See also* Coastal waters
Marginal seas. *See also* Coastal waters
  Arabian Sea, 367–368
  Atlantic Ocean, 358–363
  Bay of Bengal, 368
  Bering Sea, 364–366
  Caribbean Sea, 361
  in general, 358
  Gulf of California, 364
  Gulf of Mexico, 361–363
  Indian Ocean, 366–368
  Mediterranean Sea, 358–361
  Pacific Ocean, 364–366
  Red Sea, 366–367
Mariana Basin, 483
Mariana Trench, 12, 22, 92, 124, 126, 530
Mariana Trench/Island Arc subduction system, 125
Mariculture, 500–503
Marine deposition coast, 333, 335
Marine dust, 144
Marine ecology, 24
Marine environment. *See also* Marine organisms; Oceanic environment
  adaptation to 379–388
    dissolved gases, 385
    physical support, 379–380
    pressure, 387–388
    salinity, 382–385
    temperature, 381–382
    water's high temperature, 385–387
    water's viscosity, 380–381
  distribution of life in, 378–379
  divisions of, 388–392
    benthic environment, 391–392
    pelagic environment, 388–391
  in general, 372
  microbes, 421–422
Marine iguanas (*Amblyrhynchus cristatus*), 252
Marine mammals. *See also* Marine organisms
  Carnivora, 441–442
  Cetacea, 442–454
  in general, 440–441
  Sirenia, 442
Marine Mammals Protection Act, 497
Marine organisms. *See also* Coral reef; Life origin; Marine environment; Marine mammals
  algae. *See* Algae
  benthos, 376–378
  classification of, 374–378
    nekton, 375–376, 377
    plankton, 374–375
  diatoms. *See* Diatoms
  dinoflagellates, 404–409
  echinoderms, 468–469
  foraminifera, 145, 148, 383, 427, 428
  marine iguanas, 252
  oil spill effects, 521, 522, 523, 524
  organism size, 380–381
  photosynthetic, 402–405
  plankton. *See* Plankton
  predicting effects of pollution on, 520
  rocky shore. 459–465

sandy beach. *See* Beach, marine organisms
  sediment-covered shores, 466
  taxonomic classification, 372–374
Marine Protected Areas (MPAs), 539
Marine Protection, Research, and Sanctuaries Act (1972), 539
Marine provinces, 117. *See also* Continental margins; Deep-ocean basins; Mid-ocean ridges
Marine reserves, 539, 544
Marine resources
  chemical resources, 513–515
  ecosystems and fisheries, 493–500
  energy resources, 503–509
  in general, 492
  geologic resources, 509–513
  mariculture, 500–503
Marine sanctuaries, 549, 540, 544
Marine science, 3. *See also* Oceanography
Marine sediments, ophiolites and, 98
Marine terrace, 327, 335, 339
MarineLab Undersea Laboratory, 38
MARPOL Convention (1973), 539
Marquesas Islands, 13
Mars, 48, 52
Marsh grass, 333
Marshall Islands, 14
Marsupials, 69
Martin, John, 224
Mass balance, 50–51
Mass extinctions, 52
Matthews, Drummond, 74
Maturity, sediment, 140
Mauna Kea, 12
Maury, Matthew F., 26
Maverick's, 264, 265
Maxim Gorky, 219
Maximum sustainable yield (MSY), 493–494
Mean Lower Low Water (MLLW), 310
Meanders, 246
Mediterranean circulation, 360
Mediterranean Intermediate Water, 258, 359, 360
Mediterranean Sea, 107
  circulation, 358–360
  exploration of, 13, 17
  pollution in, 539
Mega Borg, 525
Megaplume, 130
Megapterids, 450
Meiofauna, 469
Melanesia, 13
Melon, 447
Melting, latent heat of, 169, 173
Mendocino Escarpment, 130
Mendocino Fracture Zone, 130
Mercator, Gerhardus, 20
Mercury, 531–533. *See also* Pollution
Meroplankton, 375, 376. *See also* Plankton
Merry-go-round, Coriolis effect, 200–201
Mesopelagic zone, 389, 390
Mesosaurus, 68–69
Mesosphere, 79, 80, 198
Metal sulfides, 128, 154, 511
Meteor, 155
Meteor, 28, 111
Meteorites
  ALH 84001, 52
  debris, 155
  early solar system, 49
  extinction of dinosaurs and, 156–157
  protoearth and, 44
Meteorological Equator, 241
Methane, 53, 221, 222
Methane hydrate, 510
Michel, Helen, 156
Microbes
  in deep biosphere, 488

in marine environment, 421–422
Microbial loop, 421
Microbial mats, 482, 486
Micronesia, 13
Microscopic biogenous sediment, 144, 146, 148
Mid-Atlantic Ridge. *See also* Oceanic ridges
 hydrothermal vents, 483–484
 ocean formation, 86, 88–89, 103
 rift valley, 88
Middle Ages, ocean exploration and, 16–20
Middle tide zone, 459, 462–463
Mid-ocean ridges
 formation of seamounts and tablemounts, 96–97
 fracture zones, 130–132
 in general, 74, 117, 125–130
 relationships, 77
 transform faults, 130–132
Milkfish, 501
Milky Way galaxy, 41, 42
Miller, Stanley, 53, 54
Minamata disease, 531, 532–533
Minas Basin, 313, 315, 316
Minerals. *See also specific minerals*
 deep-ocean, 493
 defined, 138
 evaporite, 154, 155
 in lithogenous sediment, 138
 in manganese nodules, 154, 511, 512–513
 metal sulfide, 154, 511
 phosphates, 154
 phosphorite, 511
 salt, 514–515
 sand and gravel deposits, 511
Ming Dynasty, ocean exploration during, 18
Missile paths, Coriolis effect, 201–202
Mississippi River Delta, 330, 331, 338
Mitch, Hurricane, 212
Mitochondria, 56, 57
Mix interference, 277, 278
Mixed surface layer, 188
Mixed tidal pattern, 310, 311
Mixotrophs, 474
MLLW. *See* Mean Lower Low Water
Moho, 99
Mollusks, 466–467
Monera, 372
Monsoon, 254, 256. *See also* Climate
 winds, 205, 367, 368
Monthly tidal cycle, 302–304
Moon. *See also* Solar system; Tides
 declination of, 305–306
 navigation and, 14
 orbit
  apogee, 306
  perigee, 306
  origin, 44
 phases of, 302, 304
 size, 301
 quadrature, 303
Moraines, 333, 338, 352
Moriarty, Jay, 264
Motu Iti, 102
Mount Everest, 12
Mount St. Helens, 92
Mountains, continental, 66–67
MPAs. *See* Marine Protected Areas
MSY. *See* Maximum sustainable yield
Mud flats, 333, 465, 466, 469
Mud lumps, 333
Mullet, 501
Multibeam echo sounders, 112
Munk, Walter, 225
Murray, Sir John, 26
Museau du singe, 447
Musée Océanographique, 27
Museum of Comparative Zoology, 27
Mussel *(Mytilus),* 460, 461, 462, 502

in hydrothermal vents, 129, 484
 red tide and, 406
Mutualism, 415
Myoglobin, 438
Mysticeti, 447, 449–451

N

Nadir, 296
Nannoplankton, 145
Nansen, Fridtjof, 24, 235, 236
Nansen bottle, 24
Nantucket Island, oil spill, 523–524
NAO. *See* North Atlantic Oscillation
Narwhal *(Monodon monocerus),* 445
NASA Scatterometer, 30
National Academy of Sciences, 345
National Biological Service, 29
National Flood Insurance Program (NFIP), 343
National Ocean Service, 29
National Oceanic and Atmospheric Administration (NOAA), 29, 38, 116
National Oceanographic Data Center, 29
National Park Service, 345
National Science Foundation, 28, 136
National Sea Grant Office, 29
National Undersea Research Program, 38
National (ship), 24
Natural Energy Laboratory of Hawaii Authority, 508
Natural selection, 57
Nautilus, 426
Navigation tools, 14–15, 20
Nazca Plate, 88, 117
Neap tide, 303
Nearshore, 320
Nebula, 43, 45
Nebular hypothesis, 43–44, 45
Negative tide, 310
Nektobenthos, 376
Nekton. *See also* Marine organisms
 deep-water, 435–437
 fin designs, 433–435
 in general, 375–376, 377, 432–433
 sharks, 437
Nematath, 96, 116
Nematocysts, 427, 463
Neritic deposits
 beach deposits, 140
 carbonate deposits, 146, 149
 continental shelf deposits, 140
 distribution of, 158, 159
 glacial deposits, 142
 in general, 140–142, 146
 turbidite deposits, 140, 142
Neritic province, 388
Net primary production, 396, 409
Neutral, 182
Neutrons, 164
New Carissa, 525
New Ireland, 13
New moon, 302, 303
New production, 396
New Siberian Islands, 24
New York Bight Sludge Site, 527–528
New Zealand, 13
Newfoundland, 17, 18
 fisheries, 497, 499
Newfoundland herring, 420–421
Newton, Sir Isaac, 20, 295
NFIP. *See* National Flood Insurance Program
NGC 4414 galaxy, 41
Nile River Delta, 330, 331
Nimbus 7, 30
Nitrates, 396, 397
Nitrogen
 atmospheric, 180
 diving and, 32, 446–447
 in dry air, 197
 fisheries and, 494
 as nutrient, 397

in ocean, 180
Nitrogen cycle, 416–418. *See also* Biogeochemical cycling
Nitrogen-fixing bacteria, 418
Nitrogen narcosis, 32, 446–447
NMFS. *See* U.S. National Marine Fisheries Service
NOAA. *See* National Oceanic and Atmospheric Administration
Nodes, 284
Nonconservative constituents, 180
Non-native species, 535
Non-point source pollution, 533
Nor'easters, 338. *See also* Weather
North American Plate, 94
North Atlantic Central Surface Water, 258
North Atlantic Current, 234, 236, 245, 247
North Atlantic Deep Water, 258, 259, 361, 410
North Atlantic Drift, 247
North Atlantic Gyre, 234–235, 236, 244, 245
North Atlantic Ocean, formation of, 103
North Atlantic Oscillation (NAO), 254
North Atlantic Sea, 11
North Atlantic Water, 361
North Equatorial Current, 236, 244, 245, 247, 248, 257, 367
North Pacific Current, 236, 248
North Pacific Gyre, 234–235, 236, 247–248
North Pacific Sea, 11
North Pacific Subtropical Gyre, 237
Northeast monsoon, 367
Northeast trade wind, 203, 204
Northern elephant seal *(Mirounga angustirostris),* 533
Northern Hemisphere, Coriolis effect and, 201
Northern right whale *(Eubalaena glacialis),* 444
Northwest Atlantic, fisheries, 497, 499
Norwegian Current, 247
Nucleotides, 53
Nucleus
 atomic, 164
 galaxy, 42
Nurse shark, 377
Nutrients
 availability of, 396–397, 398, 399
 coral reefs and, 476, 478
 Pfiesteria outbreak, 408

O

Obduction, 98, 100
Observations, 6
Ocean, 4–5. *See also* Salinity; Seawater
 color, 401
 vs. continents, 11–12
 depth, 11–12
 development of salinity, 49
 gas exchange with atmosphere, 180–181
 geography, 9–12
 layering, 28
 light transmission in, 398–402
 living in, 34, 37–38
 margins, 398
 on other planets, 48–49
 origin of, 47–49
 role in reducing greenhouse effect, 224
 size, 11
 source of water, 47, 48, 49–50
 stresses from human population, 5
 surface gravity, 115–116
 vertical distribution of life, 21
Ocean basins. *See also* Deep-ocean basins
 age of, 75–76
 development of, 87
 evolution of, 107

in general, 74–75
Ocean circulation. *See also* Atmospheric circulation
 Antarctic, 242–244
  wind drift, 243–244
 Atlantic Ocean, 244–247
  climatic effects of North Atlantic Currents, 247
  Gulf Stream, 244–247
  North and South Atlantic Gyres, 244
 deep currents, 230, 257–261
  origin of thermocline circulation, 257–258
  sources of deep water, 258–259
  worldwide deep-water circulation, 259–261
 downwelling, 239–242
 Indian Ocean, 256–257
 measuring currents, 230–233
 Pacific Ocean, 247–256
  effects of El Ninos and La Ninas, 253–254
  El Nino-Southern Oscillation conditions, 250, 252
  ENSO cool phase, 250–251, 253
  ENSO warm phase, 250
  examples from recent El Ninos, 254–256
  frequency of El Nino events, 253
  normal conditions, 247–250
  predicting El Nino events, 256
  surface currents, 230, 233–239
  upwelling, 239–242
Ocean climate patterns, 214–215
Ocean Drilling Program (ODP), 136, 156
Ocean dumping, 544
 laws on, 539, 540–541
Ocean Dumping Ban Act (1991), 539
Ocean exploration
 autonomous underwater vehicles, 33, 34, 36
 in Middle Ages, 16–20
 of Pacific Ocean, 12–13
 Phoenicians, Greeks, Romans, 13, 14, 16
 remotely operated vehicles, 33, 34, 36
 during Renaissance, 18–20
 submersibles, 31–33
 Vikings, 17–18
Ocean floor, 469–470. *See also* Abyssal plains; Bathymetry; Benthic environment; Benthos; Oceanic crust
 age of, 75–76
 coral reefs, 473–479
 features, 111
 historical record, 136
 paleomagnetism and, 73
 rocky bottoms, 470–473
Ocean sampling, 26–27
Ocean science. *See also* Oceanography
 American contribution, 25–27
 beginnings, 20–27
 European contribution, 20–24
Ocean thermal energy conversion (OTEC), 507–509
Ocean trenches, 12, 89–94. *See also specific trenches*
 dimensions, 89
 earthquakes associated with, 93–94
 location of, 124
 subduction zones, 74
Oceanic Common Water, 259
Oceanic-continental convergence, 90–92
Oceanic crust. *See also* Continental crust; Ocean floor; Sea floor spreading; Spreading centers
 composition, 79, 81
 density, 46, 82, 84
 formation, 81
 mantle convection, 81
 movement, 84
Oceanic environment. *See also* Marine environment

abyssopelagic zone, 389, 390–391
aphotic zone, 389
bathypelagic zone, 389, 390–391
disphotic zone, 389
epipelagic zone, 389–390
euphotic zone, 389
in general, 388
mesopelagic zone, 389, 390
Oceanic heat flow, 197
Oceanic-oceanic convergence, 91, 92–93
Oceanic province, 388–389
Oceanic ridges. See also Mid-Atlantic ridges; Sea floor spreading
discussed, 88
magnetic polarity reversals, 73
oceanic crust formation, 81
vs. oceanic rises, 90
slopes, 129
Oceanic rises, 88, 90, 129
Oceanic transform fault, 94
Oceanography. See also Ocean science
defined, 2–3
institutionalization of, 28–29
international cooperation, 29–30
satellite, 30
twentieth century, 27–30
Oceans Act of 2000, 539
Oceanus, 17
Odontoceti, 447–449
ODP. See Ocean Drilling Program
Office of Wetland Protection (OWP), 357
Offshore, 320, 321
Offshore wind power system (OWPS), 503, 504
Oil Pollution Act, 525
Oligotrophic area, 401
OML. See Oxygen minimum layer
Omnivores, 414
Oncaea conifera, 429
Ontong Java Plateau, 123
Oolites, 139, 154
Oozes, 145
calcereous, 145, 148, 149, 151–153, 157, 159–160
diatom, 145, 148
ostracod, 151
pteropod, 151
radiolarian, 149
siliceous, 139, 145, 146, 148, 149, 150, 152, 153, 159, 160
Ophiolites, 98–100
Open exposure, 338
Orbital waves, 267, 268
Orchestoidea, 466, 467–468
Organic substances, creation of, 53, 54
Organisms. See also Marine organisms
distribution of, 68–69
first, 54
viscosity and size of, 380
water content, 164
Origin of Species, The (Darwin), 23
Origins of Continents and Oceans, The (Wegener), 69
Orthogonal lines, 284
Osmosis, 175, 383–384
example, 384–385
Osmotic pressure, 384
Ostracod oozes, 151
OTEC. See Ocean thermal energy conversion
Outer Banks, 328, 329
Outer core, 79, 80
Outer sublittoral zone, 391
Outgassing, 46, 47
Overfishing, 493–494, 544
Owen Fracture Zone, 130, 131
OWP. See Office of Wetland Protection
OWPS. See Offshore wind power system
Oxygen. See also Gases; Respiration
concentration in atmosphere, 57, 180
dissolved in deep water, 261

in dry air, 197
eutrophication, 535–536
free, 53
importance to life, 51, 53
in marine zones, 390–391
in ocean, 4, 180–181
ocean currents and, 230
Oxygen crisis, 56–57
Oxygen minimum layer (OML), 390
Oyashio Current, 236, 248
Oyster-drill snail, 473
Oysters, 406, 470, 472–473, 502
Ozone, 51, 53
depletion of, 537–538

P
P waves, 80–81
Pacific coast, 338, 340
Pacific Decadal Oscillation (PDO), 253
Pacific Ocean, 10, 107
age of floor, 75–76
circulation, 247–256
normal conditions, 247–250. See also
El-Nino-Southern Oscillation
events; La Nina
exploration of, 12–13
marginal seas, 364–366
Bering Sea, 364–366
Gulf of California, 364
pelagic sediment within, 160
Pacific Plate, 88, 94, 95
Pacific Ring of Fire, 124, 290
Pacific Tsunami Warning Center (PTWC), 290–291
Pacific warm pool, 250
Pacillopora verrucosa, 475
Pack ice, 24, 217
Padre Island, 328, 358
PAHs (polycyclic aromatic hydrocarbons), 52
Paleoceanography, 102–106
Paleogeography, 102
Paleomagnetism. See also Dating
magnetic anomalies, 73
magnetic polarity reversals, 73
plate tectonics, 70–73
polar wandering, 71–72
seafloor, 73
Pancake ice, 216
Pangaea, 66, 68, 102, 103, 105
Panthalassa, 66
Panulirus argus, 470
Panulirus interruptus, 470, 472, 502
Paralytic shellfish poisoning, 406
Parasitism, 415
Particle size, beach slope and, 321
Parts per thousand, 174
Patagonia, 23
PCBs (polychlorinated biphenyls), 530, 531
PDO. See Pacific Decadal Oscillation
PDR. See Precision depth recorder
Peat deposits, 328, 330, 355
Pectoral fins, 433, 435
Pelagic deposits. See also Sediment
abyssal clay, 144
calcareous ooze, 149, 151–153
distribution of, 158–161
in general, 140, 149
siliceous ooze, 149, 150
Pelagic environment. See also Marine environment; Marine mammals
adaptations for seeking prey, 438–439
adaptations to avoid being prey, 439–440
discussed, 388–391, 426
gas containers, 426–427
nekton, 432–437
percent species in, 379
zooplankton, 427–432
Pelagophycus, 404
Pelvic fins, 433

Perfluorocarbon, 38
Peridotite, 98, 99, 100, 125
Perigee, 306
Perihelion, 306
Periwinkle snail (Littorina), 459, 460, 461, 462
Persian Gulf War, oil spills, 526, 527
Persistent organic pollutants (POPs), 530
Peru Current, 236, 248
Peru-Chile Trench, 91, 94, 125, 488
Petitcodiac River, 312
Petroleum
as pollution source, 518, 520–525
Argo Merchant spill of Nantucket Island, 523–524
cleaning oil spills, 524–525
Exxon Valdez oil spill, 520, 525, 526
Florida spill in West Falmouth Harbor, 522–523
largest oil spills, 527
preventing oil spills, 525
sources of in ocean, 522
as resource, 509–510, 511
Pfiesteria, 408–409
pH scale, 182
variation of pH in seawater, 183
Philadelphia Sludge Site, 527, 528
Phoenicians, 13
Phosphates, 154
Phosphorite, 511
Phosphorus, 396, 397
Phosphorus cycle, 416, 418
Photophores, 390, 435
Photosynthesis. See also
Biogeochemical cycling
biogeochemical cycling and, 415, 416, 417
compensation depth for, 396, 397, 398
discussed, 55–56
greenhouse gases and, 224
oxygen and, 181
productivity and, 395–396
Photosynthetic marine organisms, 402–405. See also Marine organisms
macroscopic algae, 403–404
microscopic algae, 404–405
seed-bearing plants, 402–403
Physical Geography of the Sea, The (Maury), 26
Physical oceanography, 3
Physical weathering, 50
Phytoplankton. See also Plankton
as base of food web, 364, 422
biomass, 395, 399
exudates, 421
frictional resistance and, 380
in general, 374, 375, 404–405
munchate, 421
ozone depletion and production of, 537–538
silica extraction and, 383
spring and fall blooms, 413
Piccard, Jacques, 12
Picoplankton, 374. See also Plankton
Pillow basalts, 98, 99, 127, 128
Pillow lavas, 127
Pinnipeds, 441
Pioneer Seamount, 226
Piranhas, 435
Pitcairn Island, 13
Planets, 41
oceans on, 48–49
Plankton. See also Algae; Marine organisms; Phytoplankton
bacterioplankton, 374
in general, 374–375
holoplankton, 375
macroplankton, 374
meroplankton, 375, 376
origin of term, 24
picoplankton, 374
upwelling and, 364
zooplankton. See Zooplankton

Plankton nets, 396
Planktonic, 145
Planktoniella sol, 381
Plantae, 372, 373
Plants. See also Algae
effect on environment, 57–58
seed-bearing, 402–403
Plasma, 167
Plastics, pollution, 533, 534, 544, 545
Plastids, 56, 57
Plate boundaries. See also Plate tectonics
convergent, 89–94
divergent, 84–89
earthquakes and, 84, 88–89
faults, 94, 95, 130–132
seismic waves, 46, 80–81
transform, 94
fracture zones, 130–132
Plate tectonics, 65. See also Continental crust; Earth; Plate boundaries
acceptance of, 79
applications
coral reef development, 100–102
hotspots, 94–96
mantle plumes, 94–96
paleoceanography, 102–106
seamounts, 96–97
tablemounts, 96–97
Deep Sea Drilling Project and, 29
evidence for
age of the ocean floor, 75–76
continental paleomagnetism, 71
earth's magnetic field and, 70–71, 72
heat flow, 77
hypsographic curve, 116–117
in general, 70
polar wandering, 71–72
seafloor paleomagnetism, 73
seafloor spreading, 74–75
worldwide earthquakes, 77–78
future predictions, 106
Laurasia and Gondwanaland, 103–106
objections to, 69
ophiolites, 98–100
predicting, 107
processes, 74
rate of movement, 76
sea floor spreading and, 75
shore interaction
Atlantic-type coasts, 338
emergent/submergent feature, 335–337
eustatic sea level change, 336
Pacific-type coasts, 338
tectonic/isostatic movements, 335–336
Platinum, 511
Pleistocene glaciers, sea level change, 336
Pleurobranchia, 431, 432
Plunging breaker, 282
Pneumatocysts, 470
Pneumatophores, 429
Poikilothermic, 438–439
Poison runoff, 533
Polar air masses, 207, 208
Polar bears, 441
Polar cell, 203, 205
Polar easterly wind belts, 203, 204, 206
Polar front, 204, 206
Polar highs, 203, 205, 206
Polar ice, 217. See also Ice formation
melting of, 222–223
Polar latitudes, 197
Polar oceanography, 24–25
Polar oceans, productivity in, 410–411
Polar regions, 215
Polar wandering curves, 71–72
Polarity, 73, 165, 177
Pollution
coastal population increase and, 538
DDT, 530–531
eutrophication, 535–537
in general, 518–519

mercury, 531–533
non-native species, 535
non-point-source, 533–535
ozone depletion, 537–538
PCBs, 530
petroleum, 518, 520–525, 526
predicting effects on marine
  organisms, 520
prevention of, 544–545
radioactive waste, 528, 530
sewage sludge, 525, 527–528, 529
waste heat, 535
Pollution control
international efforts, 539–540
of ocean dumping, 540–541
in United States, 539
Polynesia, 13
Polynesians, 14
Polynyas, 217
Polyps, 473
POPs. See Persistent organic pollutants
Population increase, coastal, 538
Porpoises, 442, 447
Porter, John, 475
Portuguese man-of-war, 427, 430
Posidonius, 111
Potassium, 397
Potential world fishery, 494, 495
Prawn, 502
Precession, 305
Precipitate, 128–129, 153
Precipitation latitudes, 171
Precipitation, 170. See also Water;
  Weather
affecting seawater salinity, 183, 184, 185
weather fronts and, 209
Precision depth recorder (PDR), 112
Pressure
adaptation to in marine environment,
  387
atmospheric, 198–199
  in circulation cells, 203, 205
deep-ocean, 479
relation to density, 187
Prestige, 518
Prevailing westerly wind belts, 203,
  204, 206, 234, 272
Primary coasts, 333, 334
Primary consumers, 427
Primary crystalline rocks, 50
Primary productivity. See also
  Productivity; Regional productivity
effect on fisheries, 494
in general, 395
measurement of, 396
photosynthetic productivity, 395–396
values for, 409
"Primordial soup," 53
Principle of constant proportions, 177
Problematum (Aristotle), 32, 35
Producers, 414
Productivity. See also Primary
  productivity; Regional productivity
defined, 146
ocean, 224
upwelling and downwelling and, 239,
  241
world, 402
Progressive waves, 267
Prokaryotic cells, 57
Proteins, 53
Protoctista, 372
Protoearth, 44, 46
Protons, 164
Protoplanets, 44
Protoplasm, 379
Protozoa, 372
Protozoans, 145
Proxigean, 306
Proxima Centauri, 41, 42
Pseudopterosins, 515
Pteropod oozes, 151
Ptolemy, Claudius, 16, 17, 18

PTWC. See Pacific Tsunami Warning
  Center
Puffer (Arothron), 478, 500
Purse seine net, 496
Pycnocline, 188–189, 233, 265, 355,
  409
Pyrosoma, 431
Pytheas, 13

O
Quadrature, 303
Quarter moon, 302
Quartz, 138, 139, 143
  shocked, 156
QuikSCAT/SeaWinds, 30

R
R/V Ballena, 278
Radar altimeter, 230, 233
Radiant energy, 194
Radioactive waste, 528, 530
Radioactivity, 46
Radiolarian ooze, 149
Radiolarians, 145, 146, 151, 383, 427,
  428. See also Marine organisms
Radiometric age dating, 58–59
Radionuclides, 528
Rafted ice, 216
Raiatea, 102
Rain, 170. See also Precipitation;
  Water; Weather
Rainbow trout, 72
Ralfsia, 404
Recreational beach, 321
Red algae (Lithothamnion), 403, 404,
  461, 464
Red clays, 144
Red Sea, 107
  bathymetry and circulation, 367
  brine pools, 367
  circulation patterns, 366
  in general, 366
  as linear sea, 87
  Suez Canal and, 358
Red snapper, 408
Red tide, 404, 405, 406–407
Redfield, Alfred, 416
Redfield ratio, 416
Redshift, 42–43
Reefs. See Coral reefs
Reef sharks (Carcharhinus perezi), 425
Regenerated production, 396
Regional productivity. See also Primary
  productivity; Productivity
in general, 405, 409–410
in polar oceans, 410–411
in temperate oceans, 412–413
in tropical oceans, 411
Relict sediments, 140
Relocation, 344
Remnant arc, 125, 126
Remotely operated vehicles (ROVs),
  33, 34, 36, 479
Renaissance
ocean exploration during, 18–20
scientific achievements during, 20
Reservoirs, 50
Resident time, 179
Resonance tide, 313. See also Tides
Respiration, 55–56, 395. See also
  Oxygen
Resultant forces, 297–298, 299
Reverse osmosis, 514
Reversing current, 316
Rhizophora, 403
Richards, Ellen Swallow, 28
Ridge Inter-Disciplinary Global
  Experiments (RIDGE/RIDGE), 29
RIDGE/RIDGE. See Ridge Inter-
  Disciplinary Global Experiments
Rift valley, 84, 87, 88, 127
Rifting, 87
Right whales, 450–451
Rings, 246

Rip current, 322, 323
Rip tide, 322
Ripples, 272
Rip-rap, 340
River Seine, 312
River-borne sediment, 136, 141
Rivers. See also Estuaries
sediment discharge, 142
tides in, 311–312
Rivers and Harbors Act (1899), 539
RMS Titanic, 31, 33, 36, 217–218, 317
RNA, 53
Rock. See also Minerals
affected by magnetic field, 71
age of continental, 76
continental, 66–67
crystalline, 50, 51
dating techniques, 58–61
igneous, 71
in ophiolites, 98–99
sedimentary, 50, 51, 71, 135
Rock fish, 386
Rock louse (Ligia), 459, 460, 461
Rock weed (Fucus filiformes), 461,
  462, 472
Rocky shores. See also Shore
high tide zone, 459, 462
in general, 459
low tide zone, 459, 463–465
middle tide zone, 459, 462–463
spray zone, 459–462
Rogue waves, 278
yachting and, 279–280
Romans, ocean exploration, 16
Rorqual whales, 450, 451
Rosenstiel School of Marine and
  Atmospheric Sciences, 28, 29, 136
Ross, Sir James Clark, 21
Ross, Sir John, 21
Ross Ice Shelf, 218, 219
Rotary current, 316
Rough keyhole limpet (Diodora
  aspera), 461
Rounded fins, 433, 434
ROVs. See Remotely operated vehicles
Runoff
peaks, 349
as source of nutrients, 396–397
stream, 177–179, 351

S
S waves, 80–81
Saffir-Simpson scale, 209
Salinity. See also Salt
Arabian Sea, 368
Bay of Bengal, 368
change through time, 179
coastal water, 349, 350
decreasing seawater, 183
depth variation in, 185, 187
determining, 176–177
development of ocean, 49
freezing point and, 171, 172
in general, 174
halocline, 187
hypersaline seep, 485
ice, 171, 172
increasing seawater, 183–184
lagoon, 357, 358
organism adaptation to in marine
  environment, 382–385
Red Sea, 366
runoff effects, 349
sources of, 50
surface variation in, 185, 186
surface water density and, 187–188
variations in, 174–176
viscosity and, 380
Salinometer, 177
Salmon, mariculture of, 500–501
Salps, 431 Salpa, 431
Salt. See also Salinity
boiling point and, 168

desalination, 513–514
distillation, 513
electrolysis, 514
freezing, 514
as mineral commodity, 514
reverse osmosis, 514
solar humidification, 513–514
Salt domes, 333
Salt marshes, 328, 355, 356, 465
Salt wedge estuary, 352, 353
Salts, 154, 155
evaporative, 514–515
Saltwater zone, 357
San Andreas Fault, 94, 95, 118
San Francisco Bay, 353
Sand. See also Beach
hard stabilization and, 340–344
movement of on beach, 321–325
as resource, 510–511
Sand beach, 333, 334
Sand crabs (Emerita), 468
Sand flats, 333
Sand star (Astropecten), 466, 467, 468
Sandwell, David, 116
Santa Barbara Harbor/breakwater,
  342–343
Sapphirina auronitens, 429
Sardine, 377
Sargasso Sea, 11, 245, 534
Sargassum, 374, 403, 404, 472
Satellite oceanography, 30
Satellites
bathymetric maps, 112, 115, 116
detecting plate motion with, 102
navigation and, 15
view of worldwide productivity, 401,
  402
Scanner, 30
Schooling, 439–440
Science, goal of, 5
Scientific fact, 6
Scientific inquiry, 5–7
Scientific method, 6
Scientific names, 373–374
Scripps Institution of Oceanography,
  28, 29, 136, 225
Scuba, 32, 35
Sculpins, 434
Scurvy, 21
Scyphozoan, 429–430
Sea anemones, 460, 462–463, 484
Sea arches, 326
Sea area, 272, 273
Sea breezes, 207. See also Wind
Sea caves, 326
Sea Cliff II, 31, 34
Sea cucumber, 460
Sea fan (Pseudopterogorgia
  elisabethae), 515
Sea floor movement, waves and, 266
Sea floor spreading. See also Oceanic
  crust; Oceanic ridges; Plate tectonics
in general, 74–75
asthenosphere, 79, 80, 83
lithosphere, 80, 81–83
spreading centers, 74
subduction, 74
mechanism for, 74–75
convection cells, 74
ocean trenches, 74
oceanic rises, 88, 90
sea level changes, 336
sediment accumulation and, 152
spreading rate, 76
Sea Grant program, 28
Sea hares (Aplysia californica), 471
Sea ice. See also Ice formation
near Antarctica, 217
in Arctic Ocean, 217
density, 187
discussed, 183, 215–216
formation, 216–217
salinity and, 184, 185

Sea lettuce (Ulva), 404, 460
Sea level
    Atlantic coast, 338
    Bangladesh and, 369
    changes in
        estuary formation, 351–352
        eustatic, 336
        discussed, 335–336
        fluctuation of, 118
        greenhouse effect and, 336–337
Sea lions, 441, 442, 443
Sea Marc (Sea Mapping and Remote
    Characterization), 112
Sea otters, 441, 526
Sea slug, 460
Sea stacks, 326
Sea star (Pisaster), 460, 462, 476
Sea urchins (Echinus), 392, 460, 464, 471
Sea waves, 272–275
SeaBeam, 112
Seafood choices, 499–500
Seaknolls, 123
Sealab, 37–38
Seals, 441–442, 443
Seamounts, 96–97, 115, 123, 125, 126.
    See also Volcanoes
    tablemounts, 96–97, 123
Search for Extraterrestrial Intelligence
    in the Universe (SETI) Institute, 52
Seas
    Beaufort Wind Scale and, 273
    defined, 11
    fully developed, 273
    linear, 87, 88
    tides in, 313
Seasat A, 30
Seasons, 195–197, 205
SeaStar/SeaWiFS, 30
Seawalls, 343, 344, 345
Seaward migration, 124
Seawater. See also Coastal waters; Water
    acidity and alkalinity of, 181–183
        buffering and alkalinity, 182–183
        pH, 182
    biogeochemical cycles
        carbon cycle, 416, 417
        in general, 415–416
        nitrogen cycle, 416–418
        phosphorus cycle, 416, 418
        silica cycle, 418
    color indications, 401
    density, 187–189, 190
    depth and latitude variations, 11–12
    dissolved components, 174, 175,
        177–179
    dissolved gases, 180–181
    freezing point, 167
    in general, 174
    hydrologic cycle, 184–185
    pH variation, 183
    vs. pure water, 190
    resident time, 179
    salinity, 174
        determining, 176–177
        processes affecting, 183–185
        variations, 174–176, 179
    viscosity, 380–381
Seaweed, 376
SeaWIFS, 396
Secchi disk, 401
Secondary coasts, 333, 334
Sediment. See also Erosion; Weathering
    biogenous, 158
        composition, 145
        distribution, 145–153
            neritic, 145–149
            pelagic, 149–153
        in general, 139, 144
        origin, 144–145
    cosmogenous, 139, 155
    deposition rate, 160
    distribution, 158–161
        neritic deposits, 158

pelagic deposits, 158–161
        in general, 135
    hydrogenous
        in general, 139, 153
        origin, 153
        composition and distribution, 153–154
    lithogenous, 158
        distribution, 140–144
            neritic, 140–142
            pelagic, 144
        in general, 135
        origin, 136–137
        composition, 138
        texture, 138–140
        maturity, 140
        mixtures, 157–158
        relict, 140
        on shores, 465–466
    sources
        biogenous particles, 139
        cosmogenous particles, 139
        glaciation, 139, 140
        hydrogenous deposits, 139
        lithogenous particles from rock,
            136–137, 139
        volcanogenic particles, 135, 139
    texture, 135, 138–140
    transport
        longshore drift, 322, 325
        prevented by dams, 331
    turbidity, 137, 140
Sediment terrigenous, 135
Sedimentary rocks, 50, 51, 71, 135
Sediment-covered shores. See also Shore
    in general, 465
    intertidal zonation, 466
    life in sediment, 466
    mud flats, 469
    sandy beaches, 466–469
    sediment, 465–466
Seed-bearing plants, 402–403
Seeps, 372, 483, 485–488
Seiches, 313
Seismic moment magnitude, 89
Seismic reflection profiles, 115–116
Seismic topography, 81
Seismic waves. See also Earthquakes;
    Volcanism
    in general, 46, 80–81
    P waves, 81
    S waves, 81
    sea, 266, 285
    Wadati-Benioff seismic zone, 94
Seismologists, 80
Semidiurnal tidal pattern, 309, 310, 311
Sepia, 426
Serpentine, 100, 125, 126
Serpulid reefs, 333
Sessile organisms, 462
Sewage sludge, 525, 527–528
    Boston Harbor sewage project, 528, 529
    disposal at sea, 527–528
Sextant, 14, 15
Shallow-water waves, 270, 271, 281.
    See also Tides; Waves
Sharks, 434, 437. See also individual
    species
Shear wave tomography, 82
Sheet dikes, 98, 99
Shelf break, 118
Shelf ice, 219
Shellfish, red tide and, 406
Shepard, Francis, 333
Shinkai 6500, 31, 33, 34
Ship passage, 493
Shoaling, 281
Shoes, drifting, 232
Shore. See also Beach; Coast;
    Sediment-covered shores
    beach, composition and slope, 321
    coast classification, 333–335
    coastal development, 340, 344
    depositional-type, 325

barrier islands, 327–328
        beach compartment, 331–333
        deltas, 328, 330–331
        in general, 327
    erosional-type
        features, 325–327
        in general, 325
        in general, 320
    geomorphic features
        barrier islands, 327–328, 329, 330,
            334, 338
        bay barrier, 327, 328, 333
        deltas, 328, 330–331, 333
        salt marsh, 328
        spits, 327, 328, 333
        tombolos, 327, 328
    longshore drift, 322, 325
    plate tectonics and
        Atlantic-type coasts, 338
        emergent/submergent features,
            335–337
        eustatic sea level changes, 336
        Pacific-type coasts, 338
        tectonic/isostatic movements, 335–336
    rip currents, 322, 323
    surf, swash, 281–283
    terminology
        backshore, 320
        beach, 321
        beach face, 321
        berm, 321
        coast, 320
        coastline, 320
        foreshore, 320
        longshore bars, 321
        longshore troughs, 321
        nearshore zone, 320–321
        offshore zone, 321
        shore, 320
        shoreline, 320
        wave-cut beach, 321
    U.S. coastal conditions, 337–340
    wave erosion
        headlands, 325–326
        sea arches, 326
        sea caves, 326
        sea stacks, 326
        wave-cut cliffs, 326
Shore crabs, 464, 465
Shoreline
    defined, 320
    emerging and submerging, 335–337
    movement of sand parallel to, 322
    movement of sand perpendicular to,
        321–322
    retreat of in U.S., 337–338
Shrimp, 390, 391, 485, 502
Side-scanning sonar, 112, 113
Silica, 145, 151, 383
Silicate minerals, 79
Siliceous ooze, 139, 145, 146, 148–150,
    152, 153, 159, 160
Siliceous sediment, 157
Silicoflagellate ooze, 149
Silicon, 397
Silicon cycle, 418
Sill, 358
Sine waves, 268
Sirenia, 442, 443
Slack water, 316
Slightly stratified estuary, 352, 353
Slocum, Joshua, 33
Slocum glider, 33, 36
Slope
    beach, 321
    continental, 117, 119–120, 144
    ocean floor, 480
Slush, 216
Small, Peter, 32
Smith, Walter, 116
Snowflakes, 172
Society Islands, 102

Sodium, 397
SOFAR (sound fixing and ranging)
    channel, 225
Solar bulges, 302
Solar day, 301
Solar energy. See also Biogeochemical
    cycling; Heat budget; Heat flow;
    Light; Sun
    distribution of, 194–195
    energy from ocean thermal, 503
    heat budget and atmospheric
        greenhouse, 219–221
    wind and, 194
Solar humidification, 513–514
Solar distillation, 513–514
Solar radiation
    availability of, 397–398
    greenhouse effect and, 220
Solar system, 41
    heliocentric theory of, 7
    origin of, 43–46
Solar wind, 46
Soldier fish, 440
Solid state, 167, 168, 170, 173
Somali Current, 236, 256, 257, 367
Somoa, 13
Sonar (Sound Navigation and Ranging),
    28, 112
    side-scanning, 112, 113
Sorting, 139–140
Soundings, 21, 22, 111
South American Plate, 117
South Atlantic Gyre, 234–235, 236, 244
South Atlantic Sea, 11
South Equatorial Current, 236, 241,
    244, 248, 256, 257, 361, 367, 368
South Georgia Islands, 21
South Pacific Gyre, 234–235, 236, 248
South Pacific Sea, 11
South Pacific Subtropical Gyre, 239
South Sandwich Islands, 21
Southeast trade wind, 203, 204
Southern Oscillation. See El
    Nino/Southern Oscillation events
Southern Hemisphere, Coriolis effect
    and, 201
Southern Ocean, 10. See also Antarctic
    Ocean
Southwest monsoon, 367, 368
Southwest Monsoon Current, 256, 257,
    367
Space dust, 155
Spanish Armada, 19
Species
    cold-water vs. warm-water, 382
    discussed, 373–374
    distribution on earth, 378–379
    diversity on deep-ocean floor, 480
    evolution and, 57
    non-native, 535
    number of marine, 378–379
Sperm whale (Physeter
    macrocephalus), 388, 444, 446, 447
Spermaceti organ, 447
Spherules, 155
Spilling breaker, 282
Spirula, 426
Spit, 327, 328, 333
Splash wave, 266, 289
Sponges, 476, 477
Spotfin butterfly fish, 387
Spotted dolphin (Stenella attnuata), 496
Spray zone, 391, 459–462
Spreading centers. See also Oceanic
    crust; Sea floor spreading
    back-arc, 124–125
    sea floor, 74, 93
Spring bloom, 413
Spring tide, 303, 306
Squid, 377, 426, 432–433
Staghorn corals (Acropora), 479
Standard laboratory bioassay, 520
Standing stock, 493

Standing waves, 284, 285, 311. *See also* Waves
State, changes of, 167, 168, 169, 170, 173
Stenella, 438
Stenohaline, 383
Stenothermal, 382
Stick chart, 14
Still water level, 268
Stingrays, 434, 435
Stomiatoid, 436
Stommel, Henry, 260
Storm, 207–209
  tropical, 209
Storm surge, 210, 212. *See also* Waves
Strabo, 16
Straightened coasts, 333
Stranded beach deposits, 335
Stratosphere, 53, 198
Stream runoff, 177–179
  affecting seawater salinity, 183, 184, 185
Streamlining, 381
Striped bass, 377
Stromatolites, 149, 150, 359
Structure and Distribution of Coral Reefs, The (Darwin), 100
Subaerial deposition coasts, 333, 335
Subduction, 74. *See also* Ocean trenches; Plate tectonics
Subduction zone, 74
Subduction zone seep biocommunity, 487–488
Sublittoral zone, 391, 470–473
Submarine, 28, 31, 35
Submarine canyons, 120, 121, 331
Submarine fans, 120
Submerged dune topography, 335
Submerging shorelines, 335–337
Submersibles, 31–33
Subneritic province, 391
Suboceanic province, 391
Subpolar gyres, 235. *See also* Gyres
Subpolar low, 203
Subpolar regions, 215
Subtropical Convergence, 237, 244
Subtropical gyres, 234–235, 236, 239. *See also* Gyres
Subtropical highs, 203
Subtropical Intermediate Water, 361, 362
Subtropical regions, 215
Subtropical Underwater, 361, 362
Suez Canal, 358
Suiko Seamount, 96
Sulfide gas, 129
Sulfur, 397
Sulfur bacteria, 55–56
Summer, productivity in temperate oceans in, 413
Summer solstice, 196
Summertime beach, 322, 324
Sun. *See also* Solar energy
  autotrophs and, 55
  coral reefs and, 474, 476
  declination of, 305–306
  diameter of, 44
  effect on tidal bulges, 301–302
  formation of solar system and, 46
  marine zones and, 389
  navigation and, 14
  size, 301
Sun-moon system, gravitational and centripetal forces in, 295–297, 298
Supernova, 43
Superwaves, 278
Supralittoral zone, 391, 459–462
Surf, 281–283. *See also* Waves
  breakers, 282
    plunging, 282
    spilling, 282
Surf grass (*Phylospadix*), 372, 403, 464, 465
Surf zone, 281
Surface area to volume ratios, 380–381
Surface currents. *See also* Currents

boundary, 234
climate and, 239
Ekman spiral and Ekman transport, 235–237
equatorial, 233–234
equatorial countercurrents, 239
in general, 230, 233
geostrophic currents, 237
gyres, 234–235
measuring, 230–233
western intensification, 237–238
wind-driven, 235
Surface salinity variation, 185, 186
Surface tension, 165–166, 173
Surface water
  converging, 241
  diverging, 241
Surfing, 282–283
Surging breaker, 282
Suspension feeding, 466, 467
Suspension settling, 121, 123
Sverdrup (Sv), 243
Swash, 321
Swell, 273, 276–278
  free and forced waves, 277
  interference patterns, 276–277
Swim bladder, 426–427
Sydney to Hobart race disaster, 279–280
Symbiosis, 57, 414–415
Syzygy, 303

T
Tablemounts, 96–97, 123. *See also* Seamounts
  upwelling and, 242, 243
Tahaa, 102
TAO. *See* Tropical Atmosphere and Ocean project
TAP. *See* Transarctic Acoustic Propagation Experiment
Tapwater, salinity, 175–176
Taxonomy, 373–374
Technology, environmental consequences, 5
Tectonic estuary, 351, 352
Tectonic movement, of Earth's crust, 335–336
Tectonic plates, detecting motion of with satellite, 102
Tektites, 155, 157
Temperate oceans, productivity in, 412–413
Temperate regions, 215
Temperature. *See also* Global warming
  Arabian Sea, 368
  atmospheric, 198
  Challenger expedition readings, 22
  coastal water, 349–350
  coral reefs and, 473
  global surface water, 261
  instrumental record, 223
  marine environment, 381–382
  Red Sea, 366
  sea surface, 255
  sea surface anomaly, 251
  surface water density and, 187–188, 189
  viscosity and, 380
  water and, 167
  world ocean surface, 240
Temperature-salinity diagram, 257, 258
Terra, 30
Terrigenous sediment, 135
Territorial sea, 492
Test, 145, 404
Testing, 6
Tethys Sea, 66, 103, 105, 358, 473
Texture, of sediment, 135
Thalassiosira eccentrica, 147
Theory, 6
  truth and, 6–7
Thermal contraction, 170
Thermal energy, 507–509
Thermal pollution, 535

Thermocline
  coastal water, 349
  Kelvin waves, 250
  relation to pycnocline, 188–189
  in subtropical gyres, 409, 410
  in tropical ocean, 44, 507
Thermohaline circulation. *See also* Ocean circulation
  climate change and, 261
  origin of, 257–258
  sources of deep water, 258–259
  worldwide deep-water circulation, 259–261
Thermostatic effects, 170
Third quarter moon, 302, 303
Thompson, C. Wyville, 22, 26
Thorvaldson, Erik "the Red," 17
Tidal bore, 311–312
Tidal bulges, 300–301, 305–306
  sun's effect, 301–302
Tidal period, 300
Tidal power plants, 506–507
Tidal range, 303
  Atlantic coast, 338
  Pacific coast, 338
Tidal waves, 286, 312. *See also* Tsunami
Tide-generating forces, 298–300, 306
Tide pools, 457
Tide zones
  high-tide, 459, 462
  intertidal zonation, 466
  low-tide, 459, 462–463
  middle-tide, 459, 462–463
Tides. *See also* Waves
  dynamic theory of, 308–309
    amphidromic points and cotidal lines, 308–309
    effect of continents, 309
  equilibrium theory of, 300–308
    declination of moon and sun, 305–306
    earth's rotation, 302
    effects of elliptical orbits, 306
    monthly tidal cycle, 302–304
    prediction of equilibrium tides, 306–309
    tidal bulges, 300–302
  in general, 266, 271, 295
  generating, 295–300
    barycenter, 295, 296, 297
    Earth-Moon system, 295–297
    gravitational attraction, 295–297, 298
    Newton's law of gravitation, 295
    rotating earth, 302
    Sun-Moon declination, 305–306
    Sun-Moon distance, 301–302
    Sun-Moon effects, 302–304
    tide-generating forces, 295–300
  lagoons and, 357
  in lakes, 313
  patterns, 309–311
  power generation from, 505–507
  tidal phenomena, 311–317
    Bay of Fundy, 313, 315–316
    coastal tidal currents, 316–317
    grunions, 314–315
    tidal bores, 312
    tides in lakes, 313
    tides in narrow basins connected to ocean, 313
  types
    diurnal, 309, 311
    equatorial, 307
    mixed, 310, 311
    neap, 303
    partial, 309
    resonance, 313
    semidiurnal, 309, 310, 311
    spring, 303, 306
    tropical, 307
Tierra del Fuego, 23
Tin, 511
Titanic, 31, 33, 36, 217–218, 317
TOGA. *See* Tropical Ocean-Global

Atmosphere program
Tombolo, 327, 328
Tonga, 13
TOPEX/Poseidon, 30
Tornadoes, El Nino and, 256
Toxic compounds, 522
Tracers, 233
Trade winds, 203, 206, 234. *See also* Wind
Transarctic Acoustic Propagation Experiment (TAP), 226
Transfer efficiency, 419–420
Transform active margins, 117–118
Transform boundaries. *See* Plate boundaries
Transform faulting, 94
Transform faults, 94, 95, 130–132
Transitional waves, 270, 271
Transitional zone, 357
Transverse waves, 267
Trent River, 312
Trieste, 12, 31, 34, 35
Tritium, 259
Trochoidal waveform, 272
Troodos Massif ophiolite, 98, 99–100
Trophic levels, 419
Tropic of Cancer, 196, 215
Tropic of Capricorn, 196, 215
Tropical air masses, 207, 208
Tropical Atmosphere and Ocean (TAO) project, 256
Tropical cyclones
  in general, 209
  historic destruction, 210, 212–214
  movement, 210
  origin, 209–210
  types of destruction, 210
Tropical depression, 209
Tropical Ocean-Global Atmosphere (TOGA) program, 256
Tropical oceans, 411
Tropical storm, 209
Tropical tides, 307
Tropics, 196, 215
Tropopause, 198
Troposphere, 198
Troughs, 268, 333
Truncate fins, 433, 434
Truth, theory and, 6–7
Tsunami
  coastal effects, 286–287
  in general, 266, 271, 285
  historic, 287–290
  origin, 286
  warning system, 290–291
Tuamotu Islands, 13
Tube worms (*Riftia*), 129, 482, 484, 485, 486
Tuna, 438–439
  incidental catch and, 496–497
Tunicates, 431
Turbidite deposits, 120, 140, 142
Turbidity, 385
Turbidity current, 120, 122. *See also* Currents
  sediment and, 137, 140, 142
  waves and, 266
Turtle grass (*Thalassia*), 469
Typhoons, 209–214

U
U-boats, 28
Ultraviolet radiation, 537–538
Undertow, 322
Underwater living, 34, 37–38
United Nations Conference on the Law of the Sea, 492
United States
  coastal characteristics, 337–340
  marine pollution control, 539
  migration to coasts in, 5
U.S. Army Corps of Engineers, 345
U.S. Coast and Geodetic Survey, 73

U.S. Coast Guard, 218
U.S. Commission on Ocean Policy, 539
U.S. Environmental Protection Agency, 520, 528
U.S. Food and Drug Administration (FDA), 532
U.S. Geological Survey, 29
U.S. National Marine Fisheries Service (NMFS), 29, 226
U.S. National Marine Sanctuaries, 539, 540, 544
Universe, origin of, 41–43
Upper water, 188
Upwelling. *See also* Coriolis effect
coastal, 241–242, 411
converging surface water, 241
diverging surface water, 241
equatorial, 411
fisheries and, 494
in general, 239–241
nutrient supply and, 398
productivity enhancement, 371
siliceous ooze and, 153
winds and, 242, 243, 361
Urchin barren, 470
USS Ramapo, 274–275

V

van der Waals forces, 167
Vapor, 167, 168, 170, 173
Vaporization, latent heat of, 169, 173
Venus, 48
Vernal equinox, 196
Vertically mixed estuary, 352, 353
Vikings, 17–18
Vinci, Leonardo de, 20
Vine, Frederick, 74
Vinland, 18
Virgo, 42
Viruses, 421–422
Viscosity
of asthenosphere, 83
of water, 380–381
Visible light, 398, 400–401
Volcanic arc, 90, 124
Volcanic coasts, 333, 334, 335
Volcanism. *See also* Basalt; Convection cells; Earthquakes; Hydrothermal vents; Magma
elements origin, 46–47
gases, 48, 49, 50
mantle plumes, 94–96
ophiolites, 98–100
paleomagnetism, 71
volcanogenic particles, 135, 138
water origin, 47
Volcanoes. *See also* Hot spots; Seamounts; Seismic waves
coral reef development and, 100–102
cycling and, 49
eruption of, 130, 178
formation of atmosphere and, 48
Hawaii, 94, 95–96, 97
North American, 92
in oceans, 4
Vortexes, 246

W

Wadati-Benioff seismic zone, 94
Wake, 266
Walker, Gilbert T., 249
Walker Circulation Cell, 249, 250
Wallace, Alfred Russel, 23
Walruses, 441, 442
Walsh, Don, 12
Waning crescent, 304
Waning gibbous, 304
Warm front, 207, 208
Warm-blooded organisms, 439
Warm-core rings, 246–247
Warm-water species, vs. cold-water, 382
Warm-water vents, 128
Waste disposal, 518–519

Waste heat, 535
Water. *See also* Ice; Ocean; Salinity; Seawater
boiling point, 167–168
capillarity, 166
as commodity, 513–514
conductivity, 177
density of, 170–173
dissolving power, 166–167
evaporation, 169
freezing point, 167
fresh, 513–514
heat capacity, 168, 173
heat capacity, 168, 173
hydrogen bonds, 170
latent heats, 169
light transmission, 398–402
mixing in estuaries, 352–353
origins, 47, 48, 49–50
as physical support, 379–380
polarity, 177
recycling, 184–185
vs. seawater, 190
sources, 47
states of, 167, 168, 170, 173
surface tension, 165–166, 173
thermal properties, 167–170, 173
transparency of, 385–387
viscosity, 380–381
wetland role in cleaning, 356–357
Water cycle. *See also* Biogeochemical cycling
in general, 177–179
resident time, 179
Water hyacinth (*Eichhornia crassipes*), 535
Water masses
Caribbean Sea, 361
Gulf of Mexico, 363
Water molecule
in general, 164–165
geometry, 165
interconnections, 165–166
polarity, 165
as universal solvent, 166–167, 173
Water vapor
atmospheric, 198
formation of atmosphere and, 47, 48
greenhouse effect and, 221
Watling Island, 18
Wave-cut beach, 320, 321
Wave-cut cliffs, 326, 327
Wave erosion coast, 334, 335
Waveform, 267, 272
Waves. *See also* Tides
cause of, 265–266
characteristics
base, 269
circular orbital motion, 268–269
crest, 268
in general, 268
height, 268, 272–275, 281
length, 268, 281
longitudinal, 267
orbital, 268–269
period, 268
steepness, 268
transverse, 267
trough, 268
deep-water, 269–271
dispersion, 276
drift, 269
forced standing, 313
frequency, 268
interaction with shore
standing waves, 284, 285
surf, 281–283
wave diffraction, 284
wave reflection, 284–285
wave refraction, 265, 283–284
internal, 265, 266
Kelvin, 250
movement of, 267–268
navigation and direction of, 72

power generation, 503–505
rogue waves, 278
yachting in, 279–280
"sea," 272–275
shallow-water, 271
speed, 269–271, 281
storm surges, 210, 212
surf, 281–283
swell, 273, 276–278
free/forced waves, 277
interference patterns, 276–277
wave dispersion, 276
wave train, 276
transitional, 271
tsunami, 285–291
historic, 287–290
warning system, 290–291
wavelength, 268, 271
wind-generated, 272
capillary waves, 272
duration, 272
fetch, 272
gravity waves, 272
ripples, 272
swell, 273, 276–278
whitecaps, 272
Waxing crescent, 304
Waxing gibbous, 304
Weather. *See also* Climate; El Nino/Southern Oscillation events; Hurricanes; Precipitation
cold front, 207, 208
defined, 205
influence of oceans on, 4
jet stream, 207, 209
storms, 207–209
warm front, 207, 208
winds, 206–207
Weathering, 136, 140. *See also* Climate; Erosion
"The Wedge," 284, 285
Wegener, Alfred, 65–66
Wentworth scale of grain size, 138, 143
West Australian Current, 236, 256, 257
West Falmouth Harbour, oil spill, 522–523
West Greenland Current, 217
West Indies, 18
West Mariana Ridge, 125
West Wind Drift, 234, 236, 242–243, 244, 247, 248, 256, 257
Western boundary currents, 233–234, 239
Western intensification, 237–238
Wetlands
coastal, 355–357
loss of, 357
role in cleaning water, 356–357
Wetting, 166
Whale shark (*Rhincodon typus*), 437
Whales. *See also* Cetacea
baleen, 444, 447, 449–451
breathing, 446
intelligence of toothed, 448–449
navigation, 72, 226
toothed, 444, 447–449
world population, 452
Whirlpools, 237
White Cliffs of Dover, 145, 149
White smokers, 128
Whitecaps, 272
Whittaker, Robert H., 372
"Wild Coast," 278
Wilson, John Tuzo, 107
Wilson cycle, 107
Wind. *See also* Coriolis effect; Ekman transport; Weathering
defined, 199
discussed, 206–207
doldrums, 203–204, 206, 214
monsoon, 205
particle transport, 136–137, 138, 141, 142
power generation, 503

salinity and, 349
solar, 46
trade, 203, 206, 234
waves and, 194, 265, 272
Wind belts, 203, 206, 230
Wind deposition coast, 334
Wind drift, 243–244
Wind-driven surface currents, 234–235
Winter, productivity in temperate oceans in, 412–413
Winter solstice, 196
Wintertime beach, 322, 324
WOCE. *See* World Ocean Circulation Experiment
Woods Hole Oceanographic Institute, 28, 29, 31, 33, 136, 367
World fishery, 494–495
World ocean, 9
pelagic sediment within, 160
surface temperature, 240
World Ocean Circulation Experiment (WOCE), 30
World Wars, oceanography and, 28
Worms, 129, 432, 466, 467, 482, 484, 485, 486. *See also* Marine organisms
Wrangell Island, 24
Wrasse, 434
Wust, George, 28

Y

Yachting, rogue waves and, 279–280
Yellowfin tuna (*Thunnus albacares*), 438, 500
Yellowstone National Park, 94
Yellowtail (*Seriola*), 438

Z

Zebra mussel (*Dreissena polymorpha*), 535
Zenith, 296
Zinc, 397
Zooplankton. *See also* Plankton
biomass, 410, 422
discussed, 427–432
excretions, 421
in general, 374, 375
macroscopic, 427–432
microscopic, 427
tropic level, 419, 422
Zooxanthellae, 473, 474, 475, 477
Zostera, 372, 402, 469, 470